ORDINARY DIFFERENTIAL EQUATIONS

ORDINARY DIFFERENTIAL EQUATIONS

BY

E. L. INCE

PROFESSOR OF PURE MATHEMATICS IN THE EGYPTIAN UNIVERSITY

WITH DIAGRAMS

DOVER PUBLICATIONS, INC.

NEW YORK

Published in Canada by General Publishing Company, Ltd., 30 Lesmill Road, Don Mills, Toronto, Ontario.

Published in the United Kingdom by Constable and Company, Ltd., 10 Orange Street, London WC 2.

This Dover edition, first published in 1956, is an unabridged and unaltered republication of the work originally published in 1926. It is reprinted by special arrangement with Longmans, Green and Co.

International Standard Book Number: 0-486-60349-0
Library of Congress Catalog Card Number: 58-12618

Manufactured in the United States of America
Dover Publications, Inc.
180 Varick Street
New York, N.Y. 10014

PREFACE

In accordance with the tradition which allows an author to make his preface serve rather as an epilogue, I submit that my aim has been to introduce the student into the field of Ordinary Differential Equations, and thereafter to guide him to this or that standpoint from which he may see the outlines of unexplored territory. Naturally, I have not covered the whole domain of the subject, but have chosen a path which I myself have followed and found interesting. If the reader would pause at any point where I have hurried on, or if he would branch off into other tracks, he may seek guidance in the footnotes. In the earlier stages I ask for little outside knowledge, but for later developments I do assume a growing familiarity with other branches of Analysis.

For some time I have felt the need for a treatise on Differential Equations whose scope would embrace not merely that body of theory which may now be regarded as classical, but which would cover, in some aspects at least, the main developments which have taken place in the last quarter of a century. During this period, no comprehensive treatise on the subject has been published in England, and very little work in this particular field has been carried out; while, on the other hand, both on the Continent and in America investigations of deep interest and fundamental importance have been recorded. The reason for this neglect of an important branch of Analysis is that England has but one school of Pure Mathematics, which implies a high development in certain fields and a comparative neglect of others. To spread the energies of this school over the whole domain of Pure Mathematics would be to scatter and weaken its forces; consequently its interests, which were at no time particularly devoted to the subject of Differential Equations, have now turned more definitely into other channels, and that subject is denied the cultivation which its importance deserves. The resources of those more fortunate countries, in which several schools of the first rank flourish, are adequate to deal with all branches of Mathematics. For this reason, and because of more favourable traditions, the subject of Differential Equations has not elsewhere met with the neglect which it has suffered in England.

In a branch of Mathematics with a long history behind it, the prospective investigator must undergo a severer apprenticeship than in a field more recently opened. This applies in particular to the branch of Analysis which lies before us, a branch in which the average worker cannot be certain of winning an early prize. Nevertheless, the beginner who has taken the pains to acquire a sound knowledge of the broad outlines of the subject will find manifold opportunities for original work in a special branch. For instance, I may draw attention to the need for an intensive study of the groups of functions defined by classes of linear equations which have a number of salient features in common.

Were I to acknowledge the whole extent of my indebtedness to others, I should transfer to this point the bibliography which appears as an appendix. But passing over those to whom I am indebted through their published work, I feel it my duty, as it is my privilege, to mention two names in particular. To the late Professor George Chrystal I owe my introduction to the subject; to Professor E. T. Whittaker my initiation into research and many acts of kind encouragement. And also I owe to a short period of study spent in Paris, a renewal of my interest in the subject and a clarifying of the ideas which had been dulled by war-time stagnation.

In compiling this treatise, I was favoured with the constant assistance of Mr. B. M. Wilson, who read the greater part of the manuscript and criticised it with helpful candour. The task of proof-correction had hardly begun when I was appointed to my Chair in the Egyptian University at Cairo, and had at once to prepare for the uprooting from my native country and transplanting to a new land. Unassisted I could have done no more than merely glance through the proof-sheets, but Mr. S. F. Grace kindly took the load from my shoulders and read and re-read the proofs. These two former colleagues of mine have rendered me services for which I now declare myself deeply grateful. My acknowledgments are also due to those examining authorities who have kindly allowed me to make use of their published questions; it was my intention to add largely to the examples when the proof stage was reached, but the circumstances already mentioned made this impossible. And lastly, I venture to record my appreciation of the consideration which the publishers, Messrs. Longmans, Green and Co., never failed to show, a courtesy in harmony with the traditions of two hundred years.

If this book is in no other respect worthy of remark, I can claim for it the honour of being the first to be launched into the world by a member of the Staff of the newly-founded Egyptian University. In all humility I trust that it will be a not unworthy forerunner of an increasing stream of published work bearing the name of the Institution which a small band of enthusiasts hopes soon to make a vigorous outpost of scientific enquiry.

E. L. INCE.

HELIOPOLIS,
December, 1926.

CONTENTS

PART I

DIFFERENTIAL EQUATIONS IN THE REAL DOMAIN

PART II

DIFFERENTIAL EQUATIONS IN THE COMPLEX DOMAIN

APPENDICES

PART I

DIFFERENTIAL EQUATIONS IN THE REAL DOMAIN

CHAPTER I

INTRODUCTORY

1·1. Definitions.—The term *æquatio differentialis* or differential equation was first used by Leibniz in 1676 to denote a relationship between the differentials dx and dy of two variables x and y.* Such a relationship, in general, explicitly involves the variables x and y together with other symbols $a, b, c, \ldots$ which represent constants.

This restricted use of the term was soon abandoned ; differential equations are now understood to include any algebraical or transcendental equalities which involve either differentials or differential coefficients. It is to be understood, however, that the differential equation is not an identity.†

Differential equations are classified, in the first place, according to the number of variables which they involve. An *ordinary* differential equation expresses a relation between an independent variable, a dependent variable and one or more differential coefficients of the dependent with respect to the independent variable. A *partial* differential equation involves one dependent and two or more independent variables, together with partial differential coefficients of the dependent with respect to the independent variables. A *total* differential equation contains two or more dependent variables together with their differentials or differential coefficients with respect to a single independent variable which may, or may not, enter explicitly into the equation.

The *order* of a differential equation is the order of the highest differential coefficient which is involved. When an equation is polynomial in all the differential coefficients involved, the power to which the highest differential coefficient is raised is known as the *degree* of the equation. When, in an ordinary or partial differential equation, the dependent variable and its derivatives occur to the first degree only, and not as higher powers or products, the equation is said to be *linear*. The coefficients of a linear equation are therefore either constants or functions of the independent variable or variables.

Thus, for example,

$$\frac{d^2y}{dx^2}+y=x^3$$

is an ordinary linear equation of the second order ;

$$(x+y)^2\frac{dy}{dx}=1$$

is an ordinary non-linear equation of the first order and the first degree ;

* A historical account of the early developments of this branch of mathematics will be found in Appendix A.

† An example of a differential identity is :

$$\left(\frac{dx}{dy}\right)^2\cdot\frac{d^3y}{dx^3}+\left(\frac{dy}{dx}\right)^2\cdot\frac{d^3x}{dy^3}+3\frac{d^2y}{dx^2}\cdot\frac{d^2x}{dy^2}=0\,;$$

this is, in fact, equivalent to :

$$\frac{dy}{dx}\cdot\frac{dx}{dy}=1.$$

$$\left\{1+\left(\frac{dy}{dx}\right)^2\right\}^{\frac{3}{2}}=3\frac{d^2y}{dx^2}$$

is an ordinary equation of the second order which when rationalised by squaring both members is of the second degree ;

$$x\frac{\partial z}{\partial x}+y\frac{\partial z}{\partial y}-z=0$$

is a linear partial differential equation of the first order in two independent variables ;

$$\frac{\partial^2 V}{\partial x^2}+\frac{\partial^2 V}{\partial y^2}+\frac{\partial^2 V}{\partial z^2}=0$$

is a linear partial differential equation of the second order in three independent variables ;

$$\frac{\partial^2 z}{\partial x^2}\cdot\frac{\partial^2 z}{\partial y^2}-\left(\frac{\partial^2 z}{\partial x\partial y}\right)^2=0$$

is a non-linear partial differential equation of the second order and the second degree in two independent variables ;

$$udx+vdy+wdz=0,$$

where u, v, and w are functions of x, y and z, is a total differential equation of the first order and the first degree, and

$$x^2dx^2+2xydxdy+y^2dy^2-z^2dz^2=0$$

is a total differential equation of the first order and the second degree.

In the case of a total differential equation any one of the variables may be regarded as independent and the remainder as dependent, thus, taking x as independent variable, the equation

$$udx+vdy+wdz=0$$

may be written

$$u+v\frac{dy}{dx}+w\frac{dz}{dx}=0,$$

or an auxiliary variable t may be introduced and the original variables regarded as functions of t, thus

$$u\frac{dx}{dt}+v\frac{dy}{dt}+w\frac{dz}{dt}=0.$$

1·2. Genesis of an Ordinary Differential Equation.—Consider an equation

(A) $$f(x, y, c_1, c_2, \ldots, c_n)=0,$$

in which x and y are variables and $c_1, c_2, \ldots, c_n$ are arbitrary and independent constants. This equation serves to determine y as a function of x ; strictly speaking, an n-fold infinity of functions is so determined, each function corresponding to a particular set of values attributed to $c_1, c_2, \ldots, c_n$. Now an ordinary differential equation can be formed which is satisfied by every one of these functions, as follows.

Let the given equation be differentiated n times in succession, with respect to x, then n new equations are obtained, namely,

$$\frac{\partial f}{\partial x}+\frac{\partial f}{\partial y}y'=0,$$

$$\frac{\partial^2 f}{\partial x^2}+2\frac{\partial^2 f}{\partial x\partial y}y'+\frac{\partial^2 f}{\partial y^2}y'^2+\frac{\partial f}{\partial y}y''=0,$$

$$\cdot \quad \cdot \quad \cdot \quad \cdot \quad \cdot$$

$$\frac{\partial^n f}{\partial x^n}+\ldots+\frac{\partial f}{\partial y}y^{(n)}=0,$$

where

$$y'=\frac{dy}{dx},\ y''=\frac{d^2y}{dx^2},\ \ldots,\ y^{(n)}=\frac{d^ny}{dx^n}.$$

Each equation is manifestly distinct from those which precede it; * from the aggregate of $n+1$ equations the n arbitrary constants $c_1, c_2, \ldots, c_n$ can be eliminated by algebraical processes, and the eliminant is the differential equation of order n:

$$F(x, y, y', y'', \ldots, y^{(n)})=0.$$

It is clear from the very manner in which this differential equation was formed that it is satisfied by every function $y=\phi(x)$ defined by the relation (A). This relation is termed the *primitive* of the differential equation, and every function $y=\phi(x)$ which satisfies the differential equation is known as a *solution*.† A solution which involves a number of essentially distinct arbitrary constants equal to the order of the equation is known as the *general solution*.‡ That this terminology is justified, will be seen when in Chapter III. it is proved that one solution of an equation of order n and one only can always be found to satisfy, for a specified value of x, n distinct conditions of a particular type. The possibility of satisfying these n conditions depends upon the existence of a solution containing n arbitrary constants. The general solution is thus essentially the same as the primitive of the differential equation.

It has been assumed that the primitive actually contains n distinct constants $c_1, c_2, \ldots, c_n$. If there are only apparently n constants, that is to say if two or more constants can be replaced by a single constant without essentially modifying the primitive, then the order of the resulting differential equation will be less than n. For instance, suppose that the primitive is given in the form

$$f\{x, y, \phi(a, b)\}=0,$$

then it *apparently* depends upon two constants a and b, but *in reality* upon one constant only, namely $c=\phi(a, b)$. In this case the resulting differential equation is of the first and not of the second order.

Again, if the primitive is reducible, that is to say if $f(x, y, c_1, \ldots, c_n)$ breaks up into two factors, each of which contains y, the order of the resulting differential equation may be less than n. For if neither factor contains all the n constants, then each factor will give rise to a differential equation of order less than n, and it may occur that these two differential equations are identical, or that one of them admits of all the solutions of the other, and therefore is satisfied by the primitive itself. Thus let the primitive be

$$y^2-(a+b)xy+abx^2=0;$$

it is reducible and equivalent to the two equations

$$y-ax=0, \quad y-bx=0,$$

each of which, and therefore the primitive itself, satisfies the differential equation

$$y-xy'=0.$$

1·201. The Differential Equation of a Family of Confocal Conics.—Consider the equation

$$\frac{x^2}{a^2+\lambda}+\frac{y^2}{b^2+\lambda}=1,$$

where a and b are definite constants, and λ an arbitrary parameter which can assume all real values. This equation represents a family of confocal conics. The

* Needless to say, it is assumed that all the partial differential coefficients of f exist, and that $\dfrac{\partial f}{\partial y}$ is not identically zero.

† Originally the terms *integral* (James Bernoulli, 1689) and *particular integral* (Euler, *Inst. Calc. Int.* 1768) were used. The use of the word *solution* dates back to Lagrange (1774), and, mainly through the influence of Poincaré, it has become established. The term *particular integral* is now used only in a very restricted sense, cf. Chap. VI. *infra*.

‡ Formerly known as the *complete integral* or *complete integral equation* (*æquatio integralis completa*, Euler). The term *integral equation* has now an utterly different meaning (cf. § 3·2, *infra*), and its use in any other connection should be abandoned.

differential equation of which it is the primitive is obtained by eliminating λ between it and the derived equation

$$\frac{2x}{a^2+\lambda}+\frac{2yy'}{b^2+\lambda}=0.$$

From the primitive and the derived equation it is found that

$$a^2+\lambda=\frac{x^2y'-xy}{y'}, \quad b^2+\lambda=y^2-xyy',$$

and, eliminating λ,

$$a^2-b^2=\frac{x^2y'-xy}{y'}-y^2+xyy',$$

and therefore the required differential equation is

$$xyy'^2+(x^2-y^2-a^2+b^2)y'-xy=0;$$

it is of the first order and the second degree.

When an equation is of the first order it is customary to represent the derivative y' by the symbol p. Thus the differential equation of the family of confocal conics may be written :

$$xy(p^2-1)+(x^2-y^2-a^2+b^2)p=0.$$

1·21. Formation of Partial Differential Equations through the Elimination of Arbitrary Constants.—Let $x_1, x_2, \ldots, x_m$ be independent variables, and let z, the dependent variable, be defined by the equation

$$f(x_1, x_2, \ldots, x_m; z; c_1, c_2, \ldots, c_n)=0,$$

where $c_1, c_2, \ldots, c_n$ are n arbitrary constants. To this equation may be adjoined the m equations obtained by differentiating partially with respect to each of the variables $x_1, x_2, \ldots, x_m$ in succession, namely,

$$\frac{\partial f}{\partial x_1}+\frac{\partial f}{\partial z}\cdot\frac{\partial z}{\partial x_1}=0, \ldots, \frac{\partial f}{\partial x_m}+\frac{\partial f}{\partial z}\cdot\frac{\partial z}{\partial x_m}=0.$$

If $m \geqslant n$, sufficient equations are now available to eliminate the constants $c_1, c_2, \ldots, c_n$. If $m<n$ the $\frac{1}{2}m(m+1)$ second derived equations are also adjoined; they are of the forms

$$\frac{\partial^2 f}{\partial x_r^2}+2\frac{\partial^2 f}{\partial x_r \partial z}\cdot\frac{\partial z}{\partial x_r}+\frac{\partial^2 f}{\partial z^2}\cdot\left(\frac{\partial z}{\partial x_r}\right)^2+\frac{\partial f}{\partial z}\cdot\frac{\partial^2 z}{\partial x_r^2}=0 \qquad (r=1, 2, \ldots, m),$$

$$\frac{\partial^2 f}{\partial x_r \partial x_s}+\frac{\partial^2 f}{\partial x_r \partial z}\cdot\frac{\partial z}{\partial x_s}+\frac{\partial^2 f}{\partial x_s \partial z}\cdot\frac{\partial z}{\partial x_r}+\frac{\partial^2 f}{\partial z^2}\cdot\frac{\partial z}{\partial x_r}\cdot\frac{\partial z}{\partial x_s}+\frac{\partial f}{\partial z}\cdot\frac{\partial^2 z}{\partial x_r \partial x_s}=0$$

$$(r, s=1, 2, \ldots, m; r \neq s).$$

This process is continued until enough equations have been obtained to enable the elimination to be carried out. In general, when this stage has been reached, there will be more equations available than there are constants to eliminate and therefore the primitive may lead not to one partial differential equation but to a system of simultaneous partial differential equations.

1·211. The Partial Differential Equations of all Planes and of all Spheres.—As a first example let the primitive be the equation

$$z=ax+by+c,$$

in which a, b, c are arbitrary constants. By a proper choice of these constants, the equation can be made to represent any plane in space except a plane parallel to the z-axis. The first derived equations are :

$$\frac{\partial z}{\partial x}=a, \quad \frac{\partial z}{\partial y}=b.$$

These are not sufficient to eliminate a, b, and c, and therefore the second derived equations are taken, namely,

$$\frac{\partial^2 z}{\partial x^2}=0, \quad \frac{\partial^2 z}{\partial x \partial y}=0, \quad \frac{\partial^2 z}{\partial y^2}=0.$$

They are free of arbitrary constants, and are therefore the differential equations required. It is customary to write

$$p=\frac{\partial z}{\partial x},\quad q=\frac{\partial z}{\partial y},\quad r=\frac{\partial^2 z}{\partial x^2},\quad s=\frac{\partial^2 z}{\partial x \partial y},\quad t=\frac{\partial^2 z}{\partial y^2}.$$

Thus any plane in space which is not parallel to the z-axis satisfies simultaneously the three equations

$$r=0,\quad s=0,\quad t=0.$$

In the second place, consider the equation satisfied by the most general sphere; it is

$$(x-a)^2+(y-b)^2+(z-c)^2=r^2,$$

where a, b, c and r are arbitrary constants. The first derived equations are

$$(x-a)+(z-c)p=0,\quad (y-b)+(z-c)q=0,$$

and the second derived equations are

$$1+p^2+(z-c)r=0,$$
$$pq+(z-c)s=0,$$
$$1+q^2+(z-c)t=0.$$

When $z-c$ is eliminated, the required equations are obtained, namely,

$$\frac{1+p^2}{r}=\frac{pq}{s}=\frac{1+q^2}{t}.$$

Thus there are two distinct equations. Let λ be the value of each of the members of the equations, then

$$\lambda^2(rt-s^2)=1+p^2+q^2>0.$$

Consequently, if the spheres considered are real, the additional condition

$$rt>s^2$$

must be satisfied.

1·22. A Property of Jacobians.—It will now be shown that the natural primitive of a single partial differential equation is a relation into which enter arbitrary functions of the variables. The investigation which leads up to this result depends upon a property of functional determinants or Jacobians.

Let $u_1, u_2, \ldots, u_m$ be functions of the independent variables $x_1, x_2, \ldots, x_n$, and consider the set of partial differential coefficients arranged in order thus:

$$\begin{matrix} \frac{\partial u_1}{\partial x_1}, & \frac{\partial u_1}{\partial x_2}, & \ldots, & \frac{\partial u_1}{\partial x_n} \\ \frac{\partial u_2}{\partial x_1}, & \frac{\partial u_2}{\partial x_2}, & \ldots, & \frac{\partial u_2}{\partial x_n} \\ \cdot & \cdot & \cdot & \cdot \\ \frac{\partial u_m}{\partial x_1}, & \frac{\partial u_m}{\partial x_2}, & \ldots, & \frac{\partial u_m}{\partial x_n} \end{matrix}$$

Then the determinant of order p whose elements are the elements common to p rows and p columns of the above scheme is known as a *Jacobian*.* Let all the different possible Jacobians be constructed, then *if a Jacobian of order p, say*

$$\begin{vmatrix} \frac{\partial u_1}{\partial x_1}, & \ldots, & \frac{\partial u_1}{\partial x_p} \\ \cdot & \cdot & \cdot \\ \frac{\partial u_p}{\partial x_1}, & \ldots, & \frac{\partial u_p}{\partial x_p} \end{vmatrix}$$

is not zero for a chosen set of values $x_1=\xi_1, \ldots, x_n=\xi_n$, but if every Jacobian of order $p+1$ is identically zero, then the functions $u_1, u_2, \ldots, u_p$ are

* Scott and Mathews, *Theory of Determinants*, Chap. XIII.

independent, but the remaining functions $u_{p+1}, \ldots, u_m$ *are expressible in terms of* $u_1, \ldots, u_p$.

Suppose that, for values of $x_1, \ldots, x_n$ in the neighbourhood of $\xi_1, \ldots, \xi_n$, the functions $u_1, \ldots, u_p$ are not independent, but that there exists an identical relationship,

$$\phi(u_1, \ldots, u_p)=0.$$

Then the equations

$$\frac{\partial\phi}{\partial u_1}\cdot\frac{\partial u_1}{\partial x_1}+\ldots+\frac{\partial\phi}{\partial u_p}\cdot\frac{\partial u_p}{\partial x_1}=0,$$

$$\cdot \quad \cdot \quad \cdot \quad \cdot \quad \cdot \quad \cdot$$

$$\frac{\partial\phi}{\partial u_1}\cdot\frac{\partial u_1}{\partial x_p}+\ldots+\frac{\partial\phi}{\partial u_p}\cdot\frac{\partial u_p}{\partial x_p}=0,$$

are satisfied identically, and therefore

$$\frac{\partial(u_1, \ldots, u_p)}{\partial(x_1, \ldots, x_p)}\equiv\begin{vmatrix}\dfrac{\partial u_1}{\partial x_1}, \ldots, \dfrac{\partial u_1}{\partial x_p} \\ \cdot \quad \cdot \quad \cdot \quad \cdot \\ \dfrac{\partial u_p}{\partial x_1}, \ldots, \dfrac{\partial u_p}{\partial x_p}\end{vmatrix}=0$$

identically in the neighbourhood of $\xi_1, \ldots, \xi_n$, which is contrary to the hypothesis. Consequently, the first part of the theorem, namely, that $u_1, \ldots, u_p$ are independent, is true.

In $u_{p+1}, \ldots, u_m$ let the variables $x_1, \ldots, x_p, x_{p+1}, \ldots, x_n$ be replaced by the new set of independent variables $u_1, \ldots, u_p, x_{p+1}, \ldots, x_n$. It will now be shown that if u_r is any of the functions $u_{p+1}, \ldots, u_m$, and x_s any one of the variables $x_{p+1}, \ldots, x_n$, then u_r is explicitly independent of x_s, that is

$$\frac{\partial u_r}{\partial x_s}=0.$$

Let

$$u_1=f_1(x_1, \ldots, x_n), \ldots, u_m=f_m(x_1, \ldots, x_n),$$

and let $x_1, \ldots, x_p$ be replaced by their expressions in terms of the new independent variables $u_1, \ldots, u_p, x_{p+1}, \ldots, x_n$, then differentiating both sides of each equation with respect to x_s,

$$0=\frac{\partial f_1}{\partial x_1}\cdot\frac{\partial x_1}{\partial x_s}+\ldots+\frac{\partial f_1}{\partial x_p}\cdot\frac{\partial x_p}{\partial x_s}+\frac{\partial f_1}{\partial x_s},$$

$$\cdot \quad \cdot \quad \cdot \quad \cdot \quad \cdot \quad \cdot \quad \cdot$$

$$0=\frac{\partial f_p}{\partial x_1}\cdot\frac{\partial x_1}{\partial x_s}+\ldots+\frac{\partial f_p}{\partial x_p}\cdot\frac{\partial x_p}{\partial x_s}+\frac{\partial f_p}{\partial x_s},$$

$$\frac{\partial u_r}{\partial x_s}=\frac{\partial f_r}{\partial x_1}\cdot\frac{\partial x_1}{\partial x_s}+\ldots+\frac{\partial f_r}{\partial x_p}\cdot\frac{\partial x_p}{\partial x_s}+\frac{\partial f_r}{\partial x_s},$$

$(r=p+1, \ldots, m)$.

The eliminant of $\dfrac{\partial x_1}{\partial x_s}, \ldots, \dfrac{\partial x_p}{\partial x_s}$ is

$$\begin{vmatrix}\dfrac{\partial f_1}{\partial x_1}, \ldots, \dfrac{\partial f_1}{\partial x_p}, \dfrac{\partial f_1}{\partial x_s} \\ \cdot \quad \cdot \quad \cdot \quad \cdot \quad \cdot \\ \dfrac{\partial f_p}{\partial x_1}, \ldots, \dfrac{\partial f_p}{\partial x_p}, \dfrac{\partial f_p}{\partial x_s} \\ \dfrac{\partial f_r}{\partial x_1}, \ldots, \dfrac{\partial f_r}{\partial x_p}, \dfrac{\partial f_r}{\partial x_s}-\dfrac{\partial u_r}{\partial x_s}\end{vmatrix}=0,$$

or

$$\frac{\partial(f_1, \ldots, f_p, f_r)}{\partial(x_1, \ldots, x_p, x_s)} = \frac{\partial u_r}{\partial x_s} \cdot \frac{\partial(f_1, \ldots, f_p)}{\partial(x_1, \ldots, x_p)}.$$

But since, by hypothesis,

$$\frac{\partial(f_1, \ldots, f_p, f_r)}{\partial(x_1, \ldots, x_p, x_s)} = 0, \quad \frac{\partial(f_1, \ldots, f_p)}{\partial(x_1, \ldots, x_p)} \neq 0,$$

it follows that

$$\frac{\partial u_r}{\partial x_s} = 0 \qquad (r=p+1, \ldots, m; \; s=p+1, \ldots, n).$$

Consequently each of the functions $u_{p+1}, \ldots, u_m$ is expressible in terms of the functions $u_1, \ldots, u_p$ alone, as was to be proved.

1·23. Formation of a Partial Differential Equation through the Elimination of an Arbitrary Function.—Let the dependent variable z be related to the independent variables $x_1, \ldots, x_n$ by an equation of the form

$$F(u_1, u_2, \ldots, u_n) = 0,$$

where F is an arbitrary function of its arguments $u_1, u_2, \ldots, u_n$ which, in turn, are given functions of $x_1, \ldots, x_n$ and z. When for z is substituted its value in terms of $x_1, \ldots, x_n$, the equation becomes an identity. If therefore $D_r u_s$ represents the partial derivative of u_s with respect to x_r when z has been replaced by its value, then

$$\begin{vmatrix} D_1 u_1, & \ldots, & D_n u_1 \\ \cdot & \cdot & \cdot \\ D_1 u_n, & \ldots, & D_n u_n \end{vmatrix} = 0.$$

But

$$D_r u_s = \frac{\partial u_s}{\partial x_r} + \frac{\partial u_s}{\partial z} \cdot \frac{\partial z}{\partial x_r},$$

and therefore the partial differential equation satisfied by z is

$$\begin{vmatrix} \frac{\partial u_1}{\partial x_1} + \frac{\partial u_1}{\partial z} \cdot \frac{\partial z}{\partial x_1}, & \ldots, & \frac{\partial u_1}{\partial x_n} + \frac{\partial u_1}{\partial z} \cdot \frac{\partial z}{\partial x_n} \\ \cdot & \cdot & \cdot \\ \frac{\partial u_n}{\partial x_1} + \frac{\partial u_n}{\partial z} \cdot \frac{\partial z}{\partial x_1}, & \ldots, & \frac{\partial u_n}{\partial x_n} + \frac{\partial u_n}{\partial z} \cdot \frac{\partial z}{\partial x_n} \end{vmatrix} = 0.$$

1·231. The Differential Equation of a Surface of Revolution.—The equation

$$F(z, x^2+y^2) = 0$$

represents a surface of revolution whose axis coincides with the z-axis. In the notation of the preceding section,

$$x_1 = x, \quad x_2 = y, \quad u_1 = z, \quad u_2 = x^2 + y^2,$$

and therefore z satisfies the partial differential equation :

$$\begin{vmatrix} \frac{\partial z}{\partial x}, & \frac{\partial z}{\partial y} \\ 2x, & 2y \end{vmatrix} = 0$$

or

$$y\frac{\partial z}{\partial x} - x\frac{\partial z}{\partial y} = 0.$$

Conversely, this equation is satisfied by

$$z = \phi(x^2+y^2),$$

where ϕ is an arbitrary function of its argument, and is therefore the differential equation of all surfaces of revolution which have the common axis $x=0$, $y=0$.

1·232. Euler's Theorem on Homogeneous Functions.—Let

$$z=\phi(x, y),$$

where $\phi(x, y)$ is a homogeneous function of x and y of degree n. Then, since $\phi(x, y)$ can be written in the form

$$x^n\psi\left(\frac{y}{x}\right),$$

it follows that

$$x^{-n}z=\psi\left(\frac{y}{x}\right).$$

In the notation of § 1·23,

$$x_1=x, \quad x_2=y, \quad u_1=x^{-n}z, \quad u_2=\psi\left(\frac{y}{x}\right),$$
$$F(u_1, u_2)=u_1-u_2,$$

and therefore z satisfies the partial differential equation :

$$\begin{vmatrix} -nx^{-n-1}z+x^{-n}\dfrac{\partial z}{\partial x}, & x^{-n}\dfrac{\partial z}{\partial y} \\ -yx^{-2}\psi', & x^{-1}\psi' \end{vmatrix}=0,$$

and this equation reduces to

$$x\frac{\partial z}{\partial x}+y\frac{\partial z}{\partial y}=nz.$$

Similarly, if u is a homogeneous function of the three variables x, y and z, of degree n,

$$x\frac{\partial u}{\partial x}+y\frac{\partial u}{\partial y}+z\frac{\partial u}{\partial z}=nu.$$

This theorem can be extended to any number of variables.

1·24. Formation of a Total Differential Equation in Three Variables.—The equation

$$\phi(x, y, z)=c$$

represents a family of surfaces, and it will be supposed that to each value of c corresponds one, and only one, surface of the family. Now let (x, y, z) be a point on a particular surface and $(x+\delta x, y+\delta y, z+\delta z)$ a neighbouring point on the same surface, then

$$\phi(x+\delta x, y+\delta y, z+\delta z)-\phi(x, y, z)=0.$$

Assuming that the partial derivatives

$$\frac{\partial\phi}{\partial x}, \quad \frac{\partial\phi}{\partial y}, \quad \frac{\partial\phi}{\partial z}$$

exist and are continuous, this equation may be written in the form

$$\left\{\frac{\partial\phi(x, y, z)}{\partial x}+\epsilon_1\right\}\delta x+\left\{\frac{\partial\phi(x, y, z)}{\partial y}+\epsilon_2\right\}\delta y+\left\{\frac{\partial\phi(x, y, z)}{\partial z}+\epsilon_3\right\}\delta z=0,$$

where $\epsilon_1, \epsilon_2, \epsilon_3 \to 0$, as $\delta x, \delta y, \delta z \to 0$.

Now let ϵ_1, ϵ_2 and ϵ_3 be made zero and let dx, dy and dz be written for δx, δy and δz respectively. Then there results the total differential equation

$$\frac{\partial\phi}{\partial x}dx+\frac{\partial\phi}{\partial y}dy+\frac{\partial\phi}{\partial z}dz=0,$$

which has been derived from the primitive by a consistent and logical process.

If the three partial derivatives have a common factor μ, and if

$$\frac{\partial\phi}{\partial x}=\mu P, \quad \frac{\partial\phi}{\partial y}=\mu Q, \quad \frac{\partial\phi}{\partial z}=\mu R,$$

then if the factor μ is removed, the equation takes the form

$$Pdx+Qdy+Rdz=0.$$

That there is no inconsistency in the above use of the differentials dx, etc., may be verified by considering a particular equation in two variables, namely,

$$y-f(x)=c.$$

The above process gives rise to the total differential equation

$$dy-f'(x)dx=0,$$

and thus the quotient of the differentials dy, dx is in fact the differential coefficient dy/dx.

Example.—The primitive

$$\frac{(x+z)(y+z)}{x+y}=c$$

gives rise to the total differential equation

$$\frac{y^2-z^2}{(x+y)^2}dx+\frac{x^2-z^2}{(x+y)^2}dy+\frac{2z+x+y}{x+y}dz=0,$$

which, after multiplication by $(x+y)^2$, becomes

$$(y^2-z^2)dx+(x^2-z^2)dy+(2z+x+y)(x+y)dz=0.$$

1·3. The Solutions of an Ordinary Differential Equation.—When an ordinary differential equation is known to have been derived by the process of elimination from a primitive containing n arbitrary constants, it is evident that it admits of a solution dependent upon n arbitrary constants. But since it is not evident that any ordinary differential equation of order n can be derived from such a primitive, it does not follow that if the differential equation is given *a priori* it possesses a general solution which depends upon n arbitrary constants. In the formation of a differential equation from a given primitive it is necessary to assume certain conditions of differentiability and continuity of derivatives. Likewise in the inverse problem of integration, or proceeding from a given differential equation to its primitive, corresponding conditions must be assumed to be satisfied. From the purely theoretical point of view the first problem which arises is that of obtaining a set of conditions, as simple as possible, which when satisfied ensure the existence of a solution. This problem will be considered in Chapter III., where an *existence theorem*, which for the moment is assumed, will be proved, namely, that when a set of conditions of a comprehensive nature is satisfied an equation of order n does admit of a unique solution dependent upon n arbitrary initial conditions. From this theorem it follows that the most general solution of an ordinary equation of order n involves n, and only n, arbitrary constants.

It must not, however, be concluded that no solution exists which is not a mere particular case of the general solution. To make this point clear, consider the differential equation obtained by eliminating the constant c from between the primitive,

$$\phi(x, y, c)=0,$$

and the derived equation,

$$\frac{\partial\phi}{\partial x}+\frac{\partial\phi}{\partial y}p=0 \qquad \left(p\equiv\frac{dy}{dx}\right).$$

The derived equation in general involves c; let the primitive be solved for c and let this value of c be substituted in the derived equation. The derived equation then becomes the differential equation

$$\left[\frac{\partial\phi}{\partial x}\right]+\left[\frac{\partial\phi}{\partial y}\right]p=0,$$

where the brackets indicate the fact of the elimination of c. In its total form, this equation can be written

$$\left[\frac{\partial\phi}{\partial x}\right]dx+\left[\frac{\partial\phi}{\partial y}\right]dy=0.$$

Now let x, y and c vary simultaneously, then

$$\frac{\partial\phi}{\partial x}dx+\frac{\partial\phi}{\partial y}dy+\frac{\partial\phi}{\partial c}dc=0.$$

When c is eliminated as before this equation becomes

$$\left[\frac{\partial\phi}{\partial x}\right]dx+\left[\frac{\partial\phi}{\partial y}\right]dy+\left[\frac{\partial\phi}{\partial c}\right]dc=0,$$

and therefore, in view of the previous equation,

$$\left[\frac{\partial\phi}{\partial c}\right]dc=0.$$

There are thus two alternatives : either c is a constant, which leads back to the primitive,

$$\phi(x, y, c)=0,$$

or else

$$\left[\frac{\partial\phi}{\partial c}\right]=0.$$

The latter relation between x and y may or may not be a solution of the differential equation ; if it is a solution, and is not a particular case of the general solution, it is known as a *singular solution.*

Consider, for instance, the primitive

$$c^2+2cy+a^2-x^2=0,$$

where c is an arbitrary, and a a definite, constant. The derived equation is

$$cdy-xdx=0,$$

which, on eliminating c, becomes the differential equation

$$[-y+(x^2+y^2-a^2)^{\frac{1}{2}}]dy-xdx=0.$$

The total differential equation obtained by varying x, y and c simultaneously is

$$(c+y)dc+cdy-xdx=0$$

or, on eliminating c,

$$(x^2+y^2-a^2)^{\frac{1}{2}}dc+[-y+(x^2+y^2-a^2)^{\frac{1}{2}}]dy-xdx=0.$$

Thus, apart from the general solution there exists the singular solution,

$$x^2+y^2=a^2,$$

which obviously satisfies the differential equation.

A differential equation of the first order may be regarded as being but one stage removed from its primitive. An equation of higher order is more remote from its primitive and therefore its integration is in general a step-by-step process in which the order is successively reduced, each reduction of the order by unity being accompanied by the introduction of an arbitrary constant. When the given equation is of order n, and by a process of integration an equation of order $n-1$ involving an arbitrary constant is obtained, the latter is known as the *first integral* of the given equation.

Thus when the given equation is

$$y''=f(y),$$

where $f(y)$ is independent of x, the equation becomes integrable when both members are multiplied by $2y'$, thus

$$2y'y''=2f(y)y',$$

and its first integral is

$$y'^2 = c + 2\int f(y)\,dy,$$

where c is the arbitrary constant of integration.

1·4. Geometrical Significance of the Solutions of an Ordinary Differential Equation of the First Order.—Since the primitive of an ordinary differential equation of the first order is a relation between the two variables x and y and a parameter c, the differential equation is said to represent a one-parameter family of plane curves. Each curve of the family is said to be an *integral-curve* of the differential equation.

Let the equation be

$$\frac{dy}{dx} = f(x, y);$$

let D be a domain in the (x, y)-plane throughout which $f(x, y)$ is single-valued and continuous, and let (x_0, y_0) be a point lying in the interior of D. Then the equation associates with (x_0, y_0) the corresponding value of dy/dx, say p_0, and thus defines a *line-element* * (x_0, y_0, p_0) issuing from the point (x_0, y_0). Choose an adjacent point (x_1, y_1) on this line-element and construct the line-element (x_1, y_1, p_1). By continuing this process a broken line is obtained which may be regarded as an approximation to the integral-curve which passes through (x_0, y_0).

This method of approximating to the integral-curves of a differential equation is illustrated in a striking manner by the iron filings method of mapping out the lines of force due to a bar magnet. Iron filings are dusted over a thin card placed horizontally and immediately above the magnet. Each iron filing becomes magnetised and tends to set itself in the direction of the resultant force at its mid-point, and if the arrangement of the filings is aided by gently tapping the card, the filings will distribute themselves approximately along the lines of force. Thus each individual filing acts as a line-element through its mid-point.

Let the bar magnet consist of two unit poles of opposite polarity situated at A and B and let P be any point on the card. Then if the co-ordinates of A, B and P are respectively $(-a, 0)$, $(a, 0)$, (x, y), if r and s are respectively the lengths of AP and BP, and if X, Y are the components of the magnetic intensity at P,

$$Y = \frac{y}{r^3} - \frac{y}{s^3}, \quad X = \frac{x+a}{r^3} - \frac{x-a}{s^3}.$$

The direction of the resultant force at P is

$$\frac{dy}{dx} = \frac{Y}{X} = \frac{y}{x + a\dfrac{r^3+s^3}{r^3-s^3}},$$

and this is the differential equation of the lines of force. Its solution is

$$\frac{x+a}{r} - \frac{x-a}{s} = \text{const.}$$

By giving appropriate values to the constant the field of force may be mapped out. The integral-curves are the lines of force approximated to by the iron filings.

Since it has been assumed that $f(x, y)$ is continuous and one-valued at every point of D, through every point there will pass one and only one integral-curve. Outside D there may be points at which $f(x, y)$ ceases to be continuous or single-valued; at such points, which are known as *singular points*, the behaviour of the integral-curves may be exceptional.

* The line-element may be defined with sufficient accuracy as the line which joins the points (x_0, y_0) and $(x_0+\delta x, y_0+\delta y)$ where δx and δy are small and $\delta y/\delta x = p_0$.

Similarly, if an equation of the second order can be written in the form

$$y''=f(x, y, y'),$$

where $f(x, y, y')$ is continuous and single-valued for a certain range of values of its arguments, the value of y' at the point (x_0, y_0) can be chosen arbitrarily within certain limits, and thus through the point (x_0, y_0) passes a one-fold infinity of integral-curves. The general solution involves two arbitrary constants, and therefore the aggregate of integral-curves forms a two-parameter family.

In general the integral-curves of an ordinary equation of order n form an n-parameter family, and through each non-singular point there passes in general an $(n-1)$-fold infinity of integral-curves.

1·5. Simultaneous Systems of Ordinary Differential Equations.—Problems occasionally arise which lead not to a single differential equation but to a system of simultaneous equations in one independent and several dependent variables. Thus, for instance, suppose that

$$\phi(x, y, z, c_1, c_2)=0,$$
$$\psi(x, y, z, c_1, c_2)=0$$

are two equations in x, y, z each containing the two arbitrary constants c_1, c_2. Then between these two equations and the pair of equations obtained by differentiating with respect to x, the constants c_1 and c_2 can be eliminated and there results a pair of simultaneous ordinary differential equations of the first order,

$$\Phi(x, y, y', z, z')=0,$$
$$\Psi(x, y, y', z, z')=0.$$

It is possible, by introducing a sufficient number of new variables, to replace either a single equation of any order, or any system of simultaneous equations, by a simultaneous system such that each equation contains a single differential coefficient of the first order. This theorem will be proved in the most important case, namely that where the equation to be considered is of the form *

$$\frac{d^ny}{dx^n}=F\left(x, y, \frac{dy}{dx}, \ldots, \frac{d^{n-1}y}{dx^{n-1}}\right).$$

In this case new variables $y_1, y_2, \ldots, y_n$ are introduced such that

$$\frac{dy_1}{dx}=y_2, \frac{dy_2}{dx}=y_3, \ldots, \frac{dy_{n-1}}{dx}=y_n,$$

where $y_1=y$. These equations, together with

$$\frac{dy_n}{dx}=F(x, y_1, y_2, \ldots, y_n),$$

form a system of n simultaneous equations, each of the first order, equivalent to the original equation. In particular it is evident that if the original equation is linear, the equations of the equivalent system are likewise linear.

Miscellaneous Examples.

1. Find the ordinary differential equations, satisfied by the following primitives :

(i) $y=Ax^m+Bx^n$;
(ii) $y=Ae^{mx}+Be^{nx}$;
(iii) $y=A\cos nx+B\sin nx$;
(iv) $y=e^{mx}(A\cos nx+B\sin nx)$;
(v) $y=A\cosh(x/A)$;
(vi) $y=x^n(A+B\log x)$;
(vii) $y=e^{mx}(A+Bx)$;
(viii) $y=(A+Bx)\cos nx+(C+Dx)\sin nx$;
(ix) $y=e^{mx}\{(A+Bx)\cos nx+(C+Dx)\sin nx\}$;
(x) $y=Ax\cos(n/x+B)$,

where A, B, C, D are arbitrary constants and m and n are fixed constants.

* D'Alembert, *Hist. Acad. Berlin*, 4 (1748), p. 289.

2. Prove that if
$$y=\frac{ax+b}{cx+d},$$
then
$$2y'y'''=3y''^2,$$
and that if $a+d=0$, then
$$(y-x)y''=2y(1+y').$$ [Math. Tripos I. 1911.]

3. Prove that if $y^3-3ax^2+x^3=0$, then
$$\frac{d^2y}{dx^2}+\frac{2a^2x^2}{y^5}=0.$$
Show that the curve given by the above equation is everywhere concave to the x-axis, and that there is a point of inflexion where $x=3a$. [Math. Tripos I. 1912.]

4. Show that if
$$x(1-x)\frac{d^2y}{dx^2}-(4-12x)\frac{dy}{dx}-36y=0$$
then
$$x(1-x)\frac{d^{n+2}y}{dx^{n+2}}-\{4-n-(12-2n)x\}\frac{d^{n+1}y}{dx^{n+1}}-(4-n)(9-n)\frac{d^ny}{dx^n}=0.$$
Hence prove by Maclaurin's theorem that the value of y which vanishes when $x=0$ and is such that its 5th differential coefficient is unity when $x=0$ is
$$\frac{4!}{9!}\{126x^5-84x^6+36x^7-9x^8+x^9\}.$$ [Math. Tripos I. 1915.]

5. Show that the differential equation of all circles in one and the same plane is
$$\frac{d^3y}{dx^3}\left\{1+\left(\frac{dy}{dx}\right)^2\right\}-3\frac{dy}{dx}\left(\frac{d^2y}{dx^2}\right)^2=0.$$

6. Any conic section which has not an asymptote parallel to the y-axis may be written in the form
$$(y-\alpha x-\beta)^2=ax^2+2bx+c.$$
Hence show that the differential equation of all such conic sections is
$$\frac{d^3}{dx^3}\left\{\left(\frac{d^2y}{dx^2}\right)^{-\frac{2}{3}}\right\}=0$$
or
$$40\left(\frac{d^3y}{dx^3}\right)^3-45\frac{d^2y}{dx^2}\cdot\frac{d^3y}{dx^3}\cdot\frac{d^4y}{dx^4}+9\left(\frac{d^2y}{dx^2}\right)^2\frac{d^5y}{dx^5}=0.$$
In particular, show that the differential equation of all coplanar parabolæ is
$$\frac{d^2}{dx^2}\left\{\left(\frac{d^2y}{dx^2}\right)^{-\frac{2}{3}}\right\}=0$$
or
$$5\left(\frac{d^3y}{dx^3}\right)^2-3\frac{d^2y}{dx^2}\cdot\frac{d^4y}{dx^4}=0.$$ [Halphen.]

7. Verify that if
$$z=3xy-y^2+(y^2-2x)^{\frac{3}{2}},$$
then
$$\frac{\partial^2z}{\partial x\partial y}=\frac{\partial^2z}{\partial y\partial x},\quad \frac{\partial^2z}{\partial x^2}\cdot\frac{\partial^2z}{\partial y^2}=\left(\frac{\partial^2z}{\partial x\partial y}\right)^2.$$

8. Prove the following extension of Euler's theorem: If f is a function homogeneous and of degree m in x_1, x_2 and homogeneous and of degree n in y_1, y_2 then
$$\left(y_1\frac{\partial}{\partial x_1}+y_2\frac{\partial}{\partial x_2}\right)\left(x_1\frac{\partial f}{\partial y_1}+x_2\frac{\partial f}{\partial y_2}\right)-\left(x_1\frac{\partial}{\partial y_1}+x_2\frac{\partial}{\partial y_2}\right)\left(y_1\frac{\partial f}{\partial x_1}+y_2\frac{\partial f}{\partial x_2}\right)=(n-m)f.$$

9. Prove that if the family of integral-curves of the linear differential equation of the first order
$$\frac{dy}{dx}+p(x)y=q(x)$$
is cut by the line $x=\xi$, the tangents at the points of intersection are concurrent.

For curves satisfying the equation
$$\frac{dy}{dx}-\frac{y}{x}=-\frac{1}{x^3},$$
prove that for varying ξ the locus of the point of concurrence is a straight line.

CHAPTER II

ELEMENTARY METHODS OF INTEGRATION

2·1. Exact Equations of the First Order and of the First Degree.—An ordinary differential equation of the first order and of the first degree may be expressed in the form of a total differential equation,

$$Pdx+Qdy=0,$$

where P and Q are functions of x and y and do not involve p. If the differential $Pdx+Qdy$ is immediately, that is without multiplication by any factor, expressible in the form du, where u is a function of x and y, it is said to be *exact*.

If the equation

$$Pdx+Qdy=0$$

is exact and its primitive is *

$$u=c,$$

the two expressions for du, namely,

$$Pdx+Qdy \quad \text{and} \quad \frac{\partial u}{\partial x}dx+\frac{\partial u}{\partial y}dy$$

must be identical, that is,

$$P=\frac{\partial u}{\partial x}, \quad Q=\frac{\partial u}{\partial y}.$$

Then

$$\text{(A)} \qquad \frac{\partial P}{\partial y}=\frac{\partial Q}{\partial x},$$

provided that the equivalent expression $\dfrac{\partial^2 u}{\partial x \partial y}$ is continuous. The *condition of integrability* (A) is therefore necessary. It remains to show that the condition is sufficient, that is to say, if it is satisfied the equation is exact and its primitive can be found by a quadrature.

Let $u(x, y)$ be defined by

$$u=\int_{x_0}^{x} P(x, y)dx+\phi(y),$$

where x_0 is an arbitrary constant, and $\phi(y)$ is a function of y alone which, for the moment, is also arbitrary. Then $u=c$ will be a primitive of

$$Pdx+Qdy=0$$

if

$$\frac{\partial u}{\partial x}=P, \quad \frac{\partial u}{\partial y}=Q.$$

The first condition is satisfied; the second determines $\phi(y)$ thus:

$$Q(x, y)=\frac{\partial u}{\partial y}$$

* Throughout this Chapter the letter c or C generally denotes a constant of integration. Any other use of these letters will be evident from the context.

$$=\int_{x_0}^{x}\frac{\partial P}{\partial y}dx+\phi'(y)$$
$$=\int_{x_0}^{x}\frac{\partial Q}{\partial x}dx+\phi'(y)$$
$$=Q(x, y)-Q(x_0, y)+\phi'(y),$$

and therefore

$$\phi(y)=\int_{y_0}^{y}Q(x_0, y)dy,$$

where y_0 is arbitrary.

The condition is therefore sufficient, for the equation is exact and has the primitive

$$\int_{x_0}^{x}P(x, y)dx+\int_{y_0}^{y}Q(x_0, y)dy=c.$$

The constants x_0 and y_0 may be chosen as is convenient, there are not, in all, three arbitrary constants but only one, for a change in x_0 or in y_0 is equivalent to adding a constant to the left-hand member of the primitive. This is obvious as far as y_0 is concerned, and as regards x_0, it is a consequence of the condition of integrability.

As an example, consider the equation

$$\frac{2x-y}{x^2+y^2}dx+\frac{2y+x}{x^2+y^2}dy=0.$$

The condition of integrability is satisfied. The primitive therefore is

$$\int_{x_0}^{x}\frac{2x-y}{x^2+y^2}dx+\int_{y_0}^{y}\frac{2y+x_0}{x_0{}^2+y^2}dy=c.$$

It is evidently an advantage to take $x_0=0$; as the second integral then involves $\log y$, y_0 may be taken to be 1. Thus

$$\int_{0}^{x}\frac{2x-y}{x^2+y^2}dx+2\int_{1}^{y}\frac{dy}{y}=c,$$

that is

$$\left[\log(x^2+y^2)-\text{arc tan}\frac{x}{y}\right]_{x=0}^{x=x}+2\log y=c,$$

which reduces to

$$\log(x^2+y^2)-\text{arc tan}\frac{x}{y}=c.$$

2·11. Separation of Variables.—A particular instance of an exact equation occurs when P is a function of x alone and Q a function of y alone. In this case X may be written for P and Y for Q. The equation

$$Xdx+Ydy$$

is then said to have *separated variables*. Its primitive is

$$\int Xdx+\int Ydy=c.$$

When the equation is such that P can be factorised into a function X of x alone and Y_1 a function of y alone, and Q can similarly be factorised into X_1 and Y, the variables are said to be *separable*, for the equation

(I) $$XY_1dx+X_1Ydy=0$$

may be written in the separated form

(II) $$\frac{X}{X_1}dx+\frac{Y}{Y_1}dy=0.$$

It must be noticed, however, that a number of solutions are lost in the division of the equation by X_1Y_1. If, for example, $x=a$ is a root of the equation $X_1=0$, it would furnish a solution of the equation (I) but not necessarily of the equation (II).

Example.—

$$(x^2+1)(y^2-1)dx+xydy=0.$$

The variables are separable thus:

$$\frac{x^2+1}{x}dx+\frac{y}{y^2-1}dy=0.$$

Integrating:

$$x^2+\log x^2+\log (y^2-1)=c$$

or if $c=\log C$,

$$y^2=1+C\frac{e^{-x^2}}{x^2}.$$

In addition $x=0$, $y=1$, $y=-1$ are real solutions of the given equation. The two latter, but not the former are included in the general solution.

2·12. Homogeneous Equations.—If P and Q are homogeneous functions of x and y of the same degree n, the equation is reducible by the substitution * $y=vx$ to one whose variables are separable. For

$$P(x, y)=x^nP(1, v), \quad P(x, y)=x^nQ(1, v),$$

and therefore

$$P(x, y)dx+Q(x, y)dy=0$$

becomes

$$\{P(1, v)+vQ(1, v)\}dx+xQ(1, v)dv=0$$

or

$$\frac{dv}{\phi(v)}+\frac{dx}{x}=0,$$

where

$$\phi(v)=v+\frac{P(1, v)}{Q(1, v)}.$$

The solution is

$$\int\frac{dv}{\phi(v)}=\log\frac{c}{x}.$$

Example.—

$$(y^4-2x^3y)dx+(x^4-2xy^3)dy=0.$$

Let $y=vx$, then

$$(v^4+v)dx-(1-2v^3)xdv=0$$

or

$$\frac{dx}{x}=\frac{1-2v^3}{v+v^4}dv$$

$$=\left(\frac{1}{v}-\frac{3v^2}{1+v^3}\right)dv,$$

whence

$$\log x=\log v-\log (1+v^3)+\log c$$

or

$$x(1+v^3)=cv.$$

Thus the primitive is

$$x^3+y^3=cxy.$$

When the equation

$$Pdx+Qdy=0$$

is both homogeneous and exact, it is immediately integrable without the

* This device was first used by Leibniz in 1691.

introduction of a quadrature, provided that its degree of homogeneity n is not -1. Its primitive is, in fact,

$$Px+Qy=c.$$

For let $u=Px+Qy$, then

$$\frac{\partial u}{\partial x}=P+x\frac{\partial P}{\partial x}+y\frac{\partial Q}{\partial x}$$

$$=P+x\frac{\partial P}{\partial x}+y\frac{\partial P}{\partial y}=(n+1)P,$$

by Euler's theorem (§ 1·232), and similarly

$$\frac{\partial u}{\partial y}=(n+1)Q.$$

Consequently

$$du=\frac{\partial u}{\partial x}dx+\frac{\partial u}{\partial y}dy$$

$$=(n+1)(Pdx+Qdy),$$

and therefore

$$Pdx+Qdy=\frac{d(Px+Qy)}{n+1}.$$

Hence if $n \neq 1$, the primitive is

$$Px+Qy=c.$$

Example.—

$$x(x^2+3y^2)dx+y(y^2+3x^2)dy=0.$$

Solution:

$$x^4+6x^2y^2+y^4=c.$$

When $n=-1$ the integration in general involves a quadrature. It is a noteworthy fact that the homogeneous equation

$$\frac{Pdx+Qdy}{Px+Qy}=0$$

is exact, for the condition of integrability, namely

$$\frac{\partial}{\partial y}\left(\frac{P}{Px+Qy}\right)=\frac{\partial}{\partial x}\left(\frac{Q}{Px+Qy}\right),$$

reduces to

$$Q\left(x\frac{\partial P}{\partial x}+y\frac{\partial P}{\partial y}\right)=P\left(x\frac{\partial Q}{\partial x}+y\frac{\partial Q}{\partial y}\right),$$

which is true, by Euler's theorem, since P and Q are homogeneous and of the same degree. Thus any homogeneous equation may be made exact by introducing the *integrating factor* $1/(Px+Qy)$. The degree of homogeneity of this exact equation is, however, -1, so that the integration of a homogeneous equation in general involves a quadrature.

An equation of the type

$$\frac{dy}{dx}=F\left(\frac{Ax+By+C}{ax+by+c}\right),$$

in which A, B, C, a, b, c are constants such that $Ab-aB\neq 0$, may be brought into the homogeneous form by a linear transformation of the variables, for let

$$x=h+\xi, \quad y=k+\eta,$$

where ξ, η are new variables and h, k are constants such that

$$Ah+Bk+C=0,$$
$$ah+bk+c=0.$$

The equation becomes

$$\frac{d\eta}{d\xi}=F\left(\frac{A\xi+B\eta}{a\xi+b\eta}\right),$$

so that F is a homogeneous function of ξ, η of degree zero. The constants h, k are determinate since $Ab-aB\neq 0$.

When $Ab-aB=0$, let η be a new dependent variable defined by

$$\eta=x+By/A=x+by/a,$$

then

$$\frac{d\eta}{dx}=1+\frac{b}{a}F\left(\frac{A\eta+C}{a\eta+c}\right).$$

The variables are now separable.

Example.—

$$(3y-7x+7)dx+(7y-3x+3)dy=0.$$

The substitution

$$x=\xi+1,\ y=\eta$$

reduces the equation to

$$(3\eta-7\xi)d\xi+(7\eta-3\xi)d\eta=0.$$

It is now homogeneous ; the transformation $\eta=v\xi$ changes it into

$$(7v-3)\xi dv+(7v^2-7)d\xi=0$$

or

$$\left(\frac{2}{v-1}+\frac{5}{v+1}\right)dv+\frac{7}{\xi}d\xi=0,$$

whence

$$(v-1)^2(v+1)^5\xi^7=c,$$

where c is the constant of integration, that is

$$(\eta-\xi)^2(\eta+\xi)^5=c.$$

The primitive therefore is

$$(y-x+1)^2(y+x-1)^5=c.$$

2·13. Linear Equations of the First Order.—The most general linear equation of the first order is of the type

$$\frac{dy}{dx}+\phi y=\psi,$$

where ϕ and ψ are functions of x alone. Consider first of all the homogeneous linear equation *

$$\frac{dy}{dx}+\phi y=0.$$

Its variables are separable, thus :

$$\frac{dy}{y}+\phi dx=0,$$

and the solution is

$$y=ce^{-\int\phi dx},$$

where c is a constant.

Now substitute in the non-homogeneous equation, the expression

$$y=ve^{-\int\phi dx},$$

* The term *homogeneous* is applied to a linear equation when it contains no term independent of y and the derivatives of y. This usage of the term is to be distinguished from that of the preceding section in which an equation (in general non-linear) was said to be homogeneous when P and Q were homogeneous functions of x and y of the same degree. There should be no confusion between the two usages of the term.

in which v, a function of x, has replaced the constant c. The equation becomes

$$\frac{dv}{dx}e^{-\int\phi dx}=\psi,$$

whence

$$v=C+\int\psi e^{\int\phi dx}dx.$$

The solution of the general linear equation is therefore

$$y=Ce^{-\int\phi dx}+e^{-\int\phi dx}\int\psi e^{\int\phi dx}dx,$$

and involves two quadratures.

The method here adopted of finding the solution of an equation by regarding the parameter, or constant of integration c of the solution of a simpler equation, as variable, and so determining it that the more general equation is satisfied, is a particular case of what is known as the method of *variation of parameters*.*

It is to be noted that the general solution of the linear equation is linearly dependent upon the constant of integration C. Conversely the differential equation obtained by eliminating C between any equation

$$y=Cf(x)+g(x),$$

and the derived equation

$$y'=Cf'(x)+g'(x),$$

is linear.

If any particular solution of the linear equation is known, the general solution may be obtained by one quadrature. For let y_1 be a solution, then the relation

$$\frac{dy_1}{dx}+\phi y_1=\psi$$

is satisfied identically. By means of this relation, ψ can be eliminated from the given equation, which becomes

$$\frac{d}{dx}(y-y_1)+\phi(y-y_1)=0.$$

The equation is now homogeneous in $y-y_1$, and has the solution

$$y-y_1=Ce^{-\int\phi dx},$$

where C is the constant of integration.

If two distinct particular solutions are known, the general solution may be expressed directly in terms of them. For it is known that the general solution has the form

$$y=Cf(x)+g(x),$$

and any two particular solutions y_1 and y_2 are obtained by assigning definite values C_1 and C_2 to the arbitrary constant C, thus

$$y_1=C_1f(x)+g(x),$$
$$y_2=C_2f(x)+g(x),$$

and therefore

$$\frac{y-y_1}{y_2-y_1}=\frac{C-C_1}{C_2-C_1}.$$

Examples.—(i) $y'-ay=e^{mx}$ (a and m constants, $m\neq a$).

The solution of the homogeneous equation

$$y'-ay=0$$

* *Vide* § 5·23. The application of the method to the linear equation of the first order is due to John Bernoulli, *Acta Erud.*, 1697, p. 113, but the solution by quadratures was known to Leibniz several years earlier.

is $y=ce^{ax}$. In the original equation, let

$$y=ve^{ax},$$

where v is a function of x, then

$$v'e^{ax}=e^{mx}$$

or

$$v=C+\frac{e^{(m-a)x}}{m-a}.$$

Thus the general solution is

$$y=Ce^{ax}+\frac{e^{mx}}{m-a}.$$

(ii) $$y'-ay=e^{ax}$$

Solution: $$y=Ce^{ax}+xe^{ax}.$$

(iii) $$y'-\frac{2x}{x^2+1}y=2x(x^2+1)$$

Solution: $$y=C(x^2+1)+(x^2+1)^2.$$

(iv) $$y'\cos x+y\sin x=1$$

Solution: $$y=C\cos x+\sin x.$$

2·14. The Equations of Bernoulli and Jacobi.—The equation

$$\frac{dy}{dx}+\phi y=\psi y^n,$$

in which ϕ and ψ are functions of x alone, is known as the *Bernoulli equation.** It may be brought into the linear form by a change of dependent variable. Let

$$z=y^{1-n},$$

then

$$\frac{dz}{dx}=(1-n)y^{-n}\frac{dy}{dx},$$

and thus if the given equation is written in the form

$$y^{-n}\frac{dy}{dx}+\phi y^{1-n}=\psi,$$

it becomes

$$\frac{dz}{dx}+(1-n)\phi z=(1-n)\psi,$$

and is linear in z.

The *Jacobi equation,*†

$$(a_1+b_1x+c_1y)(xdy-ydx)-(a_2+b_2x+c_2y)dy+(a_3+b_3x+c_3y)dx=0,$$

in which the coefficients a, b, c are constants, is closely connected with the Bernoulli equation. Make the substitution

$$x=X+\alpha,\quad y=Y+\beta,$$

where α, β are constants to be determined so as to make the coefficients of $XdY-YdX$, dY and dX separately homogeneous in X and Y. When this substitution is made, the equation is so arranged that the coefficient of $XdY-YdX$ is homogeneous and of the first degree, thus

$$(b_1X+c_1Y)(XdY-YdX)$$
$$-\{A_2+b_2X+c_2Y-\alpha(A_1+b_1X+c_1Y)-A_1X\}dY$$
$$+\{A_3+b_3X+c_3Y-\beta(A_1+b_1X+c_1Y)-A_1Y\}dX=0,$$

* James Bernoulli, *Acta Erud.* 1695, p. 553 [*Opera* 1, p. 663]. The method of solution was discovered by Leibniz, *Acta Erud.* 1696, p. 145 [*Math. Werke* 5, p. 329].

† *J. für Math.* 24 (1842), p. 1 [*Ges. Werke*, 4, p. 256]. See also the Darboux equation, § 2·21, *infra*.

where

$$A_r = a_r + b_r\alpha + c_r\beta \qquad (r=1, 2, 3).$$

The coefficients of dY and dX also become homogeneous if α and β are so chosen that

$$A_2 - \alpha A_1 = 0, \qquad A_3 - \beta A_1 = 0,$$

or, more symmetrically, if

$$A_1 = \lambda, \qquad A_2 = \alpha\lambda, \qquad A_3 = \beta\lambda,$$

that is if

$$(\Lambda) \qquad a_1 - \lambda + b_1\alpha + c_1\beta = a_2 + (b_2-\lambda)\alpha + c_2\beta = a_3 + b_3\alpha + (c_3-\lambda)\beta = 0.$$

Thus λ is determined by the cubic equation

$$\begin{vmatrix} a_1-\lambda, & b_1, & c_1 \\ a_2, & b_2-\lambda, & c_2 \\ a_3, & b_3, & c_3-\lambda \end{vmatrix} = 0,$$

and when λ is so determined, α and β are then the solutions of any two of the consistent equations (Λ).

The equation may now be written in the form

$$XdY - YdX - \Phi\left(\frac{Y}{X}\right)dY + \Psi\left(\frac{Y}{X}\right)dX = 0.$$

The substitution $Y = Xu$ brings it into the form of a Bernoulli equation,

$$\frac{dX}{du} + U_1X + U_2X^2 = 0,$$

where U_1 and U_2 are functions of u alone.

It will be shown in a later section (§ 2·21) that if the three roots of the equation in λ are λ_1, λ_2, λ_3 and are distinct,* the general solution of the Jacobi equation is

$$U^{\lambda_2-\lambda_3}V^{\lambda_3-\lambda_1}W^{\lambda_1-\lambda_2} = \text{const}.$$

where U, V, W are linear expressions in x and y.

2·15. The Riccati Equation.—The equation

$$\frac{dy}{dx} + \psi y^2 + \phi y + \chi = 0,$$

in which ψ, ϕ and χ are functions of x, is known as the *generalised Riccati equation*.† It is distinguished from the previous equations of this chapter in that it is not, in general, integrable by quadratures. It therefore defines a family of transcendental functions which are essentially distinct from the elementary transcendents.‡

When any particular solution $y = y_1$ is known, the general solution may be obtained by means of two successive quadratures. Let

$$y = y_1 + z,$$

* The case in which they are not distinct is discussed by Serret, *Calc. Diff. et Int.* 2, p. 431.

† Riccati, *Acta Erud. Suppl.*, VIII. (1724), p. 73, investigated the equation $y' + ay^2 = bx^m$, with which his name is usually associated. The generalised equation was studied by d'Alembert, *vide infra*, § 12·51.

‡ The elementary transcendents are functions which can be derived from algebraic functions by integration, and the inverses of such functions. Thus the logarithmic function is defined as $\int_1^x x^{-1}dx$; its inverse is the exponential function. From the exponential function the trigonometrical and the hyperbolic functions are derived by rational processes, and such functions as the error-function by integration.

then the equation becomes

$$\frac{dy_1}{dx}+\frac{dz}{dx}+\psi(y_1^2+2y_1z+z^2)+\phi(y_1+z)+\chi=0,$$

and since $y=y_1$ is a solution, it reduces to

$$\frac{dz}{dx}+(2y_1\psi+\phi)z+\psi z^2=0.$$

This is a case of the Bernoulli equation ; it is reduced to the linear form by the substitution

$$z=1/u,$$

from which the theorem stated follows immediately.

Let y_1, y_2, y_3 be three distinct particular solutions of the Riccati equation and y its general solution. Then

$$u=\frac{1}{y-y_1}, \qquad u_1=\frac{1}{y_2-y_1}, \qquad u_2=\frac{1}{y_3-y_1}$$

satisfy one and the same linear equation, and consequently

$$\frac{u-u_1}{u_2-u_1}=C,$$

where C is a constant. When u, u_1 and u_2 are replaced by their expressions in terms of y, y_1 and y_2 this relation may be written

$$\frac{y-y_2}{y-y_1}=C\frac{y_3-y_2}{y_3-y_1}.$$

This formula shows that the general solution of the Riccati equation is expressible rationally in terms of any three distinct particular solutions, and also that the anharmonic ratio of any four solutions is constant. It also shows that the general solution is a rational function of the constant of integration. Conversely any function of the type

$$y=\frac{Cf_1+f_2}{Cf_3+f_4},$$

where f_1, f_2, f_3, f_4 are given functions of x and C an arbitrary constant, satisfies a Riccati equation, as may easily be proved by eliminating C between the expressions for y and the derived expression for y'.

When ψ is identically zero, the Riccati equation reduces to the linear equation ; when ψ is not zero, the equation may be transformed into a linear equation of the second order. Let v be a new dependent variable defined by

$$y=v/\psi,$$

then the equation becomes

$$\frac{dv}{dx}+v^2+Pv+Q=0,$$

where

$$P=\phi-\frac{\psi'}{\psi}, \quad Q=\psi\chi.$$

The substitution

$$v=u'/u$$

now brings the equation into the proposed form, namely,

$$\frac{d^2u}{dx^2}+P\frac{du}{dx}+Qu=0.$$

In particular, the original equation of Riccati, namely,

$$\frac{dy}{dx}+ay^2=bx^m,$$

where a and b are constants, becomes *

$$\frac{d^2u}{dx^2}-abx^mu=0.$$

2·16. The Euler Equation.—An important type of equation with separated variables is the following: †

$$\frac{dx}{X^{\frac{1}{2}}}+\frac{dy}{Y^{\frac{1}{2}}}=0,$$

in which

$$X=a_0x^4+a_1x^3+a_2x^2+a_3x+a_4,$$
$$Y=a_0y^4+a_1y^3+a_2y^2+a_3y+a_4.$$

Consider first of all the particular equation

$$\frac{dx}{\sqrt{(1-x^2)}}+\frac{dy}{\sqrt{(1-y^2)}}=0,$$

one solution is ‡

$$\text{arc sin } x+\text{arc sin } y=c,$$

but the equation has also the solution

$$x\sqrt{(1-y^2)}+y\sqrt{(1-x^2)}=C.$$

Since, as will be proved in Chapter III., the differential equation has but one distinct solution, the two solutions must be related to one another in a definite way. This relation is expressed by the equation

$$C=f(c)$$

Now let

$$x=\sin u, \quad y=\sin v,$$

then

$$u+v=c,$$
$$\sin u\cos v+\sin v\cos u=f(c)$$
$$=f(u+v).$$

Let $v=0$, then

$$\sin u=f(u)$$

and therefore

$$\sin u\cos v+\sin v\cos u=\sin(u+v).$$

Thus the addition formula for the sine-function is established.

In the same way, the differential equation

$$\frac{dx}{(1-x^2)^{\frac{1}{2}}(1-k^2x^2)^{\frac{1}{2}}}+\frac{dy}{(1-y^2)^{\frac{1}{2}}(1-k^2y^2)^{\frac{1}{2}}}=0$$

has the solution

$$\text{arg sn } x+\text{arg sn } y=c,$$

* The solution of this equation may be expressed in terms of Bessel functions (§ 7·31).

† Euler, *Inst. Calc. Int.*, 1, Chaps. V., VI.

‡ The function arc sin x is defined as $\int_0^x(1-t^2)^{-\frac{1}{2}}dt$; sin x is defined as the inverse of arc sin x, so that $\sin 0=0$; and cos x is defined as $(1-\sin^2 x)^{\frac{1}{2}}$ with the condition that $\cos 0=1$. No further properties of the trigonometrical functions are assumed.

where arg sn x is the inverse Jacobian elliptic function* defined by

$$\arg \operatorname{sn} x = \int_0^x \frac{dt}{(1-t^2)^{\frac{1}{2}}(1-k^2t^2)^{\frac{1}{2}}}.$$

Let

$$x = \operatorname{sn} u, \quad y = \operatorname{sn} v,$$

then

$$u+v=c.$$

A second and equivalent solution may be found as follows. By definition

$$\frac{dx}{du} = (1-x^2)^{\frac{1}{2}}(1-k^2x^2)^{\frac{1}{2}},$$

and therefore

$$\frac{d^2x}{du^2} = -(1+k^2)x + 2k^2x^3.$$

Similarly

$$\frac{dy}{du} = -\frac{dy}{dv} = -(1-y^2)^{\frac{1}{2}}(1-k^2y^2)^{\frac{1}{2}},$$

$$\frac{d^2y}{du^2} = \frac{d^2y}{dv^2} = -(1+k^2)y + 2k^2y^3,$$

from which it follows that

$$x\frac{d^2y}{du^2} - y\frac{d^2x}{du^2} = 2k^2xy(y^2-x^2),$$

$$x^2\left(\frac{dy}{du}\right)^2 - y^2\left(\frac{dx}{du}\right)^2 = (x^2-y^2)(1-k^2x^2y^2).$$

Hence

$$\frac{x\dfrac{d^2y}{du^2} - y\dfrac{d^2x}{du^2}}{x\dfrac{dy}{du} - y\dfrac{dx}{du}} = -\left(x\frac{dy}{du} + y\frac{dx}{du}\right)\frac{2k^2xy}{1-k^2x^2y^2}.$$

This equation is immediately integrable ; the solution is

$$\log\left(x\frac{dy}{du} - y\frac{dx}{du}\right) = \text{const.} + \log(1-k^2x^2y^2)$$

or

$$x\frac{dy}{dv} + y\frac{dx}{du} = C(1-k^2x^2y^2),$$

that is

$$\operatorname{sn} u \operatorname{sn}' v + \operatorname{sn} v \operatorname{sn}' u = f(c)(1-k^2 \operatorname{sn}^2 u \operatorname{sn}^2 v).$$

By putting $v=0$ it is found that $f(u) = \operatorname{sn} u$, and therefore

$$\operatorname{sn}(u+v) = \frac{\operatorname{sn} u \operatorname{sn}' v + \operatorname{sn} v \operatorname{sn}' u}{1-k^2 \operatorname{sn}^2 u \operatorname{sn}^2 v}.$$

This is the addition formula for the Jacobian elliptic function sn$u$.

The same process of integration may be applied to the general Euler equation.† In particular it may be noted that when $a_0=0$ a linear transformation brings the equation into the form

$$\frac{dx}{\sqrt{(4x^3-g_2x-g_3)}} + \frac{dy}{\sqrt{(4y^3-g_2y-g_3)}} = 0.$$

* Whittaker and Watson, *Modern Analysis*, Chap. XXII.
† Cayley, *Elliptic Functions*, Chap. XIV.

If $\wp(z)$ is the Weierstrassian elliptic function defined by

$$z=\int_{\wp(z)}^{\infty}(4t^3-g_2t-g_3)^{-\frac{1}{2}}dt,$$

and $x=\wp(u)$, $y=\wp(v)$, the general solution of the equation is

$$u+v=c.$$

An equivalent general solution is

$$\{(4x^3-g_2x-g_3)^{\frac{1}{2}}-(4y^3-g_2y-g_3)^{\frac{1}{2}}\}^2=(x-y)^2(C+4x+4y)$$

It may thus be shown that the addition-formula for the $\wp$-function is

$$\wp(u+v)=-\wp(u)-\wp(v)+\tfrac{1}{4}\left\{\frac{\wp'(u)-\wp'(v)}{\wp(u)-\wp(v)}\right\}^2$$

2·2. The Integrating Factor.—Let

$$Pdx+Qdy=0$$

be a differential equation which is not exact. The theoretical method of integrating such an equation is to find a function $\mu(x, y)$ such that the expression

$$\mu(Pdx+Qdy)$$

is a total differential du. When μ has been found the problem reduces to a mere quadrature.

The main question which arises is as to whether or not integrating factors exist. It will be proved that on the assumption that the equation itself has one and only one solution,* which depends upon one arbitrary constant, *there exists an infinity of integrating factors.*

Let the general solution be written in the form

$$\phi(x, y)=c,$$

where c is the arbitrary constant. Then, taking the differential,

$$\frac{\partial\phi}{\partial x}dx+\frac{\partial\phi}{\partial y}dy=0$$

or, as it may be written,

$$\phi_x dx+\phi_y dy=0.$$

Since, therefore,

$$\phi(x, y)=c$$

is the general solution of

$$Pdx+Qdy=0,$$

the relation

$$\frac{\phi_x}{P}=\frac{\phi_y}{Q}$$

must hold identically, whence it follows that a function μ exists such that

$$\phi_x=\mu P, \qquad \phi_y=\mu Q.$$

Consequently

$$\mu(Pdx+Qdy)=d\phi,$$

that is to say an integrating factor μ exists.

Let $F(\phi)$ be any function of ϕ, then the expression

$$\mu F(\phi)\{Pdx+Qdy\}=F(\phi)d\phi$$

is exact. If, therefore, μ is any integrating factor, giving rise to the solution $\phi=c$, then $\mu F(\phi)$ is an integrating factor. Since $F(\phi)$ is an arbitrary-function of ϕ, there exists an infinity of integrating factors.

* This assumption will be justified in the following chapter.

Since the equation

$$\mu(Pdx+Qdy)=0$$

is exact, the integrating factor satisfies the relation

$$\frac{\partial(\mu P)}{\partial y}=\frac{\partial(\mu Q)}{\partial x}$$

or

$$P\frac{\partial\mu}{\partial y}-Q\frac{\partial\mu}{\partial x}+\mu\left(\frac{\partial P}{\partial y}-\frac{\partial Q}{\partial x}\right)=0.$$

Thus μ satisfies a partial differential equation of the first order. In general, therefore, the direct evaluation of μ depends upon an equation of a more advanced character than the ordinary linear equation under consideration. It is, however, to be noted that any particular solution, and not necessarily the general solution of the partial differential equation is sufficient to furnish an integrating factor. Moreover, in many particular cases, the partial differential equation has an obvious solution which gives the required integrating factor.

As an instance, suppose that μ is a function of x alone, then

$$\frac{1}{\mu}\frac{d\mu}{dx}=\frac{1}{Q}\left(\frac{\partial P}{\partial y}-\frac{\partial Q}{\partial x}\right).$$

It is therefore necessary that the right-hand member of this equation should be independent of y. When this is the case, μ is at once obtainable by a quadrature. Now suppose also that Q is unity, then P must be a linear function of y. The equation is therefore of the form

$$dy+(py-q)dx=0,$$

where p and q are functions of x alone. The equation is therefore linear; the integrating factor, determined by the equation

$$\frac{d\mu}{dx}=p\mu$$

is

$$\mu=e^{\int pdx}.$$

(cf. § 2·13).

An example of an equation in which an integrating factor can readily be obtained is

$$\alpha xdy+\beta ydx+x^m y^n(axdy+bydx)=0.$$

Consider first of all the expression $\alpha xdy+\beta ydx$; an integrating factor is $x^{\beta-1}y^{\alpha-1}$ and since

$$x^{\beta-1}y^{\alpha-1}(\alpha xdy+\beta ydx)=d(x^\beta y^\alpha),$$

the more general expression

$$x^{\beta-1}y^{\alpha-1}\Phi(x^\beta y^\alpha)$$

is also an integrating factor. In the same way

$$x^{b-m-1}y^{a-n-1}F(x^b y^a)$$

is an integrating factor for $x^m y^n(axdy+bydx)$. If, therefore, Φ and F can be so determined that

$$x^{\beta-1}y^{\alpha-1}\Phi(x^\beta y^\alpha)=x^{b-m-1}y^{a-n-1}F(x^b y^a),$$

an integrating factor for the original equation will have been obtained. Let

$$\Phi(z)=z^\rho,\ F(z)=z^r,$$

then $x^\lambda y^\mu$ will be an integrating factor if

$$\lambda=(\rho+1)\beta-1=(r+1)b-m-1,$$
$$\mu=(\rho+1)\alpha-1=(r+1)a-n-1.$$

These equations determine ρ and r, and consequently λ and μ if only $a\beta - b\alpha \neq 0$. If, on the other hand, $a = k\alpha$, $b = k\beta$, the original equation is

$$(1 + kx^m y^n)(\alpha x dy + \beta y dx) = 0.$$

The integrating factor is now

$$\frac{x^{\beta-1} y^{\alpha-1}}{1 + kx^m y^n}.$$

2·21. The Darboux Equation.—A type of equation which was investigated by Darboux is the following: *

$$-Ldy + Mdx + N(xdy - ydx) = 0.$$

where L, M, N are polynomials in x and y of maximum degree m.

It will be shown that when a certain number of particular solutions of the form

$$f(x, y) = 0,$$

in which $f(x, y)$ is an irreducible polynomial, are known, the equation may be integrated.

Let the general solution be

$$u(x, y) = \text{const.}$$

then the given equation is equivalent to

$$\frac{\partial u}{\partial x} dx + \frac{\partial u}{\partial y} dy = 0,$$

and therefore

$$L\frac{\partial u}{\partial x} + M\frac{\partial u}{\partial y} - N\left(x\frac{\partial u}{\partial x} + y\frac{\partial u}{\partial y}\right) = 0$$

Replace x by $\frac{x}{z}$, y by $\frac{y}{z}$, where z is a third independent variable, then $u\left(\frac{x}{z}, \frac{y}{z}\right)$ is a homogeneous rational function of x, y, z of degree zero, and by Euler's Theorem (§ 1·232)

$$x\frac{\partial u}{\partial x} + y\frac{\partial u}{\partial y} + z\frac{\partial u}{\partial z} = 0.$$

Moreover $u\left(\frac{x}{z}, \frac{y}{z}\right)$ satisfies the relation

$$A(u) \equiv L\frac{\partial u}{\partial x} + M\frac{\partial u}{\partial y} + N\frac{\partial u}{\partial z} = 0,$$

in which L, M, N are homogeneous polynomials in x, y, z of degree m.

The theory depends on the fact that if

$$u(x, y) = \text{const.}$$

is a solution of the given equation, $u\left(\frac{x}{z}, \frac{y}{z}\right)$ is homogeneous and of degree zero, and satisfies the relation $A(u) = 0$. The converse is clearly also true.

Now let

$$f(x, y) = 0$$

be any particular solution, where $f(x, y)$ is an irreducible polynomial of degree h, and let

$$g(x, y, z) = z^h f\left(\frac{x}{z}, \frac{y}{z}\right).$$

* *Bull. Sc. Math.* (2), 2 (1878), p. 72.

Then, since g is homogeneous and of degree h,

$$x\frac{\partial g}{\partial x}+y\frac{\partial g}{\partial y}+z\frac{\partial g}{\partial z}=hg.$$

Also

$$\begin{aligned}A(g)&\equiv L\frac{\partial g}{\partial x}+M\frac{\partial g}{\partial y}+N\frac{\partial g}{\partial z}\\ &=z^h\left(L\frac{\partial f}{\partial x}+M\frac{\partial f}{\partial y}+N\frac{\partial f}{\partial z}\right)+hz^{h-1}fN\\ &=hz^{-1}Ng,\end{aligned}$$

since $f=0$ is a solution. This relation may be written in the form

$$A(g)=Kg,$$

since $A(g)$ is a polynomial of degree $m+h-1$ and g is a polynomial of degree h, K is a polynomial of degree $m-1$.

The operator A has the property that if F is any function of $u, v, w, \ldots,$ where $u, v, w, \ldots$ are themselves functions of x, y, z,

$$A(F)=\frac{\partial F}{\partial u}A(u)+\frac{\partial F}{\partial v}A(v)+\frac{\partial F}{\partial w}A(w)+\ldots$$

Let

$$f_1(x, y)=0, \quad f_2(x, y)=0, \quad \ldots, \quad f_p(x, y)=0$$

be particular solutions of the given equation, where $f(x, y)$ is an irreducible polynomial of degree h_r. Let

$$g_r(x, y, z)=z^{h_r}f\left(\frac{x}{z}, \frac{y}{z}\right) \qquad (r=1, 2, \ldots, p),$$

and consider the function

$$u(x, y, z)=\prod_{r=1}^{p}(g_r)^{\alpha_r},$$

where $\alpha_1, \alpha_2, \ldots, \alpha_r$ are constants to be determined. Now

$$\begin{aligned}A(u)&=\sum_{r=1}^{p}\frac{\partial u}{\partial g_r}A(g_r)\\ &=\sum\alpha_r g_1^{\alpha_1}\ldots g_r^{\alpha_r-1}\ldots g_p^{\alpha_p}\ldots K_r g_r\\ &=u\sum\alpha_r K_r,\end{aligned}$$

where K_r is, for every value of r, a polynomial of degree $m-1$. Also $u(x, y, z)$ is a polynomial in x, y, z of degree $h_1\alpha_1+h_2\alpha_2+\ldots+h_p\alpha_p$. If $u(x, y, z)$ is to furnish the required solution when $z=1$, it must be a polynomial in x, y, z of degree zero, and must satisfy the relation $A(u)=0$, whence

$$\begin{aligned}h_1\alpha_1+h_2\alpha_2+\ldots+h_p\alpha_p&=0,\\ K_1\alpha_1+K_2\alpha_2+\ldots+K_p\alpha_p&=0.\end{aligned}$$

Each polynomial K_r contains at most $\frac{1}{2}m(m+1)$ terms, so that the last equation, being an identity in x, y, z, is equivalent to not more than $\frac{1}{2}m(m+1)$ relations between the constants $\alpha_1, \alpha_2, \ldots, \alpha_p$. There are, therefore, in all, at most

$$\tfrac{1}{2}m(m+1)+1$$

equations between the p unknown constants α. Suitable values can therefore be given to these constants if the number p exceeds the number of equations, that is if

$$p\geqslant\tfrac{1}{2}m(m+1)+2.$$

If, therefore, $\frac{1}{2}m(m+1)+2$ *particular solutions are known, the general solution can be obtained without quadratures.*

If $p=\frac{1}{2}m(m+1)+1$ and the discriminant of the equations is zero, the same result holds. Let $p=\frac{1}{2}m(m+1)+1$ and let the discriminant be not zero. In this case, let the constants be determined by the equations

$$h_1\alpha_1+h_2\alpha_2+\ .\ .\ .\ +h_p\alpha_p=-m-2,$$

$$K_1\alpha_1+K_2\alpha_2+\ .\ .\ .\ +K_p\alpha_p=-\frac{\partial L}{\partial x}-\frac{\partial M}{\partial y}-\frac{\partial N}{\partial z}.$$

There are now $\frac{1}{2}m(m+1)+1$ non-homogeneous equations which determine the constants α. This determination of the constants gives rise to a function $u(x, y, z)$ such that

$$x\frac{\partial u}{\partial x}+y\frac{\partial u}{\partial y}+z\frac{\partial u}{\partial z}=-(m+2)u,$$

$$A(u)\equiv L\frac{\partial u}{\partial x}+M\frac{\partial u}{\partial y}+N\frac{\partial u}{\partial z}=-\left(\frac{\partial L}{\partial x}+\frac{\partial M}{\partial y}+\frac{\partial N}{\partial z}\right)u.$$

Eliminate $\frac{\partial u}{\partial z}$ between these equations, then

$$\left\{L\frac{\partial u}{\partial x}+M\frac{\partial u}{\partial y}+\left(\frac{\partial L}{\partial x}+\frac{\partial M}{\partial y}+\frac{\partial N}{\partial z}\right)u\right\}z-\left\{x\frac{\partial u}{\partial x}+y\frac{\partial u}{\partial y}+(m+2)u\right\}N=0.$$

But since N is homogeneous and of degree m,

$$x\frac{\partial N}{\partial x}+y\frac{\partial N}{\partial y}+z\frac{\partial N}{\partial z}=mN,$$

and therefore, eliminating $\frac{\partial N}{\partial z}$,

$$(Lz-Nx)\frac{\partial u}{\partial x}+(Mz-Ny)\frac{\partial u}{\partial y}+\left(z\frac{\partial L}{\partial x}+z\frac{\partial M}{\partial y}-x\frac{\partial N}{\partial x}-y\frac{\partial N}{\partial y}-2N\right)u=0.$$

Let $z=1$, then $u(x, y)$ satisfies the equation

$$(L-Nx)\frac{\partial u}{\partial x}+(M-Ny)\frac{\partial u}{\partial y}+\left\{\frac{\partial(L-Nx)}{\partial x}+\frac{\partial(M-Ny)}{\partial y}\right\}u=0.$$

But this is precisely the condition that $u(x, y)$ should be an integrating factor for the equation

$$-Ldy+Mdx+N(xdy-ydx)=0.$$

If, therefore, $\frac{1}{2}m(m+1)+1$ *particular solutions are known, an integrating factor can be obtained.*

To return to the Jacobi equation (§ 2·14),

$$(a_1+b_1x+c_1y)(xdy-ydx)-(a_2+b_2x+c_2y)dy+(a_3+b_3x+c_3y)dx=0.$$

In this case $m=1$. The equation will have a solution of the linear form

$$\alpha x+\beta y+\gamma=\text{const.}$$

$$A(f)\equiv(a_2z+b_2x+c_2y)\frac{\partial f}{\partial x}+(a_3z+b_3x+c_3y)\frac{\partial f}{\partial y}+(a_1z+b_1x+c_1y)\frac{\partial f}{\partial z}=\lambda f,$$

where λ is a constant and $f=\alpha x+\beta y+\gamma z$. This leads to three equations between $\alpha, \beta, \gamma, \lambda$, namely,

$$\gamma(a_1-\lambda)+\alpha a_2+\beta a_3=0,\ \gamma b_1+\alpha(b_2-\lambda)+\beta b_3=0,\ \gamma c_1+\alpha c_2+\beta(c_3-\lambda)=0,$$

whence

$$\begin{vmatrix} a_1-\lambda, & a_2, & a_3 \\ b_1, & b_2-\lambda, & b_3 \\ c_1, & c_2, & c_3-\lambda \end{vmatrix}=0.$$

It will be assumed that this equation has three distinct roots, $\lambda_1, \lambda_2, \lambda_3$, to which correspond three values of f, namely, U, V, W. Then

$$U^i V^j W^k = \text{const.}$$

will be the general solution, when z is made equal to unity, if

$$i+j+k=0,$$
$$\lambda_1 i+\lambda_2 j+\lambda_3 k=0.$$

It is sufficient to take $i=\lambda_2-\lambda_3$, $j=\lambda_3-\lambda_1$, $k=\lambda_1-\lambda_2$. The general solution is therefore

$$U^{\lambda_2-\lambda_3} V^{\lambda_3-\lambda_1} W^{\lambda_1-\lambda_2}=\text{const.}$$

2·3. Orthogonal Trajectories.—The equation

$$\Phi(x, y, c)=0,$$

in which c is a parameter, represents a family of plane curves. To this family of curves there is related a second family, namely, the family of *orthogonal trajectories* or curves which cut every curve of the given family at right angles. To return to the instance given in § 1·4, the first family of curves may be considered as the lines of force due to a given plane magnetic or electrostatic distribution. The family of orthogonal trajectories will then represent the equipotential lines in the given plane.

Let

$$F(x, y, p)=0$$

be the differential equation of the given family of curves; it determines the gradient p of any curve of the family which passes through the point (x, y). The gradient ϖ of the orthogonal curve through (x, y) is connected with p by the relation

$$p\varpi=-1,$$

and consequently the differential equation of the family of orthogonal trajectories is

$$F\left(x, y, -\frac{1}{p}\right)=0.$$

Since the differential equation of the given family is obtained by eliminating c between the two equations

$$\Phi=0, \quad \frac{\partial \Phi}{\partial x}+p\frac{\partial \Phi}{\partial y}=0,$$

the differential equation of the orthogonal trajectories arises through the elimination of c between the equations

$$\Phi=0, \quad p\frac{\partial \Phi}{\partial x}-\frac{\partial \Phi}{\partial y}=0.$$

Examples.—(i) The family of parabolas,

$$y^2=4cx,$$

where c is a parameter, are integral-curves of the differential equation

$$2xp=y.$$

The differential equation of the orthogonal trajectories is therefore

$$2x+py=0,$$

and the trajectories themselves are the curves

$$2x^2+y^2=c^2;$$

they compose a family of similar ellipses whose axes lie along the co-ordinate axes.

(ii) The family of confocal conics,

$$\frac{x^2}{a^2-\lambda}+\frac{y^2}{b^2-\lambda}=1,$$

where λ is the parameter, are integral-curves of the differential equation

$$(x+py)(y-px)+(a^2-b^2)p=0.$$

This equation is unaltered by the substitution of $-p^{-1}$ for p. The family is therefore self-orthogonal.

2·31. Oblique Trajectories.—An oblique trajectory is a curve which cuts the curves of a family at a given angle. Let the given angle be arc tan m. Then if p and ϖ are respectively the gradients of a curve of the given family and the trajectory at a point where they intersect,

$$\varpi=\frac{p-m}{1+mp}.$$

If the differential equation of the given family is

$$F(x, y, p)=0,$$

that of the family of oblique trajectories will be

$$F\left(x, y, \frac{p-m}{1+mp}\right)=0.$$

Example.—Consider the family of concentric circles,

$$x^2+y^2+c^2;$$

their differential equation is

$$x+yp=0.$$

The family of curves which cut the circles at the angle arc tan m is therefore

$$x+\frac{p-m}{1+mp}y=0$$

or

$$(mx+y)p+x-my=0.$$

This equation is homogeneous : its solution is

$$\log\sqrt{(x^2+y^2)}+m\arctan\frac{y}{x}=\text{const.}$$

In polar co-ordinates, the equation of the trajectories is

$$r=Ce^{-m\theta},$$

the curves are therefore equiangular spirals.

2·32. Conformal Representation of a Surface on a Plane.—Another important application of differential equations of the first order is to the conformal representation of an algebraic surface upon a plane. The real quadratic form

$$dS^2=Edu^2+2Fdudv+Gdv^2 \qquad (EG-F^2\neq 0)$$

represents an element of surface. Since it is essentially positive, its linear factors,

$$adu+bdv, \quad a'du+b'dv$$

are such that a and b are, in general, complex functions of u and v, and a' and b' are respectively the conjugate complex functions.

Let $\mu(u, v)$ be an integrating factor for $adu+bdv$, then the conjugate μ' will be an integrating factor for $a'du+b'dv$. If

$$\mu(adu+bdv)=dV, \; \mu'(a'du+b'dv)=dV'$$

then V and V' will be conjugate complexes, and

$$\mu\mu'dS^2=dVdV'.$$

Define x and y as new variables by the equations

$$V=x+iy, \quad V'=x-iy$$

and let

$$\lambda^2=\mu\mu',$$

then

$$dS^2=\lambda^2(dx^2+dy^2)$$
$$=\lambda^2 ds^2.$$

Thus the surface (u, v) is conformally represented on the plane (x, y).*

Example.—Consider the representation of the sphere

$$dS^2=a^2du^2+a^2 \sin^2u\, dv^2$$

on the plane.

$$dS^2=a^2(du+i \sin u\, dv)(du-i \sin u\, dv)$$
$$=a^2 \sin^2 u(\operatorname{cosec} u\, du+idv)(\operatorname{cosec} u\, du-idv).$$

Let

$$\operatorname{cosec} u\, du=dy, \quad dv=dx,$$

that is

$$y=\log \tan \tfrac{1}{2}u, \quad x=v.$$

Then

$$dS^2=4a^2 \operatorname{sech}^2 y\, (dx^2+dy^2).$$

This correspondence between the sphere and the plane is Mercator's projection † Meridians on the sphere are represented by lines parallel to the y-axis in the plane, and parallels of latitude by lines parallel to the x-axis. The whole sphere is represented by that strip of the plane which lies between $x=-\pi$ and $x=+\pi$. Any straight line in the plane represents a *loxodrome* on the sphere, that is a curve which cuts all the meridians at a constant angle.

2·4. Equations of the First Order but not of the First Degree.—An equation of the first order and of degree m may be written

$$\text{(A)} \qquad F\left(x, y, \frac{dy}{dx}\right)\equiv\left(\frac{dy}{dx}\right)^m+P_1\left(\frac{dy}{dx}\right)^{m-1}+\ .\ .\ .\ +P_{m-1}\frac{dy}{dx}+P_m=0,$$

where $P_1, \ldots, P_m$ are functions of x and y. Theoretically, the equation may be brought into the factorised form,

$$\left(\frac{dy}{dx}-p_1\right)\left(\frac{dy}{dx}-p_2\right)\ .\ .\ .\ \left(\frac{dy}{dx}-p_m\right)=0,$$

where $p_1, p_2, \ldots, p_m$ are functions of x and y.

Let

$$\phi_r(x, y, c_r)=0$$

be the general solution of the equation

$$\frac{dy}{dx}-p_r=0\,;$$

it will also be a solution of the given equation. Conversely if

$$\Phi(x, y, C)=0$$

is a solution of the given equation, it must satisfy one or other of the equations

$$\frac{dy}{dx}-p_r=0 \qquad (r=1, 2, \ldots, m).$$

* For the general theory of conformal representation, see Forsyth, *Theory of Functions*, Chap. XIX.

† Gerhard Kremer (*latine* Mercator) published his map of the world in 1538. The underlying mathematical principles were first explained by Edward Wright in 1594.

It follows that every solution of (A) will be included in the solution

$$\phi_1(x, y, c)\phi_2(x, y, c) \ . \ . \ . \ \phi_m(x, y, c)=0,$$

which is therefore the general solution. The one arbitrary constant c is sufficient for complete generality, for a particular solution is obtained explicitly by solving one or other of the equations

$$\phi_r(x, y, c)=0,$$

in which c has any numerical value.

Example.— $$\left(\frac{dy}{dx}\right)^2-(1+y^2)=0.$$

In the factorised form the equation is

$$\left\{\frac{dy}{dx}-\sqrt{(1+y^2)}\right\}\left\{\frac{dy}{dx}+\sqrt{(1+y^2)}\right\}=0,$$

the two factors give rise to solutions

$$y=\sinh(c\pm x)$$

respectively, where c is a constant. The general solution therefore is

$$\begin{aligned} y^2&=\tfrac{1}{4}(e^{c+x}-e^{-c-x})(e^{c-x}-e^{-c+x}) \\ &=\tfrac{1}{4}(e^{2c}+e^{-2c}-e^{2x}-e^{-2x}) \\ &=\tfrac{1}{2}(C-\cosh 2x), \end{aligned}$$

where $C=\cosh 2c$.

2·41. Geometrical Treatment.—The theory of the differential equation

$$F\left(x, y, \frac{dy}{dx}\right)=0$$

may also be approached from a geometrical point of view. Replace $\frac{dy}{dx}$ by z and regard z as the third rectangular co-ordinate in space. Then the equation

$$F(x, y, z)=0$$

represents a surface S.

Let

$$y=\phi(x)$$

be any solution of the differential equation, then the pair of equations

$$y=\phi(x),\ z=\phi'(x)$$

represents a space-curve Γ which, since

$$F\{x, \phi(x), \phi'(x)\}=0$$

identically, lies upon the surface S. There is not a solution of the differential equation corresponding to every curve which lies on S, but only to those curves at all points of which the differential relation

$$dy-zdx=0$$

is satisfied.

Let

$$x=x(t), \quad y=y(t), \quad z=z(t)$$

be the parametric representation of a curve Γ upon S for which the relation

$$dy-zdx=0$$

is satisfied. The projection of Γ upon the (x, y)-plane will be the curve C

$$x=x(t),\ y=y(t)$$

or

$$y=\phi(x).$$

Since at all points of the curve Γ the equation

$$F(x, y, z)=0$$

becomes

$$F\{x, \phi(x), \phi'(x)\}=0,$$

the curve C, or

$$y=\phi(x)$$

is an integral-curve of the equation

$$F(x, y, y')=0.$$

Let the parametric representation of the surface S be

$$x=f(u, v), \quad y=g(u, v), \quad z=h(u, v),$$

then the relation

$$dy-z dx=0$$

becomes

$$\frac{\partial g}{\partial u}du+\frac{\partial g}{\partial v}dv-h\left\{\frac{\partial f}{\partial u}du+\frac{\partial f}{\partial v}dv\right\}=0$$

or, say,

$$\frac{dv}{du}=k(u, v).$$

Any solution of this differential equation is a relation between u and v which defines a curve Γ on the surface S such that the projection of this curve on the (x, y)-plane is an integral-curve of the differential equation.

Consider, as an example, an equation which can be written in the form

$$y-g(x, p)=0.$$

The corresponding surface S is then representable parametrically as

$$x=x, \quad y=g(x, p), \quad z=p,$$

and the relation $dy-z dx=0$ becomes

$$p=\frac{\partial g}{\partial x}+\frac{\partial g}{\partial p}p'.$$

This is a differential equation of the form

$$\frac{dp}{dx}=k(x, p);$$

let its general solution be

$$l(x, p, c)=0.$$

Then the integral-curves are the projections on the (x, y)-plane of the intersection of the surface

$$y-g(x, z)=0$$

with the family of cylindrical surfaces

$$l(x, z, c)=0.$$

The general solution of the given equation is therefore obtained by eliminating p between the two equations

$$y=g(x, p), \quad l(x, p, c)=0.$$

2·42. Equations in which x or y does not explicitly occur.—When an equation of either of the forms

$$F(x, p)=0, \quad F(y, p)=0$$

can be solved for p, the equation can be integrated by quadratures. On the other hand it may occur that the equation is more readily soluble for x (or y as the case may be) in terms of p. Let

$$x=f(p)$$

be the solution, then, on differentiating with respect to y,

$$\frac{1}{p}=f'(p)\frac{dp}{dy},$$

whence

$$y=c+\int pf'(p)dp$$
$$=c+g(p),$$

say. Then the equations

$$x=f(p),\quad y=c+g(p)$$

may be regarded as a parametric representation of the solution, which is obtained explicitly by eliminating p between the two equations.

If the equation does not involve x, it is solved for y and then differentiated with respect to x. The solution is then obtained in the parametric form

$$y=f(p),\ x=c+g(p),$$

where

$$g(p)=\int p^{-1}f'(p)dp.$$

More generally, it may be possible to express the equation

$$F(x,\ p)=0$$

parametrically in the form

$$x=u(t),\quad p=v(t),$$

then, on differentiating the former with respect to t,

$$\frac{1}{p}\cdot\frac{dy}{dt}=u'(t),$$

whence

$$y=c+\int v(t)u'(t)dt.$$

The solution is then obtained by eliminating t between the expressions for x and y. The equation

$$F(y,\ p)=0,$$

if expressible in the form

$$y=u(t),\quad p=v(t),$$

is solved by eliminating t between

$$y=u(t)\quad\text{and}\quad x=c+\int\frac{u'(t)}{v(t)}dt.$$

Example.—Consider the equation

$$p^3-p^2+y^2=0.$$

It may be represented parametrically as

$$y=t-t^3,\ p=1-t^2.$$

Differentiate the first equation with respect to t, then

$$p\frac{dx}{dt}=1-3t^2,$$

whence

$$x=c+\int\frac{1-3t^2}{1-t^2}dt=c+3t+\log\frac{t-1}{t+1}.$$

Thus x and y are expressed in terms of the parameter t.

2·43. Equations homogeneous in x and y.—An equation which is homogeneous and of degree m in x and y may be written

$$x^mF\left(\frac{y}{x},\ p\right)=0.$$

If it is soluble for p, equations of the type

$$p=f\left(\frac{y}{x}\right)$$

already considered (§ 2·12) will arise. This case, therefore, presents no new features of interest. Consider, however, the case in which the equation is soluble for $\frac{y}{x}$; thus

$$\frac{y}{x}=f(p)$$

or

$$y=xf(p).$$

Differentiate this equation with respect to x, then

$$p=f(p)+xf'(p)\frac{dp}{dx}.$$

Let p be taken as dependent variable, then in this equation the variables are separable, and it has the solution

$$\log cx=\int\frac{f'(p)dp}{p-f(p)},$$

or, say,

$$cx=g(p).$$

The simultaneous equations

$$y=xf(p), \quad cx=g(p)$$

furnish the general solution of the equation.

Example.—
$$y=yp^2+2px.$$

Solve for x, thus

$$2x=y\left(\frac{1}{p}-p\right);$$

differentiate with respect to y, then

$$\frac{2}{p}=\frac{1}{p}-p-y\left(\frac{1}{p^2}+1\right)\frac{dp}{dy}$$

or

$$\frac{dp}{dy}=-\frac{p}{y},$$

whence

$$py=c.$$

Eliminating p from the original equation gives the required solution

$$y^2=2cx+c^2.$$

2·44. Equations linear in x and y.—A general type of equation whose solution can be obtained in a parametric form by differentiation is the following: [*]

$$y=x\phi(p)+\psi(p).$$

The derived equation is

$$p=\phi(p)+\{x\phi'(p)+\psi'(p)\}\frac{dp}{dx};$$

[*] The equations appear to have been integrated by John Bernoulli before the year 1694. Its singular solutions were studied by d'Alembert, *Hist. Acad. Berlin* 4 (1748), p. 275.

if x is regarded as dependent variable, and p as independent variable the equation may, when $p-\phi(p)\neq 0$, be written

$$\frac{dx}{dp}-\frac{\phi'(p)}{p-\phi(p)}x=\frac{\psi'(p)}{p-\phi(p)}$$

and is then a linear equation in the ordinary sense. Its solution in general involves two quadratures; let it be

$$x=cf(p)+g(p),$$

then x may be eliminated from the original equation, giving an expression for y in the form

$$y=cf_1(p)+g_1(p).$$

The general solution is thus expressed parametrically in terms of p.

Consider now those particular values of p, say $p_1, p_2, \ldots$, for which

$$p-\phi(p)=0;$$

for those values of p,

$$\frac{dp}{dx}=0.$$

Thus there arises a certain set of isolated integral curves such as

$$y=x\phi(p_1)+\psi(p_1).$$

They are straight lines such that if an integral curve of the general family meets one of them, it will have, in general, an inflexion at the common point. The straight lines furnish an example of singular solutions, that is of solutions of the equation which are not included in the general family of integral curves, and not obtainable from the general solution by attributing a special value to the constant of integration.

Example.— $$y=2px-p^2.$$

The derived equation is

$$p=2p+2(x-p)\frac{dp}{dx},$$

whence, if $p\neq 0$,

$$\frac{dx}{dp}+\frac{2x}{p}=2.$$

The solution of this linear equation is

$$x=\frac{c}{p^2}+\frac{2}{3}p,$$

which, combined with the original equation, gives the required solution.

On the other hand, when $p=0$, there is a solution

$$y=0.$$

2·45. The Clairaut Equation.—The Clairaut equation,*

$$y=px+\psi(p),$$

is not included in the class of equations studied in the preceding section because, in the notation of that section,

$$\phi(p)=p$$

identically, and therefore the method adopted fails.

The derived equation is

$$p=p+\{x+\psi'(p)\}\frac{dp}{dx};$$

* *Hist. Acad. Paris* (1734), p. 209.

it can be satisfied either by $p=c$, a constant, or by

$$x+\psi'(p)=0.$$

The first possibility, $p=c$, leads to the general solution

$$y=cx+\psi(c).$$

The second possibility leads to a particular solution obtained by eliminating p between the two equations

$$y=px+\psi(p), \quad x+\psi'(p)=0.$$

It contains no arbitrary constant, and is not a particular case of the general solution; it is therefore a singular solution.

Now the envelope of the family of straight lines

$$y=cx+\psi(c)$$

is obtained by eliminating c between this equation and

$$0=x+\psi'(c),$$

and is identical with the curve furnished by the singular solution. In the case of the Clairaut equation, therefore, the singular solution represents the envelope of the family of integral-curves.

Conversely, the family of tangents to a curve

$$y=f(x)$$

satisfies an equation of the Clairaut form, for if

$$y=\alpha x+\beta$$

is a tangent, then

$$\alpha x+\beta=f(x), \quad \alpha=f'(x).$$

The elimination of x between these equations gives rise to a relation

$$\beta=\psi(\alpha),$$

and since, on the tangent, $\alpha=p$, the tangents satisfy the equation

$$y=px+\psi(p).$$

Example.— $$y=px+1/p.$$

Differentiating,

$$p=p+(x-1/p^2)\frac{dp}{dx},$$

whence either $p=c$, giving the general solution

$$y=cx+1/c,$$

or else

$$p^2=1/x.$$

The singular solution is found by eliminating p between

$$p^2=1/x \quad \text{and} \quad y=px+1/p$$

and is

$$y^2=4x.$$

2·5. The Principle of Duality.—There exists a certain transformation, due to Legendre, by which a dual relationship can be set up between one equation of the first order and another of the same order. Let X and Y be new variables defined by the relations

$$X=p, \quad Y=xp-y,$$

and let

$$P=\frac{dY}{dX}.$$

Now, assuming that $\dfrac{dp}{dx} \neq 0$,

$$dX = dp, \quad dY = xdp + pdx - dy$$
$$= xdp,$$

and therefore

$$P = x.$$

Also

$$y = xp - Y$$
$$= XP - Y.$$

Thus the transformation

$$X = p, \quad Y = xp - y$$

is equivalent to

$$x = P, \quad y = XP - Y.$$

They are therefore reciprocally related to one another.*

By means of this substitution, either of the equations

$$F(x, y, p) = 0, \quad F(P, XP - Y, X) = 0$$

may be transformed into the other, and in this sense a dual relationship exists between them. When one of the equations is integrable, the other may be integrated by purely algebraical processes.

For instance, let

$$\phi(X, Y) = 0$$

be a solution of the second equation, then on differentiating with respect to X,

$$\frac{\partial \phi}{\partial X} + \frac{\partial \phi}{\partial Y} P = 0.$$

Now X, Y, P may be eliminated between these two equations and

$$x = P, \quad y = XP - Y,$$

thus giving a solution of the equation

$$F(x, y, p) = 0.$$

In particular, an equation of the form

$$\Phi(xp - y) = x\Psi(p)$$

would become

$$\Phi(Y) = P\Psi(X).$$

The variables X and Y are now separable, and the equation is integrable by quadratures.

Example.— $(y - px)x = y.$

The transformed equation is

$$P = \frac{Y}{Y + X};$$

it is homogeneous, and has the solution,

$$\log Y - \frac{X}{Y} = \text{const.}$$

Differentiate with respect to X, then

$$\frac{P}{Y} - \frac{Y - XP}{Y^2} = 0,$$

* If (x, y) and (X, Y) are regarded as points in the plane of the variables u, v, the locus of (x, y) is the polar reciprocal of the locus of (X, Y) with respect to the parabola $u^2 = 2v$, and conversely.

whence

$$Y=\frac{Y-XP}{P}=-\frac{y}{x},$$

and consequently

$$\frac{X}{Y}=\frac{1}{P}-1=\frac{1}{x}-1$$

Hence the solution of the original equation is

$$\log\left(-\frac{y}{x}\right)-\frac{1}{x}=\text{const.}$$

or

$$y=cxe^{1/x}.$$

NOTE.—In the case of the Clairaut equation the condition that $\frac{dp}{dx}\neq 0$ is violated for the general solution ; this method therefore leads only to the singular solution.

2·6. Equations of Higher Order than the First.—The simplest of all differential equations of general order n is the following:

$$\frac{d^n y}{dx^n}=f(x).$$

Its integration is simply the process of n-ple integration and may be carried out in successive stages as follows. Let x_0 be a constant, chosen at random, then

$$\frac{d^{n-1}y}{dx^{n-1}}=\int_{x_0}^{x} f(x)dx+C_0,$$

$$\frac{d^{n-2}y}{dx^{n-2}}=\int_{x_0}^{x} dx\int_{x_0}^{x} f(x)dx+C_0(x-x_0)+C_1,$$

$$\cdot\quad\cdot\quad\cdot\quad\cdot\quad\cdot\quad\cdot\quad\cdot\quad\cdot\quad\cdot\quad\cdot$$

$$y=\int_{x_0}^{x} dx\int_{x_0}^{x} dx\ \ldots\int_{x_0}^{x} f(x)dx+C_0\frac{(x-x_0)^{n-1}}{(n-1)!}+\ldots+C_{n-1},$$

where $C_0, C_1, \ldots, C_{n-1}$ are arbitrary constants.

The multiple integral may, however, be replaced by a single integral. Let

$$Y=\frac{1}{(n-1)!}\int_{x_0}^{x}(x-t)^{n-1}f(t)dt,$$

then

$$\frac{dY}{dx}=\frac{1}{(n-2)!}\int_{x_0}^{x}(x-t)^{n-2}f(t)dt,$$

$$\cdot\quad\cdot\quad\cdot\quad\cdot\quad\cdot\quad\cdot$$

$$\frac{d^{n-1}Y}{dx^{n-1}}=\int_{x_0}^{x} f(t)dt,$$

whence, finally,

$$\frac{d^n Y}{dx^n}=f(x).$$

Y is therefore a solution of the equation which, together with its first $(n-1)$ derivatives vanishes when $x=x_0$. It is therefore identified with the multiple integral. The general solution of the equation is therefore

$$y=\frac{1}{(n-1)!}\int_{x_0}^{x}(x-t)^{n-1}f(t)dt+C_0\frac{(x-x_0)^{n-1}}{(n-1)!}+\ldots+C_{n-1}.$$

Apart from this simple case, and the case of linear equations with constant coefficients, which will be dealt with in Chapter VI., there are but few equations of order higher than the first which yield to an elementary treatment. In a number of very special cases, however, the order of an equation can be lowered by means of a suitable transformation of the variables, combined with one or more quadratures. The main cases of this kind which can arise will be dealt with in the three following sections.

2·61. Equations which do not explicitly involve the Dependent Variable.—Consider the equation

$$F\left(x, \frac{d^k y}{dx^k}, \frac{d^{k+1}y}{dx^{k+1}}, \ldots, \frac{d^n y}{dx^n}\right)=0,$$

in which y and its first $k-1$ derivatives do not appear. The transformation

$$v=\frac{d^k y}{dx^k}$$

reduces the equation to an equation in v of order $n-k$. If this equation can be integrated and its solution is $v=v(x)$, it only remains to integrate the equation

$$\frac{d^k y}{dx^k}=v(x),$$

which is of the type dealt with in the preceding section.

More generally, however, the reduced equation has a solution of the form

$$\phi(x, v)=0,$$

which is not readily soluble for v. For the method to be practicable it is necessary to express x and v in terms of a parameter t, thus

$$y^{(k)}=v(t), \quad x=x(t).$$

Then

$$dy^{(k-1)}=v(t)dx=v(t)x'(t)dt,$$

which, on integration, gives $y^{(k-1)}$. The process is repeated, k times in all, until the explicit solution is reached.

An important particular case is that of equations of the form

$$\frac{d^n y}{dx^n}=f\left(\frac{d^{n-1}y}{dx^{n-1}}\right);$$

such equations are integrable by quadratures.

2·62. Equations which do not explicitly involve the Independent Variable.—When an equation has the form

$$F\left(y, \frac{dy}{dx}, \frac{d^2y}{dx^2}, \ldots, \frac{d^n y}{dx^n}\right)=0,$$

its order may be reduced to $n-1$ by a change of variables. Let y be taken as a new independent variable, and p as the dependent variable. The formulæ by means of which this transformation is effected are

$$\frac{dy}{dx}=p, \quad \frac{d^2y}{dx^2}=p\frac{dp}{dy}, \quad \frac{d^3y}{dx^3}=p\frac{d}{dy}\left(p\frac{dp}{dy}\right), \ldots.$$

The given equation is thus reduced to one of the form

$$\Phi\left(y, p, \frac{dp}{dy}, \ldots, \frac{d^{n-1}p}{dy^{n-1}}\right)=0.$$

Let it be supposed that this equation can be integrated, and that its solution is expressible in the parametric form

$$y=f(t),\ p=g(t),$$

where f and g are functions of the auxiliary variable t, and depend also on $n-1$ constants of integration. Then x is obtained, in terms of t, by a quadrature, thus:

$$x=\int\frac{dy}{p}$$

$$=\int\frac{f'(t)dt}{g(t)},$$

In particular, an equation of the second order, which does not explicitly involve x, namely

$$F\left(y,\ \frac{dy}{dx},\ \frac{d^2y}{dx^2}\right)=0,$$

is transformed into the equation

$$F\left(y,\ p,\ p\frac{dp}{dy}\right)=0,$$

which is of the first order.

An equation of the form

$$\frac{d^ny}{dx^n}=f\left(\frac{d^{n-2}y}{dx^{n-2}}\right)$$

is reduced, by the substitution

$$\frac{d^{n-2}y}{dx^{n-2}}=v,$$

to

$$\frac{d^2v}{dx^2}=f(v).$$

If $\dfrac{dv}{dx}=p$, this last equation becomes

$$p\frac{dp}{dv}=f(v),$$

whence

$$p^2=c+\int f(v)dv,$$

and therefore

$$x=\int\{c+\int f(v)dv\}^{-\frac{1}{2}}dv.$$

In order that y may be obtained, v must be expressed in terms of x; the solution is then completed by $n-2$ quadratures.

2·63. Equations exhibiting a Homogeneity of Form.—Two classes of equations will be discussed, the first class being that of equations which are homogeneous in $y, y', y'', \ldots, y^{(n)}$, and which may also involve x explicitly. An equation of this class may, if m is the degree of homogeneity, be written

$$y^mF\left(x,\ \frac{y'}{y},\ \frac{y''}{y},\ \ldots,\ \frac{y^{(m)}}{y}\right)=0.$$

Let u be a new dependent variable, defined by the relation

$$y=e^{\int u\,dx},$$

then

$$y'=ue^{\int u\,dx},\quad y''=(u'+u^2)e^{\int u\,dx},\ \ldots,$$

and in general

$$y^{(n)}=U_n e^{\int u\,dx},$$

where U_n is a polynomial in $u, u', \ldots, u^{(n-1)}$. The change of dependent variable from y to u therefore reduces the order of the equation from n to $n-1$.

The second class includes those equations which are homogeneous in $y, xy', x^2y'', \ldots, x^ny^{(n)}$ and do not otherwise involve x. Let

$$F(y, xy', x^2y'', \ldots, x^ny^{(n)})=0$$

be the typical equation. Change the independent variable by the substitution

$$x=e^t,$$

then

$$x\frac{dy}{dx}=\frac{dy}{dt},\quad x^2\frac{d^2y}{dx^2}=\frac{d^2y}{dt^2}-\frac{dy}{dt},\ \ldots$$

and, in general,

$$x^r\frac{d^ry}{dx^r}=\frac{d}{dt}\left(\frac{d}{dt}-1\right)\ldots\left(\frac{d}{dt}-r+1\right)y.$$

Thus the transformed equation is of the form

$$\Phi\left(y, \frac{dy}{dt}, \frac{d^2y}{dt^2}, \ldots, \frac{d^ny}{dt^n}\right)=0$$

and does not explicitly involve x. It thus comes under the heading of § 2·62.

An equation which comes under the last class, but which can be integrated by a simpler method is the following: *

$$F(y'',\quad y'-xy'',\quad y-xy'+\tfrac{1}{2}x^2y'')=0.$$

The derived equation is simply

$$y'''.(F_1-xF_2+\tfrac{1}{2}x^2F_3)=0,$$

where F_1, F_2, F_3 are the partial derivatives of F with respect to its first, second, and third arguments respectively. It is satisfied by $y'''=0$, or

$$y=A+Bx+\tfrac{1}{2}Cx^2,$$

where A, B, C are arbitrary constants. This will be the general solution of the original equation provided that

$$F(C, B, A)=0.$$

2·7. Simultaneous Systems in Three Variables.—Before the general theory of the integration of simultaneous systems of differential equations is attacked, it will be convenient to dispose of a simple case in which the equations are integrable by the methods which were detailed in the earlier sections of the chapter.

Consider the system

$$\frac{dx}{\xi}=\frac{dy}{\eta}=\frac{dz}{\zeta};$$

ξ, η and ζ are, in general, functions of x, y and z. A very special, but

* Dixon, *Phil. Trans. R. S.* (A) 186 (1894), p. 563. The generalisation to any order is obvious. See also Raffy, *Bull. Soc. Math. France*, 25 (1897), p. 71.

important case is that in which ξ and η are independent of z. In this case the equation

$$\frac{dx}{\xi}=\frac{dy}{\eta}$$

involves only x and y; it will be supposed that this equation can be integrated and that its solution is

$$\Phi(x, y, \alpha)=0,$$

where α is the constant of integration. Let this equation be solved for y, thus

$$y=\phi(x, \alpha),$$

and let ξ_1 and ζ_1 be what ξ and ζ become when y is replaced therein by $\phi(x, \alpha)$. Then the equation

$$\frac{dx}{\xi_1}=\frac{dz}{\zeta_1}$$

does not involve y. Its solution will be of the form

$$\Theta(x, z, \alpha, \beta)=0,$$

where β is the constant of integration. Now let α be eliminated between the two solutions

$$\Phi(x, y, \alpha)=0, \ \Theta(x, z, \alpha, \beta)=0;$$

the solutions then take the form

$$\Phi(x, y, \alpha)=0, \ \Psi(x, y, z, \beta)=0.$$

2·701. Integration of a Simultaneous Linear System with Constant Coefficients.—The system

$$\frac{dx}{\xi}=\frac{dy}{\eta}=\frac{dz}{\zeta},$$

where

$$\xi=a_1x+b_1y+c_1z+d_1,$$
$$\eta=a_2x+b_2y+c_2z+d_2,$$
$$\zeta=a_3x+b_3y+c_3z+d_3,$$

is not of the form dealt with in the preceding section. It can, however, be dealt with in a similar manner after a linear transformation of the variables has been made. To simplify the working a new variable t is introduced such that

$$\frac{dx}{\xi}=\frac{dy}{\eta}=\frac{dz}{\zeta}=\frac{dt}{t},$$

then, whatever constants l, m, n may be,

$$\frac{dt}{t}=\frac{ldx+mdy+ndz}{l\xi+m\eta+n\zeta}.$$

Let l, m, n be so chosen that

$$la_1+ma_2+na_3=l\rho,$$
$$lb_1+mb_2+nb_3=m\rho,$$
$$lc_1+mc_2+nc_3=n\rho,$$

then

$$\frac{dt}{t}=\frac{d(lx+my+nz)}{\rho(lx+my+nz+r)},$$

where $r\rho=ld_1+md_2+nd_3$. This choice of l, m, n is possible if ρ is a root of the equation

$$\begin{vmatrix} a_1-\rho, & a_2, & a_3 \\ b_1, & b_2-\rho, & b_3 \\ c_1, & c_2, & c_3-\rho \end{vmatrix}=0.$$

Let the roots of this equation, supposed distinct be $\frac{1}{\lambda_1}, \frac{1}{\lambda_2}, \frac{1}{\lambda_3}$, and let the corresponding values of l, m, n, r be

$$l_i, m_i, n_i, r_i \qquad (i=1, 2, 3),$$

then

$$\frac{dt}{t}=\frac{\lambda_i d(l_i x+m_i y+n_i z)}{l_i x+m_i y+n_i z+r_i},$$

whence

$$t=C_i(l_i x+m_i y+n_i z+r_i)^{\lambda_i}.$$

The solution of the system is therefore

$$C_1(l_1x+m_1y+n_1z+r_1)^{\lambda_1}=C_2(l_2x+m_2y+n_2z+r_2)^{\lambda_2}=C_3(l_3x+m_3y+n_3z+r_3)^{\lambda_3}$$

and contains three constants of integration, C_1, C_2, C_3, of which two are arbitrary.

2·71. The Equivalent Partial Differential Equation.—Let x and y be regarded as independent variables, and z as a dependent variable. Let p and q be the partial derivatives of z with respect to x and y respectively, then

$$\xi p+\eta q=\zeta$$

is a linear partial differential equation of the first order and is known as the Lagrange linear equation. If

$$z=f(x, y)$$

is a solution of the equation, then

$$\xi\frac{\partial f}{\partial x}+\eta\frac{\partial f}{\partial y}=\zeta,$$

for all values of x, y. This solution represents a surface, known as an *integral-surface* of the partial differential equation. Since the direction cosines of the normal to a surface $z=f(x, y)$ are proportional to

$$\frac{\partial f}{\partial x}, \frac{\partial f}{\partial y}, -1,$$

the differential equation expresses a distinguishing property of the tangent plane to the integral-surface.

Now consider the system of simultaneous ordinary differential equations

$$\frac{dx}{\xi}=\frac{dy}{\eta}=\frac{dz}{\zeta},$$

and let its solutions be solved for the constants of integration, thus

$$u(x, y, z)=\alpha, \quad v(x, y, z)=\beta.$$

These solutions represent a two-parameter family of curves in space, which are known as the *characteristics* of the system. If ξ, η, ζ exist and are one-valued at a point (x_0, y_0, z_0), and at least one of them is not zero at (x_0, y_0, z_0), one and only one characteristic passes through that point.

It will now be shown that the characteristics of the simultaneous differential system bear an intimate relationship to the integral-surface of the partial differential equation. In the first place it will be proved that, *if an integral-surface passes through* (x_0, y_0, z_0), *it contains the characteristic through that point.* Let the integral surface through (x_0, y_0, z_0) be

$$z=f(x, y)$$

and, supposing that ξ does not vanish at (x_0, y_0, z_0), consider the differential equation

$$\frac{dy}{dx}=\frac{\eta}{\xi},$$

in which z has been replaced by $f(x, y)$. The equation defines y as a function of x and is therefore the differential equation of a family of cylinders whose generators are parallel to the axis of z. The cylinder through $(x_0, y_0, 0)$ intersects the integral-surface in a curve through (x_0, y_0, z_0). Along this curve

$$\frac{dx}{\xi}=\frac{dy}{\eta}=\frac{pdx+qdy}{p\xi+q\eta}=\frac{dz}{\zeta}.$$

The curve so defined is therefore a characteristic, and the theorem is proved. An immediate consequence of this theorem is the fact that every integral surface is a locus of characteristics. In particular if any non-characteristic curve in space is drawn, the characteristics which pass through the points of this curve build up an integral-surface.

In the second place, the converse of this theorem will be shown to be true, namely, that *in general every surface which arises as a locus of characteristic curves is an integral-surface of the partial differential equation.** The tangent line to the characteristic at any arbitrary point (x_0, y_0, z_0) is

$$\frac{x-x_0}{\xi_0}=\frac{y-y_0}{\eta_0}=\frac{z-z_0}{\zeta_0},$$

where ξ_0, η_0, ζ_0 are the values of ξ, η, ζ at (x_0, y_0, z_0). The equation of the tangent plane at (x_0, y_0, z_0) to the surface which envelopes the characteristics will be

$$(x-x_0)p_0+(y-y_0)q_0=z-z_0,$$

where p_0 and q_0 are respectively the values of $\dfrac{\partial z}{\partial x}, \dfrac{\partial z}{\partial y}$ on the surface at (x_0, y_0, z_0). Since the characteristic lies in the surface, the tangent line lies in the tangent plane, and therefore

$$\xi_0 p_0+\eta_0 q_0=\zeta_0.$$

But (x_0, y_0, z_0) is any point on the surface ; the latter is therefore an integral-surface of the partial differential equation

$$\xi p+\eta q=\zeta.$$

2·72. Formation of the Integral-Surface.—The aggregate of characteristics form a two-parameter family or congruence of curves. Just as a plane curve is formed by selecting, according to a definite law, a one-fold infinity of the two-fold infinity of points in a plane, so an integral-surface is formed by selecting a one-fold infinity of curves of the congruence. Let

$$u(x, y, z)=\alpha, \quad v(x, y, z)=\beta$$

be the aggregate of characteristics from which a one-fold infinity is chosen by setting up a relationship between α and β, say

$$\Omega(\alpha, \beta)=0.$$

The equation to the integral-surface is therefore

$$\Omega(u, v)=0,$$

and this equation, in which the function Ω is arbitrary, is the general solution of the partial differential equation.

In the theory of ordinary differential equations of the first order, it is often required to find that integral-curve which passes through a given point of the plane. The corresponding problem in the case of partial

* The exceptional case arises when the surface has a tangent plane parallel to the z-axis, for then p and q become infinite and the proof fails.

differential equation is to find that integral-surface which passes through a given (non-characteristic) *base-curve* in space. This problem, in its general form, is known as Cauchy's problem.

Let

$$\phi(x, y, z)=0, \quad \psi(x, y, z)=0$$

represent the base-curve, and let

$$u(x, y, z)=\alpha, \quad v(x, y, z)=\beta$$

be the characteristics. If, between these four equations, x, y, z are eliminated, there remains a relation between α and β which expresses the condition that the characteristics and the base-curve have points in common. Let this relation be

$$\Phi(\alpha, \beta)=0,$$

then

$$\Phi(u, v)=0$$

is the required integral-surface.

Example.—Consider the partial differential equation

$$(cy-bz)\frac{\partial z}{\partial x}+(az-cx)\frac{\partial z}{\partial y}=bx-ay.$$

The subsidiary differential system is

$$\frac{dx}{cy-bz}=\frac{dy}{az-cx}=\frac{dz}{bx-ay}.$$

This system is equivalent to

$$\begin{cases} adx+bdy+cdz=0, \\ xdx+ydy+zdz=0, \end{cases}$$

and therefore the equations of the characteristics are

$$\begin{cases} ax+by+cz=\alpha, \\ x^2+y^2+z^2=\beta, \end{cases}$$

where α and β are arbitrary constants. The characteristics are the intersections of all spheres whose centre is at the origin with all planes which are parallel to the straight line

(l) $$\frac{x}{a}=\frac{y}{b}=\frac{z}{c},$$

that is to say, they are the aggregate of circles whose planes are perpendicular to, and whose centres lie on, this line.

The integral-surfaces have the equation

$$x^2+y^2+z^2=f(ax+by+cz),$$

and are surfaces of revolution which have the line (l) as axes of symmetry.

Now consider that particular integral-surface which contains the y-axis; it is built up of those characteristic curves which pass through the y-axis. The characteristics are those for which α and β are such that the equations

$$ax+by+cz=\alpha, \quad x^2+y^2+z^2=\beta, \quad x=0, \; z=0$$

are consistent. The condition that they are consistent is obtained by eliminating y from

$$by=\alpha, \quad y^2=\beta$$

and therefore is

$$b^2\beta=\alpha^2$$

The required integral-surface is

$$b^2(x^2+y^2+z^2)=(ax+by+cz)^2$$

or

$$(a^2-b^2)x^2+(c^2-b^2)z^2+2abxy+2bcyz+2cazx=0.$$

2·73. The Homogeneous Linear Partial Differential Equation.—When ζ is identically zero, the equation has the so-called homogeneous form

$$\xi p+\eta q=0.$$

The equations of the characteristics then become

$$\frac{dx}{\xi}=\frac{dy}{\eta}=\frac{dz}{0}.$$

The last equation gives at once

$$z=\alpha,$$

and therefore the characteristics are plane curves whose planes are perpendicular to the z-axis.

The most important case is that in which ξ and η are independent of z; the equation of the characteristics is then

$$z=\alpha, \quad u(x, y)=\beta,$$

and the equation of the integral-surface may be written in the form

$$z=f(u).$$

Now consider the equation

$$\xi\frac{\partial f}{\partial x}+\eta\frac{\partial f}{\partial y}+\zeta\frac{\partial f}{\partial z}=0,$$

where ξ, η, ζ are functions of x, y, z and do not involve f. If

$$f(x, y, z)=c,$$

where c is a constant, is a solution of the partial differential equation, then

$$df\equiv\frac{\partial f}{\partial x}dx+\frac{\partial f}{\partial y}dy+\frac{\partial f}{\partial z}dz=0,$$

and therefore $f(x, y, z)=c$ is a solution of the simultaneous system

$$\frac{dx}{\xi}=\frac{dy}{\eta}=\frac{dz}{\zeta}.$$

The converse is also true, for if

$$u(x, y, z)=\alpha$$

is any solution of the simultaneous system, then

$$du\equiv\frac{\partial u}{\partial x}dx+\frac{\partial u}{\partial y}dy+\frac{\partial u}{\partial z}dz=0.$$

and therefore

$$\xi\frac{\partial u}{\partial x}+\eta\frac{\partial u}{\partial y}+\zeta\frac{\partial u}{\partial z}=0.$$

Let

$$v(x, y, z)=\beta$$

be a second, and distinct, solution of the simultaneous system; it will also be a solution of the partial differential equation, so that

$$\xi\frac{\partial v}{\partial x}+\eta\frac{\partial v}{\partial y}+\zeta\frac{\partial v}{\partial z}=0.$$

If any other solution

$$w(x, y, z)=\gamma$$

exists, then

$$\xi\frac{\partial w}{\partial x}+\eta\frac{\partial w}{\partial y}+\zeta\frac{\partial w}{\partial z}=0,$$

and, eliminating ξ, η, ζ,

$$\frac{\partial(u, v, w)}{\partial(x, y, z)} \equiv \begin{vmatrix} \frac{\partial u}{\partial x}, & \frac{\partial u}{\partial y}, & \frac{\partial u}{\partial z} \\ \frac{\partial v}{\partial x}, & \frac{\partial v}{\partial y}, & \frac{\partial v}{\partial z} \\ \frac{\partial w}{\partial x}, & \frac{\partial w}{\partial y}, & \frac{\partial w}{\partial z} \end{vmatrix} = 0$$

identically. Consequently w is a function of u and v,* and therefore the partial differential equation admits of two and only two distinct solutions.

From the three equations

$$u(x, y, z) = \alpha, \quad v(x, y, z) = \beta, \quad w(x, y, z) = \gamma,$$

two of the variables, say x and y, may be eliminated, and the eliminant can be expressed in the form

$$w = \phi(u, v, z).$$

Now

$$0 = \frac{\partial(u, v, w)}{\partial(x, y, z)} = \frac{\partial(u, v, \phi)}{\partial(u, v, z)} \cdot \frac{\partial(u, v, z)}{\partial(x, y, z)}.$$

The first determinant on the right is simply $\frac{\partial \phi}{\partial z}$, the second is $\frac{\partial(u, v)}{\partial(x, y)}$. The second of these is not zero, since u and v are supposed to be independent. Consequently

$$\frac{\partial \phi}{\partial z} = 0,$$

that is to say, ϕ is explicitly independent of z, or in other words w is a function of u and v alone.

The general solution of the partial differential equation

$$\xi \frac{\partial f}{\partial x} + \eta \frac{\partial f}{\partial y} + \zeta \frac{\partial f}{\partial z} = 0$$

is therefore

$$\Omega(u, v) = \text{const.},$$

where Ω is an arbitrary function of its arguments, and

$$u = \alpha, \; v = \beta$$

are any two independent solutions of the subsidiary system

$$\frac{dx}{\xi} = \frac{dy}{\eta} = \frac{dz}{\zeta}.$$

The extension to the case of n variables is obvious. An exceptional case occurs when ξ, η, ζ have a common factor; the result of equating this factor to zero provides a special solution of the partial differential equation which may or may not be included in the general solution.

As an example consider the equation

$$x^2 \frac{\partial f}{\partial x} + xy \frac{\partial f}{\partial y} + z^2 \frac{\partial f}{\partial z} = 0.$$

The subsidiary system

$$\frac{dx}{x^2} = \frac{dy}{xy} = \frac{dz}{z^2}$$

* § 1·22.

has the two distinct solutions

$$\frac{y}{x}=\alpha, \quad \frac{z-x}{xz}=\beta.$$

The general solution is

$$\Omega\left(\frac{y}{x}, \frac{z-x}{xz}\right)=\text{const.}$$

2·8. Total Differential Equations.—An algebraic equation in three variables, of the form

$$\phi(x, y, z)=c,$$

where c is a constant, leads to the total differential equation

$$\frac{\partial\phi}{\partial x}dx+\frac{\partial\phi}{\partial y}dy+\frac{\partial\phi}{\partial z}dz=0.$$

If $\dfrac{\partial\phi}{\partial x}, \dfrac{\partial\phi}{\partial y}, \dfrac{\partial\phi}{\partial z}$ have a common factor μ, and if

$$\frac{\partial\phi}{\partial x}=\mu P, \quad \frac{\partial\phi}{\partial y}=\mu Q, \quad \frac{\partial\phi}{\partial z}=\mu R,$$

the total differential equation may be written in the form

$$Pdx+Qdy+Rdz=0.$$

On the other hand, if P, Q, and R are arbitrarily-assigned functions of x, y, z, the total differential equation does not necessarily correspond to a primitive of the form

$$\phi(x, y, z)=c.$$

For if such a primitive exists, P, Q, R are respectively proportional to the three partial differential coefficients of a function $\phi(x, y, z)$, which is not in general true. The problem therefore arises, to find a necessary and sufficient condition that a given total differential equation should be integrable, that is to say, derived from a primitive of the form considered.

It is first of all necessary that functions $\phi(x, y, z)$ and $\mu(x, y, z)$ exist such that the conditions

$$\frac{\partial\phi}{\partial x}=\mu P, \quad \frac{\partial\phi}{\partial y}=\mu Q, \quad \frac{\partial\phi}{\partial z}=\mu R$$

are satisfied. Then *

$$\frac{\partial}{\partial y}(\mu P)=\frac{\partial^2\phi}{\partial y\partial x}$$

$$=\frac{\partial^2\phi}{\partial x\partial y}=\frac{\partial}{\partial x}(\mu Q),$$

that is

$$\mu\left\{\frac{\partial P}{\partial y}-\frac{\partial Q}{\partial x}\right\}=Q\frac{\partial\mu}{\partial x}-P\frac{\partial\mu}{\partial y},$$

and similarly

$$\mu\left\{\frac{\partial Q}{\partial z}-\frac{\partial R}{\partial y}\right\}=R\frac{\partial\mu}{\partial y}-Q\frac{\partial\mu}{\partial z},$$

$$\mu\left\{\frac{\partial R}{\partial x}-\frac{\partial P}{\partial z}\right\}=P\frac{\partial\mu}{\partial z}-R\frac{\partial\mu}{\partial x}.$$

* It is, of course, assumed that the change of order of differentiation is valid.

The unknown μ is eliminated from these three equations by multiplying respectively by R, P, Q and adding. The resulting equation

$$P\left\{\frac{\partial Q}{\partial z}-\frac{\partial R}{\partial y}\right\}+Q\left\{\frac{\partial R}{\partial x}-\frac{\partial P}{\partial z}\right\}+R\left\{\frac{\partial P}{\partial y}-\frac{\partial Q}{\partial x}\right\}=0$$

is a necessary condition for integrability.*

It is obvious from the above demonstration, and may easily be verified independently, that if λ is a function of x, y, z and

$$P_1=\lambda P, \quad Q_1=\lambda Q, \quad R_1=\lambda R,$$

the condition for integrability is satisfied by P_1, Q_1, R_1.

It will now be proved that the condition of integrability is a sufficient condition, that is to say, when it is satisfied, there exists a solution involving an arbitrary constant. The proof incidentally furnishes a method of obtaining the solution when the condition for integrability is satisfied.

Let one of the variables be, for the moment, regarded as a constant. If the variable chosen is z, the equation reduces to

$$Pdx+Qdy=0,$$

where P and Q are to be regarded as functions of x and y into which z enters as a parameter. This equation has a solution

$$u(x, y, z)=\text{const.}$$

where, if $\lambda(x, y, z)$ is the integrating factor,

$$\frac{\partial u}{\partial x}=\lambda P=P_1, \quad \frac{\partial u}{\partial y}=\lambda Q=Q_1,$$

but, of course, it does not follow that

$$\frac{\partial u}{\partial z}=\lambda R=R_1.$$

Let

$$R_1=\lambda R=\frac{\partial u}{\partial z}+S,$$

then since, by hypothesis,

$$P_1\left\{\frac{\partial Q_1}{\partial z}-\frac{\partial R_1}{\partial y}\right\}+Q_1\left\{\frac{\partial R_1}{\partial x}-\frac{\partial P_1}{\partial z}\right\}+R_1\left\{\frac{\partial P_1}{\partial y}-\frac{\partial Q_1}{\partial x}\right\}=0,$$

it follows that

$$\frac{\partial S}{\partial x}\cdot\frac{\partial u}{\partial y}-\frac{\partial S}{\partial y}\cdot\frac{\partial u}{\partial x}=0.$$

This relation is not satisfied in virtue of

$$u(x, y, z)=\text{const.},$$

it is therefore an identity. Consequently S and u, regarded as functions of x and y are functionally dependent upon one another. The functional relationship between them, however, involves also the third variable z, and thus S is expressible in terms of u and z alone.

Now

$$\lambda(Pdx+Qdy+Rdz)=\frac{\partial u}{\partial x}dx+\frac{\partial u}{\partial y}dy+\frac{\partial u}{\partial z}dz+Sdz$$
$$=du+Sdz.$$

* Euler, *Inst. Calc. Int.* 3 (1770), p. 1.

The original equation is therefore equivalent to

$$du+Sdz=0\,;$$

let $\mu(u, z)$ be an integrating factor, then

$$\lambda\mu(Pdx+Qdy+Rdz)=\mu(du+Sdz)$$

is an exact differential $d\psi$. The primitive is

$$\psi(u, z)=c\,;$$

and if u is replaced by its expression in x, y, z the primitive takes the form

$$\phi(x, y, z)=c.$$

Similarly it may be proved that a necessary and sufficient condition that the equation in n variables

$$X_1dx_1+X_2dx_2+\ .\ .\ .\ +X_ndx_n=0$$

should have a primitive of the form

$$\phi(x_1, x_2, \ldots, x_n)=c$$

is that the set of equations

$$X\left\{\frac{\partial X_\mu}{\partial x_\lambda}-\frac{\partial X_\lambda}{\partial x_\mu}\right\}+X_\mu\left\{\frac{\partial X_\lambda}{\partial x_\nu}-\frac{\partial X_\nu}{\partial x_\lambda}\right\}+X_\lambda\left\{\frac{\partial X_\nu}{\partial x_\mu}-\frac{\partial X_\mu}{\partial x_\nu}\right\}=0$$

$$(\lambda, \mu, \nu=1, 2, \ldots, n).$$

are satisfied simultaneously and identically. The total number of such equations is $\frac{1}{6}n(n-1)(n-2)$; of these $\frac{1}{2}(n-1)(n-2)$ are independent.

The main lines upon which the integration proceeds is illustrated by the following example :

$$yz(y+z)dx+zx(z+x)dy+xy(x+y)dz=0.$$

In this case

$$P=yz(y+z),\ Q=zx(z+x),\ R=xy(x+y),$$

and the condition for integrability is satisfied.

When z is regarded as a constant the equation reduces to

$$yz(y+z)dx+zx(z+x)dy=0,$$

and this reduced equation has the solution

$$u\equiv\frac{(z+x)(z+y)}{xy}=\text{const.}$$

Now

$$\frac{\partial u}{\partial x}=-\frac{z(z+y)}{x^2y}=-\frac{1}{x^2y^2}P,$$

$$\frac{\partial u}{\partial y}=-\frac{z(z+x)}{xy^2}=-\frac{1}{x^2y^2}Q,$$

so that

$$\lambda=-\frac{1}{x^2y^2}.$$

Also

$$S=\lambda R-\frac{\partial u}{\partial z}$$

$$=-\frac{x+y}{xy}-\frac{2z+x+y}{xy}=-2\frac{x+y+z}{xy}$$

$$=-2\frac{u-1}{z},$$

and therefore

$$\lambda(Pdx+Qdy+Rdz)=du-2\frac{u-1}{z}\,dz.$$

An integrating factor is $\mu = z^{-2}$, and

$$\lambda\mu(Pdx+Qdy+Rdz)=\frac{du}{z^2}-\frac{2(u-1)dz}{z^3}$$
$$=d\left\{\frac{u-1}{z^2}\right\}$$

The primitive therefore is

$$\frac{u-1}{z^2}=c$$

or, replacing u by its expression in terms of x, y, z,

$$\frac{x+y+z}{xyz}=c.$$

2·81. Geometrical Interpretation.—When R is not zero, the total differential equation may be written

$$dz=-\frac{P}{R}dx-\frac{Q}{R}dy$$

or

$$dz=Udx+Vdy.$$

Since

$$dz=pdx+qdy,$$

the total differential equation is equivalent to the two simultaneous partial differential equations

$$p=U(x, y, z), \quad q=V(x, y, z).$$

The equation of the tangent plane at (x_0, y_0, z_0) to the integral-surface which passes through (x_0, y_0, z_0) is therefore,

$$z-z_0=U_0(x-x_0)+V_0(y-y_0),$$

where U_0 and V_0 are respectively the values of U and V at (x_0, y_0, z_0).

The problem of integration is therefore equivalent to finding a surface such that the direction cosines of its normal at every point (x, y, z) are proportional to

$$U(x, y, z), V(x, y, z), -1.$$

This problem is, in general, insoluble; in order that it may be soluble the condition for integrability, which reduces to

$$\frac{\partial U}{\partial y}+V\frac{\partial U}{\partial z}=\frac{\partial V}{\partial x}+U\frac{\partial V}{\partial z},$$

must be satisfied.

The general solution of each of the partial differential equations

$$\frac{\partial z}{\partial x}=U, \quad \frac{\partial z}{\partial y}=V$$

represents a family of surfaces, such that through every curve in space there passes, in general, one and only one surface of each family.* Their common solution represents a family of space-curves

$$u(x, y, z)=\alpha, \quad v(x, y, z)=\beta,$$

depending upon the two parameters α and β, and such that through each point in space then passes one and only one integral-curve.

An integral-surface of the total differential equation cuts every curve of

* This depends upon the fact that a partial differential equation possesses, in general, a unique solution satisfying assigned initial conditions. The truth of the underlying existence-theorem is assumed.

this family orthogonally, that is the tangent plane at any point P of an integral-surface must contain the normals at P of the two surfaces $u=\alpha$, $v=\beta$ which pass through P. Hence

$$p\frac{\partial u}{\partial x}+q\frac{\partial u}{\partial y}-\frac{\partial u}{\partial z}=0,$$

$$p\frac{\partial v}{\partial x}+q\frac{\partial v}{\partial y}-\frac{\partial v}{\partial z}=0.$$

These two equations determine

$$p=U(x, y, z), \quad q=V(x, y, z).$$

These are consistent if, and only if

$$\frac{\partial p}{\partial y}=\frac{\partial q}{\partial x},$$

that is, if the condition for integrability

$$\frac{\partial U}{\partial y}+V\frac{\partial U}{\partial z}=\frac{\partial V}{\partial x}+U\frac{\partial V}{\partial z}$$

is satisfied.

2·82. Mayer's Method of Integration.—The method of integration developed in § 2·8 depends upon the integration of two successive differential equations in two variables. In Mayer's method * only one integration is necessary. Let (x_0, y_0) be any chosen pair of values of (x, y) and let z_0 be an arbitrary value of z such that the four differential coefficients

$$\frac{\partial U}{\partial y}, \quad \frac{\partial U}{\partial z}, \quad \frac{\partial V}{\partial x}, \quad \frac{\partial V}{\partial z}$$

exist and are continuous in the neighbourhood of (x_0, y_0, z_0). Then if the equation is integrable, its solution will be completely determined by the initial value z_0. The value of z at (x, y) can therefore be obtained by following the variation of z from its initial value z_0 as a point P moves in a straight line in the (x, y)-plane from (x_0, y_0) to (x, y).

There is no loss in generality in supposing that the point (x_0, y_0) is the origin, and this will be assumed. On the straight line joining the origin to (x, y),

$$y=\kappa x, \; dy=\kappa dx,$$

where κ is constant. The equation therefore becomes

$$dz=(U_1+\kappa V_1)dx,$$

where U_1 and V_1 are what U and V become when y is replaced by κx. This equation, in the two variables x and z, has a solution of the form

$$\phi(x, z, \kappa)=\text{const.}$$

or, since $z=z_0$ when $x=0$,

$$\phi(x, z, \kappa)=\phi(0, z_0, \kappa).$$

On replacing κ by y/x, the solution

$$\phi(x, z, y/x)=\phi(0, z_0, y/x)$$

is obtained in a form which indicates its dependence upon the arbitrary constant z_0.

* *Math. Ann.*, 5 (1872), p. 448.

Example —Consider the equation

$$dz = \frac{1+yz}{1+xy}dx + \frac{x(z-x)}{1+xy}dy;$$

the coefficients of dx and dy are continuous in the neighbourhood of $x=0$, $y=0$, $z=z_0$, and so are their partial differential coefficients.

Let

$$y=\kappa x, \quad dy=\kappa dx,$$

then the equation reduces to

$$\frac{dz}{dx}=\frac{2\kappa x}{1+\kappa x^2}z+\frac{1-\kappa x^2}{1+\kappa x^2};$$

it is now linear, and has the solution

$$z=x+z_0(1+\kappa x^2).$$

The solution of the given equation is therefore

$$z=x+z_0(1+xy).$$

2·83. Pfaff's Problem.—When the condition for integrability is not satisfied, the total differential equation is not derivable from a single primitive. On this account such an equation was at one time regarded as meaningless.* Further consideration, however, brought to light the fact that the total differential equation is equivalent to a pair of algebraic equations † known as its *integral equivalents*. In general, when the equations for integrability are not all satisfied, a total differential equation in $2n$ or $2n-1$ variables is equivalent to a system of not more than n algebraic equations.‡ The problem of determining the integral equivalents of any given total differential equation is known as Pfaff's Problem. A sketch of the method of procedure, in the case of three variables, will now be given.§

The first step consists in showing that the differential expression

$$Pdx+Qdy+Rdz$$

can be reduced to the form

$$du+vdw,$$

where u, v, w are functions of x, y, z. The two forms are identical if

$$\text{(A)} \qquad P=\frac{\partial u}{\partial x}+v\frac{\partial w}{\partial x}, \quad Q=\frac{\partial u}{\partial y}+v\frac{\partial w}{\partial y}, \quad R=\frac{\partial u}{\partial z}+v\frac{\partial w}{\partial z}.$$

Let

$$P'=\frac{\partial Q}{\partial z}-\frac{\partial R}{\partial y}, \quad Q'=\frac{\partial R}{\partial x}-\frac{\partial P}{\partial z}, \quad R'=\frac{\partial P}{\partial y}-\frac{\partial Q}{\partial x},$$

then

$$P'=\frac{\partial v}{\partial z}\cdot\frac{\partial w}{\partial y}-\frac{\partial v}{\partial y}\cdot\frac{\partial w}{\partial z},$$

$$Q'=\frac{\partial v}{\partial x}\cdot\frac{\partial w}{\partial z}-\frac{\partial v}{\partial z}\cdot\frac{\partial w}{\partial x},$$

$$R'=\frac{\partial v}{\partial y}\cdot\frac{\partial w}{\partial x}-\frac{\partial v}{\partial x}\cdot\frac{\partial w}{\partial y}.$$

* Euler, *Inst. Calc. Int.*, 3 (1770), p. 5.

† Monge, *Mém. Acad. Sc. Paris* (1784), p. 535.

‡ Pfaff, *Abh. Akad. Wiss. Berlin* (1814), p. 76.

§ An extended treatment in the general case is given in Forsyth, *Theory of Differential Equations*, Part I., and in Goursat, *Leçons sur le Problème de Pfaff*.

It follows that

$$P'\frac{\partial v}{\partial x}+Q'\frac{\partial v}{\partial y}+R'\frac{\partial v}{\partial z}=0,$$

$$P'\frac{\partial w}{\partial x}+Q'\frac{\partial w}{\partial y}+R'\frac{\partial w}{\partial z}=0.$$

Thus v and w are solutions of one and the same linear partial differential equation; the equivalent simultaneous system is

$$\frac{dx}{P'}=\frac{dy}{Q'}=\frac{dz}{R'}.$$

Let

$$\alpha(x, y, z)=\text{const.},\quad \beta(x, y, z)=\text{const.}$$

be two independent solutions of the simultaneous system, then v and w are functions of α and β.

Now return to the variable u; since

$$\left\{P-\frac{\partial u}{\partial x}\right\}P'+\left\{Q-\frac{\partial u}{\partial y}\right\}Q'+\left\{R-\frac{\partial u}{\partial z}\right\}R'=v\left\{P'\frac{\partial w}{\partial x}+Q'\frac{\partial w}{\partial y}+R'\frac{\partial w}{\partial z}\right\}=0,$$

it follows that

$$P'\frac{\partial u}{\partial x}+Q'\frac{\partial u}{\partial y}+R'\frac{\partial u}{\partial z}=PP'+QQ'+RR'.$$

But the condition

$$PP'+QQ'+RR'=0$$

is the condition for integrability; since it is supposed not to be satisfied, u does not satisfy the same partial differential equation as v and w.

Now w may be any function of α and β; for simplicity let

$$w=\alpha.$$

Then if the relation

$$\alpha(x, y, z)=a,$$

where a is a constant, is set up between the variables x, y, z, the differential form $Pdx+Qdy+Rdz$ reduces to du, and therefore becomes a perfect differential. Thus the relation $\alpha(x, y, z)=a$ is used to express any variable, say z, and its differential dz in terms of the other two variables and their differentials, and when these expressions are substituted for z and dz in $Pdx+Qdy+Rdz$, the latter becomes a total differential $d\phi(x, y, a)$. When a is replaced by $\alpha(x, y, z)$ this differential becomes du. Thus u is obtained, and since u and w are known, v may be deduced algebraically from any one of the equations (A). The total differential equation

$$Pdx+Qdy+Rdz=0$$

is thus reduced to the *canonical form*

$$du+vdw=0.$$

The canonical equation may be satisfied in various ways, as follows:

(i) $u=\text{const.},\quad w=\text{const.}$ (ii) $u=\text{const.},\quad v=0.$

More generally, if $\psi(u, w)$ is any arbitrary function of u and w, an integral equivalent is

$$\text{(iii)}\ \psi(u, w)=0,\quad v\frac{\partial\psi}{\partial u}-\frac{\partial\psi}{\partial w}=0;$$

(iii) includes (ii) but not (i). In each case, *the integral equivalent consists of a pair of algebraic equations.*

As an example, consider the equation

$$ydx+zdy+xdz=0.$$

In this case

$$P=y,\ Q=z,\quad R=x,\quad P'=Q'=R'=1,$$

and thus

$$PP'+QQ'+RR'\neq 0,$$

that is, the condition for integrability is not satisfied.

The simultaneous system is

$$dx=dy=dz\,;$$

one solution is

$$\alpha\equiv x-y=a.$$

Let $w=\alpha$, and eliminate x from the given equation, which becomes

$$(y+z)dy+(y+a)dz=0.$$

This reduced equation is immediately integrable and its solution is

$$\phi\equiv\tfrac{1}{2}y^2+yz+az=\text{const.}$$

When a is replaced by $x-y$, ϕ becomes u, thus

$$u=\tfrac{1}{2}y^2+yz+(x-y)z$$
$$=\tfrac{1}{2}y^2+xz.$$

Finally v is obtained as follows :

$$v\frac{\partial w}{\partial x}=P-\frac{\partial u}{\partial x},$$

that is

$$v=y-z.$$

Thus

$$ydx+zdy+xdz=du+vdw,$$

where

$$u=\tfrac{1}{2}y^2+xy,\quad v=y-z,\quad w=x-y.$$

Integral equivalents are therefore

(i) $\tfrac{1}{2}y^2+xz=\text{const.},\quad x-y=\text{const.},$

(ii) $\tfrac{1}{2}y^2+xz=\text{const.},\quad y-z=0.$

(iii) $\psi(u,w)=0,\quad v\dfrac{\partial\psi}{\partial u}-\dfrac{\partial\psi}{\partial w}=0.$

Other integral equivalents are obtained by permuting x, y, z, cyclically.

2·84. Reduction of an Integrable Equation to Canonical Form.—The foregoing reduction to canonical form may equally well be performed in the case of an integrable equation, but since, in this case,

$$PP'+QQ'+RR'=0,$$

identically, u satisfies the same partial differential equation as v and w and therefore u, v and w are functions of α and β.

It follows that

$$du+vdw=Ad\alpha+Bd\beta,$$

where A and B are functions of α and β alone. When α and β have been determined, A and B are derivable algebraically from any two of the three consistent equations,

$$P=A\frac{\partial\alpha}{\partial x}+B\frac{\partial\beta}{\partial x},\quad Q=A\frac{\partial\alpha}{\partial y}+B\frac{\partial\beta}{\partial y},\quad R=A\frac{\partial\alpha}{\partial z}+B\frac{\partial\beta}{\partial z}.$$

Thus the total differential equation is transformed into an ordinary equation in the two variables α and β.

This leads to a practical method of solving an integrable equation, as is shown by the following example (cf. § 2·8):

$$yz(y+z)dx+zx(z+x)dy+xy(x+y)dz=0.$$

Here

$$P'=2x(z-y),\quad Q'=2y(x-z),\quad R'=2z(y-x),$$

and the condition for integrability is satisfied. The simultaneous system

$$\frac{dx}{x(z-y)}=\frac{dy}{y(x-z)}=\frac{dz}{z(y-x)}$$

is equivalent to

$$d(x+y+z)=0,\quad \frac{dx}{x}+\frac{dy}{y}+\frac{dz}{z}=0,$$

and has the solution

$$\alpha\equiv x+y+z=\text{const.},\quad \beta\equiv xyz=\text{const.}$$

Thus the given equation reduces to

$$Ad\alpha+Bd\beta=0,$$

where

$$yz(y+z)=A+Byz,$$
$$zx(z+x)=A+Bzx,$$
$$xy(x+y)=A+Bxy.$$

Hence

$$A=-xyz,\quad B=x+y+z,$$

that is to say, the equation becomes

$$\alpha d\beta-\beta d\alpha=0,$$

and has the solution

$$\frac{\alpha}{\beta}\equiv\frac{x+y+z}{xyz}=\text{const.}$$

Miscellaneous Examples.

1. Integrate the following equations:

(i) $(1-x^2)^{\frac{1}{2}}dx+(1-y^2)^{\frac{1}{2}}dy=0$;
(ii) $x(1+y^2)^{\frac{1}{2}}dx+y(1+x^2)^{\frac{1}{2}}dy=0$;
(iii) $(x^2+2xy-y^2)dx-(x^2-2xy-y^2)dy=0$;
(iv) $(y^2-xy)dx+(x^2-xy)dy=0$;
(v) $x^2ydx+(x^3-y^3)dy=0$;
(vi) $(x+y)dx-(2x-y-1)dy=0$;
(vii) $(x+2y+1)dx-(2x+4y+3)dy=0$;
(viii) $(2x^2+6xy+y^2)dx+(3x^2+2xy+4y^2)dy=0$;
(ix) $(x^2+y^2)dx+xydy=0$;
(x) $(1+x^2)p+xy=1$;
(xi) $p+y\tan x=\sin 2x$;
(xii) $p+y\cos x=e^{2x}$;
(xiii) $xp-ay=x^n$;
(xiv) $xp-y=x^2\sin x$;
(xv) $p+2xy=xe^{-x^2}$;
(xvi) $p\sin x\cos x-y=\sin^3 x$;
(xvii) $(y^2-xy)dx+(x+1)dy=0$;
(xviii) $(y-xp)^2=4p$;
(xix) $y-px+p(p-1)=0$;
(xx) $xyp^2+(x^2+y^2)p+xy=0$;
(xxi) $yp^2+2px-y=0$;
(xxii) $y-(x+5)p+p^3=0$;
(xxiii) $(x+1)p^2-(x+y)p+y=0$;
(xxiv) $y-2px+p^2=0$.

2. Determine n so that the equation

$$\frac{ax^2+2bxy+cy^2}{(x^2+y^2)^n}(ydx-xdy)=0$$

is exact.

3. Show that the equation

$$(y^4-2y^2)dx+(3xy^3-4xy+y)dy=0$$

has an integrating factor which is a function of xy^2, and solve the equation.

4. Show that $\cos x \cos y$ is an integrating factor for

$$(2x \tan y \sec x + y^2 \sec y)dx + (2y \tan x \sec y + x^2 \sec x)dy$$

and integrate the resulting product. [Edinburgh, 1915.]

5. From the relation

$$A(x^2+y^2)-2Bxy+C=0$$

derive the differential equation

$$\frac{dx}{\sqrt{(x^2-c^2)}}+\frac{dy}{\sqrt{(y^2-c^2)}}=0,$$

where $c^2=AC(B^2-A^2)$. Deduce the addition theorem for the hyperbolic cosine.

6. Verify that a solution of

$$\frac{dx}{\sqrt{(1+x^3)}}=\frac{dy}{\sqrt{(1+y^3)}}$$

is

$$x^2y^2+2axy(x+y)+a^2(x-y)^2-4(x+y)+4a=0,$$

where a is an arbitrary constant. In what way is this result connected with the theory of elliptic functions ?

7. Find the curves for which

(i) The subnormal is constant and equal to $2a$;

(ii) The subtangent is equal to twice the abscissa at the point of contact ;

(iii) The perpendicular from the origin upon the tangent is equal to the abscissa at the point of contact ;

(iv) The subtangent is the arithmetical mean of the abscissa and the ordinate ;

(v) The intercept of the normal upon the x-axis is equal to the radius vector ;

(vi) The intercept of the tangent upon the y-axis is equal to the radius vector.

8. P is a point (x, y) on a plane curve, C is the corresponding centre of curvature, and T the point in which the tangent at P meets the x-axis. If the line drawn through T parallel to the y-axis bisects PC prove that

$$2yy''=y'^2(1+y'^2),$$

and hence prove that the curve is a cycloid. [Paris, 1914.]

9. Prove that every curve whose ordinate, considered as a function of its abscissa, satisfies the differential equation

$$(xy'-y)^2=a(1+y'^2)\sqrt{(x^2+y^2)},$$

where a is a constant, has the following property. If H is the foot of the perpendicular from the origin O upon the tangent at any point P of the curve and Q is the foot of the perpendicular from H upon OP, then P lies upon the circle of centre O and radius a.

Change the variables by the substitution

$$x=r \cos \theta,\ y=r \sin \theta$$

and integrate the equation thus obtained. [Paris, 1917.]

CHAPTER III

THE EXISTENCE AND NATURE OF SOLUTIONS OF ORDINARY DIFFERENTIAL EQUATIONS

3·1. Statement of the Problem.—The equations of the type

$$\frac{dy}{dx}=f(x, y),$$

whose solutions were found, in the preceding chapter, by the application of elementary processes, are integrable on account of the fact that they belong to certain simple classes. In general, however, an equation of the type in question is not amenable to so elementary a treatment, and in many cases the investigator is obliged to have recourse to a method of numerical approximation. The theoretical question therefore arises as to whether a solution does exist, either in general or under particular restrictions. Researches into this question have brought to light a group of theorems known as existence-theorems, the more important of which will be studied in the present chapter.*

Let (x_0, y_0) be a particular pair of values assigned to the real variables

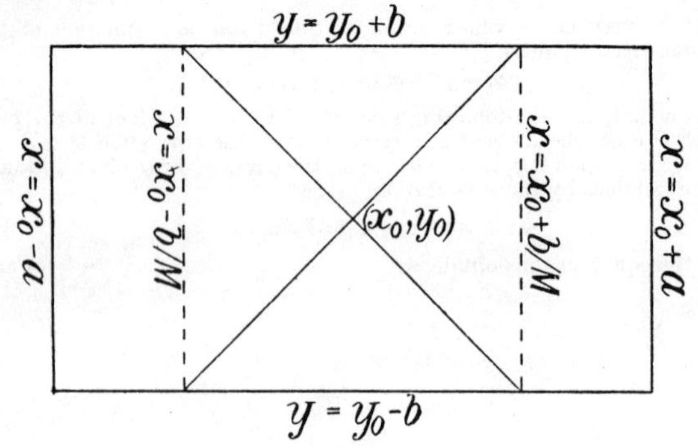

FIG. 1.

(x, y) such that within a rectangular domain D surrounding the point (x_0, y_0) and defined by the inequalities

$$|x-x_0| \leqslant a, \quad |y-y_0| \leqslant b,$$

$f(x, y)$ is a one-valued continuous † function of x and y.

* See also Chap. XII., where the question is discussed from the point of view of the theory of functions of a complex variable.

† $f(x, y)$ is a continuous function of x and y in D if, given an arbitrarily small positive number ϵ, a number δ can be determined such that $|f(x+h, y+k)-f(x, y)| < \epsilon$, provided that (x, y) and $(x+h, y+k)$ are in D and $|h| < \delta$, $|k| < \delta$. It is important to note that h and k vary independently.

Let M be the upper bound of $|f(x, y)|$ in D and let h be the smaller of a and b/M. If $h<a$, the more stringent restriction

$$|x-x_0| \leqslant h$$

is imposed upon x. (Fig. 1.)

Yet another condition must be satisfied by $f(x, y)$, namely that, if (x, y) and (x, Y) be two points within D, of the same abscissa, then

$$|f(x, Y)-f(x, y)| < K|(Y-y)|,$$

where K is a constant. This is known as the *Lipschitz condition.**

Then, these conditions being satisfied, *there exists a unique continuous function of x, say $y(x)$, defined for all values of x such that $|x-x_0|<h$, which satisfies the differential equation and reduces to y_0 when $x=x_0$.*

Two entirely distinct proofs of this existence theorem will now be given, known respectively as the *Method of Successive Approximations* and the *Cauchy-Lipschitz Method.*

3·2. The Method of Successive Approximations.†—Suppose for the moment that a solution $y(x)$ is known, which reduces to y_0 when $x=x_0$; this solution evidently satisfies the relation

$$y(x)=y_0+\int_{x_0}^{x} f\{t, w(t)\}dt.$$

This relation is, in reality, an *integral equation*,‡ involving the dependent variable under the integral sign. Let the function $y(x)$ be now regarded as unknown; the integral equation may then be solved by a method of successive approximation in the following manner.

Let x lie in the interval § (x_0, x_0+h) and consider the sequence of functions $y_1(x), y_2(x), \ldots, y_n(x)$ defined as follows:

$$y_1(x)=y_0+\int_{x_0}^{x} f\{t, y_0\}dt,$$

$$y_2(x)=y_0+\int_{x_0}^{x} f\{t, y_1(t)\}dt,$$

$$\cdot \quad \cdot \quad \cdot \quad \cdot \quad \cdot \quad \cdot$$

$$y_n(x)=y_0+\int_{x_0}^{x} f\{t, y_{n-1}(t)\}dt.$$

It will now be proved

(*a*) that, as n increases indefinitely, the sequence of functions $y_n(x)$ tends to a limit which is a continuous function of x,

(*b*) that the limit-function satisfies the differential equation, and

(*c*) that the solution thus defined assumes the value y_0 when $x=x_0$ and is the only continuous solution which does so.

* It will be seen, as the theory develops, that it is only necessary that the Lipschitz condition should hold in the smaller region $|x-x_0|<h$, $|y-y_0|<M|x-x_0|$.

† This method, though probably known to Cauchy, appears to have been first published by Liouville, *J. de Math.* (1) 2 (1838), p. 19; (1) 3 (1838), p. 565, who applied it to the case of the homogeneous linear equation of the second order. Extensions to the linear equation of order n are given by Caqué, *J. de Math.* (2) 9 (1864), p. 185; Fuchs, *Annali di Mat.* (2) 4 (1870), p. 36 [*Ges. Werke*, I. p. 295]; and Peano, *Math. Ann.* 32 (1888), p. 450. In its most general form it has been developed by Picard, *J. de Math.* (4) 9 (1893), p. 217; *Traité d'Analyse*, 2, p. 301; (2nd ed.) 2, p. 340; and Bôcher, *Am. J. Math.* 24 (1902), p. 311.

‡ Bôcher, *Introduction to the Theory of Integral Equations*; Whittaker and Watson, *Modern Analysis*, Chap. XI.

§ This restriction is a matter of convenience, not of necessity, and will shortly be removed.

In the first place, it will be proved by induction that, when x lies in the interval considered, $|y_n(x)-y_0|\leqslant b$. Suppose then that $|y_{n-1}(x)-y_0|\leqslant b$; it follows that $|f\{t, y_{n-1}(t)\}|\leqslant M$, and consequently

$$\begin{aligned}|y_n(x)-y_0| &\leqslant \int_{x_0}^{x} |f\{t, y_{n-1}(t)\}|\,dt \\ &\leqslant M(x-x_0) \\ &\leqslant Mh \\ &\leqslant b.\end{aligned}$$

But evidently

$$|y_1(x)-y_0|\leqslant b;$$

it is therefore true that

$$|y_n(x)-y_0|\leqslant b,$$

for all values of n. It follows that $f\{x, y_n(x)\}\leqslant M$ when $x_0<x<x_0+h$.

It will now be proved, in a similar way, that

$$|y_n(x)-y_{n-1}(x)|<\frac{MK^{n-1}}{n\,!}(x-x_0)^n.$$

For suppose it to be true that, when $x_0\leqslant x\leqslant x_0+h$,

$$|y_{n-1}(x)-y_{n-2}(x)|<\frac{MK^{n-2}}{(n-1)\,!}(x-x_0)^{n-1},$$

then

$$\begin{aligned}|y_n(x)-y_{n-1}(x)| &\leqslant \int_{x_0}^{x} |f\{t, y_{n-1}(t)\}-f\{t, y_{n-2}(t)\}|\,dt \\ &< \int_{x_0}^{x} K\,|y_{n-1}(t)-y_{n-2}(t)|\,dt,\end{aligned}$$

by the Lipschitz condition, so that

$$\begin{aligned}|y_n(x)-y_{n-1}(x)| &< \frac{MK^{n-1}}{(n-1)\,!}\int_{x_0}^{x} |t-x_0|^{n-1}dt \\ &= \frac{MK^{n-1}}{n\,!}|x-x_0|^n.\end{aligned}$$

But the inequality is clearly true when $n=1$, it is therefore true for all values of n. In the same way it can be proved to hold when $x_0-h\leqslant x\leqslant x_0$, it is therefore true for $|x-x_0|\leqslant h$.

It follows that the series

$$y_0+\sum_{r=1}^{\infty}\{y_r(x)-y_{r-1}(x)\}$$

is absolutely and uniformly convergent when $|x-x_0|\leqslant h$ and moreover each term is a continuous function of x. But

$$y_n(x)=y_0+\sum_{r=1}^{n}\{y_r(x)-y_{r-1}(x)\};$$

consequently the limit-function

$$y(x)=\lim_{n\to\infty} y_n(x)$$

*exists and is a continuous function of x in the interval (x_0-h, x_0+h).**

* Bromwich, *Theory of Infinite Series*, § 45.

Now if it is true that

$$\lim_{n\to\infty} y_n(x)=y_0+\lim_{n\to\infty}\int_{x_0}^{x} f\{t,\, y_{n-1}(t)\}dt$$
$$=y_0+\int_{x_0}^{x} \lim_{n\to\infty} f\{t,\, y_{n-1}(t)\}dt,$$

it will follow that $y(x)$ is a solution of the integral equation

$$y(x)=y_0+\int_{x_0}^{x} f\{t,\, y(t)\}dt.$$

That the inversion of the order of integration and procedure to the limit is legitimate may be proved as follows :

$$\left|\int_{x_0}^{x} [f\{t,\, y(t)\}-f\{t,\, y_{n-1}(t)\}]dt\right| < K\int_{x_0}^{x} |\, y(t)-y_{n-1}(t)\,|\, dt$$
$$< K\epsilon_n\,|\,x-x_0\,| < K\epsilon_n h,$$

where ϵ_n is independent of x and tends to zero as n tends to infinity.

The function $f\{t,\, y(t)\}$ is continuous in the interval $x_0-h\leqslant t\leqslant x_0+h$; consequently

$$\frac{dy(x)}{dx}=\frac{d}{dx}\int_{x_0}^{x} f\{t,\, y(t)\}dt$$
$$=f\{x,\, y(x)\}.$$

The limit-function $y(x)$ therefore satisfies the differential equation ; it also reduces to y_0 when x assumes the value x_0.

It remains to prove that this solution $y(x)$ is unique. Suppose $Y(x)$ to be a solution distinct from $y(x)$, satisfying the initial condition $Y(x_0)=y_0$, and continuous in an interval $(x_0,\, x_0+h')$ where $h'\leqslant h$ and h' is such that the condition

$$|\, Y(x)-y_0\,| < b$$

is satisfied for this interval. Then, since $Y(x)$ is a solution of the given equation, it satisfies the integral equation

$$Y(x)=y_0+\int_{x_0}^{x} f\{t,\, Y(t)\}dt,$$

and consequently

$$Y(x)-y_n(x)=\int_{x_0}^{x} [f\{t,\, Y(t)\}-f\{t,\, y_{n-1}(t)\}]dt.$$

Let $n=1$, then

$$Y(x)-y_1(x)=\int_{x_0}^{x} [f\{t,\, Y(t)\}-f\{t,\, y_0\}]dt,$$

and it follows from the Lipschitz condition that

$$|\, Y(x)-y_1\,(x)\,| < Kb(x-x_0).$$

Similarly, when $n=2$,

$$|\, Y(x)-y_2(x)| < \left|\int_{x_0}^{x} [f\{t,\, Y(t)\}-f\{t,\, y_1\,(t)\}]dt\right|$$
$$< K\int_{x_0}^{x} |\, Y(t)-y_1\,(t)\,|\, dt$$
$$< K\int_{x_0}^{x} Kb(t-x_0)dt=\tfrac{1}{2}K^2b(x-x_0)^2,$$

and in general

$$|Y(x)-y_n(x)| < \frac{K^n b(x-x_0)^n}{n!},$$

whence

$$Y(x) = \lim_{n\to\infty} y_n(x) = y(x)$$

for all values of x in the interval (x_0, x_0+h'), and therefore the new solution is identical with the old. *There is therefore one and only one continuous solution of the differential equation which satisfies the initial conditions.*

3·21. Observations on the Method of Successive Approximation.—The two main assumptions which were made regarding the behaviour of the function $f(x, y)$ in the domain D, namely the assumption of continuity and that of the Lipschitz condition are quite independent of one another. The question arises as to the necessity of these assumptions; it is therefore well to look a little more closely into them and to enquire whether or not they may be unduly restrictive.

In the first place, it will be seen that the continuity of $f(x, y)$ is not necessary for the existence of a continuous solution; in fact all that the previous investigation demands is that $f(x, y)$ be bounded, and that all integrals of the type

$$\int_{x_0}^{x} |f\{t, y_n(t)\}|\, dt$$

exist. In particular, $f(x, y)$ may admit of a limited number of finite discontinuities.*

Thus, for instance, the differential equation

$$\frac{dy}{dx} = y(1-2x) \text{ when } x>0,$$
$$= y(2x-1) \text{ when } x<0$$

admits of a continuous solution satisfying the initial condition $y=1$ when $x=1$. This solution is

$$y=e^{x-x^2} \text{ when } x \geqslant 0,$$
$$=e^{x^2-x} \text{ when } x \leqslant 0,$$

and the solution is valid for all real values of x, moreover it is unique.

On the other hand, the Lipschitz condition, or a condition of a similar character, must be imposed in order to ensure the uniqueness of the solution. It is not difficult to construct an equation for which the Lipschitz condition is not satisfied, and which admits of more than one continuous solution fulfilling the initial conditions.†

Thus, for instance, in the equation

$$\frac{dy}{dx} = \sqrt{|y|},$$

the Lipschitz condition is violated in any region which includes the line $y=0$. The

* These may be discrete points or lines parallel to the y-axis; any other lines of discontinuity imply a violation of the Lipschitz condition throughout an interval of finite dimensions. Mie, *Math. Ann.* 43 (1893), p. 553, has shown that solutions exist whenever $f(x, y)$ is continuous in y and discontinuous but integrable (in Riemann's sense) with respect to x.

† Peano, *Math. Ann.* 37 (1890), p. 182; Mie, *loc. cit., ante*; Perron, *Math. Ann.* 76 (1915), p. 471.

equation admits of two real continuous solutions satisfying the initial conditions $x=0$, $y=0$, viz.

$$(1^0)\quad y=0,$$
$$(2^0)\quad y=\tfrac{1}{4}x^2 \text{ when } x\geqslant 0,$$
$$= -\tfrac{1}{4}x^2 \text{ when } x\leqslant 0.$$

Another example is given by the equation

$$\frac{dy}{dx}=f(x, y),$$

where

$$f(x, y) = \frac{4x^3y}{x^4+y^2} \text{ when } x \text{ and } y \text{ are not both zero,}$$
$$= 0 \text{ when } x=y=0.$$

It is easily proved that $f(x, y)$ is a continuous function of x and y. On the other hand

$$f(x, Y)-f(x, y) = \frac{4x^3(x^4-yY)}{(x^4+y^2)(x^4+Y^2)}(Y-y).$$

If $y=px^2$, $Y=qx^2$,

$$|f(x, Y)-f(x, y)| = 4\left|\frac{1-pq}{(1+p^2)(1+q^2)}\right|\frac{|Y-y|}{|x|},$$

and therefore the Lipschitz condition is not satisfied throughout any region containing the origin.

The equation admits of the solution

$$y=c^2-\sqrt{(x^4+c^4)},$$

c being an arbitrary real constant, and thus there is an infinity of solutions satisfying the initial conditions $x=0$, $y=0$.

The question has been placed on a firm basis by Osgood,* who proved that, if $f(x, y)$ be continuous in the neighbourhood of (x_0, y_0), there exists in general a one-fold infinity of solutions satisfying the initial conditions. These solutions lie entirely within the area bounded by two extremal solutions

$$y=Y_1(x),\quad y=Y_2(x).$$

A necessary and sufficient condition that there be a unique solution is that $Y_1(x)$ and $Y_2(x)$ be identical. This is the case when the Lipschitz condition is satisfied, but it is also true when the Lipschitz condition is replaced by one or other of the less restrictive conditions

$$|f(x, Y)-f(x, y)| < K_1|Y-y|\log\frac{1}{|Y-y|},$$

$$|f(x, Y)-f(x, y)| < K_2|Y-y|\log\frac{1}{|Y-y|}\log\log\frac{1}{|Y-y|},$$

. . . .

in which K_1, K_2, . . . are constants.

The constant K which occurs in the Lipschitz condition determines, for any given value of x, the rapidity with which the comparison series

$$\sum\frac{MK^{n-1}}{n!}|x-x_0|^n$$

converges, and therefore gives an indication of the utility of the series

$$y_0+\sum_{r=1}^{n}\{y_r(x)-y_{r-1}(x)\}$$

as an approximation to the limit-function $y(x)$. Thus if K were small,

* *Monatsh. Math. Phys.* 9 (1898), p. 331.

$y_n(x)$ would tend to the limit $y(x)$ more rapidly than if K were large. Now in most cases occurring in practice K is the upper bound of

$$\frac{\partial f(x, y)}{\partial y}$$

in the domain D. To make use of this fact, consider the family of curves

$$f(x, y)=C,$$

for all values of the constant C. The typical curve of this family is such that it intersects each integral curve in a point at which the gradient of the latter curve is C. For this reason the curves are known as the *isoclinal lines*.* Let the isoclinal lines be plotted for a succession of discrete equally-spaced (*e.g.* integral) values of C, and let a line be drawn parallel to the y-axis. Then the intervals along this line in which the points of intersection with the isoclinal lines are densely packed correspond to large values of K, whereas those intervals in which the intersections are more widely spaced correspond to smaller values of K. This brings out the fact that the regions in which the method of successive approximations may most successfully be applied as a practical method of computation are those in which the isoclinal lines tend to run more or less parallel to the y-axis.†

The method of successive approximations leads to a solution which was shown to converge in the interval $|x-x_0| \leqslant h$, where h is the least of a and b/M. But, as was remarked in passing, the assumption originally made that certain conditions are satisfied throughout the region $|x-x_0| \leqslant a$, $|y-y_0| \leqslant b$ was unnecessarily restrictive. If a region $|x-x_0| \leqslant k$, $|y-y_0| \leqslant M|x-x_0|$ can be found such that $f(x, y)$ satisfies the necessary conditions in that region, and M is the upper bound of $|f(x, y)|$, then k will certainly not be less, and may quite conceivably be greater, than h. Several writers have succeeded in thus extending the range in which the solution can be proved to converge,‡ but no general method of determining the exact boundaries of the interval of convergence has yet been discovered.

3·22. Variation of the Initial Conditions.—Let the given initial condition that $y=y_0$ when $x=x_0$ be replaced by the new condition $y=y_0+\eta$ when $x=x_0$, where $(x_0, y_0+\eta)$ is a point within the domain D such that $|\eta| \leqslant \delta$. Then, in place of the sequence of functions

$$y_1(x), y_2(x), \ldots, y_n(x),$$

as defined in § 3·2, there now arises the sequence

$$Y_1(x), Y_2(x), \ldots, Y_n(x),$$

defined as follows:

$$Y_1(x)=y_0+\eta+\int_{x_0}^{x} f\{t, y_0+\eta\}dt,$$

$$Y_2(x)=y_0+\eta+\int_{x_0}^{x} f\{t, Y_1(t)\}dt,$$

$$\cdots\cdots\cdots$$

$$Y_n(x)=y_0+\eta+\int_{x_0}^{x} f\{t, Y_{n-1}(t)\}dt.$$

* The term is due to Chrystal, see Wedderburn, *Proc. Roy. Soc. Edin.* 24 (1902), p. 400.

† Practical methods of approximate computation based upon the method of successive approximations have been devised by Severini, *Rend. Ist. Lombard.* (2) 31 (1898), pp. 657, 950; Cotton, *C. R. Acad. Sc. Paris*, 140 (1905), p. 494; 141 (1905), p. 177; 146 (1908), pp. 274, 510; *Math. Ann.* 31 (1908), p. 107.

‡ Lindelöf, *C. R. Acad. Sc. Paris*, 118 (1894), p. 454; *J. de Math.* (4) 10 (1894), p. 117. See Picard, *Traité d'Analyse*, 3, p. 88; (2nd ed.) 2, p. 340; and also § 3·41 below.

The existence and uniqueness of the solution

$$Y(x)=\lim Y_n(x)$$

then follow as before. Now

$$| Y_1(x)-y_1(x) | \leqslant \delta+\left| \int_{x_0}^{x} [f\{t, y_0+\eta\}-f\{t, y_0\}]dt \right|$$
$$< \delta+K\delta\, | x-x_0 |,$$

$$| Y_2(x)-y_2(x) | \leqslant \delta+\left| \int_{x_0}^{x} [f\{t, Y_1(t)\}-f\{t, y_1(t)\}dt \right|$$
$$< \delta+K\delta\, | x-x_0 | +\tfrac{1}{2}K^2\delta\, | x-x_0 |^2,$$

and, by induction,

$$| Y_n(x)-y_n(x) | < \delta+K\delta\, | x-x_0 |+ \ldots +\frac{1}{n!}K^n\delta\, | x-x_0 |^n$$
$$< \delta e^{K|x-x_0|},$$

so that, in the limit,

$$| Y(x)-y(x) | \leqslant \delta e^{K|x-x_0|}.$$

Consequently, when $|x-x_0|\leqslant h$, the solution is uniformly continuous in the initial value y_0. To bring out this fact, it may be written in either of the forms

$$y(x, y_0) \quad \text{and} \quad y(x-x_0, y_0).$$

Moreover,

$$\left|\frac{y_n(x, y_0+\eta)-y_n(x, y_0)}{\eta}\right| < 1+K\, | x-x_0 | + \ldots +\frac{1}{n!}K^n\, | x-x_0 |^n,$$

and consequently

$$\left|\frac{\partial y_n(x, y_0)}{\partial y_0}\right| \leqslant 1+K\, | x-x_0 | + \ldots +\frac{1}{n!}K^n\, | x-x_0 |^n,$$

from which it may be deduced that the series

$$\frac{\partial y(x, y_0)}{\partial y_0}=1+\sum_{n=1}^{\infty}\frac{\partial\{y_n(x, y_0)-y_{n-1}(x, y_0)\}}{\partial y_0}$$

is absolutely and uniformly convergent. Therefore $y(x, y_0)$ is uniformly differentiable with respect to y_0 when $|x-x_0|\leqslant h$.

A proof proceeding on similar lines to the above shows that if the differential equation involves a parameter λ, that is to say if

$$\frac{dy}{dx}=f(x, y\,;\ \lambda),$$

where $f(x, y\,;\ \lambda)$ is single-valued and continuous and satisfies the Lipchitz condition uniformly in D when $\Lambda_1\leqslant\lambda\leqslant\Lambda_2$, then the solution depends continuously upon λ, and in fact is uniformly differentiable with respect to λ when $|x-x_0|\leqslant h$.

3·23. Singular Points.—A singular point may be defined as a point of the (x, y)-plane at which one or other of the conditions necessary for the establishment of the existence theorem ceases to hold. In fact if for the initial value-pair (x_0, y_0) the solution

(*a*) is discontinuous, (*b*) is not unique, or (*c*) does not exist,

then the point (x_0, y_0) is a singular point of the equation. As illustrations of the diverse ways in which the solutions of an equation may behave at or in the neighbourhood of a singular point, the following examples may be taken.

$$(1^0) \qquad \frac{dy}{dx}=\frac{y}{x}.$$

The conditions requisite for the existence of a unique and continuous solution are fulfilled except in the neighbourhood of $x=0$. The solution corresponding to the initial value-pair (x_0, y_0) is

$$y=\frac{y_0}{x_0}x,$$

when $x_0 \neq 0$. If $x_0=0$ and $y_0 \neq 0$, the solution reduces to

$$x=0.$$

The only exceptional case is when $x_0=y_0=0$; the only singular point in the finite part of the (x, y)-plane is the origin. Now *every* integral-curve passes through the origin, which is a *node* of the integral-curves.

$$(2^0) \qquad \frac{dy}{dx}=m\frac{y}{x}.$$

In this case also, the only singular point is the origin. To any other point (x_0, y_0) corresponds the solution

$$y=y_0\left(\frac{x}{x_0}\right)^m.$$

The family of integral curves corresponding to all possible values of (x_0, y_0) touch the x-axis at the origin if $m>1$ and the y-axis at the origin if $0<m<1$. Thus if $m>0$, every integral-curve passes through the origin.

On the other hand, if $m<0$, say $m=-p$, the solution is

$$yx^p=y_0x_0{}^p.$$

The family of integral-curves is asymptotic to the x- and y-axes. The degenerate curve

$$yx^p=0$$

passes through the origin, but no other integral-curve does so. The origin is a *saddle-point*, for in its neighbourhood the integral-curves resemble the contour lines around a mountain pass.

$$(3^0) \qquad \frac{dy}{dx}=\frac{x+y}{x}.$$

The origin is the only singular point; to any other point (x_0, y_0) corresponds the solution

$$y=\frac{y_0}{x_0}x+x\log\left|\frac{x}{x_0}\right|.$$

The origin is a *node* of the integral-curves.

$$(4^0) \qquad \frac{dy}{dx}=-\frac{x}{y}.$$

The solution is, in general,

$$x^2+y^2=x_0{}^2+y_0{}^2.$$

No real integral-curve, except the degenerate curve $x^2+y^2=0$ passes through the origin, which is a *focal point.*

$$(5^0) \qquad \frac{dy}{dx}=\frac{x+y}{x-y}.$$

This equation is most effectively dealt with by means of a transformation to polar co-ordinates

$$x=r\cos\theta,\ y=r\sin\theta.$$

It then becomes

$$\frac{dr}{d\theta}=r;$$

the integral-curves are the family of logarithmic spirals

$$r=ce^\theta.$$

One curve of the family goes through each point of the plane except the origin. No integral-curve passes through the origin, which is a *focal point* of every curve of the family.

It will be noticed that all these examples are particular cases of the general form

$$\frac{dy}{dx}=\frac{ax+by}{cx+dy},$$

which may be integrated by the method of § 2·12. It will be found that, from the point of view of the behaviour of the integral-curves in the neighbourhood of the origin, the equation is of one or other of three main types according as

$$\text{I. } (b-c)^2+4ad>0,$$
$$\text{II. } (b-c)^2+4ad<0,$$
$$\text{III. } (b-c)^2+4ad=0$$

In Case I. the origin is a node of $ad-bc<0$, and a saddle-point if $ad-bc>0$; in Case II. the origin is a focal point, and in Case III. a node.

3·3. Extension of the Method of Successive Approximation to a System of Equations of the First Order.—Let the system of equations be

$$\frac{dy_1}{dx}=f_1(x, y_1, y_2, \ldots, y_m),$$
$$\frac{dy_2}{dx}=f_2(x, y_1, y_2, \ldots, y_m),$$
$$\cdot \quad \cdot \quad \cdot \quad \cdot \quad \cdot \quad \cdot \quad \cdot$$
$$\frac{dy_m}{dx}=f_m(x, y_1, y_2, \ldots, y_m),$$

then, under conditions which will be stated, *there exists a unique set of continuous solutions of this system of equations which assume given values* y_1^0, $y_2^0 \ldots y_m^0$ *when* $x=x_0$. A bare outline of the proof will be given; the method follows exactly on the lines of the preceding section.

The functions $f_1, f_2, \ldots f_m$ are supposed to be single-valued and continuous in their $m+1$ arguments when these arguments are restricted to lie in the domain D defined by

$$| x-x_0 | \leqslant a, | y_1-y_1^0 | \leqslant b_1, \ldots, | y_m-y_m^0 | \leqslant b_m.$$

Let the greatest of the upper bounds of $f_1, f_2, \ldots, f_m$ in this domain be M; if h is the least of a, $b_1/M, \ldots, b_m/M$, let x be further restricted, if necessary, by the condition $| x-x_0 | <h$.

The Lipschitz condition to be imposed is

$$|f_r(x, Y_1, Y_2, \ldots, Y_m)-f_r(x, y_1, y_2, \ldots, y_m)| <K_1 | Y_1-y_1 | +K_2 | Y_2-y_2 | + \ldots +K_m | Y_m-y_m |,$$

for $r=1, 2, \ldots, m$.

Now define the functions $y_1^n(x), y_2^n(x), \ldots, y_m^n(x)$ by the relations

$$y_r^n(x)=y_r^0+\int_{x_0}^{x} f_r[t, y_1^{n-1}(t), y_2^{n-1}(t), \ldots, y_m^{n-1}(t)]dt,$$

then it can be proved by induction that

$$| y_r^n(x)-y_r^{n-1}(x) | <\frac{M(K_1+K_2+ \ldots +K_m)^{n-1}}{n!}|x-x_0|^n,$$

and the existence, continuity, and uniqueness of the set of solutions follow immediately.

Since the differential equation of order m

$$\frac{d^my}{dx^m}=f\left(x, y, \frac{dy}{dx}, \ldots, \frac{d^{m-1}y}{dx^{m-1}}\right)$$

is equivalent to the set of m equations of the first order

$$\frac{dy}{dx}=y_1, \quad \frac{dy_1}{dx}=y_2, \quad . \; . \; ., \quad \frac{dy_{m-2}}{dx}=y_{m-1},$$

$$\frac{dy_{m-1}}{dx}=f(x, y, y_1, \; . \; . \; ., y_{m-1}),$$

it follows that if f is continuous and satisfies a Lipschitz condition in a domain D, *the equation admits of a unique continuous solution which, together with its first m—1 derivatives, which are also continuous, will assume an arbitrary set of initial conditions for the initial value $x=x_0$.*

3·31. Application to a System of Linear Equations.—Consider the set of m linear equations

$$\frac{dy_i}{dx}=p_{i1}y_1+p_{i2}y_2+ \; . \; . \; . \; +p_{im}y_m+r_i \qquad (i=1, 2, \; . \; . \; ., m),$$

in which the coefficients p_{ij} and r_i are continuous functions of x in the interval $a \leqslant x \leqslant b$. The right-hand member of the equation is therefore continuous for all values of $y_1, y_2, \; . \; . \; ., y_m$ when x lies in the interval (a, b). No further restrictions are necessary; the set of continuous solutions

$$y_1(x), y_2(x), \; . \; . \; ., y_m(x)$$

exists and is unique in the interval (a, b).

If, moreover, the coefficients are continuous for all positive and negative values of x, then the set of solutions will be continuous for all real values of x. This is the case, for instance, when all the functions p_{ij} and r_i are polynomials in x.

Suppose now that the coefficients p_{ij} and r_i, in addition to being continuous functions of x in (a, b), are analytic * functions of a parameter λ in a domain Λ. The moduli $| p_{ij} |$ are therefore bounded; let K (a number independent of λ) be their upper bound.

Now the integrals such as

$$y_i^n(x, \lambda)=y_i^0+\int_x^x \{p_{i1}(t)y_1^{n-1}(t, \lambda)+ \; . \; . \; . \; +p_{im}(t)y_m^{n-1}(t, \lambda)+r_i(t)\}dt$$

are continuous in x and analytic in λ. Also

$$| y_i^n(x, \lambda)-y_i^{n-1}(x, \lambda) | < \frac{M(mK)^{n-1}}{n!} | x-x_0 |^n.$$

Thus the comparison of the series

$$y_i^0+\sum_{n=1}^{\infty}\{y_i^n(x, \lambda)-y_i^{n-1}(x, \lambda)\}$$

with the power series

$$M+\sum_{n=1}^{\infty}\frac{M(mK)^{n-1}}{n!} | x-x_0 |^n$$

shows that the functions $y_i^n(x, \lambda)$ tend respectively to their limits $y_i(x, \lambda)$ uniformly in (x, λ), when $a \leqslant x \leqslant b$ and λ is in Λ. Consequently the solutions $y_i(x, \lambda)$ are continuous in x and analytic in λ. In particular, if the coefficients

* It is inexpedient to restrict the discussion to real values of λ, as it so frequently happens that imaginary or complex values have to be considered. Let λ, then, be a complex number restricted to such a region Λ of the Argand diagram (or λ-plane) that the coefficients are analytic in λ, that is to say, they are single-valued, continuous, and admit of a unique derivative (*i.e.* a derivative independent of the direction of approach), at each point of the domain Λ.

are integral functions (or polynomials) of λ, the solutions $y_i(x, \lambda)$ will themselves be integral functions of λ, and may be written in the form

$$y_i(x, \lambda)=u_{i0}+u_{i1}\lambda+ \ . \ . \ . \ +u_{ir}\lambda^r+ \ . \ . \ . \qquad (i=1, 2, \ldots, m)$$

uniformly convergent for all values of λ when $a\leqslant x\leqslant b$. If the initial conditions do not themselves involve the parameter λ, u_{i0} must alone satisfy the appropriate initial conditions, whilst each u_{ij} $(j>0)$ reduces to zero for the initial value of x.

Frequently a convenient method of obtaining a series-solution of an equation, or set of equations, involving a parameter λ is to assume a solution of this form and then to proceed by a method of undetermined coefficients.*

3·32. The Existence Theorem for a Linear Differential Equation of Order n.— It has already been pointed out (§ 1·5) that the linear differential equation

$$p_0(x)\frac{d^ny}{dx^n}+p_1(x)\frac{d^{n-1}y}{dx^n}+ \ . \ . \ . \ p_{n-1}(x)\frac{dy}{dx}+p_n(x)y=r(x)$$

is equivalent to the system of n linear equations of the first order

$$\begin{cases}\dfrac{dy}{dx}=y_1, \quad \dfrac{dy_1}{dx}=y_2, \quad \ldots \quad \dfrac{dy_{n-2}}{dx}=y_{n-1}, \\[2ex] \dfrac{dy_{n-1}}{dx}=\dfrac{r(x)}{p_0(x)}-\dfrac{p_n(x)}{p_0(x)}y-\dfrac{p_{n-1}(x)}{p_0(x)}y_1- \ . \ . \ . \ -\dfrac{p_1(x)}{p_0(x)}y_{n-1}.\end{cases}$$

It follows from the preceding section that *if $p_0(x)$, $p_1(x)$, . . ., $p_n(x)$ and $r(x)$ are continuous functions of x in the interval $a\leqslant x\leqslant b$ and $p_0(x)$ does not vanish at any point of that interval, the differential equation admits of a unique solution which, together with its first $(n-1)$ derivatives, is continuous in (a, b) and satisfies the following initial conditions :*

$$y(x_0)=y_0, \ y'(x_0)=y_0', \ . \ . \ ., \ y^{(n-1)}(x_0)=y_0^{(n-1)}$$

where x_0 is a point of (a, b).

A direct proof of this theorem will now be given, but in order to abbreviate the work, it will be restricted to the equation of the second order

$$\frac{d^2y}{dx^2}+p\frac{dy}{dx}+qy=r,$$

associated with the initial conditions

$$y(c)=\gamma, \quad y'(c)=\gamma',$$

where c is an internal point of the interval (a, b) in which p, q and r are continuous.

As a preliminary, consider the equation

$$\frac{d^2y}{dx^2}=v(x) ;$$

a solution which satisfies the initial conditions is

$$y=\int_c^x(x-t)v(t)dt+\gamma'(x-c)+\gamma,$$

and this solution is unique.

Let $y_0(x)$ be any continuous function of x such that $y_0'(x)$ is also continuous in (a, b), and form the equation

$$\frac{d^2y}{dx^2}=r(x)-qy_0(x)-py_0'(x).$$

* See Poincaré, *Les Méthodes nouvelles de la Mécanique céleste*, I., Chap. II.

Let $y=y_1(x)$ be the solution of this equation which satisfies the initial conditions $y_1(c)=\gamma$, $y_1'(c)=\gamma'$, and form the equation

$$\frac{d^2y}{dx^2}=r(x)-qy_1(x)-py_1'(x),$$

of which the solution which satisfies the initial condition will be denoted by $y_2(x)$.

By proceeding in this way, a sequence of functions

$$y_1(x),\ y_2(x),\ \ldots,\ y_n(x),\ \ldots$$

continuous and differentiable in (a, b) and such that

$$y_n(c)=\gamma,\quad y_n'(c)=\gamma'$$

is obtained. It will now be proved that this sequence has a limit, and that the limit function is the solution required. Write

$$u_n(x)=y_n(x)-y_{n-1}(x),$$

then

$$\frac{d^2u_n(x)}{dx^2}=-qu_{n-1}(x)-pu'_{n-1}(x),$$

and since

$$u_n(c)=0,\quad u_n'(c)=0,$$

it follows that

$$u_n(x)=\int_c^x\{-q(t)u_{n-1}(t)-p(t)u'_{n-1}(t)\}(x-t)dt,$$

$$u_n'(x)=\int_c^x\{-q(t)u_{n-1}(t)-p(t)u'_{n-1}(t)\}dt.$$

The coefficients $p(x)$ and $q(x)$ are finite in (a, b), so that

$$|p(x)|+|q(x)|\leqslant M,$$

also, a number A exists such that

$$|u_1(x)|\leqslant A,\quad |u_1'(x)|\leqslant A,$$

Let L be the greater of 1 and $b-a$. Then it follows by induction that

$$|u_n(x)|\leqslant\frac{AM^{n-1}L^{2n-2}}{(n-1)!},$$

and $|u_n'(x)|$ satisfies the same inequality.

The series

$$y_0(x)+\{y_1(x)-y_0(x)\}+\ \ldots\ +\{y_n(x)-y_{n-1}(x)\}+\ \ldots$$

and

$$y_0'(x)+\{y_1'(x)-y_0'(x)\}+\ \ldots\ +\{y_n'(x)-y'_{n-1}(x)\}+\ \ldots$$

are therefore absolutely and uniformly convergent in (a, b). Consequently

$$y(x)=\lim y_n(x),\quad y_n'(x)=\lim y_n'(x)$$

exist and are continuous in (a, b). Now

$$\begin{aligned}q(x)y(x)+p(x)y'(x)&=q(x)y_0(x)+p(x)y_0'(x)+\sum_{n=1}^{\infty}\{q(x)u_n(x)+p(x)u_n'(x)\}\\&=r(x)-y_1''(x)-\sum_{n=2}^{\infty}u_n''(x)\\&=r(x)-y''(x),\end{aligned}$$

since the series which represents $y''(x)$ is uniformly convergent in (a, b).

The limit-function $y(x)$ therefore satisfies the differential equation; it remains to show that it is the only solution which fulfils all the conditions specified.

Suppose that two such solutions $y(x)$ and $Y(x)$ exist, and let

$$v(x)=Y(x)-y(x).$$

Then $v(x)$ would satisfy the homogeneous differential equation

$$\frac{d^2v}{dx^2}+p(x)\frac{dv}{dx}+q(x)v=0$$

together with the initial conditions

$$v(c)=0, \quad v'(c)=0.$$

Now this is impossible, for if $v_1(x)$ and $v_2(x)$ are any two distinct solutions of the homogeneous equation, then

$$v_1(x)\{v_2''(x)+p(x)v_2'(x)+q(x)v(x)\}-v_2(x)\{v_1''(x)+p(x)v_1'(x)+q(x)v_1(x)\}=0,$$

whence

$$\frac{d}{dx}\{v_1(x)v_2'(x)-v_2(x)v_1'(x)\}+p(x)\{v_1(x)v_2'(x)-v_2(x)v_1'(x)\}=0,$$

a linear differential equation of the first order whose general solution is

$$v_1(x)v_2'(x)-v_2(x)v_1'(x)=Ce^{-\int_c^x p(x)dx},$$

where C is a constant determined by the initial values of $v_1(x)$, $v_2(x)$, $v_1'(x)$, $v_2'(x)$. This is known as the *Abel identity*.*

Now let $v_1(x)$ be the solution which satisfies the initial conditions

$$v_1(c)=v_1'(c)=0,$$

then $C=0$ and

$$v_1(x)v_2'(x)-v_2(x)v_1'(x)=0$$

identically.

If $v_1(x)$ is not identically zero, this identity may be written

$$\frac{v_2'(x)}{v_2(x)}=\frac{v_1'(x)}{v_1(x)},$$

which implies that $v_2(x)$ is a constant multiple of $v_1(x)$, or that the solutions $v_1(x)$ and $v_2(x)$ are not distinct. This contradiction proves that $v_1(x)$ is identically zero, and therefore the solution $y(x)$ is unique.

If the coefficients $p(x)$, $q(x)$ and $r(x)$ depend upon a real parameter λ, and are continuous for all values of x in (a, b) when λ ranges between Λ_1 and Λ_2, then $y(x)$ can be proved to depend continuously upon λ when λ lies within a closed interval interior to (Λ_1, Λ_2). For it is sufficient to assign such a value to the number M that the inequality

$$|p(x)|+|q(x)|\leqslant M$$

holds for all values of λ in (Λ_1, Λ_2). Then the subsequent inequalities prove the uniform convergence of the series

$$y_0(x)+\{y_1(x)-y_0(x)\}+ \dots +\{y_n(x)-y_{n-1}(x)\}+ \dots$$

and of its derivative for all values of x in $a\leqslant x\leqslant b$ and for any closed interval of λ in (Λ_1, Λ_2). The existence and uniform continuity of the limit-functions $y(x)$ and $y'(x)$ follow immediately. By a slight change of wording the theorem may be extended to cover the case of a complex parameter λ.

3·4. The Cauchy-Lipschitz Method.—This method of proving the existence of solutions of a differential equation or system of equations is essentially

* Abel, *J. für Math.* 2 (1827), p. 22 [*Œuvres complètes* (1839) 1, p. 93; (1881) 1, p. 251].

distinct from the method of successive approximations. It is in reality a refinement of the primitive existence theorem invented by Cauchy.*

Let (x_0, y_0) be the initial pair of values to be satisfied by the solution of

$$\frac{dy}{dx}=f(x, y);$$

dividing the interval (x_0, x) into n subdivisions

$$(x_0, x_1), (x_1, x_2), \ldots, (x_{n-1}, x)$$

such that

$$x_0<x_1<x_2 \ldots <x_{n-1}<x,$$

consider the sequence $y_0, y_1, y_2, \ldots, y_{n-1}, y_n$ defined as follows :

$$y_1=y_0+f(x_0, y_0)(x_1-x_0),$$
$$y_2=y_1+f(x_1, y_1)(x_2-x_1),$$
$$\cdot \quad \cdot \quad \cdot \quad \cdot \quad \cdot \quad \cdot \quad \cdot$$
$$y_n=y_{n-1}+f(x_{n-1}, y_{n-1})(x-x_{n-1}).$$

Then the sum

$$y_n=y_0+f(x_0, y_0)(x_1-x_0)+f(x_1, y_1)(x_2-x_1)+ \ldots +f(x_{n-1}, y_{n-1})(x-x_{n-1})$$

offers a close analogy to the sum which leads to Cauchy's definition of the definite integral. This sum will now be generalised in a way which exhibits the closest possible analogy with the more general Riemann definition.†

Consider the triangle ABC (Fig. 2) formed by the three straight lines

$$X=x_0+h, \quad Y=y_0+M(X-x_0), \quad Y=y_0-M(X-x_0),$$

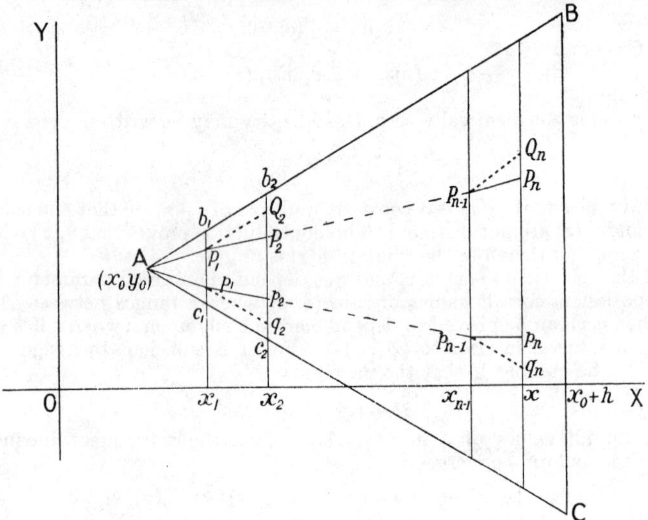

FIG. 2.

* The original method was developed by Cauchy in his lectures at the École polytechnique between the years 1820 and 1830 ; it is summarised in a memoir, *Sur l'intégration des équations différentielles*, lithographed Prague, 1835, reprinted *Exercises d'Analyse*, 1840, p. 327 [*Œuvres completès*, (2) 11, p. 399]. In a fuller form, it was preserved by Cauchy's pupil, l'abbé Moigno, *Leçons de calcul*, 2 (1844), pp. 385, 513. The essence of the method, however, goes back to Euler, *Inst. Calc. Int.* 1 (1768), p. 493. The improvement due to Lipschitz was given in *Bull. Sc. Math.* 10 (1876), p. 149.

† This generalisation is due to Goursat, *Cours d'Analyse*, 2 (2nd ed.), p. 375. A generalisation on different lines is given by Cotton, *Acta Math.* 31 (1908), p. 107.

where h is as defined in § 3·2. Then if a continuous integral-curve passing through the vertex A exists, this curve will lie below AB and above AC, because for any x such that $x_0 \leqslant x \leqslant x_0+h$ the gradient of the integral-curve is less than that of AB and greater than that of AC. Now let the triangle be divided up into strips by the lines $X=x_1$, $X=x_2$, . . ., $X=x$, parallel to BC. The first of these strips is the triangle Ab_1c_1, the second the trapezium $c_1b_1b_2c_2$, and so on.

In the triangle Ab_1c_1 let the upper and lower bounds of $f(x, y)$ be M_1 and m_1, then

$$-M \leqslant m_1 < M_1 \leqslant M.$$

Let P_1 and p_1 be the points on the line $X=x_1$ whose ordinates are respectively $Y_1=y_0+M_1(x_1-x_0)$ and $y_1=y_0+m_1(x_1-x_0)$. Draw P_1Q_2 and p_1q_2 parallel to AB and AC respectively, to meet the line $X=x_2$ in Q_2 and q_2. Let M_2 and m_2 be the upper and lower bounds of $f(x, y)$ in the trapezium $p_1P_1Q_2q_2$, then since this trapezium lies entirely within the trapezium $c_1b_1b_2c_2$ it follows that $-M \leqslant m_2 < M_2 \leqslant M$. Let P_2 and p_2 be points on the line $X=x_2$ of ordinates $Y_2=Y_1+M_2(x_2-x_1)$ and $y_2=y_1+m_2(x_2-x_1)$ respectively. The process is continued from one trapezium to the next until points P_n and p_n on the $X=x$ are reached, whose ordinates are

$$Y_n=Y_{n-1}+M_n(x-x_{n-1}) \quad \text{and} \quad y_n=y_{n-1}+m_n(x-x_{n-1}).$$

Thus two polygonal arcs $AP_1P_2 \ldots P_n$ and $Ap_1p_2 \ldots p_n$ are defined and lie entirely within the angle CAB.

The sums

$$Y_n=y_0+M_1(x_1-x_0)+M_2(x_2-x_1)+ \ldots +M_n(x-x_{n-1})$$

and

$$y_n=y_0+m_1(x_1-x_0)+m_2(x_2-x_1)+ \ldots +m_n(x-x_{n-1})$$

are exactly analogous to the sums S_n and s_n in the classical Riemann theory of integration.* To take full advantage of the analogy, S_n will be written for Y_n and s_n for y_n. If, then, S_ν and s_ν are the corresponding sums arising from a new mode of subdivision of the same range (x_0, x) into ν intervals,

$$S_n > s_\nu; \quad S_\nu > s_n.$$

As the number, n or ν, of subdivisions increases by the addition of new points of subdivision, the existing points being retained, S_n and S_ν do not increase, nor do s_n and s_ν decrease. Let the lower bound of S_n and the upper bound of s_n be Y and y respectively, then

$$S_n \geqslant Y, \quad s_n \leqslant y, \quad Y \geqslant y.$$

Now

$$S_n-s_n=(S_n-Y)+(Y-y)+(y-s_n),$$

and each of the three bracketed terms is positive or zero. If, therefore, it is proved that, as $n \to \infty$,

$$S_n-s_n \to 0,$$

it will follow that

$$S_n \to Y, \quad s_n \to y, \quad Y=y,$$

since Y and y are independent of n. Hence

$$\lim S_n \quad \text{and} \quad \lim s_n$$

will both exist and will be equal.

* For a full explanation of the steps which are here merely outlined, see Whittaker and Watson, *Modern Analysis*, § 4·11.

It is therefore sufficient to prove that, ϵ being assigned, N can be determined such that

$$S_n - s_n < \epsilon \quad \text{when } n > N.$$

This is true if, in ABC,

(i) $f(x, y)$ is a uniformly continuous function of x, *i.e.* given λ, arbitrarily small, a number σ, independent of x and y, may be found such that

$$|f(x', y) - f(x'', y)| < \lambda \quad \text{when } |x' - x''| < \sigma.$$

It will be supposed that the subdivision of (x_0, x) has been carried so far that the length of every interval $x_{r-1}x_r$ is less than σ.

(ii) The Lipschitz condition,

$$|f(x, y') - f(x, y'')| < K|y' - y''|,$$

is satisfied for all pairs of points in the triangle ABC which lie on lines parallel to BC.

In any given mode of subdivision with a pre-assigned value of λ, let

$$\delta_r = Y_r - y_r,$$

then

$$\delta_r = \delta_{r-1} + (M_r - m_r)(x_r - x_{r-1}).$$

But

$$\begin{aligned} M_r - m_r &= f(x_r', y_r') - f(x_r'', y_r'') \\ &= \{f(x_r', y_r') - f(x_r'', y_r')\} + \{f(x_r'', y_r') - f(x_r'', y_r'')\}, \end{aligned}$$

where (x_r', y_r') and (x_r'', y_r'') are the co-ordinates of two particular points in the trapezium $p_{r-1}P_{r-1}Q_rq_r$. Hence

$$M_r - m_r < \lambda + K|y_r' - y_r''|.$$

But

$$|y_r' - y_r''| \leqslant \delta_r \leqslant \delta_{r-1} + 2M(x_r - x_{r-1}),$$

and therefore

$$M_r - m_r < \lambda + 2MK(x_r - x_{r-1}) + K\delta_{r-1}.$$

Let the intervals be taken so small that

$$2MK(x_r - x_{r-1}) < \lambda$$

for $r = 1, 2, \ldots, n$, then

$$M_r - m_r < 2\lambda + K\delta_{r-1},$$

whence

$$\delta_r < \delta_{r-1}\{1 + K(x_r - x_{r-1})\} + 2\lambda(x_r - x_{r-1}),$$

and therefore

$$\begin{aligned} \delta_r + \frac{2\lambda}{K} &< \left(\delta_{r-1} + \frac{2\lambda}{K}\right)\left\{1 + K(x_r - x_{r-1})\right\} \\ &< \left(\delta_{r-1} + \frac{2\lambda}{K}\right)e^{K(x_r - x_{r-1})} \\ &< \left(\delta_{r-2} + \frac{2\lambda}{K}\right)e^{K(x_r - x_{r-2})} \\ &\quad \cdot \quad \cdot \quad \cdot \quad \cdot \\ &< \frac{2\lambda}{K}e^{K(x_r - x_0)}. \end{aligned}$$

Consequently

$$\delta_n + \frac{2\lambda}{K} < \frac{2\lambda}{K}e^{K(x - x_0)},$$

that is

$$S_n - s_n = \delta_n < \frac{2\lambda}{K}\left\{e^{K(x-x_0)} - 1\right\}$$
$$< \frac{2\lambda}{K}\left\{e^{Kh} - 1\right\} < \epsilon,$$

provided that λ, and therefore σ, is sufficiently small. Now λ is quite arbitrary; if therefore n is sufficiently large, and each interval sufficiently small,

$$S_n - s_n < \epsilon.$$

But σ is independent of x and consequently a number N, independent of x, exists such that this inequality holds for $n > N$ and for all x in the interval (x_0, x_0+h). *The expressions* S_n *and* s_n *therefore tend uniformly to a common limit* $F(x)$.

Let the two polygonal arcs $AP_1P_2 \ldots P_n$ and $Ap_1p_2 \ldots p_n$ be continued right up to the line BC, and let $P(x)$ be the ordinate of a point on the upper, and $Q(x)$ the ordinate of the corresponding point on the lower arc. Then

$$P(x) - Q(x) < \epsilon.$$

The two polygonal arcs therefore tend uniformly to a limit-curve Γ, namely the curve

$$y = F(x).$$

But $P(x)$ and $Q(x)$ are continuous, therefore $F(x)$ is continuous and Γ is a continuous curve.

Now any other continuous polygonal arc which lies below $AP_1P_2 \ldots$ and above $Ap_1p_2 \ldots$ has the same limit-curve Γ. In particular the polygonal arc Λ, the angular points of which have ordinates defined by the relation

$$z_r = z_{r-1} + f(x_{r-1}, z_{r-1})(x_r - x_{r-1}),$$

is so situated (Fig. 3) and its limit is the curve Γ. If therefore (x_r', y_r') is

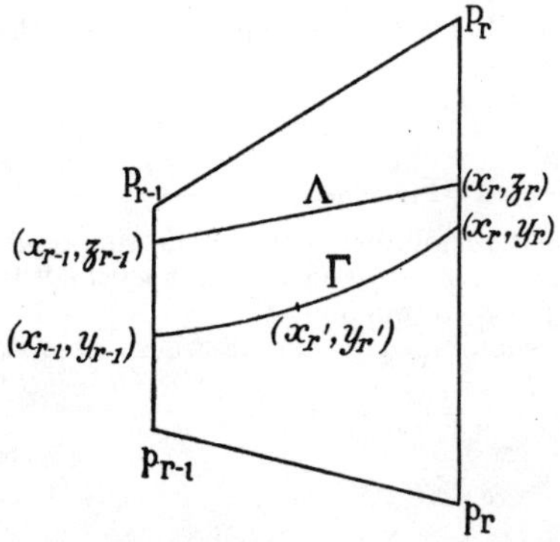

FIG. 3.

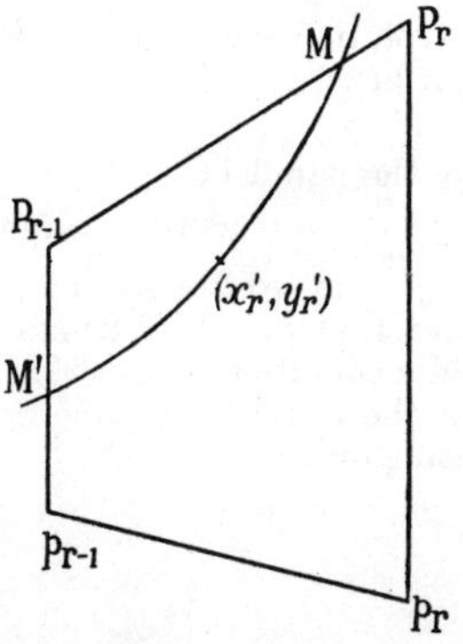

FIG. 4.

any point on the curve Γ lying in the trapezium $p_{r-1}\, P_{r-1}P_rp_r$, then the differences

$$x_r' - x_{r-1}, \quad |y_r' - z_{r-1}|$$

may be made arbitrarily small by assigning a sufficiently small upper bound to

$$x_1 - x_0,\ x_2 - x_1,\ \ldots,\ x_r - x_{r-1},$$

and therefore

$$|f(x_r', y_r')-f(x_{r-1}, z_{r-1})|$$

may be made arbitrarily small. Consequently the gradient of Γ at (x', y') is $f(x', y')$ and therefore Γ is an integral-curve of the differential equation. Moreover Γ passes through the point (x_0, y_0). Thus *the limit-function*

$$y=F(x)$$

is a solution of the differential equation and satisfies the initial conditions.

The integral-curve Γ is the only continuous integral-curve which passes through the point A. For if another such integral-curve existed, the subdivision of the interval (x_0, x_0+h) could be carried to such a degree of fineness that this integral-curve would pass across one or other of the polygonal arcs corresponding to this mode of subdivision. Suppose, for instance, that it crosses the arc $P_{r-1}P_r$ at the point M, and let M' be the point in which it cuts $p_{r-1}P_{r-1}$ (Fig. 4). Then the gradient of the chord $M'M$ is equal to the gradient of the curve at a point (x_r', y_r') of the arc $M'M$. But the gradient of the integral-curve at (x_r', y_r') is $f(x_r', y_r')$ which is by definition less than the gradient of $P_{r-1}P_r$, thus leading to a contradiction.

Consequently *there exists one and only one continuous solution of the differential equation which satisfies the initial conditions.*

3·41. Extended Range of the Cauchy-Lipschitz Method.—The method of successive approximation and the Cauchy-Lipschitz method lead to a demonstration of the existence and uniqueness of a continuous solution in the minimum interval (x_0, x_0+h). The ideal method would be one which leads to a solution which converges uniformly throughout any greater interval (x_0, x_0+k) in which the solution, defined by the assigned initial conditions, is continuous. The advantage of the Cauchy-Lipschitz method is that it does actually furnish a solution which converges in a maximum interval.

To show that such is the case, let

$$y=F(x)$$

be the solution such that $y_0=F(x_0)$. Let S be the strip bounded by the two straight lines

$$x=x_0, \quad x=x_0+k,$$

and by the parallel curves

$$y=F(x)-\eta, \quad y=F(x)+\eta$$

where η is an arbitrarily small positive number. It will be supposed that k is such that $F(x)$ is continuous in (x_0, x_0+k) and that η is so small that a Lipschitz condition is satisfied by $f(x, y)$ throughout S.

Let the interval (x_0, x_0+k) be subdivided by points whose abscissæ, in increasing order, are

$$x_0, x_1, \ldots, x_{n-1}, x_n,$$

where

$$x_n=x_0+k;$$

let

$$y_0, y_1, \ldots, y_{n-1}, y_n$$

be the corresponding ordinates of the integral-curve Γ, and let

$$y_0, z_1, \ldots, z_{n-1}, z_n$$

be the corresponding angular points of the polygonal line Λ defined by the recurrence formulæ

$$z_r=z_{r-1}+f(x_{r-1}, z_{r-1})(x_r-x_{r-1}),$$

with $z_0=y_0$ (Fig. 3).

It will now be proved that if the subdivision of the interval (x_0, x_0+k) is sufficiently fine, then the polygonal line Λ will be wholly within the strip S, and if $d_r = |z_r - y_r|$, then $d_r < \epsilon$ where ϵ is arbitrarily small. Let it be supposed that the angular points up to and including the point (x_{r-1}, z_{r-1}) are within the strip S. Then, by the mean-value theorem,

$$y_r = y_{r-1} + f(x_r', y_r')(x_r - x_{r-1}),$$

where (x_r', y_r') is a point of Γ lying between the points (x_{r-1}, y_{r-1}) and (x_r, y_r).

Consequently

$$z_r - y_r = z_{r-1} - y_{r-1} + \{f(x_{r-1}, z_{r-1}) - f(x_r', y_r')\}(x_r - x_{r-1}).$$

But

$$f(x_{r-1}, z_{r-1}) - f(x_r', y_r') = \{f(x_{r-1}, z_{r-1}) - f(x_{r-1}, y_{r-1})\} + \{f(x_{r-1}, y_{r-1}) - f(x_r', y_r')\},$$

and by the Lipschitz condition, since (x_{r-1}, z_{r-1}), (x_{r-1}, y_{r-1}) are both in S,

$$|f(x_{r-1}, z_{r-1}) - f(x_{r-1}, y_{r-1})| < K d_{r-1}.$$

Also since $f(x, y)$ is continuous in S, it is a continuous function of x along Γ, and therefore, if λ is arbitrarily assigned, σ may be chosen sufficiently small that

$$|f(x_{r-1}, y_{r-1}) - f(x_r', y_r')| < 2\lambda \quad \text{if} \quad |x_r - x_{r-1}| < \sigma.$$

Thus if the sub-interval (x_r, x_{r-1}) is sufficiently small,

$$d_r < d_{r-1} + (x_r - x_{r-1})(2\lambda + K d_{r-1}),$$

whence, as in the preceding section,

$$\delta_r < \frac{2\lambda}{K}\{e^{K(x_r - x_0)} - 1\}.$$

If, therefore, λ is so chosen that

$$2\lambda(e^{Kk} - 1) < K\eta,$$

then it follows by induction that

$$|d_1| < \eta, \ldots, |d_n| < \eta,$$

that is to say all the angular points of Λ lie within the strip S.

Let Λ' denote the polygonal line formed by joining the successive points of abscissæ $x_0, x_1, \ldots, x_n$ of the integral-curve Γ; let $P(x)$ be the ordinate of any point of Λ and $Q(x)$ be the ordinate of the corresponding point of Λ'. Then, if the difference between the greatest and least values of $F(x)$ in each sub-interval (x_{r-1}, x_r) is less than $\frac{1}{2}\epsilon$,

$$|Q(x) - F(x)| < \tfrac{1}{2}\epsilon.$$

Now η is arbitrary; let $\eta < \frac{1}{2}\epsilon$, then

$$|P(x) - Q(x)| < \tfrac{1}{2}\epsilon,$$

and since

$$P(x) - F(x) = \{P(x) - Q(x)\} + \{Q(x) - F(x)\},$$

it follows that, throughout the interval (x_0, x_0+k),

$$|P(x) - F(x)| < \epsilon.$$

Thus *if the equation possesses a solution*

$$y = F(x)$$

continuous in the interval (x_0, x_0+k), *and* ϵ *is an arbitrary positive number, the Cauchy-Lipschitz method will, for a sufficient fineness of subdivision of the interval, define a function* $P(x)$ *such that*

$$|P(x) - F(x)| < \epsilon$$

for

$$x_0 \leqslant x \leqslant x_0 + k.$$

3·5. Discussion of the Existence Theorem for an Equation not of the First Degree.—Consider a differential equation of the form

$$F\left(x, y, \frac{dy}{dx}\right)=0,$$

in which F is a polynomial in $\frac{dy}{dx}$, and is single-valued in x and y. Let (x_0, y_0) be any initial pair of values of (x, y). Then if the equation

$$F(x, y, p)=0$$

has a non-repeated root $p=p_0$ when $x=x_0$, $y=y_0$, it will have one and only one root

$$p=f(x, y),$$

which reduces to p_0 when $x=x_0$, $y=y_0$, and $f(x, y)$ will be single-valued in the neighbourhood of (x_0, y_0).

Now if $f(x, y)$ is continuous and satisfies a Lipschitz condition throughout a rectangle surrounding the point (x_0, y_0), the equation

$$\frac{dy}{dx}=f(x, y)$$

will possess a unique solution, continuous for values of x sufficiently near to x_0, and satisfying the assigned initial conditions. This solution clearly satisfies the original equation for the same range of values of x, and thus in this case the problem presents no new features.

On the other hand, when the given equation

$$F(x, y, p)=0$$

has a multiple root $p=p_0$ for $x=x_0$, $y=y_0$, then p is a non-uniform function of (x, y) in any domain including the point (x_0, y_0) and therefore the existence theorem is not applicable.

If $p=p_0$ is a root of multiplicity μ at (x_0, y_0), then

$$\frac{\partial F}{\partial p_0}=\ldots=\frac{\partial^{\mu-1}F}{\partial p_0^{\mu-1}}=0, \quad \frac{\partial^{\mu}F}{\partial p_0^{\mu}}\neq 0,$$

so that if

$$x=x_0+X, \quad y=y_0+Y, \quad p=p_0+P,$$

the equation $F(x, y, p)=0$ takes the form

$$\frac{\partial F}{\partial x_0}X+\frac{\partial F}{\partial y_0}Y+\frac{\partial^{\mu}F}{\partial p_0^{\mu}}\cdot\frac{P^{\mu}}{\mu\,!}+\ldots=0.$$

Let

$$Y=p_0X+Y_1,$$

then

$$p_0+P=p=\frac{dy}{dx}=\frac{dY}{dX}=p_0+\frac{dY_1}{dX},$$

and therefore

$$P=\frac{dY_1}{dX}.$$

Since X and P are small, Y_1 is of a higher order than X. Thus, retaining only terms of lowest order,

$$\left(\frac{\partial F}{\partial x_0}+p_0\frac{\partial F}{\partial y_0}\right)X+\frac{\partial^{\mu}F}{\partial p_0^{\mu}}\cdot\frac{P^{\mu}}{\mu\,!}+\ldots=0,$$

from which, with the assumption that

$$\frac{\partial F}{\partial x_0}+p_0\frac{\partial F}{\partial y_0}\neq 0,$$

it follows that

$$P^{\mu}=\left(\frac{dY_1}{dX}\right)^{\mu}=KX+\ldots,$$

where K is a constant, not zero, whence

$$Y_1=K_1X^{1+\frac{1}{\mu}}+\ldots,$$

where K_1 depends upon K and μ and is not zero. Thus *when the equations*

$$F(x, y, p)=0, \quad \frac{\partial F}{\partial p}=0, \ldots, \quad \frac{\partial^{\mu-1}F}{\partial p^{\mu-1}}=0, \quad \frac{\partial^{\mu}F}{\partial^{\mu}p}\neq 0, \quad \frac{\partial F}{\partial x}+p\frac{\partial F}{\partial y}\neq 0$$

are simultaneously satisfied for

$$x=x_0, \quad y=y_0, \quad p=p_0,$$

the solution which assumes the value y_0 when $x=x_0$ is, in the neighbourhood of (x_0, y_0), of the form

$$y=y_0+p_0(x-x_0)+K_1(x-x_0)^{1+\frac{1}{\mu}}+\ldots,$$

and is a function having μ values which become equal when $x=x_0$.

The most general case in which $F=0$, $F_p=0$ are satisfied simultaneously is when $F=0$ has a *double* root $p=p_0$ for $x=x_0$, $y=y_0$, and therefore $\mu=2$. In this case the solution is of the form

$$\{y-y_0-p_0(x-x_0)\}^2=A(x-x_0)^3+\ldots$$

and therefore *in the most general case the integral-curve has a cusp at (x_0, y_0).*

3·51. The p-discriminant and its locus.—A triad (x_0, y_0, p_0) for which

$$F=0, \quad F_p=0$$

is said to be a *singular line-element.* The corresponding pair of values (x_0, y_0) must satisfy the equation obtained by eliminating p between

$$F(x, y, p)=0, \quad F_p(x, y, p)=0.$$

The eliminant * is termed the *p-discriminant* † of the differential equation and is denoted by

$$\Delta_pF(x, y, p);$$

the curve which the equation

$$\Delta_pF(x, y, p)=0$$

in general defines is known as the *p-discriminant locus.*

Assuming for the moment that $x_0=0$, $y_0=0$, the differential equation can be written

$$F(x, y, p)\equiv U_0+U_1(p-p_0)+\ldots+U_m(p-p_0)^m=0,$$

where the coefficients are developable in series of ascending integral powers of x and y, and since $F(x, y, p)$ is to be of the second order in $p-p_0$ when $x=0$, $y=0$, U_0 and U_1 must be of the forms

$$U_0=\alpha_0x+\beta_0y+\ldots, \quad U_1=\alpha_1x+\beta_1y+\ldots.$$

Then the approximation to the p-discriminant at the origin is $p=p_0$ or

$$\alpha_0x+\beta_0y=0.$$

* It should be observed that in the process of elimination no variable factor is to be discarded. The use of a general method such as Sylvester's dialytic method of elimination (Scott and Mathews, *Theory of Determinants*, Chap. X., § 10) is therefore to be recommended.

† For references, see § 3·6.

But the integral curve has the equation

$$(y-p_0x)^2=Ax^3+\ .\ .\ .$$

and is therefore not, in general, tangential to the p-discriminant locus (Fig. 5). *In general the p-discriminant locus is the locus of cusps on the integral-curves of the differential equation.*

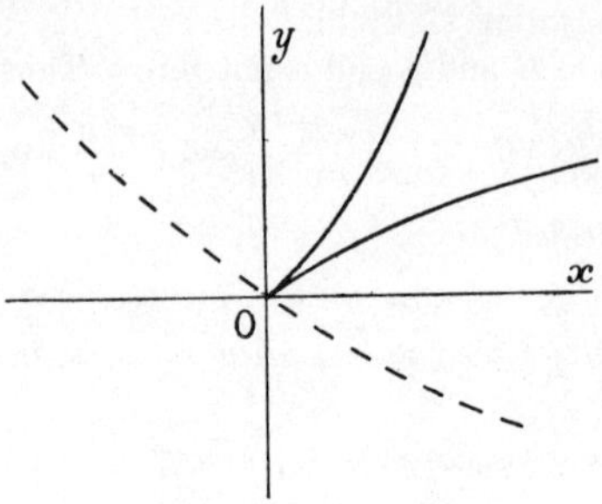

FIG. 5.

[The p-discriminant is the broken line, the integral-curve which meets the p-discriminant at the origin is the full line.]

At a point on the p-discriminant locus, the equation

$$F(x, y, p)=0$$

has at least two equal roots in p. This is in general owing to the presence of a cusp of an integral-curve at the point in question. When more than two roots in p become equal, there is in general a multiple point with coincident tangents. The preceding theorem thus becomes still more general if the term *locus of cusps* is understood to mean *locus of multiple points with coincident tangents.*

But the p-discriminant locus is not necessarily only a locus of cusps, because equal roots in p may occur through circumstances other than the presence of a cusp. The most important case of all is when consecutive members of the family of integral-curves have the same tangent, that is to say at points on the envelope of the family of integral-curves. The p-discriminant therefore includes the envelope in all cases in which an envelope exists. Moreover, the envelope is an integral-curve, for the line-elements of the envelope coincide with the line-elements of the integral-curves at the points of contact, and thus the envelope is built up of continuous line-elements which satisfy the differential equation. But the line-elements on the p-discriminant are, by definition, singular; the envelope is therefore said to be a *singular integral-curve.* An example of an envelope singular solution has been met with in the Clairaut equation (§ 2·44).

A singular integral-curve is not, however, necessarily an envelope; the exceptional case arises when the singular integral-curve touches every member of the family of integral-curves at a point which is the same for all curves. In this case the singular integral-curve is a member of the general family of integral-curves and is obtained by assigning a particular value to the parameter of the family. It is generally known as a *particular curve.*

As an example, consider the equation

$$(2x-p)^2+x(y-x)^2(2x-p)-(y-x^2)^3=0,$$

whose general solution is $$y=x^2+\frac{c^2}{1+cx}.$$

The p-discriminant of the equation, as well as the c-discriminant of its solution, contain the factor $y-x^2$, and yet the curve $y=x^2$ is not an envelope. In fact this curve does not have any finite point in common with any integral-curve for which $c\neq 0$. It is therefore a particular curve, and corresponds to $c=0$.

There remains one other possibility, namely that two non-consecutive integral-curves have the same tangent at a point on the p-discriminant locus. Such a point is said to be a *tac-point*; the locus of tac-points is a *tac-locus.* In general the common tangent to the integral-curves is not a tangent to the p-discriminant locus, and therefore the tac-locus, like the cusp-locus, is not, except in very special cases, an integral-curve of the differential equation.

3·52. The c-discriminant.—When the differential equation can be integrated, and its solution is

$$\Phi(x, y, c)=0,$$

the envelope, if it exists, is given by the c-discriminant equation

$$\Delta_c\Phi(x, y, c)=0,$$

obtained by eliminating c between the two equations

$$\Phi=0, \quad \frac{\partial\Phi}{\partial c}=0.$$

But, as will now be proved, the c-discriminant does not furnish the envelope alone.

Let the equations $\Phi=0$, $\Phi_c=0$ be solved for x and y, thus giving the c-discriminant in the parametric form

$$x=\phi(c), \quad y=\psi(c),$$

then the direction of tangent at any point of the c-discriminant locus is

$$\psi'(c)/\phi'(c).$$

Since

$$\frac{\partial\Phi}{\partial x}dx+\frac{\partial\Phi}{\partial y}dy=0,$$

the tangent at any point of the integral-curve $c=c_0$ has the direction

$$-\frac{\partial\Phi(x, y, c_0)}{\partial x}\bigg/\frac{\partial\Phi(x, y, c_0)}{\partial y},$$

Let (x_0, y_0) be the co-ordinates of a point of intersection of the two curves

$$\Phi(x, y, c_0)=0, \quad \frac{\partial\Phi}{\partial c_0}=0,$$

and if the functions ϕ and ψ are many-valued let them be so determined that

$$\phi(c_0)=x_0, \quad \psi(c_0)=y_0.$$

Then the parametric equations

$$x=\phi(c), \quad y=\psi(c)$$

represent a branch of the c-discriminant locus through (x_0, y_0).

Now at any point of the c-discriminant locus

$$\frac{\partial\Phi}{\partial x}\cdot\frac{\partial x}{\partial c}+\frac{\partial\Phi}{\partial y}\cdot\frac{\partial y}{\partial c}=0,$$

and therefore, at (x_0, y_0),

$$\left(\frac{\partial\Phi}{\partial x}\right)_0 \phi'(c_0)+\left(\frac{\partial\Phi}{\partial y}\right)_0 \psi'(c_0)=0.$$

Thus the integral-curve through (x_0, y_0) and the c-discriminant locus have a common tangent unless

$$\left(\frac{\partial\Phi}{\partial x}\right)_0=0, \quad \left(\frac{\partial\Phi}{\partial y}\right)_0=0,$$

that is to say, unless the integral-curve has a singular point at (x_0, y_0).

Thus the branch of the c-discriminant locus through (x_0, y_0) is either an envelope or a locus of singular points. In general the c-discriminant locus breaks up into two distinct parts, of which one furnishes the envelope, whilst the other furnishes the locus or loci of singular points. In the most general case the singular points are cusps and nodes, so that the c-discriminant locus includes the cusp- and node-loci. As in the example of the preceding section, a particular curve may also be included. The c-discriminant and p-discriminant loci, therefore, have in common the envelope and cusp-locus and possibly also a particular curve.

It is not always possible to obtain the explicit general solution of an equation and therefore it is necessary to investigate criteria for the discrimination of the various curves which may occur in the p-discriminant locus without having recourse to the solution. These criteria will be obtained after the foregoing discussion has been illustrated by examples.

3·521. Examples of Discriminant-loci.—

(i) The curves of the family

$$(y+c)^2=x(x-\alpha)(x-\beta),$$

where c is the parameter of the family and α and β are constants $(\beta>\alpha>0)$, are integral-curves of the differential equation

$$4p^2x(x-\alpha)(x-\beta)=\{3x^2-2(\alpha+\beta)x+\alpha\beta\}^2.$$

The p-discriminant equation is

$$x(x-\alpha)(x-\beta)\{3x^2-2(\alpha+\beta)x+\alpha\beta\}^2=0,$$

and the c-discriminant equation is

$$x(x-\alpha)(x-\beta)=0.$$

The three lines

$$x=0, \quad x=\alpha, \quad x=\beta$$

are common to both discriminant loci, each line touches every member of the family, and therefore the three lines form the envelope. The remaining part of the p-discriminant locus breaks up into two pairs of coincident straight lines

$$3x=\alpha+\beta+\sqrt{(\alpha^2-\alpha\beta+\beta^2)},$$
$$3x=\alpha+\beta-\sqrt{(\alpha^2-\alpha\beta+\beta^2)}.$$

These are tac-loci ; the former is the locus of imaginary, and the latter of real, points of contact of non-consecutive curves of the family.

(ii) Now let $\beta=\alpha>0$; the differential equation of the family

$$(y+c)^2=x(x-\alpha)^2$$

is

$$4p^2x=(3x-\alpha)^2.$$

The p-discriminant equation is

$$x(3x-\alpha)^2=0,$$

and the c-discriminant equation is

$$x(x-\alpha)^2=0.$$

The common ocus $x=0$ is the envelope. The p-discriminant locus also contains the line $x=\frac{1}{3}\alpha$ which is the tac-locus, and the c-discriminant locus contains the line $x=\alpha$ which is the node locus.

(iii) Finally, let $\beta=\alpha=0$; the differential equation of the family

$$(y+c)^2=x^3$$

is

$$4p^2=9x.$$

The p-discriminant locus is $x=0$ and the c-discriminant locus is $x^3=0$. Every member of the family of integral-curves has a cusp on the y-axis, which is therefore a cusp-locus.

3·6. Singular Solutions.—When a continuous succession of singular line-elements build up an integral-curve of the equation, that integral-curve is singular, and the corresponding solution is known as a *singular solution.** Since singular line-elements exist, by definition, only at points on the p-discriminant locus, a singular integral-curve must be a branch of the p-discriminant locus.

To obtain the direction of the tangent at any point of the p-discriminant locus, differentiate the equation

$$F(x, y, p)=0$$

with respect to x, thus

$$\frac{\partial F}{\partial x}+\frac{\partial F}{\partial y}\cdot\frac{dy}{dx}+\frac{\partial F}{\partial p}\cdot\frac{dp}{dx}=0.$$

But at any point on the p-discriminant locus

$$\frac{\partial F}{\partial p}=0,$$

and therefore the direction of the tangent is given by

$$\frac{\partial F}{\partial x}+\frac{\partial F}{\partial y}\cdot\frac{dy}{dx}=0.$$

But since the tangent to the p-discriminant locus now coincides with the tangent to an integral-curve,

$$\frac{dy}{dx}=p,$$

and therefore a *necessary condition for the existence of a singular solution is that the three equations*

$$F(x, y, p)=0,$$

$$\frac{\partial F(x, y, p)}{\partial p}=0,$$

$$\frac{\partial F(x, y, p)}{\partial x}+p\frac{\partial F(x, y, p)}{\partial y}=0$$

should be satisfied simultaneously for a continuous set of values of (x, y).

* The first examples of singular solutions were given by Brook Taylor in 1715 (see Appendix A). The earlier attempts at a systematic treatment of the subject, such as Lagrange, *Mém. Acad. Sc. Berlin*, 1774 [*Œuvres*, 4, p. 5] ; De Morgan, *Trans. Camb. Phil. Soc.* 9 (1851), p. 107 ; Darboux, *C. R. Acad. Sc. Paris*, 70 (1870), p. 1331 ; 71, p. 267 ; *Bull. Sc. Math.* 4 (1873), p. 158 ; Mansion, *Bull. Acad. Sc. Belg.* 34 (1872), p. 149 ; Cayley, *Mess. Math.* 2 (1873), p. 6 ; 6 (1877), p. 23 [*Coll. Math. Papers*, 8, p. 529 ; 10, p. 19] ; Glaisher, *ibid.* 12 (1882), p. 1 ; Hamburger, *J. für Math.* 112 (1893), p. 205, are not altogether satisfactory. The first complete direct treatment of the p-discriminant is due to Chrystal, *Trans. Roy. Soc. Edin.* 38 (1896), p. 803. Other noteworthy papers are : Hill, *Proc. London Math. Soc.* (1) 19 (1888), p. 561 ; 22 (1891), p. 216 ; Hudson, *ibid.* 33 (1901), p. 380 ; Petrovitch, *Math. Ann.* 50 (1898), p. 103. See also Bateman, *Differential Equations*, Chap. IV. The theory has been extended to equations with transcendental coefficients by Hill, *Proc. London Math. Soc.* (2) 17 (1918), p. 149.

Conversely suppose that the equations

$$F(x, y, \lambda)=0,$$
$$\frac{\partial F(x, y, \lambda)}{\partial \lambda}=0,$$
$$\frac{\partial F(x, y, \lambda)}{\partial x}+\lambda\frac{\partial F(x, y, \lambda)}{\partial y}=0,$$

where λ is a parameter, represent a curve. Then by differentiating the first equation and simplifying the derived equation by means of the second, the direction, p, of the tangent at any point of the curve is given by

$$\frac{\partial F(x, y, \lambda)}{\partial x}+p\frac{\partial F(x, y, \lambda)}{\partial y}=0,$$

and therefore, in view of the third equation,

$$(p-\lambda)\frac{\partial F(x, y, \lambda)}{\partial y}=0.$$

Consequently, if F_y is not zero at all points of the curve,

$$\lambda=p,$$

and therefore the curve is an integral-curve of the differential equation

$$F(x, y, p)=0.$$

Thus *the conditions*

$$F=0, \quad F_p=0, \quad F_x+pF_y=0,$$

together with the condition $F_y \neq 0$, *are sufficient for the existence of a singular solution.**

3·61. Conditions for a Tac-locus.—It was seen in § 3·5 that if

$$\frac{\partial F}{\partial x}+p\frac{\partial F}{\partial y}\neq 0$$

at all but a finite number of points of a branch of the p-discriminant locus, that branch is a cusp-locus or locus of multiple points. At any point at which

$$\frac{\partial F}{\partial x}+p\frac{\partial F}{\partial y}=0,$$

two distinct integral-curves touch one another. If, in the notation of the preceding section, $\lambda \neq p$, the integral-curves do not touch, and therefore are both distinct from, the p-discriminant locus, or in other words a tac-point occurs. Necessary conditions for a tac-point are therefore

$$\frac{\partial F}{\partial x}=0, \quad \frac{\partial F}{\partial y}=0,$$

which implies that at a tac-point a double-point of the p-discriminant locus occurs.

In order that the p-discriminant may furnish a tac-locus it is necessary that every point of some particular branch should be a double-point, which is impossible unless that branch is a double-line. The p-discriminant must therefore contain (as in § 3·521, (i) and (ii)) a squared factor, which, equated to zero, gives the equation of the tac-locus.

It follows that *a necessary condition that the p-discriminant should furnish a tac-locus is that the four equations*

$$F(x, y, p)=0, \quad F_p=0, \quad F_x=0, \quad F_y=0$$

should be satisfied for a continuous set of values of (x, y).

* The examples of § 3·521 show that an envelope may exist when $F_y=0$.

Since these equations are satisfied at every point of a tac-locus

$$F_{pp}dp+F_{px}dx+F_{py}dy=0,$$
$$F_{px}dp+F_{xx}dx+F_{xy}dy=0,$$
$$F_{py}dp+F_{xy}dx+F_{yy}dy=0,$$

and thus the condition for a tac-locus becomes

$$\begin{vmatrix} F_{pp}, & F_{px}, & F_{py} \\ F_{px}, & F_{xx}, & F_{xy} \\ F_{py}, & F_{xy}, & F_{yy} \end{vmatrix}=0.$$

3·611. A Deduction from the Symmetry of the Condition for a Tac-locus.—It appears from the symmetry of the conditions for a tac-locus that if the p-discriminant of the equation $F(x, y, p)=0$ furnishes a tac-locus, the same is true, in general, with regard to the equations

$$F(y, x, p)=0, \quad F(x, p, y)=0, \quad F(y, p, x)=0, \quad F(p, x, y)=0, \quad F(p, y, x)=0.$$

In particular cases, however, the tac-locus may reduce to a tac-point.

Consider, for example, the equation *

$$F(x, y, p)\equiv(x^2-a^2)p^2-2xyp-x^2=0.$$

The conditions for a tac-locus are

$$xp^2-yp-x=0, \quad xp=0, \quad (x^2-a^2)p-xy=0,$$

whence

$$x=0, \quad y=y, \quad p=0.$$

The tac-locus is $x=0$. In the case of the equation

$$F(y, x, p)\equiv(y^2-a^2)p^2-2xyp-y^2=0,$$

the conditions are

$$x=x, \quad y=0, \quad p=0,$$

and the tac-locus is $y=0$. But in the equation

$$F(x, p, y)=(x^2-a^2)y^2-2xpy-x^2=0,$$

the conditions are

$$x=0, \quad y=0, \quad p=p,$$

and there is no tac-locus, but a tac-point at the origin.

3·62. The Locus of Inflexions.—An integral-curve may be regarded either as the locus of its points or as the envelope of its tangents. Now the analytical conditions for a cusp, in point-co-ordinates, are formally identical with the analytical conditions for an inflexion in line-co-ordinates. Since, therefore, the family of integral-curves has in general a cusp-locus, it will have, in general, also an inflexion locus.

Since

$$\frac{\partial F}{\partial x}+\frac{\partial F}{\partial y}p+\frac{\partial F}{\partial p}\cdot\frac{dp}{dx}=0,$$

and at an inflexion $\frac{dp}{dx}=0$, *the inflexion locus is furnished by the p-eliminant of the equations*

$$F(x, y, p)=0, \quad F_x+pF_y=0.$$

In the general case $\frac{d^2p}{dx^2}$ is finite on the inflexion locus. But

$$\frac{\partial^2 F}{\partial x^2}+2p\frac{\partial^2 F}{\partial x \partial y}+p^2\frac{\partial^2 F}{\partial y^2}+\frac{\partial F}{\partial p}\cdot\frac{d^2p}{dx^2}=0,$$

* Glaisher, *Mess. Math.* 12 (1882), p. 6.

and therefore it is necessary that

$$F_p \neq 0.$$

3·7. Discussion of a Special Differential Equation.—The equation

$$F(x, y, p) \equiv ay + \beta x^2 + \gamma xp + p^2 = 0$$

will now be considered.* It will first of all be proved that when the equation has an envelope singular-solution, its integral-curves are algebraic. When the equation is solved for p,

$$p = -\tfrac{1}{2}\{\gamma x \pm \sqrt{(\gamma^2 x^2 - 4\beta x^2 - 4ay)}\}.$$

Let $y = vx^2$, so that

$$xv' + 2v = -\tfrac{1}{2}\{\gamma \pm \sqrt{(\gamma^2 - 4\beta - 4av)}\},$$

and, assuming that $a \neq 0$, write

$$u^2 = \gamma^2 - 4\beta - 4av.$$

The equation is now rational, and its variables are separable, thus

$$\frac{udu}{a\gamma + \gamma^2 - 4\beta \pm au - u^2} - \frac{dx}{x} = 0.$$

The conditions $F_p = 0$, $F_x + pF_y = 0$, for a singular solution, are respectively

$$\gamma x + 2p = 0, \quad 2\beta x + \gamma p + pa = 0,$$

whence, eliminating p,

$$a\gamma + \gamma^2 - 4\beta = 0.$$

With this condition the equation in u and x is reduced to

$$\frac{du}{u \pm a} + \frac{dx}{x} = 0,$$

and has the general solution

$$x(u \pm a) = \text{const.},$$

or

$$ax \pm \sqrt{(-a\gamma x^2 - 4ay)} = c,$$

where c is the parameter of the family of integral-curves. In its rationalised form the solution is

$$(ax - c)^2 + a(\gamma x^2 + 4y) = 0,$$

and the integral-curves constitute a family of parabolæ, whose envelope is the parabola

$$4y + \gamma x^2 = 0.$$

Thus, when there exists an envelope singular-solution, the integral-curves are algebraic. The converse is not, however, true. In order to obtain a condition that the general solution be algebraic, express the equation in the form

$$\frac{udu}{(u-\lambda)(u-\mu)} + \frac{dx}{x} = 0,$$

where

$$(u-\lambda)(u-\mu) = u^2 \pm au - a\gamma - \gamma^2 + 4\beta.$$

Let

$$2\lambda = \pm a + \sqrt{k}, \quad 2\mu = \pm a - \sqrt{k},$$

* This equation is effectively the first approximation in the neighbourhood of the origin to the equation $F(x, y, p) = 0$, when the axes are so chosen that an integral-curve touches the x-axis at the origin. The investigation here reproduced is due to Chrystal, *Trans. Roy. Soc. Edin.* 38 (1896), p. 813.

then

$$a^2-k=4(-a\gamma-\gamma^2+4\beta),$$

that is

$$k=(a+2\gamma)^2-16\beta.$$

The solution now is

$$\left(\pm\frac{a}{\sqrt{k}}+1\right)\log(u-\lambda)-\left(\pm\frac{a}{\sqrt{k}}-1\right)\log(u-\mu)+\log x^2=\log c^2,$$

whence

$$\left\{\frac{u-\lambda}{u-\mu}\right\}^{\pm a/\sqrt{k}}(u-\lambda)(u-\mu)x^2=c^2,$$

where

$$u=\pm\sqrt{(\gamma^2-4\beta-4ayx^{-2})}.$$

Thus, assuming that a, β and γ are rational numbers, a necessary and sufficient condition that the general solution be algebraic is that k, or

$$(a+2\gamma)^2-16\beta,$$

be the square of a rational number.

But when this condition is satisfied, the condition for an envelope singular-solution, namely

$$a\gamma+\gamma^2-4\beta=0,$$

is not necessarily satisfied. On the other hand, when this condition is satisfied,

$$(a+2\gamma)^2-16\beta=(a+2\gamma)^2-4(a\gamma+\gamma^2)$$
$$=a^2,$$

and the general solution is algebraic.

The equation

$$3y+\tfrac{1}{9}x^2-\tfrac{2}{3}xp+p^2=0$$

has an algebraic primitive, namely,

$$(x^2+12y)c^2-2x(x^2+9y)c+(x^2+3y)^2=0.$$

The c- and p-discriminants are effectively y^3 and y respectively. The negative half of the y-axis is a locus of real cusps. There is no true envelope because the point of ultimate intersection of consecutive curves is the same, namely the origin, for all curves of the family.

Miscellaneous Examples.

1. Modify the method of successive approximations so as to prove the following existence theorem. If x_0, y_0, a, b, and K have the meanings attributed to them in § 3·1, but M now signifies the upper bound of $|f(x, y_0)|$ for values of x in the interval (x_0, x_0+a), then there exists a unique solution of the equation

$$y'=f(x, y),$$

which reduces to y_0 when $x=x_0$ and is continuous in the interval $(x_0, x_0+\rho)$, where ρ is the smaller of the two numbers a and $K^{-1}\log(1+KbM^{-1})$.

[Lindelöf, *J. de Math.* (4) 10 (1894), p. 117.]

2. Investigate the behaviour, near the origin, of solutions of

(i) $y'=y^2$; (ii) $x^2y'=y$;

(iii) $y'=-\dfrac{x+2x^3}{y+2y^3}$; (iv) $y'=\dfrac{4x^3y}{x^4+y^2}$;

(v) $xy'+y^2=0$; (vi) $x^2y'+y=x$.

3. Discuss the p- and c-discriminants of the equations :

(i) $3xy=2px^2-2p^2$, Primitive : $(3y+2c)^2=4cx^3$;
(ii) $p^3-4xyp+8y^2=0$, Primitive : $y=c(x-c)^2$;
(iii) $xp^2-2yp+4x=0$, Primitive : $cy=c^2x^2+1$;
(iv) $p^2(2-3y)^2=4(1-y)$, Primitive : $y^2-y^3=(x-c)^2$;
(v) $yp^2-4xp+y=0$, Primitive : $y^6-3x^2y^4+2cx(3y^2-8x^2)+c^2=0$;
(vi) $8p^3x=y(12p^2-9)$, Primitive : $3cy^2=(x+c)^3$.

4. Integrate the equation $(y+px)^2=4x^2p$ and discuss the discriminants.

5. Show that the equation

$$(1-x^2)p^2=1-y^2$$

represents a family of conics touching the four sides of a square.

6. Let $\Phi(x, y, c)=0$ be a general family of integral-curves. Then $\Phi(x, y, z)=0$ represents a surface, and the c-discriminant locus is the orthogonal projection on the (x, y)-plane of the curve of intersection of the two surfaces

$$\Phi(x, y, z)=0, \quad \frac{\partial \Phi}{\partial z}=0.$$

By considering the section of $\Phi=0$ by a plane parallel to the z-axis, prove that in general

$$\Delta_c \Phi(x, y, c)=EN^2C^3,$$

where $E=0$ is the envelope, $N=0$ is the node-locus and $C=0$ is the cusp-locus.

[Cayley, Hill, Hudson ; see Salmon, *Higher Plane Curves*, 3rd ed., p. 54.]

7. Show that the locus of inflexions on the orthogonal trajectories of $F(x, y, p)=0$ is a branch of the curve

$$F(x, y, p)=0, \quad pF_x-F_y=0.$$

Discuss the case in which this curve has a branch in common with

$$F(x, y, p)=0, \quad F_x+pF_y=0.$$

[Chrystal.]

8. Show that an irreducible differential equation of the first order, polynomial in x, y and p, whose degree in x, y and p collectively does not exceed the second, can have no tac-locus. [Chrystal.]

CHAPTER IV

CONTINUOUS TRANSFORMATION-GROUPS

4·1. Lie's Theory of Differential Equations.—The earliest researches in the subject of differential equations were devoted to the problem of integration in the crude sense, that is to say to finding devices by which particular equations or classes of equations could be forced to yield up their solutions directly, or be reduced to a more tractable form. The next stage was the investigation of existence theorems, which served as criteria to settle, in a rigorous manner, the question of the existence of solutions of those equations which were not found to be integrable by elementary methods. Thus on the one hand, there exists a number of apparently disconnected methods of integration, each adapted only to one particular class of equations, whilst on the other hand, the existence theorems show that, except possibly for certain very unnatural equations, every equation has one or more solutions.

This heterogeneous mass of knowledge was co-ordinated in a very striking way by means of the theory of continuous groups.* The older methods of integration were shown to depend upon one general principle, which in its turn proved to be a powerful instrument for breaking new ground. In the following sections this co-ordinating method will be explained in its simplest aspects and with reference only to equations of the first order in one independent and one dependent variable.

4·11. The Transformation-Group of One Parameter.—Consider a transformation

(T) $$x_1=\phi(x, y), \quad y_1=\psi(x, y),$$

by means of which the point (x, y) is transferred to the new position (x_1, y_1) in the same plane and referred to the same pair of rectangular axes. If the equations which represent the transformation are solved for x and y in terms of x_1 and y_1, thus

$$x=\Phi(x_1, y_1), \quad y=\Psi(x_1, y_1),$$

they represent the *inverse* transformation (T_1), namely the operation of transferring the point (x_1, y_1) back to its original position (x, y). The result of performing the transformations T and T_1 in succession, in either order, is the *identical* transformation

$$x_1=x, \quad y_1=y.$$

Now consider the aggregate of the transformations included in the family

$$x_1=\phi(x, y\,;\, a), \quad y_1=\psi(x, y\,;\, a),$$

* Klein and Lie, *Math. Ann.* 4 (1871), p. 80; Lie, *Forhand. Vid.-Selsk. Christiania* (1874), p. 198; (1875), p. 1; *Math. Ann.* 9 (1876), p. 245; 11 (1877), p. 464; 24 (1884), p. 537; 25 (1885), p. 71 [Lie's *Ges. Abhandlungen*, iii. iv.] See also Lie-Scheffers, *Vorlesungen über Differentialgleichungen mit Bekannten Infinitesimalen Transformationen* (1891) and Page, *Ordinary Differential Equations* (1896).

where a is a parameter which can vary continuously over a given range.* Any particular transformation of the family is obtained by assigning a particular value to a. Now, in general, the result of applying two successive transformations of the family is not identical with the result of applying a third transformation of the family, for in general a_3 cannot be found such that

$$\phi(x, y\,;\, a_3)=\phi\{\phi(x, y\,;\, a_1),\quad \psi(x, y\,;\, a_1)\,;\, a_2\},$$

or in particular, taking

$$\phi\,(x, y;\, a)=a-x,\quad \psi(x, y\,;\, a)=y$$

it is not true that a_3 can be so chosen that for all values of x,

$$a_3-x=a_2-(a_1-x).$$

When, however, any two successive transformations of the family are equivalent to a single transformation of the family, the transformations are said to form a *finite continuous group*. It will be assumed that every group considered contains the inverse of each of its transformations, and therefore also the identical transformation. Since the transformations which form the group depend upon a single parameter a, the group will be referred to as a G_1 or *group of one parameter*.

4·111. *Examples of G_1.*—

(*a*) The group of translations parallel to the x-axis

$$x_1=x+a,\quad y_1=y.$$

The result of performing in succession the transformations of parameters a_1 and a_2 is

$$x_1=x+a_1+a_2,\quad y_1=y,$$

and is the transformation of parameter a_1+a_2. The inverse of the transformation of parameter a_1 is

$$x_1=x-a_1,\quad y_1=y\,;$$

its parameter is $-a_1$

(*b*) The group of rotations about the origin

$$x_1=x\cos a-y\sin a,\quad y_1=x\sin a+y\cos a.$$

The result of performing successive transformations of parameters a_1 and a_2 is

$$\begin{aligned}x_1&=(x\cos a_2-y\sin a_2)\cos a_1-(x\sin a_2+y\cos a_2)\sin a_1\\&=x\cos(a_1+a_2)-y\sin(a_1+a_2),\\y_1&=(x\cos a_2-y\sin a_2)\sin a_1+(x\sin a_2+y\cos a_2)\cos a_1\\&=x\sin(a_1+a_2)+y\cos(a_1+a_2),\end{aligned}$$

and is the transformation of parameter a_1+a_2. The inverse of the transformation of parameter a_1 is

$$x_1=x\cos a_1+y\sin a_1,\quad y_1=-x\sin a_1+y\cos a_1\,;$$

its parameter is $-a_1$.

(*c*) The group

$$x_1=ax,\quad y_1=a^2y.$$

The transformations of parameters a_1 and a_2 applied in succession are equivalent to the transformation of parameter a_1a_2. The inverse of the transformation of parameter a_1 is the transformation of parameter $1/a_1$.

4·12. Infinitesimal Transformations.—Let a_0 be that value of the parameter a which corresponds to the identical transformation, so that

$$\phi(x, y\,;\, a_0)=x,\quad \psi(x, y\,;\, a_0)=y,$$

* It will be assumed that ϕ and ψ are differentiable with respect to a in the given range.

then if ϵ is small, the transformation

$$x_1=\phi(x, y; a_0+\epsilon), \quad y_1=\psi(x, y; a_0+\epsilon)$$

will be such that x_1 differs only infinitesimally from x, and y_1 from y. This transformation therefore differs only infinitesimally from the identical transformation, and is said to be an *infinitesimal transformation.* It will now be shown that *every G_1 contains an infinitesimal transformation.**

Let α be any fixed value of the parameter a, and β the parameter of the corresponding inverse transformation. Thus

$$x_1=\phi(x, y; \alpha), \quad y_1=\psi(x, y; \alpha),$$
$$x=\phi(x_1, y_1; \beta), \quad y=\psi(x_1, y_1; \beta).$$

Let δt be small, and consider the transformation

$$x'=\phi\{\phi(x,y; \alpha), \ \psi(x,y; \alpha); \beta+\delta t\}, \ \ y'=\psi\{\phi(x,y; \alpha), \ \psi(x,y; \alpha); \beta+\delta t\}.$$

By the mean-value theorem, if θ_1 and θ_2 are positive and less than unity,

$$\begin{aligned}x' &=\phi\{\phi(x, y; \alpha), \psi(x, y; \alpha); \beta\}+\frac{\partial\phi\{\phi(x, y; \alpha), \psi(x, y; \alpha); \beta+\theta_1\delta t\}}{\partial\beta}\delta t\\ &=x+\xi(x, y; \alpha)\,\delta t,\\ y' &=\psi\{\phi(x, y; \alpha), \psi(x, y; \alpha); \beta\}+\frac{\partial\psi\{\phi(x, y; \alpha), \psi(x, y; \alpha); \beta+\theta_2\delta t\}}{\partial\beta}\delta t\\ &=y+\eta(x, y; \alpha)\delta t,\end{aligned}$$

where $\xi(x, y; \alpha)$, $\eta(x, y; \alpha)$ do not in general vanish identically and are independent of δt if terms of the second and higher orders are neglected. These equations represent an infinitesimal transformation. Every G_1 in two variables contains therefore an infinitesimal transformation; the method is evidently applicable to the case of any number of variables, with the same result.

Geometrically, this infinitesimal transformation represents a small displacement of length

$$\sqrt{\{(x'-x)^2+(y'-y)^2\}}=\sqrt{(\xi^2+\eta^2)}\,\delta t$$

in the direction θ where

$$\cos\theta=\xi/\sqrt{(\xi^2+\eta^2)}, \quad \sin\theta=\eta/\sqrt{(\xi^2+\eta^2)}.$$

Two transformation groups are said to be *similar* when they can be derived from one another by a change of variables and parameter. It will be shown that every G_1 in two variables is similar to the group of translations. To prove this theorem, write the equations of the infinitesimal transformation in the form

$$\delta x=\xi(x, y)\delta t, \quad \delta y=\eta(x, y)\delta t,$$

then the finite equations of the group are found by integrating the equations

$$\frac{dx}{\xi(x, y)}=\frac{dy}{\eta(x, y)}=dt.$$

The solutions are expressible in the form

$$F_1(x, y)=C_1, \quad F_2(x, y)=C_2+t,$$

where C_1 and C_2 are constants. Let $t=0$ correspond to the identical transformation, then

$$F_1(x_1, y_1)=F_1(x, y), \quad F_2(x_1, y_1)=F_2(x, y)+t.$$

Let $u=F_1(x, y)$, $v=F_2(x, y)$ be taken as new variables, then

$$u_1=u, \quad v_1=v+t.$$

* It will be proved later that no G_1 contains more than one infinitesimal transformation.

Thus the given group has been reduced to the group of translations. It is clear that this group has one and only one infinitesimal transformation, namely $\delta u=0$, $\delta v=\delta t$. Since ξ and η are uniquely determined in terms of u and v it follows that the original G_1 has *only one infinitesimal transformation.*

4·121. Examples.—(*a*) The identical transformation of the rotation group defined by

$$x_1=x \cos a-y \sin a, \quad y_1=x \sin a+y \cos a$$

is given by $a=0$. The infinitesimal transformation is therefore

$$x_1=x \cos \delta t-y \sin \delta t, \quad y_1=x \sin \delta t+y \cos \delta t,$$

or to the first order of small quantities

$$x_1=x-y\delta t, \quad y_1=y+x\delta t.$$

This transformation represents a rotation, in the positive sense, through the small angle δt.

(*b*) The equations

$$x_1=ax, \quad y_1=a^2y$$

define a group ; the identical transformation corresponds to $a=1$. The infinitesimal transformation therefore is

$$x_1=(1+\delta t)x, \quad y_1=(1+\delta t)^2y$$

or, to the first order of small quantities,

$$x_1=x+x\delta t, \quad y_1=y+2y\delta t.$$

To reduce the group to the translation group it is necessary to solve the equations

$$\frac{dx}{x}=\frac{dy}{2y}=dt,$$

whence

$$\frac{y_1}{x_1^2}=\frac{y}{x^2}, \quad \log x_1=\log x+t.$$

The required new variables are therefore

$$u=y/x^2, \quad v=\log x.$$

4·13. Notation for an Infinitesimal Transformation.—Consider the variation undergone by any given function $f(x, y)$ when the variables x, y are subjected to the infinitesimal transformation

$$x_1=x+\delta x=x+\xi(x, y)\delta t,$$
$$y_1=y+\delta y=y+\eta(x, y)\delta t.$$

The change in the value of $f(x, y)$ is :

$$\begin{aligned}\delta f(x, y)&=f(x_1, y_1)-f(x, y)\\&=f(x+\delta x, y+\delta y)-f(x, y)\\&=\left\{\xi(x, y)\frac{\partial f(x, y)}{\partial x}+\eta(x, y)\frac{\partial f(x, y)}{\partial y}\right\}\delta t,\end{aligned}$$

retaining only small quantities of the first order. Conversely, if the increment $\delta f(x, y)$ is known, which a given function $f(x, y)$ assumes under the infinitesimal transformation of a G_1, then $\xi(x, y)$ and $\eta(x, y)$ are known and therefore the infinitesimal transformation itself is known. Thus *the infinitesimal transformation is completely represented by the symbol*

$$Uf\equiv\xi(x, y)\frac{\partial f}{\partial x}+\eta(x, y)\frac{\partial \eta}{\partial y}.$$

Thus, for example, the symbol

$$-y\frac{\partial f}{\partial x}+x\frac{\partial f}{\partial y}$$

represents the infinitesimal rotation

$$x_1=x-y\delta t,\quad y_1=y+x\delta t.$$

In particular

$$Ux=\xi(x, y),\quad Uy=\eta(x, y),$$

so that

$$Uf=Ux\cdot\frac{\partial f}{\partial x}+Uy\cdot\frac{\partial f}{\partial y}.$$

It is obvious that if in a G_1 operating on variables x, y, these variables are replaced by x', y', where x', y' are any functions of x, y, the group property is maintained. Now since

$$\frac{\partial f(x', y')}{\partial x}=\frac{\partial f(x', y')}{\partial x'}\cdot\frac{\partial x'}{\partial x}+\frac{\partial f(x', y')}{\partial y'}\cdot\frac{\partial y'}{\partial x},$$

$$\frac{\partial f(x', y')}{\partial y}=\frac{\partial f(x', y')}{\partial x'}\cdot\frac{\partial x'}{\partial y}+\frac{\partial f(x', y')}{\partial y'}\cdot\frac{\partial y'}{\partial y},$$

it follows that

$$\begin{aligned}Uf(x', y') &= \xi(x', y')\left\{\frac{\partial f}{\partial x'}\cdot\frac{\partial x'}{\partial x}+\frac{\partial f}{\partial y'}\cdot\frac{\partial y'}{\partial x}\right\}+\eta(x', y')\left\{\frac{\partial f}{\partial x'}\cdot\frac{\partial x'}{\partial y}+\frac{\partial f}{\partial y'}\cdot\frac{\partial y'}{\partial y}\right\}\\ &= Ux'\cdot\frac{\partial f}{\partial x'}+Uy'\cdot\frac{\partial f}{\partial y'}.\end{aligned}$$

Now, let the finite equations of the G_1 be

$$x_1=\phi(x, y\,;t),\quad y_1=\psi(x, y\,;t),$$

and let $t=0$ give the identical transformation. The function $f(x_1, y_1)$ may be regarded as a function of x, y and t; regard x and y as fixed and let the function be expanded as a Maclaurin series in t, thus

$$f(x_1, y_1)=f_0+f_0't+\tfrac{1}{2}f_0''t^2+\ldots,$$

where

$$\begin{aligned}f_0&=[f(x_1, y_1)]_{t=0}=f(x, y),\\ f_0'&=\left[\frac{df(x_1, y_1)}{dt}\right]_{t=0}=\left[\frac{\partial f}{\partial x_1}\cdot\frac{dx_1}{dt}+\frac{\partial f}{\partial y_1}\cdot\frac{dy_1}{dt}\right]_{t=0}\\ &=\left[\frac{\partial f}{\partial x_1}\xi(x_1, y_1)+\frac{\partial f}{\partial y_1}\eta(x_1, y_1)\right]_{t=0}\\ &=\frac{\partial f}{\partial x}\xi(x, y)+\frac{\partial f}{\partial y}\eta(x, y)=Uf(x, y),\\ f_0''&=\left[\frac{d^2f(x_1, y_1)}{dt^2}\right]_{t=0}=\left[\left\{\xi(x_1, y_1)\frac{\partial}{\partial x_1}+\eta(x_1, y_1)\frac{\partial}{\partial y_1}\right\}^2 f(x_1, y_1)\right]_{t=0}\\ &=U^2f(x, y),\end{aligned}$$

. . . .

Consequently the expansion of $f(x_1, y_1)$ is

$$f(x_1, y_1)=f(x, y)+\frac{t}{1\,!}Uf+\frac{t^2}{2\,!}U^2f+\ldots,$$

where U^nf symbolises the result of operating n times in succession on $f(x, y)$ by the operator

$$U\equiv\xi(x, y)\frac{\partial}{\partial x}+\eta(x, y)\frac{\partial}{\partial y}.$$

In particular,

$$x_1 = x + \frac{t}{1!}Ux + \frac{t^2}{2!}U^2x + \ldots,$$

$$y_1 = y + \frac{t}{1!}Uy + \frac{t^2}{2!}U^2y + \ldots,$$

and these are the finite equations of the group. It may easily be verified that, regarding x and y as fixed numbers to which x_1 and y_1 reduce when $t=0$, these equations furnish the solution of the simultaneous system

$$\frac{dx_1}{\xi(x_1, y_1)} = \frac{dy_1}{\eta(x_1, y_1)} = dt.$$

The infinitesimal transformation therefore defines the group, which may therefore be spoken of as the group Uf.

4·131. Examples of the Deduction of the Finite Equations from the Infinitesimal Transformation.—

(*a*) Given the infinitesimal transformation

$$Uf \equiv -y\frac{\partial f}{\partial x} + x\frac{\partial f}{\partial y},$$

to find the corresponding G_1. It may be verified that

$$\begin{array}{ll} Ux = -y, & Uy = x, \\ U^2x = -x, & U^2y = -y, \\ U^3x = y, & U^3y = -x, \\ U^4x = x, & U^4y = y, \\ \ldots & \ldots \end{array}$$

Thus U is a cyclic operation of period 4, with respect to x and y. It follows that

$$x_1 = x - \frac{t}{1!}y - \frac{t^2}{2!}x + \frac{t^3}{3!}y + \frac{t^4}{4!}x - \ldots$$

$$= x\left(1 - \frac{t^2}{2!} + \frac{t^4}{4!} - \ldots\right) - y\left(\frac{t}{1!} - \frac{t^3}{3!} + \ldots\right)$$

$$= x\cos t - y\sin t,$$

$$y_1 = y + \frac{t}{1!}x - \frac{t^2}{2!}y - \frac{t^3}{3!}x + \frac{t^4}{4!}y + \ldots$$

$$= x\left(\frac{t}{1!} - \frac{t^3}{3!} + \ldots\right) + y\left(1 - \frac{t^2}{2!} + \frac{t^4}{4!} - \ldots\right)$$

$$= x\sin t + y\cos t.$$

Thus the corresponding G_1 is the rotation-group.

(*b*) In the same way, if

$$Uf \equiv x\frac{\partial f}{\partial x} + y\frac{\partial f}{\partial y},$$

it is found that

$$x_1 = x + \frac{t}{1!}x + \frac{t^2}{2!}x + \frac{t^3}{3!}x + \ldots = xe^t,$$

$$y_1 = y + \frac{t}{1!}y + \frac{t^2}{2!}y + \frac{t^3}{3!}y + \ldots = ye^t.$$

If e^t is replaced by the new parameter a, the equations become

$$x_1 = ax, \quad y_1 = ay,$$

and define the group of uniform magnifications.

4·2. Functions Invariant under a Given Group.—As before, let the finite equations of the group,

$$x_1=\phi(x, y; t), \quad y_1=\psi(x, y; t),$$

be such that the identical transformation corresponds to $t=0$, and let

$$Uf\equiv \xi(x, y)\frac{\partial f}{\partial x}+\eta(x, y)\frac{\partial f}{\partial y}$$

denote the infinitesimal transformation of the group.

A function $\Omega(x, y)$ is said to be invariant if, when x_1, y_1 are derived from x, y by operations of the given group,

$$\Omega(x_1, y_1)=\Omega(x, y),$$

for all values of t.

Now the expansion of $\Omega(x_1, y_1)$ in powers of t may be written in the form

$$\Omega(x, y)+\frac{t}{1!}U\Omega+\frac{t^2}{2!}U\{U\Omega\}+\frac{t^3}{3!}U^2\{U\Omega\}+\ldots$$

If, therefore, $\Omega(x, y)$ is invariant under the group, this expression must be equal to $\Omega(x, y)$ for all values of t in a given range. For this *it is necessary and sufficient that $U\Omega$ should be identically zero*, that is,

$$\xi\frac{\partial\Omega}{\partial x}+\eta\frac{\partial\Omega}{\partial y}\equiv 0.$$

The function $z=\Omega(x, y)$ is therefore a solution of the partial differential equation

$$\xi\frac{\partial z}{\partial x}+\eta\frac{\partial z}{\partial y}=0,$$

and consequently,

$$\Omega(x, y)=\text{constant}$$

is a solution of the equivalent ordinary differential equation

$$\frac{dx}{\xi}=\frac{dy}{\eta}.$$

Since this equation has one, and only one, solution depending upon a single arbitrary constant, it follows that every G_1 in two variables has one and only one independent invariant. In other words, there exists one invariant in terms of which all other invariants may be expressed.

4·201. The Invariants of the Group of Rotations.—The infinitesimal transformation of the G_1 of rotations is

$$Uf\equiv -y\frac{\partial f}{\partial x}+x\frac{\partial f}{\partial y}.$$

The equation to determine Ω is

$$\frac{dx}{y}+\frac{dy}{x}=0,$$

and has the solution $x^2+y^2=\text{const.}$ Hence

$$\Omega(x, y)=x^2+y^2.$$

It is, of course, geometrically evident that circles whose centres are at the origin are invariant under the group. To verify this fact analytically, note that the finite equations of the group are

$$x_1=x\cos t-y\sin t, \quad y_1=x\sin t+y\cos t.$$

Then

$$\begin{aligned}\Omega(x_1, y_1)=x_1{}^2+y_1{}^2&=(x\cos t-y\sin t)^2+(x\sin t+y\cos t)^2\\&=x^2+y^2=\Omega(x, y),\end{aligned}$$

whatever value t may have. The invariance of x^2+y^2 is therefore established. Any other invariant under the group must be a function of x^2+y^2, and conversely, any function of x^2+y^2 is invariant under the group.

4·21. Invariant Points, Curves and Families of Curves.—If, for any point in the (x, y)-plane, $\xi(x, y)$ and $\eta(x, y)$ are both zero, that point is a fixed point under the infinitesimal transformation

$$Uf=\xi\frac{\partial f}{\partial x}+\eta\frac{\partial f}{\partial y},$$

and is consequently fixed under all transformations of the group. Such points are said to be *absolutely invariant* under the group.

A point (x_0, y_0) which is not invariant under the group is transferred, by the infinitesimal transformation, into a neighbouring point $(x_0+\delta x, y_0+\delta y)$ such that

$$\frac{\delta y}{\delta x}=\frac{\eta}{\xi}.$$

If the infinitesimal transformation is repeated indefinitely, the point P, originally situated at (x_0, y_0), will trace out a curve which will be one of the integral-curves of the equation

$$\frac{dy}{dx}=\frac{\eta}{\xi}.$$

The family of integral-curves

$$\Omega(x, y)=\text{const.}$$

is such that each curve is invariant under the group.

But a family of curves may also be invariant in the sense that each curve is transformed into another curve of the same family by the operations of the group. Thus the family of curves may be invariant as a whole although the individual curves of the family are not invariant under the group. Let

$$\Omega(x, y)=\text{const.}$$

be such a family of curves. If, under any transformation of the group, (x, y) becomes (x_1, y_1), then

$$\Omega(x_1, y_1)=\text{const.}$$

must represent the same family of curves. But

$$\Omega(x_1, y_1)=\Omega(x, y)+\frac{t}{1\,!}U\Omega+\frac{t^2}{2\,!}U^2\Omega+\ \dots$$

and therefore, if the two families of curves

$$\Omega(x_1, y_1)=\text{const.},\quad \Omega(x, y)=\text{const.}$$

are identical, the expression

$$\frac{t}{1\,!}U\Omega+\frac{t^2}{2\,!}U^2\Omega+\ \dots$$

must be a constant for every fixed value of t, that is, for every curve of the family

$$U\Omega=\text{const.}$$

Thus a necessary and sufficient condition that $\Omega(x, y)=\text{const.}$ should represent a family of curves invariant, as a whole, under the group, is that $U\Omega=\text{const.}$ should represent the same family of curves, that is $U\Omega$ should be a certain function of Ω, say $F(\Omega)$. When $F(\Omega)$ is zero, the individual curves of the family are invariant curves.

Thus, for instance, under the rotation group

$$x_1=x\cos t-y\sin t,\quad y_1=y\cos t+x\sin t,$$

the family of straight lines $\dfrac{y}{x}=\alpha$

becomes $\dfrac{y}{x}=\beta,$

where α and β are the parameters of the families. But

$$\frac{y_1}{x_1}=\frac{y\cos t+x\sin t}{x\cos t-y\sin t}=\frac{y}{x}+\left(1+\frac{y^2}{x^2}\right)t+\left(\frac{y}{x}+\frac{y^3}{x^3}\right)t^2+\ \ldots$$

If the family $\dfrac{y}{x}=\alpha$ is invariant, the family

$$1+\frac{y^2}{x^2}=\gamma$$

must be identical with it. This is, in fact, the case ; the parameters α and γ being connected by the relation

$$\gamma=\alpha^2+1.$$

Now

$$\Omega=\frac{y}{x},\quad U\Omega=-y\frac{\partial\Omega}{\partial x}+x\frac{\partial\Omega}{\partial y}=\frac{y^2}{x^2}+1=\Omega^2+1.$$

This is the form which the condition $U\Omega=F(\Omega)$ takes in this case.

4·3. Extension to n Variables.—The G_1 in n variables $x_1, x_2, \ldots, x_n$ defined by the transformations

$$x_i'=\phi_i(x_1, x_2, \ldots, x_n\,;\,a)\qquad (i=1, 2, \ldots, n)$$

may be proved as above to admit of, and to be equivalent to, a unique infinitesimal transformation

$$Uf(x_1, x_2, \ldots, x_n)=\xi_1(x_1, x_2, \ldots, x_n)\frac{\partial f}{\partial x_1}+\ \ldots\ +\xi_n(x_1, x_2, \ldots, x_n)\frac{\partial f}{\partial x_n}.$$

Let t be the parameter of the group such that the infinitesimal transformation is

$$x_i'=x_i+\xi_i(x_1, x_2, \ldots, x_n)\delta t\qquad (i=1, 2, \ldots, n),$$

then if $F(x_1, x_2, \ldots, x_n)$ is a function which can be differentiated any number of times with respect to its arguments,

$$F(x_1', x_2', \ldots, x_n')=F(x_1, x_2, \ldots, x_n)+\frac{t}{1\,!}UF+\frac{t^2}{2\,!}U^2F+\ \ldots$$

Let $(x_1, x_2, \ldots, x_n)$ be considered as the co-ordinates of a point in space of n dimensions, and t as a parameter independent of these co-ordinates; t may, for instance, be regarded as a measure of time. As t varies, the point $(x_1', x_2', \ldots, x_n')$ describes a trajectory starting from the point $(x_1, x_2, \ldots, x_n)$. Every trajectory is evidently an invariant curve under the group.

As before, a necessary and sufficient condition that $\Omega(x_1, x_2, \ldots, x_n)$ should be an invariant function is that $U\Omega$ be identically zero. A curve $\Omega=0$ is a trajectory and therefore an invariant curve if $U\Omega=0$. The family of curves

$$\Omega=\text{const.}$$

is, as a family, invariant if $U\Omega$ is a definite function of Ω above.

Lastly, an equation

$$\Omega(x_1, x_2, \ldots, x_n)=0,$$

is invariant if $U\Omega$ is zero, whether identically or by virtue of the equation

$\Omega=0$. In the former case the equation $\Omega=a$ is invariant, and in the latter case not invariant, for all values of the constant a.

Examples of invariant equations are as follows :

(*a*) the equation $\Omega\equiv x^2+y^2-c^2=0$, where c is any constant, is invariant under the rotation-group, for

$$U\Omega=\left(-y\frac{\partial}{\partial x}+x\frac{\partial}{\partial y}\right)(x^2+y^2-c^2)=-2yx+2xy=0.$$

(*b*) the equation $\Omega\equiv y-x=0$ is invariant under the group

$$Uf=x\frac{\partial f}{\partial x}+y\frac{\partial f}{\partial y},$$

for

$$U\Omega=\left(x\frac{\partial}{\partial x}+y\frac{\partial}{\partial y}\right)(y-x)=-x+y=\Omega.$$

On the other hand, the equation $y-x+c=0$, where c is any constant not zero, is not invariant under the group.

4·4. Determination of all Equations which admit of a given Group.—An equation is said to *admit of* a given group when it is invariant under that group. Let the group be

$$Uf\equiv\xi_1(x_1, x_2, \ldots, x_n)\frac{\partial f}{\partial x_1}+\ldots+\xi_n(x_1, x_2, \ldots, x_n)\frac{\partial f}{\partial x_n}.$$

Let

$$\Omega(x_1, x_2, \ldots, x_n)=0$$

be an equation which admits of the group so that $U\Omega=0$. It will be supposed that Ω is not a factor common to all of the functions $\xi_1, \ldots, \xi_n$; let ξ_n, for instance, be not zero when $\Omega=0$. Then if

$$Vf\equiv\frac{\xi_1}{\xi_n}\cdot\frac{\partial f}{\partial x_1}+\ldots+\frac{\xi_{n-1}}{\xi_n}\cdot\frac{\partial f}{\partial x_{n-1}}+\frac{\partial f}{\partial x_n},$$

$V\Omega=0$, and therefore Ω is invariant under the group Vf.

Let $y_1, y_2, \ldots, y_{n-1}$ be an independent set of solutions of the partial differential equation

$$Vf=0;$$

since they are also solutions of $Uf=0$ they are functions of the original variables $x_1, x_2, \ldots, x_n$. Now adjoin to $y_1, y_2, \ldots, y_{n-1}$ the function x_n; the functions of the set thus formed are also independent, for if not there would be a relation of the form

$$x_n=W(y_1, y_2, \ldots, y_{n-1}),$$

and therefore x_n would be a solution of the linear partial differential equation $Vf=0$, which is manifestly untrue.

On the other hand $x_1, x_2, \ldots, x_{n-1}$ are expressible in terms of the n variables $y_1, y_2, \ldots, y_{n-1}$ and x_n. When this change of variables is effected let the invariant equation $\Omega=0$ become

$$\Psi(y_1, y_2, \ldots, y_{n-1}, x_n)=0.$$

Apparently Ψ involves x_n; in reality it does not. For if a is any constant,

$$V\Psi(y_1, y_2, \ldots, y_{n-1}, x_n)$$
$$=V\Psi(y_1, y_2, \ldots, y_{n-1}, a)+\frac{\partial}{\partial x_n}\Psi(y_1, y_2, \ldots, y_{n-1}, x_n).$$

Since $V\Psi(y_1, y_2, \ldots, y_{n-1}, a)$ is identically zero and $V\Psi(y_1, y_2, \ldots, y_{n-1}, x_n)$ is zero either identically or because of the equation $\Psi=0$, it follows that

$$\frac{\partial \Psi}{\partial x_n}=0,$$

that is to say, Ψ is effectively independent of x_n. Thus Ψ and consequently Ω is expressible in terms of $y_1, y_2, \ldots, y_{n-1}$ alone.

Thus if the equation $\Omega=0$ is invariant, Ω *must be expressible in terms of the* $n-1$ *independent solutions of the partial differential equation* $Uf=0$. In other words, every invariant equation $\Omega=0$ is a particular integral of the equation $Uf=0$.

In particular, if u and v are two independent solutions of the equation

$$Uf(x, y, z) \equiv \xi\frac{\partial f}{\partial x}+y\frac{\partial f}{\partial y}+\zeta\frac{\partial f}{\partial z}=0,$$

the most general equation, invariant under the group Uf has the form

$$\Omega(u, v)=0,$$

or the equivalent form

$$v-F(u)=0.$$

This result is the foundation upon which most of the following work will be based.

4·5. The Extended Group.—Let

$$x_1=\phi(x, y\,;\, a) \quad y_1=\psi(x, y\,;\, a)$$

define a G_1 in two variables. Consider the differential coefficient p as a third variable which under the group becomes p_1 where

$$p_1=\frac{dy_1}{dx_1}=\frac{d\psi}{d\phi}$$

$$=\frac{\dfrac{\partial\psi}{\partial x}+\dfrac{\partial\psi}{\partial y}p}{\dfrac{\partial\phi}{\partial x}+\dfrac{\partial\phi}{\partial y}p}=\chi(x, y, p\,;\, a).$$

Let α and β be two particular values of a, such that

$$x_1=\phi(x, y\,;\, \alpha), \quad y_1=\psi(x, y\,;\, \alpha),$$
$$x_2=\phi(x_1, y_1\,;\, \beta), \quad y_2=\psi(x_1, y_1\,;\, \beta),$$

then the resultant transformation

$$x_2=\phi(x, y\,;\, \gamma), \quad y_2=\psi(x, y\,;\, \gamma)$$

is the result of eliminating x_1, y_1 between the equations of the two component transformations. In the same way,

$$p_2=\frac{d\psi(x, y\,;\, \gamma)}{d\phi(x, y\,;\, \gamma)}$$

is the result of eliminating y_1 between

$$p_1=\frac{d\psi(x, y\,;\, \alpha)}{d\phi(x, y\,;\, \alpha)} \quad \text{and} \quad p_2=\frac{d\psi(x_1, y_1\,;\, \beta)}{d\phi(x_1, y_1\,;\, \beta)}.$$

Thus in general the transformations

$$x_1=\phi(x, y\,;\, a), \quad y_1=\psi(x, y\,;\, a), \quad p_1=\chi(x, y, p\,;\, a),$$

acting on the line-element (x, y, p), form a group. This group is known as the *extended group* of the given group.

The finite equations of the given group may be developed in the form

$$x_1=x+\frac{t}{1!}\xi(x, y)+\frac{t^2}{2!}\left(\xi\frac{\partial\xi}{\partial x}+\eta\frac{\partial\xi}{\partial y}\right)+\ldots,$$

$$y_1=y+\frac{t}{1!}\eta(x, y)+\frac{t^2}{2!}\left(\xi\frac{\partial\eta}{\partial x}+\eta\frac{\partial\eta}{\partial y}\right)+\ldots,$$

and hence

$$p_1=\frac{dy+t\left(\frac{\partial\eta}{\partial x}dx+\frac{\partial\eta}{\partial y}dy\right)+\ldots}{dx+t\left(\frac{\partial\xi}{\partial x}dx+\frac{\partial\xi}{\partial y}dy\right)+\ldots}$$

$$=p+t\left\{\frac{\partial\eta}{\partial x}+\left(\frac{\partial\eta}{\partial y}-\frac{\partial\xi}{\partial x}\right)p-\frac{\partial\xi}{\partial y}p^2\right\}+\ldots$$

$$=p'+t\zeta(x, y, p)+\ldots$$

When $\zeta(x, y, p)$ is thus defined, the infinitesimal transformation of the extended group is

$$U'f\equiv\xi\frac{\partial f}{\partial x}+\eta\frac{\partial f}{\partial y}+\zeta\frac{\partial f}{\partial p}.$$

The group can be further extended in a similar way by considering as new variables the higher differential coefficients $y'', \ldots, y^{(n)}$.

4·6. Integration of a Differential Equation of the First Order in Two Variables.—It has been proved (§ 2·1) that an exact differential equation of the first order in two variables is immediately integrable by means of a quadrature. When an equation is not exact, the first step towards its integration is the determination of an integrating factor by means of which the equation is made exact. It will now be shown that when an equation is invariant under a known group, an integrating factor may, at least theoretically, be found, and the equation integrated by a quadrature.

Let it be supposed, then, that the differential equation

$$F(x, y, p)=0$$

is invariant under the extended group

$$U'f\equiv\xi\frac{\partial f}{\partial x}+\eta\frac{\partial f}{\partial y}+\zeta\frac{\partial f}{\partial p}$$

derived from

$$Uf\equiv\xi\frac{\partial f}{\partial x}+\eta\frac{\partial f}{\partial y}.$$

Then the necessary and sufficient condition for this invariant property is satisfied, namely that $U'F$ is zero either *per se* or in virtue of the equation $F=0$.

It is proposed to determine, and to integrate, the most general differential equation which admits of the given group $U'f$. The problem is therefore to determine two independent solutions of the partial differential equation

$$U'f\equiv\xi\frac{\partial f}{\partial x}+\eta\frac{\partial f}{\partial y}+\zeta\frac{\partial f}{\partial p}=0,$$

and this, in turn, depends upon finding two distinct solutions of the simultaneous system

$$\frac{dx}{\xi}=\frac{dy}{\eta}=\frac{dp}{\zeta}.$$

Let $u=\alpha$ be the solution of

$$\frac{dx}{\xi}=\frac{dy}{\eta},$$

then since ξ and η are independent of p, u is independent of p. Let $v=\beta$ be a solution of

$$\frac{dx}{\xi}=\frac{dy}{\eta}=\frac{dp}{\zeta}$$

distinct from $u=\alpha$; v must necessarily involve p. Then if $H(u)$ is an arbitrary function of u, $f=v-H(u)$ satisfies the partial differential equation $U'f=0$, that is

$$U'\{v-H(u)\}=0.$$

Consequently,

$$v-H(u)=0$$

is the most general ordinary differential equation of the first order invariant under U'.

It will now be shown that, when u is known, v can be determined by a quadrature. It has been proved that any group is reducible to a translation group. Let the change of variables from (x, y) to (x_1, y_1) reduce Uf to the group of translations parallel to the y_1-axis, namely U_1f. Then

$$U_1f\equiv U(x_1)\frac{\partial f}{\partial x_1}+U(y_1)\frac{\partial f}{\partial y_1}=\frac{\partial f}{\partial y_1},$$

from which it follows that

$$U(x_1)=0, \quad U(y_1)=1.$$

Thus x_1, y_1 are determined as functions of x, y by the equations

$$\xi\frac{\partial x_1}{\partial x}+\eta\frac{\partial x_1}{\partial y}=0, \quad \xi\frac{\partial y_1}{\partial x}+\eta\frac{\partial y_1}{\partial y}=1.$$

The first equation has the solution

$$x_1=u(x, y),$$

the second equation is equivalent to the simultaneous system

$$\frac{dx}{\xi}=\frac{dy}{\eta}=\frac{dy_1}{1}.$$

One solution of this system is

$$u(x, y)=\alpha;$$

if this solution is used to eliminate x from the equation

$$\frac{dy_1}{dy}=\eta(x, y),$$

y_1 is obtainable, in terms of x and α, by a quadrature. By eliminating α, y_1 is obtained in terms of x and y. Thus the necessary change of variables has been found.

It is easily verified that $U_1'f$, the extended group of U_1f, is identical with U_1f itself. The most general equation invariant under $U_1'f$ is found by solving the simultaneous system

$$\frac{dx_1}{0}=\frac{dy_1}{1}=\frac{dp_1}{0}.$$

Since two solutions of this system are

$$x_1=\text{const.}, \quad p_1=\text{const.},$$

the most general invariant differential equation of the first order may be written in the form

$$p_1 = \Phi(x_1),$$

and is integrable by quadratures. In the original variables this equation has the form

$$v = H(u).$$

But since $x_1 = u$, p_1 is necessarily a function of v alone, and since H is arbitrary there is no loss of generality in taking $p_1 = v$.

Thus *when one solution of the equation*

$$\frac{dx}{\xi} = \frac{dy}{\eta}$$

is known, the most general differential equation of the first order invariant under the group $Uf \equiv \xi \frac{\partial f}{\partial x} + \eta \frac{\partial f}{\partial y}$ *can be constructed, and this equation is integrable by quadratures.*

4·61. Integration of a Differential Equation invariant under G_1.—Let the given differential equation be

$$\frac{dx}{P(x, y)} = \frac{dy}{Q(x, y)},$$

and let

$$\phi(x, y) = c$$

be its solution. Then $\phi(x, y)$ is an integral of the partial differential equation

$$P\frac{\partial f}{\partial x} + Q\frac{\partial f}{\partial y} = 0.$$

It will be assumed that, for at least one value of c, the integral-curve $\phi(x, y) = c$ is not invariant under the group. As a family, however, the integral curves are invariant, so that

$$U\phi(x, y) \equiv \xi\frac{\partial \phi}{\partial x} + \eta\frac{\partial \phi}{\partial y} = F(\phi),$$

where $F(\phi)$ is a definite function of ϕ not identically zero. Now if Φ is a function of ϕ alone, the family of curves $\Phi = C$ is identical with the family $\phi = c$. Let

$$\Phi = \int \frac{d\phi}{F(\phi)},$$

then

$$U\Phi = U\phi \cdot \frac{d\Phi}{d\phi} = 1.$$

Thus Φ is an integral of the two partial differential equations

$$P\frac{\partial \Phi}{\partial x} + Q\frac{\partial \Phi}{\partial y} = 0,$$

$$\xi\frac{\partial \Phi}{\partial x} + \eta\frac{\partial \Phi}{\partial y} = 1,$$

from which it is found that

$$\frac{\partial \Phi}{\partial x} = \frac{-Q}{P\eta - Q\xi}, \quad \frac{\partial \Phi}{\partial y} = \frac{P}{P\eta - Q\xi}$$

and therefore

$$d\Phi=\frac{\partial\Phi}{\partial x}dx+\frac{\partial\Phi}{\partial y}dy$$
$$=\frac{Pdy-Qdx}{P\eta-Q\xi}.$$

Consequently $\dfrac{1}{P\eta-Q\xi}$ *is an integrating factor for the differential equation*

$$Pdy-Qdx=0.$$

The solution of the equation

$$\frac{dx}{P}=\frac{dy}{Q}$$

is therefore

$$\int\frac{Pdy-Qdx}{P\eta-Q\xi}=K,$$

where K is a constant.

When every individual integral-curve is invariant under the group, $U\phi$ is identically zero, that is

$$\xi\frac{\partial\phi}{\partial x}+\eta\frac{\partial\phi}{\partial y}=0,$$

and therefore

$$P\eta-Q\xi=0.$$

The infinitesimal transformation then takes the form

$$Uf\equiv\rho(x,y)\left\{P\frac{\partial f}{\partial x}+Q\frac{\partial f}{\partial y}\right\},$$

and conversely, when it is of this form it does not furnish an integrating factor of the equation

$$Pdy-Qdx=0.$$

Such an integrating factor is said to be *trivial* with respect to the equation in question.

4·62. Differential Equations of the First Order invariant under a Translation Group.—It is now proposed to investigate the most general differential equations which are invariant under particular groups of an elementary character. To begin with consider the G_1 of translations parallel to the x-axis,

$$Uf\equiv\frac{\partial f}{\partial x}.$$

In this case the extended group $U'f$ is identical with Uf. The simultaneous system to be considered is therefore

$$\frac{dx}{1}=\frac{dy}{0}=\frac{dp}{0};$$

it possesses the solutions

$$y=\text{const.},\quad p=\text{const.}$$

The most general differential equation invariant under the group is therefore

$$p=F(y),$$

where F is arbitrary.

Similarly, the most general equation invariant under

$$Uf\equiv\frac{\partial f}{\partial y}$$

is

$$p=F(x).$$

In these two cases the variables are separable.

The general translation group is

$$Uf\equiv\frac{1}{a}\cdot\frac{\partial f}{\partial x}-\frac{1}{b}\cdot\frac{\partial f}{\partial y},$$

where a and b are constants. $U'f$ is again identical with Uf. The simultaneous system is

$$adx=-bdy=\frac{dp}{0},$$

and hence the most general differential equation invariant under the group is

$$p=F(ax+by).$$

It is integrated by taking $ax+by$ as a new dependent variable.

4·63. Differential Equations of the First Order invariant under the Affine Group.*—

$$Uf\equiv x\frac{\partial f}{\partial x}.$$

In this case $\xi=x$, $\eta=0$, $\zeta=-p$, and hence the extended group is

$$U'f\equiv x\frac{\partial f}{\partial x}-p\frac{\partial f}{\partial p}.$$

The simultaneous system

$$\frac{dx}{x}=\frac{dy}{0}=\frac{dp}{-p},$$

has solutions

$$xp=\text{const.}\quad y=\text{const.}$$

The most general equation which admits of the group is therefore

$$xp=F(y).$$

Similarly it is found that the general differential equation which admits of the affine group

$$Uf\equiv y\frac{\partial f}{\partial y}$$

is

$$p=yF(x).$$

In both cases the variables are separable.

4·64. Differential Equations of the First Order invariant under the Magnification Group.†—

$$Uf\equiv x\frac{\partial f}{\partial x}+y\frac{\partial f}{\partial y}.$$

* An *affine transformation* is a projective collineation which transforms the Euclidean plane into itself. It preserves the parallelism of straight lines and may be represented by

$$x_1=ax+by+c,\quad y_1=a'x+b'y+c'. \qquad (ab'-a'b\neq 0).$$

An *affine group* is a group of such transformations, and is a one-parameter group if a, b, c, a', b', c' are functions of a single parameter (Euler, 1748 ; Klein, *Erlanger Programm*, 1872).

† Or group of perspective transformations.

Here $p=0$, and $U'f$ is identical with Uf. The simultaneous system

$$\frac{dx}{x}=\frac{dy}{y}=\frac{dp}{0}$$

has solutions

$$p=\text{const.}, \quad \frac{y}{x}=\text{const.}$$

The invariant differential equation of general form is therefore

$$p=F\left(\frac{y}{x}\right);$$

it is of the type known as *homogeneous* (§ 2·12).

If the equation is written in the form

$$dy-F\left(\frac{y}{x}\right)dx=0,$$

it has for integrating factor the reciprocal of

$$y-xF\left(\frac{y}{x}\right).$$

Example.— $$(y^4-2x^3y)dx+(x^4-2xy^3)dy=0.$$

The integrating factor is the reciprocal of

$$(y^4-2x^3y)x+(x^4-2xy^3)y=-(x^4y+xy^4).$$

Now

$$\frac{(y^4-2x^3y)dx+(x^4-2xy^3)dy}{x^4y+xy^4}=\frac{d(x^4y+xy^4)}{x^4y+xy^4}-2\frac{d(x^3+y^3)}{x^3+y^3}.$$

The solution therefore is

$$\frac{x^4y+xy^4}{(x^3+y^3)^2}=\text{const.}$$

or

$$x^3+y^3=cxy.$$

Consider now the more general group

$$Uf\equiv\frac{x}{a}\cdot\frac{\partial f}{\partial x}+\frac{y}{b}\cdot\frac{\partial f}{\partial y}.$$

The extended group is

$$U'f\equiv\frac{x}{a}\cdot\frac{\partial f}{\partial x}+\frac{y}{b}\cdot\frac{\partial f}{\partial y}+\frac{a-b}{ab}p\frac{\partial f}{\partial p}.$$

The simultaneous system

$$\frac{adx}{x}=\frac{bdy}{y}=\frac{abdp}{(a-b)p}$$

has the solutions

$$y^b=\alpha x^a, \quad p=\beta x^{\frac{a}{b}-1},$$

where α and β are constants. The typical invariant differential equation is therefore

$$p=x^{\frac{a}{b}-1}F\left(\frac{y^b}{x^a}\right).$$

Particular examples are:

(i) $Uf \quad x\frac{\partial f}{\partial x}-y\frac{\partial f}{\partial y}$; Equation: $xdy=F(xy)ydx$,
Integrating factor: xy.

(ii) $Uf \equiv 2x\dfrac{\partial f}{\partial x}+y\dfrac{\partial f}{\partial x}$; Equation : $ydy=F\left(\dfrac{y^2}{x}\right)dx$,

Integrating factor : $\dfrac{1}{y^2-xF\left(\dfrac{y^2}{x}\right)}$.

(iii) $Uf \equiv x\dfrac{\partial f}{\partial x}+2y\dfrac{\partial f}{\partial x}$; Equation : $dy=xF\left(\dfrac{y}{x^2}\right)dx$,

Integrating factor : $\dfrac{1}{y-x^2F\left(\dfrac{y}{x^2}\right)}$.

Similar types :

(iv) $Uf \equiv \dfrac{\partial f}{\partial x}+\dfrac{y}{x}\cdot\dfrac{\partial f}{\partial y}$; Equation : $p=\dfrac{y}{x}+xF\left(\dfrac{y}{x}\right)$,

Integrating factor: $dx/xF\left(\dfrac{y}{x}\right)$.

(v) $Uf \equiv x^2\dfrac{\partial f}{\partial x}+xy\dfrac{\partial f}{\partial y}$; Equation : $xp-y=F\left(\dfrac{y}{x}\right)$,

Integrating factor : $dx/x^2F\left(\dfrac{y}{x}\right)$.

4·65. Differential Equations of the First Order invariant under the Rotation Group.—

$$Uf \equiv -y\frac{\partial f}{\partial x}+x\frac{\partial f}{\partial y}.$$

The extended group is

$$U'f \equiv -y\frac{\partial f}{\partial x}+x\frac{\partial f}{\partial y}+(1+p^2)\frac{\partial f}{\partial p}.$$

The first equation of the simultaneous system

$$\frac{dx}{-y}=\frac{dy}{x}=\frac{dp}{1+p^2}$$

has the solution

$$x^2+y^2=a^2,$$

where a is a constant. The last equation may therefore be written

$$\frac{dy}{\sqrt{(a^2-y^2)}}=\frac{dp}{1+p^2};$$

its solution is

$$\text{arc sin}\frac{y}{a}-\text{arc tan } p=\beta,$$

where β is a second constant. This solution is equivalent to

$$\text{arc tan}\frac{y}{\sqrt{(a^2-y^2)}}-\text{arc tan } p=\beta$$

or to

$$\text{arc tan } \frac{y}{x}-\text{arc tan } p=\beta$$

and therefore may be written

$$\frac{\dfrac{y}{x}-p}{1+\dfrac{y}{x}p}=\tan\beta.$$

The most general differential equation which admits of the group is therefore

$$\frac{xp-y}{x+yp}=F(x^2+y^2).$$

When this equation is written in the form

$$(x-yF)dy-(y+xF)dx=0,$$

it admits of the integrating factor

$$\frac{1}{x^2+y^2}.$$

Similar examples are:

(i) $Uf\equiv y\frac{\partial f}{\partial x}$; Equation: $\frac{xp-y}{p}=F(y),$

or $\{x-F(y)\}dy-ydx=0,$

Integrating factor: $\frac{1}{y^2}.$

(ii) $Uf\equiv x\frac{\partial f}{\partial y}$; Equation: $xp-y=F(x),$

or $xdy-\{y+F(x)\}dx=0,$

Integrating factor: $\frac{1}{x^2}.$

4·66. Differential Equations of the First Order invariant under the Group.—

$$Uf\equiv e^{\int\phi(x)dx}\frac{\partial f}{\partial y}.$$

The extended group is

$$U'f\equiv e^{\int\phi(x)dx}\left\{\frac{\partial f}{\partial y}+\phi(x)\frac{\partial f}{\partial p}\right\}.$$

The simultaneous system is virtually

$$\frac{dx}{0}=\frac{dy}{1}=\frac{dp}{\phi(x)};$$

one solution is

$$x=a,$$

where a is a constant. In view of this solution the last equation becomes

$$\frac{dy}{1}=\frac{dp}{\phi(a)},$$

whence

$$p-y\phi(a)=\beta,$$

where β is a second constant. The invariant equation is therefore

$$p-y\phi(x)=F(x),$$

that is, the general linear equation of the first order. When it is written in the form

$$dy-\{y\phi(x)+F(x)\}dx=0,$$

it has the integrating factor

$$e^{-\int\phi(x)dx}.$$

4·7. Integral-Curves which are Invariant under a Group of the Equation.— The family of integral-curves is invariant, as a whole, under any group which the differential equation admits, but unless the group is trivial all individual curves of the family are not invariant under the group. It may, however, occur that particular integral-curves are invariant, and it is important to note the special properties which these curves enjoy.

If

$$\Omega(x, y, p)=0$$

is a differential equation invariant under the group

$$Uf\equiv\xi\frac{\partial f}{\partial x}+\eta\frac{\partial f}{\partial y},$$

and if any integral-curve is invariant under the group its gradient at any point (x, y) will be η/ξ. Hence any such integral-curve is found by substituting η/ξ for p in the differential equation itself; all such curves, if any exist, are included in the equation

$$\Omega\left(x, y, \frac{y}{x}\right)=0.$$

But this equation may include curves which are invariant under the group and have equations which are solutions of the differential equation, but are not particular integral-curves. An instance arises when the integral-curves have an envelope; the envelope itself is invariant under the group which transforms the family of integral-curves into itself, has an equation which satisfies the differential equation, but is not in general a particular integral-curve. The equation of such a curve is a *singular solution* of the differential equation.

Example.—The differential equation

$$p^3-4xyp+8y^2=0$$

admits of the group

$$Uf\equiv x\frac{\partial f}{\partial x}+3y\frac{\partial f}{\partial y}.$$

If a singular solution exists, it is obtained by replacing $3y/x$ for p in the differential equation, which becomes

$$27y^3-4x^3y^2=0,$$

whence either $y=0$ or $27y=4x^3$.
The general solution of the equation is

$$y=c(x-c)^2,$$

and thus $y=0$ is a particular solution. On the other hand, $27y=4x^3$ is an envelope singular solution.

Miscellaneous Examples.

1. Find the general differential equations of the first order invariant under the groups:

(i) $Uf\equiv\frac{\partial x}{\partial f}+x\frac{\partial f}{\partial y}$; (iii) $Uf\equiv ax\frac{\partial f}{\partial x}+y\frac{\partial f}{\partial y}$,

(ii) $Uf\equiv y\frac{\partial f}{\partial x}+\frac{\partial f}{\partial y}$; (iv) $Uf\equiv x\frac{\partial f}{\partial x}+ay\frac{\partial f}{\partial y}$,

and determine the corresponding integrating factors.

2. Show that each of the equations

(i) $2xyp+x-y^2=0$; (iv) $p^2-x^2-y=0$;
(ii) $xp-y-x^m=0$; (v) $p^4-4y(xp-2y)^2=0$;
(iii) $y+xp-x^4p^2=0$; (vi) $p^2+2x^3p-4x^2y=0$

admits of a group of the form

$$Uf\equiv ax\frac{\partial f}{\partial x}+by\frac{\partial f}{\partial y}.$$

Integrate the equations and examine them for singular solutions.

3. Show that if

$$U^{(n)}f \equiv \xi\frac{\partial f}{\partial x}+\eta\frac{\partial f}{\partial y}+\eta'\frac{\partial f}{\partial y'}+\ .\ .\ .\ +\eta^{(n)}\frac{\partial f}{\partial y^{(n)}}$$

is the n times extended group of $Uf\equiv\xi\frac{\partial f}{\partial x}+\eta\frac{\partial f}{\partial y}$, then

$$\eta^{(r)}=\frac{d\eta^{(r-1)}}{dx}-y^{(r)}\frac{d\xi}{dx} \qquad (r=1, 2, \ .\ .\ ., n).$$

4. Prove that if $u=\alpha$, $v=\beta$, $w=\gamma$ (α, β, γ constants) are distinct solutions of the system

$$\frac{dx}{\xi}=\frac{dy}{\eta}=\frac{dy'}{\eta'}=\frac{dy''}{d\eta''}$$

such that u involves only x and y, v involves y' but not y'', and w involves y'', then the most general differential equation of the second order invariant under the twice extended group

$$U''f\equiv\xi\frac{\partial f}{\partial x}+\eta\frac{\partial f}{\partial y}+\eta'\frac{\partial f}{\partial y'}+\eta''\frac{\partial f}{\partial y''}$$

is of the form

$$w=\Phi(u, v),$$

where Φ is an arbitrary function of its arguments.

Show that there is no loss of generality in taking $w\equiv dv/du$, and that therefore the second order equation $w=\Phi(u, v)$ is equivalent to the first order equation

$$\frac{dv}{du}=\Phi(u, v).$$

Verify this theorem in the case of the following groups and corresponding invariant differential equations, and in each case show how a first integral may be obtained:

(i) $Uf\equiv\frac{\partial f}{\partial x}$, $\qquad y''=F(y, y')$;

(ii) $Uf\equiv\frac{\partial f}{\partial y}$, $\qquad y''=F(x, y')$;

(iii) $Uf\equiv x\frac{\partial f}{\partial x}$, $\qquad xy''=y'F(y, xy')$;

(iv) $Uf\equiv y\frac{\partial f}{\partial y}$, $\qquad y''=yF(x, y'/y)$;

(v) $Uf\equiv x\frac{\partial f}{\partial x}+y\frac{\partial f}{\partial y}$, $\qquad xy''=F(y/x, y')$;

(vi) $Uf\equiv\phi(x)\frac{\partial f}{\partial y}$, $\qquad y''=p(x)y'+q(x)y+r(x)$.

[Page, *Ordinary Differential Equations*, Chap. IX.]

CHAPTER V

THE GENERAL THEORY OF LINEAR DIFFERENTIAL EQUATIONS

5·1. Properties of a Linear Differential Operator.—The most general linear differential equation is of the type

$$p_0(x)\frac{d^ny}{dx^n}+p_1(x)\frac{d^{n-1}y}{dx^{n-1}}+\ldots+p_{n-1}(x)\frac{dy}{dx}+p_n(x)y=r(x),$$

which may be written symbolically as *

$$\text{(A)}\qquad L(y)\equiv\{p_0D^n+p_1D^{n-1}+\ldots+p_{n-1}D+p_n\}y=r(x).$$

It will be assumed that the coefficients $p_0, p_1, \ldots, p_n$, and the function $r(x)$ are continuous one-valued functions of x throughout an interval $a \leqslant x \leqslant b$, and that p_0 does not vanish at any point of that interval. Then the fundamental existence theorem of § 3 proves that there exists a unique continuous solution $y(x)$ which assumes a given value y_0 at a point x_0 within (a, b), and whose first $n-1$ derivatives are continuous and assume respectively the values $y_0', y_0'' \ldots y_0^{(n-1)}$ at x_0.

The expression

$$L\equiv p_0D^n+p_1D^{n-1}+\ldots+p_{n-1}D+p^n$$

is known as a *linear differential operator of order n*. The differential equation

$$\text{(B)}\qquad L(u)=0$$

is said to be the *homogeneous equation* corresponding to (A). It is so called because $L(u)$ is a homogeneous linear form in $u, u', \ldots, u^{(n)}$. It is also known as the *reduced equation.*

The following elementary theorems bring out clearly the nature of the operator L:

I. *If $u=u_1$ is a solution of the homogeneous equation* (B), *then $u=Cu_1$ is also a solution, where C is any arbitrary constant.*

This follows from the fact that

$$D^rCu_1=CD^ru_1.$$

For then,

$$L(C_1u_1)=\sum_{r=0}^{n}p_rD^{n-r}Cu_1$$

$$=C\sum_{r=0}^{n}p_rD^{n-r}u_1=CL(u_1)=0.$$

II. *If $u=u_1, u_2, \ldots, u_m$ are m solutions of the homogeneous equation* (B), *then $u=C_1u_1+C_2u_2+\ldots+C_mu_m$ is a solution, where $C_1, C_2, \ldots, C_m$ are arbitrary constants.*

* The notion of a symbolic operator has been traced back to Brisson, *J. Éc. Polyt.* (1) Cah. 14 (1808), p. 197. Its use was extended by Cauchy.

This follows in a similar way from the fact that

$$D^r\{C_1u_1+C_2u_2+\ldots+C_mu_m\}=C_1D^ru_1+C_2D^ru_2+\ldots+C_mD^ru_m.$$

If n linearly-distinct solutions $u_1, u_2, \ldots, u_n$, of the homogeneous equation are known,* then the solution

$$u(x)=C_1u_1+C_2u_2+\ldots+C_nu_n,$$

containing n arbitrary constants, is the complete primitive of the homogeneous equation. The constants $C_1, C_2, \ldots, C_n$ may be chosen, and in one way only, so that

$$\text{(C)} \qquad u(x_0)=y_0, \quad u'(x_0)=y_0', \ldots, \quad u^{(n-1)}(x_0)=y_0^{(n-1)}.$$

III. *Let $y=y_0(x)$ be any solution of the non-homogeneous equation* (A), *then if $u(x)$ is the complete primitive of* (B), *$y=y_0(x)+u(x)$ will be the most general solution of* (A). †

Since the operator D^r is distributive, L is distributive, that is to say,

$$L\{y_0(x)+u(x)\}=L\{y_0(x)\}+L\{u(x)\}=r(x),$$

for

$$L\{y_0(x)\}=r(x), \quad L\{u(x)\}=0.$$

But the solution

$$y=y_0(x)+u(x)$$

involves n arbitrary constants; it is therefore the most general solution of (A).

If $u(x)$ is chosen so as to satisfy the conditions (C), and $y_0(x)$ is such that

$$y_0(x_0)=y_0'(x_0)=\ldots=y_0^{(n-1)}(x_0)=0,$$

which is possible provided that $r(x)$ is not identically zero, then the solution

$$y=y_0(x)+u(x)$$

also satisfies the conditions

$$y(x_0)=y_0, \quad y'(x_0)=y_0', \ldots, \quad y^{(n-1)}(x_0)=y_0^{(n-1)}.$$

This general solution of (A) may be considered as consisting of two parts, viz.

(1°). The complete primitive of the corresponding homogeneous equation, which is of the form

$$u(x)=C_1u_1+C_2u_2+\ldots+C_nu_n,$$

containing n arbitrary constants—this is known as the *Complementary Function.*

(2°). The *Particular Integral*, which is any particular solution of (A), and contains no arbitrary constant. It may, for definiteness, be that solution of (A) which, together with its first $n-1$ derivatives, vanishes at a point x_0 in the interval (a, b).

Thus, for instance, if the equation considered is

$$\frac{d^2y}{dx^2}+y=x,$$

then the complementary function is $A \cos x+B \sin x$, in which A and B are arbitrary constants; the particular integral may be taken as $y=x$. The general solution is therefore

$$y=A \cos x+B \sin x+x.$$

Any special solution is obtained by assigning to A and B definite numerical values.

* Conditions for linear independence follow in § 5·2.

† d'Alembert, *Misc. Taur.*, 3 (1762–65), p. 381.

5·2. The Wronskian.—Let $u_1, u_2, \ldots, u_n$ be n solutions of the homogeneous equation of degree n,

$$L(u)=0,$$

then the most general solution or complete primitive of this equation is

$$u=C_1u_1+C_2u_2+\ldots+C_nu_n,$$

provided that the solutions $u_1, u_2, \ldots, u_n$ are linearly independent, that is to say, such that it is impossible so to choose constants $C_1, C_2, \ldots, C_n$ not all zero so that the expression

$$C_1u_1+C_2u_2+\ldots+C_nu_n$$

is identically zero. Conditions that the n functions

$$u_1(x), \quad u_2(x), \ldots, \quad u_n(x),$$

which are supposed to be differentiable $n-1$ times in (a, b), be linearly independent will now be obtained.

In the first place, if these n functions are not linearly independent, then constants $C_1, C_2, \ldots, C_n$ may be determined so that

$$C_1u_1+C_2u_2+\ldots+C_nu_n=0$$

identically. Since this relation is satisfied identically in the interval (a, b) it may be differentiated any number of times up to $n-1$ in that interval, thus

$$C_1u_1'+C_2u_2'+\ldots+C_nu_n'=0,$$
$$\cdots\cdots\cdots$$
$$C_1u_1^{(n-1)}+C_2u_2^{(n-1)}+\ldots+C_nu_n^{(n-1)}=0.$$

There are thus n equations to determine the constants $C_1, C_2, \ldots, C_n$; if these equations are consistent then

$$\Delta(u_1, u_2, \ldots, u_n)\equiv\begin{vmatrix} u_1, & u_2, \ldots & u_n \\ u_1', & u_2', \ldots, & u_n' \\ \cdot & \cdot & \cdot \\ u_1^{(n-1)}, & u_2^{(n-1)}, \ldots, & u_n^{(n-1)} \end{vmatrix}=0.$$

The determinant is known as the *Wronskian* * of the functions $u_1, u_2, \ldots, u_n$. Its identical vanishing is a necessary condition for the linear dependence of $u_1, u_2, \ldots, u_n$; therefore *its non-vanishing is sufficient for the linear independence of* $u_1, u_2, \ldots, u_n$.

Conversely suppose that Δ is identically zero. It may happen that the Wronskian of a lesser number of the functions, say $u_1, u_2, \ldots, u_k$, is also identically zero. It will then be proved that, when $u_1, u_2, \ldots, u_k$ are solutions of the differential equation, they are linearly dependent.

In the first place, let $u_1(x), u_2(x), \ldots, u_k(x)$ be functions whose first $k-1$ derivatives are finite in the interval $a\leqslant x\leqslant b$, and such that their Wronskian vanishes identically in (a, b). Then if the Wronskian of $u_1(x), u_2(x), \ldots, u_{k-1}(x)$ does not vanish identically there is an identical relationship of the form

$$u_k(x)=c_1u_1(x)+c_2u_2(x)+\ldots+c_{k-1}u_{k-1}(x),$$

* After H. Wronski (*c.* 1821). The identical vanishing of the Wronskian is not a sufficient condition for the linear dependence of the n functions. See Peano, *Mathesis*, 6 (1889), pp. 75, 110; *Rend. Accad. Lincei* (5), 6 (1897), p. 413; Bortolotti, *ibid.* 7 (1898), p. 45; Vivanti, *ibid.* 7, p. 194; Bôcher, *Trans. Am. Math. Soc.* 2 (1901), p. 139; Curtiss, *Math. Ann.* 65 (1908), p. 282.

where $c_1, c_2, \ldots, c_{k-1}$ are constants.* To prove this theorem, denote by $U_1, U_2, \ldots, U_k$ the minors of the elements in the last line of the Wronskian

$$\begin{vmatrix} u_1, & u_2, \ldots, & u_k \\ u_1{}', & u_2{}', \ldots, & u_k{}' \\ \cdot & \cdot \quad \cdot & \cdot \\ u_1^{(k-1)}, & u_2^{(k-1)}, \ldots, & u_k^{(k-1)} \end{vmatrix}$$

then there follow the k identities

$$U_1u_1^{(r)}+U_2u_2^{(r)}+\ldots+U_ku_k^{(r)} \qquad (r=0, 1, \ldots, k-1).$$

If each of the first $k-1$ of these identities is differentiated and the next identity subtracted from the result, it follows that

$$U_1{}'u_1^{(r)}+U_2{}'u_2^{(r)}+\ldots+U_k{}'u_k^{(r)}=0 \qquad (r=0, 1, \ldots, k-2).$$

Multiply the rth of these $k-1$ identities by the co-factor of $u_1^{(r-1)}$ in the determinant U_k, and add the products. Then

$$U_1{}'U_k-U_k{}'U_1=0,$$

and since U_k is not identically zero in (a, b) it follows that

$$U_1=-c_1U_k.$$

In the same way it may be proved that

$$U_2=-c_2U_k,$$

$$\cdot \quad \cdot \quad \cdot$$

$$U_{k-1}=-c_{k-1}U_k.$$

From the identity

$$U_1u_1+U_2u_2+\ldots+U_ku_k=0,$$

it therefore follows that

$$U_k\{-c_1u_1-c_2u_2-\ldots-c_{k-1}u_{k-1}+u_k\}=0,$$

which proves the theorem.

Now let $u_1, u_2, \ldots, u_n$ be such that their first $n-1$ derivatives are finite in (a, b), and such that no non-zero expression of the form

$$g_1u_1+g_2u_2+\ldots+g_nu_n,$$

where $g_1, g_2, \ldots, g_n$ are constants, vanishes together with its first $n-1$ derivatives at any point of (a, b). Then if the Wronskian of $u_1, u_2, \ldots, u_n$ vanishes at any point p of (a, b), these functions are linearly dependent.† For the vanishing of the Wronskian for $x=p$ implies that constants $c_1, c_2, \ldots, c_n$, not all zero, can be found such that

$$c_1u_1^{(r)}(p)+c_2u_2^{(r)}(p)+\ldots+c_nu_n^{(r)}(p)=0 \qquad (r=0, 1, \ldots, n-1),$$

that is to say the function

$$c_1u_1(x)+c_2u_2(x)+\ldots+c_nu_n(x)$$

vanishes, together with its first $n-1$ derivatives, at $x=p$, and is therefore identically zero. Thus the theorem is proved.

Now let $u_1, u_2, \ldots, u_k$ be functions of x which at every point of the interval (a, b) have finite derivatives of the first $n-1$ orders $(n>k)$ and which are such that no non-zero function of the form

$$g_1u_1+g_2u_2+\ldots+g_ku_k$$

* Frobenius, *J. für Math.* 76 (1873), p. 238.

† This and the following theorem are due to Bôcher, *loc. cit.*

(where $g_1, g_2, \ldots, g_k$ are constants) vanishes together with its first $n-1$ derivatives at any point of (a, b). Then if the Wronskian of $u_1, u_2, \ldots, u_k$ vanishes identically, the functions are linearly dependent.

To prove this theorem, consider first the case in which the Wronskian of $u_1, u_2, \ldots, u_{k-1}$ does not vanish identically in (a, b). Let p be a point of the interval in which it does not vanish. Then since the Wronskian is continuous, it will not vanish in the immediate neighbourhood of p, and from what has already been proved it follows that constants $c_1, c_2, \ldots, c_k$ will exist such that the function

$$c_1u_1+c_2u_2+\ldots+c_ku_k$$

is zero in the neighbourhood of p. The first $n-1$ derivatives of this function therefore also vanish in the neighbourhood of p, and thus, by hypothesis, the function must be identically zero.

Now consider the general case, and let the Wronskian of $u_1, u_2, \ldots, u_m$ vanish identically $(1<m<k)$, whilst that of $u_1, u_2, \ldots, u_{m-1}$ does not vanish identically in (a, b). Then it follows that $u_1, u_2, \ldots, u_m$ are linearly dependent and the theorem is proved.

These theorems may now be applied to the solutions

$$u_1(x), \quad u_2(x), \ldots, \quad u_n(x)$$

of the differential equation. Since any solution which, together with its first $n-1$ derivatives, vanishes at any point of the interval (a, b) is identically zero, it follows that:

I. *If the Wronskian of $u_1, u_2, \ldots, u_n$ vanishes at any point of (a, b) these n solutions are linearly dependent.*

II. *If the Wronskian of the k solutions $u_1, u_2, \ldots, u_k$ $(k<n)$ vanishes identically in (a, b), these k solutions are linearly dependent.*

If $v_1, v_2, \ldots, v_n$ are derived from $u_1, u_2, \ldots, u_n$ by the linear transformation

$$v_r=a_{r1}u_1+a_{r2}u_2+\ldots+a_{rn}u_n \qquad (r=1, 2, \ldots, n),$$

then it is easily verified that

$$\Delta(v_1, v_2, \ldots, v_n)=A\Delta(u_1, u_2, \ldots, u_n),$$

where A is the determinant $|a_{rs}|$. Consequently $\Delta(v_1, v_2, \ldots, v_n)$ is not zero, and therefore $v_1, v_2, \ldots, v_n$ are linearly independent provided (1°) that the determinant A is not zero, that is to say, the transformation is ordinary, (2°) that $u_1, u_2, \ldots, u_n$ are linearly independent.

Let $u_1, u_2, \ldots, u_n$ be n linearly independent solutions of the equation

$$L(u)=0,$$

then the Wronskian $\Delta(u_1, u_2, \ldots, u_n)$ is expressible in a simple form which will now be obtained. In the first place,

$$\frac{d\Delta}{dx}=\begin{vmatrix} u_1, & u_2, \ldots, & u_n \\ u_1', & u_2', \ldots, & u_n' \\ \cdot & \cdot & \cdot \\ u_1^{(n-2)}, & u_2^{(n-2)}, \ldots, & u_n^{(n-2)} \\ u_1^{(n)}, & u_2^{(n)}, \ldots, & u_n^{(n)} \end{vmatrix}$$

for all the other determinants which arise in the differentiation have two rows alike, and therefore vanish. Then since

$$p_0u_r^{(n)}=-p_1u_r^{(n-1)}-\ldots-p_{n-1}u_r'-p_nu_r,$$

it follows, after a slight reduction, that

$$\frac{d\Delta}{dx} = -\frac{p_1}{p_0}\Delta,$$

or

$$\Delta = \Delta_0 \exp\left\{-\int_{x_0}^{x} \frac{p_1}{p_0}\,dx\right\},$$

where Δ_0 is the value to which Δ reduces when $x = x_0$. This relation is known as the *Abel identity* (cf. § 3·32).

This shows that if $p_0(x)$ does not vanish in the interval (a, b), then if Δ vanishes at x_0, Δ will be identically zero. If Δ_0 is not zero, then Δ will not be zero except at a singular point, that is to say, at a point in which p_1/p_0 becomes infinite. Such points are excluded by the supposition that the coefficients in $L(u)$ are continuous and p_0 does not vanish in (a, b).

5·21. Fundamental Sets of Solutions.—Any linearly independent set of n solutions $u_2, u_2, \ldots, u_n$ of the equation

$$L(u) = 0$$

is said to form a *fundamental set* or *fundamental system*.* Conversely, the condition that any given set of n solutions should be a fundamental set is that the Wronskian of the n solutions is not zero. The general solution of the equation will be †

$$u = C_1 u_1 + C_2 u_2 + \ldots + C_n u_n,$$

which cannot vanish identically unless the constants $C_1, C_2, \ldots C_n$ be all zero.

There is clearly an infinite number of possible fundamental sets of solutions, but one particular set is of importance on account of its simplicity.

Let $u_1(x)$ be such that

$$u_1(x_0) = 1, \quad u_1'(x_0) = u_1''(x_0) = \ldots = u_1^{(n-1)}(x_0) = 0,$$

and define $u_r(x)$, where $r = 2, 3, \ldots, n$, as that particular solution which satisfies the initial conditions

$$u(x_0) = u'(x_0) = \ldots = u^{(r-2)}(x_0) = 0,$$
$$u^{(r-1)}(x_0) = 1,$$
$$u^{(r)}(x_0) = u^{(r+1)}(x_0) = \ldots = u^{(n-1)}(x_0) = 0.$$

Then $u_1(x), u_2(x), \ldots, u_n(x)$ form a fundamental set; the value of their Wronskian when $x = x_0$ is unity.

The unique solution of

$$L(u) = 0$$

which satisfies the initial conditions

$$u(x_0) = y_0, \quad u'(x_0) = y_0', \ldots, \quad u^{(n-1)}(x_0) = y_0^{(n-1)}$$

is

$$u(x) = y_0 u_1(x) + y_0' u_2(x) + \ldots + y_0^{(n-1)} u_n(x).$$

Any fundamental set of solutions

$$u_1(x), \quad u_2(x), \ldots, \quad u_n(x)$$

* The term *fundamental system* is due to Fuchs, *J. für Math.* 66 (1866), p. 126 [*Ges. Math. Werke*, 1, p. 165].

† Lagrange, *Misc. Taur.*, 3 (1762-65), p. 181 [*Œuvres*, 1, p. 473].

may be re-written in the form

$$u_1(x)=v_1,\quad u_2(x)=v_1\int v_2dx,\quad u_3(x)=v_1\int v_2\int v_3(dx)^2,$$

and in general

$$u_r(x)=v_1\int v_2\int \ldots \int v_r(dx)^{r-1},$$

where

$$v_r=\frac{d}{dx}\left[\frac{1}{v_{r-1}}\cdot\frac{d}{dx}\left\{\frac{1}{v_{r-2}}\cdot\frac{d}{dx}\ \cdots\ \frac{1}{v_2}\cdot\frac{d}{dx}\left(\frac{u_r}{u_1}\right)\right\}\right].$$

Now the homogeneous differential equation which has as a fundamental set of solutions the n functions

$$u_1,\quad u_2,\ \ldots,\ u_n$$

is obtained by eliminating the n arbitrary constants C from between the $n+1$ equations

$$\begin{aligned} u&=C_1u_1+C_2u_2+\ldots+C_nu_n,\\ u'&=C_1u_1'+C_2u_2'+\ldots+C_nu_n',\\ &\cdot\quad\cdot\quad\cdot\quad\cdot\quad\cdot\\ u^{(n)}&=C_1u_1^{(n)}+C_2u_2^{(n)}+\ldots+C_nu_n^{(n)}\end{aligned}$$

and is therefore

$$\Delta(u,\ u_1,\ u_2,\ \ldots,\ u_n)=0,$$

where Δ is the Wronskian of $u,\ u_1,\ u_2,\ \ldots,\ u_n$. In its development, the coefficient of $u^{(n)}$ will be $\Delta(u_1,\ u_2,\ \ldots,\ u_n)$ which is not zero since $u_1,\ u_2,\ \ldots,\ u_n$ form a fundamental set.

It is convenient to write

$$L(u)\equiv\frac{\Delta(u,\ u_1,\ u_2,\ \ldots,\ u_n)}{\Delta(u_1,\ u_2,\ \ldots,\ u_n)}=0,$$

so that the coefficient of $u^{(n)}$ in $L(u)$ is unity.

Then the equation is

$$L(u)\equiv\frac{d^nu}{dx^n}+p_1\frac{d^{n-1}u}{dx^{n-1}}+\ldots+p_{n-1}\frac{du}{dx}+p_nu=0,$$

where

$$p_r=-\Delta_r/\Delta$$

and Δ_r is obtained from Δ by replacing $u_1^{(n-r)}$ by $u_1^{(n)}$, $u_2^{(n-r)}$ by $u_2^{(n)}$ and so on.

In order to express the operator L as the product of n operators of the first order, write

$$\begin{aligned} U_r&=\Delta(u,\ u_1,\ u_2,\ \ldots,\ u_r),\\ \Delta_r&=\Delta(u_1,\ u_2,\ \ldots,\ u_r). \end{aligned}\qquad (\Delta_0=1.)$$

Then *

$$\begin{aligned} L(u)&=\frac{U_n}{\Delta_n}=\frac{\Delta_n}{\Delta_{n-1}}\cdot\frac{\Delta_{n-1}U_n}{\Delta_n^2}\\ &=-\frac{\Delta_n}{\Delta_{n-1}}\cdot\frac{d}{dx}\left(\frac{U_{n-1}}{\Delta_n}\right)=-\frac{\Delta_n}{\Delta_{n-1}}\cdot\frac{d}{dx}\left(\frac{\Delta_{n-1}}{\Delta_n}\cdot\frac{U_{n-1}}{\Delta_{n-1}}\right).\end{aligned}$$

By repeating the reduction it is found that

$$L(u)=(-1)^n\frac{\Delta_n}{\Delta_{n-1}}\cdot\frac{d}{dx}\cdot\frac{\Delta^2_{n-1}}{\Delta_n\Delta_{n-2}}\cdot\frac{d}{dx}\ \cdots\ \frac{d}{dx}\cdot\frac{\Delta_1^2}{\Delta_2\Delta_0}\cdot\frac{d}{dx}\cdot\frac{\Delta_0}{\Delta_1}u,$$

where each differential operator acts on all that follows it.

* The essential step that $U_r\Delta_{r-1}=U_{r-1}\Delta_r'-U'_{r-1}\Delta_r$ is proved by partially expanding the determinants.

When the fundamental system is taken in the form

$$u_1=v_1,\ u_2=v_1\int v_2dx,\ \ldots,\qquad u_n=v_1\int v_2\int\ \ldots\ \int v_n(dx)^{n-1},$$

the equation becomes *

$$\frac{d}{dx}\cdot\frac{d}{v_ndx}\cdot\frac{d}{v_{n-1}dx}\cdot\ \ldots\ \cdot\frac{d}{v_2dx}\cdot\frac{u}{v_1}=0.$$

Symbolically, the equation $L(u)=0$ may be written in the form †

$$L_nL_{n-1}\ \ldots\ L_2L_1(u)=0,$$

where the symbol L_i represents the operator $D-a_i$, in which

$$a_i=\frac{d}{dx}\log\frac{\Delta_i}{\Delta_{i-1}}=\frac{d}{dx}\log\,(v_1\,v_2\ \ldots\ v_i).$$

This follows from the fact that

$$\frac{\Delta_i}{\Delta_{i-1}}\cdot\frac{d}{dx}\left(\frac{\Delta_{i-1}z}{\Delta_i}\right)=\frac{dz}{dx}-z\frac{d}{dz}\log\frac{\Delta_i}{\Delta_{i-1}}.$$

It is to be noted that the order in which the factors $(D-a_i)$ occur must in general be preserved, for it is not true that for any two suffixes i and j

$$(D-a_i)(D-a_j)=(D-a_j)(D-a_i).$$

In other words, the factors of the differential operator are not in general permutable.

5·22. Depression of the Order of an Equation.—If r independent solutions of the equation of order n,

$$L(u)=0,$$

are known, then the order of the equation may be reduced to $n-r$. For let

$$u_1,\ u_2,\ \ldots,\ u_r$$

be the known solutions, and let

$$v_1=u,\ v_2=\frac{d}{dx}\left(\frac{v_1}{u_2}\right),\ v_3=\frac{d}{dx}\left\{\frac{1}{v_2}\cdot\frac{d}{dx}\left(\frac{u_3}{v_1}\right)\right\},$$

and so on as before. Then since the equation is known to be ultimately of the form

$$\frac{d}{dx}\cdot\frac{d}{v_ndx}\cdot\ \ldots\ \cdot\frac{d}{v_{r+1}dx}\cdot\frac{d}{v_rdx}\cdot\ \ldots\ \cdot\frac{d}{v_2dx}\cdot\frac{u}{v_1}=0,$$

it may be written as

$$P(v)=0,$$

where

$$\text{(A)}\qquad v=\frac{d}{v_rdx}\cdot\frac{d}{v_{r-1}dx}\cdot\ \ldots\ \cdot\frac{d}{v_2dx}\cdot\frac{u}{v_1},$$

and P is a linear operator of order $n-r$.

If any solution of $P(v)=0$ is obtainable, the corresponding value of u may be obtained directly from (A) by r quadratures.

* Frobenius, *J. für Math.* 76 (1873), p. 264 ; 77 (1874), p. 256.

† Floquet, *Ann. Éc. Norm.* (2) 8 (1879), suppl. p. 49.

The actual way of carrying out the process may be illustrated by the case of the equation of the second degree

(B) $$\frac{d^2y}{dx^2}+p\frac{dy}{dx}+qy=0.$$

Suppose that one solution of this equation is known; let it be denoted by y_1 and write

$$y=y_1\int u dx.$$

Then

$$y_1''\int u dx+2y_1'u+y_1u'+p\{y_1'\int u dx+y_1u\}+qy_1\int u dx=0,$$

and this reduces to

$$y_1u'+(2y_1'+py_1)u=0.$$

This is a linear equation of the first order in u whose solution is

$$u=Cy_1^{-2}e^{-\int p dx},$$

and therefore the two distinct solutions of (B) are

$$y_1 \quad \text{and} \quad y_1\int\{y_1^{-2}e^{-\int p dx}\}dx.$$

5·23. Solution of the Non-homogeneous Equation.—Consider now the general equation

(A) $$L(y)=r(x),$$

it being supposed that a fundamental set of solutions $u_1(x), u_2(x), \ldots, u_n(x)$ of the reduced equation

$$L(u)=0,$$

are known.

Then the general solution of the reduced equation is

$$u=C_1u_1+C_2u_2+\ldots+C_nu_n,$$

in which $C_1, C_2, \ldots, C_n$ are arbitrary constants. Now just as in the case of linear equations of the first order (§ 2·13), so here also the method of *variation of parameters* * can be applied to determine the general solution of the equation under consideration. Let

$$y=V_1u_1+V_2u_2,+\ldots+V_nu_n,$$

in which $V_1, V_2, \ldots, V_n$ are undetermined functions of x, be assumed to satisfy the equation (A). The problem is to determine the functions V explicitly. Since the differential equation itself is equivalent to a single relation between the functions V and $r(x)$, it is clear that $n-1$ other relations may be set up provided only that these relations are consistent with one another. The set of $n-1$ relations which will actually be chosen is

(B) $$\begin{cases} V_1'u_1+V_2'u_2+\ldots+V_n'u_n=0, \\ V_1'u_1'+V_2'u_2'+\ldots+V'_nu_n'=0, \\ \cdot \quad \cdot \quad \cdot \quad \cdot \quad \cdot \quad \cdot \quad \cdot \\ V_1'u_1^{(n-2)}+V_2'u_2^{(n-2)}+\ldots+V_n'u_n^{(n-2)}=0. \end{cases}$$

As a consequence of these relations it follows that

$$\begin{aligned} y'&=V_1u_1'+V_2u_2'+\ldots+V_nu_n', \\ y''&=V_1u_1''+V_2u_2''+\ldots+V_nu_n'', \\ &\cdot \quad \cdot \quad \cdot \quad \cdot \quad \cdot \quad \cdot \\ y^{(n-1)}&=V_1u_1^{(n-1)}+V_2u_2^{(n-1)}+\ldots+V_nu_n^{(n-1)}, \end{aligned}$$

* Lagrange, *Nouv. Mém. Acad. Berlin*, 5 (1774), p. 201; 6 (1775), p. 190 [*Œuvres*, 4, pp. 9, 159].

whilst

$$y^{(n)}=V_1u_1^{(n)}+V_2u_2^{(n)}+\ldots+V_nu_n^{(n)}$$
$$+V_1'u_1^{(n-1)}+V_2'u_2^{(n-1)}+\ldots+V_n'u_n^{(n-1)}.$$

Thus the expression

$$y=V_1u_1+V_2u_2+\ldots+V_nu_n$$

satisfies the differential equation

$$L(y)=r(x),$$

in which the coefficient of $y^{(n)}$ is supposed to be unity, provided that

$$\text{(C)}\qquad V_1'u_1^{(n-1)}+V_2'u_2^{(n-1)}+\ldots+V_n'u_n^{(n-1)}=r(x).$$

Since the solutions $u_1(x), u_2(x), \ldots, u_n(x)$ form a fundamental set the n equations in (B) and (C) are sufficient to determine $V_1', V_2', \ldots, V_n'$ uniquely in terms of $u_1, u_2, \ldots, u_n$ and $r(x)$. Then $V_1, V_2, \ldots, V_n$ are obtained by quadratures.

In particular, if the equation is of the second degree,

$$V_1=-\int\frac{u_2(x)}{\Delta(u_1, u_2)}r(x)dx, \quad V_2=\int\frac{u_1(x)}{\Delta(u_1, u_2)}r(x)dx,$$

where $\Delta(u_1, u_2)$ is the Wronskian of u_1 and u_2.

5·3. The Adjoint Equation.—The conception of an integrating factor, which plays so important a part in the theory of linear equations of the first order, may be brought into use in the theory of linear equations of higher order, and leads to results of supreme importance. Let

$$\text{(A)}\qquad L(u)\equiv p_0\frac{d^nu}{dx^n}+p_1\frac{d^{n-1}u}{dx^{n-1}}+\ldots+p_{n-1}\frac{du}{dx}+p_nu,$$

and let a function v be supposed to exist such that $vL(u)dx$ is a perfect differential. Then the formula

$$U^{(r)}=V\frac{d}{dx}\{U^{(r-1)}V-U^{(r-2)}V'+\ldots+(-1)^{r-1}UV^{(r-1)}\}+(-1)^rUV^{(r)}$$

applied to $vL(u)$ in its extended form gives

$$\begin{aligned}\text{(B)}\quad vL(u)=&\frac{d}{dx}\{u^{(n-1)}p_0v-u^{(n-2)}(p_0v)'+\ldots+(-1)^{n-1}u(p_0v)^{(n-1)}\}\\ &+\frac{d}{dx}\{u^{(n-2)}p_1v-u^{(n-3)}(p_1v)'+\ldots+(-1)^{n-2}u(p_1v)^{(n-2)}\}\\ &+\quad\cdot\quad\cdot\quad\cdot\quad\cdot\quad\cdot\\ &+\frac{d}{dx}\{u'p_{n-2}v-u(p_{n-2}v)'\}\\ &+\frac{d}{dx}(up_{n-1}v)+u\bar{L}(v),\end{aligned}$$

where

$$\text{(C)}\quad \bar{L}(v)=(-1)^n\frac{d^n(p_0v)}{dx^n}+(-1)^{n-1}\frac{d^{n-1}(p_1v)}{dx^{n-1}}+\ldots-\frac{d(p_{n-1}v)}{dx}+p_nv.$$

The differential expression $\bar{L}(v)$ is said to be adjoint to $L(u)$, and the equation

$$\bar{L}(v)=0$$

is the *adjoint equation* * corresponding to

$$L(u)=0.$$

The relation (B) may be expressed in the form

$$vL(u)-u\bar{L}(v)=\frac{d}{dx}\{P(u,\ v)\},$$

and is known as the *Lagrange identity*. The expression $P(u, v)$, which is linear and homogeneous in

$$u,\ u',\ \ldots,\ u^{(n-1)},$$

as well as in

$$v,\ v',\ \ldots,\ v^{(n-1)},$$

is known as the *bilinear concomitant*.

In order that v may be an integrating factor for $L(u)$ it is necessary and sufficient that v should satisfy the adjoint equation $\bar{L}(v)=0$. If v is taken to be a solution of this equation, then the equation

$$L(u)=0$$

reduces to the linear equation of order $n-1$,

$$P(u,\ v)=C,$$

where C is an arbitrary constant.

If r distinct solutions of the adjoint equation are known, for example,

$$v_1,\ v_2,\ \ldots,\ v_r,$$

then there will be the r distinct equations

$$P(u,\ v_1)=C_1,\quad P(u,\ v_2)=C_2,\ \ldots,\quad P(u,\ v_r)=C_r,$$

each of order $n-1$. Between these r equations, the $r-1$ quantities $u^{(n-1)}$, $u^{(n-2)}, \ldots, u^{(n-r+1)}$ may be eliminated; the eliminant will be a linear equation of order $n-r$ whose coefficients involve the r arbitrary constants $C_1, C_2, \ldots, C_r$. In particular, if $r=n$ all the derivatives $u^{(n-1)}, u^{(n-2)}, \ldots, u'$ may be eliminated; and the result is an explicit expression of u in terms of $v_1, v_2, \ldots, v_n$ and $C_1, C_2, \ldots, C_n$. In other words, the equation is then completely integrable.

It will now be proved that the relation between $L(u)$ and $\bar{L}(v)$ is a reciprocal one, that is to say if $\bar{L}(v)$ be adjoint to $L(u)$, then $L(u)$ is adjoint to $\bar{L}(v)$. For if not, let $L_1(u)$ be adjoint to $\bar{L}(v)$ Then there exists a function $P_1(u, v)$ such that

$$vL_1(u)-u\bar{L}(v)=\frac{d}{dx}\{P_1(u,\ v)\}.$$

But

$$vL(u)-u\bar{L}(v)=\frac{d}{dx}\{P(u,\ v)\},$$

and therefore

$$v\{L_1(u)-L(u)\}=\frac{d}{dx}\{P_1(u,\ v)-P(u,\ v)\}.$$

Now $P_1(u, v)-P(u, v)$ is homogeneous and linear in $v, v', \ldots, v^{(n-1)}$. But $v\{L_1(u)-L(u)\}$ does not involve $v^{(n)}$ and therefore the coefficient of $v^{(n-1)}$ in $P(u, v)-P_1(u, v)$ is zero. The argument may be repeated, and proves that

* Lagrange *Misc. Taur.* 3 (1762–65), p. 179 [*Œuvres*, 1, p. 471]. The term *adjoint* is due to Fuchs, *J. für Math.* 76 (1873), p. 183 [*Ges. Math. Werke*, 1, p. 422].

$P_1(u, v)-P(u, v)$ is identically zero, and therefore that $L_1(u)$ is identical with $L(u)$.

When an equation is identical with its adjoint it is said to be *self-adjoint*.*

Now let $L(u)$ be factorised after the manner of the preceding section, thus let

$$L(u)\equiv \frac{d}{v_{n+1}dx}\cdot\frac{d}{v_ndx}\cdot\ \ldots\ \cdot\frac{d}{v_2dx}\cdot\frac{u}{v_1}.$$

Then since †

$$\int vL(u)dx=\frac{v}{v_{n+1}}\cdot\frac{d}{v_ndx}\cdot\frac{d}{v_{n-1}dx}\cdot\ \ldots\ \cdot\frac{d}{v_2dx}\cdot\frac{u}{v_1}$$
$$-\int\left(\frac{d}{dx}\cdot\frac{v}{v_{n+1}}\right)\left(\frac{d}{v_ndx}\cdot\frac{d}{v_{n-1}dx}\cdot\ \ldots\ \cdot\frac{d}{v_2dx}\cdot\frac{u}{v_1}\right)dx.$$

$$\int\left(\frac{d}{dx}\cdot\frac{v}{v_{n+1}}\right)\left(\frac{d}{v_ndx}\cdot\frac{d}{v_{n-1}dx}\cdot\ \ldots\ \cdot\frac{d}{v_2dx}\cdot\frac{u}{v_1}\right)dx$$
$$=\left(\frac{d}{v_ndx}\cdot\frac{v}{v_{n+1}}\right)\left(\frac{d}{v_{n+1}dx}\cdot\frac{d}{v_{n-2}dx}\cdot\ \ldots\ \cdot\frac{d}{v_2dx}\cdot\frac{u}{v_1}\right)$$
$$-\int\left(\frac{d}{dx}\cdot\frac{d}{v_ndx}\cdot\frac{v}{v_{n+1}}\right)\left(\frac{d}{v_{n-1}dx}\cdot\frac{d}{v_{n-2}dx}\cdot\ \ldots\ \cdot\frac{d}{v_2dx}\cdot\frac{u}{v_1}\right)dx,$$

and so on; if

$$P(u, v)=\frac{v}{v_{n+1}}\cdot\frac{d}{v_ndx}\cdot\frac{d}{v_{n-1}dx}\cdot\ \ldots\ \cdot\frac{d}{v_2dx}\cdot\frac{u}{v_1}$$
$$-\left(\frac{d}{v_ndx}\cdot\frac{v}{v_{n+1}}\right)\left(\frac{d}{v_{n-1}dx}\cdot\frac{d}{v_{n-2}dx}\cdot\ \ldots\ \cdot\frac{d}{v_2dx}\cdot\frac{u}{v_1}\right)$$
$$+\ \cdot\ \cdot\ \cdot\ \cdot\ \cdot\ \cdot\ \cdot\ \cdot$$
$$+(-1)^{n-2}\left(\frac{d}{v_3dx}\cdot\frac{d}{v_4dx}\cdot\ \ldots\ \cdot\frac{d}{v_ndx}\cdot\frac{v}{v_{n+1}}\right)\left(\frac{d}{v_2dx}\cdot\frac{u}{v_1}\right)$$
$$+(-1)^{n-1}\left(\frac{d}{v_2dx}\cdot\frac{d}{v_3dx}\cdot\ \ldots\ \cdot\frac{d}{v_ndx}\ \frac{v}{v_{n+1}}\right)\frac{u}{v_1},$$

and

$$\bar{L}(v)=(-1)^n\frac{d}{v_1dx}\cdot\frac{d}{v_2dx}\cdot\ \ldots\ \cdot\frac{d}{v_ndx}\cdot\frac{v}{v_{n+1}},$$

then

$$vL(u)=\frac{d}{dx}\{P(u, v)\}+u\bar{L}(v).$$

In particular, if the expression $L(u)$ is self-adjoint, then

$$v_{n+1}=\pm v_1,$$
$$v_n=\pm v_2,$$
$$\cdot\quad\cdot\quad\cdot$$

Thus if $L(u)=0$ is a self-adjoint linear differential equation of even order $2m$. it may be written as ‡

$$\frac{d}{v_1dx}\cdot\frac{d}{v_2dx}\cdot\ \ldots\ \cdot\frac{d}{v_mdx}\cdot\frac{d}{v_{m+1}dx}\cdot\frac{d}{v_mdx}\cdot\ \ldots\ \cdot\frac{d}{v_2dx}\cdot\frac{u}{v_1}=0$$

* An early example of a self-adjoint equation is given by Jacobi, *J. für Math.* 17 (1837), p. 71 [*Werke*, 4, p. 44], who proved that when the order is $2m$ the operator is of the form $P\bar{P}$, where P and $\bar{P}$ are adjoint operators of order m. See also Jacobi, *J. für Math.* 32 (1846), p. 189 (*Werke*, 2, p. 127], Hesse, *ibid.* 54 (1857), p. 230.

† Frobenius, *J. für Math.* 76 (1873), p. 264; Thomé, *ibid.* 76, p. 277. See also Frobenius *ibid.* 77 (1874), p. 257; 80 (1875), p. 328.

‡ Frobenius, *ibid.* 85 (1878), p. 192.

or

$$P\bar{P}(u)=0,$$

where P is the differential operator

$$\frac{d}{v_1dx}\cdot\frac{d}{v_2dx}\cdot\ \cdot\ \cdot\ \cdot\frac{d}{v_mdx}\cdot\frac{1}{v_{m+1}^{\frac{1}{2}}},$$

and $\bar{P}$ is its adjoint.

Similarly, if the equation $L(u)=0$ is self-adjoint, and of odd order $2m-1$, it may be written as *

$$\frac{d}{v_1dx}\cdot\frac{d}{v_2dx}\cdot\ \cdot\ \cdot\ \cdot\frac{d}{v_mdx}\cdot\frac{d}{v_mdx}\cdot\ \cdot\ \cdot\ \cdot\frac{d}{v_2dx}\cdot\frac{u}{v_1}=0$$

or

$$P\frac{d}{dx}\bar{P}(u)=0.$$

where P is the operator

$$\frac{d}{v_1dx}\cdot\frac{d}{v_2dx}\cdot\ \cdot\ \cdot\ \cdot\frac{d}{v_{m-1}dx}\cdot\frac{1}{v_m},$$

and $\bar{P}$ is its adjoint.

5·4. Solutions common to two Linear Differential Equations.—If it is known, *a priori*, that the equation

$$L(u)=0,$$

of order n, has solutions in common with another homogeneous linear equation, of lower order, then the order of the first equation may be depressed, even though the common solutions may not explicitly be given. Let

$$L\equiv p_0D^n+p_1D^{n-1}+\ .\ .\ .\ +p_{n-1}D+p_n,$$

and let

$$L_1\equiv q_0D^m+q_1D^{m-1}+\ .\ .\ .\ +q_{m-1}D+q_m$$

be an operator of order m, less than n. Consider a third operator

$$R_1=r_0D^{n-m}+r_1D^{n-m-1}+\ .\ .\ .\ +r_{n-m-1}D+r_{n-m},$$

in which the coefficients r are to be determined in such a manner as to depress the order of the operator

$$L-R_1L_1$$

as far as possible. By choosing the coefficients r so as to satisfy the relations

$$p_0=r_0q_0,$$
$$p_1=r_1q_0+r_0\{(n-m)q_0'+q_1\},$$
$$p_2=r_2q_0+r_1\{(n-m-1)q_0'+q_1\}+r_0\{\tfrac{1}{2}(n-m)(n-m-1)q_0''+(n-m)q_1'+q_2\},$$
$$.\quad .\quad .\quad .\quad .\quad .\quad .\quad .\quad .\quad .\quad .\quad .\quad .$$
$$p_{n-m}=r_{n-m}q_0+r_{n-m-1}\{q_0'+q_1\}+r_{n-m-2}\{q_0''+2q_1'+q_2\}+\ .\ .\ .$$

it is possible to clear the operator $L-R_1L_1$ of terms in D^m, D^{m+1}, . . ., D^n. Now these relations are sufficient to determine in succession r_0, r_1, . . ., r_{n-m}, and when these coefficients have been so determined, the operator $L-R_1L_1$ is reduced to the order $m-1$ at most.

It should be noticed, in passing, that the functions r are derived from the functions p and q by the rational processes of addition, subtraction, multipli-

* Darboux, *Théorie des Surfaces*, 2, p. 127.

cation, division, and differentiation. If, therefore, the coefficients of L and L_1 are rational functions of x, then so also are the coefficients of R_1.

Thus

$$L = R_1 L_1 + L_2,$$

where L_2 is an operator similar to L and L_1 but of order not exceeding $m-1$. Consider the case in which the equations

$$L(u)=0, \quad L_1(u)=0$$

have a solution in common. Then this solution will also satisfy the equation

$$L_2(u)=0.$$

If every solution of $L_1(u)=0$ were a solution of $L(u)=0$, and L_2 were not identically zero, the equation $L_2(u)=0$, whose order is at most $m-1$, would be satisfied by the m solutions of $L_1(u)=0$, which is impossible. L_2 would therefore be identically zero, and L would be decomposable into the product $R_1 L_1$. The converse is also true.

Suppose, on the other hand, that $L_1(u)=0$ has solutions which do not belong to $L(u)=0$, then L_2 would not be identically zero. Then operators R_2 and L_3, where the order of L_3 is less than that of L_2, exist, such that

$$L_1 = R_2 L_2 + L_3,$$

and so on until, finally,

$$L_{\nu-1} = R_\nu L \;\; + L_{\nu+1}.$$

In this last equation $L_{\nu+1}$ is either identically zero, or else an operator of order zero, for in any other circumstance the process could be advanced a stage further.

In the first case, every solution of $L_\nu(u)=0$ is a solution of $L_{\nu-1}(u)=0$, and therefore also of

$$L_{\nu-2}(u)=0, \; . \; . \; ., \quad L_1(u)=0, \quad L(u)=0.$$

Then

$$\begin{aligned} L &= R_1(R_2 L_2 + L_3) + L_2 \\ &= (R_1 R_2 + 1) L_2 + R_1 L_3 \\ &= (R_1 R_2 R_3 + R_1 + R_3) L_3 + (R_1 R_2 + 1) L_4 \\ &\quad . \quad . \quad . \quad . \quad . \quad . \quad . \quad . \\ &= \mathbf{R} L_\nu, \end{aligned}$$

and thus L has been decomposed by rational processes into the product of two operators.

If, therefore, the change of dependent variable

$$v = L_\nu(u)$$

is made, the equation $L(u)=0$ becomes

$$\mathbf{R}(v)=0,$$

where $\mathbf{R}$ is an operator of order $n-k$, if k is the order of L_ν.

Let $v=V$ be the most general solution (involving $n-k$ arbitrary constants) of $\mathbf{R}(v)=0$, then the general solution of

$$L(u)=0$$

is obtained by solving completely the non-homogeneous equation

$$L_\nu(u)=V;$$

this solution will contain, in all, n arbitrary constants.

In the second case, $L_{\nu+1}$ is either a function of x or a constant, not zero, which shows that

$$L_{\nu-1}(u)=0, \quad L_\nu(u)=0$$

have no common solution, not identically zero.

When an equation with rational coefficients has no solution in common with any other equation of lower order than itself, whose coefficients are also rational, it is said to be *irreducible.* This idea may be extended very considerably by appealing to the concept of a field of rationality. The independent variable x and certain irrational or transcendental functions of x are taken as the elements or base of a field $[R]$. Then any function which is derived by rational processes * from these elements is said to be rational in the field $[R]$. If an equation whose coefficients are rational in $[R]$ has no solution in common with an equation of lower order, whose coefficients are also rational in $[R]$, that equation is said to be irreducible in the field $[R]$.

5·5. Permutable Linear Operators—Any differential equation of the type

$$\frac{d^2y}{dx^2}+2p\frac{dy}{dx}+qy=0$$

may be expressed in a factorised form as

$$\{D+a_2(x)\}\{D+a_1(x)\}y=0,$$

for it is only necessary to determine the functions a_1 and a_2 by the equations

$$a_1(x)+a_2(x)=2p, \quad a_1(x)a_2(x)+a_1'(x)=q.$$

which may, at least theoretically, be solved.† The given equation is therefore satisfied by the general solution of

$$\{D+a_1(x)\}y=0,$$

but not, except in a very special case, by the general solution of

$$\{D+a_2(x)\}y=0.$$

It will be satisfied by the general solution of the latter equation as well as by that of the former if, and only if, the two operators

$$D+a_1(x) \quad\text{and}\quad D+a_2(x)$$

are *permutable* or *commutative*, that is to say if

$$\{D+a_1(x)\}\{D+a_2(x)\}u=\{D+a_2(x)\}\{D+a_1(x)\}u,$$

whatever differentiable function u may be. A necessary and sufficient condition that the operators be commutative is that

$$a_2'(x)=a_1'(x)$$

or

$$a_2(x)=a_1(x)+A,$$

where A is an arbitrary constant. The differential equation is therefore of the form

$$(P+A)Py=0,$$

where P represents the operator $D+a_1(x)$. Also, the equation

$$\frac{d^2y}{dx^2}+2p\frac{dy}{dx}+(p^2+p'-a^2)y=0,$$

where a is an arbitrary constant, may be factorised into

$$(D+p-a)(D+p+a)y=0,$$

and is completely integrable.

* The rational processes include *differentiation.*

† Cayley, *Quart. J. Math.* 21 (1886), p. 331. [*Coll. Math. Papers*, 12, p. 403.]

It is not difficult to prove that the operator

$$D+a(x)$$

is permutable with the operator of the second order

$$D^2+2pD+q$$

when, and only when, the latter is expressible in the form

$$\{D+a(x)+A_1\}\{D+a(x)+A_2\},$$

where A_1 and A_2 are constants. In general, if P and Q are operators of orders m and n respectively, P and Q are commutative if

$$P=\{D+a(x)+A_1\}\ldots\{D+a(x)+A_m\},$$
$$Q=\{D+a(x)+A_{m+1}\}\ldots\{D+a(x)+A_{m+n}\},$$

but this condition, though clearly sufficient, is far from necessary. Thus, for instance, the operators

$$D^2-2x^{-2}$$

and

$$D^3-3x^{-2}D+3x^{-3}$$

are commutative, but cannot be expressed in the above product-form. This at once suggests the problem of determining a necessary and sufficient condition that two operators P and Q be permutable, when these two operators are not themselves expressible as polynomials in a differential operator R of lower order.

5·51. The Condition for Permutability.*—Let P and Q be linear operators of orders m and n, then if P and Q are permutable, and h is an arbitrary constant

$$(P-h)Q=Q(P-h).$$

Consequently, if

$$y_1,\ y_2,\ \ldots,\ y_m$$

is a fundamental set of solutions of the equation

$$\text{(A)}\qquad P(y)-hy=0,$$

then

$$Q(y_1),\ Q(y_2),\ \ldots,\ Q(y_n)$$

are likewise solutions of (A), and there exist relations of the form

$$Q(y_1)=a_{11}y_1+a_{12}y_2+\ldots+a_{1m}y_m,$$
$$Q(y_2)=a_{21}y_1+a_{22}y_2+\ldots+a_{2m}y_m,$$
$$\cdots\cdots\cdots$$
$$Q(y_m)=a_{m1}y_1+a_{m2}y_2+\ldots+a_{mm}y_m.$$

Now let

$$Y=c_1y_1+c_2y_2+\ldots+c_my_m,$$

then

$$Q(Y)=kY,$$

provided that k and the constants c satisfy the equations

$$kc_r=a_{r1}c_1+a_{r2}c_2+\ldots+a_{rm}c_m\quad (r=1,2,\ldots,m).$$

In order that these equations may be consistent it is necessary that k be determined by the relation

$$\begin{vmatrix} a_{11}-k, & a_{12}, & \ldots, & a_{1m} \\ a_{21}, & a_{22}-k, & \ldots, & a_{2m} \\ \cdot & \cdot & \cdot & \cdot \\ a_{m1}, & a_{m2}, & \ldots, & a_{mm}-k \end{vmatrix}=0.$$

* Burchnall and Chaundy, *Proc. London Math. Soc.* (2) 21 (1922), p. 420.

Thus corresponding to each h there exist m values of the constant k (not necessarily all distinct) such that the equations

$$\text{(A)} \qquad P(y)-hy=0,$$

$$\text{(B)} \qquad Q(y)-ky=0$$

have a common solution.

Similarly, corresponding to each k of (B) there exist n values of h in (A) such that (A) and (B) have a common solution. Thus, when (A) and (B) have a common solution, h and k are related by an algebraic equation

$$F(h, k)=0,$$

of degree n in h and m in k. This expression may be obtained explicitly by eliminating

$$y, y', \ldots, y^{(m+n-1)}$$

between the $m+n$ equations

$$\begin{array}{cc} P(y)-hy=0, & Q(y)-ky=0, \\ DP(y)-hy'=0, & DQ(y)-ky'=0, \\ \cdot \quad \cdot \quad \cdot \quad \cdot & \cdot \quad \cdot \quad \cdot \quad \cdot \\ D^{n-1}P(y)-hy^{(n-1)}=0, & D^{m-1}Q(y)-ky^{(m-1)}=0. \end{array}$$

Now since

$$P(y)-hy=0, \quad Q(y)-ky=0,$$

it follows that

$$F(P, Q)y=F(h, k)y=0,$$

and therefore y is a solution of the equation

$$L(y)\equiv F(P, Q)y=0,$$

which is of order mn.

Now let the numbers

$$h_1, \quad h_2, \ldots, \quad h_r$$

be all distinct, and let

$$Y_1, Y_2, \ldots, Y_r$$

be common solutions of

$$P(y)-hy=0, \qquad Q(y)-ky=0,$$

for these values of h and corresponding values of k. These functions $Y_1, Y_2, \ldots, Y_r$ are linearly distinct, for if there existed an identical relation of the type

$$C_1Y_1+C_2Y_2+\ldots+C_rY_r=0,$$

then by operation on the left-hand member of this identity by $P, P^2, \ldots, P^{r-1}$, further relations

$$C_1h_1Y_1+C_2h_2Y_2+\ldots+C_rh_rY_r=0,$$

$$\cdot \quad \cdot \quad \cdot \quad \cdot \quad \cdot \quad \cdot \quad \cdot \quad \cdot$$

$$C_1h_1^{r-1}Y_1+C_2h_2^{r-1}Y_2+\ldots+C_rh_r^{r-1}Y_r=0$$

are obtained. But these relations are inconsistent unless $C_1, C_2, \ldots, C_r$ are all zero. This is true no matter how many distinct numbers h are chosen.

Thus there exists an unlimited set of linearly distinct functions $Y_1, Y_2, \ldots$ all of which satisfy the equation

$$F(P, Q)y=0.$$

But the order of this equation is mn, and therefore it cannot possess more than mn linearly distinct solutions. It follows that

$$F(P, Q)=0$$

identically.

This leads to the fundamental theorem that *if P and Q are permutable operators of orders m and n respectively, they satisfy identically an algebraic relation of the form*

$$F(P, Q)=0$$

of degree n in P and of degree m in Q.

Thus, for instance, if

$$P\equiv D^2-2x^{-2},$$

$$Q\equiv D^3-3x^{-2}D+3x^{-3},$$

then

$$P^3\equiv Q^2,$$

and the equations

$$(P-h)y=0, \quad (Q-k)y=0$$

have common solutions if

$$h^3-k^2=0.$$

Miscellaneous Examples.

1. If the equation
$$\frac{d^2y}{dx^2}+P\frac{dy}{dx}+Qy=0$$
is transformed by the substitution $x=Q(\xi)$ into
$$\frac{d^2y}{d\xi^2}+p\frac{dy}{d\xi}+qy=0,$$
prove that
$$\left\{2PQ+\frac{dQ}{dx}\right\}Q^{-\frac{3}{2}}=\left\{2pq+\frac{dq}{dx}\right\}q^{-\frac{3}{2}}.$$
Hence integrate the equation
$$(x^3-x)\frac{d^2y}{dx^2}+\frac{dy}{dx}+n^2x^3y=0.$$

2. Verify that x^3 and x^{-2} are solutions of
$$\frac{d^2y}{dx^2}-\frac{6}{x^2}y=0,$$
and obtain a particular integral of
$$\frac{d^2y}{dx^2}-\frac{6}{x^2}y=x\log x.$$

3. Integrate the equation
$$x^2(1-x)\frac{d^2y}{dx^2}+2x(2-x)\frac{dy}{dx}+2(1+x)y=x^2$$
given that the reduced equation has a particular solution of the form $y=x^n$.

4. Prove that any homogeneous self-adjoint equation of order $2m$ may be written in the form
$$\frac{d^m}{dx^m}\left\{A_0y^{(m)}\right\}+\frac{d^{m-1}}{dx^{m-1}}\left\{A_1y^{(m-1)}\right\}+\ldots+A_my=0.$$
Investigate the corresponding theorem for the equation of order $2m+1$.

[Bertand, Hesse.]

5. Prove that if the general solution $u=u(x)$ of the equation

$$\frac{d^2u}{dx^2}=\{\phi(x)+h\}u$$

is known for all values of h, and that any particular solution for the particular value $h=h_1$ is $u=f(x)$, then the general solution of the equation

$$\frac{d^2y}{dx^2}=\left\{f(x)\frac{d^2}{dx^2}\left(\frac{1}{f(x)}\right)+h-h_1\right\}y$$

for $h\neq h_1$ is

$$y=u'(x)-u(x)\frac{f'(x)}{f(x)}.$$

[Darboux, *C. R. Acad. Sc. Paris*, 94 (1882), p. 1456; *Théorie des Surfaces*, II. p. 210.]

6. By considering, as the initial equation,

$$\frac{d^2u}{dx^2}=hu.$$

with $h_1=0$, integrate the equation

$$\frac{d^2y}{dx^2}=\left\{\frac{2}{x^2}+h\right\}y.$$

By repeating the process, integrate

$$\frac{d^2y}{dx^2}=\left\{\frac{m(m-1)}{x^2}+h\right\}y,$$

where m is an integer. [Darboux.]

7. By considering the same initial equation, but taking $h_1=-1$, integrate

$$\frac{d^2y}{dx^2}=\left\{\frac{m(m-1)}{\sin^2 x}+\frac{n(n-1)}{\cos^2 x}+h\right\}y$$

where m and n are integers. [Darboux.]

CHAPTER VI

LINEAR EQUATIONS WITH CONSTANT COEFFICIENTS

6·1. The Linear Operator with Constant Coefficients.—The homogeneous linear differential equation with constant coefficients

$$\text{(A)} \qquad A_0\frac{d^ny}{dx^n}+A_1\frac{d^{n-1}y}{dx^{n-1}}+\ \ldots\ +A_{n-1}\frac{dy}{dx}+A_ny=0$$

was the first equation of a general type to be completely solved.* But apart from its historical interest, the equation has important practical applications and is of theoretical interest because of the simplicity of its general solution. The corresponding non-homogeneous equation †

$$\text{(B)} \qquad A_0\frac{d^ny}{dx^n}+A_1\frac{d^{n-1}y}{dx^{n-1}}+\ \ldots\ +A_{n-1}\frac{dy}{dx}+A_ny=f(x),$$

has also many important applications.

It is assumed that A_0 is not zero ; the remaining coefficients may or may not be zero. Equation (A), which may be written as ‡

$$F(D)y\equiv(A_0D^n+A_1D^{n-1}+\ \ldots\ +A_{n-1}D+A_n)y=0,$$

has an operator which may be factorised thus :

$$A_0(D-a_1)(D-a_2)\ \ldots\ (D-a_n).$$

But now $a_1, a_2, \ldots, a_n$ are constants, namely the roots of the algebraic equation

$$\text{(C)} \qquad A_0\xi^n+A_1\xi^{n-1}+\ \ldots\ +A_{n-1}\xi+A_n=0,$$

and therefore the factors

$$D-a_1,\quad D-a_2,\ \ldots,\quad D-a_n$$

are permutable. It follows that the given homogeneous equation is satisfied by the solution of each of n equations of the first order, namely,

$$(D-a_1)y=0,\ (D-a_2)y=0,\ \ldots,\ (D-a_n)y=0.$$

6·11. Solution of the Homogeneous Equation.—Let y_r be the general solution of

$$(D-a_r)y=0,$$

* It appears that the solution was known to Euler and to Daniel Bernoulli about the year 1739. The first published account was given by Euler, *Misc. Berol.* 7 (1743), p. 193 ; see also *Inst. Calc. Int.* 2, p. 375.

† D'Alembert, *Misc. Taur.* 3 (1762–65), p. 381.

‡ The symbolic notation $F(D)$ is due to Cauchy, see *Exercises math.* 2 (1827), p. 159 [*Œuvres* (2) 7, p. 198].

then

$$y_r = C_r e^{a_r x},$$

and therefore the general solution of (A) is

$$y = C_1 e^{a_1 x} + C_2 e^{a_2 x} + \ldots + C_n e^{a_n x},$$

where $C_1, C_2, \ldots, C_n$ are arbitrary constants. It has been tacitly assumed that $a_1, a_2, \ldots, a_n$ are unequal; the case in which the algebraic equation (C) has equal roots will be set aside for the moment.

Now let it be assumed that the coefficients $A_0, A_1, \ldots, A_n$ are real numbers, so that $a_1, \ldots, a_n$ are either real or conjugate imaginaries. The preceding solution is, as it stands, appropriate to the case in which $a_1, \ldots, a_n$ are real, but requires a slight modification when one or more pairs of conjugate complex quantities are included. For instance, let a_r and a_s be conjugate complex numbers, say

$$a_r = \alpha + i\beta, \quad a_s = \alpha - i\beta.$$

Then the terms

$$C_r e^{a_r x} + C_s e^{a_s x}$$

may be written as

$$\begin{aligned} C_r e^{(\alpha+i\beta)x} + C_s e^{(\alpha-i\beta)x} &= e^{\alpha x}\{C_r(\cos\beta x + i\sin\beta x) + C_s(\cos\beta x - i\sin\beta x)\} \\ &= e^{\alpha x}\{C_r'\cos\beta x + C_s'\sin\beta x\}, \end{aligned}$$

where

$$C_r' = C_r + C_s, \quad C_s' = i(C_r - C_s).$$

The number of arbitrary constants therefore remains as before.

As an example, consider the equation

$$\frac{d^3y}{dx^3} + \frac{d^2y}{dx^2} - 7\frac{dy}{dx} - 15y = 0.$$

The roots of

$$\xi^3 + \xi^2 - 7\xi - 15 = 0$$

are

$$3, \; -2+i, \; -2-i,$$

and therefore the general solution is

$$y = C_1 e^{3x} + e^{-2x}(C_2\cos x + C_3\sin x).$$

6·12. Repeated Factors.—Now let the operator

$$A_0D^n + A_1D^{n-1} + \ldots + A_{n-1}D + A_n$$

have a repeated factor, for example, the factor

$$(D-a)^p.$$

Then the general solution of

$$(D-a)^p y = 0$$

will be included in the general solution of (A). One solution corresponding to this factor is known, namely,

$$y = Ce^{ax},$$

where C is a constant; to determine all the general solution, the method of variation of parameters is applied. Write

$$y = e^{ax}v,$$

where v is a function to be determined, then

$$
\begin{aligned}
(D-a)^p e^{ax} v &= (D-a)^{p-1} e^{ax} Dv \\
&= (D-a)^{p-2} e^{ax} D^2 v \\
&= \ldots = e^{ax} D^p v.
\end{aligned}
$$

Consequently $y=e^{ax}v$ is a solution of

$$(D-a)^p y=0,$$

provided that v is a solution of

$$D^p v=0,$$

and hence v is an arbitrary polynomial in x of degree $p-1$. Thus the solution required is

$$y=(C_1+C_2x+\ldots+C_px^{p-1})e^{ax},$$

and contains, as it theoretically must contain, p arbitrary constants.

Lastly, if two conjugate imaginaries occur, each in a factor repeated p times, for example,

$$(D-\alpha+i\beta)^p(D-\alpha-i\beta)^p,$$

the solution which corresponds will be of the form

$$y=(C_1+C_2x+\ldots+C_px^{p-1})e^{\alpha x}\cos\beta x+(C_1'+C_2'x+\ldots+C_p'x^{p-1})e^{\alpha x}\sin\beta x,$$

with the correct number, $2p$, of arbitrary constants.

For example, the general solution of

$$\frac{d^4y}{dx^4}+2\frac{d^2y}{dx^2}+y=0$$

is

$$y=(C_1+C_2x)\cos x+(C_3+C_4x)\sin x.$$

6·13. The Complementary Function.—The complementary function of any linear equation has been defined as the general solution of the corresponding homogeneous equation. Now that all possible cases which may arise when the coefficients are constants have been discussed, it is important to determine whether or not the solution obtained is the most general solution.

Consider, first of all, the case in which

$$y=C_1e^{a_1x}+C_2e^{a_2x}+\ldots+C_ne^{a_nx},$$

and the numbers $a_1, a_2, \ldots, a_n$, which may be real or complex, are distinct. In this case, if

$$s=a_1+a_2+\ldots+a_n,$$

the value of the Wronskian of the solution is

$$e^{sx}\begin{vmatrix} 1, & 1, & \ldots, & 1 \\ a_1, & a_2, & \ldots, & a_n \\ \cdot & \cdot & \cdot & \cdot \\ a_1^{n-1}, & a_{2n-1}, & \ldots, & a_n^{n-1} \end{vmatrix} = e^{sx}\Pi(a_i-a_j),$$

and cannot be zero since $a_i \neq a_j$. The n functions

$$e^{a_1x},\quad e^{a_2x},\ \ldots,\quad e^{a_nx}$$

are therefore linearly distinct, and

$$y=C_1e^{a_1x}+C_2e^{a_2x}+\ldots+C_ne^{a_nx}$$

is the general solution.

In the next place consider the extreme case in which the numbers a are all equal. Then

$$y=(C_1+C_2x+\ .\ .\ .\ +C_nx^{n-1})e^{ax}.$$

If, for any particular values of the constants C, y is identically zero, then $C_1+C_2x+\ .\ .\ .\ +C_nx^{n-1}$ will be identically zero, which is impossible unless $C_1=C_2=\ .\ .\ .\ =C_n=0$. In this case also the solution is general.

In any other case the solution would be of the form

$$P_1e^{a_1x}+P_2e^{a_2x}+\ .\ .\ .\ +P_me^{a_mx},$$

where $P_1, P_2, \ldots, P_m$ are polynomials in x, and the numbers $a_1, a_1, \ldots, a_m$ are distinct. It will be shown that a function of this kind cannot be identically zero unless the polynomials P are themselves identically zero. Assume then that

$$P_1e^{a_1x}+P_2e^{a_2x}+\ .\ .\ .\ +P_me^{a_mx}=0$$

identically. Let

$$b_k=a_k-a_1,$$

then the identity may be written

$$P_1+P_2e^{b_2x}+\ .\ .\ .\ +P_me^{b_mx}=0.$$

Let r_1 be the degree of the polynomial P_1, then if the identity is differentiated r_1+1 times it takes the form

$$Q_2e^{b_2x}+\ .\ .\ .\ +Q_me^{b_mx}=0,$$

where $Q_2, \ldots, Q_m$ are polynomials whose degrees are the same as the degrees of $P_2, \ldots, P_m$ respectively and the numbers $b_2, \ldots, b_m$ are unequal. If this process is continued, a stage is arrived at in which

$$R_me^{r_mx}=0$$

identically, where R_m is a polynomial whose degree is equal to that of P_m. Hence R_m must vanish identically, which is impossible. If follows that

$$P_1e^{a_1x}+P_2e^{a_2x}+\ .\ .\ .\ +P_me^{a_mx}$$

is not identically zero.

The investigation of the complementary function may therefore be regarded as complete.

6·14. The Case of Repeated Factors regarded as a Limiting Case.—A very powerful method of attacking the case in which the operator

$$A_0D^n+A_1D^{n-1}+\ .\ .\ .\ +A_{n-1}D+A_n$$

has a repeated factor is due to d'Alembert.* As the scope of the method extends beyond the case in which the coefficients are constants, it will be convenient to suppose for the moment that the equation is of the form

$$p_0\frac{d^ny}{dx^n}+p_1\frac{d^{n-1}y}{dx^{n-1}}+\ .\ .\ .\ +p_{n-1}\frac{dy}{dx}+p_ny=0,$$

where $p_0, p_1, \ldots\ p_n$ depend upon certain parameters

$$a_1,\ a_2,\ \ldots,\ a_\mu$$

and possibly also upon x. Let $f(x, r)$ be a function which for certain values

* *Hist. Acad. Berlin* 1748, p. 283.

of r, depending upon the parameters $\alpha_1, \alpha_2, \ldots, \alpha_\mu$, satisfies the equation. Let

$$r_1, \quad r_2, \ldots, \quad r_\nu$$

be such a set of values of r, so chosen that functions

$$f(x, r_1), \quad f(x, r_2), \ldots, \quad f(x, r_\nu)$$

are in general distinct. The functions are thus a set of ν particular solutions of the equation. For particular values of $\alpha_1, \alpha_2, \ldots, \alpha_\mu$, however, two or more of the quantities r, say r_1 and r_2, and the corresponding functions $f(x, r_1)$ and $f(x, r_2)$ become equal, and therefore the number of solutions of the equation represented by the functions $f(x, r)$ is reduced. In such a case, however, the limiting value of

$$\frac{f(x, r_2)-f(x, r_1)}{r_2-r_1},$$

when that limit exists, is a solution of the equation. But this limit is

$$\left[\frac{\partial f(x, r)}{\partial r}\right]_{r=r_1}$$

The case in which r_1, r_2 and r_3 become equal may be treated in the same way. The function

$$\left\{\frac{f(x, r_3)-f(x, r_2)}{r_3-r_2}-\frac{f(x, r_2)-f(x, r_1)}{r_2-r_1}\right\} \Big/ (r_2-r_1)$$

satisfies the equation, and if its limit exists, this limit, namely

$$\left[\frac{\partial^2 f(x, r)}{\partial r^2}\right]_{r=r_1},$$

is a solution of the equation.

In general if, for particular values of the parameters $\alpha_1, \alpha_2, \ldots, \alpha_\mu$,

$$r_1=r_2= \ldots =r_\mu,$$

the equation has the μ solutions

$$f(x, r_1), \quad \left[\frac{\partial f(x, r)}{\partial r}\right]_{r=r_1}, \ldots, \left[\frac{\partial^{\mu-1} f(x, r)}{\partial r^{\mu-1}}\right]_{r=r_1}.$$

Consider, as an example, the equation

$$(D^2+1)^2 y=0;$$

replace it by the more general equation

$$(D^2+\alpha^2)(D^2+\beta^2)y=0.$$

The latter equation, when $\alpha^2 \neq \beta^2$, has the general solution

$$y=A_1 \cos \alpha x+A_2 \cos \beta x+A_3 \sin \alpha x+A_4 \sin \beta x.$$

When $\alpha=\beta=1$ this solution ceases to be general, and reduces to

$$y=C_1 \cos x+C_3 \sin x.$$

But the functions

$$\left[\frac{\partial}{\partial \alpha} \cos \alpha x\right]_{\alpha=1}=-x \sin x, \quad \left[\frac{\partial}{\partial \alpha} \sin \alpha x\right]_{\alpha=1}=x \cos x$$

are particular solutions not obtainable by attributing particular values to C_1 and C_3. The general solution of the given equation is thus

$$y=(C_1+C_2 x) \cos x+(C_3+C_4 x) \sin x.$$

6·2. Discussion of the Non-Homogeneous Equation.—The determination of a particular integral of the non-homogeneous equation depends upon the properties of operators inverse to D, $D-a$, etc., for the problem really amounts to attributing a value to the expression *

$$(D-a_1)^{-1} \ldots (D-a_n)^{-1}f(x).$$

The operator inverse to D is D^{-1} and is the operation of simple indefinite integration; similarly D^{-p} is the operation of p-ple integration. A significance must now be given to the operators $(D-a)^{-1}$ and $(D-a)^{-p}$ where a is a non-zero constant.

In order to make these operators as definite as possible, it will be stipulated that the arbitrary element which they introduce is to be discarded. Just as the operation D^{-1} introduces an arbitrary additive constant, and, more generally, the operation D^{-p} introduces an arbitrary element $C_1+C_2x+\ldots+C_px^{p-1}$, so also $(D-a)^{-1}$ brings in an arbitrary element Ce^{ax}, and $(D-a)^{-p}$ introduces $e^{ax}(C_1+C_2x+\ldots+C_px^{p-1})$. These expressions are already accounted for by the complementary function; they are therefore discarded in determining the particular integral.

When $f(x)$ is a function of a simple type the effect of operating upon $f(x)$ by $(D-a)^{-1}$ or by $(D-a)^{-p}$ is as follows:

1°. Let

$$f(x)=e^{kx}, \quad k \text{ a constant.}$$

Operating upon both sides of the identity

$$(D-a)e^{kx}=(k-a)e^{kx}$$

by $(k-a)^{-1}(D-a)^{-1}$ gives

$$(D-a)^{-1}e^{kx}=(k-a)^{-1}e^{kx},$$

provided that $k\neq a$; this exceptional case is treated below.

Similarly

$$(D-a_1)^{-1} \ldots (D-a_p)^{-1}e^{kx}=(k-a_1) \ldots (k-a_p)e^{kx},$$

provided that $a_1, \ldots, a_p$ are distinct from k. In particular

$$(D-a)^{-p}e^{kx}=(k-a)^{-p}e^{kx}.$$

Thus if $F(D)$ is a polynomial in D such that $F(k)\neq 0$, then

$$F^{-1}(D)e^{kx}=\frac{e^{kx}}{F(k)},$$

where $F^{-1}(D)$ is the operator inverse to $F(D)$.

2°. Let

$$f(x)=e^{kx}\phi(x).$$

In the identity

$$(D-a)e^{kx}X=e^{kx}(D+k-a)X,$$

where X is an arbitrary function of x, write

$$(D+k-a)X=\phi(x),$$

then

$$(D-a)e^{kx}(D+k-a)^{-1}\phi(x)=e^{kx}\phi(x),$$

and hence

$$(D-a)^{-1}e^{kx}\phi(x)=e^{kx}(D+k-a)^{-1}\phi(x).$$

* Lobatto, *Théorie des caractéristiques* (Amsterdam, 1837); Boole, *Camb. Math. J.*, 2 (1841), p. 114.

Similarly it may be proved that in general

$$(D-a_1)^{-1}\ldots(D-a_p)^{-1}e^{kx}\phi(x)=e^{kx}(D+k-a_1)^{-1}\ldots(D+k-a_p)^{-1}\phi(x),$$

or

$$F^{-1}(D)e^{kx}\phi(x)=e^{kx}F^{-1}(D+k)\phi(x).$$

In particular, taking $k=a_1=a_2=\ldots=a_p=a$, $\phi(x)=1$, it follows that

$$\begin{aligned}(D-a)^{-p}e^{ax}&=e^{ax}D^{-p}.1\\&=e^{ax}\frac{x^p}{p!},\end{aligned}$$

and thus the exceptional case left over from above is accounted for.

3°. Let

$$f(x)=\sin ax.$$

If $F(D)$ is an even polynomial in D, write $F(D)=\Phi(D^2)$ so that $\Phi(D^2)$ is a polynomial in D^2. Then

$$\Phi(D^2)\sin ax=\Phi(-a^2)\sin ax,$$

and hence

$$F^{-1}(D)\sin ax=\frac{\sin ax}{\Phi(-a^2)}.$$

In the most general case, the polynomial $F(D)$ is not even; if it has an even polynomial factor $G(D)$, let $F(D)=G(D)H(D)$. Then

$$\begin{aligned}F^{-1}(D)\sin ax&=\frac{1}{G(D)H(D)}\sin ax\\&=\frac{H(-D)}{G(D)H(D)H(-D)}\sin ax.\end{aligned}$$

Now $G(D)H(D)H(-D)$ is an even polynomial in D and may be written $K(D^2)$, thus

$$\begin{aligned}F^{-1}(D)\sin ax&=\frac{H(-D)}{K(D^2)}\sin ax\\&=\frac{H(-D)}{K(-a^2)}\sin ax,\end{aligned}$$

and thus $F^{-1}(D)\sin ax$ and similarly $F^{-1}(D)\cos ax$ may be evaluated provided that $K(-a^2)\neq 0$.

By combining this case with the previous case, particular integrals of the form

$$F^{-1}(D)e^{kx}\sin ax,\quad F^{-1}(D)e^{kx}\cos ax$$

may be evaluated.

Example.—

$$(3D^2+2D-8)y=5\cos x.$$

A particular integral is

$$\begin{aligned}y&=\frac{5}{3D^2+2D-8}\cos x\\&=\frac{5(3D^2-2D-8)}{(3D^2+2D-8)(3D^2-2D-8)}\cos x\\&=\frac{5(3D^2-2D-8)}{9D^4-52D^2+64}\cos x\\&=\frac{5(3D^2-2D-8)}{9+52+64}\cos x\\&=\tfrac{1}{25}\{(3D^2-8)-2D\}\cos x=\tfrac{1}{25}(2\sin x-11\cos x).\end{aligned}$$

4°. Let
$$f(x) = x^n.$$
Then
$$(D-a)x^n = nx^{n-1} - ax^n,$$
$$(D-a)\frac{n}{a}x^{n-1} = \frac{n(n-1)}{a}x^{n-2} - nx^{n-1},$$
$$(D-a)\frac{n(n-1)}{a^2}x^{n-2} = \frac{n(n-1)(n-2)}{a^2}x^{n-3} - \frac{n(n-1)}{a}x^{n-2},$$
$$\cdot \quad \cdot \quad \cdot \quad \cdot \quad \cdot \quad \cdot \quad \cdot \quad \cdot$$
$$(D-a)\frac{n(n-1)\ldots 3.2}{a^{n-1}}x = \frac{n!}{a^{n-1}} - \frac{n(n-1)\ldots 3.2}{a^{n-2}}x,$$
$$(D-a)\frac{n!}{a^n} = -\frac{n!}{a^{n-1}}.$$
Hence, by addition,
$$(D-a)\left(x^n + \frac{n}{a}x^{n-1} + \ldots + \frac{n!}{a^n}\right) = -ax^n,$$
and consequently
$$(D-a)^{-1}x^n = -\sum_{r=0}^{n} \frac{n!}{r!\, a^{n-r+1}} x^r.$$

This result is the same as would have been obtained by formally expanding the operator $(D-a)^{-1}$ in ascending powers of D and performing the differentiations. It follows that if X is a polynomial in x of degree n,
$$F^{-1}(D)X = (a_0 + a_1 D + \ldots + a_n D^n)X,$$
where
$$a_0 + a_1 D + \ldots + a_n D^n$$
is the expansion of $F^{-1}(D)$ to $n+1$ terms.

The inverse differential operator $F^{-1}(D)$ may be decomposed into partial fractions in precisely the same way as the reciprocal of a polynomial, for if this process is formally carried out, the resulting expression will be reduced to unity by the operator $F(D)$. Consequently the material which has been accumulated is sufficient to determine the particular integral in cases where the function $f(x)$ is a sum of terms or products of terms of the form x^n, e^{kx}, $\sin ax$ and $\cos ax$. Sine and cosine terms may equally well be dealt with by expressing them in the exponential form.

6·21. Determination of a Particular Integral by Quadratures.—If the function $f(x)$ is such that $f(x)$ and $e^{-ax}f(x)$ are integrable, a particular integral may be determined by quadratures. Suppose in the first place that $F(D)$ has no repeated factors. Then $F^{-1}(D)$ can be decomposed into simple partial fractions thus
$$F^{-1}(D) = \sum_{r=1}^{n} \frac{\alpha_r}{D - a_r}.$$
A particular integral is then
$$\begin{aligned} y &= \sum_{r=1}^{n} \frac{\alpha_r}{D-a_r} f(x) \\ &= \sum \alpha_r (D-a_r)^{-1} e^{a_r x}\{e^{-a_r x} f(x)\} \\ &= \sum \alpha_r e^{a_r x} D^{-1}\{e^{-a_r x} f(x)\} \\ &= \sum \alpha_r e^{a_r x} \int e^{-a_r x} f(x)dx \\ &= \sum_{r=1}^{n} \alpha_r \int^{x} e^{a_r(x-t)} f(t)dt. \end{aligned}$$

The lower limit of integration may be arbitrarily fixed, for the term which proceeds from a constant lower limit of integration is a constant multiple of $e^{a_r x}$, and is therefore included in the Complementary Function.

Consider now the case in which $F(D)$ contains the factor $(D-a)^p$. The part of the expression of $F^{-1}(D)$ in partial fractions which corresponds to this repeated factor is

$$\sum_{r=1}^{p}\frac{\beta_r}{(D-a)^r},$$

and the corresponding contribution to the particular integral is

$$\begin{aligned}\sum_{r=1}^{p}\frac{\beta_r}{(D-a)^r}f(x) &= \sum\beta_r(D-a)^{-r}e^{ax}\{e^{-ax}f(x)\}\\ &= \sum\beta_r e^{ax}D^{-r}\{e^{-ax}f(x)\}\\ &= \sum_{r=1}^{p}\beta_r\int^x\int^t\ldots\int^t e^{a(x-t)}f(t)dt^r.\end{aligned}$$

Example.—

$$\frac{d^2y}{dx^2}-4y=16x^2e^{2x^2}.$$

The Complementary Function is $Ae^{2x}+Be^{-2x}$; the Particular Integral may be written as

$$\begin{aligned}y&=4\left(\frac{1}{D-2}-\frac{1}{D+2}\right)x^2e^{2x^2}\\ &=4e^{2x}\int_{-\frac{1}{2}}^{x}t^2e^{2t^2-2t}dt-4e^{-2x}\int_{\frac{1}{2}}^{x}t^2e^{2t^2+2t}dt.\end{aligned}$$

The lower limits of integration are so chosen as to make the particular integral as simple as possible. By integration by parts it is found that

$$y=e^{2x^2}.$$

6·3. The Euler Linear Equation.—The equation of the type

$$A_0x^n\frac{d^ny}{dx^n}+A_1x^{n-1}\frac{d^{n-1}y}{dx^{n-1}}+\ldots+A_{n-1}x\frac{dy}{dx}+A_ny=f(x),$$

in which $A_0, A_1, \ldots, A_n$ are constants, is known as the Euler equation.* It may be transformed into a linear equation with constant coefficients by means of the substitution $x=e^z$, for

$$x\frac{dy}{dx}=\frac{dy}{dz}=Dy,$$

where D now signifies $\frac{d}{dz}$, and similarly

$$x^2\frac{d^2y}{dx^2}=\frac{d^2y}{dz^2}-\frac{dy}{dz}=D(D-1)y,$$

$$\cdot\quad\cdot\quad\cdot\quad\cdot\quad\cdot\quad\cdot\quad\cdot$$

$$x^n\frac{d^ny}{dx^n}=D(D-1)\ldots(D-n+1)y,$$

and thus the equation is brought into the form

$$F(D)y\equiv(A_0D^n+A'_1D^{n-1}+\ldots+A'_{n-1}D+A_n)y=f(e^z).$$

and may be solved by the foregoing methods.

* Its general solution was, however, known to John Bernoulli at least as early as the year 1700. Euler's work on the equation was done about the year 1740, published *Inst. Calc. Int.* (1769), 2, p. 483. Later work was done by Cauchy; see also Malmsten, *J. für Math.* 39 (1850), p. 99.

A simple factor $(D-a)$ of the operator $F(D)$ leads to a term in the Complementary Function of the form

$$Ce^{az}=Cx^a,$$

whilst a repeated factor $(D-a)^p$ leads to

$$\begin{aligned} y &= e^{az}(C_1+C_2z+\ldots+C_pz^{p-1}) \\ &= x^a\{C_1+C_2\log x+\ldots+C_p(\log x)^{p-1}\}. \end{aligned}$$

This solution should be particularly noted. It might equally well have been arrived at by the application of d'Alembert's method. For since $y=x^\alpha$ is a solution of the homogeneous equation, corresponding to a p-ple factor in $F(D)$,

$$y_1=\left\{\frac{d}{d\alpha}x^\alpha\right\}_{\alpha=a}=\left\{\frac{d}{d\alpha}e^{\alpha\log x}\right\}_{\alpha=a}=x^a\log x,$$

$$\cdot\quad\cdot\quad\cdot\quad\cdot\quad\cdot\quad\cdot\quad\cdot$$

$$y_{p-1}=\left\{\frac{d^{p-1}}{d\alpha^{p-1}}x^\alpha\right\}_{\alpha=a}=x^a(\log x)^{p-1}$$

are also solutions of the homogeneous equation.

In the same way, equations of the type *

$$\sum_{r=0}^{n}A_r(ax+b)^{n-r}\frac{d^{n-r}y}{dx^{n-r}}=f(x),$$

where a, b, A_r are constants and $a\neq 0$ can be dealt with by the substitution $ax+b=e^z$.

A particular integral of the non-homogeneous equation may be obtained by quadratures in a manner analogous to that adopted in the case of the equation with constant coefficients.

Let ϑ denote the operator $x\dfrac{d}{dx}$, then since †

$$x^m\frac{d^m}{dx^m}=\vartheta(\vartheta-1)\ldots(\vartheta-m+1),$$

the operator

$$A_0x^n\frac{d^n}{dx^n}+A_1x^{n-1}\frac{d^{n-1}}{dx^{n-1}}+\ldots+A_{n-1}x\frac{d}{dx}+A_n$$

may be written

$$F(\vartheta)\equiv A_0\vartheta^n+A'_1\vartheta^{n-1}+\ldots+A'_{n-1}\vartheta+A_n,$$

and $F(\vartheta)$ may be decomposed into permutable linear factors as follows:

$$F(\vartheta)=A_0(\vartheta-a_1)(\vartheta-a_2)\ldots(\vartheta-a_n).$$

Now

$$x^m\frac{d^m}{dx^m}x^\mu=\mu(\mu-1)\ldots(\mu-m+1)x^\mu,$$

so that

$$F(\vartheta)x^\mu=x^\mu F(\mu).$$

If therefore a is such that $F(a)=0$,

$$y=Ax^a$$

is a solution of the homogeneous equation

$$F(\vartheta)y=0.$$

* Lagrange, *Misc. Taur.* 3 (1762–65), p. 190 [*Œuvres*, 1, p. 481].

† Note that if $x=e^z$, $\vartheta=x\dfrac{d}{dx}$ and $D=\dfrac{d}{dz}$, then $\vartheta\equiv D$.

Also if X is a function of x,

$$\vartheta x^\mu X = x(x^\mu X' + \mu x^{\mu-1} X)$$
$$= x^\mu(\vartheta+\mu)X,$$

and in general,

$$\vartheta^m x^\mu X = x^\mu(\vartheta+\mu)^m X,$$

from which it follows that

$$F(\vartheta)x^\mu X = x^\mu F(\vartheta+\mu)X.$$

Write $\phi(x)$ for $F(\vartheta+\mu)X$ and operate on both sides of this identity by $F^{-1}(\vartheta)$, then

$$F^{-1}(\vartheta)x^\mu\phi(x) = x^\mu F^{-1}(\vartheta+\mu)\phi(x).$$

When $F(\vartheta)$ has no repeated factors, the inverse operator $F^{-1}(\vartheta)$ may be decomposed into simple partial fractions thus

$$F^{-1}(\vartheta) = \sum_{r=1}^{n} \frac{\alpha_r}{\vartheta - a_r},$$

for $\sum \alpha_r(\vartheta - a_r)^{-1}$ is reduced to unity by the direct operator $F(\vartheta)$. A particular solution of the non-homogeneous equation is therefore

$$y = \sum_{r=1}^{n} \frac{\alpha_r}{\vartheta - a_r} f(x)$$
$$= \sum \frac{\alpha_r}{\vartheta - a_r} x^{a_r}\left\{x^{-a_r} f(x)\right\}$$
$$= \sum \alpha_r x^{a_r} \vartheta^{-1}\{x^{-a_r} f(x)\}$$
$$= \sum_{r=1}^{n} \alpha_r x^{a_r} \int^x t^{-1-a_r} f(t)\,dt.$$

If $F(\vartheta)$ has a repeated factor, say $(\vartheta - a)^p$, the corresponding part of the partial fraction representation of $F^{-1}(\vartheta)$ will be of the form

$$\sum_{r=1}^{p} \frac{\beta_r}{(\vartheta - a)^r}.$$

The corresponding terms in the particular integral are

$$\sum_{r=1}^{p} \frac{\beta_r}{(\vartheta - a)^r} f(x) = \sum \frac{\beta_r}{(\vartheta - a)^r} x^a\{x^{-a} f(x)\}$$
$$= \sum \beta_r x^a \vartheta^{-r}\{x^{-a} f(x)\}$$
$$= \sum \beta_r x^a \vartheta^{-r+1} \int^x t^{-1-a} f(t)\,dt$$
$$= \sum \beta_r x^a \vartheta^{-r+2} \int^x t^{-1} \int^t t^{-1-a} f(t)\,dt^2$$
$$= \sum_{r=1}^{p} \beta_r x^a \int^x t^{-1} \int^t \ldots t^{-1} \int^t t^{-1-a} f(t)\,dt^r.$$

Example.—

$$x^2 \frac{d^2y}{dx^2} - 3x\frac{dy}{dx} + 4y = 2x^2.$$

This may be written

$$(\vartheta - 2)^2 y = 2x^2.$$

The complementary function is

$$y = Ax^2 + Bx^2 \log x.$$

The particular integral is

$$y=\frac{1}{(\vartheta-2)^2}2x^2=2x^2\vartheta^{-2}.1$$

$$=2x^2\vartheta^{-1}\int_1^x t^{-1}dt=2x^2\vartheta^{-1}\log x$$

$$=2x^2\int_1^x t^{-1}\log t\,dt=(x\log x)^2.$$

6·4. Systems of Simultaneous Linear Equations with Constant Coefficients.—It was remarked in a previous section (§ 1·5) that a single linear differential equation may be replaced by a system of simultaneous equations each of lower order than n, and in particular by a system of n simultaneous equations of the first order. The converse question now suggests itself, namely, given a system of simultaneous linear equations, is this system equivalent to a single linear equation, in the sense that the general solution of the system contains the same number of arbitrary constants as does the complete solution of the simultaneous system ? This question will now be discussed with the assumption that the equations considered have constant coefficients.* Such equations appear in many dynamical problems ; their importance is therefore both practical and theoretical.

The germ of the problem to be considered can be made clear by considering a system of three homogeneous linear equations between three dependent variables, namely,

$$F_{11}(D)y_1+F_{12}(D)y_2+F_{13}(D)y_3=0,$$
$$F_{21}(D)y_1+F_{22}(D)y_2+F_{23}(D)y_3=0,$$
$$F_{31}(D)y_1+F_{32}(D)y_2+F_{33}(D)y_3=0,$$

where $F_{rs}(D)$ are polynomials in the operator D, with constant coefficients, and the independent variable is x.

The variable y_3 may be eliminated from these equations by first of all operating on the first by $F_{23}(D)$, and on the second by $F_{13}(D)$ and subtracting, and then by operating on the second by $F_{33}(D)$, and on the third by $F_{23}(D)$ and subtracting. Then y_2 may be eliminated in the same way between the resulting two equations, leaving an equation in y_1 only. This process is identical with that of algebraic elimination, and is formally carried out as if the operators $F_{rs}(D)$ were constants. The result is therefore

$$\begin{vmatrix} F_{11}(D), & F_{12}(D), & F_{13}(D) \\ F_{21}(D), & F_{22}(D), & F_{23}(D) \\ F_{31}(D), & F_{32}(D), & F_{33}(D) \end{vmatrix} y_1=0$$

or, say,

$$F(D)y_1=0.$$

This equation exists if $F(D)$ is not identically zero, that is to say when the three equations of the given set are really distinct from one another. In the same way

$$F(D)y_2=0, \quad F(D)y_3=0.$$

$F(D)$ may be a constant, in which case the only possible solution is

$$y_1=y_2=y_3=0,$$

* The original discussion of the problem, by Jacobi, *J. für Math.* 60 (1865), p. 297 [*Ges Werke* 5, p. 193], was defective ; a rigorous investigation into the equivalence of two systems of simultaneous linear equations was made by Chrystal, *Trans. Roy. Soc. Edin.* 38 (1895), p. 163. The account here given is based upon Chrystal's memoir.

or in other words the equations are inconsistent, but in general $F(D)$ is a polynomial in D; let its degree be m. Let the factors of $F(D)$ be

$$D-a_1, \quad D-a_2, \quad \ldots, \quad D-a_m,$$

and suppose that $a_1, a_2, \ldots, a_m$ are all distinct. Then the solution of $F(D)y_1=0$ will be

$$y_1=C_{11}e^{a_1x}+C_{12}e^{a_2x}+\ldots+C_{1m}e^{a_mx},$$

and similarly the solutions of $F(D)y_2=0$ and $F(D)y_3=0$ will be

$$y_2=C_{21}e^{a_1x}+C_{22}e^{a_2x}+\ldots+C_{2m}e^{a_mx},$$
$$y_3=C_{31}e^{a_1x}+C_{32}e^{a_2x}+\ldots+C_{3m}e^{a_mx},$$

respectively. In all $3m$ constants enter into these solutions, but these constants are not all independent. For since y_1, y_2, y_3 must satisfy the given system, the constants are connected by the relations

$$C_{1r}F_{11}(a_r)+C_{2r}F_{12}(a_r)+C_{3r}F_{13}(a_r)=0,$$
$$C_{1r}F_{21}(a_r)+C_{2r}F_{22}(a_r)+C_{3r}F_{23}(a_r)=0,$$
$$C_{1r}F_{31}(a_r)+C_{2r}F_{32}(a_r)+C_{3r}F_{33}(a_r)=0,$$

$(r=1, 2, \ldots m)$, and if these equations are sufficient to determine all the ratios $C_{1r} : C_{2r} : C_{3r}$, the number of constants is effectively reduced to m. But although it is true that the order of the system, which is equal to the number of independent constants in its general solution, is always the same as the order of the *characteristic determinant* $F(D)$, the assumptions which have been made are not always valid. The difficulty arises from the fact that even when y_1, y_2, y_3 form a general solution of the system, it may happen that no one of the functions y_1, y_2, y_3 satisfies the *characteristic equation*

$$F(D)y=0.$$

A rigorous proof of the theorem that *the order of the system is equal to the order of the characteristic equation* will now be given; the first step consists in establishing a necessary and sufficient condition for the equivalence of two systems of linear equations, not necessarily homogeneous with constant coefficients.

The following example is illustrative. Consider the system:

$$U\equiv(D^2+1)y_1+(D^2+D+1)y_2=x,$$
$$V\equiv \qquad Dy_1+ \qquad (D+1)y_2=e^x.$$

Its characteristic determinant reduces to a constant, the natural inference from which is that the solution of the system involves no arbitrary constants. Consider the derived system:

$$U-DV=x-e^x,$$
$$DU-(D^2+1)V=1-2e^x.$$

This system reduces to

$$y_1+y_2=x-e^x,$$
$$-y_2=1-2e^x,$$

whence

$$y_1=1+x-3e^x, \quad y_2=2e^x-1.$$

This is a solution of the given system. The investigation which follows shows that when the determinant of the *multiplier system*, which is here

$$\begin{vmatrix} 1, & -D \\ D, & -D^2-1 \end{vmatrix},$$

is a constant, the given system and the derived system are *equivalent*, and have the same general solution. In this case, therefore, the general solution has no arbitrary constants.

6·41. Conditions for the Equivalence of Two Systems of Linear Equations.—Let

$$F_{r1}(D)y_1+F_{r2}(D)y_2+\ \dots\ +F_{rn}(D)y_n=f_r(x) \qquad (r=1, 2, \dots, m),$$

$$G_{r1}(D)y_1+G_{r2}(D)y_2+\ \dots\ +G_{rn}(D)y_n=g_r(x) \qquad (r=1, 2, \dots, m),$$

be two systems of linear equations in the n dependent variables

$$y_1,\ y_2,\ \dots,\ y_n,$$

where $n \geqslant m$. The m equations of each set are supposed to be linearly distinct, and the operators $F(D)$ and $G(D)$ are polynomials in D with constant coefficients. These systems may be written respectively

(U) $\qquad U_1=0,\ U_2=0,\ \dots,\ U_m=0,$

(V) $\qquad V_1=0,\ V_2=0,\ \dots,\ V_m=0.$

The system (V) is said to be *derived from* the system (U) when every solution of (U) satisfies (V). When this is the case, any equation of (V) can be obtained by operating upon the equations of (U) by polynomials in D and adding the results together. Thus

$$V_1=\delta_{11}U_1+\ \dots\ +\delta_{1m}U_m,$$
$$\cdot \quad \cdot \quad \cdot \quad \cdot \quad \cdot$$
$$V_m=\delta_{m1}U_1+\ \dots\ +\delta_{mm}U_m.$$

The set of operators δ_{rs} is known as the *multiplier system* by means of which the system (V) is derived from the system (U), and the determinant

$$\Delta=\begin{vmatrix} \delta_{11}, & \dots, & \delta_{1m} \\ \cdot & \cdot & \cdot \\ \delta_{m1}, & \dots, & \delta_{mm} \end{vmatrix}$$

is its *modulus*. Δ cannot be zero since the equations of (V) are linearly independent.

If, when (V) is derived from (U), every solution of (V) satisfies (U), the systems are said to be *equivalent*. It will now be proved that *a necessary and sufficient condition that the two systems be equivalent is that the modulus is a constant.*

Let

$$\begin{vmatrix} \Delta_{11}, & \dots, & \Delta_{1m} \\ \cdot & \cdot & \cdot \\ \Delta_{m1}, & \dots, & \Delta_{mm} \end{vmatrix}$$

be the reciprocal of Δ, then $U_1, \dots, U_m$ are expressed in terms of $V_1, \dots, V_m$ as follows: *

$$\Delta U_1=\Delta_{11}V_1+\ \dots\ +\Delta_{m1}V_m,$$
$$\cdot \quad \cdot \quad \cdot \quad \cdot \quad \cdot \quad \cdot \quad \cdot$$
$$\Delta U_m=\Delta_{1m}V_1+\ \dots\ +\Delta_{mm}V_m.$$

Hence every solution of (V) satisfies the system

$$\Delta U_1=0,\ \dots,\ \Delta U_m=0.$$

If, therefore, Δ is a constant, every solution of (V) satisfies (U). The condition stated is therefore sufficient. To prove that it is necessary, suppose that the system (U) is derived from the system (V). Then there will exist a set of polynomials in D, say δ'_{rs}, such that

$$U_1=\delta'_{11}V_1+\ \dots\ +\delta'_{1m}V_m,$$
$$\cdot \quad \cdot \quad \cdot \quad \cdot \quad \cdot \quad \cdot \quad \cdot$$
$$U_m=\delta'_{m1}V_1+\ \dots\ +\delta'_{mm}V_m.$$

* See Scott and Mathews, *Theory of Determinants*, Chaps. VI. and XI.

By substituting the values of $V_1, \ldots, V_m$ in terms of $U_1, \ldots, U_m$ it is found that

$$U_r = \delta'_{r1}(\delta_{11}U_1 + \ldots + \delta_{1m}U_m) + \ldots + \delta'_{rm}(\delta_{m1}U_1 + \ldots + \delta_{mm}U_m) \qquad (r=1, 2, \ldots, m).$$

But $U_1, \ldots, U_m$ are linearly independent, and therefore

$$\delta'_{r1}\delta_{1s} + \ldots + \delta'_{rm}\delta_{ms} = 0 \qquad (r \neq s),$$
$$\delta'_{r1}\delta_{1r} + \ldots + \delta'_{rm}\delta_{mr} = 1.$$

There are, for each value of r, m equations to determine $\delta'_{r1}, \ldots, \delta'_{rm}$; their solution is

$$\delta'_{r1} = \Delta_{1r}/\Delta, \ldots, \delta'_{rm} = \Delta_{mr}/\Delta,$$

and, therefore, if Δ' is the modulus of the multiplier system δ'_{rs},

$$\Delta' = \begin{vmatrix} \delta'_{11}, & \ldots, & \delta'_{1m} \\ \cdot & \cdot & \cdot \\ \delta'_{m1}, & \ldots, & \delta'_{mm} \end{vmatrix} = \begin{vmatrix} \Delta_{11}, & \ldots, & \Delta_{m1} \\ \cdot & \cdot & \cdot \\ \Delta_{1m}, & \ldots, & \Delta_{mm} \end{vmatrix} \div \Delta^m$$
$$= \Delta^{m-1}/\Delta^m = 1/\Delta.$$

But both Δ and Δ' are polynomials in D. The identity

$$\Delta\Delta' = 1$$

cannot therefore be satisfied unless Δ and Δ' are both independent of D, that is to say Δ and Δ' are constants. The condition is therefore necessary as well as sufficient.

6·42. An Alternative Form of the Equivalence Conditions.—The above form of the equivalence theorem explicitly involves the multiplier system; a second form of the theorem, and one which does not require the direct calculation of the multiplier system can be derived as follows. Since

$$U_r \equiv F_{r1}(D)y_1 + \ldots + F_{rn}(D)y_n - f_r(x)$$

and

$$V_r \equiv G_{r1}(D)y_1 + \ldots + G_{rn}(D)y_n - g_r(x),$$

and since

$$V_r = \delta_{r1}U_1 + \ldots + \delta_{rm}U_m,$$

it follows, on equating the operators on $y_1, \ldots, y_n$ and writing F_{rs} and G_{rs} in short for $F_{rs}(D)$ and $G_{rs}(D)$, that

$$G_{r1} = \delta_{r1}F_{11} + \ldots + \delta_{rm}F_{m1},$$
$$\cdot \quad \cdot \quad \cdot \quad \cdot \quad \cdot \quad \cdot$$
$$G_{rn} = \delta_{r1}F_{1n} + \ldots + \delta_{rm}F_{mn},$$
$$g_r(x) = \delta_{r1}f_1(x) + \ldots + \delta_{rm}f_m(x).$$

From these equations it follows that

$$\begin{pmatrix} G_{11}, & \ldots, & G_{1n}, & g_1 \\ \cdot & \cdot & \cdot & \cdot \\ G_{m1}, & \ldots, & G_{mn}, & g_m \end{pmatrix} = \Delta \begin{pmatrix} F_{11}, & \ldots, & F_{1n}, & f_1 \\ \cdot & \cdot & \cdot & \cdot \\ F_{m1}, & \ldots, & F_{mn}, & f_m \end{pmatrix}$$

in the sense that every determinant * of order m whose columns are columns of the first matrix is equal to the corresponding determinant of the second matrix, multiplied by the constant Δ. This condition is both necessary and sufficient for the equivalence of the systems.

* It must be noted that in evaluating determinants containing $f(x)$ and $g(x)$ the operators F and G are multiplied by, and do not operate on, $f(x)$ and $g(x)$. Thus a typical term of the expansion of a determinant of the first matrix is $g_1(x)\, G_{21}\, G_{34}\, G_{42} \ldots$ and not $G_{21}\, G_{31}\, G_{42} \ldots g_1(x)$.

In particular, if there are as many equations as dependent variables, namely n, then

$$\begin{vmatrix} G_{11}, & \cdots, & G_{1n} \\ \cdot & \cdot & \cdot \\ G_{n1}, & \cdots & G_{nn} \end{vmatrix} = \Delta \begin{vmatrix} F_{11}, & \cdots, & F_{1n} \\ \cdot & \cdot & \cdot \\ F_{n1}, & \cdots, & F_{nn} \end{vmatrix}$$

Therefore *a necessary and sufficient condition that two homogeneous systems of n equations in n dependent variables be equivalent is that the determinants of the operators of the two systems are constant multiples of one another.* This condition is also necessary when the two systems are non-homogeneous; the remaining conditions requisite for a sufficient set of conditions are easily supplied.

6·5. Reduction of a System of Linear Equations to the Equivalent Diagonal System.—A system of linear equations of the forms

$$\begin{aligned} H_{11}(D)y_1+H_{12}(D)y_2+H_{13}(D)y_3+\ \cdots\ &+H_{1n}(D)y_n=h_1(x), \\ H_{22}(D)y_2+H_{23}(D)y_3+\ \cdots\ &+H_{2n}(D)y_n=h_2(x), \\ H_{33}(D)y_3+\ \cdots\ &+H_{3n}(D)y_n=h_3(x), \\ \cdot\quad \cdot\quad \cdot\quad &\cdot\quad \cdot\quad \cdot \\ &H_{nn}(D)y_n=h_n(x), \end{aligned}$$

in which the first equation involves y_1, the second equation involves y_2 but does not involve y_1, the third equation involves y_3 but not y_1 or y_2 and so on, until the last involves y_n only is called a *Diagonal System.* The operators $H_{11}(D)$, $H_{22}(D)$, . . ., $H_{nn}(D)$ are known as its *diagonal coefficients.* Each dependent variable is associated with one, and only one, diagonal coefficient; the mode of this association is known as the *diagonal order.*

It will now be proved that *every determinate system of linear equations with constant coefficients can be reduced to an equivalent diagonal system in which the dependent variables have any assigned diagonal order.* For definiteness it will be supposed that the diagonal order is that of increasing suffixes, as in the scheme above.

As a preliminary lemma it will be proved that if

$$\begin{aligned} U_1 &\equiv F_{11}(D)y_1+\ \cdots\ +F_{1n}(D)y_n-f_1(x)=0, \\ U_2 &\equiv F_{21}(D)y_1+\ \cdots\ +F_{2n}(D)y_n-f_2(x)=0 \end{aligned}$$

are two equations both containing any particular dependent variable, say y_1, they can be replaced by an equivalent pair of equations, one of which does not contain y_1. If such an equivalent pair exists, it will be of the form

$$\begin{aligned} LU_1+MU_2&=0, \\ L'U_1+M'U_2&=0, \end{aligned}$$

where L, M, L', M' are polynomials in D with constant coefficients such that

$$LM'-L'M$$

is a constant. Let Γ be the highest common factor of $F_{11}(D)$ and $F_{21}(D)$, then

$$F_{11}(D)=\Gamma\Phi, \quad F_{21}(D)=\Gamma\Psi,$$

and Φ and Ψ are polynomials in D having no common factor. Let

$$L=\Psi, \quad M=-\Phi,$$

then there will be no term in y_1 in

$$LU_1+MU_2.$$

But since L and M are relatively prime with respect to D, two polynomials in D, namely L' and M', can be determined * so that

$$LM'-L'M$$

s a constant, not zero. The lemma is therefore established.

Now let the given system be

$$U_1=0, \ldots, U_n=0,$$

so arranged that any equations which do not contain y_1 are placed at the end of the system. Let $U_r=0$ be the last equation which contains y_1, then $U_{r-1}=0$ and $U_r=0$ can be replaced by an equivalent pair of equations $U'_{r-1}=0$ and $U'_r=0$ of which the second does not contain y_1. Similarly $U_{r-2}=0$ and $U'_{r-1}=0$ can be replaced by the equivalent pair $U'_{r-2}=0$ and $U''_{r-1}=0$ of which the second does not involve y_1. This process may be repeated until an equivalent system, say,

$$V_1=0, \ldots, V_n=0$$

is reached in which the V_1 alone involves y_1. V_1 itself must involve y_1, since the original system is determinate. Then setting $V_1=0$ aside, the remaining system

$$V_2=0, \ldots, V_n=0,$$

which involves all of the remaining variables $y_2, \ldots y_n$, is dealt with in the same way with respect to y_2, and reduced to a system

$$W_2=0, \ldots, W_n=0,$$

in which W_2 alone involves y_2. The process is repeated, until finally the diagonal system is reached.

6·501. Example of a Reduction to a Diagonal System.—Consider the homogeneous system

$$\begin{aligned}(D+1)y_1+D^2y_2+(D+1)y_3&=0,\\ (D-1)y_1+D\,y_2+(D-1)y_3&=0,\\ y_1+y_2+Dy_3&=0.\end{aligned}$$

By means of the multiplier system

$$\begin{pmatrix}-1, & D\\ -1, & D-1\end{pmatrix}$$

the last two equations may be replaced by an equivalent pair of equations, one of which does not contain y_1, and thus the system becomes

$$\begin{aligned}(D+1)y_1+D^2y_2+(D+1)y_3&=0,\\ y_1+0y_2+(D^2-D+1)y_3&=0,\\ -y_2+(D^2-2D+1)y_3&=0.\end{aligned}$$

Next by operating on the first two equations of this equivalent set by the multiplier system

$$\begin{pmatrix}-1, & D\\ -1, & D+1\end{pmatrix}$$

the set of equations becomes

$$\begin{aligned}-y_1-D^2y_2+(D^3-D^2-1)y_3&=0,\\ -D^2y_2+(D^3-D)y_3&=0,\\ -y_2+(D^2-2D+1)y_3&=0.\end{aligned}$$

Lastly, by applying the multiplier system

$$\begin{pmatrix}0, & 1\\ -1, & D^2\end{pmatrix}$$

to the last pair of equations, the diagonal system is obtained, namely

$$\begin{aligned}-y_1-D^2y_2+(D^3-D^2-1)y_3&=0,\\ -y_2+(D^2-2D+1)y_3&=0,\\ (D^4-3D^3+D^2+D)y_3&=0.\end{aligned}$$

The last equation is easily solved for y_3, the second and first equations then give, in turn, y_2 and y_1.

* Chrystal, *Algebra*, i., Chap. VI., § 11.

6·51. Properties of a Diagonal System. Proof of the Fundamental Theorem.—Let $U_1=0, \ldots, U_n=0$, be a system in the diagonal form; its determinant is clearly the product of its diagonal coefficients. This product is therefore equal to, or a constant multiple of, the determinant of any other system to which the diagonal system is equivalent.

Now a diagonal system can be solved by a continued application of the methods given in the earlier sections of this chapter for the solution of single linear equations with constant coefficients. Let ω_r be the degree in D of the diagonal coefficient of y_r. The last equation of the system gives a general value for y_n, with a definite number ω_n of arbitrary constants. If this value for y_n is substituted in the last equation but one, and that equation solved for y_{n-1}, a number ω_{n-1} of additional arbitrary constants are introduced. The process is repeated; in general the expression for y_r will introduce ω_r *new* arbitrary constants in addition to some or all of the arbitrary constants which enter into the equation for y_r owing to the fact that that equation may involve the expressions for $y_{r+1}, \ldots, y_n$ previously obtained.

Since the ω_r constants introduced by the process of integrating the equation for y_r are essentially new constants, altogether distinct from the constants which $y_{r+1}, \ldots, y_n$ involve, the general solution of the system involves $\omega_1+\omega_2+\ldots+\omega_n$ constants, none of which are superfluous. The total number of distinct arbitrary constants which occur in the complete solution of the system is therefore equal to the degree of its determinant.

From this follows the main theorem which it was the aim of this investigation to establish, namely, that *the order of any determinate system of linear equations with constant coefficients is equal to the order of its characteristic equation.*

6·52. Equivalent Diagonal Systems.—Let $L_{r+1}, \ldots, L_n$ be polynomials in D with constant coefficients, then any set of solutions of $U_{r+1}=0, \ldots, U_n=0$ will satisfy the equation

$$L_{r+1}U_{r+1}+\ldots+L_nU_n=0.$$

If, therefore, an expression of the form $L_{r+1}U_{r+1}+\ldots+L_nU_n$ is added to the left-hand member of any equation $U_r=0$, the modified system will have all the solutions of the old system. But in the resulting system the diagonal coefficients are precisely those of the original system. The equivalence of a diagonal system is consequently not affected by this process, but on the other hand a gain in simplicity may be attained.

Thus when the diagonal order of the dependent variables is assigned, the diagonal coefficients are uniquely determined, but the non-diagonal coefficients are not so determined. Moreover, the diagonal coefficient of any variable is uniquely determined if the aggregate of the variables which follow in the diagonal order is known. Thus let the variable y_r be followed, in any order, by the $n-r$ variables $y_{r+1}, \ldots, y_n$. Let the diagonal coefficient of y_r and the succeeding variables be

$$K_r, \quad K_{r+1}, \quad \ldots, \quad K_n$$

in one order and

$$K'_r, \quad K'_{r+1}, \quad \ldots, \quad K'_n$$

in another order. Then since the two systems are equivalent,

$$K_rK_{r+1}\ldots K_n=K'_rK'_{r+1}\ldots K'_n.$$

But in the two cases the last $n-r$ equations, between the variables $y_{r+1}, \ldots, y_n$, form equivalent systems, and therefore

$$K_{r+1}\ldots K_n=K'_{r+1}\ldots K'_n,$$

whence it follows that

$$K_r = K'_r,$$

that is to say, the diagonal coefficient of y_r is unaltered if the aggregate of variables which follow y_r is unchanged.

In the complete solution let y_r involve ν_r arbitrary constants, then if the diagonal system is so arranged that y_r occurs in the last equation, the diagonal coefficient of y_r will be of degree ν_r in D. Now let the system be transformed so that y_r occurs as the diagonal term in the last equation but one. Then, since in the complete solution y_r still involves ν_r arbitrary constants, the degree of the diagonal coefficient of y_r will not exceed ν_r; in fact it may be less than ν_r by the degree of the diagonal coefficient of the last equation of the system. The degree of the diagonal coefficient of y_r may be diminished still further by so transforming the system that y_r occurs in the diagonal term of the last equation but two, and so on. Thus the diagonal coefficient for any given variable is least when that variable occurs first in diagonal order; it may increase but cannot diminish as the variable advances in diagonal order, and is greatest when the variable is last in diagonal order.

When the variable is last in diagonal order, the degree of its diagonal coefficient is equal to the total number of arbitrary constants in the complete expression for that variable; when the variable is first in diagonal order, the degree of its diagonal coefficients is equal to the number of arbitrary constants which enter into it but not into any other variable. The diagonal coefficients in these two extreme cases are therefore important; a set of rules for calculating the diagonal coefficients of any particular variable will therefore be given, when that variable occupies the first or the last place in diagonal order.

Let

(U) $$U_1=0, \quad U_2=0, \quad \ldots, \quad U_n=0,$$

where

$$U_r = F_{r1}y_1 + \ldots + F_{rn}y_n - f_r(x),$$

be the given system, and let

(V) $$V_1=0, \quad V_2=0, \quad \ldots, \quad V_n=0,$$

where

$$V_r = H_{rr}y_r + \ldots + H_{rn}y_n - h_r(x),$$

be an equivalent diagonal system. Let

$$\begin{pmatrix} \delta_{11}, & \ldots, & \delta_{1n} \\ \cdot & \cdot & \cdot \\ \delta_{n1}, & \ldots, & \delta_{nn} \end{pmatrix}, \quad \begin{pmatrix} \delta'_{11}, & \ldots, & \delta'_{1n} \\ \cdot & \cdot & \cdot \\ \delta'_{n1}, & \ldots, & \delta'_{nn} \end{pmatrix}$$

be the multiplier systems which transform (U) into (V) and (V) into (U) respectively, then, since

$$V_1 = \delta_{11}U_1 + \ldots + \delta_{1n}U_n,$$

it follows by comparing coefficients of y_1 that

$$H_{11} = \delta_{11}F_{11} + \ldots + \delta_{1n}F_{n1},$$

and since

$$U_r = \delta'_{r1}V_1 + \ldots + \delta'_{rn}V_n \qquad (r=1, 2, \ldots, n),$$

it follows similarly that

$$F_{r1} = \delta'_{r1}H_{11} \qquad (r=1, 2, \ldots, n).$$

Hence H_{11} must be a common factor in D of

$$F_{11}, \ldots, F_{n1},$$

and the highest common factor of these quantities must be a divisor of H_{11}. Consequently, apart from a constant multiplier, H_{11} must be the highest common factor of $F_{11}, \ldots, F_{n1}$. This is the rule for calculating the diagonal coefficient of y_1 when y_1 is first in diagonal order. If y_r were to be first in diagonal order, its diagonal coefficient would then be a constant multiple of the highest common factor γ_r of

$$F_{1r}, \ldots, F_{nr}.$$

The rule for calculating H_{nn}, that is to say the diagonal coefficient of y_n, when y_n is last in diagonal order, is as follows. Since

$$V_n = \delta_{n1}U_1 + \ldots + \delta_{nn}U_n,$$

it follows by comparing the coefficients of $y_1, \ldots, y_{n-1}, y_n$ that

$$\begin{aligned} \delta_{n1}F_{11} + \ldots + \delta_{nn}F_{n1} &= 0, \\ \cdot \quad \cdot \quad \cdot \quad \cdot \quad \cdot \quad \cdot & \\ \delta_{n1}F_{1n-1} + \ldots + \delta_{nn}F_{n,n-1} &= 0, \\ \delta_{n1}F_{1n} + \ldots + \delta_{nn}F_{nn} &= H_{nn}. \end{aligned}$$

Let G_{rs} be the co-factor of F_{rs} in the characteristic determinant

$$F = \begin{vmatrix} F_{11}, & \ldots, & F_{n1} \\ \cdot & \cdot & \cdot \\ F_{1n}, & \ldots, & F_{nn} \end{vmatrix},$$

let Γ_n be the highest common factor of $G_{1n}, \ldots, G_{nn}$, and let

$$G_{1n} = G'_{1n}\Gamma_n, \ldots, G_{nn} = G'_{nn}\Gamma_n.$$

Therefore

$$\delta_{n1} = \lambda G'_{1n}, \ldots, \delta_{nn} = \lambda G'_{nn},$$

where λ is defined by the relation

$$F\lambda = \Gamma_n H_{nn},$$

and since $G'_{1n}, \ldots, G'_{nn}$ are relatively prime, λ is either a constant or a polynomial in D.

Now since the two systems are equivalent, the modulus

$$\begin{vmatrix} \delta_{11}, & \ldots, & \delta_{1n} \\ \cdot & \cdot & \cdot \\ \delta_{n1}, & \ldots, & \delta_{nn} \end{vmatrix}$$

must be a constant. But this determinant clearly has the factor λ, therefore λ is a constant. Hence

$$\begin{aligned} H_{nn} &= \lambda(F_{1n}G'_{1n} + \ldots + F_{nn}G'_{1n}) \\ &= \frac{\lambda}{\Gamma_n}(F_{1n}G_{1n} + \ldots + F_{nn}G_{nn}) = \frac{\lambda F}{\Gamma_n}. \end{aligned}$$

More generally, when y_r is last in diagonal order, its diagonal coefficient is a constant multiple of F/Γ_r, where F is the characteristic determinant of the system and Γ_r is the highest common factor of

$$G_{1r}, \ldots, G_{nr},$$

and G_{mr} is the minor of F_{mr} in the characteristic determinant.

Finally, the differential equations which determine $y_1, \ldots, y_n$ separately are

$$\begin{aligned} \frac{F}{\Gamma_1}y_1 &= \frac{G_{11}}{\Gamma_1}f_1(x) + \ldots + \frac{G_{n1}}{\Gamma_1}f_n(x), \\ \cdot \quad \cdot \quad \cdot \quad & \cdot \quad \cdot \quad \cdot \quad \cdot \quad \cdot \\ \frac{F}{\Gamma_n}y_n &= \frac{G_{1n}}{\Gamma_n}f_1(x) + \ldots + \frac{G_{nn}}{\Gamma_n}f_n(x). \end{aligned}$$

But it is to be noted that although this set of equations fully determines each of $y_1, \ldots y_n$ yet, considered as a system, it is not necessarily equivalent to the given system. For the aggregate of the arbitrary constants in the solutions of this set of equations may, and in general does, exceed the order of the given system.

6·53. Simple Diagonal Systems : Prime Systems.—It may happen that of the total number of dependent variables, certain variables are wholly determined by non-differential equations, and therefore involve no arbitrary constants. If this is the case, it may be supposed that the variables in question are removed from the system by being replaced wherever they occur, by their actual values. The system then involves no dependent variable which can be determined without integrating a differential equation.

Suppose that in a solution thus restricted there occurs only one differential equation. This equation must be the last equation of the system, for otherwise the last dependent variable in diagonal order would be determined by a non-differential equation. Let the last variable be y_n, then y_n is determined by the equation,

$$H_{nn}y_n=0.$$

whose order is equal to the order of the system, so that the expression for y_n involves all the arbitrary constants of the complete solution of the system. The remaining diagonal coefficients $H_{n-1,\,n-1}, \ldots, H_{11}$ are constants; the corresponding dependent variables $y_{n-1}, \ldots, y_1$ depend upon some or all of the constants which enter into the expression for y_n, but do not involve any other arbitrary constants than these. Such a system is known as a *Simple Diagonal System.* Conditions in which a given system is reducible to a simple diagonal system will now be investigated.

If F is the determinant of the given system, then

$$F=H_{11}H_{22} \ldots H_{nn},$$

and since the operations by which $H_{11}, H_{22}, \ldots, H_{nn}$ are obtained from the coefficients of the original system are rational operations, it follows that H_{11}, $H_{22}, \ldots, H_{nn}$ are rational in the operator coefficients F_{rs} of the original system. If therefore F has no factor of lower order in D than F itself, which is a polynomial in the coefficients F_{rs}, then $H_{11}, \ldots, H_{n-1,\,n-1}$ must reduce to constants, and the equivalent diagonal system is simple.

As before, let G_{rs} denote the co-factor of F_{rs} in the characteristic determinant F. Consider the matrix

$$\begin{pmatrix} G_{11}, & \ldots, & G_{1n} \\ \cdot & \cdot & \cdot \\ G_{n1}, & \ldots, & G_{nn} \end{pmatrix},$$

and suppose that the constituents of any one column, say the rth, are relatively prime. Then, in the notation of the previous section, Γ_r is a constant, and consequently if y_r is taken as the dependent variable last in order in the equivalent diagonal system, the coefficient of y_r in the last equation of the diagonal system is a constant multiple of F itself. The diagonal system thus obtained is simple. Thus, *for every prime column in the reciprocal matrix of a given system an equivalent simple diagonal system can be formed in which the corresponding variable is last in diagonal order.*

In particular, if every column of the reciprocal matrix is prime, then every equivalent diagonal system will be simple and the expression for each dependent variable will contain all the arbitrary constants.

A system every column of whose characteristic determinant is prime is known as a *prime system.* Any given system may be transformed into a prime system, for if γ_r is the highest common factor of $F_{1r}, \ldots, F_{nr}$, it is only necessary to introduce new dependent variables $u_1, u_2, \ldots, u_n$, where

$$u_1=\gamma_1 y_1, \ldots, u_n=\gamma_n y_n.$$

The characteristic property of a prime system is that, in any equivalent diagonal system, the first equation is non-differential.

The homogeneous system

$$\begin{aligned}(D^2-1)y_1+D^3y_2+(D+1)y_3&=0,\\ (D-1)^2y_1+D^2y_2+(D-1)y_3&=0,\\ (D-1)y_1+Dy_2+Dy_3&=0\end{aligned}$$

is reduced, by the transformation

$$u_1=(D-1)y_1,\quad u_2=Dy_2,\quad u_3=y_3$$

into the prime system

$$\begin{aligned}(D+1)u_1+D^2u_2+(D+1)u_3&=0,\\ (D-1)u_1+Du_2+(D-1)u_3&=0,\\ u_1+u_2\quad+Du_3&=0.\end{aligned}$$

This is the system whose reduction to equivalent diagonal form was effected in § 6·501.

Example.—

$$\begin{aligned}(2D-2)y_1+(D^3-D+2)y_2&=e^{-x},\\ (D^3+3D^2+5D-1)y_1+(-3D^2-4D+1)y_2&=0.\end{aligned}$$

The characteristic determinant is

$$\begin{vmatrix} 2D-2 & D^3-D+2 \\ D^3+3D^2+5D-1, & -3D^2-4D+1\end{vmatrix}$$

and the characteristic equation is

$$D(D+1)^3(D^2+1)y=0.$$

The reciprocal matrix is

$$\begin{pmatrix}-3D^2-4D+1, & -D^3-3D^2-5D+1\\ -D^3+D-2\;, & 2D-2\end{pmatrix};$$

its columns are both prime. Thus, in any equivalent diagonal system the first equation is non-differential, and as there are only two equations in the system, the system is therefore simple. The multiplier system

$$\begin{pmatrix} L & , & M\\ D^3+3D^2+5D-1, & & -2D+2\end{pmatrix}$$

will transform the given system into an equivalent diagonal system in which y_2 is last in diagonal order provided that L and M are so chosen that

$$L(2D-2)+M(D^3+3D^2+5D-1)$$

is a constant.

L and *M* are readily determined as follows : let

$$u=2D-2,\quad v=D^3+3D^2+5D-1,$$

then, eliminating D^3,

$$D^2u-2v=-8D^2-10D+2.$$

Next, eliminating D^2 between the expressions for D^2u-2v and u,

$$(D^2+4D)u-2v=-18D+2,$$

and finally, eliminating D between this expression and the expression for u.

$$(D^2+4D+9)u-2v=-16.$$

Thus, suitable values for L and M are

$$L=D^2+4D+9,\quad M=-2.$$

The required multiplier system

$$\begin{pmatrix} D^2+4D+9 & , & -2\\ D^3+3D^2+5D-1, & & -2D+2\end{pmatrix}$$

reduces the given system to the equivalent diagonal system.

$$\begin{aligned}-16y_1+(D^5+4D^4+8D^3+4D^2+7D+16)y_2&=6e^{-x},\\ D(D+1)^3(D^2+1)y_2&=-4e^{-x}.\end{aligned}$$

The general solution of the equation for y_2 is

$$y_2=C_1+(C_2+C_3x+C_4x^2)e^{-x}+C_5\cos x+C_6\sin x+\tfrac{1}{3}x^3e^{-x}.$$

Since the equation for y_1 is non-differential, the expression for y_1 is

$$y_1=C_1+\tfrac{1}{4}\{(2C_2+3C_3-3C_4)+(2C_3+6C_4)x+2C_4x^2\}e^{-x}$$
$$+C_5\cos x+C_6\sin x-\tfrac{1}{8}(1+6x-6x^2-\tfrac{4}{3}x^3)e^{-x}.$$

6·6. Behaviour at Infinity of Solutions of a Linear Differential System with bounded Coefficients.—It is convenient at this stage to enlarge the scope of the investigation in order to consider the behaviour, for large values of the independent variable, of solutions of systems whose coefficients are not necessarily constant, but are bounded.*

The following lemma will be assumed. Let $f(x)$ be a function which is finite when $x_0<x<\infty$, and let λ_1 and λ_2 be two real numbers such that $e^{\lambda_1 x}f(x)$ tends to zero and $e^{\lambda_2 x}f(x)$ does not tend to zero as $x\to\infty$. Then there will exist in the interval (λ_1, λ_2) a number $\lambda_0\leqslant\lambda_2$ such that, if ϵ is a small positive number, $e^{(\lambda_0-\epsilon)x}f(x)$ tends to zero and $e^{(\lambda_0+\epsilon)x}f(x)$ to infinity as $x\to\infty$.† Similarly, if

$$f_1(x),\ f_2(x),\ \ldots,\ f_n(x)$$

are functions defined in the range (x_0, ∞) and λ_1 and λ_2 are such that each product $e^{\lambda_1 x}f_r(x)$ tends to zero, whereas the products $e^{\lambda_2 x}f_r(x)$ do not all tend to zero as $x\to\infty$, then there will exist a number λ_0 such that $\lambda_1<\lambda_0\leqslant\lambda_2$ and such that each product $e^{(\lambda_0-\epsilon)x}f_r(x)$ tends to zero, but one at least of the products $e^{(\lambda_0+\epsilon)x}f_r(x)$ is unbounded. The number λ_0 is said to be *characteristic* for the system of functions in question.

Now consider the system

$$\frac{dy_1}{dx}=a_{11}y_1+a_{12}y_2+\ \ldots\ +a_{1n}y_n,$$
$$\frac{dy_2}{dx}=a_{21}y_1+a_{22}y_2+\ \ldots\ +a_{2n}y_n,$$
$$\cdots\cdots\cdots\cdots$$
$$\frac{dy_n}{dx}=a_{n1}y_1+a_{n2}y_2+\ \ldots\ +a_{nn}y_n,$$

where all the coefficients a are real functions of x, bounded in the range (x_0, ∞). Let

$$y_r=e^{\lambda x}v_r,$$

where λ is an arbitrary real number, then

$$\frac{dv_1}{dx}=(a_{11}-\lambda)v_1+a_{12}v_2+\ \ldots\ +a_{1n}v_n,$$
$$\frac{dv_2}{dx}=a_{21}v_1+(a_{22}-\lambda)v_2+\ \ldots\ +a_{2n}v_n,$$
$$\cdots\cdots\cdots\cdots$$
$$\frac{dv_n}{dx}=a_{n1}v_1+a_{n2}v_2+\ \ldots\ +(a_{nn}-\lambda)v_n.$$

When these equations are multiplied respectively by $v_1, v_2, \ldots v_n$, and then added together the resulting equation is

$$\frac{1}{2}\frac{d(v_1^2+v_2^2+\ \ldots\ +v_n^2)}{dx}=(a_{11}-\lambda)v_1^2+(a_{22}-\lambda)v_2^2+\ \ldots\ +(a_{nn}-\lambda)v_n^2$$
$$+\sum a_{rs}v_rv_s.$$

* Liapounov, *Comm. Math. Soc. Kharkov* (1892) ; *Ann. Fac. Sc. Toulouse* (2), 9 (1908).

† This theorem is proved by repeated subdivision of the interval (λ_1, λ_2).

Now, if λ is sufficiently large, the quadratic form which stands on the right-hand side of this equation is definite and negative, and therefore, if for λ a sufficiently large positive number α is taken,

$$\frac{d(v_1^2+v_2^2+\ .\ .\ .\ +v_n^2)}{dx}<0,$$

for all values of x in the interval (x_0, ∞). Thus, the positive function $v_1^2+v_2^2+\ .\ .\ .\ +v_n^2$ diminishes as x increases, and therefore $v_1, v_2, \ .\ .\ , v_n$ are severally bounded. It follows that

$$y_1e^{-\alpha x}, \quad y_2e^{-\alpha x}, \quad .\ .\ ., \quad y_ne^{-\alpha x}$$

are bounded in the interval $x_0<x<\infty$, and it is obvious that α can be so chosen that the limiting value of each product is zero.

Similarly if $\lambda=-\beta$, where β is a sufficiently large positive number,

$$\frac{d(v_1^2+v_2^2+\ .\ .\ .\ +v_n^2)}{dx}>0,$$

and therefore the limiting value of $v_1^2+v_2^2+\ .\ .\ .\ +v_n^2$ is not zero.

Consequently, one at least of

$$y_1e^{\beta x}, \quad y_2e^{\beta x}, \quad .\ .\ ., \quad y_ne^{\beta x}$$

does not tend to zero as $n\rightarrow\infty$.

It follows, therefore, that *any system of solutions*

$$y_1, \quad y_2, \ .\ .\ ., \quad y_n$$

not identically zero admits of a characteristic number λ_0.

An immediate consequence of this theorem is that *there exists a real number* κ *such that*

$$y_1e^{\kappa x}, \quad y_2e^{\kappa x}, \quad .\ .\ ., \quad y_ne^{\kappa x}$$

tend simultaneously to zero as $x\rightarrow\infty$.

The corresponding theorem in the case of the single linear differential equation of order n is that *if the coefficients* p_r *in the equation*

$$\frac{d^ny}{dx^n}+p_1\frac{d^{n-1}y}{dx^{n-1}}+\ .\ .\ .\ +p_{n-1}\frac{dy}{dx}+p_ny=0$$

are bounded in the interval $(0, \infty)$, *there exists a number* κ *such that, if* y *is any solution of the equation,*

$$ye^{\kappa x}, \quad y'e^{\kappa x}, \quad .\ .\ ., \quad y^{(n-1)}e^{\kappa x}$$

all tend to zero as $n\rightarrow\infty$.

Miscellaneous Examples.

1. Integrate the following equations:

(i) $\frac{d^2y}{dx^2}-2\frac{dy}{dx}+10y=\sin x$;

(ii) $\frac{d^2y}{dx^2}-2\frac{dy}{dx}+10y=e^x\cos 3x$;

(iii) $\frac{d^3y}{dx^3}-y=\cosh x$;

(iv) $\frac{d^2y}{dx^2}-3\frac{dy}{dx}+2y=x+2e^x$;

(v) $\frac{d^4y}{dx^4}+4y=e^x\sin 2x\cos x$;

(vi) $\frac{d^4y}{dx^4}+2\frac{d^2y}{dx^2}+y=\sin x$;

(vii) $\frac{d^2y}{dx^2}+m^2y=x\cos mx$;

(viii) $\frac{d^3y}{dx^3}-5\frac{d^2y}{dx^2}+8\frac{dy}{dx}-4y=e^{2x}+e^{3x}$;

(ix) $\frac{d^3y}{dx^3}-2\frac{dy}{dx}+4y=e^x\cos x$;

(x) $x^3\frac{d^3y}{dx^3}+3x^2\frac{d^2y}{dx^2}+x\frac{dy}{dx}=24x^2$;

(xi) $x^2\frac{d^2y}{dx^2}+x\frac{dy}{dx}+n^2y=x^m$;

(xii) $x^3\frac{d^3y}{dx^3}+3x^2\frac{d^2y}{dx^2}-6x\frac{dy}{dx}+6y=x(1-x)$

2. Find the solution of

$$\frac{d^4y}{dx^4}+m^4y=0$$

which satisfies the conditions

$$y=\frac{dy}{dx}=0 \text{ when } x=0, \qquad y=\frac{d^2y}{dx^2}=0 \text{ when } x=l.$$

3. Prove that a particular solution of the equation

$$\frac{d^2y}{dx^2}+m^2y=m^2f(x)$$

is

$$y=m \sin mx\int_0^x f(x)\cos mx\, dx-m \cos mx\int_0^x f(x) \sin mx\, dx.$$

[Fourier.]

4. Integrate the systems

(i) $\dfrac{dx}{dt}+ax-by=e^t, \quad \dfrac{dy}{dt}-ay+bx=e^t \quad (a^2-b^2=1)$;

(ii) $\dfrac{d^2x}{dt^2}+n^2y=0, \quad \dfrac{d^2y}{dt^2}-n^2x=0$;

(iii) $\dfrac{d^2x}{dt^2}+2k\dfrac{dy}{dt}+n^2x=0, \quad \dfrac{d^2y}{dt^2}-2k\dfrac{dx}{dt}+n^2y=0$;

(iv) $\dfrac{d^2x}{dt^2}+n\dfrac{dy}{dt}=a \cos nt, \quad \dfrac{d^2y}{dt^2}-n\dfrac{dx}{dt}=0.$

5. Solve the system

$$\frac{d^2x}{dt^2}-3x-4y+3=0, \quad \frac{d^2y}{dt^2}+x+y+5=0$$

subject to the condition that, when $t=0$,

$$x=y=\frac{dx}{dt}=\frac{dy}{dt}=0.$$

6. Integrate the system

$$x\frac{d^2x}{dt^2}-y\frac{d^2y}{dt^2}=0, \quad \frac{d^2x}{dt^2}+\frac{d^2y}{dt^2}+x+y=0.$$

[Edinburgh, 1909.]

7. Reduce to diagonal systems and integrate

(i) $\begin{cases}(D^2-1)x+Dy=e^t,\\ Dx+(D^2+1)y=0;\end{cases}$

(ii) $\begin{cases}(D^2-1)x+2(D+1)y+(D+1)z=2e^t,\\ (D-1)^2x+4Dy+(D-3)z=0,\\ (3D-D^3)x-2Dy-(D-1)z=0,\end{cases}$

where $D\equiv\dfrac{d}{dt}$.

CHAPTER VII

THE SOLUTION OF LINEAR DIFFERENTIAL EQUATIONS IN AN INFINITE FORM

7·1. Failure of the Elementary Methods.—Apart from equations with constant coefficients, and such equations as can be derived therefrom by a change of the independent variable, there is no known type of linear equation of general order n which can be fully and explicitly integrated in terms of elementary functions. When an equation arises which can not be reduced to one or other of the general types discussed in Chapter VI., it is almost invariably the case that the solution has to be expressed in an infinite form, that is to say as an infinite series, an infinite continued fraction, or a definite integral. Thus, in the great majority of cases, equations which arise out of problems of applied mathematics and which are not reducible to equations with constant coefficients, have as their solutions new transcendental functions. It may, perhaps, be not without profit to emphasise the fact that transcendental functions may be divided into two classes, namely those which, like the Bessel functions, are solutions of ordinary differential equations, and those which, like the Riemann-Zeta function, do not satisfy any ordinary differential equation of finite order.

The present chapter will deal in the main with the process of expressing the solution of linear differential equations as infinite series; continued fractions will briefly be mentioned, and the problem of expressing solutions in the form of definite integrals will be postponed to the following chapter.

It was proved in Chapter III. that if the coefficients of the equation

$$L(y) \equiv p_0(x)\frac{d^ny}{dx^n} + p_1(x)\frac{d^{n-1}y}{dx^{n-1}} + \cdots + p_{n-1}(x)\frac{dy}{dx} + p_n(x)y = 0$$

are all finite, one-valued and continuous throughout an interval $a \leqslant x \leqslant b$, the only singular points which can occur within that interval are the zeros of the leading coefficient $p_0(x)$. All other points of the interval are ordinary points.

From the point of view of the problem of developing the solutions of the equation as infinite series, the distinction between ordinary and singular points is fundamental. The following sections aim at making clear the distinction between solutions relative to an ordinary point and those appropriate to a singular point.

7·2. Solutions relative to an Ordinary Point.—The fundamental existence theorems show that if x_0 is a non-singular point of the differential equation, then there exists a unique solution $y(x)$ such that $y(x)$ and its first $n-1$ derivatives assume a set of arbitrarily-assigned values,

$$y_0, y_0', \ldots, y_0^{(n-1)}$$

when $x=x_0$, and such that $y(x)$ may be developed as a Taylor's series convergent in a certain interval (x_0-h, x_0+h). It has also been seen that if

$Y_1(x), Y_2(x), \ldots, Y_n(x)$ are the particular solutions defined by the conditions

$$\begin{array}{lll} Y_1(x_0)=1, & Y_1'(x_0)=0, \ldots, & Y_1^{(n-1)}(x_0)=0, \\ Y_2(x_0)=0, & Y_2'(x_0)=1, \ldots, & Y_2^{(n-1)}(x_0)=0, \\ \cdot \quad \cdot & \cdot \quad \cdot \quad \cdot & \cdot \quad \cdot \quad \cdot \\ Y_n(x_0)=0, & Y_n'(x_0)=0, \ldots, & Y_n^{(n-1)}(x_0)=1, \end{array}$$

then

$$y(x)=y_0Y_1(x)+y_0'Y_2(x)+ \ldots +y_0^{(n-1)}Y_n(x).$$

Thus, in order to arrive at either the general solution of the equation, or a particular solution satisfying pre-assigned conditions, it is sufficient to have derived the n fundamental solutions $Y_1(x), Y_2(x), \ldots, Y_n(x)$.

It is characteristic of $Y_{r+1}(x)$ that its leading term is $(x-x_0)^r/r!$ and that no terms in $(x-x_0)^{r+1}, (x-x_0)^{r+2}, \ldots, (x-x_0)^{n-1}$ are present. In practice, however, it is more convenient to take the coefficient of the leading term to be unity and to endeavour to satisfy the equation by series of the form

$$y_{r+1}(x)=(x-x_0)^r\{1+a_{r1}(x-x_0)+ \ldots +a_{r\nu}(x-x_0)^\nu+ \ldots\}.$$

Since

$$y_{r+1}^{(s)}(x_0)=0 \qquad (s<r),$$
$$y_{r+1}^{(r)}(x_0)=r!,$$

the Wronskian of the set of solutions $y_1(x), y_2(x), \ldots, y_n(x)$ does not vanish ; the set is therefore fundamental.

The actual method of solution as carried out in practice is to substitute the series in the left-hand member of the differential equation, to arrange the resulting expression in ascending powers of $x-x_0$ and then to equate to zero the coefficients of successive powers of $x-x_0$. There results a set of linear algebraic relations between the coefficients $a_{r1}, a_{r2}, \ldots a_{r\nu}, \ldots$, known as the recurrence-relations ; thus the coefficients are determined by algebraical processes.

7·201. The Weber Equation.—In the case of the Weber equation *

$$\frac{d^2y}{dx^2}+(n+\tfrac{1}{2}-\tfrac{1}{4}x^2)y=0,$$

the point $x=0$ is an ordinary point and the two fundamental solutions may be expressed in ascending series of powers of x. But it is more advantageous to make the preliminary transformation

$$y=e^{-\frac{1}{4}x^2}v,$$

when the new dependent variable is found to satisfy the equation

$$\frac{d^2v}{dx^2}-x\frac{dv}{dx}+nv=0.$$

Now assume the solution

$$v=a_0+a_1x+a_2x^2+ \ldots +a_rx^r+ \ldots ;$$

the two fundamental solutions v_1 and v_2 are obtained by assigning the initial conditions

$$\text{(i) } a_0=1, \quad a_1=0, \qquad \text{(ii) } a_0=0, \quad a_1=1.$$

The recurrence-relation which the coefficients must satisfy is

$$(r+1)(r+2)a_{r+2}=(r-n)a_r, \qquad (r=0, 1, 2, \ldots)$$

* Weber, *Math. Ann.* 1 (1869), p. 29. The equation in v was previously studied by Hermite, *C. R. Acad. Sc. Paris,* 58 (1864), pp. 93, 266 [*Œuvres* II., p. 293]. The functions defined by the equation were standardised by Whittaker, *Proc. London Math. Soc.* (1) 35 (1903), p. 417. See also Whittaker and Watson, *Modern Analysis*, §§ 16·5–16·7.

and thus

$$v_1=1-\frac{n}{2!}x^2+\frac{n(n-2)}{4!}x^4-\frac{n(n-2)(n-4)}{6!}x^6+\ldots,$$

$$v_2=x-\frac{n-1}{3!}x^3+\frac{(n-1)(n-3)}{5!}x^5-\frac{(n-1)(n-3)(n-5)}{7!}x^7+\ldots$$

The ordinary tests show that these series converge for all finite values of $|x|$.

7·21. Solutions relative to a Singular Point.—Let the point x_0, which for the purposes of the argument will be taken to be the origin, be not an ordinary point. Then a natural hypothesis to make is that there is nevertheless a solution of the form

$$y=x^r(a_0+a_1x+\ldots+a_\nu x^\nu+\ldots) \qquad (a_0\neq 0),$$

though perhaps in this case r may not be a positive integer.

To investigate the possible existence of such solutions, substitute the series for y in $L(y)$ and equate to zero the coefficient of the dominant term, namely the term of lowest degree in x. This coefficient will be either independent of r or a polynomial $P(r)$ in r whose degree will not exceed the order of the equation.* In the former case no solution of the type in question exists, and the singularity, $x=0$, is said to be *irregular*. In the latter case, if $P(r)$ is of degree n, the singularity is said to be *regular*; if the degree of $P(r)$ is less than n the singularity is again said to be *irregular*. For the present the singularity will be assumed to be regular; then the equation

$$P(r)=0,$$

which is known as the *indicial equation*, will have n roots some or all of which may be equal. If, for the moment, the equation is reduced to the form

$$\frac{d^ny}{dx^n}+p_1\frac{d^{n-1}y}{dx^{n-1}}+\ldots+p_{n-1}\frac{dy}{dx}+p_ny=0,$$

then in order that $P(r)$ may be of degree n it is necessary and sufficient that †

$$p_r=O(x^{-r}) \qquad (r=1, 2, \ldots, n).$$

The roots of the indicial equation are known as the *exponents* relative to the singular point in question. It will now be stated as a general principle, which will be proved at a later stage with the aid of the theory of the complex variable,‡ that if the exponents are distinct, and no two of them differ by an integer, then there are n linearly-distinct solutions of the type contemplated. If, on the other hand, two or more of the exponents are equal, or differ by an integer, then the number of solutions of the type in question in general falls short of n, and the remaining solutions of a fundamental set are of a less simple character.

7·22. The Point at Infinity as a Regular Singular Point.—The question as to whether any finite singularity is regular or irregular can be settled almost at a glance; the nature of the point at infinity can be determined with little extra trouble. The transformation

$$x=z^{-1}$$

carries the point at infinity to the origin, and the criteria for an ordinary

* It is obvious that $P(r)$ will be independent of the coefficients $a_1, a_2, \ldots$, and will involve a_0 as a multiplicative factor.

† The *ordo-symbol* $O(x^{-r})$ will frequently be used in the following pages. Its definition is as follows: if a function $f(x)$ is such that as $x\to 0$ (or ∞), $|x^rf(x)|<K$, where K is a positive number independent of x or zero, then $f(x)$ is said to be *of the order of* x^{-r} or $f(x)=O(x^{-r})$. It will generally be clear from the context whether the limiting process is for $x\to 0$ or for $x\to\infty$. If $\lim |x^rf(x)|=0$ the state of affairs is indicated by writing $f(x)=o(x^{-r})$.

A rigorous proof that $p_r=O(x^{-r})$ is a necessary and sufficient condition for a regular singularity will be given later (§ 15·3).

‡ See Chap. XV.

point, a regular singularity, and an irregular singularity may then be applied directly.

Consider the equation of the second order

$$\frac{d^2y}{dx^2}+p(x)\frac{dy}{dx}+q(x)y=0\,;$$

when transformed by the substitution $x=z^{-1}$ it becomes

$$\frac{d^2y}{dz^2}+\left\{\frac{2}{z}-\frac{p(z^{-1})}{z^2}\right\}\frac{dy}{dz}+\frac{q(z^{-1})}{z^4}y=0.$$

If the original equation has an ordinary point at infinity, the transformed equation will have an ordinary point at the origin, and therefore the conditions

$$\frac{2}{z}-\frac{p(z^{-1})}{z^2}=O(1),$$

$$\frac{q(z^{-1})}{z^4}=O(1)$$

must hold as $z\to 0$. The corresponding conditions for the original equation are that

$$p(x)=\frac{2}{x}+O(x^{-2}),$$

$$q(x)=O(x^{-4}),$$

as $x\to\infty$.

The conditions for a regular singularity are

$$\frac{2}{z}-\frac{p(z^{-1})}{z^2}=O\left(\frac{1}{z}\right),$$

$$\frac{q(z^{-1})}{z^4}=O\left(\frac{1}{z^2}\right),$$

as $z\to 0$, that is

$$p(x)=O(x^{-1}),$$

$$q(x)=O(x^{-2}),$$

as $x\to\infty$. Let

$$p(x)=p_0x^{-1}+O(x^{-2}),$$

$$q(x)=q_0x^{-2}+O(x^{-3}),$$

then the indicial equation relative to the singularity $z=0$ will be

$$r^2+(1-p_0)r+q_0=0.$$

Let its roots be α and β. Then, in the general case, when α and β are unequal and do not differ by an integer, there will exist two solutions of the original equation, relative to the singularity $x=\infty$, namely,

$$y_1=x^{-\alpha}(1+a_1x^{-1}+a_2x^{-2}+\ .\ .\ .),$$

$$y_2=x^{-\beta}(1+b_1x^{-1}+b_2x^{-2}+\ .\ .\ .\ ),$$

and these developments will converge for sufficiently large values of $|x|$. It is to be noted that the exponents relative to the point at infinity are α, β and not $-\alpha$, $-\beta$.

The foregoing general principles will now be illustrated by considering an equation of particular importance, known as the hypergeometric equation.

7·23. The Hypergeometric Equation.—The hypergeometric equation *

$$x(1-x)\frac{d^2y}{dx^2}+\{\gamma-(\alpha+\beta+1)x\}\frac{dy}{dx}-\alpha\beta y=0$$

* Gauss, *Comm. Gott.* 2 (1813) [*Werke*, 3, pp. 123, 207]. A detailed study of the hypergeometric function, with references, is given in Whittaker and Watson, *Modern Analysis*, Chap. XIV.

has three singular points, namely, $x=0$, $x=1$, and $x=\infty$. The exponents relative to $x=0$ are 0 and $1-\gamma$, those relative to $x=1$ are 0 and $\gamma-\alpha-\beta$, and those relative to $x=\infty$ are α and β. To express this fact the most general solution of the equation is written in the symbolic form,

$$y=P\begin{Bmatrix} 0 & \infty & 1 & \\ 0 & \alpha & 0 & x \\ 1-\gamma & \beta & \gamma-\alpha-\beta & \end{Bmatrix},$$

and the entity which stands on the right-hand side of this relation is known as the Riemann P-function.*

The solution relative to the singularity $x=0$ and exponent 0 is developable in the series

$$1+\frac{\alpha\beta}{1!.\gamma}x+\frac{\alpha(\alpha+1)\beta(\beta+1)}{2!.\gamma(\gamma+1)}x^2+\frac{\alpha(\alpha+1)(\alpha+2)\beta(\beta+1)(\beta+2)}{3!.\gamma(\gamma+1)(\gamma+2)}x^3+\ldots$$

and is denoted by $F(\alpha, \beta; \gamma; x)$. It may be verified that the series converges when $|x|<1$ for all finite values of α and β, and for all finite values of γ except negative integer values, and diverges when $|x|>1$. If α, β and γ are real, the series converges when $x=1$ if $\gamma>\alpha+\beta$, and diverges if $\gamma\leqslant\alpha+\beta$; it converges when $x=-1$ if $\gamma+1>\alpha+\beta$, and diverges if $\gamma+1\leqslant\alpha+\beta$.

Now consider the solution relative to the singularity $x=0$, with exponent $1-\gamma$; assuming the series-solution

$$y=x^{1-\gamma}(1+a_1x+a_2x^2+\ldots+a_\nu x^\nu+\ldots),$$

it is found that

$$(\nu+1)(\nu-\gamma+2)a_{\nu+1}=(\nu+\alpha-\gamma+1)(\nu+\beta-\gamma+1)a_\nu$$

for $\nu=0, 1, 2, \ldots$, with $a_0=1$. Thus

$$y=x^{1-\gamma}F(\alpha-\gamma+1, \beta-\gamma+1; 2-\gamma; x).$$

It may be found in the same way that two solutions appropriate to the singularity $x=1$ are

$$y=F(\alpha, \beta; \alpha+\beta-\gamma+1; 1-x),$$
$$y=(1-x)^{\gamma-\alpha-\beta}F(\gamma-\alpha, \gamma-\beta; \gamma-\alpha-\beta+1; 1-x),$$

and that two solutions appropriate to the point at infinity are

$$y=x^{-\alpha}F(\alpha, \alpha-\gamma+1; \alpha-\beta+1; x^{-1}),$$
$$y=x^{-\beta}F(\beta, \beta-\gamma+1; \beta-\alpha+1; x^{-1}).$$

The interval of convergence for the series in $1-x$ is $0<x<2$, and for the series in x^{-1} it is $|x|>1$. Thus six solutions have been obtained; † since not more than two solutions are linearly distinct, linear relations must exist between them. An example of this linear relationship will now be given.

7·231. Linear Relationship between the Series-Solutions.—It will first of all be proved that, when $\gamma>\alpha+\beta$, and γ is not a negative integer,

$$F(\alpha, \beta; \gamma; 1)=\frac{\Gamma(\gamma)\Gamma(\gamma-\alpha-\beta)}{\Gamma(\gamma-\alpha)\Gamma(\gamma-\beta)}.$$

Since, when $0\leqslant x\leqslant 1$, $F(\alpha, \beta; \gamma; x)$ satisfies the identity

$$\{\gamma-(\alpha+\beta+1)x\}F'(\alpha, \beta; \gamma; x)=\alpha\beta F(\alpha, \beta; \gamma; x)-x(1-x)F''(\alpha, \beta; \gamma; x),$$

and since, as may be verified from the series itself, $F''(\alpha, \beta; \gamma; 1)$ is finite, it follows that

$$(\gamma-\alpha-\beta-1)F'(\alpha, \beta; \gamma; 1)=\alpha\beta F(\alpha, \beta; \gamma; 1).$$

* Riemann, *Abh. Ges. Wiss. Gött.* 7 (1857) [*Math. Werke*, 2nd ed, p 67].

† Kummer, *J. für Math.* 15 (1836), pp. 39, 127. See also Whittaker and Watson, *loc. cit.*

It may also be verified by comparing the coefficients of like terms that

$$F(\alpha, \beta ; \gamma+1 ; x)-F(\alpha, \beta ; \gamma ; x)= -\frac{\alpha\beta x}{\gamma(\gamma+1)}F(\alpha+1, \beta+1 ; \gamma+2 ; x)$$

$$= -\frac{x}{\gamma}F'(\alpha, \beta ; \gamma+1 ; x),$$

and therefore

$$F(\alpha, \beta ; \gamma+1 ; 1)-F(\alpha, \beta ; \gamma ; 1)= -\frac{1}{\gamma}F'(\alpha, \beta ; \gamma+1 ; 1)$$

$$= -\frac{\alpha\beta}{\gamma(\gamma-\alpha-\beta)}F(\alpha, \beta ; \gamma+1 ; 1).$$

Consequently

$$F(\alpha, \beta ; \gamma ; 1) = \frac{(\gamma-\alpha)(\gamma-\beta)}{\gamma(\gamma-\alpha-\beta)}F(\alpha, \beta ; \gamma+1 ; 1).$$

By repeated use of this formula it is found that

$$F(\alpha, \beta ; \gamma ; 1) = \lim_{n\to\infty}\left\{\prod_{r=0}^{n-1}\frac{(\gamma-\alpha+r)(\gamma-\beta+r)}{(\gamma+r)(\gamma-\alpha-\beta+r)}F(\alpha, \beta ; \gamma+n ; 1)\right\}.$$

But, by a well-known theorem,* the limiting value of the infinite product is

$$\frac{\Gamma(\gamma)\Gamma(\gamma-\alpha-\beta)}{\Gamma(\gamma-\alpha)\Gamma(\gamma-\beta)},$$

and since

$$F(\alpha, \beta ; \gamma+n ; 1)=1+\frac{\alpha\beta}{\gamma+n}U_n,$$

where U_n is a convergent series and is positive and decreases as n increases,

$$\lim F(\alpha, \beta ; \gamma+n ; 1)=1,$$

and the theorem is proved.

Now, since any solution is linearly expressible in terms of two independent solutions, there will be an identical relationship of the form

$$F(\alpha, \beta ; \gamma ; x)=AF(\alpha, \beta ; \alpha+\beta-\gamma+1 ; 1-x)$$
$$+B(1-x)^{\gamma-\alpha-\beta}F(\gamma-\alpha, \gamma-\beta ; \gamma-\alpha-\beta+1 ; 1-x),$$

where A and B are constants to be determined.

In order that all series may converge throughout the common interval $0\leqslant x\leqslant 1$ it is assumed that †

$$1>\gamma>\alpha+\beta.$$

Then, putting in succession $x=1$ and $x=0$, it is found that

$$F(\alpha, \beta ; \gamma ; 1)=A,$$

$$1=AF(\alpha, \beta ; \alpha+\beta-\gamma+1 ; 1)+BF(\gamma-\alpha, \gamma-\beta ; \gamma-\alpha-\beta+1 ; 1).$$

From these two equations the values of A and B are obtained. The resulting relationship is

$$F(\alpha, \beta ; \gamma ; x) = \frac{\Gamma(\gamma)\Gamma(\gamma-\alpha-\beta)}{\Gamma(\gamma-\alpha)\Gamma(\gamma-\beta)}F(\alpha, \beta ; \alpha+\beta-\gamma+1 ; 1-x)$$
$$+\frac{\Gamma(\gamma)\Gamma(\alpha+\beta-\gamma)}{\Gamma(\alpha)\Gamma(\beta)}(1-x)^{\gamma-\alpha-\beta}F(\gamma-\alpha, \gamma-\beta ; \gamma-\alpha-\beta+1 ; 1-x).$$

7·232. The Case of Integral Exponent-Difference.—The two solutions appropriate to the singularity $x=0$, namely,

$$y_1=F(\alpha, \beta ; \gamma ; x), \qquad y_2=x^{1-\gamma}F(\alpha-\gamma+1, \beta-\gamma+1 ; 2-\gamma ; x),$$

are distinct when the exponent-difference $1-\gamma$ is not zero or a negative integer. When $\gamma=1$, the two solutions become identical ; when $\gamma=2, 3, 4, \ldots$, the solution

* Whittaker and Watson, *Modern Analysis*, § 12·13.

† This severe restriction is not essential to the result, it is merely inherent to the method followed.

y_2 becomes illusory through the vanishing of the denominator in the coefficients of an infinite number of terms of the series. Nevertheless, the solution y_2 can be made significant when $\gamma=m$, a positive integer, by multiplying it by an appropriate constant factor. Consider the solution

$$\frac{(2-\gamma)\ldots(m-\gamma)\,.\,(m-1)\,!}{(\alpha-\gamma+1)\ldots(\alpha-\gamma+m-1)(\beta-\gamma+1)\ldots(\beta-\gamma+m-1)}y_2.$$

This solution remains finite when γ is made equal to m, the first $m-1$ terms of the series-development vanish and there remains the solution

$$1+\frac{\alpha\beta}{1\,!\,m}x+\frac{\alpha(\alpha+1)\beta(\beta+1)}{2\,!\,m(m+1)}x^2+\ldots=F(\alpha,\beta\,;\,m\,;\,x).$$

Thus, when γ is a positive integer or zero the two solutions y_1 and y_2 are effectively the same. The general method * of obtaining another solution which is essentially distinct from the one considered will be investigated in a later chapter (Chapter XVI). A simple example which illustrates the general case is the following:

Consider the equation

$$\frac{d^2y}{dx^2}+\left\{-\frac{1}{4}+\frac{1}{4x^2}\right\}y=0\,;$$

the origin is a regular singular point to which corresponds the indicial equation

$$(r-\tfrac{1}{2})^2=0,$$

whose roots are equal. One solution is obtainable directly, namely,

$$y_1=x^{\frac{1}{2}}\left(1+\frac{x^2}{4^2}+\frac{x^4}{4^2\,.\,8^2}+\ldots\right);$$

the second solution is now arrived at by making the substitution

$$y=y_1v,$$

where v is a new dependent variable. The equation for v, namely

$$y_1v''+2y_1'v'=0.$$

has the solution

$$v=\int\frac{dx}{\{y_1(x)\}^2}=\int\frac{dx}{x\{1+\frac{1}{8}x^2+O(x^4)\}}$$

$$=\int\{x^{-1}-\tfrac{1}{8}x+O(x^3)\}dx=\log x-\tfrac{1}{16}x^2+O(x^4).$$

The second solution y_2 is therefore of the form

$$y_2=y_1\log x-x^{\frac{1}{2}}\{\tfrac{1}{16}x^2+O(x^4)\}.$$

Thus the logarithmic case arises, just as it arose in similar circumstances in the case of the Euler equation (§ 6·3).

7·24. The Legendre Equation.—The differential equation

$$(1-x^2)\frac{d^2y}{dx^2}-2x\frac{dy}{dx}+n(n+1)y=0,$$

known as the Legendre equation, is of great importance in physical problems; its solutions are known as Legendre Functions.† The equation has regular singularities at the points ±1 and at infinity, and is defined by the scheme

$$y=P\begin{Bmatrix}-1 & \infty & +1 & \\ 0 & n+1 & 0 & x\\ 0 & -n & 0 & \end{Bmatrix}$$

or by the equivalent scheme

$$y=P\begin{Bmatrix}0 & \infty & 1 & \\ 0 & n+1 & 0 & \frac{1}{2}-\frac{1}{2}x\\ 0 & -n & 0 & \end{Bmatrix}.$$

* See Lindelöf, *Acta Soc. Sc. Fenn.* 19 (1893), p. 15.

† Legendre, *Mém. Acad. Sc. Paris*, 10 (1785); see Whittaker and Watson, *Modern Analysis*, Chap. XV.

The most manageable expansion for the solution is that which proceeds in descending powers of x, and is therefore appropriate to the singularity at infinity. It may easily be verified that the equation is satisfied by the two series

$$y_1 = x^n - \frac{n(n-1)}{2.(2n-1)} x^{n-2} + \frac{n(n-1)(n-2)(n-3)}{2.4.(2n-1)(2n-3)} x^{n-4} - \ldots,$$

$$y_2 = x^{-n-1} + \frac{(n+1)(n+2)}{2.(2n+3)} x^{-n-3} + \frac{(n+1)(n+2)(n+3)(n+4)}{2.4.(2n+3)(2n+5)} x^{-n-5} + \ldots$$

both of which are convergent when $|x|>1$.

In the first place, let n be an integer; moreover, as no further essential restriction is thereby introduced, n will be regarded as a positive integer.* Then the solution y_1 is a polynomial of degree n and after multiplication by the factor

$$\frac{(2n)!}{2^n(n!)^2}$$

will be denoted by $P_n(x)$. This particular choice of multiplying factor is made so that, for all values of n, $P_n(1)=1$. The polynomials so defined are known as the *Legendre Polynomials*; they play the central part in the theory of *Spherical Harmonics*.

The first six Legendre polynomials are:

$$P_0(x)=1; \; P_1(x)=x; \; P_2(x)=\tfrac{1}{2}(3x^2-1); \; P_3(x)=\tfrac{1}{2}(5x^3-3x);$$
$$P_4(x)=\tfrac{1}{8}(35x^4-30x^2+3); \; P_5(x)=\tfrac{1}{8}(63x^5-70x^3+15x).$$

It may be proved directly that if n is a positive integer,

$$P_n(x) = \frac{1}{2^n.n!}\cdot\frac{d^n}{dx^n}(x^2-1)^n.$$

This result is known as the *Rodrigues formula*.

Now consider the second series y_2; since this series does not terminate when $n>-1$ there is no point in restricting n to be an integer. This series-solution, when multiplied by the factor †

$$\frac{\pi^{\frac{1}{2}}\Gamma(n+1)}{2^{n+1}\Gamma(n+\frac{3}{2})},$$

is denoted by $Q_n(x)$. It may be verified, by comparing the series y_2 with the hypergeometric series in x^{-2} that, when $x>1$,

$$Q_n(x) = \frac{\pi^{\frac{1}{2}}\Gamma(n+1)}{2^{n+1}\Gamma(n+\frac{3}{2})} x^{-n-1}F(\tfrac{1}{2}n+\tfrac{1}{2},\; \tfrac{1}{2}n+1;\; n+\tfrac{3}{2};\; x^{-2}).$$

The function $Q_n(x)$, thus defined, may be taken as one standard solution of the Legendre equation, and is known as the *Legendre function of the second kind.*

The series y_1 ceases to be essentially distinct from y_2 when $2n$ assumes the value -1 or any positive odd integral value, and is therefore unsuitable as a standard solution. Now it follows immediately from the second of the

* In general, n being real, it is sufficient to consider values of n such that $n \geqslant -\frac{1}{2}$.

† On account of the duplication-formula for the Gamma-function, namely,

$$2^{2z-1}\Gamma(z)\Gamma(z+\tfrac{1}{2})=\pi^{\frac{1}{2}}\Gamma(2z)$$

this multiplier can be written $2^n\{\Gamma(n+1)\}^2/\Gamma(2n+2)$, and when n is a positive integer, has the value

$$\frac{2^n.\{n!\}^2}{(2n+1)!}.$$

The reason for this choice will appear later.

two schemes by which the Legendre equation may be defined that the hypergeometric series

$$F(n+1, -n; 1; \tfrac{1}{2}-\tfrac{1}{2}x)$$

satisfies the Legendre equation and assumes the value 1 when $x=1$. Moreover it is a polynomial when n is a positive integer and since, when $n \geqslant 0$, only one solution, namely $P_n(x)$, is a polynomial, it follows that

$$P_n(x)=F(n+1, -n; 1; \tfrac{1}{2}-\tfrac{1}{2}x).$$

There is no value of n for which this solution ceases to be significant; it is therefore taken as standard. As the hypergeometric function has only been defined as a series, convergent when $-1<\tfrac{1}{2}-\tfrac{1}{2}x<1$, it follows that when n is not an integer the series-development of $P_n(x)$ is only valid in the range $-1<x<3$. Thus the series-solutions $P_n(x)$ and $Q_n(x)$ have the common range of validity $1<x<3$.*

7·241. The Second Solution when n is an Integer.—Since the exponents relative to the singularities $x=\pm 1$ are equal, it is to be expected that the companion solution to $y=P_n(x)$ is of a form which involves logarithmic terms. Let

$$y=uP_n(x)-v$$

be assumed as a tentative solution, then

$$\{(1-x^2)u''-2xu'\}P_n(x)+2(1-x^2)u'P_n'(x)-\{(1-x^2)v''-2xv'+n(n+1)v\}=0.$$

Let u be so chosen that

$$(1-x^2)u''-2xu''=0$$

or

$$(1-x^2)u'=-1.$$

The choice of the number -1 as the constant of integration is made so as to facilitate the subsequent identification of the solution which will be obtained. Then

$$u=\frac{1}{2}\log\frac{x+1}{x-1},$$

and v is determined by the equation

$$(1-x^2)v''-2xv'+n(n+1)v=2P_n'(x).$$

Now it may be verified directly that

$$P_n'(x)-P_{n-2}'(x)=(2n-1)P_{n-1}(x),$$

and therefore

$$P_n'(x)=(2n-1)P_{n-1}(x)+(2n-5)P_{n-3}(x)+(2n-9)P_{n-5}(x)+\ldots;$$

the last term of the series is $3P_1(x)$ or $P_0(x)$ according as n is even or odd. Consequently v is to be determined by the equation

$$\frac{d}{dx}\{(1-x^2)v'\}+n(n+1)v=2\sum_{r=1}^{N}(2n-4r+3)P_{n-2r+1}(x),$$

where $N=\tfrac{1}{2}n$ or $\tfrac{1}{2}(n+1)$ according as n is even or odd. But a particular solution of the equation

$$\frac{d}{dx}\{(1-x^2)w'\}+n(n+1)w=2(2n-4r+3)P_{n-2r+1}(x)$$

is

$$w=\frac{2n-4r+3}{(2r-1)(n-r+1)}P_{n-2r+1}(x)$$

and consequently

$$v=\sum_{r=1}^{N}\frac{2n-4r+3}{(2r-1)(n-r+1)}P_{n-2r+1}(x).$$

* Extended ranges of validity are obtainable by expressing the solutions in the form of definite integrals.

Thus the solution sought for is

$$\tfrac{1}{2}P_n(x)\log\frac{x+1}{x-1}-\left\{\frac{2n-1}{1\,.\,n}P_{n-1}(x)+\frac{2n-5}{3\,.\,(n-1)}P_{n-3}(x)+\frac{2n-9}{5\,.\,(n-2)}P_{n-5}(x)+\ldots\right\};$$

the last term is

$$\frac{3}{(n-1)(\tfrac{1}{2}n+1)}P_1(x)\quad\text{or}\quad\frac{1}{\tfrac{1}{2}n(n+1)}P_0(x)$$

according as n is even or odd. The solution is obviously valid for all values of x such that $|x|>1$.

Let the solution obtained be denoted, for the moment, by $S_n(x)$, then since $P_n(x)$ and $Q_n(x)$ are distinct solutions,

$$S_n(x)=AP_n(x)+BQ_n(x),$$

where A and B are constants. Now for large values of $|x|$,

$$P_n(x)=O(x^n),\quad Q_n(x)=O(x^{-n-1}),$$

and since

$$\tfrac{1}{2}\log\frac{1+x}{1-x}=\frac{1}{x}+\frac{1}{3x^3}+\frac{1}{5x^5}+\ldots,$$

$S_n(x)$ is at most $O(x^{n-1})$. Consequently $A=0$ and $S_n(x)$ is a mere multiple of $Q_n(x)$. Thus

$$\begin{aligned}BQ_n(x)&=S_n(x)\\&=\tfrac{1}{2}P_n(x)\log\frac{x+1}{x-1}-R_n(x),\end{aligned}$$

where $R_n(x)$ is a polynomial of degree $n-1$. Divide both sides of the equation by $P_n(x)$ and differentiate with respect to x, then

$$B\frac{d}{dx}\left\{\frac{Q_n(x)}{P_n(x)}\right\}=\frac{-1}{x^2-1}-\frac{T_n(x)}{\{P_n(x)\}^2},$$

where $T_n(x)$ is a polynomial of degree $2n-2$ at most.

Now since

$$(1-x^2)\frac{d^2P_n(x)}{dx^2}-2x\frac{dP_n(x)}{dx}+n(n+1)P_n(x)=0,$$

$$(1-x^2)\frac{d^2Q_n(x)}{dx^2}-2x\frac{dQ_n(x)}{dx}+n(n+1)Q_n(x)=0,$$

it is found, by multiplying the first equation by $Q_n(x)$ and the second by $P_n(x)$ and subtracting, that

$$(x^2-1)\left\{Q_n(x)\frac{d^2P_n(x)}{dx^2}-P_n(x)\frac{d^2Q_n(x)}{dx}\right\}+2x\left\{Q_n(x)\frac{dP_n(x)}{dx}-P_n(x)\frac{dQ_n(x)}{dx}\right\}=0,$$

whence, by integration,

$$(x^2-1)\left\{Q_n(x)\frac{dP_n(x)}{dx}-P_n(x)\frac{dQ_n(x)}{dx}\right\}=C,$$

where C is a constant to be determined. Now, since the leading terms in $P_n(x)$ and $Q_n(x)$ are respectively

$$\frac{(2n)!}{2^n(n!)^2}x^n\quad\text{and}\quad\frac{2^n(n!)^2}{(2n+1)!}x^{-n-1},$$

it is found that $C=1$. Therefore

$$Q_n(x)\frac{dP_n(x)}{dx}-P_n(x)\frac{dQ_n(x)}{dx}=\frac{1}{x^2-1},$$

or

$$\frac{d}{dx}\left\{\frac{Q_n(x)}{P_n(x)}\right\}=\frac{1}{(1-x^2)\{P_n(x)\}^2}.$$

Thus it follows that

$$\frac{B}{(1-x^2)\{P_n(x)\}^2}=\frac{1}{1-x^2}-\frac{T_n(x)}{\{P_n(x)\}^2},$$

or

$$B=\{P_n(x)\}^2+(x^2-1)T_n(x).$$

Let $x=1$, then, since $P_n(1)=1$ and $T_n(1)$ is finite, it follows that $B=1$.

Consequently

$$Q_n(x)=\tfrac{1}{2}P_n(x)\log\frac{x+1}{x-1}-\sum_{r=1}^{N}\frac{2n-4r+3}{(2r-1)(n-r+1)}P_{n-2r+1}(x).$$

In particular,

$$Q_0(x)=\tfrac{1}{2}\log\frac{x+1}{x-1};\quad Q_1(x)=\tfrac{1}{2}x\log\frac{x+1}{x-1}-1;$$

$$Q_2(x)=\tfrac{1}{2}P_2(x)\log\frac{x+1}{x-1}-\tfrac{3}{2}x;\quad Q_3(x)=\tfrac{1}{2}P_3(x)\log\frac{x+1}{x-1}-\tfrac{5}{2}x^2+\tfrac{2}{3}.$$

7·3. The Point at Infinity as an Irregular Singular Point.—Equations whose solutions are irregular at infinity are of frequent occurrence; linear equations with constant coefficients furnish a case in point. To study the behaviour of solutions of such equations for numerically large values of x is therefore a problem of some importance, a problem, however, which cannot be fully treated except with the aid of the theory of functions of a complex variable.*

It is, however, possible to give some rather crude indications of the behaviour of solutions which are irregular at infinity, which, crude as they are, will be found to be not without value in their applications.

Consider the equation of the second order,

$$\frac{d^2y}{dx^2}+p(x)\frac{dy}{dx}+q(x)y=0,$$

in which at least one of the conditions for a regular singularity at infinity, namely,

$$p(x)=O(x^{-1}),\quad q(x)=O(x^{-2})$$

as $x\rightarrow\infty$, is violated. It will be supposed that the coefficients $p(x)$ and $q(x)$ can be developed as series of descending powers of x, thus

$$p(x)=p_0x^{\alpha}+\ldots,\quad q(x)=q_0x^{\beta}+\ldots,$$

then since the point at infinity is irregular, one or both of the inequalities

$$\alpha>-1,\quad \beta>-2$$

must be satisfied.

Now consider the possibility of satisfying the equation by a function which, for large values of x, is of the form

$$x^{\sigma}e^{P(x)}v(x),$$

where $P(x)$ is a polynomial in x and $v(x)=O(1)$ as $x\rightarrow\infty$. Let λx^{ν} be the leading term in $P(x)$, then on substituting the above expression in the equation and extracting the dominant part of each term it is found that

$$\lambda^2\nu^2x^{2\nu-2}+p_0\lambda\nu x^{\nu+\alpha-1}+q_0x^{\beta}=0.$$

Thus ν is given by

$$\nu=\alpha+1\quad\text{or}\quad 2\nu=\beta+2,$$

whichever furnishes the greater value of ν. Thus 2ν is a positive integer, for simplicity it will be supposed that ν is a positive integer also.

Then a solution of the form

$$y=e^{\lambda x^{\nu}+\mu x^{\nu-1}+\ldots+\varpi x}x^{\sigma}v(x)$$

* See Chaps. XVII.–XIX.

is assumed, where

$$v(x)=1+\frac{a_1}{x}+\frac{a_2}{x^2}+\ldots,$$

and the constants λ, μ, . . . , ϖ, σ, a_1, a_2, . . . determined in succession.

When a solution of this type exists, it is said to be normal and of rank ν. Unfortunately, however, when the series $v(x)$ does not terminate, it diverges in general, and therefore the solution is illusory. Nevertheless it can be shown that the series, though divergent, is *asymptotic,** and therefore is of value in practical computation. It will now be shown, by an application of the process of successive approximation, how it is that the divergent series are of practical value, and an illustration will be taken from the theory of Bessel functions.

7·31. Asymptotic Development of Solutions.—Consider the linear equation of the second order

$$\frac{d^2y}{dx^2}+p(x)\frac{dy}{dx}+q(x)y=0,$$

in which p and q are real and finite at infinity; let p and q be developed in the convergent series

$$p(x)=p_0+p_1x^{-1}+p_2x^{-2}+\ldots,$$
$$q(x)=q_0+q_1x^{-1}+q_2x^{-2}+\ldots$$

The substitution $y=e^{\lambda x}v$ transforms the equation into

$$\frac{d^2v}{dx^2}+(2\lambda+p)\frac{dv}{dx}+(\lambda^2+\lambda p+q)v=0;$$

if λ is a root of the equation

$$\lambda^2+\lambda p_0+q_0=0,$$

the constant term in the coefficient of v disappears and the equation takes the form

$$\frac{d^2v}{dx^2}+(\varpi_0+\varpi_1x^{-1}+\ldots)\frac{dv}{dx}+(\rho_1x^{-1}+\rho_2x^{-2}+\ldots)v=0.$$

Now let

$$v=x^\sigma u,$$

then if

$$\varpi_0\sigma+\rho_1=0,$$

the term in x^{-1} in the coefficient of v disappears.

The leading term in the coefficient of $\frac{dv}{dx}$ is ϖ_0 and is real if λ is real. It will be supposed that ϖ_0 is negative,† then multiplication of the independent variable by the positive number $(-\varpi_0)^{-1}$ replaces ϖ_0 by -1.

The equation thus becomes

$$\frac{d^2u}{dx^2}+\left\{-1+\frac{a_1}{x}+\frac{a_2}{x^2}+\ldots\right\}\frac{du}{dx}+\left\{\frac{b_2}{x^2}+\frac{b_3}{x^3}+\ldots\right\}u=0;$$

a solution will be found which assumes the value η when $x=+\infty$. Let $u_1=\eta$ and define the sequence of functions (u_n) by the relations

$$\frac{d^2u_2}{dx^2}-\frac{du_2}{dx}=-\left\{\frac{a_1}{x}+\frac{a_2}{x^2}+\ldots\right\}\frac{du_1}{dx}-\left\{\frac{b_2}{x^2}+\frac{b_3}{x^3}+\ldots\right\}u_1,$$

. . . .

$$\frac{d^2u_n}{dx^2}-\frac{du_n}{dx}=-\left\{\frac{a_1}{x}+\frac{a_2}{x^2}+\ldots\right\}\frac{du_{n-1}}{dx}-\left\{\frac{b_2}{x^2}+\frac{b_3}{x^3}+\ldots\right\}u_{n-1},$$

. . . .

* Whittaker and Watson, *Modern Analysis*, Chap. VIII.

† The case in which ϖ_0 is positive and that in which λ is imaginary may be left to the reader. An example of the latter circumstance is given in the following section.

Then *

$$u_n=\eta+\int_x^\infty (e^{x-t}-1)\left\{\frac{a_1}{t}+\frac{a_2}{t^2}+\ .\ .\ .\ \right\}\frac{du_{n-1}(t)}{dt}\,dt$$

$$+\int_x^\infty (e^{x-t}-1)\left\{\frac{b_2}{t^2}+\frac{b_3}{t^3}+\ .\ .\ .\ \right\}u_{n-1}(t)dt$$

$$=\eta+\int_x^\infty e^{x-t}\left\{\frac{\alpha_1}{t}+\frac{\alpha_2}{t^2}+\ .\ .\ .\ \right\}u_{n-1}(t)dt+\int_x^\infty\left\{\frac{\beta_2}{t^2}+\frac{\beta_3}{t^3}+\ .\ .\ .\ \right\}u_{n-1}(t)dt,$$

where $\alpha_1, \alpha_2, \ldots, \beta_2, \beta_3, \ldots$ are expressible in terms of $a_1, a_2, \ldots, b_2, b_3, \ldots$.

It follows that

$$u_n-u_{n-1}=\int_x^\infty e^{x-t}\left\{\frac{\alpha_1}{t}+\frac{\alpha_2}{t^2}+\ .\ .\ .\ \right\}\left\{u_{n-1}(t)-u_{n-2}(t)\right\}dt$$

$$+\int_x^\infty\left\{\frac{\beta_2}{t^2}+\frac{\beta_3}{t^3}+\ .\ .\ .\ \right\}\left\{u_{n-1}(t)-u_{n-2}(t)\right\}dt.$$

Let it be supposed that $| u_{n-1}-u_{n-2} |$ is bounded for $x>a$, and that its upper bound is M_{n-1}. Then $| u_n-u_{n-1} |$ is bounded in the same range and its upper bound M_n satisfies the inequality

$$M_n<\frac{K}{x}M_{n-1},$$

where K is a constant, independent of n. Now M_2 is bounded for sufficiently large values of x; consequently the inequality holds for all values of n. It follows by comparison that the series

$$u=u_1+(u_2-u_1)+\ .\ .\ .\ +(u_n-u_{n-1})+\ .\ .\ .$$

is convergent for sufficiently large values of x. Moreover its sum is a solution of the differential equation in u.

Now

$$u_2-u_1=\int_x^\infty e^{x-t}\left\{\frac{\alpha_1}{t}+\frac{\alpha_2}{t^2}+\ .\ .\ .\ \right\}\eta dt+\int_x^\infty\left\{\frac{\beta_2}{t^2}+\frac{\beta_3}{t^3}+\ .\ .\ .\ \right\}\eta dt$$

$$=\frac{A^1{}_1}{x}+\frac{A^1{}_2}{x^2}+\ .\ .\ .\ +\frac{A^1{}_{m-1}}{x^{m-1}}+\frac{A^1{}_m+\epsilon_1}{x^m},$$

where $\epsilon_1\to 0$ as $x\to\infty$.

Similarly

$$u_3-u_2=\frac{A^2{}_2}{x^2}+\ .\ .\ .\ +\frac{A^2{}_{m-1}}{x^{m-1}}+\frac{A^2{}_m+\epsilon_2}{x^m},$$

and finally, if $m>n$,

$$u_n-u_{n-1}=\frac{A^{n-1}{}_{n-1}}{x^{n-1}}+\ .\ .\ .\ +\frac{A^{n-1}{}_{m-1}}{x^{m-1}}+\frac{A^{n-1}{}_m+\epsilon_{n-1}}{x^m},$$

where $\epsilon_{n-1}\to 0$ as $x\to\infty$.

Consequently,

$$u_1+(u_2-u_1)+\ .\ .\ .\ +(u_n-u_{n-1})$$

$$=\eta+\frac{C_1}{x}+\frac{C_2}{x^2}+\ .\ .\ .\ +\frac{C_{m-1}}{x^{m-1}}+\frac{C_m+\epsilon}{x^m},$$

* The solution of

$$\frac{d^2u}{dx^2}-\frac{du}{dx}=-f(x)$$

which reduces to η when $x=+\infty$ is

$$u=\eta+\int_x^\infty(e^{x-t}-1)f(t)dt,$$

provided that the integral exists.

where $\epsilon \to 0$ as $x \to \infty$. On the other hand

$$|(u_{n+1}-u_n)+(u_{n+2}-u_{n+1})+\ldots| < M_n\left(\frac{K}{x}+\frac{K^2}{x^2}+\ldots\right) < \frac{H}{x^n},$$

where H is a constant, for sufficiently large values of x.

It follows that

$$u=\eta+\frac{C_1}{x}+\frac{C_2}{x^2}+\ldots+\frac{C_{n-1}}{x^{n-1}}+\frac{C_n+\gamma_n}{x^n},$$

where $\gamma_n \to 0$ as $x \to \infty$.

Consequently the given differential equation admits of a solution of the form

$$y=e^{\lambda x}x^{\sigma}\left\{\eta+\frac{C_1}{x}+\frac{C_2}{x^2}+\ldots+\frac{C_{n-1}}{x^{n-1}}+\frac{C_n+\gamma_n}{x^n}\right\}.$$

The series $\sum C_r x^{-r}$ may terminate, in which case the representation is exact. But when the series does not terminate, it in general diverges.* Nevertheless if m is fixed, and S_m denotes the sum of the series

$$e^{\lambda x}x^{\sigma}\left\{\eta+\frac{C_1}{x}+\ldots+\frac{C_m}{x^m}\right\},$$

then if ϵ is arbitrarily small,

$$|x^m(y-S_m)|<\epsilon$$

for sufficiently large values of $|x|$. Consequently the series furnishes an asymptotic representation of the solution, and the sign of equality is replaced by the sign of asymptotic equivalence, thus:

$$y \sim e^{\lambda x}x^{\sigma}\left\{\eta+\frac{C_1}{x}+\ldots+\frac{C_n}{x^n}+\ldots\right\}.$$

7·32. The Bessel Equation.—When n is not an integer, the Bessel equation †

$$x^2\frac{d^2y}{dx^2}+x\frac{dy}{dx}+(x^2-n^2)y=0$$

is satisfied by the two distinct solutions

$$y_1=J_n(x), \quad y_2=J_{-n}(x),$$

where

$$J_n(x)=\frac{x^n}{2^n\Gamma(n+1)}\left\{1-\frac{x^2}{2^2.1!.(n+1)}+\frac{x^4}{2^4.2!.(n+1)(n+2)}-\ldots\right\}.$$

When n is an integer these two solutions cease to be independent. The second solution, when n is an integer, is of the logarithmic type.‡

Now consider solutions appropriate to the irregular singularity at infinity.§ The substitution

$$y=x^{-\frac{1}{2}}u$$

removes the second term from the equation, which becomes

$$\frac{d^2u}{dx^2}+\left\{1+\frac{\frac{1}{4}-n^2}{x^2}\right\}u=0$$

* This can be verified by considering the simple equation

$$\frac{d^2y}{dx^2}-\frac{dy}{dx}+\frac{\beta}{x^2}y=0.$$

† Bessel, *Abh. Akad. Wiss. Berlin*, 1824, p. 34. An account of the early history of this and allied equations is given by Watson, *Bessel Functions*, Chap. I.

‡ This solution will be given explicitly in a later section (§ 16·32).

§ For a complete discussion of the problem, see Watson, *Bessel Functions*, Chap. VII.

For large values of $|x|$ this equation becomes effectively $u''+u=0$, which suggests the substitution *

$$u=e^{ix}v.$$

The equation now becomes

$$\frac{d^2v}{dx^2}+2i\frac{dv}{dx}+\frac{\frac{1}{4}-n^2}{x^2}v=0.$$

This equation is formally satisfied by a series of descending powers of x, namely

$$1-\frac{\frac{1}{4}-n^2}{2x}i-\frac{(\frac{1}{4}-n^2)(\frac{9}{4}-n^2)}{2^2.2!.x^2}+\frac{(\frac{1}{4}-n^2)(\frac{9}{4}-n^2)(\frac{25}{4}-n^2)}{2^3.3!.x^3}i$$
$$+\frac{(\frac{1}{4}-n^2)(\frac{9}{4}-n^2)(\frac{25}{4}-n^2)(\frac{49}{4}-n^2)}{2^4.4!.x^4}-\dots$$

This series is divergent for all values of x, but it is of asymptotic type. In fact, if $|x|$ is large, the earlier terms diminish rapidly with increasing rank, and as will be seen later the series furnishes a valuable method for computing $J_n(x)$ when x is large.

By combining the series obtained with that obtained by changing i into $-i$ two asymptotic relations are obtained, namely

$$y_1 \sim x^{-\frac{1}{2}}(U\cos x+V\sin x),$$
$$y_2 \sim x^{-\frac{1}{2}}(U\sin x-V\cos x),$$

where U and V stand respectively for the even and odd series

$$1-\frac{(\frac{1}{4}-n^2)(\frac{9}{4}-n^2)}{2^2.2!.x^2}+\frac{(\frac{1}{4}-n^2)(\frac{9}{4}-n^2)(\frac{25}{4}-n^2)(\frac{49}{4}-n^2)}{2^4.4!\,x^4}-\dots$$

and

$$\frac{\frac{1}{4}-n^2}{2x}-\frac{(\frac{1}{4}-n^2)(\frac{9}{4}-n^2)(\frac{25}{4}-n^2)}{2^3.3!.x^3}+\dots.$$

The connection between the function $J_0(x)$ and the corresponding asymptotic series may be derived from the relation,†

$$\pi J_0(x)=\int_0^\pi \cos(x\cos\theta)d\theta.$$

Let

$$J_0(x)=Ay_1+By_2,$$

then as $x\to\infty$,

$$\lim x^{\frac{1}{2}}J_0(x)=A\cos x+B\sin x,$$
$$\lim x^{\frac{1}{2}}J_0'(x)=-A\sin x+B\cos x.$$

Thus

$$\begin{aligned}A&=\lim x^{\frac{1}{2}}\{J_0(x)\cos x-J_0'(x)\sin x\}\\&=\lim\frac{x^{\frac{1}{2}}}{\pi}\int_0^\pi\{\cos x\cos(x\cos\theta)+\sin x\cos\theta\sin(x\cos\theta)\}d\theta\\&=\lim\frac{x^{\frac{1}{2}}}{\pi}\int_0^\pi\cos(2x\sin^2\tfrac{1}{2}\theta)\cos^2\tfrac{1}{2}\theta d\theta\\&\quad+\lim\frac{x^{\frac{1}{2}}}{\pi}\int_0^\pi\cos(2x\cos^2\tfrac{1}{2}\theta)\sin^2\tfrac{1}{2}\theta d\theta.\end{aligned}$$

Let

$$\sqrt{(2x)}\sin\tfrac{1}{2}\theta=\phi,$$

* For an alternative method of procedure when $n=0$, see Stokes, *Trans. Camb. Phil. Soc.* 9 (1850), p. 182; [*Math. and Phys. Papers*, 2, p. 350].

† An equivalent relation will be established in the following chapter, § 8·22.

then

$$\lim \frac{x^{\frac{1}{2}}}{\pi}\int_0^{\pi} \cos (2x \sin^2\tfrac{1}{2}\theta) \cos^2\tfrac{1}{2}\theta d\theta = \lim \frac{2^{\frac{1}{2}}}{\pi}\int_0^{\sqrt{(2x)}} \left(1-\frac{\phi^2}{2x}\right)^{\frac{1}{2}} \cos \phi^2 d\phi$$

$$= \frac{2^{\frac{1}{2}}}{\pi}\int_0^{\infty} \cos \phi^2 d\phi = \tfrac{1}{2}\pi^{-\frac{1}{2}}.$$

The second integral has the same limit and therefore

$$A = \pi^{-\frac{1}{2}}.$$

Similarly $B = \pi^{-\frac{1}{2}}$, and thus

$$J_0(x) \sim \left(\frac{2}{\pi x}\right)^{\frac{1}{2}}\left\{\left(1 - \frac{1^2 . 3^2}{2^6 . 2\,! . x^2} + \frac{1^2 . 3^2 . 5^2 . 7^2}{2^{12} . 4\,! . x^4} - \ldots\right) \cos (x - \tfrac{1}{4}\pi)\right.$$

$$\left. + \left(\frac{1^2}{2^3 . x^3} - \frac{1^2 . 3^2 . 5^2}{2^9 . 3\,! . x^3} + \ldots\right) \sin (x - \tfrac{1}{4}\pi).\right.$$

7·321. Use of the Asymptotic Series in Numerical Calculations.—The value of the asymptotic series may be illustrated by computing particular values of $J_0(x)$. If the ascending series

$$J_0(x) = 1 - \frac{x^2}{2^2} + \frac{x^4}{2^6} - \frac{x^6}{2^8 . 3^2} + \frac{x^8}{2^{10} . 3^2 . 4^2} - \frac{x^{10}}{2^{12} . 3^2 . 4^2 . 5^2} + \ldots$$

is used to evaluate $J_0(2)$, and the last term taken is that in x^{16}, the value

$$J_0(2) = 0{\cdot}223\ 890\ 779\ 14$$

correct to eleven places is obtained. But if $x = 6$, and terms up to and including that in x^{20} are taken, the value obtained is

$$J_0(6) = 0{\cdot}15067,$$

which is correct to four places only; in fact the last term used has the value 0·00026 which affects the fourth decimal place. Thus for even comparatively small values of x the ascending series is useless for practical calculations.

Now consider the asymptotic representation of $J_0(6)$; it is found that

$$J_0(6) = \frac{1}{\sqrt{(3\pi)}}\{(\sin 6 + \cos 6)U + (\sin 6 - \cos 6)V\},$$

where

$$U = 1 - \frac{1^2 . 3^2}{2^6 . 2\,! . 6^2} + \frac{1^2 . 3^2 . 5^2 . 7^2}{2^{12} . 4\,! . 6^4} - \frac{1^2 . 3^2 . 5^2 . 7^2 . 9^2 . 11^2}{2^{18} . 6\,! . 6^6} + \ldots$$

$$= 1 - 0{\cdot}00195 + 0{\cdot}00009 - 0{\cdot}00001 + \ldots$$

$$= 0{\cdot}99812,$$

and

$$V = \frac{1^2}{2^3 . 6} - \frac{1^2 . 3^2 . 5^2}{2^9 . 3\,! . 6^3} + \frac{1^2 . 3^2 . 5^2 . 7^2 . 9^2}{2^{15} . 5\,! . 6^5} - \ldots$$

$$= 0{\cdot}02083 - 0{\cdot}00034 + 0{\cdot}00003$$

$$= 0{\cdot}02052.$$

Since $2\pi - 6 = 0{\cdot}28318$, it is found from Burrau's tables that

$$\sin 6 = -0{\cdot}27941, \quad \cos 6 = 0{\cdot}96017,$$

and therefore

$$J_0(6) = 0{\cdot}23033\,(0{\cdot}67948 - 0{\cdot}02544)$$

$$= 0{\cdot}15064,$$

correct to five places of decimals. Thus by the use of the asymptotic series a more correct result is obtained with far less labour than in using the convergent ascending series.

7·322. The Large Zeros of the Bessel Functions.—It may be proved, as in § 7·32, that

$$J_n(x) \sim \left(\frac{2}{\pi x}\right)^{\frac{1}{2}}\{U_n \cos (x - \tfrac{1}{2}n\pi - \tfrac{1}{4}\pi) + V_n \sin (x - \tfrac{1}{2}n\pi - \tfrac{1}{4}\pi)\},$$

where

$$U_n = 1 - \frac{(\frac{1}{4}-n^2)(\frac{9}{4}-n^2)}{8x^2} + \ldots,$$

$$V_n = \frac{\frac{1}{4}-n^2}{2x} - \ldots.$$

If, therefore, ξ is a zero of $J_n(x)$, ξ is given by the relation

$$\cot(\xi - \tfrac{1}{2}n\pi - \tfrac{1}{4}\pi) \sim \frac{n^2-\frac{1}{4}}{2\xi} - \ldots.$$

Consequently if ξ is a zero of large absolute value and n is not very large, ξ is approximately given by the equation

$$\cot(\xi - \tfrac{1}{2}n\pi - \tfrac{1}{4}\pi) = 0,$$

or

$$\xi = (\tfrac{1}{2}n \pm m - \tfrac{1}{4})\pi,$$

where m is large.*

An immediate consequence of this result is that the large zeros of consecutive Bessel functions separate one another,† that is between two consecutive large zeros of $J_n(x)$ lies one and only one zero of $J_{n+1}(x)$.

7·323. Further Illustration of the Use of an Asymptotic Series.—The differential equation

$$\frac{dy}{dx} + y = \frac{1}{x}$$

is formally satisfied by the series

$$\frac{1}{x} + \frac{1!}{x^2} + \frac{2!}{x^3} + \ldots + \frac{n!}{x^{n+1}} + \ldots$$

but the series is obviously divergent for all values of x.

Now the equation possesses the particular integral

$$y = e^{-x}\int_{-\infty}^{x} x^{-1}e^x dx,$$

which is convergent when x is negative.

By repeated integration by parts it is found that

$$e^{-x}\int_{-\infty}^{x} x^{-1}e^x dx = \frac{1}{x} + \frac{1!}{x^2} + \frac{2!}{x^3} + \ldots + \frac{n!}{x^{n+1}} + R_n,$$

where

$$R_n = (n+1)!\,e^{-x}\int_{-\infty}^{x} x^{-n-2}e^x dx,$$

Now when $x<0$,

$$|R_n| < (n+1)!\,e^{-x}\,|x^{-n-2}|\left|\int_{-\infty}^{x} e^x dx\right|$$
$$= \frac{(n+1)!}{|x^{n+2}|}.$$

Consequently the error committed in taking the first n terms of the series is numerically less than the $(n+1)^{\text{th}}$ term. The series is therefore asymptotic and may be used for computing the integral.

The function defined by the integral

$$\int_{-\infty}^{x} \frac{e^x}{x} dx$$

is known as the exponential-integral function and is denoted by $Ei(x)$.

* The method is due to Stokes, *Trans. Camb. Phil. Soc.* 9 (1850), p. 184; [*Math. and Phys. Papers*, 2, p. 352]. For its full development see Watson, *Bessel Functions*, § 15·53.

† This theorem is in fact true of all the zeros. The general problem of the distribution of the zeros of solution of a linear differential equation of the second order is treated in Chap. X.

7·4. Equations with Periodic Coefficients ; the Mathieu Equation.—When the coefficients of a differential equation are one-valued, continuous, and periodic, say with period π, the general solution does not necessarily also possess the period π. In fact the equation may not, and in general does not, admit of such a periodic solution.

Thus the equation

$$\frac{dy}{dx}+(a+b\cos 2x)y=0$$

has no periodic solution unless $a=0$, and although the equation

$$\frac{d^2y}{dx^2}+n^2y=0$$

has always a periodic general solution, the period is not π unless n is an even integer.

The consideration of the general case will be deferred to a later chapter,* but a particular equation which has some important applications, namely the Mathieu equation †

$$\frac{d^2y}{dx^2}+(a-2\theta\cos 2x)y=0,$$

will be considered. This equation has no finite singular points and therefore its solutions are valid for all finite values of x. Moreover if $G(x)$ is a solution which is neither even nor odd, then $G(-x)$ is a distinct solution and

$$\tfrac{1}{2}\{G(x)+G(-x)\}$$

is an even solution, not identically zero, and

$$\tfrac{1}{2}\{G(x)-G(-x)\}$$

is an odd solution, not identically zero. Thus it is sufficient to consider only even or odd solutions. Now if the equation possessed two distinct even solutions, a solution satisfying the initial conditions

$$y(0)=0, \quad y_0'(0)=1$$

would not exist, which is in contradiction to the fact that the origin is an ordinary point. Thus two distinct even solutions, and likewise two distinct odd solutions, cannot exist. Thus one fundamental solution is even and the other odd.

Now assume that an even periodic solution with period 2π exists, and admits of the development ‡

$$C_0(x)=\sum_{r=0}^{\infty}c_r\cos(2r+1)x.$$

By substituting this series in the equation and equating the coefficients of like terms, a set of recurrence-relations connecting the coefficients c_r is obtained, namely

$$(a-1-\theta)c_0-\theta c_1=0,$$
$$\{(2r+1)^2-a\}c_r+\theta(c_{r+1}+c_{r-1})=0 \qquad (r=1, 2, 3, \ldots).$$

* See Chap. XV.

† Mathieu, *J. de Math.* (2) 13 (1868), p. 146 ; Whittaker and Watson, *Modern Analysis*, Chap. XIX. ; Humbert, *Fonctions de Lamé et Fonctions de Mathieu.*

‡ The differential equation has no finite singular point, and therefore (§§ 3·32, 12·22) its solution has no finite singularity, and the development converges for all values of x. See also Whittaker and Watson, *Modern Analysis*, § 9·11.

Now these equations must be consistent; the condition for their consistency is

$$\Delta(a,\theta)\equiv\begin{vmatrix} a-1-\theta, & -\theta, & 0, & 0, & \dots \\ -\theta, & a-9, & -\theta, & 0, & \dots \\ 0, & -\theta, & a-25, & -\theta, & \dots \\ 0, & 0, & -\theta, & a-49, & \dots \\ \cdot & \cdot & \cdot & \cdot & \cdot \end{vmatrix}=0.$$

Thus, in order that a periodic solution of the type considered may exist, the constant a must have one of the values determined by the determinantal equation *

$$\Delta(a,\theta)=0.$$

These values of a are known as the *characteristic values*; when a has been determined, the coefficients c_r may be obtained from the recurrence-relations, and are determined uniquely, apart from a constant factor.

Let a_n be that root of the determinantal equation which reduces to n^2 when $\theta=0$. Then it may be verified that †

$$a_1=1+\theta+O(\theta^2),$$

$$a_3=9+\frac{\theta^2}{16}+O(\theta^3),$$

$$a_n=n^2+\frac{\theta^2}{2(n^2-1)}+O(\theta^4) \qquad (n=5,7,9,\dots).$$

It may also be verified that if $a=a_{2n+1}$ and $c_n=1$,

$$c_{n-1}=\frac{\theta}{8n}+O(\theta^2),\quad c_{n-2}=\frac{\theta^2}{64n(2n-1)}+O(\theta^3),$$

$$c_{n-r}=\frac{(2n-r)!\,\theta^r}{2^{2r}r!\,(2n)!}+O(\theta^{r+1}),$$

$$c_{n+1}=-\frac{\theta}{8(n+1)}+O(\theta^2),\quad c_{n+2}=\frac{\theta^2}{64(n+1)(2n+3)}+O(\theta^3),$$

$$c_{n+r}=(-1)^r\frac{(2n+1)!\,\theta^r}{2^{2r}r!\,(2n+r+1)!}+O(\theta^{r+1}),$$

which, at least for small values of $|\theta|$, confirms the convergence of the series

In the same way, a solution of the type

$$S_o(x)=\sum_{r=0}^{\infty}c'_r\sin(2r+1)x$$

exists, where a is a root of the determinantal equation

$$\Delta(a,-\theta)=0.$$

The recurrence-relations from which the coefficients c'_r are determined are

$$(a-1+\theta)c'_0-\theta c'_1=0,$$

$$\{(2r+1)^2-a\}c'_r+\theta(c'_{r+1}+c'_{r-1})=0 \qquad (r=1,2,3,\dots).$$

There also exist, for appropriate values of a, solutions of period π, of the form

$$C_e(x)=\sum_{r=0}^{\infty}c_r\cos 2rx,$$

$$S_e(x)=\sum_{r=1}^{\infty}c'_r\sin 2rx.$$

* As it stands, the determinant is not convergent; it may, however, be made absolutely convergent by multiplying each row by an appropriate factor. See Whittaker and Watson, *Modern Analysis*, § 2·81.

† The verification is most easily affected by expressing a and $C_0(x)$ as ascending series in θ and determining the first two or three coefficients.

The recurrence-relations in these cases are respectively

$$ac_0 - \theta c_1 = 0,$$
$$(4r^2 - a)c_r + \theta(c_{r+1} + c_{r-1}) = 0 \qquad (r=1, 2, 3, \ldots),$$
$$(a-4)c'_1 - \theta c'_2 = 0,$$
$$(4r^2 - a)c'_r + \theta(c'_{r+1} + c'_{r-1}) = 0, \qquad (r=2, 3, 4, \ldots).$$

Thus there are four distinct types of solution of the Mathieu equation, having a period π or 2π; these solutions, multiplied by appropriate factors, are known as the *Mathieu Functions*. The Mathieu Function which reduces to $\cos mx$ when $\theta=0$ and in which the coefficient of $\cos mx$ is unity is denoted by $ce_m(x)$. Similarly the function which reduces to $\sin mx$ when $\theta=0$ and in which the coefficient of $\sin mx$ is unity is denoted by $se_m(x)$. Thus

$$ce_{2n+1}(x) \text{ is of type } C_o(x),$$
$$ce_{2n}(x) \text{ is of type } C_e(x),$$
$$se_{2n+1}(x) \text{ is of type } S_o(x),$$
$$se_{2n}(x) \text{ is of type } S_e(x).$$

7·41. The Non-Existence of Simultaneous Periodic Solutions.—Let a be such that Mathieu's equation has a periodic solution of type $C_o(x)$. Then the question arises as to whether in any circumstances the second solution, and therefore the general solution, can be periodic. Since if y_1 and y_2 are distinct solutions of the equation,

$$y_1 \frac{d^2y_2}{dx^2} - y_2 \frac{d^2y_1}{dx^2} = 0,$$

and therefore

$$y_1 \frac{dy_2}{dx} - y_2 \frac{dy_1}{dx} = \text{constant},$$

it follows that if y_1 is of type $C_o(x)$, y_2 is of type $S_o(x)$ and not of type $S_e(x)$. If the equation admits both of a solution $C_o(x)$ and of a solution $S_o(x)$ the equations

$$(a-1-\theta)c_0 - \theta c_1 = 0,$$
$$(a-1+\theta)c'_0 - \theta c'_1 = 0,$$
$$\{(2r+1)^2 - a\}c_r + \theta(c_{r+1} + c_{r-1}) = 0,$$
$$\{(2r+1)^2 - a\}c'_r + \theta(c'_{r+1} + c'_{r-1}) = 0,$$

$(r=1, 2, 3, \ldots)$ must be satisfied simultaneously. It will be shown that this is impossible.

From the first two equations it is found, on eliminating a, that

$$c_0c'_1 - c'_0c_1 = 2c_0c'_0$$

or

$$\begin{vmatrix} c_0, & c_1 \\ c'_0, & c'_1 \end{vmatrix} = 2c_0c'_0.$$

Similarly the last two equations give

$$c_r(c'_{r+1} + c'_{r-1}) = c'_r(c_{r+1} + c_{r-1})$$

or

$$\begin{vmatrix} c_r, & c_{r+1} \\ c'_r, & c'_{r+1} \end{vmatrix} = \begin{vmatrix} c_{r-1}, & c_r \\ c'_{r-1}, & c'_r \end{vmatrix}$$

whence, for all values of r,

$$\begin{vmatrix} c_r, & c_{r+1} \\ c'_r, & c'_{r+1} \end{vmatrix} = 2c_0c'_0.$$

But if c_0 is zero and θ is not zero, the remaining coefficients c_n are zero and the solution is identically zero. Therefore c_0 is not zero, and similarly c'_0 is not zero. But in order that the series may converge it is necessary that

$$c_r \to 0 \quad \text{as} \quad r \to \infty,$$

which leads to a contradiction. Thus, except when $\theta=0$, solutions of types $C_o(x)$ and $S_o(x)$ cannot exist simultaneously. In the same way it may be proved that solutions of types $C_e(x)$ and $S_e(x)$ do not co-exist.

7·42. The Nature of the Second Solution.—It has thus been proved that if one solution y_1 has the period π, or 2π, the second solution y_2 is definitely aperiodic. An indication of the general character of this second solution will now be given. Since

$$y_1y'_2-y_2y'_1=C,$$

where C is a constant,

$$y_2=Cy_1\int\frac{dx}{y_1{}^2}.$$

Now let

$$y_1=C_o(x)=\sum_{r=0}^{\infty}c_r\cos(2r+1)x,$$

then

$$y_1{}^2=\sum_{r=0}^{\infty}e_r\cos 2rx,$$

and since y_1 is not zero when $x=0$,

$$y_1{}^{-2}=\sum_{r=0}^{\infty}g_r\cos 2rx.$$

The last series is convergent at least for sufficiently small values of x.

Consequently

$$y_2=C\left\{\sum_{r=0}^{\infty}c_r\cos(2r+1)x\right\}\left\{g_0x+\sum_{r=1}^{\infty}h_r\sin 2rx\right\},$$

where since y_2 is known not to be periodic, g_0 is not zero, and therefore, with an appropriate choice of C,

$$y_2=xC_o(x)+S_o'(x),$$

where $S_o'(x)$ is a series of the same type as $S_o(x)$.

Thus $y_2(x)$ is not periodic, but quasi-periodic, and

$$y_2(x+2\pi)=y_2(x)+2\pi y_1(x).$$

The nature of the second solution, when the first solution is of type $S_o(x)$, $C_e(x)$, $S_e(x)$ may be investigated in the same way.*

7·5. A connexion between Differential Equations and Continued Fractions.—The particular method of dealing with differential equations which will now be outlined has the advantage that it is direct and not so artificial as the method of solution in series. It suffers on the other hand that it is applicable only to linear equations of the second order and admits of no obvious extension to equations of higher order.†

The equation to be considered may, without loss in generality, be assumed to be of the form

$$y=Q_0y'+P_1y'',$$

* The general solution when a is not a characteristic number may be exhibited in a variety of forms. See, for example, Whittaker, *Proc. Edin. Math. Soc.* 32 (1914), p. 75.

† The method was originally applied by Euler to the Riccati differential equation.

where Q_0 and P_1 are functions of x. The equation is differentiated and becomes

$$y'=Q_1y''+P_2y''',$$

where

$$Q_1=\frac{Q_0+P_1'}{1-Q_0'},\quad P_2=\frac{P_1}{1-Q_0'}.$$

This process is repeated indefinitely, and a set of relations

$$y^{(n)}=Q_ny^{(n+1)}+P_{n+1}y^{(n+2)}$$

is obtained, where $n=1, 2, 3, \ldots$, and

$$Q_n=\frac{Q_{n-1}+P_n'}{1-Q'_{n-1}},\quad P_{n+1}=\frac{P_n}{1-Q'_{n-1}}.$$

Then

$$\begin{aligned}\frac{y}{y'}&=Q_0+P_1\Big/\frac{y'}{y''}\\ &=Q_0+\cfrac{P_1}{Q_1+P_2\Big/\frac{y''}{y'''}}\\ &=\ldots\ldots\\ &=Q_0+\frac{P_1}{Q_1+}\,\frac{P_2}{Q_2+}\,\cdots\,\frac{P_n}{+Q_n+R_n},\end{aligned}$$

where

$$R_n=P_{n+1}\Big/\frac{y^{(n+1)}}{y^{(n+2)}}.$$

It is therefore natural to consider the continued fraction *

$$\text{(A)}\qquad \frac{1}{Q_0+}\,\frac{P_1}{Q_1+}\,\frac{P_2}{Q_2+}\,\cdots\,\frac{P_n}{+Q_n+}\cdots;$$

if it terminates it will represent the logarithmic derivative of a solution of the equation; if it does not terminate the problem of its convergence arises. This question is settled by the following theorem, which is fundamental in the theory of continued fractions.† *The continued fraction* (A) *converges and has the value y'/y if $y\neq 0$ and* (i) $P_n\to P,\ Q_n\to Q$ *as* $n\to\infty$, (ii) *the roots* ρ_1 *and* ρ_2 *of the equation* $\rho^2=Q\rho+P$ *are of unequal modulus, and* (iii) *if* $|\rho_2|<|\rho_1|$ *then*

$$\lim |y^{(n)}|^{\frac{1}{n}}<|\rho_2|^{-1}$$

provided that $|\rho_2|\neq 0$.

When $|\rho_2|=0$ the last condition is replaced by the condition that the limit is finite.

7·501. An example of a terminating Continued Fraction.—In the case of the equation

$$y=\frac{x}{m}y'+\frac{1}{m}y'',$$

where m is a positive integer, the derived equations are

$$y^{(n)}=\frac{x}{m-n}y^{(n+1)}+\frac{1}{m-n}y^{(n+2)}\qquad (n=1, 2, \ldots m-1),$$
$$0=xy^{(m+1)}+y^{(m+2)}.$$

* A similar continued fraction may frequently be obtained by integrating instead of differentiating.

† A proof of this theorem will be found in Perron, *Die Lehre von den Kettenbrüchen*, § 57.

It follows that

$$\frac{y'}{y}=\frac{m}{x+}\,\frac{m-1}{x+}\,\frac{m-2}{x+}\cdots\frac{1}{+x}.$$

Since the continued fraction terminates, it may explicitly be evaluated by calculating its successive convergents,* and it is found that

$$\frac{y'}{y}=\frac{mx^{m-1}+(m-2)a_1x^{m-3}+(m-4)a_2x^{m-5}+\ldots}{x^m+a_1x^{m-2}+a_2x^{m-4}+\ldots},$$

where

$$a_r=\frac{n!}{2^r r!(n-2r)!}.$$

Thus, as may be verified directly, the equation has the polynomial solution

$$y=x^m+a_1x^{m-2}+a_2x^{m-4}+\ldots.$$

7·51. The Function, ${}_1F_1(\alpha;\gamma;x)$ and the associated Continued Fraction.—The function †

$${}_1F_1(\alpha;\gamma;x)=1+\sum_{r=1}^{\infty}\frac{(\alpha)_r}{r!(\gamma)_r}x^r,$$

where

$$(\alpha)_r=\alpha(\alpha+1)\ldots(\alpha+r-1),$$

is a solution of the equation

$$\alpha y=(\gamma-x)y'+xy'';$$

when γ is not an integer an independent second solution is

$$x^{1-\gamma}{}_1F_1(\alpha-\gamma+1;2-\gamma;x).$$

The series terminates when α is zero or a negative integer; this case is of no new interest and will be put aside. When the series is multiplied by $1/\Gamma(\gamma)$ its coefficients are always finite, and the function vanishes only when $\gamma-\alpha$ as well as α is zero or a negative integer. This case also will be excluded.

Now let

$$Y=\frac{{}_1F_1(\alpha;\gamma;x)}{\Gamma(\gamma)},$$

then

$$Y^{(n)}=\frac{\gamma-x+n}{\alpha+n}Y^{(n+1)}+\frac{x}{\alpha+n}Y^{(n+2)}\qquad(n=1,2,3,\ldots).$$

All the derivatives $Y^{(n)}$ cannot vanish, for if $Y^{(m+1)}$ and $Y^{(m+2)}$ were to vanish when $x=x_0$, it would follow from the above relation that $Y^{(m)}$, $Y^{(m-1)}$, and finally Y itself would vanish when $x=x_0$. Thus Y would vanish identically, which except in the excluded cases is not true.

It may be verified directly that

$$Y'(\alpha;\gamma;x)=\alpha Y(\alpha+1;\gamma+1;x),$$

* Chrystal, *Algebra*, II., Chap. XXXIV.

† This function was first considered by Kummer, *J. für Math.* 15 (1836), p. 139; the notation is due to Barnes, *Trans. Camb. Phil. Soc.* 20 (1906), p. 253. The confluent hypergeometric functions are closely allied; in Whittaker's notation

$$M_{k,m}(x)=x^{\frac{1}{2}+m}e^{-\frac{1}{2}x}\cdot{}_1F_1(\tfrac{1}{2}+m-k;\ 2m+1;\ x);$$

see Whittaker and Watson, *Modern Analysis*, Chap. XVI. The Bessel functions are particular cases, in fact

$$J_n(x)=\frac{(\frac{1}{2}x)^n e^{-ix}}{\Gamma(n+1)}\,{}_1F_1(n+\tfrac{1}{2};\ 2n+1;\ 2ix).$$

and in general,

$$Y^{(n)}(\alpha;\gamma;x)=(\alpha)_n Y(\alpha+n;\gamma+n;x)$$
$$=\frac{(\alpha)_n}{\Gamma(\gamma+n)}\left\{1+\frac{\alpha+n}{\gamma+n}\cdot\frac{x}{1!}+\frac{(\alpha+n)(\alpha+n+1)}{(\gamma+n)(\gamma+n+1)}\cdot\frac{x^2}{2!}+\dots\right\}$$

Let m be a positive integer such that

$$m>|\alpha|,\quad m>2|\gamma|,$$

then, if $n\geqslant m$,

$$\left|\frac{\alpha+n}{\gamma+n}\right|\leqslant\frac{n+|\alpha|}{n-|\gamma|}$$
$$<\frac{n+m}{n-\frac{1}{2}m}<\frac{2n}{\frac{1}{2}n}=4,$$

and, *a fortiori*, if $r\geqslant 1$

$$\left|\frac{\alpha+n+r}{\gamma+n+r}\right|<4.$$

Consequently, when $n\geqslant m$,

$$|Y^{(n)}|=\left|\frac{(\alpha)_m}{\Gamma(\gamma+m)}\right|\cdot\left|\frac{(\alpha+m)\dots(\alpha+n-1)}{(\gamma+m)\dots(\gamma+n-1)}\right|\cdot\left|1+\frac{\alpha+n}{\gamma+n}\cdot\frac{x}{1!}+\dots\right|$$
$$<\left|\frac{(\alpha)_m}{\Gamma(\gamma+m)}\right|\cdot 4^{n-m}\left(1+\frac{|4x|}{1!}+\frac{|4x|^2}{2!}+\dots\right)$$
$$=\left|\frac{(\alpha)_m}{\Gamma(\gamma+m)}\right|4^{n-m}e^{4|x|},$$

and therefore $|Y^{(n)}|^{1/n}$ is finite. But the equation for ρ is

$$\rho^2=\rho,$$

and $\rho_2=0$. It follows that the continued fraction *

$$\frac{1}{\frac{\gamma-x}{\alpha}}+\frac{\frac{x}{\alpha}}{\frac{\gamma-x+1}{\alpha+1}}+\frac{\frac{x}{\alpha+1}}{\frac{\gamma-x+2}{\alpha+2}}+\dots$$

or

$$\frac{\alpha}{\gamma-x+}\,\frac{(\alpha+1)x}{\gamma-x+1+}\,\frac{(\alpha+2)x}{\gamma-x+2+}\dots$$

converges and has the value

$$\frac{d}{dx}\{\log {}_1F_1(\alpha;\gamma;x)\}$$

for all values of x for which the latter function is finite.

The hypergeometric equation may be treated in a somewhat similar way, but the results obtained are by no means as simple as in the above case. The main result is that, for real values of x, the continued fraction

$$\frac{\alpha\beta}{\gamma-(\alpha+\beta+1)x+}\,\frac{(\alpha+1)(\beta+1)x(1-x)}{\gamma+1-(\alpha+\beta+3)x+}\dots\frac{(\alpha+r)(\beta+r)x(1-x)}{+\gamma+r-(\alpha+\beta+2r+1)x+}\dots$$

converges to the value $\frac{d}{dx}\log F(\alpha,\beta;\gamma;x)$ when $x<\frac{1}{2}$, and to the value $\frac{d}{dx}\log F(\alpha,\beta;\alpha+\beta-\gamma+1;1-x)$ when $x>\frac{1}{2}$.†

* Perron, *Rend. Circ. Mat. Palermo*, 29 (1910), p. 124.

† Ince, *Proc. London Math. Soc.* (2), 18 (1919), p. 236.

7·511. Continued Fractions and Legendre Functions.—It may be verified that, if y_n is a Legendre function $Q_n(x)$ of degree n,*

$$y_1 - x\,y_0 + 1 = 0,$$
$$(n+2)y_{n+2} - (2n+3)xy_{n+1} + (n+1)y_n = 0 \qquad (n=0, 1, 2, 3, \ldots).$$

These recurrence relations lead to the infinite continued fraction

$$y_0 = \frac{1}{x-}\,\frac{1^2}{3x-}\,\frac{2^2}{5x-}\,\frac{3^2}{7x-}\,\cdots$$

the convergence and significance of which will now be investigated.

Since, as $n \to \infty$,

$$\frac{2n+3}{n+1}x \to 2x, \qquad \frac{n+2}{n+1} \to 1,$$

the equation in ρ is

$$\rho^2 = 2x\rho - 1$$

and

$$\rho_2 = x - \sqrt{(x^2-1)}.$$

The continued fraction will therefore converge and have the value y_0 if

$$\lim\, |\,y_n\,|^{\frac{1}{n}} < \left|\frac{1}{x-\sqrt{(x^2-1)}}\right| = |\,x+\sqrt{(x^2-1)}\,|.$$

Now, since

$$Q_n(x) = \frac{2^n(n\,!)^2}{(2n+1)\,!}x^{-n-1} + O(x^{-n-3}),$$

$$\lim\, Q_{n+1}(x)\,/\,Q_n(x) = \frac{1}{2x}$$

and therefore †

$$\lim\, |\,Q_n(x)\,|^{\frac{1}{n}} = \frac{1}{2\,|\,x\,|}.$$

Thus when

$$\left|x+\sqrt{(x^2-1)}\right| > \frac{1}{2\,|\,x\,|},$$

or at least when $|\,x\,| > 1$, y_0 can be identified with $Q_0(x)$, and therefore

$$Q_0(x) = \frac{1}{x-}\,\frac{1^2}{3x-}\,\frac{2^2}{5x-}\,\frac{3^2}{7x-}\,\cdots.$$

Now (§ 7·241), since

$$Q_n(x) = \tfrac{1}{2}P_n(x)\log\frac{x+1}{x-1} - R_n(x)$$
$$= P_n(x)Q_0(x) - R_n(x),$$

where R_n is a polynomial of degree $n-1$,

$$\frac{R_n(x)}{P_n(x)} = Q_0(x) - \frac{Q_n(x)}{P_n(x)}$$
$$= Q_0(x) + O(x^{-2n-1}).$$

It follows that the convergents of the continued fraction for $Q_0(x)$ are

$$\frac{R_1(x)}{P_1(x)},\ \frac{R_2(x)}{P_2(x)},\ \ldots,\ \frac{R_n(x)}{P_n(x)},\ \ldots$$

This result furnishes a practical method of evaluating the polynomials $R_n(x)$

* These recurrence-relations are also satisfied by $y_n = P_n(x)$ except the first, which is evidently not satisfied.

† Bromwich, *Infinite Series*, Appendix I., p. 421.

Miscellaneous Examples.

1. Find a series which satisfies the differential equation

$$(1+x)\frac{dy}{dx}=my.$$

Prove from the differential equation that if $f(m)$ is the solution which reduces to unity when $x=0$ then, for all values of x,

$$f(m_1)f(m_2)=f(m_1+m_2).$$

2. Show that the function

$$O_n(x)=\frac{2^{n-1}n!}{x^{n+1}}\left\{1+\frac{x^2}{2.(2n-2)}+\frac{x^4}{2.4.(2n-2)(2n-4)}+\ldots\right\}$$

satisfies the equation

$$\frac{d^2y}{dx^2}+\frac{3}{x}\cdot\frac{dy}{dx}+\left\{1-\frac{n^2-1}{x^2}\right\}y=\frac{1}{x} \text{ when } n \text{ is an even positive integer,}$$

$$=\frac{n}{x^2} \text{ when } n \text{ is an odd positive integer.}$$

[Edinburgh, 1912.]

3. Find two independent series of ascending powers of x which satisfy the differential equation

$$\frac{d^2y}{dx^2}+xy=0.$$

Show that the equation is also satisfied by an asymptotic expansion of the form

$$e^{\mu}x^{-\frac{1}{4}}v$$

where $\mu=\frac{2}{3}ix^{\frac{3}{2}}$ and v is a series of descending powers of $x^{\frac{3}{2}}$. [Edinburgh, 1914.]

4. Show that the following functions satisfy the hypergeometric equation

$$\text{(i)}\ (1-x)^{\gamma-\alpha-\beta}F(\gamma-\alpha,\ \gamma-\beta\ ;\ \gamma\ ;\ x),$$

$$\text{(ii)}\ x^{1-\gamma}(1-x)^{\gamma-\alpha-\beta}F(1-\alpha,\ 1-\beta\ ;\ 2-\gamma\ ;\ x).$$

Transform the equation by taking in succession as new independent variables

$$z=1-x,\quad z=1/x,\quad z=1/(1-x),\quad z=x/(x-1),\quad z=(x-1)/x$$

and write down four solutions in each of the new variables. Show that the aggregate of twenty-four solutions may be grouped into six classes, such that the members of each class are equal or are constant multiples of one another. [Kummer.]

5. Prove that, when m is a positive integer and $-1<x<1$, the associated Legendre equation

$$(1-x^2)\frac{d^2y}{dx^2}-2x\frac{dy}{dx}+\left\{n(n+1)-\frac{m^2}{1-x^2}\right\}y=0$$

is satisfied by the *associated Legendre functions*

$$P_n^m(x)=(1-x^2)^{\frac{1}{2}m}\frac{d^mP_n(x)}{dx^m},\quad Q_n^m(x)=(1-x^2)^{\frac{1}{2}m}\frac{d^mQ_n(x)}{dx^m}.$$ [Ferrers.]

Obtain and identify descending series which satisfy the equation.

6. Show that if $C_\nu^\mu(x)$ is the coefficient of h^ν in the expansion of $(1-2xh+h^2)^{-\mu}$ in ascending powers of h, then $C_\nu^\mu(x)$ satisfies the differential equation

$$\frac{d^2y}{dx^2}+\frac{(2\mu-1)x}{x^2-1}\cdot\frac{dy}{dx}-\frac{\nu(\nu+2\mu)}{x^2-1}y=0,$$

and express $C_\nu^\mu(x)$ as an associated Legendre function.

7. Show that the differential equation for $C_\nu^\mu(x)$ is defined by the scheme

$$P\left\{\begin{matrix} -1 & \infty & 1 & \\ 0 & \nu+2\mu & 0 & x \\ \frac{1}{2}-\mu & -\nu & \frac{1}{2}-\mu & \end{matrix}\right\}.$$

8. Prove that the differential equation

$$(1-x)x^2\frac{d^3y}{dx^3}+\{\theta+\epsilon+1-(\alpha+\beta+\gamma+3)x\}x\frac{d^2y}{dx}$$
$$+\{\theta\epsilon-(\alpha\beta+\beta\gamma+\gamma\alpha+\alpha+\beta+\gamma+1)x\}\frac{dy}{dx}-\alpha\beta\gamma=0$$

is satisfied by a function ${}_3F_2(\alpha,\ \beta,\ \gamma;\ \theta,\ \epsilon;\ x)$ whose series development is

$$1+\frac{\alpha\beta\gamma}{1!\,\theta\epsilon}x+\frac{\alpha(\alpha+1)\,.\,\beta(\beta+1)\,.\,\gamma(\gamma+1)}{2!\,\theta(\theta+1)\,.\,\epsilon(\epsilon+1)}x^2+\ .\ .\ .$$

9. Prove that, when n is not an integer,

$$\frac{d}{dx}\left\{\frac{J_{-n}(x)}{J_n(x)}\right\}=\frac{-2\sin n\pi}{\pi x\{J_n(x)\}^2},$$

$$J_n(x)J_{1-n}(x)+J_{n-1}(x)J_{-n}(x)=\frac{2\sin n\pi}{\pi x}.$$ [Lommel.]

10. Show that when n is half an odd integer, the Bessel equation admits of solution in a finite form, and that

$$J_{\frac{1}{2}}(x)=\left(\frac{2}{\pi x}\right)^{\frac{1}{2}}\sin x,$$

$$J_{k+\frac{1}{2}}(x)=\frac{(-1)^k(2x)^{k+\frac{1}{2}}}{\pi^{\frac{1}{2}}}\cdot\frac{d^k}{d(x^2)^k}\left(\frac{\sin x}{x}\right),$$

and obtain the general solution in each case.

11. Show that the general solution of the equation

$$4\frac{d^2y}{dx^2}+9xy=0$$

may be written in the form

$$y=Ax^{\frac{1}{2}}J_{\frac{1}{3}}(x^{\frac{3}{2}})+Bx^{\frac{1}{2}}J_{-\frac{1}{3}}(x^{\frac{3}{2}}).$$

12. Show that the equation

$$\frac{d^2y}{dx^2}-\left\{n^2-\frac{m(m+1)}{x^2}\right\}y=0,$$

is integrable in terms of Bessel functions, and that, when m is a positive integer, it admits of the following general solution :

$$y=x^{m+1}\left\{x^{-1}\frac{d}{dx}\right\}^m\left\{\frac{Ae^{nx}+Be^{-nx}}{x}\right\},$$

where A and B are arbitrary constants.

13. Find ascending and descending (asymptotic) series-solutions for the confluent hypergeometric equation

$$\frac{d^2y}{dx^2}+\left\{-\frac{1}{4}+\frac{k}{x}+\frac{\frac{1}{4}-m^2}{x^2}\right\}y=0,$$

and show that, when $k=0$, a solution is

$$y=x^{\frac{1}{2}}J_n(\tfrac{1}{2}ix).$$

14. Show that if $W_{k,m}(x)$ is a solution of the confluent hypergeometric equation, the function

$$x^{\frac{1}{2}}W_{\frac{1}{2}n+\frac{1}{4},\ -\frac{1}{4}}(\tfrac{1}{2}x^2)$$

satisfies the Weber equation

$$\frac{d^2y}{dx^2}+(n+\tfrac{1}{2}-\tfrac{1}{4}x^2)y=0.$$

Solutions of this equation are known as the Weber-Hermite or parabolic-cylinder functions and are denoted by $D_n(x)$. Verify the asymptotic relationship

$$D_n(x)\sim e^{-\frac{1}{4}x^2}x^n\left\{1-\frac{n(n-1)}{2x^2}+\frac{n(n-1)(n-2)(n-3)}{2\,.\,4x^4}-\ .\ .\ .\right\}$$

and show that $D_{-n-1}(ix)$ is an independent solution.

15. By considering the differential equation

$$\alpha y = -xy' + y'',$$

show that when $\alpha > 0$, $x > 0$,

$$\frac{\alpha}{x+}\,\frac{\alpha+1}{x+}\,\frac{\alpha+2}{x+}\,\frac{\alpha+3}{x}+\cdots=\frac{\int_0^\infty e^{-xt-\frac{1}{2}t^2}t^\alpha\,dt}{\int_0^\infty e^{-xt-\frac{1}{2}t^2}t^{\alpha-1}\,d}$$

16. Show that the substitution $y = e^x u$ transforms the equation

$$\alpha y = (\gamma - x)y' + xy''$$

into

$$(\alpha - \gamma)u = (\gamma + x)u' + xu'',$$

and hence prove that

$$\frac{\alpha x}{\gamma - x+}\,\frac{(\alpha+1)x}{\gamma - x + 1+}\,\frac{(\alpha+2)x}{\gamma - x + 2}+\cdots$$

$$= x - \frac{(\gamma-\alpha)x}{\gamma + x -}\,\frac{(\gamma-\alpha-1)x}{\gamma + x + 1 -}\,\frac{(\gamma-\alpha-2)x}{\gamma + x + 2 -}\cdots$$

[Perron.]

17. Show that, if $D_n(x)$ is the Weber-Hermite function,

$$\frac{D_n'(x)}{D_n(x)} = \frac{n}{x-}\,\frac{n-1}{x-}\,\frac{n-2}{x-}\cdots,$$

and that

$$\frac{D'_{-n-1}(ix)}{D_{-n-1}(ix)} = x - \frac{n+1}{x-}\,\frac{n+2}{x-}\,\frac{n+3}{x-}\cdots.$$

CHAPTER VIII

THE SOLUTION OF LINEAR DIFFERENTIAL EQUATIONS BY DEFINITE INTEGRALS

8·1. The General Principle.—The object which is now in view is to obtain a definite integral of the form

(A) $$y(x)=\int_{\alpha}^{\beta} K(x, t)v(t)dt,$$

wherein x enters as a parameter, to satisfy the given linear differential equation

(B) $$L_x(y)=0.$$

There are three distinct elements in the definite integral which have to be chosen as circumstances demand, namely :

(i) the function $K(x, t)$, which will be known as the *nucleus* of the definite integral,

(ii) the function $v(t)$,

(iii) the limits of integration, α and β.

Now let it be supposed that the nucleus $K(x, t)$ can be found to satisfy a partial differential equation of the form *

(C) $$L_x(K)=M_t(K),$$

where M_t is a linear differential operator involving only t and $\frac{\partial}{\partial t}$.

Then, if it is permissible to apply the operator L_x to the definite integral $y(x)$,†

$$\begin{aligned} L_x\{y(x)\} &= \int_{\alpha}^{\beta} L_x\{K(x, t)\}v(t)dt \\ &= \int_{\alpha}^{\beta} M_t\{K(x, t)\}v(t)dt. \end{aligned}$$

Let $\overline{M}_t$ be the operator adjoint to M_t, then from the Lagrange identity (§ 5·3) which is here of the form ‡

$$v(t)M_t\{K(x, t)\}-K(x, t)\overline{M}_t\{v(t)\}=\frac{\partial}{\partial t}P\{K, v\},$$

it follows that

$$L_x\{y(x)\}=\int_{\alpha}^{\beta} K(x, t)\overline{M}_t(v)dt+\Big[P\{K, v\}\Big]_{t=\alpha}^{t=\beta}$$

In order that the integral (A) may be a solution of the equation (B), the right-hand member of this last equation must be zero. Such is the case if, in the first place, $v(t)$ is a solution of the equation

$$\overline{M}_t(v)=0,$$

* Bateman, *Trans. Camb. Phil. Soc.* 21 (1909), p. 171.

† This assumption will be made throughout the present chapter.

‡ The bilinear concomitant $P\{K, v\}$ here involves x as a parameter.

and secondly, if the limits of integration are so chosen that

$$\Big[P\{K, v\}\Big]_{t=\alpha}^{t=\beta}=0$$

identically.

This method admits of considerable generalisation. Thus, for instance, let it be supposed, not that the nucleus $K(x, t)$ satisfies the partial differential equation (C), but merely that two functions $K(x, t)$ and $\kappa(x, t)$ can be found such that

$$L_x\{K(x, t)\}=M_t\{\kappa(x, t)\},$$

then

$$L_x\{y(x)\}=\int_\alpha^\beta \kappa(x, t)\bar{M}_t(v)dt+\Big[P\{\kappa, v\}\Big]_{t=\alpha}^{t=\beta}$$

and it is now necessary to find the function $v(t)$ and the limits of integration α and β as before.

8·2. The Laplace Transformation.—If, in the operator L_x, each coefficient is of degree m at most, and the operator itself is of order n, L_x may be written in the extended form

(A) $$L_x\equiv\sum_{r=0}^{n}\sum_{s=0}^{m} a_{rs}x^s\frac{\partial^r}{\partial x^r},$$

in which the coefficients a_{rs} are constants.

Consider, together with L_x, the operator

(B) $$M_t\equiv\sum_{r=0}^{n}\sum_{s=0}^{m} a_{rs}t^r\frac{\partial^s}{\partial t^s},$$

then

$$L_x(e^{xt})\equiv M_t(e^{xt}),$$

for each member of this identity is

$$e^{xt}\sum_{r=0}^{n}\sum_{s=0}^{m} a_{rs}x^s t^r.$$

Consequently the equation

$$L_x(y)=0,$$

is satisfied by the definite integral

(C) $$y(x)=\int_\alpha^\beta e^{xt}v(t)dt,$$

provided that $v(t)$ satisfies the differential equation

(D) $$\bar{M}_t(v)=0,$$

and that the limits of integration are so chosen that

$$\Big[P\{e^{xt}, v\}\Big]_{t=\alpha}^{t=\beta}=0$$

identically.

The equation (D) is known as the *Laplace-transform* of $L_x(y)=0$, and e^{xt} as the nucleus of the transformation from $v(t)$ into $y(x)$. The success of the method as a means of obtaining an explicit solution of the given equation depends primarily upon the readiness with which a solution of (D) is obtainable. In the particular and very special case in which $m=1$, that is to say, when the coefficients of the given equation are linear in x, the Laplace-transform is a linear equation of the first order and may therefore be integrated by quadratures.*

* See Example 1, p. 201.

An important reciprocal relationship * exists between the equations $L_x(y)=0$ and $\bar{M}_t(v)=0$, namely, that the former is the Laplace-transform of the latter, the nucleus of the transformation being e^{-xt}. This follows at once from the identity

$$L_x(e^{-xt})=\bar{M}_t(e^{-xt}).$$

Since

$$L_x(u)\equiv\sum_{r=0}^{n}\sum_{s=0}^{m}(-1)^r a_{s\,r}\frac{d^r(x^s u)}{dx^r},$$

$$\bar{M}_t(v)\equiv\sum_{r=0}^{n}\sum_{s=0}^{m}(-1)^s a_{s\,r}\frac{d^s(t^r v)}{dt^s},$$

it is sufficient to prove that

$$(-1)^r\frac{d^r(x^s e^{-xt})}{dx^r}=(-1)^s\frac{d^s(t^r e^{-xt})}{dt^s},$$

and this is true since each member of the equation is equal to

$$e^{-xt}\Big\{x^s t^r-rsx^{s-1}t^{r-1}+\frac{r(r-1)s(s-1)}{2\,!}x^{s-2}t^{r-2}$$
$$-\frac{r(r-1)(r-2)s(s-1)(s-2)}{3\,!}x^{s-3}t^{r-3}+\dots\Big\}.$$

It follows that, if γ and δ are appropriately chosen,

(E) $$v(t)=\int_\gamma^\delta e^{-xt}y(x)dx$$

is a solution of (D). The relationship between (C) and (E) furnishes an example of the *inversion* of a definite integral, that is to say the determination of an unknown function $v(t)$ in the integrand, so that the definite integral may represent the function $y(x)$ which is now supposed to be known.

8·201. Example illustrating the Laplace Transformation.—Let

$$L_x(y)\equiv x\frac{d^2y}{dx^2}+(p+q+x)\frac{dy}{dx}+py=0,$$

then

$$M_t(u)\equiv t(t+1)\frac{du}{dt}+\{p+(p+q)t\}u,$$

$$\bar{M}_t(v)\equiv -t(t+1)\frac{dv}{dt}+\{p-1+(p+q-2)t\}v,$$

and

$$vM_t(u)-u\bar{M}_t(v)=\frac{d}{dt}[t(t+1)uv].$$

The equation $\bar{M}_t(v)=0$ possesses the solution

$$v(t)=t^{p-1}(t+1)^{q-1},$$

and therefore an integral of the type

$$y(x)=\int_\alpha^\beta e^{xt}t^{p-1}(t+1)^{q-1}dt$$

will satisfy the equation $L_x(y)=0$ provided that α and β can be so chosen that

$$\Big[e^{xt}t^p(t+1)^q\Big]_{t=\alpha}^{t=\beta}=0$$

identically.

* Petzval, *Integration der linearen Differentialgleichungen*, 1 (Vienna, 1851), p. 472.

It is convenient to write $-t$ for t. Then the integral

$$y(x)=\int_{\alpha}^{\beta} e^{-xt}t^{p-1}(1-t)^{q-1}dt$$

satisfies $L_x(y)=0$ if α and β are such that

$$\Big[e^{-xt}t^{p}(1-t)^{q}\Big]_{t=\alpha}^{t=\beta}$$

vanishes identically. Appropriate pairs of values are

$$\begin{array}{llll} \text{(i)} & \alpha=0, & \beta=1 & (p>0,\quad q>0), \\ \text{(ii)} & \alpha=0, & \beta=\infty & (x>0,\quad p>0), \\ \text{(iii)} & \alpha=1, & \beta=\infty & (x>0,\quad q>0), \\ \text{(iv)} & \alpha=-\infty, & \beta=0 & (x<0,\quad p>0), \\ \text{(v)} & \alpha=-\infty, & \beta=1 & (x<0,\quad q>0). \end{array}$$

Thus required values of α and β exist in all cases except when p and q are both negative. In particular when p, q and x are all positive, the general solution of $L_x(y)=0$ can be written

$$y=A\int_0^1 e^{-xt}t^{p}(1-t)^{q}dt+B\int_1^{\infty} e^{-xt}t^{p}(1-t)^{q}dt,$$

where A and B are arbitrary constants.

8·21. Determination of the Limits of Integration.—The equation $\overline{M}(v)=0$, which serves to determine $v(t)$, is of order m; its general solution is of the form

$$v=C_1v_1(t)+C_2v_2(t)+\ .\ .\ .\ +C_mv_m(t),$$

where $v_1, v_2, \ldots, v_m$ form a fundamental set of solutions and the constants $C_1, C_2, \ldots, C_m$ are arbitrary. These constants and the limits of integration α and β have to be so determined that the expression

$$\Big[P\{K, v\}\Big]_{t=\alpha}^{t=\beta}$$

vanishes identically.

Now it will be seen from the form of the bilinear concomitant (§ 5·3) that it is sufficient to determine the constants $C_1, \ldots, C_m$, α and β so that

$$v(t),\quad v'(t),\ \ldots,\quad v^{(m-1)}(t)$$

vanish when $t=\alpha$ and $t=\beta$. Such cannot be the case unless α and β are singular points of $\overline{M}(v)=0$. But if α and β are singular points, and a solution $v(t)$ exists such that the exponent relative to each of these points is greater than $m-1$, the bilinear concomitant vanishes at α and at β and therefore the limits of integration may be taken to be α and β. This case is of practical importance, and is illustrated by the example of the preceding section. Every distinct pair of limits, if distinct pairs exist, leads to a distinct particular solution of the equation. In some cases a sufficient number of definite integrals is available to build up the general solution, in others only a partial solution is attained.

8·22. Definite-Integral Expressions for the Bessel Functions.—A function which may be taken, instead of e^{xt} as the nucleus of a definite integral is

$$K(x, t)=e^{\frac{1}{2}x(t-t^{-1})}.$$

Now the two functions $e^{\frac{1}{2}xt}$ and $e^{-\frac{1}{2}xt^{-1}}$ may be expanded respectively in ascending powers of xt and xt^{-1} which converge absolutely for all values of x and all non-zero values of t. The double series which represents their product therefore converges for the same values of x and t, and is as follows :

$$e^{\frac{1}{2}x(t-t^{-1})}=\sum_{r=0}^{\infty}\sum_{s=0}^{\infty}\frac{(-1)^s x^{r+s}t^{r-s}}{2^{r+s}\,r!\,s!}.$$

When $n>0$, the coefficient of t^n is obtained by selecting those terms of the double series, for which $r=n+s$. These terms form a singly infinite series, namely,

$$\sum_{s=0}^{\infty}\frac{(-1)^s x^{n+2s}}{2^{n+2s}(n+s)!\;\;s!}=J_n(x),$$

where $J_n(x)$ is the Bessel function of order n. Similarly, the coefficient of t^{-n} is $(-1)^n J_n(x)$. Thus

$$e^{\frac{1}{2}x(t-t^{-1})}=J_0(x)+\sum_{n=1}^{\infty}\{t^n+(-1)^n t^{-n}\}J_n(x).$$

Now let $t=e^{i\theta}$, and this relation becomes

$$e^{ix\sin\theta}=J_0(x)+2\sum_{m=1}^{\infty}J_{2m}(x)\cos 2m\theta+2i\sum_{m=1}^{\infty}J_{2m-1}(x)\sin(2m-1)\theta.$$

By separating real and imaginary parts, the following two expressions are obtained :

$$\cos(x\sin\theta)=J_0(x)+2\sum_{m=1}^{\infty}J_{2m}(x)\cos 2m\theta,$$

$$\sin(x\sin\theta)=2\sum_{m=1}^{\infty}J_{2m-1}(x)\sin(2m-1)\theta.$$

By changing θ into $\frac{1}{2}\pi-\theta$ it follows that

$$\cos(x\cos\theta)=J_0(x)+2\sum_{m=1}^{\infty}(-1)^m J_{2m}(x)\cos 2m\theta,$$

$$\sin(x\cos\theta)=2\sum_{m=1}^{\infty}(-1)^{m+1}J_{2m-1}(x)\cos(2m-1)\theta.$$

From the first of these four relations it follows that

$$\int_0^{\pi}\cos(x\sin\theta)\cos n\theta\,d\theta=\pi J_n(x)\quad\text{when } n \text{ is even},$$
$$=0\qquad\text{when } n \text{ is odd},$$

and from the second it follows that

$$\int_0^{\pi}\sin(x\sin\theta)\sin n\theta\,d\theta=\pi J_n(x)\quad\text{when } n \text{ is odd},$$
$$=0\qquad\text{when } n \text{ is even}.$$

By addition it follows that when n is any positive integer, or zero,

$$\int_0^{\pi}\cos(n\theta-x\sin\theta)d\theta=\pi J_n(x).$$

Thus the ordinary Bessel function with integer suffix is expressed as a definite integral.*

8·3. The Nucleus $K(x-t)$.—Consider the possibility of satisfying a linear differential equation of the Laplace type

$$L_x(y)\equiv\left\{xF\left(\frac{d}{dx}\right)+G\left(\frac{d}{dx}\right)\right\}y=0$$

* Bessel, *Abh. Akad. Wiss. Berlin*, 1824, p. 34.

by a definite integral of the form *

$$y(x)=\int_{a}^{\beta}K(x-t)v(t)dt.$$

It is clear that $K(x-t)$ will satisfy a partial differential equation of the form

$$\text{(A)}\qquad \left\{xF\left(\frac{\partial}{\partial x}\right)+G\left(\frac{\partial}{\partial x}\right)\right\}K=\left\{tF\left(-\frac{\partial}{\partial t}\right)+H\left(-\frac{\partial}{\partial t}\right)\right\}K$$

provided that $K(z)$, regarded as a function of the single variable z, satisfies the ordinary linear equation

$$\left\{zF\left(\frac{d}{dz}\right)+G\left(\frac{d}{dz}\right)-H\left(\frac{d}{dz}\right)\right\}K(z)=0.$$

If, therefore, $v(t)$ is a solution of the equation

$$F\left(\frac{d}{dt}\right)tv+H\left(\frac{d}{dt}\right)v=0,$$

the left-hand member of which is the adjoint expression of the right-hand member of (A), and if the limits of integration can be suitably chosen, the given equation has a solution expressible as a definite integral of the specified type.

8·31. The Euler Transformation.—A frequently-occurring instance of a nucleus of the type studied in the preceding section is

$$K(x-t)=(x-t)^{-\nu-1}.$$

The transformation of which $(x-t)^{-\nu-1}$ is the nucleus is adaptable to any linear differential equation in which the coefficient of $y^{(r)}$ is a polynomial in x of degree r. Such an equation may always be written in the form

$$\begin{aligned}L_x(y)\equiv G_0(x)\frac{d^ny}{dx^n}&-\mu G_0'(x)\frac{d^{n-1}y}{dx^{n-1}}+\frac{\mu(\mu+1)}{1\,.\,2}G_0''(x)\frac{d^{n-2}y}{dx^{n-2}}-\ldots\\&-G_1(x)\frac{d^{n-1}y}{dx^{n-1}}+(\mu+1)G_1'(x)\frac{d^{n-2}y}{dx^{n-2}}-\ldots\\&+G_2(x)\frac{d^{n-2}y}{dx^{n-2}}-\ldots\\&-\ldots=0,\end{aligned}$$

or

$$\Gamma_0(y)-\Gamma_1(y)+\Gamma_2(y)-\ldots\pm\Gamma_p(y)=0,$$

where

$$\Gamma_r(y)=\sum_{s=0}^{n-r}(-1)^s\frac{(\mu+r)\ldots(\mu+r+s-1)}{s\,!}G_r^{(s)}(x)\frac{d^{n-r-s}y}{dx^{n-r-s}}.$$

In these expressions G_r is a polynomial of degree $n-r$ and μ is a constant It is supposed that the $p+1$ polynomials $G_0\ \ldots\ G_p$ suffice.

Now, writing $-\nu=n+\mu$ in the nucleus $K(x-t)$,

$$\begin{aligned}\Gamma_r\{(x-t)^{n+\mu-1}\}&=\sum_{s=0}^{n-r}(-1)^s\frac{(n+\mu-1)(n+\mu-2)\ldots(\mu+r)}{s\,!}G_r^{(s)}(x)(x-t)^{\mu+r+s-1}\\&=(n+\mu-1)\ldots(\mu+r)(x-t)^{\mu+r-1}\sum_{s=0}^{n-r}(-1)^s\frac{(x-t)^sG_r^{(s)}(x)}{s\,!}\\&=(n+\mu-1)\ldots(\mu+r)(x-t)^{\mu+r-1}G_r(t),\end{aligned}$$

* Cailler, *Bull. Sc. Math.* 34 (1899), p. 26; see also Mellin, *Acta Soc. Sc. Fenn.* 21 (1896), No. 6.

and therefore

$$L_x\{(x-t)^{n+\mu-1}\}=A\sum_{r=0}^{p}(-1)^r\frac{(x-t)^{\mu+r-1}}{\mu(\mu+1)\ .\ .\ .\ (\mu+r-1)}G_r(t),$$

where

$$A=(n+\mu-1)(n+\mu-2)\ .\ .\ .\ (\mu+1)\mu.$$

Now if

$$M_t(u)\equiv G_0(t)\frac{d^pu}{dt^p}+G_1(t)\frac{d^{p-1}u}{dt^{p-1}}+\ .\ .\ .\ +G_p(t)u,$$

then

$$M_t\{(x-t)^{p+\mu-1}\}=\sum_{r=0}^{p}(-1)^{p-r}(p+\mu-1)(p+\mu-2)\ .\ .\ .\ (\mu-r)(x-t)^{\mu+r-1}G_r(t)$$

$$=B\sum_{r=0}^{p}(-1)^r\frac{(x-t)^{\mu+r-1}}{\mu(\mu+1)\ .\ .\ .\ (\mu+r-1)}G_r(t),$$

where

$$B=(-1)^p(p+\mu-1)(p+\mu-2)\ .\ .\ .\ (\mu+1)\mu.$$

Consequently

$$L_x\{(x-t)^{n+\mu-1}\}=CM_t\{(x-t)^{p+\mu-1}\},$$

where $C=A/B$.

If, therefore,

$$y(x)=\int_a^\beta(x-t)^{n+\mu-1}v(t)dt,$$

then

$$L_x(y)=\int_a^\beta L_x\{(x-t)^{n+\mu-1}\}v(t)dt$$

$$=C\int_a^\beta M_t\{(x-t)^{p+\mu-1}\}v(t)dt$$

and now, as in the general case, $v(t)$ has to be chosen so that the integrand is a perfect differential, and thereafter the limits of integration have to be fixed. The determination of $v(t)$ involves the solution of the equation

$$\overline{M}_t(v)=0,$$

which is known as the *Euler-transform* of $L_x(y)=0$. When $p=1$ the Euler transform is a linear equation of the first order, and $v(t)$ can then be determined explicitly.*

8·311. An Example of the Euler Transformation.—Take, as an illustration, the case of the Legendre equation (§ 7·24),

$$(1-x^2)\frac{d^2y}{dx^2}-2x\frac{dy}{dx}+n(n+1)y=0.$$

In the notation of the preceding section,

$$G_0(x)=1-x^2,$$

$$\mu G_0'(x)+G_1(x)=2x,$$

$$\tfrac{1}{2}\mu(\mu+1)G_0''(x)+(\mu+1)G_1'(x)+G_2(x)=n(n+1).$$

These relations are satisfied by

$$G_1(x)=2(\mu+1)x,\quad G_2(x)=0,$$

provided that

$$\mu=n-1\quad\text{or}\quad\mu=-n-2.$$

* The full discussion involves the use of the complex variable and is postponed to § 18·4.

So in this case $p=1$, and the equation $M_t(u)=0$ becomes

$$(1-t^2)\frac{du}{dt}+2(\mu+1)tu=0.$$

The adjoint equation is

$$\overline{M}_t(v)\equiv(1-t^2)\frac{dv}{dt}-2(\mu+2)tv=0,$$

and has the solution $v(t)=(1-t^2)^{-\mu-2}$.

The limits of integration α and β are to be so chosen that

$$\Big[(x-t)^{\mu}(1-t^2)^{-\mu-1}\Big]_{t=\alpha}^{t=\beta}=0$$

identically. When $\mu=-n-2$, $n+1>0$ and $|x|>1$ this condition is satisfied by taking $\alpha=-1$, $\beta=+1$. Hence the definite integral

$$y(x)=\int_{-1}^{+1}(x-t)^{-n-1}(1-t^2)^n dt$$

satisfies the Legendre equation. In fact, if $Q_n(x)$ is the Legendre function of the second kind,*

$$Q_n(x)=\frac{1}{2^{n+1}}\int_{-1}^{1}(x-t)^{-n-1}(1-t^2)^n dt.$$

8·32. The Laplace Integrals.—It is possible, by modifying the path of integration, to obtain an integral expression for the Legendre function $P_n(x)$ similar to that which in the preceding example was stated to represent $Q_n(x)$. This cannot, however, be carried out without making use of the complex variable, and will be postponed to a later chapter.† In view of the importance of the Legendre polynomials, however, it is well at this point to interpolate a simple method by which they may be expressed as definite integrals.

Consider that branch of the function

$$(1-2xh+h^2)^{-\frac{1}{2}}$$

which has the value $+1$ when $h=0$. When $|h|$ is less than the smaller of $|x+(x^2-1)^{\frac{1}{2}}|$ and $|x-(x^2-1)^{\frac{1}{2}}|$, the function can be expanded as a power series in h, namely

$$P_0(x)+hP_1(x)+h^2P_2(x)+\ldots,$$

where $P_0(x)$, $P_1(x)$, $P_2(x)$, . . . are polynomials in x which will be proved to be the Legendre polynomials.

Now the equation

$$v=x+\tfrac{1}{2}h(v^2-1)$$

has a root

$$v=\frac{1-\sqrt{(1-2xh+h^2)}}{h}$$

which reduces to x when $h=0$ and which, when $|h|$ is sufficiently small, is developable in the form of the series

$$v=x+\sum_{n=1}^{\infty}\frac{h^n}{n!}\cdot\left(\frac{\partial^n v}{\partial h^n}\right)_0.$$

It is easily verified that

$$\frac{\partial v}{\partial h}=\tfrac{1}{2}(v^2-1)\frac{\partial v}{\partial x},$$

* Whittaker and Watson, *Modern Analysis*, § 15·3.

† § 18·5.

and that if $\phi(v)$ is any function of v,

$$\frac{\partial}{\partial h}\left\{\phi(v)\frac{\partial v}{\partial x}\right\}=\frac{\partial}{\partial x}\left\{\phi(v)\frac{\partial v}{\partial h}\right\}.$$

Let it be supposed that, for a certain integral value of n,

$$\frac{\partial^n v}{\partial h^n}=\frac{\partial^{n-1}}{\partial x^{n-1}}\left\{\frac{1}{2^n}(v^2-1)^n\frac{\partial v}{\partial x}\right\},$$

then

$$\begin{aligned}\frac{\partial^{n+1}v}{\partial h^{n+1}}&=\frac{\partial^n}{\partial x^{n-1}\partial h}\left\{\frac{1}{2^n}(v^2-1)^n\frac{\partial v}{\partial x}\right\}\\&=\frac{\partial^n}{\partial x^n}\left\{\frac{1}{2^n}(v^2-1)^n\frac{\partial v}{\partial h}\right\}\\&=\frac{\partial^n}{\partial x^n}\left\{\frac{1}{2^{n+1}}(v^2-1)^{n+1}\frac{\partial v}{\partial x}\right\}\end{aligned}$$

Thus, since the relation holds when $n=1$, it holds for all values of n.

Now let $h=0$, so that $v=x$, then

$$\left[\frac{\partial^n v}{\partial h^n}\right]_0=\frac{d^{n-1}}{dx^{n-1}}\left\{\frac{1}{2^n}(x^2-1)^n\right\}$$

and consequently,

$$v=\frac{1-\sqrt{(1-2xh+h^2)}}{h}=x+\sum_{n=1}^{\infty}\frac{h^n}{n!}\cdot\frac{d^{n-1}}{dx^{n-1}}\left\{\frac{1}{2^n}(x^2-1)^n\right\}.$$

Thus

$$\frac{dv}{dx}=(1-2xh+h^2)^{-\frac{1}{2}}=1+\sum_{n=1}^{\infty}\frac{h^n}{n!}\cdot\frac{d^n}{dx^n}\left\{\frac{1}{2^n}(x^2-1)^n\right\},$$

and therefore

$$P_n(x)=\frac{1}{2^n n!}\cdot\frac{d^n}{dx^n}(x^2-1)^n,$$

and $P_n(x)$ is identified, on account of the Rodrigues formula (§ 7·24), with the Legendre polynomial.

Now since, when $|b|<|a|$,

$$\int_0^\pi\frac{dx}{a+b\cos x}=\frac{\pi}{\sqrt{(a^2-b^2)}},$$

it follows that

$$\begin{aligned}\sum_{n=0}^{\infty}h^nP_n(x)&=\frac{1}{\sqrt{(1-2xh+h^2)}}\\&=\frac{1}{\pi}\int_0^\pi\frac{dt}{1-hx-h\sqrt{(x^2-1)}\cos t}.\end{aligned}$$

This integral is absolutely and uniformly convergent for sufficiently small values of $|h|$; by developing the integrand as a series of ascending powers of h and comparing the coefficients of h^n it is found that

$$P_n(x)=\frac{1}{\pi}\int_0^\pi\{x+\sqrt{(x^2-1)}\cos t\}^n dt.$$

This is the Laplace integral for the Legendre polynomial $P_n(x)$; the choice of the determination of $\sqrt{(x^2-1)}$ is immaterial.

Similar integrals are

$$Q_n(x)=\int_0^\infty \{x+\sqrt{(x^2-1)}\cosh t\}^{-n-1}dt,$$

$$P_n{}^m(x)=\frac{(n+1)(n+2)\ldots(n+m)}{\pi}\int_0^\pi \{x+(\sqrt{x^2-1})\cos t\}^n \cos mt\,dt,$$

$$Q_n{}^m(x)=(-1)^m n(n-1)\ldots(n-m+1)\int_0^\infty \{x+\sqrt{(x^2-1)}\cosh t\}^{-n-1}\cosh mt\,dt.$$

8·4. The Mellin Transformation.—Definite integral solutions in which the nucleus is a function of the product xt have been exhaustively studied by Mellin.* Such solutions may be obtained when the differential equation in question is of the form

$$\text{(A)}\qquad L_x(y)\equiv x^n F\left(x\frac{d}{dx}\right)y+G\left(x\frac{d}{dx}\right)y=0.$$

Let H be any polynomial of its argument and $K(z)$ any solution of the ordinary differential equation

$$z^n F\left(z\frac{d}{dz}\right)K-H\left(z\frac{d}{dz}\right)K=0,$$

then $K(xt)$ satisfies the partial differential equation

$$\left\{x^n F\left(x\frac{d}{dx}\right)+G\left(x\frac{d}{dx}\right)\right\}K=\left\{G\left(t\frac{d}{dt}\right)+t^{-n}H\left(t\frac{d}{dt}\right)\right\}K$$

or

$$L_xK=M_tK.$$

The integral

$$y=\int_\alpha^\beta K(xt)v(t)dt$$

satisfies (A) provided that $v(t)$ is a solution of

$$\overline{M}_t(v)=0,$$

where $\overline{M}_t$ is the operator adjoint to M_t, and provided that appropriate limits of integration α and β are taken.

8·41. Application of the Mellin Transformation to the Hypergeometric Equation.—The example taken for illustration will be the hypergeometric equation

$$\text{(A)}\qquad x(1-x)\frac{d^2y}{dx^2}+\{c-(a+b+1)x\}\frac{dy}{dx}-aby=0$$

which, after multiplication by x, may be written in the form

$$L_x(y)\equiv x\left\{\left(x\frac{d}{dx}\right)^2+(a+b)x\frac{d}{dx}+ab\right\}y-\left\{\left(x\frac{d}{dx}\right)^2+(c-1)x\frac{d}{dx}\right\}y=0.$$

Let

$$M_t\equiv t(1-t)\frac{d^2}{dt^2}+(e-c)\frac{d}{dt}$$

$$=-\left\{\left(t\frac{d}{dt}\right)^2+(c-1)t\frac{d}{dt}\right\}+t^{-1}\left\{\left(t\frac{d}{dt}\right)^2+(e-1)t\frac{d}{dt}\right\},$$

where the constant e is arbitrary. Then the partial differential equation

$$L_x(K)=M_t(K)$$

* *Acta Soc. Sc. Fenn.* 21 (1896), No. 6, p. 39.

is satisfied by $K(xt)$ provided that $u=K(z)$ is a solution of

$$z(1-z)\frac{d^2u}{dz^2}+\{e-(a+b+1)z\}\frac{du}{dz}-abu=0.$$

Now the equation

$$\overline{M}_t(v)=0$$

is satisfied by

$$v(t)=t^{e-1}(1-t)^{c-e-1},$$

and limits of integration are to be determined so that

$$\left[t^e(1-t)^{c-e}\frac{\partial u}{\partial t}\right]_a^\beta$$

vanishes identically. If $u=F(a,\ b\ ;\ e\ ;\ xt)$ this condition is satisfied when $a=0$, $\beta=1$ provided that $e>0$, $c>e$. Under these conditions, then

$$y(x)=\int_0^1 F(a,\ b\ ;\ e\ ;\ xt)t^{e-1}(1-t)^{c-e-1}dt$$

satisfies (A). Now

$$y(0)=\int_0^1 t^{e-1}(1-t)^{c-e-1}dt=\frac{\Gamma(e)\Gamma(c-e)}{\Gamma(c)}$$

$$y'(0)=\frac{ab}{e}\int_0^1 t^e(1-t)^{c-e-1}dt=\frac{ab}{e}\cdot\frac{\Gamma(e+1)\Gamma(c-e)}{\Gamma(c+1)}$$

$$=\frac{ab}{c}\cdot\frac{\Gamma(e)\Gamma(c-e)}{\Gamma(c)}.$$

But these initial conditions determine the unique solution

$$\frac{\Gamma(e)\Gamma(c-e)}{\Gamma(c)}F(a,\ b\ ;\ c\ ;\ x),$$

and consequently

$$\int_0^1 F(a,\ b\ ;\ e\ ;\ xt)\,t^{e-1}(1-t)^{c-e-1}dt=\frac{\Gamma(e)\Gamma(c-e)}{\Gamma(c)}F(a,\ b\ ;\ c\ ;\ x).$$

In particular, let $e=b$, then since

$$F(a,\ b\ ;\ b\ ;\ xt)=(1-xt)^{-a},$$

it follows that

$$\int_0^1 (1-xt)^{-a}\,t^{b-1}(1-t)^{c-b-1}dt=\frac{\Gamma(b)\Gamma(c-b)}{\Gamma(c)}F(a,\ b\ ;\ c\ ;\ x)$$

provided that $b>0$, $c>b$.

8·42. Derivation of the Definite Integral from the Hypergeometric Series.—By making use of the properties of the Gamma and Beta functions it is a simple matter to transform the series expression for the hypergeometric function into the equivalent definite integral. Since

$$b(b+1)\ .\ .\ .\ (b+r-1)=\Gamma(b+r)/\Gamma(b),$$

$$F(a,\ b\ ;\ c\ ;\ x)=1+\sum_{r=1}^{\infty}\frac{a(a+1)\ .\ .\ .\ (a+r-1)\,.\,b(b+1)\ .\ .\ .\ (b+r-1)}{r!\ \ c(c+1)\ .\ .\ .\ (c+r-1)}x^r$$

$$=\frac{\Gamma(c)}{\Gamma(b)}\left\{\frac{\Gamma(b)}{\Gamma(c)}+\sum\frac{a(a+1)\ .\ .\ .\ (a+r-1)}{r!}\cdot\frac{\Gamma(b+r)}{\Gamma(c+r)}x^r\right\}.$$

Now

$$\frac{\Gamma(b+r)\Gamma(c-b)}{\Gamma(c+r)}=B(b+r,\ c-b)$$
$$=\int_0^1 t^{b+r-1}(1-t)^{c-b-1}dt,$$

provided that the real parts of $b+r$ and $c-b$ are positive, and therefore

$$F(a,\ b;\ c;\ x)=\frac{\Gamma(c)}{\Gamma(b)\Gamma(c-b)}\sum_{r=0}^{\infty}\frac{a(a+1)\ .\ .\ .\ (a+r-1)}{r!}x^r\int_0^1 t^{b+r-1}(1-t)^{c-b-1}dt$$
$$=\frac{\Gamma(c)}{\Gamma(b)\Gamma(c-b)}\int_0^1 t^{b-1}(1-t)^{c-b-1}\sum\frac{a(a+1)\ .\ .\ .\ (a+r-1)}{r!}x^r t^r dt$$
$$=\frac{\Gamma(c)}{\Gamma(b)\Gamma(c-b)}\int_0^1 t^{b-1}(1-t)^{c-b-1}(1-xt)^{-a}dt.$$

The inversion of the order of summation and integration which has been made is valid so long as the hypergeometric series remains uniformly convergent, that is to say if $|x|\leqslant\rho<1$. Nevertheless the definite integral representation of the function is valid for all values of x, but to compensate for this increase of validity, restrictions have been imposed upon b and c.

It is possible to alter the path of integration in such a way that the integral constitutes an independent solution of the differential equation.

8·5. Solution by Double Integrals.—In many cases in which attempts to satisfy a given linear differential equation by a definite integral of the type (8·1, A) fail, it is possible to solve the problem by means of a multiple integral. For instance, a method such as that based upon the Laplace transformation is practically useless unless the transformed equation is of the first order and the equation to be solved restricted accordingly. In the present section a method of expressing the solution of a differential equation by a double integral will be outlined, and in the following section a particular example will be treated in detail.

Let $L_x(y)=0$ be the given differential equation, and let it be supposed that a function $K(x;\ s,\ t)$ can be found such that

(A) $$L_xK(x;\ s,\ t)=M_{s,\,t}K(x;\ s,\ t),$$

where $M_{s,\,t}$ is a partial differential operator of the second order of the type

(B) $$M_{s,\,t}\equiv a\frac{\partial^2}{\partial s\partial t}+b\frac{\partial}{\partial s}+c\frac{\partial}{\partial t}+d,$$

where a, b, c and d are functions of s and t. Such relations as these can as a rule only be arrived at tentatively; no general method for setting them up is known.

Now consider the double integral

(C) $$y(x)=\iint K(x;\ s,\ t)w(s,\ t)dsdt,$$

where both the function $w(s,\ t)$ and the domain of integration are at present unspecified. Then, assuming the validity of differentiation under the integral sign a sufficient number of times with respect to x,

$$L_xy(x)=\iint L_xK(x;\ s,\ t)w(s,\ t)dsdt$$
$$=\iint M_{s,\,t}K(x;\ s,\ t)w(s,\ t)dsdt.$$

But, by integration by parts,

$$\iint aw\frac{\partial^2 K}{\partial s\partial t}dsdt=\left[\int aw\left(\frac{\partial K}{\partial s}ds+\frac{\partial K}{\partial t}dt\right)-Kaw\right]+\int K\int\frac{\partial^2(aw)}{\partial s\partial t}dsdt,$$

$$\iint bw\frac{\partial K}{\partial s}dsdt=\left[\int bwKdt\right]-\iint K\frac{\partial(bw)}{\partial s}dsdt,$$

$$\iint cw\frac{\partial K}{\partial t}dsdt=\left[\int cwKds\right]-\iint K\frac{\partial(cw)}{\partial t}dsdt,$$

and therefore

$$L_x y(x)=\iint K(x\,;\ s,t)\overline{M}_{s,t}(w)dsdt+[P\{K,w\}],$$

where

(D) $$\overline{M}_{s,t}\equiv\frac{\partial^2}{\partial s\partial t}a-\frac{\partial}{\partial s}b-\frac{\partial}{\partial t}c+d$$

is the partial differential operator adjoint to (B), and $P\{K, w\}$ is an expression analogous to the bilinear concomitant which may easily be written out in full.

In the first place, then, $w(s, t)$ is to be determined as a solution of the partial differential equation

(E) $$\overline{M}_{s,t}(w)=0.$$

Thus the solution of the problem appears to depend, and in fact may depend upon an appeal to a higher branch of analysis, namely the theory of partial differential equations. But in most cases of practical importance $w(s, t)$ has the particular form $u(s)v(t)$, and the single partial equation (E) is replaced by a pair of ordinary equations each of the first order:

$$\alpha\frac{du}{ds}+\beta=0,\quad \gamma\frac{dv}{dt}+\delta=0,$$

where α and β are functions of s only, and γ and δ functions of t only.

In the second place, $w(s, t)$ having been determined, it remains to choose a domain of integration such that the integral in (C) exists and the expression $[P\{K, w\}]$ vanishes identically.

8·501. Example of Solution by a Double Integral.—Consider the equation

$$L_x(y)\equiv(x^2-1)\frac{d^2y}{dx^2}+(a+b+1)x\frac{dy}{dx}+aby=0.$$

It does not yield to treatment by the simple Laplace transformation because th first coefficient is of the second degree. It can, however, be solved by a doubl integral whose nucleus is e^{xst}, a form suggested by Laplace's nucleus e^{xt}. In th case

$$L_x e^{xst}=\{x^2s^2t^2-s^2t^2+(a+b+1)xst+ab\}e^{xst}$$
$$=\left\{\left(s\frac{\partial}{\partial s}+a\right)\left(t\frac{\partial}{\partial t}+b\right)-s^2t^2\right\}e^{xst}=M_{s,t}(e^{xst}).$$

The multiplier $w(s, t)$ therefore satisfies the differential equation

$$\overline{M}_{s,t}(w)\equiv\left\{\left(s\frac{\partial}{\partial s}-a+1\right)\left(t\frac{\partial}{\partial t}-b+1\right)-s^2t^2\right\}w=0,$$

and it is sufficient to write $w(s, t)=u(s)v(t)$, where

$$s\frac{du}{ds}-(a-1)u=-s^2u,$$

whence

$$u(s)=e^{-\frac{1}{2}s^2}s^{a-1},$$

and

$$t\frac{dv}{dt}-(b-1)v=-t^2v,$$

whence

$$v(t)=e^{-\frac{1}{2}t^2}t^{b-1}.$$

The domain of integration may be taken to be the quadrant $x\geqslant 0$ $y\geqslant 0$ provided only that a and b are numbers whose real parts are positive.

It follows that

$$y_1=\int_0^\infty\int_0^\infty e^{xst-\frac{1}{2}(s^2+t^2)}s^{a-1}t^{b-1}dsdl,$$

and similarly that

$$y_2=\int_0^\infty\int_0^\infty e^{-xst-\frac{1}{2}(s^2+t^2)}s^{a-1}t^{b-1}dsdt$$

are solutions of the given equation.

8·502. Connection of the Double Integral with the Solutions in Series.—The double integrals which satisfy the differential equation of the previous section may readily be derived from the series solution by making use of the property of the Gamma function that *

$$\Gamma(z+1)=z\Gamma(z).$$

A pair of series solutions, even and odd functions of x respectively, is

$$Y_1=1+\frac{ab}{2!}x^2+\frac{a(a+2)\,.\,b(b+2)}{4!}x^4$$
$$+\frac{a(a+2)(a+4)\,.\,b(b+2)(b+4)}{6!}x^6+\dots,$$

$$Y_2=x+\frac{(a+1)(b+1)}{3!}x^3+\frac{(a+1)(a+3)\,.\,(b+1)(b+3)}{5!}x^5+\dots,$$

where the law of formation of the coefficients is sufficiently obvious.

Then

$$\Gamma(\tfrac{1}{2}a)\Gamma(\tfrac{1}{2}b)Y_1=\Gamma(\tfrac{1}{2}a)\Gamma(\tfrac{1}{2}b)+2^2\frac{\frac{1}{2}a\Gamma(\frac{1}{2}a)\,.\,\frac{1}{2}b\Gamma(\frac{1}{2}b)}{2!}x^2$$
$$+2^4\frac{\frac{1}{2}a(\frac{1}{2}a+1)\Gamma(\frac{1}{2}a)\,.\,\frac{1}{2}b(\frac{1}{2}b+1)\Gamma(\frac{1}{2}b)}{4!}x^4+\dots$$
$$=\Gamma(\tfrac{1}{2}a)\Gamma(\tfrac{1}{2}b)+\frac{2^2\Gamma(\frac{1}{2}a+1)\Gamma(\frac{1}{2}b+1)}{2!}x^2+\frac{2^4\Gamma(\frac{1}{2}a+2)\Gamma(\frac{1}{2}b+2)}{4!}x^4+\dots$$
$$=2^{2-\frac{1}{2}a-\frac{1}{2}b}\int_0^\infty\int_0^\infty e^{-\frac{1}{2}(s^2+t^2)}\left\{s^{a-1}t^{b-1}+\frac{s^{a+1}t^{b+1}x^2}{2!}+\frac{s^{a+3}t^{b+3}x^4}{4!}+\dots\right\}ds\dot{a}.$$
$$=2^{2-\frac{1}{2}a-\frac{1}{2}b}\int_0^\infty\int_0^\infty e^{-\frac{1}{2}(s^2+t^2)}s^{a-1}t^{b-1}\cosh(xst)dsdt$$
$$=2^{1-\frac{1}{2}a-\frac{1}{2}b}(y_1+y_2),$$

and in the same way it may be proved that

$$\Gamma(\tfrac{1}{2}a+\tfrac{1}{2})\Gamma(\tfrac{1}{2}b+\tfrac{1}{2})Y_2=2^{1-\frac{1}{2}a-\frac{1}{2}b}\int_0^\infty\int_0^\infty e^{-\frac{1}{2}(s^2+t^2)}s^{a-1}t^{b-1}\sinh(xst)dsdt.$$
$$=2^{-\frac{1}{2}a-\frac{1}{2}b}(y_1-y_2).$$

The series Y_1 and Y_2 converge for any values of a and b when $|x|<1$; the corresponding integrals exist for all values of x when the real parts of a and b are

* It will be remembered that $\Gamma(z)=\int_0^\infty e^{-u}u^{z-1}du$

$$=2^{1-z}\int^\infty e^{-\frac{1}{2}t^2}t^{2z-1}dt,\text{ writing } u=\tfrac{1}{2}t^2.$$

positive. Thus the increase in the range of validity of the expression for the solution is gained at the expense of a restriction on the parameters a and b.

8·6. Periodic Transformations.—It will now be supposed that, in the integral

(A) $$y(x)=\int_a^\beta K(x, t)v(t)dt,$$

the nucleus $K(x, t)$ satisfies the partial differential equation

(B) $$L_x(K)=\bar{L}_t(K).$$

Then, if the differentiation under the integral sign is valid, and if A is an arbitrary constant,

$$\begin{aligned} L_x(y)+Ay &= \int_a^\beta \{L_x(K)+AK\}v(t)dt \\ &= \int_a^\beta \{\bar{L}_t(K)+AK\}v(t)dt \\ &= \int_a^\beta K(x, t)\{L_t(v)+Av\}dt+\Big[P\{K, v\}\Big]_{t=a}^{t=\beta}. \end{aligned}$$

Thus if, for any choice of the constant A, the function $v(t)$ satisfies the differential equation

(C) $$L_t(v)+Av=0,$$

and the limits of integration are chosen so that the integrated part is identically zero, the definite integral will satisfy the equation

(D) $$L_x(y)+Ay=0$$

for the same value of A.

The solution of an equation such as (C) or (D) is often, as was seen in § 7·4, a twofold process involving not merely the formal determination of a function which satisfies the equation together with a set of initial conditions relative to a specified point, but also the determination of the constant A so that other conditions may be satisfied. Such conditions might be introduced, for instance, by supposing that the solution is purely periodic with a given period, or has a zero at a point other than that to which the initial conditions refer.

It will be supposed then, that such conditions are imposed upon the solution of (C), that such a solution can exist only for a set of discrete values of A, and when it exists is uniquely determined apart from an arbitrary constant multiplier. Precisely the same set of conditions will be imposed upon the nucleus $K(x, t)$ regarded as a function of the single variable x with t as a parameter.* Then clearly, if the relation $v_r(t)$, so determined, corresponds to the *characteristic* value A_r, then

$$y_r(x)=\int_a^\beta K(x, t)v_r(t)dt$$

satisfies (D) for the parameter A_r and satisfies all the initial conditions which were imposed upon $v_r(t)$. But $y_r(x)$ is, under these restrictions, unique, that is to say, a mere multiple of $v_r(x)$. If $v_r(x)=\lambda_r y_r(x)$, then $y_r(x)$ satisfies the homogeneous integral equation †

$$y(x)=\lambda\int_a^\beta K(x, t)y(t)dt,$$

when λ has the *characteristic* value λ_r.

* The possibility of determining $K(x, t)$ to satisfy the imposed conditions identically in t is assumed.

† Bateman, *Proc. London Math. Soc.* (2), 4 (1907), pp. 90, 461 ; *Trans. Camb. Phil. Soc.* 21 (1909), p. 187 ; Ince, *Proc. Roy. Soc. Edin.* 42 (1922), p. 43.

8·601. Example of Solution by an Integral Equation.—Let the given equation be

$$L_x(y)+Ay\equiv(1-x^2)\frac{d^2y}{dx^2}+\{n-m-(m+n)x\}\frac{dy}{dx}+\{A-p(m-n)x+p^2x^2\}y=0,$$

where m, n and p are constants and $m \geqslant 0$, $n \geqslant 0$. For certain discrete characteristic values of A there exists a solution, unique apart from a constant multiplier, which is finite in the neighbourhood of the singular points $x=\pm 1$. The nucleus $K(x, t)$ which satisfies the equation $L_x(K)=\bar{L}_t(K)$ and is finite, for all values of t except $t=\pm 1$, in the neighbourhood of $x=\pm 1$, is $e^{pxt}(1+t)^{m-1}(1-t)^{n-1}$. Now

$$\frac{\partial P\{K, v\}}{\partial t}=v(t)\bar{L}_t(K)-K(x, t)L(v)$$

$$=\frac{\partial}{\partial t}\left[(1-t^2)v\frac{\partial K}{\partial t}-K\frac{\partial}{\partial t}\{(1-t^2)v\}+\{n-m-(m+n+4)t\}Kv\right].$$

If $v(t)$ is finite in the neighbourhood of $t=\pm 1$, the expression in square brackets will vanish at those points provided $n>0$, $m>0$. Consequently solutions of the given equation will satisfy the integral equation

$$y(x)=\lambda\int_{-1}^{1}e^{pxt}(1+t)^{m-1}(1-t)^{n-1}y(t)dt.$$

Miscellaneous Examples.

1. Show that the differential equation

$$x\phi(D)y+\psi(D)y=0,$$

where ϕ and ψ are polynomials with constant coefficients, is satisfied by

$$y=\int_a^\beta e^{xt+\int\psi(t)\chi(t)dt}\chi(t)dt,$$

where $\chi(t)$ is the reciprocal of $\phi(t)$, and a and β are so chosen that for all values of x,

$$\left[e^{xt+\int\psi(t)\chi(t)dt}\right]_a^\beta=0.$$

2. Express the general solution of

$$x\frac{d^2y}{dx^2}-(a+\beta)(1+x)\frac{dy}{dx}+a\beta xy=0$$

in integral form (i) for positive, and (ii) for negative values of x. [Petzval.]

3. Show that the most general solution of

$$\frac{d^ny}{dx^n}-xy=a,$$

where a is a constant, is

$$y=\sum_{r=0}^{n}A_r\omega^r\int_0^\infty \exp\left\{\omega^r xt-\frac{t^{n+1}}{n+1}\right\}dt,$$

where $\omega^{n+1}=1$, and the constants A_r are connected by the single relation

$$\sum_{r=0}^{n}A_r=a.$$

4. Prove that the equation

$$x\frac{d^3y}{dx^3}-y=0$$

has the particular solution

$$y=\int_0^\infty \sin(x/v)e^{-\frac{1}{2}v^2}v\,dv;$$

and that the equation

$$x\frac{d^3y}{dx^3}+y=0$$

has the particular solution

$$y=\int_0^\infty e^{-x/v-\frac{1}{2}v^2}v\,dv$$

when $x>0$. What modification is required when $x<0$?

Derive the general solution of each equation. [Petzval.]

5. Show that the equation

$$x\frac{d^2y}{dx^2}+\frac{dy}{dx}+(x+a)y=0$$

has the solution, finite at the origin,

$$y=\int_0^{\frac{1}{2}\pi}\cos\,(x\cos\theta+a\log\cot\tfrac{1}{2}\theta)dt,$$

when a is real. [Sharpe, *Mess. Math.* x.]

6. Prove that

$$x\frac{d^2y}{dx^2}+2m\frac{dy}{dx}+xy=0$$

is satisfied by

$$y=\int_{-1}^{+1}(1-t^2)^{m-1}\cos xt\,dt,$$

and deduce the series-development of this solution.

7. Prove that, in the notation of Chapter VII., Example 8,

$$ {}_3F_2(\alpha,\beta,\gamma\,;\,\theta,\epsilon\,;\,x) $$
$$=\frac{\Gamma(\theta)\Gamma(\epsilon)}{\Gamma(\beta)\Gamma(\theta-\beta)\Gamma(\gamma)\Gamma(\epsilon-\gamma)}\int_0^1\int_0^1(1-xst)^{-\alpha}s^{\beta-1}(1-s)^{\theta-\beta-1}t^{\gamma-1}(1-t)^{\epsilon-\gamma-1}ds\,dt$$

and thus express the general solution of the ${}_3F_2$-equation in terms of double integrals.

8. Prove that a particular integral of

$$\left(x\frac{d}{dx}+\alpha_1\right)\left(x\frac{d}{dx}+\alpha_2\right)y=f(x)$$

is

$$y=\int_0^1\int_0^1 f(stx)s^{\alpha_1-1}t^{\alpha_2-1}\,ds\,dt,$$

and obtain the corresponding result for the equation of order n:

$$\left(x\frac{d}{dx}+\alpha_1\right)\left(x\frac{d}{dx}+\alpha_2\right)\cdots\left(x\frac{d}{dx}+\alpha_n\right)=f(x).$$

9. Prove that, if $P_m(x)$ is a Legendre polynomial, and $Q_m(x)$ the corresponding Legendre function of the second kind,

$$Q_m(x)=\tfrac{1}{2}\int_{-1}^{+1}\frac{P_m(t)}{x-t}\,dt,$$

and deduce, by induction, that if m and n are positive integers and $m\geqslant n$,

$$P_n(x)Q_m(x)=\tfrac{1}{2}\int_{-1}^{+1}\frac{P_n(t)P_m(t)}{x-t}\,dt.$$

10. Find the differential equation of the fourth order satisfied by

$$P_n(x)P_m(x),\quad P_n(x)Q_m(x),\quad P_m(n)Q_n(x),\quad Q_m(x)Q_n(x)$$

and show that it is transformed into itself by the Euler transformation

$$f(x)=\int\frac{\phi(t)}{t-x}\,dt.$$

Obtain a general type of equation of order n invariant under this transformation.

11. Show that the relation

$$f(x)=\int_0^x K(x-t)\phi(t)dt,$$

may be replaced by the three relations

$$u(s)=\int_0^\infty e^{-st}K(t)dt,\quad v(s)=\int_0^\infty e^{-st}\phi(t)dt,$$
$$u(s)v(s)=\int_0^\infty e^{-st}f(t)dt.$$

[Borel.]

Hence prove that, if $J_n(x)$ is a Bessel function

$$\int_0^x J_m(x-t)J_n(t)t^{-1}dt = n^{-1}J_{m+n}(x). \qquad \text{[Bateman.]}$$

12. Show that the nucleus $K(xt)$ satisfies the partial differential equation

$$\left\{x^n F\left(x\frac{\partial}{\partial x}\right) + x^{-n}G\left(x\frac{\partial}{\partial x}\right)\right\}K = \left\{t^n F\left(t\frac{\partial}{\partial t}\right) + t^{-n}H\left(t\frac{\partial}{\partial t}\right)\right\}$$

if $u = K(s)$ is a solution of

$$\left\{s^n F\left(s\frac{d}{ds}\right) - H\left(s\frac{d}{ds}\right)\right\}u = 0,$$

and that there is then a transformation depending upon the nucleus $K(xt)$ from

$$\left\{x^n F\left(x\frac{d}{dx}\right) + G\left(x\frac{d}{dx}\right)\right\}y = 0$$

to the adjoint equation of

$$\left\{t^n G\left(t\frac{d}{dt}\right) + H\left(t\frac{d}{dt}\right)\right\}v = 0.$$

Hence prove that

$$J_n(x) = \int_0^\infty J_{2n}\{2\sqrt{(xt)}\}J_n(t)dt,$$

where x is positive and n is real and greater than $-\frac{1}{2}$. [Bateman.]

CHAPTER IX

THE ALGEBRAIC THEORY OF LINEAR DIFFERENTIAL SYSTEMS

9·1. Definition of a Linear Differential System.—The linear differential equation

$$L(y) \equiv p_0(x)\frac{d^ny}{dx^n} + p_1(x)\frac{d^{n-1}y}{dx^{n-1}} + \ldots + p_{n-1}(x)\frac{dy}{dx} + p_n(x)y = r(x),$$

taken together with one or more supplementary conditions which are to be satisfied, for particular values of x, by y and its first $(n-1)$ derivatives, is said to form a *linear differential system.* The simplest set of supplementary conditions is that which was postulated for the fundamental existence theorem (§ 3·32), viz.:

$$y(x_0) = y_0, \quad y'(x_0) = y_0', \ldots, y^{(n-1)}(x_0) = y_0^{(n-1)},$$

$y_0, y_0', \ldots, y_0^{(n-1)}$ being n pre-assigned constants. The existence theorem reveals the fact that, when x_0 is an ordinary point of the equation, the system has one and only one solution. This particular set of supplementary conditions provides what is known as a *one-point boundary problem,* since a solution of the differential equation has to be found which satisfies the initial conditions at one specified point. Such a problem, then, has one and only one solution provided that the number of independent conditions is equal to the order of the equation.

In a *two-point boundary problem* the differential system is composed of the differential equation and a number of supplementary linear conditions of the form

$$U_i(y) \equiv \alpha_i y(a) + \alpha_i' y'(a) + \ldots + \alpha_i^{(n-1)} y^{(n-1)}(a)$$
$$+ \beta_i y(b) + \beta_i y'(b) + \ldots + \beta_i^{(n-1)}(b) = \gamma_i,$$

in which the numbers α, β and γ are given constants, and (a, b) is a definite range of variation of x. It will be supposed that m linearly-independent supplementary conditions of this type are assigned; since there cannot be more than $2n$ independent linear relations between the $2n$ quantities

$$y(a), \quad y'(a), \ldots, \quad y^{(n-1)}(a), \quad y(b), \quad y'(b), \ldots, \quad y^{(n-1)}(b),$$

it follows that $m \leqslant 2n$.

The system will be written in brief as

$$\begin{cases} L(y) = r(x), \\ U_i(y) = \gamma_i \end{cases} \qquad (i = 1, 2, \ldots, m).$$

Intimately related to the given system is the completely homogeneous system

$$\begin{cases} L(y) = 0, \\ U_i(y) = 0 \end{cases} \qquad (i = 1, 2, \ldots, m).$$

This is known as the *reduced system.*

In the case of the reduced system, there are clearly two possibilities to consider:

(i) The system may possess no solution which is not identically zero; the system is then said to be *incompatible*.

(ii) The system may have $k(\leqslant n)$ linearly independent solutions

$$y_1(x), \quad y_2(x), \ldots, \quad y_k(x).$$

Then the general solution of the reduced system may be written

$$c_1y_1(x)+c_2y_2(x)+\ldots+c_ky_k(x),$$

and depends upon the k arbitrary constants $c_1, c_2, \ldots, c_k$. The system is said, in this case, to be *k-ply compatible*; k is called the *index of compatibility*.

Similarly, in the non-homogeneous system there arise two cases:

(i) The system may admit of no solution at all, which implies that no solution of the equation $L(y)=r(x)$ can be found which satisfies the m boundary conditions $U_i(y)=\gamma_i$.

(ii) The system may be satisfied by a particular solution $y_0(x)$. Then if the index of the reduced system is k, the general solution of the non-homogeneous system is

$$y_0(x)+c_1y_1(x)+c_2y_2(x)+\ldots+c_ky_k(x),$$

where $c_1y_1(x)+c_2y_2(x)+\ldots+c_ky_k(x)$ is the general solution of the reduced system. It bears a close analogy to the complementary function (§ 5·1) of the linear differential equation, when the latter is unrestricted by boundary conditions.

The present chapter will be devoted to the general question of the compatibility or incompatibility of a linear differential system, and will show the very close resemblance which exists between the theory of linear differential systems on the one hand, and the theory of simultaneous linear algebraic equations on the other.

9·2. Analogy with the Theory of a System of Linear Algebraic Equations.—A linear differential system may be regarded as the limiting case of a system of M linear algebraic equations involving N variables, when, in the limit, M and N tend to infinity. For simplicity, the analogy will, in the first place, be developed for the case of a linear differential system of the second order,

$$\begin{cases} p_0(x)\dfrac{d^2y}{dx^2}+p_1(x)\dfrac{dy}{dx}+p_2(x)y=r(x), \\ \alpha_iy(a)+\alpha_i'y'(a)+\beta_iy(b)+\beta_i'y'(b)=\gamma_i \qquad (i\leqslant 4). \end{cases}$$

It will be supposed that $p_0(x)$, $p_1(x)$, $p_2(x)$ and $r(x)$ are continuous functions of the real variable x throughout the closed interval $a\leqslant x\leqslant b$. Let this interval be divided into s equal parts by the points

$$x_0, x_1, x_2, \ldots, x_s,$$

where $x_0=a$, $x_s=b$, and let

$$\Delta x=x_{\nu+1}-x_\nu,$$
$$\Delta y_\nu=y(x_{\nu+1})-y(x_\nu),$$
$$\Delta^2y_\nu=y(x_{\nu+2})-2y(x_{\nu+1})+y(x_\nu).$$

Then the differential equation may be regarded as the limiting form of the difference equation *

$$p_0(x_\nu)\frac{\Delta^2y_\nu}{\Delta x^2}+p_1(x_\nu)\frac{\Delta y_\nu}{\Delta x}+p_2(x_\nu)y_\nu=r(x_\nu)$$

when, in the limit, Δx tends to zero. As it stands, this difference equation holds for $\nu=0, 1, 2, \ldots, s-2$. In virtue of the expressions for Δy_ν and

* Porter, *Ann. of Math.* (2), 3 (1902), p. 55, proved that the passage to the limit from the difference equation to the differential equation may be made with complete rigour.

$\Delta^2 y_\nu$, it may be written, after both members have been multiplied by Δx^2, in the form

$$P_{0\nu}y_\nu + P_{1\nu}y_{\nu+1} + P_{2\nu}y_{\nu+2} = R_\nu \qquad (\nu = 0, 1, 2, \ldots, s-2).$$

There are thus $s-1$ equations connecting the $s+1$ unknown quantities

$$y_0, \; y_1, \; y_2, \; \ldots, \; y_s.$$

In the same way, each boundary condition

$$\alpha_i y(a) + \alpha_i' y'(a) + \beta_i y(b) + \beta_i' y'(b) = \gamma_i$$

may be expressed as the limiting form of

$$\alpha_i y_0 + \alpha_i' \frac{\Delta y_0}{\Delta x} + \beta_i y_s + \beta_i' \frac{\Delta y_{s-1}}{\Delta x} = \gamma_i$$

which, in turn, may be written as

$$A_{i0}y_0 + A_{i1}y_1 + A_{i,\,s-1}y_{s-1} + A_{is}y_s = B_i,$$

and so each boundary condition is equivalent to a linear difference equation connecting y_0, y_1, y_{s-1}, and y_s.

The ideas here involved are clearly quite general; thus a linear differential equation of order n, whose coefficients are continuous in (a, b), may be regarded as the limiting case of a family of difference equations of the type

$$P_{0\nu}y_\nu + P_{1\nu}y_{\nu+1} + \ldots + P_{n\nu}y_{\nu+n} = R_\nu \qquad (\nu = 0, 1, 2, \ldots, s-n)$$

where, as before, s is the number of equal segments into which the interval (a, b) has been subdivided. Each boundary condition, whether it relates to one, to two, or to several, points of (a, b), also leads to an equation of precisely the same type; if there are m boundary conditions, there will be, in all, $s+m-n+1$ equations between the $s+1$ unknown quantities

$$y_0, \; y_1, \; y_2, \; \ldots, \; y_s.$$

In order to emphasise the analogy which is thus seen to exist between linear differential systems and systems of linear algebraic equations, it is necessary to record the main properties which the latter are known to possess.*

9·21. Properties of a Linear Algebraic System.—Consider the set of M simultaneous linear equations

$$\begin{cases} a_{11}X_1 + a_{12}X_2 + \ldots + a_{1N}X_N = 0, \\ \cdot \quad \cdot \quad \cdot \quad \cdot \quad \cdot \quad \cdot \\ a_{M1}X_1 + a_{M2}X_2 + \ldots + a_{MN}X_N = 0, \end{cases}$$

between the N variables $X_1, X_2, \ldots, X_N$. Two cases may arise:

(i) The system may admit of no solution except

$$X_1 = X_2 = \ldots = X_N = 0;$$

that is, the system may be incompatible.

(ii) There may be several, say k, sets of solutions:

$$\begin{array}{llll} X_{11}, & X_{21}, & \ldots, & X_{N1}, \\ X_{12}, & X_{22}, & \ldots, & X_{N2}, \\ \cdot & \cdot & \cdot & \cdot \\ X_{1k}, & X_{2k}, & \ldots, & X_{Nk}. \end{array}$$

These relations are said to be linearly independent if it is impossible to determine constants c, which are not all zero, such that the N equations

$$\begin{array}{l} c_1X_{11} + c_2X_{12} + \ldots + c_kX_{1k} = 0, \\ \cdot \quad \cdot \quad \cdot \quad \cdot \quad \cdot \quad \cdot \\ c_1X_{N1} + c_2X_{N2} + \ldots + c_kX_{Nk} = 0, \end{array}$$

* See Bôcher, *Introduction to Higher Algebra*, Chap. IV.

are satisfied simultaneously. When the k sets of solutions are in fact linearly independent, then on account of homogeneity of the system, the general solution is

$$\begin{aligned} X_1&=c_1X_{11}+c_2X_{12}+ \dots +c_kX_{1k}, \\ X_2&=c_1X_{21}+c_2X_{22}+ \dots +c_kX_{2k}, \\ &\cdot \quad \cdot \quad \cdot \quad \cdot \quad \cdot \quad \cdot \quad \cdot \\ X_N&=c_1X_{N1}+c_2X_{N2}+ \dots +c_kX_{Nk}, \end{aligned}$$

in which $c_1, c_2, \dots, c_k$ are arbitrary constants. The system is, in this case, k-ply compatible.

The index of compatibility, k, of the given system is determined by the following theorem: If p is the order of the non-zero determinant of highest order which can be extracted from the matrix

$$(A) \equiv \begin{pmatrix} a_{11}, & a_{12}, & \dots, & a_{1N} \\ \cdot & \cdot & & \cdot \\ a_{M1}, & a_{M2}, & \dots, & a_{MN} \end{pmatrix},$$

then $k=N-p$. The number p is called the *rank* of the matrix (A).

Consider now the non-homogeneous system of equations

$$\begin{aligned} a_{11}X_1+a_{12}X_2+ \dots +a_{1N}X_N&=b_1, \\ \cdot \quad \cdot \quad \cdot \quad \cdot \quad \cdot& \\ a_{M1}X_1+a_{M2}X_2+ \dots +a_{MN}X_N&=b_M, \end{aligned}$$

and with it the augmented matrix

$$(B) \equiv \begin{pmatrix} a_{11}, & a_{12}, & \dots, & a_{1N}, & b_1 \\ \cdot & \cdot & & \cdot & \cdot \\ a_{M1}, & a_{M2}, & \dots, & a_{MN}, & b_M \end{pmatrix}.$$

The rank of (B) is *at least* equal to that of (A); a necessary and sufficient condition that the non-homogeneous system of equations be compatible is that the rank of (B) should be *exactly* equal to that of (A). In this case, if

$$X_{10}, \quad X_{20}, \quad \dots, \quad X_{N0}$$

is any particular solution of the non-homogeneous system, then the general solution is

$$\begin{aligned} X_1&=X_{10}+c_1X_{11}+c_2X_{12}+ \dots +c_kX_{1k}, \\ X_2&=X_{20}+c_1X_{21}+c_2X_{22}+ \dots +c_kX_{2k}, \\ &\cdot \quad \cdot \quad \cdot \quad \cdot \quad \cdot \quad \cdot \quad \cdot \quad \cdot \\ X_N&=X_{N0}+c_1X_{N1}+c_2X_{N2}+ \dots +c_kX_{Nk}. \end{aligned}$$

9·22. Determination of the Index of a Linear Differential System.—Let $y_1, y_2, \dots, y_n$ be a fundamental set of solutions of the homogeneous linear differential equation

$$L(y)=0.$$

The question as to whether or not this equation is compatible with the m homogeneous linear boundary conditions

$$U_i(y)=0 \qquad (i=1, 2, \dots, m)$$

is equivalent to the problem of investigating the possibility of determining the constants $c_1, c_2, \dots, c_n$ in the general solution

$$y=c_1y_1+c_2y_2+ \dots +c_ny_n$$

in such a way that the boundary conditions are satisfied. Everything therefore depends upon the compatibility or incompatibility of the system of m simultaneous equations

$$c_1U_1(y_1)+c_2U_1(y_2)+ \cdots +c_nU_1(y_n)=0,$$
$$\cdot \quad \cdot \quad \cdot \quad \cdot \quad \cdot \quad \cdot \quad \cdot$$
$$c_1U_m(y_1)+c_2U_m(y_2)+ \cdots +c_nU_m(y_n)=0,$$

and therefore upon the rank of the matrix

$$(U)\equiv\begin{pmatrix} U_1(y_1), & U_1(y_2), & \cdots & U_1(y_n) \\ \cdot & \cdot & \cdot & \cdot \\ U_m(y_1), & U_m(y_2), & \cdots, & U_m(y_n) \end{pmatrix}.$$

If the rank of this matrix is p, there will be $n-p$ linearly independent sets of values of $c_1, c_2, \ldots, c_n$ and corresponding to each of these sets of values there will be one solution of the differential equation which satisfies the boundary conditions. The index of the differential system is therefore $k=n-p$. Consequently, *a necessary and sufficient condition that the given system should be k-ply compatible is that the rank of the matrix* (U) *is* $n-k$. In particular, if the rank of the matrix is n (which implies the condition that $m\geqslant n$), the system will be incompatible.

Consider now the non-homogeneous system

$$\begin{cases} L(y)=r(x), \\ U_i(y)=\gamma_i \qquad\qquad (i=1, 2, \ldots, m). \end{cases}$$

If $y_1, y_2, \ldots, y_n$ form, as before, a fundamental set of solutions of the homogeneous equation, and if y_0 is a particular solution of the non-homogeneous equation, then the general solution of the latter will be

$$y=y_0+c_1y_1+c_2y_2+ \cdots +c_ny_n.$$

In order that the boundary conditions of the non-homogeneous system may be satisfied, it must be possible to determine the constants $c_1, c_2, \ldots, c_n$ from the equations

$$c_1U_1(y_1)+c_2U_1(y_2)+ \cdots +c_nU_1(y_n)=\gamma_1-U_1(y_0),$$
$$\cdot \quad \cdot \quad \cdot \quad \cdot \quad \cdot \quad \cdot \quad \cdot \quad \cdot \quad \cdot$$
$$c_1U_m(y_1)+c_2U_m(y_2)+ \cdots +c_nU_m(y_n)=\gamma_m-U_m(y_0).$$

The possibility of so doing depends upon the rank of the augmented matrix

$$(U^1)\equiv\begin{pmatrix} U_1(y_1), & U_1y_2), & \cdots, & U_1(y_n), & \gamma_1-U_1(y_0) \\ \cdot & \cdot & \cdot & \cdot & \cdot \\ U_m(y_1), & U_m(y_2), & \cdots, & U_m(y_n), & \gamma_m-U_m(y_0) \end{pmatrix}.$$

A necessary and sufficient condition that the non-homogeneous system should be compatible is that the rank of the matrix (U^1) *is equal to the rank of the matrix* (U). If p is the common rank of the matrices, the general solution of each system will depend upon $n-p$ arbitrary constants.

As an important corollary it follows that *when* $m<n$ *a necessary and sufficient condition that a non-homogeneous system should have a solution is that the corresponding reduced system is* $(n-m)$*-ply compatible*; when $m=n$ the condition is that *the reduced system is incompatible.*

9·3. Properties of a Bilinear Form.—The expression

$$\begin{aligned} &a_{11}x_1y_1+a_{12}x_1y_2+ \cdots +a_{1N}x_1y_N \\ &+a_{21}x_2y_1+a_{22}x_2y_2+ \cdots +a_{2N}x_2y_N \\ &\cdot \quad \cdot \quad \cdot \quad \cdot \quad \cdot \quad \cdot \quad \cdot \\ &+a_{N1}x_Ny_1+a_{N2}x_Ny_2+ \cdots +a_{NN}x_Ny_N \end{aligned}$$

is said to be a bilinear form in the two sets of N variables

$$x_1,\quad x_2,\quad \ldots,\quad x_N,$$
$$y_1,\quad y_2,\quad \ldots,\quad y_N,$$

for the reason that the coefficient of each x is a linear function of the variables y and conversely. A distinction has to be made between the cases when the determinant

$$A=\begin{vmatrix} a_{11}, & a_{12}, & \ldots, & a_{1N} \\ a_{21}, & a_{22}, & \ldots, & a_{2N} \\ \cdot & \cdot & \cdot & \cdot \\ a_{N1}, & a_{N2}, & \ldots, & a_{NN} \end{vmatrix}$$

which is known as the determinant of the form, is or is not zero respectively. In the former case the bilinear form is said to be *singular*, in the latter it is said to be *ordinary*. It will here be assumed that the form considered is ordinary.

Let the variables $x_1, x_2, \ldots, x_N$ be replaced by a new set of N variables $X_1, X_2, \ldots, X_N$ by means of the substitution

$$\begin{aligned} x_1&=c_{11}X_1+c_{12}X_2+\ldots+c_{1N}X_N, \\ x_2&=c_{21}X_1+c_{22}X_2+\ldots+c_{2N}X_N, \\ &\cdot\quad\cdot\quad\cdot\quad\cdot\quad\cdot\quad\cdot \\ x_N&=c_{N1}X_1+c_{N2}X_2+\ldots+c_{NN}X_N, \end{aligned}$$

such that the determinant

$$C=|\,c_{ij}\,|$$

is not zero. Since $C\neq 0$ the substitution is reversible, that is to say the variables X are uniquely determinate in terms of the variables x. The bilinear form is then expressible as

$$\sum_{i=1}^{N}\sum_{j=1}^{N} d_{ij}X_iy_j,$$

and the corresponding determinant is

$$\begin{aligned} D=|\,d_{ij}\,|&=|\,a_{ik}\,|\,|\,c_{kj}\,| \\ &=AC\neq 0. \end{aligned}$$

The form therefore remains ordinary after the substitution has been made.

Now let the variables $y_1, y_2, \ldots, y_N$ be replaced by the set $Y_1, Y_2, \ldots, Y_N$ by means of the ordinary substitution

$$\begin{aligned} Y_1&=d_{11}y_1+d_{12}y_2+\ldots+d_{1N}y_N, \\ Y_2&=d_{21}y_1+d_{22}y_2+\ldots+d_{2N}y_N, \\ &\cdot\quad\cdot\quad\cdot\quad\cdot\quad\cdot\quad\cdot \\ Y_N&=d_{N1}y_1+d_{N2}y_2+\ldots+d_{NN}y_N. \end{aligned}$$

The form is thus reduced to

$$X_1Y_1+X_2Y_2+\ldots+X_NY_N,$$

which may be regarded as the canonical representation of an ordinary bilinear form. This reduction may be carried out in an infinite number of ways because the variables $x_1, x_2, \ldots, x_N$ may be transformed into the new set $X_1, X_2, \ldots, X_N$ by any linear substitution whose determinant is not zero. But once the new set of variables $X_1, X_2, \ldots, X_N$ has been determined, the corresponding set $Y_1, Y_2, \ldots, Y_N$ is unique.

Consider then what change is introduced into the set of variables Y in consequence of a change in one or more of the variables X. In the first place suppose that to

$$X_1,\quad X_2,\quad \ldots,\quad X_M,\quad X_{M+1},\quad \ldots,\quad X_N,$$

correspond respectively

$$Y_1, \quad Y_2, \quad \ldots, \quad Y_M, \quad Y_{M+1}, \quad \ldots, \quad Y_N.$$

Let $X_1, X_2, \ldots, X_M$ remain unchanged; let $X_{M+1}, \ldots, X_N$ be replaced by the new variables $X'_{M+1}, \ldots, X'_N$ which are such that

$$X_1, \quad X_2, \quad \ldots, \quad X_M, \quad X'_{M+1}, \quad \ldots, \quad X'_N$$

form a linearly independent system, and, further, let

$$Y'_1, \quad Y'_2, \quad \ldots, \quad Y'_M, \quad Y'_{M+1}, \quad \ldots, \quad Y'_N$$

be the corresponding system. Then

$$X_1Y_1+X_2Y_2+\ldots+X_MY_M+X_{M+1}Y_{M+1}+\ldots+X_NY_N$$
$$\equiv X_1Y'_1+X_2Y'_2+\ldots+X_MY'_M+X'_{M+1}Y'_{M+1}+\ldots X'_NY'_N.$$

Since $X_1, X_2, \ldots, X_M, X'_{M+1}, \ldots, X'_N$ are linearly independent quantities derived from the variables $x_1, x_2, \ldots, x_N$ by a substitution whose determinant is not zero, it follows that a unique set of values of $x_1, x_2, \ldots, x_N$ can be found such that

$$X_1=\ldots=X_M=0, \quad X'_{M+1}=1, \quad X'_{M+2}=\ldots=X'_N=0.$$

Then, if for these values of $x_1, x_2, \ldots, x_N$, $X_{M+1}, X_{M+2}, \ldots, X_N$ become respectively $A_{M+1}, A_{M+2}, \ldots, A_N$, it follows that

$$Y'_{M+1}=A_{M+1}Y_{M+1}+A_{M+2}Y_{M+2}+\ldots+A_NY_N.$$

In the same way $Y'_{M+2}, \ldots, Y'_N$ are expressible as linear combinations of $Y_{M+1}, \ldots, Y_N$.

The quantities $Y'_1, Y'_2, \ldots, Y'_M$ may be dealt with in a similar way. In particular, let that set of values of $x_1, x_2, \ldots, x_N$ be determined for which

$$X_1=1, \quad X_2=\ldots=X_M=X'_{M+1}=\ldots=X'_N=0;$$

and for this set of values let $X_{M+1}, \ldots, X_N$ become $B_{M+1}, \ldots, B_N$ respectively. Then

$$Y'_1=Y_1+B_{M+1}Y_{M+1}+\ldots+B_NY_N,$$

and similar expressions are found for $Y'_2, \ldots, Y'_M$.

9·31. Adjoint Differential Systems.—The theory of the bilinear form, which was outlined in the previous section, finds an important application in the development of the conception of an adjoint pair of linear differential systems.* Let

$$L(u)\equiv p_0\frac{d^nu}{dx^n}+p_1\frac{d^{n-1}u}{dx^{n-1}}+\ldots+p_{n-1}\frac{du}{dx}+p_nu$$

be a linear differential expression, in which it is assumed that the coefficients p_i are continuous functions of the real variable x for $a\leqslant x\leqslant b$, that the first $n-i$ derivatives of p_i exist and are continuous, and that p_0 does not vanish at any point of the closed interval (a, b).†

Then the adjoint differential expression is

$$\overline{L}(v)\equiv(-1)^n\frac{d^n(p_0v)}{dx^n}+(-1)^{n-1}\frac{d^{n-1}(p_1v)}{dx^{n-1}}+\ldots-\frac{d(p_{n-1}v)}{dx}+p_nv,$$

* A special pair of adjoint differential systems is given by Liouville, *J. de Math.* 3 (1838), p. 604. Mason, *Trans. Am. Math. Soc.* 7 (1906), p. 337, deals with systems of the second order. Birkhoff, *ibid.* 9 (1908), p. 373, and Bôcher, *ibid.* 14 (1913), p. 403, treat the general question. Extensions to systems of differential equations have been made by Bounitzky, *J. de Math.* (6), 5 (1909), p. 65, and Bôcher, *loc. cit.*

† This implies that the equation has no singular points within the interval (a, b) or at its end-points.

and $L(u)$ and $\bar{L}(v)$ are related by the Lagrange identity (§ 5·3)

$$vL(u)-u\bar{L}(v)=\frac{d}{dx}P(u, v),$$

where $P(u, v)$ is the bilinear concomitant

$$u\Big[p_{n-1}v-\frac{d(p_{n-2}v)}{dx}+\ldots+(-1)^{n-1}\frac{d^{n-1}(p_0v)}{dx^{n-1}}\Big]$$
$$+\frac{du}{dx}\Big[p_{n-2}v-\frac{d(p_{n-3}v)}{dx}+\ldots+(-1)^{n-2}\frac{d^{n-2}(p_0v)}{dx^{n-2}}\Big]$$
$$+\ldots\ldots\ldots$$
$$+\frac{d^{n-1}u}{dx^{n-1}}p_0v.$$

The determinant of this form is

$$\Delta(x)=\begin{vmatrix} \cdots & \cdots & \cdots & \cdots & (-1)^{n-1}p_0 \\ \cdots & \cdots & \cdots & (-1)^{n-2}p_0, & 0 \\ \cdot & \cdot & \cdot & \cdot & \cdot \\ \cdots & -p_0, & \cdots & 0, & 0 \\ p_0, & 0, & \cdots & 0, & 0 \end{vmatrix}$$

The elements below the secondary diagonal are all zero, and therefore the value of the determinant is $\pm(p_0)^n$, which is not zero at any point of (a, b). The bilinear concomitant is therefore an ordinary (*i.e.* a non-singular) bilinear form in the set of variables

$$u,\ u',\ \ldots,\ u^{(n-1)},$$
$$v,\ v',\ \ldots,\ v^{(n-1)}.$$

If the Lagrange identity is integrated between the limits a and b, Green's formula

$$\int_a^b\{vL(u)-u\bar{L}(v)\}dx=\Big[P(u, v)\Big]_a^b$$

is obtained. The right-hand member is a bilinear form in the two sets of $2n$ quantities

$$u(a),\ u'(a),\ \ldots,\ u^{(n-1)}(a),\ u(b),\ u'(b),\ \ldots,\ u^{(n-1)}(b),$$
$$v(a),\ v'(a),\ \ldots,\ v^{(n-1)}(a),\ v(b),\ v'(b),\ \ldots,\ v^{(n-1)}(b);$$

its determinant is

$$\begin{vmatrix} \Delta(a), & 0 \\ 0, & \Delta(b) \end{vmatrix}=\{p_0(a)p_0(b)\}^n,$$

and is not zero. The form $\Big[P(u, v)\Big]_a^b$ is therefore ordinary, and consequently reducible to the canonical form.

Let $U_1, U_2, \ldots, U_{2n}$ be any $2n$ linearly independent homogeneous expressions of the type

$$U_i(u)\equiv\alpha_iu(a)+\alpha_i'u'(a)+\ldots+\alpha_i^{(n-1)}u^{(n-1)}(a)$$
$$+\beta_iu(b)+\beta_i'u'(b)+\ldots+\beta_i^{(n-1)}u^{(n-1)}(b),$$

where the determinant of the $4n^2$ coefficients is not zero, then there exists a unique set

$$V_1,\ V_2,\ \ldots,\ V_{2n}$$

of independent forms linear in

$$v(a),\ v'(a),\ \ldots,\ v^{(n-1)}(a),\ v(b),\ v'(b),\ \ldots,\ v^{(n-1)}(b),$$

such that

$$\Big[P(u, v)\Big]_a^b = U_1V_{2n} + U_2V_{2n-1} + \dots + U_{2n}V_1.$$

Consequently Green's formula may be written

$$\int_a^b \{vL(a) - u\bar{L}(v)\}dx = U_1V_{2n} + U_2V_{2n-1} + \dots + U_{2n}V_1.$$

If $U_1, U_2, \dots, U_m$ remain unchanged, whilst a different choice of $U_{m+1}, \dots, U_{2n}$ is made, $V_1, V_2, \dots, V_{2n-m}$ will change into a new set $V'_1, V'_2, \dots, V'_{2n-m}$ which are linear combinations of $V_1, V_2, \dots, V_{2n-m}$. Thus $V_1, V_2, \dots, V_{2n-m}$ depend in reality upon $U_1, U_2, \dots, U_m$ only.

The system

$$\begin{cases} \bar{L}(v) = 0, \\ V_i(v) = 0 \end{cases} \qquad (i = 1, 2, \dots, 2n-m)$$

is said to be the adjoint of

$$\begin{cases} L(u) = 0, \\ U_i(u) = 0 \end{cases} \qquad (i = 1, 2, \dots, m).$$

The symmetry of the formulæ brings out the fact that, conversely, the second system is the adjoint of the first.

When a homogeneous linear differential system is regarded as the analogy of the set of equations

$$\begin{aligned} a_{11}X_1 + \dots + a_{1N}X_N &= 0, \\ \cdot \quad \cdot \quad \cdot \quad \cdot \quad &\cdot \quad \cdot \\ a_{M1}X_1 + \dots + a_{MN}X_N &= 0, \end{aligned}$$

the adjoint equation is the corresponding analogy of

$$\begin{aligned} a_{11}Y_1 + \dots + a_{M1}Y_M &= 0, \\ \cdot \quad \cdot \quad \cdot \quad \cdot \quad &\cdot \quad \cdot \\ a_{1N}Y_1 + \dots + a_{MN}Y_M &= 0. \end{aligned}$$

9·32. A Property of the Solutions of a k-ply Compatible System.—The forms $U_{m+1}, \dots, U_{2n}$ are restricted only by the condition that

$$U_1, U_2, \dots U_m, U_{m+1}, \dots, U_{2n}$$

are linearly independent. They have, however, the important property that if $u_1, u_2, \dots, u_k$ form a linearly independent set of solutions of the k-ply compatible system

$$\begin{cases} L(u) = 0, \\ U_i(u) = 0 \end{cases} \qquad (i = 1, 2, \dots, m),$$

then

$$U_i(u_1), \quad U_i(u_2), \quad \dots, \quad U_i(u_k) \qquad (i = m+1, \dots, 2n)$$

are linearly independent.

For if not, then constants $c_1, c_2, \dots, c_k$ can be found so that

$$\begin{aligned} &U_i(c_1u_1 + c_2u_2 + \dots + c_ku_k) \\ &\quad = c_1U_i(u_1) + c_2U_i(u_2) + \dots + c_kU_i(u_k) = 0 \\ &\qquad\qquad (i = m+1, \dots, 2n). \end{aligned}$$

But

$$U_i(c_1u_1 + c_2u_2 + \dots + c_ku_k) = 0 \qquad (i = 1, 2, \dots, m).$$

and hence

$$U_i(u) = 0,$$

where $\quad i = 1, 2, \dots, 2n, \quad$ and $\quad u = c_1u_1 + c_2u_2 + \dots + c_ku_k.$

These $2n$ independent homogeneous equations involve the $2n$ quantities

$$u(a), \quad u'(a), \quad \dots, \quad u^{(n-1)}(a), \quad u(b), \quad u'(b), \quad \dots, \quad u^{(n-1)}(b);$$

since the determinant of the $4n^2$ coefficients is not zero, these equations are not satisfied unless each of these quantities is zero. This, however, is impossible, since then u would vanish identically. The theorem is therefore established.

9·33. The Case in which the Number of independent Boundary Conditions is equal to the Order of the Equation.—The case $m=n$ is of considerable importance, and is of rather greater simplicity than the more general case. In this case, it will be proved that *the index of compatibility of a homogeneous differential system is equal to the index of the adjoint system.*

Let the given system be

$$\begin{cases} L(u)=0 \\ U_i(u)=0. \qquad (i=1, 2, \ldots, n). \end{cases}$$

Let k be its index, and let $u_1, u_2, \ldots, u_k$ be a set of linearly independent solutions. The adjoint system is

$$\begin{cases} \bar{L}(v)=0, \\ V_i(v)=0 \qquad (i=1, 2, \ldots, n). \end{cases}$$

Let $v_1, v_2, \ldots, v_n$ be a fundamental set of solutions of the equation

$$\bar{L}(v)=0\,;$$

then Green's formula

$$\int_a^b \{vL(u)-u\bar{L}(v)\}dx=U_1V_{2n}+U_2V_{2n}+\ \ldots\ +U_{2n}V_1$$

reduces to

$$U_{n+1}(u)V_n(v_1)+\ \ldots\ +U_{2n}(u)V_1(v_1)=0,$$

$$\cdot\quad\cdot\quad\cdot\quad\cdot\quad\cdot\quad\cdot\quad\cdot\quad\cdot\quad\cdot$$

$$U_{n+1}(u)V_n(v_n)+\ \ldots\ +U_{2n}(u)V_1(v_n)=0,$$

where u denotes any solution of the set $u_1, u_2, \ldots, u_k$.

This set of equations, regarded as equations to determine $U_{n+1}, \ldots, U_{2n}$ has the k solutions

$$U_{n+1}(u_i), \ \ldots, \ U_{2n}(u_i) \qquad (i=1, 2, \ldots, k),$$

and these solutions, in virtue of the lemma of the preceding section, are linearly independent. Consequently the rank of the matrix

$$\begin{pmatrix} V_n(v_1), & \ldots, & V_1(v_1) \\ \cdot & \cdot & \cdot \\ V_n(v_n), & \ldots, & V_1(v_n) \end{pmatrix}$$

is $n-k$ at most. But this is precisely the matrix which determines the index of the adjoint system. If the index of the adjoint system is k', the rank of this matrix is $n-k'$, and hence

$$n-k'\leqslant n-k,$$

or

$$k'\geqslant k.$$

But if in this reasoning the two systems are interchanged it would follow that $k\geqslant k'$, whence finally $k'=k$ as was to be proved.

If the restriction $m=n$ is removed, the more general form of the theorem is that $k'=k+m-n$. The proof follows on the same general lines. It is first established that $k'\geqslant k+m-n$. From the reciprocity between the system and its adjoint, it is deduced that $k\geqslant k'+(2n-m)-n$, or $k'\leqslant k+m-n$, whence the theorem follows.

9·34. The Non-homogeneous System.—Let the given complete system be

$$\text{(A)}\qquad \begin{cases} L(u)=r, \\ U_i(u)=\gamma_i \qquad (i=1, 2, \ldots, n), \end{cases}$$

then *a necessary and sufficient condition that this system may have a solution is that every solution v of the homogeneous adjoint system*

$$\text{(B)} \qquad \begin{cases} \bar{L}(v)=0, \\ V_i(v)=0 \end{cases} \qquad (i=1, 2, \ldots, n)$$

satisfies the relation

$$\text{(C)} \qquad \int_a^b vrdx=\gamma_1 V_{2n}(v)+ \ldots +\gamma_n V_{n+1}(v).$$

Let k be the index of the homogeneous system (B); if $k=0$ the theorem follows from § 9·22, it will therefore be supposed that $k>0$, and that $v_1, v_2, \ldots, v_k$ form a linearly independent set of solutions.

If the given complete system has a solution u, let v be any solution of the system (B). Then if u and v so defined are substituted in Green's formula, equation (C) follows immediately. The condition is therefore *necessary*.

In order to prove the condition *sufficient*, let u_0 be any solution of the equation

$$L(u)=r,$$

then Green's theorem leads to the relation

$$\int_a^b vrdx=U_1(u_0)V_{2n}(v)+ \ldots +U_n(u_0)V_{n+1}(v),$$

where v denotes any solution of the system (B).

By subtraction from (C), it follows that

$$\text{(D)} \qquad \{U_1(u_0)-\gamma_1\}V_{2n}(v)+ \ldots +\{U_n(u_0)-\gamma_n\}V_{n+1}(v)=0.$$

Now let $u_1, u_2, \ldots, u_n$ be a fundamental system of solutions of the homogeneous equation

$$L(u)=0,$$

then, by Green's theorem,

$$\text{(E)} \qquad \begin{matrix} U_1(u_1)V_{2n}(v)+ \ldots +U_n(u_1)V_{n+1}(v)=0, \\ \cdot \quad \cdot \quad \cdot \quad \cdot \quad \cdot \quad \cdot \quad \cdot \quad \cdot \quad \cdot \quad \cdot \\ U_1(u_n)V_{2n}(v)+ \ldots +U_n(u_n)V_{n+1}(v)=0. \end{matrix}$$

Thus there are in all $n+1$ linear homogeneous equations in the n unknowns $V_{2n}(v), \ldots, V_{n+1}(v)$, and they are satisfied by the k solutions

$$V_{2n}(v_i), \ldots, \quad V_{n+1}(v_i) \quad (i=1, 2, \ldots, k)$$

which, by § 9·32, are linearly independent. The rank of the matrix of the set of $n+1$ equations (D, E) is therefore at most $n-k$, but it cannot be less than $n-k$ since the rank of the matrix of the n equations (E) is exactly $n-k$. The rank of both matrices is therefore $n-k$, from which it follows that the given complete system has a solution.

When $m \neq n$ the theorem is that a necessary and sufficient condition that the complete system

$$\begin{cases} L(u)=r, \\ U_i(u)=\gamma_i \end{cases} \qquad (i=1, 2, \ldots, m)$$

should have a solution is that every solution v of the homogeneous adjoint system

$$\begin{cases} \bar{L}(v)=0, \\ V_i(v)=0 \end{cases} \qquad (i=1, 2, \ldots, 2n-m)$$

satisfies the relation

$$\int_a^b vrdx=\gamma_1 V_{2n}(v)+ \ldots +\gamma_m V_{2n-m+1}(v).$$

The case $n \geqslant m$, $k=n-m$ is disposed of by reference to § 9·22; the proof then follows on the above lines.

9·4. The self-adjoint Linear Differential System of the Second Order.—Let

$$L(u)\equiv p_0\frac{d^2u}{dx^2}+p_1\frac{du}{dx}+p_2u$$

be a homogeneous linear differential expression of the second order. The adjoint expression is

$$\bar{L}(v)=p_0\frac{d^2v}{dx^2}+(2p_0'-p_1)\frac{dv}{dx}+(p_0''-p_1'+p_2)v.$$

A necessary and sufficient condition that $L(u)$ be identical in form with its adjoint $\bar{L}(v)$ is clearly

$$p_0'=p_1.$$

The expression may then be written

$$\frac{d}{dx}\left(p_0\frac{du}{dx}\right)+p_2u.$$

In its general form, $L(u)$ is not self-adjoint, but the expression

$$\frac{1}{p_0}e^{\int\frac{p_1}{p_0}dx}L(u)\equiv\frac{d}{dx}\left\{e^{\int\frac{p_1}{p_0}dx}\frac{du}{dx}\right\}+\frac{p_2}{p_0}e^{\int\frac{p_1}{p_0}dx}u$$

is self-adjoint. Since, therefore, any equation of the second order can be made self-adjoint by multiplying throughout by an appropriate factor (which does not vanish or become infinite in (a, b) if the assumptions of § 9·31 are maintained), there is no loss in generality in regarding as the general equation of the second order the self-adjoint equation

$$\frac{d}{dx}\left\{K\frac{du}{dx}\right\}-Gu=R,$$

which is known as the Sturm equation. In this case, let

$$L(u)\equiv\frac{d}{dx}\left\{K\frac{du}{dx}\right\}-Gu,$$

then, if u and v are any two functions of x whose first and second derivatives are continuous in (a, b),

$$vL(u)-uL(v)=\frac{d}{dx}\left[K\left(v\frac{du}{dx}-u\frac{dv}{dx}\right)\right],$$

and hence the bilinear concomitant is

$$P(u, v)=K\left(v\frac{du}{dx}-u\frac{dv}{dx}\right).$$

Green's formula reduces, in this case, to the simple form

$$\int_a^b\{vL(u)-uL(v)\}dx=\left[K\left(v\frac{du}{dx}-u\frac{dv}{dx}\right)\right]_a^b.$$

In particular, if $L(u)=0$, $L(v)=0$, it reduces to Abel's formula

$$K(b)\{v(b)u'(b)-u(b)v'(b)\}=K(a)\{v(a)u'(a)-u(a)v'(a)\}.$$

Consider, then, the homogeneous differential system

$$\begin{cases}L(u)\equiv\dfrac{d}{dx}\left\{K\dfrac{du}{dx}\right\}-Gu=0,\\ U_1(u)=\alpha_1u(a)+\alpha_2u(b)+\alpha_3u'(a)+\alpha_4u'(b)=0,\\ U_2(u)=\beta_1u(a)+\beta_2u(b)+\beta_3u'(a)+\beta_4u'(b)=0,\end{cases}$$

where it is supposed that U_1 and U_2 are linearly independent. This condition implies that, of the six determinants $\delta_{ij}=\alpha_i\beta_j-\alpha_j\beta_i$ contained in the matrix

$$\begin{pmatrix}\alpha_1, & \alpha_2, & \alpha_3, & \alpha_4\\ \beta_1, & \beta_2, & \beta_3, & \beta_4\end{pmatrix},$$

not all are zero.

Suppose, in the first place, that $\delta_{12}\neq 0$, then let U_3 and U_4 be taken in such a way that U_1, U_2, U_3 and U_4 are linearly independent. For instance, let

$$U_3(u)=u'(a), \quad U_4(u)=u'(b),$$

then if u and v are any functions of x such that $L(u)$ and $L(v)$ are continuous in (a, b),

$$\int_a^b \{vL(u)-uL(v)\}dx=\left[K\left(v\frac{du}{dx}-u\frac{dv}{dx}\right)\right]_a^b$$
$$=U_1V_4+U_2V_3+U_3V_2+U_4V_1,$$

that is

$$K(b)\{v(b)u'(b)-u(b)v'(b)\}-K(a)\{v(a)u'(a)-u(a)v'(a)\}$$
$$=\{\alpha_1u(a)+\alpha_2u(b)+\alpha_3u'(a)+\alpha_4u'(b)\}V_4$$
$$+\{\beta_1u(a)+\beta_2u(b)+\beta_3u'(a)+\beta_4u'(b)\}V_3$$
$$+u'(a)V_2+u'(b)V_1.$$

A comparison of the coefficients of $u(a)$, $u(b)$, $u'(a)$ and $u'(b)$ gives rise to the four equations

$$\alpha_1V_4+\beta_1V_3=K(a)v'(a),$$
$$\alpha_2V_4+\beta_2V_3=-K(b)v'(b),$$
$$V_2+\alpha_3V_4+\beta_3V_3=-K(a)v(a),$$
$$V_1+\alpha_4V_4+\beta_4V_3=K(b)v(b).$$

From these equations V_1, V_2, V_3 and V_4 may be obtained explicitly, viz.

$$V_1(v)=K(b)v(b)+\frac{1}{\delta_{12}}\{\delta_{24}K(a)v'(a)+\delta_{14}K(b)v'(b)\},$$

$$V_2(v)=-K(a)v(a)-\frac{1}{\delta_{12}}\{\delta_{23}K(a)v'(a)+\delta_{13}K(b)v'(b)\},$$

$$V_3(v)=-\frac{1}{\delta_{12}}\{\alpha_2K(a)v'(a)+\alpha_1K(b)v'(b)\},$$

$$V_4(v)=\frac{1}{\delta_{12}}\{\beta_2K(a)v'(a)+\beta_1K(b)v'(b)\}.$$

In order that the given system may be self-adjoint, it is necessary and sufficient that $V_1(v)$ and $V_2(v)$ should each be a linear combination of $U_1(v)$ and $U_2(v)$. Since $v(a)$ does not enter into V_1, V_1 may be obtained by eliminating $v(a)$ between $U_1(v)$ and $U_2(v)$. Hence V_1 is a multiple of

$$\delta_{12}v(b)+\delta_{13}v'(a)+\delta_{14}v'(b),$$

and thus

$$V_1(v)=K(b)v(b)+\frac{1}{\delta_{12}}\{\delta_{13}K(b)v'(a)+\delta_{14}K(b)v'(b)\}.$$

If this expression is compared with the previous expression for V_1 it is seen that the condition sought for is that

$$\delta_{24}K(a)=\delta_{13}K(b).$$

Precisely the same condition is obtained by expressing the fact that $V_2(v)$ is essentially the eliminant of $v(b)$ between $U_1(v)$ and $U_2(v)$. Thus the condition obtained is a necessary and a sufficient condition that the system dealt with may be self-adjoint.

Assume next that $\delta_{13} \neq 0$; in this case an independent set of linear expressions U may be obtained by taking

$$U_3(u)=u(b), \quad U_4(u)=u'(b).$$

It is then easily found that

$$V_1(v)=K(b)v(b)+\frac{1}{\delta_{13}}\{\delta_{14}K(a)v(a)+\delta_{34}K(a)v'(a)\},$$

$$V_2(v)=-K(b)v'(b)+\frac{1}{\delta_{13}}\{\delta_{12}K(a)v(a)-\delta_{23}K(a)v'(a)\},$$

$$V_3=-\frac{1}{\delta_{13}}\{\alpha_1K(a)v(a)+\alpha_3K(a)v'(a)\},$$

$$V_4=\frac{1}{\delta_{13}}\{\beta_1K(a)v(a)+\beta_3K(a)v'(a)\}.$$

Again, it may be inferred that

$$\delta_{24}K(a)=\delta_{13}K(b)$$

is a necessary and sufficient condition that the given system may be self-adjoint.

The remaining four cases $\delta_{14} \neq 0$, $\delta_{23} \neq 0$, $\delta_{24} \neq 0$, and $\delta_{34} \neq 0$ may be dealt with in the same way; each case leads to the same condition that the given system may be self-adjoint.

9·41. Sturm-Liouville Systems.—A system of the type

$$(A) \quad \begin{cases} \dfrac{d}{dx}k\left\{\dfrac{dy}{dx}\right\}+\{\lambda g-l\}y=0, \\ \alpha_1y(a)+\alpha_2y(b)+\alpha_3y'(a)+\alpha_4y'(b)=0, \\ \beta_1y(a)+\beta_2y(b)+\beta_3y'(a)+\beta_4y'(b)=0, \end{cases}$$

is known as a Sturm-Liouville system. In the interval $a \leqslant x \leqslant b$, k (which is everywhere positive), g and l are continuous functions of x, and λ is an arbitrary parameter. The condition that the system may be self-adjoint is $\delta_{24}k(a)=\delta_{13}k(b)$; it will be supposed that this condition is satisfied.

A special case of the system, in which the boundary conditions are

$$(B) \quad \begin{cases} y'(a)-hy(a)=0, \\ y'(b)+Hy(b)=0, \end{cases}$$

arises in the problem of the distribution of temperature in a heterogeneous bar; * the system is self-adjoint, for in this case $\delta_{24}=\delta_{13}=0$.

By eliminating in turn $y'(a)$ and $y(a)$, the boundary conditions of (A) may be brought into the form

$$\begin{cases} \delta_{13}y(a)+\delta_{23}y(b)-\delta_{34}y'(b)=0, \\ \delta_{13}y'(a)+\delta_{12}y(b)+\delta_{14}y'(b)=0. \end{cases}$$

If $\delta_{13}=0$, which since the system is self-adjoint implies that $\delta_{24}=0$, the boundary conditions reduce to (B). When $\delta_{13} \neq 0$ the system may be written

$$(C) \quad \begin{cases} \dfrac{d}{dx}\left\{k\dfrac{dy}{dx}\right\}+\{\lambda g-l\}y=0, \\ y(a)=\gamma_1y(b)+\gamma_1'y'(b), \\ y'(a)=\gamma_2y(b)+\gamma_2'y'(b), \end{cases}$$

* In this problem k represents the conductivity, g the specific heat, and l, h and H depend upon the emissivity at the surface and at the ends of the bar respectively. λ has to be determined so as to render the system compatible.

and the condition that the system may be self-adjoint becomes

$$k(b)=(\gamma_1\gamma_2'-\gamma_1'\gamma_2)k(a).$$

In particular, the system involving the so-called *periodic boundary conditions*

$$\begin{cases} y(a)=y(b), \\ y'(a)=y'(b) \end{cases}$$

is self-adjoint provided that

$$k(a)=k(b).$$

9·5. Differential Systems which involve a Parameter. The Characteristic Numbers.— It frequently happens that, in the homogeneous differential system of order n

$$\begin{cases} L(y)=0, \\ U_i(y)=0 \qquad (i=1, 2, \ldots, n), \end{cases}$$

the coefficients in the differential equation, and possibly those in the boundary conditions, depend upon a parameter λ. A case in point was met with in the preceding section. The capital question here is to determine those particular values of λ for which the system becomes compatible. Such values are known as the *characteristic numbers* of the system, the solutions which correspond to them are termed the *characteristic functions*. A later chapter (Chap. XI.) will be devoted to a closer study of the characteristic functions; the present section serves as a link between the theory which was expounded in the preceding pages and that which will be developed subsequently.

Let $y_1, y_2, \ldots, y_n$ be a fundamental set of real solutions of the equation

$$L(y)=0,$$

these are to be regarded as functions of the real variable-pair (x, λ), and as such are continuous functions * of (x, λ), and possess derivatives with respect to x up to and including the $(n-1)$th order, which are likewise continuous functions of (x, λ) when $a \leqslant x \leqslant b$ and λ lies in a certain interval, say (Λ_1, Λ_2). The condition for compatibility is that

$$\begin{vmatrix} U_1(y_1), & \ldots, & U_1(y_n) \\ \cdot & \cdot & \cdot \\ U_n(y_1), & \ldots, & U_n(y_n) \end{vmatrix} =0,$$

which may be written

$$F(\lambda)=0.$$

It will be assumed that the coefficients in U_i are continuous functions of λ, then $F(\lambda)$ will be continuous in the interval (Λ_1, Λ_2). This equation is known as the characteristic equation of the system, its roots are the characteristic numbers. For values of λ which lie in the open interval $\Lambda_1<\lambda<\Lambda_2$, the roots of the characteristic equation are isolated; † the end points Λ_1 and Λ_2 may, however, be the limit points of an infinite number of roots.

The characteristic equation is independent of the fundamental set of solutions chosen, for the effect of replacing y_i by Y_i, where

$$Y_i=c_{i1}y_1+c_{i2}y_2+\ldots+c_{in}y_n \qquad (i=1, 2, \ldots, n),$$

is to multiply the left-hand member of the characteristic equation by the

* For a definition of a continuous function of two real variables, see footnote to § 3·1. That $y_1, y_2, \ldots, y_n$ and their first $(n-1)$ derivatives with respect to x are continuous functions of (x, λ) follows from the existence theorems of §§ 3·31, 3·32.

† See § 9·6 *infra*.

determinant $|c_{ij}|$ which is not zero since $Y_1, Y_2, \ldots, Y_n$ form a fundamental set.*

By definition, each characteristic number λ_i renders the system compatible; the system will then have a certain index of compatibility, say k_i. Furthermore λ_i, regarded as a root of the characteristic equation, is of a certain multiplicity m_i. Now m_i may be unequal to k_i, but in all cases,

$$k_i \leqslant m_i.$$

(It will be remembered that $k_i \leqslant n$). To establish this inequality, it will be sufficient to prove that, if λ is any characteristic number, and k its index,

$$F'(\lambda) = F''(\lambda) = \ldots = F^{(k-1)}(\lambda) = 0.$$

Now $F^{(r)}(\lambda)$ is obtained by writing down a number of determinants, each of which contains at least $(n-r)$ columns of $F(\lambda)$ unaltered, the remaining columns being derived by differentiation from the corresponding columns of $F(\lambda)$. Let each of these determinants be developed, by Laplace's formula,† in terms of the minors contained in the $n-r$ undifferentiated columns. Since the index of λ is k, all determinants of order greater than or equal to $n-k+1$ extracted from the matrix $(U_i(y_j))$ are zero. That is to say each term in the development of $F^{(r)}(\lambda)$ will be zero, or

$$F^{(r)}(\lambda) = 0,$$

provided that $r \leqslant k-1$. Therefore the root λ is of multiplicity k *at least*, as was to be proved.

9·6. The Effect of Small Variations in the Coefficients of a Linear Differential System.—The supposition that the coefficients of the linear differential system

$$\text{(A)} \qquad \begin{cases} L(y) = 0, \\ U_i(y) = 0 \end{cases} \qquad (i = 1, 2, \ldots, n),$$

depend upon a parameter λ raises the question as to how a change in the value of λ will influence the compatibility of the system. In particular, it is important to determine whether an arbitrarily small variation in λ will raise, lower or leave unaltered the index of the system when it is known that for a given value of λ, say λ_0, the system is k-ply compatible. In its broader aspect, this question is settled by the following theorem.‡

Theorem I.—*The index of the system is not raised by any variation of the coefficients which is uniformly sufficiently small.*§

The index of the system for the characteristic number λ_0 being k, there exists within the matrix

$$\begin{pmatrix} U_1(y_1), & \ldots, & U_1(y_n) \\ \ldots & \ldots & \ldots \\ U_n(y_1), & \ldots, & U_n(y_n) \end{pmatrix}$$

at least one determinant of order $n-k$ which is not zero when $\lambda = \lambda_0$ (§ 9·22). Let λ_0 be given a small variation, then if a number δ (independent of x) exists such that, consequent on this variation, every coefficient in $L(y)$ and in $U_i(y)$ changes by an amount not greater in absolute magnitude than δ, the

* The coefficients c_{ij} may be functions of λ, but then the set $Y_1, Y_2, \ldots, Y_n$ ceases to be fundamental for any values of λ for which $|c_{ij}| = 0$. The difficulty is overcome by stipulating that $y_1, y_2, \ldots, y_n$ form a fundamental set for all values of λ in (Λ_1, Λ_2).

† Scott and Mathews, *Theory of Determinants*, p. 30.

‡ The present discussion is due to Bôcher, *Bull. Am. Math. Soc.* 21 (1914), p. 1.

§ That is to say, corresponding to each characteristic number λ_0, a number δ exists, such that in each coefficient of $L(y)$ is, in absolute magnitude, less than δ for all values of x in $(a. b)$. The variation of every coefficient in $U_i(y)$ is similarly less, in absolute magnitude, than δ.

variation in the value of the determinant will be comparable with δ. Thus a sufficiently small variation in λ_0 will not reduce the determinant to zero, which proves the theorem.

On the other hand, all determinants of order $n-k+1$ extracted from the matrix are zero when $\lambda=\lambda_0$. It is at least extremely probable that a small variation given to λ_0 would alter the value of at least one of these determinants, which would mean that the index had fallen below k. Without going into the question in its fullest aspect, an important case will be taken up, and it will be proved that by a uniformly sufficiently small variation in one coefficient alone, namely the coefficient of y in $L(y)$, the index may be reduced to zero. The proof depends upon three preliminary lemmas.

Lemma I.—*Let $y_0(x)$ be any particular solution of the given system corresponding to the characteristic number λ_0. Then there exists a function $y(x, \lambda)$, continuous in (x, λ), which satisfies the system* (A) *for values of λ in an interval Λ including λ_0, and which reduces to $y_0(x)$ when $\lambda=\lambda_0$.*

To make matters definite, let it be supposed that the determinant which does not vanish, when $\lambda=\lambda_0$, is that formed by the first $(n-k)$ rows and columns of the matrix (U). Then any solution of $L(y)=0$ which satisfies the first $(n-k)$ boundary conditions will also satisfy the remaining k conditions.

Such a solution is given by

$$\text{(B)}\quad y(x, \lambda) = \begin{vmatrix} y_1, & \ldots, & y_{n-k}, & c_1y_{n-k+1}+\ldots+c_ky_n \\ U_1(y_1), & \ldots, & U_1(y_{n-k}), & c_1U_1(y_{n-k+1})+\ldots+c_kU_1(y_n) \\ \ldots & \ldots & \ldots & \ldots \\ \ldots & \ldots & \ldots & \ldots \\ U_{n-k}(y_1), & \ldots, & U_{n-k}(y_{n-k}), & c_1U_{n-k}(y_{n-k+1})+\ldots+c_kU_{n-k}(y_n) \end{vmatrix}$$

The identical vanishing of this determinant, were it possible, would express a linear relationship between the fundamental solutions $y_1, y_2, \ldots, y_n$. Since this is contrary to hypothesis, the determinant is not identically zero. Consequently, the formula (B) represents a solution of the given system, and, being dependent upon k arbitrary constants, is its general solution.

Suppose now that there exists an interval Λ, containing λ_0, such that the system remains of index k for all values of λ within Λ. Then Λ may be taken sufficiently small to ensure that the $(n-k)$-rowed determinant which does not vanish for λ_0, is not zero for any value of λ in Λ. Consequently (B) is the general solution of the system for all values of λ in Λ, and is a continuous function of (x, λ), provided that the c_i are determined as constants or as continuous functions of λ.

Lemma II.—*Let $u(x)$ be a real solution of the system*

$$\text{(C)}\qquad \begin{cases} L(u)=gu, \\ U_i(u)=0 \qquad (i=1, 2, \ldots, n), \end{cases}$$

where g is a continuous function of x, and $v(x)$ a real solution of the system adjoint to (A)

$$\text{(D)}\qquad \begin{cases} M(v)=0, \\ V_i(v)=0 \qquad (i=1, 2, \ldots, n), \end{cases}$$

then

$$\text{(E)}\qquad \int_a^b gu(x)v(x)dx=0.$$

This lemma is a consequence of Green's Theorem.*

* For details of the proof, refer to the more general case of § 10·7 *infra*.

LEMMA III.—*If the given system* (A) *is compatible and of index* $k \geqslant 1$, *and if an arbitrarily small positive number* ϵ *is assigned, there exists a continuous real function* $g(x)$ *such that* $0 \leqslant g(x) < \epsilon$ *for which the index of the system* (C) *is less than* k.

Let $y(x)$ be a solution of the system (A) when $\lambda = \lambda_0$, and let $v(x)$ be a solution of (D) for the same value of λ. Neither $y(x)$ nor $v(x)$ can have an infinite number of zeros in (a, b).* Consequently a point c can be found in (a, b) at which the product $y(x)v(x)$ is not zero. Moreover, since $y(x)v(x)$ is a continuous function of x, the point c can be included in an interval (a', b') within which $y(x)v(x)$ does not vanish. Now define ϕ as a continuous real function of x which is zero outside (a', b') and positive, but less than ϵ, for $a' < x < b'$. From this definition, it follows that

$$\int_a^b \phi y(x)v(x)dx \neq 0.$$

Define g by the relation

$$g = \kappa\phi,$$

where κ is a constant and $0 < \kappa < 1$. Then, from (E),

$$\int_a^b \phi u(x)v(x)dx = 0.$$

Let it be assumed, for the moment, that Lemma III. is false. Then for $0 < \kappa < 1$, the system (C) is *at least* k-ply compatible, whereas, by virtue of Theorem I., its index cannot exceed k for sufficiently small values of κ. Let κ then be restricted to values sufficiently small to ensure that the index of (C) is precisely k. Then, by Lemma I., $u(x)$ is a continuous function of (x, κ) which approaches $y(x)$ uniformly as κ approaches zero through positive values, consequently

$$\int_a^b \phi u(x)v(x)dx \to \int_a^b \phi y(x)v(x)dx$$

uniformly as $\kappa \to 0$. But this is impossible since the first integral is zero for all values of k, whereas the second integral is not zero This contradiction demonstrates the truth of Lemma III.

From it follows :

THEOREM II.—*If a positive number* ϵ *is arbitrarily assigned, there exists a continuous real function* $g(x)$ *such that* $0 \leqslant g(x) < \epsilon$ *for which the system* (C) *is incompatible. The function* $g(x)$ *may be chosen as zero except in an arbitrarily small sub-interval of* (a, b).

The function g which was defined in the proof of Lemma III. lowers the index of the system (C) by at least unity. If the index is not then zero, the process may be repeated by defining a function $g_1(x)$ such that $0 \leqslant g_1 < \epsilon$, which is everywhere zero except in an interval (a'', b'') which does not overlap the interval (a', b'). Then the index of the system

$$\begin{cases} L(u) = gu + g_1 u, \\ U_1(u) = 0 \qquad (i = 1, 2, \ldots, n) \end{cases}$$

is at least one unit lower than that of (C) and therefore at least two units lower than that of (A). By continuing the process, the index may be reduced to zero. Theorem II. is therefore true.

If to the function $g(x)$, which renders the system (C) incompatible, there is added a sufficiently small function of x which is positive, but not zero, at

* If $y(x)$, for instance, had an infinite number of zeros in (a, b), these zeros would have a limit point, say c, in (a, b). Then $y(c) = y'(c) = \ldots = y^{(n-1)}(c) = 0$, which is impossible unless $y(x)$ is identically zero. See § 10·2, *infra*.

all points in (a, b), then, by Theorem I., the system remains incompatible. This consideration leads to a new theorem as follows :

THEOREM III.—*If a positive number* ϵ *be arbitrarily assigned, a continuous real function* $g(x)$ *such that* $0<g(x)<\epsilon$ *exists for which the system* (C) *is incompatible.*

MISCELLANEOUS EXAMPLES.

1. Show that the system

$$\begin{cases} y''+py'+qy=r, \\ a_1y(a)+a_2y(b)+a_3y'(a)+a_4y'(b)=A, \\ \beta_1y(a)+\beta_2y(b)+\beta_3y'(b)+\beta_4y'(b)=B \end{cases}$$

is self-adjoint if

$$a_2\beta_4-a_4\beta_2=(a_1\beta_3-a_3\beta_1)\exp\int_a^b p\,dx.$$

2. Prove that if ϵ is an arbitrary positive number and $x_1, \ldots, x_p$ are arbitrarily assigned points in (a, b), there exists a real continuous function $g(x)$ which vanishes and changes sign at each of the points x_i, but vanishes at no other point of (a, b), which satisfies the condition $|(g(x))|<\epsilon$, and which is such that the system

$$\begin{cases} L(u)=gu, \\ U_i(u)=0 \end{cases} \qquad (i=1, 2, \ldots, n)$$

is incompatible.

CHAPTER X

THE STURMIAN THEORY AND ITS LATER DEVELOPMENTS

10·1. The Purpose of the Sturmian Theory.—The present chapter deals, in the main, with equations of the type

$$L(y) \equiv \frac{d}{dx}\left\{K\frac{dy}{dx}\right\} - Gy = 0,$$

in which K and G are, throughout the closed interval $a \leqslant x \leqslant b$, continuous real functions of the real variable x. K does not vanish, and may therefore be assumed to be positive, and has a continuous first derivative throughout the interval.

The fundamental existence theorem (§ 3·32) has established the fact that this equation has one and only one continuous solution with a continuous derivative which satisfies the initial conditions

$$y(c) = \gamma_0, \quad y'(c) = \gamma_1,$$

where c is any point of the closed interval (a, b). But valuable as the existence theorem is from the theoretical point of view, it supplies little or no information as to the nature of the solution whose existence it demonstrates.

It is important from the point of view of physical applications, and not without theoretical interest, to determine the number of zeros which the solution has in the interval (a, b). This problem was first attacked by Sturm;* the theory based upon his work may now be regarded as classical. The two *Theorems of Comparison*, which form the core of the present chapter, are fundamental, and serve as the basis of a considerable body of further investigation.

10·2. The Separation Theorem.—No continuous solution of the equation can have an infinite number of zeros in (a, b) without being identically zero. For if there were an infinite number of zeros, these zeros would, by the Bolzano-Weierstrass theorem,† have at least one limit-point c. Then, not only $y(c)=0$, but also $y'(c)=0$. For

$$y(c+h) = y(c) + hy'(c+\theta h) \qquad (0 \leqslant \theta < 1)$$

and, since c is a limit-point of zeros, h may be taken so small that

$$y(c+h) = 0,$$

and therefore

$$y'(c+\theta h) = 0,$$

* *J. de Math.* 1 (1836), p. 106. The most complete account of the theory and its modern development is that given in the monograph by Bôcher: *Leçons sur les méthodes de Sturm* (Paris, 1917). See also the paper by the same author in the *Proceedings, Fifth International Congress* (Cambridge, 1912), I. p. 163.

† Whittaker and Watson, *Modern Analysis*, § 2·21.

from which, on account of the continuity of $y'(x)$, it follows that

$$y'(c)=0.$$

But the system

$$\begin{cases} L(y)=0, \\ y(c)=y'(c)=0 \end{cases}$$

has no solution not identically zero. This proves the theorem, which may be extended to the linear homogeneous equation of order n.

Now let y_1 and y_2 be any two real linearly-distinct solutions of the differential equation. It will be supposed that y_1 vanishes at least twice in (a, b); let x_1 and x_2 be two consecutive zeros of y_1 in that interval. Then *y_2 vanishes at least once in the open interval $x_1<x<x_2$.*

In the first place y_2 cannot vanish at x_1 or at x_2, for y_2 would then be a mere multiple of y_1. Suppose then that y_2 does not vanish at any point of (x_1, x_2). Now, the function y_1/y_2 is continuous and has a continuous derivative throughout the interval $x_1 \leqslant x \leqslant x_2$, and vanishes at its end-points. Its derivative must therefore vanish at not less than one internal point of the interval. But

$$\frac{d}{dx}\left\{\frac{y_1}{y_2}\right\}=\frac{y_2y_1'-y_1y_2'}{y_2^2},$$

a fraction whose numerator is the Wronskian of y_1 and y_2 and therefore cannot vanish at any point of (x_1, x_2). This contradiction proves that y_2 must have at least one zero between x_1 and x_2. It cannot have more than one such zero, for if it had two, then y_1 would have a zero between them, and x_1 and x_2 would not be consecutive zeros of y_1. The theorem which has thus been proved may be restated as follows: *the zeros of two real linearly-distinct solutions of a linear differential equation of the second order separate one another.*

This theorem does not hold if the solutions are not real. Thus, in the equation

$$y''+y=0,$$

the roots of the real solutions

$$y_1=\sin x, \quad y_2=\cos x$$

separate one another. More generally the roots of any two real solutions

$$y_1=A \sin x+B \cos x, \quad y_2=C \sin x+D \cos x$$

separate one another provided that $AD-BC\neq 0$, which is merely the condition that these two solutions are linearly independent. But the imaginary solution

$$y=\cos x+i \sin x$$

has no zero in any interval of the real variable x.

10·3. Sturm's Fundamental Theorem.—If there are two functions of x, say y_1 and y_2, defined and continuous in the interval (a, b), and if in this interval y_2 has more zeros than y_1, then y_2 is said to oscillate more rapidly than y_1. Thus, for instance, if m and n are positive integers and $m>n$, $\cos mx$ oscillates more rapidly than $\cos nx$ in the interval $(0, \pi)$ for the former has m, and the latter n zeros in that interval. The separation theorem of the previous paragraph may be stated roughly as follows: the zeros of all solutions of a given differential equation oscillate equally rapidly, by which it is implied that the number of zeros of any solution in an interval (α, β) lying in (a, b) cannot exceed the number of zeros of any independent solution in the same interval by more than one. If, in any interval, a solution has not more than one zero, it is said to be non-oscillatory in that interval.

The theorem to which this and the succeeding paragraph are devoted asserts that if the solutions of

$$\frac{d}{dx}\left\{K\frac{dy}{dx}\right\}-Gy=0$$

oscillate in the interval (a, b), they will oscillate more rapidly when K and G are diminished. In the first place, the theorem will be proved when G alone diminishes, K remaining unchanged.

Let u be a solution of

$$\frac{d}{dx}\left\{K\frac{du}{dx}\right\}-G_1u=0,$$

and v a solution of

$$\frac{d}{dx}\left\{K\frac{dv}{dx}\right\}-G_2v=0,$$

where $G_1 \geqslant G_2$ throughout (a, b), but $G_1 \neq G_2$ at all points of the interval. By multiplying the first equation throughout by v, and the second by u, and subtracting, it is found that

$$\frac{d}{dx}\{K(u'v-uv')\}=(G_1-G_2)uv,$$

whence

$$\Big[K(u'v-uv')\Big]_{x_1}^{x_2}=\int_{x_1}^{x_2}(G_1-G_2)uvdx,$$

a particular case of Green's formula.

Let the limits of integration x_1 and x_2 be taken to be consecutive zeros of u; suppose that v has no zero in the interval $x_1<x<x_2$. With no loss in generality u and v may be regarded as positive within that interval. The right-hand member of the above equation is then definitely positive. On the left-hand side u is zero at x_1 and at x_2, u_1' is positive at x_1 and negative at x_2, and v is positive at both limits. The left-hand member is, therefore, negative, which leads to a contradiction. Hence v vanishes at least once between x_1 and x_2.

In particular, if u and v are both zero at x_1, the theorem shows that v vanishes again before the consecutive zero of u appears. *Thus v oscillates more rapidly than u.*

For instance, the solutions of

$$v''+m^2v=0$$

oscillate more rapidly than those of

$$u''+n^2u=0,$$

provided that $m>n$.

10·31. The Modification due to Picone.—The more general theorem which compares the rapidity of the oscillation of the solutions of the two differential equations

$$\frac{d}{dx}\left\{K_1\frac{du}{du}\right\}-G_1u=0,$$

$$\frac{d}{dx}\left\{K_2\frac{dv}{dx}\right\}-G_2v=0,$$

wherein

$$K_1 \geqslant K_2 > 0, \quad G_1 \geqslant G_2,$$

may be attacked by means of the extended formula

$$\Big[K_1u'v-K_2uv'\Big]_{x_1}^{x_2}=\int_{x_1}^{x_2}(G_1-G_2)uvdx+\int_{x_1}^{x_2}(K_1-K_2)u'v'dx,$$

but a difficulty arises through the presence of the product $u'v'$ in the second integral. This difficulty was overcome by Picone,* who replaced the above formula by a similar one obtained as follows:

$$\frac{d}{dx}\left\{\frac{u}{v}(K_1u'v-K_2uv')\right\}$$

$$=\frac{u}{v}\left\{v\frac{d}{dx}(K_1u')-u\frac{d}{dx}(K_2v')+(K_1-K_2)u'v'\right\}+\frac{u'v-uv'}{v^2}(K_1u'v-K_2uv')$$

$$=\frac{u}{v}\{(G_1-G_2)uv+(K_1-K_2)u'v'\}+K_1u'^2-(K_1+K_2)uu'\frac{v'}{v}+K_2u^2\frac{v'^2}{v^2}$$

$$=(G_1-G_2)u^2+(K_1-K_2)u'^2+K_2\left\{u'-u\frac{v'}{v}\right\}^2.$$

Then

$$\left[\frac{u}{v}(K_1u'v-K_2uv')\right]_{x_1}^{x_2}=\int_{x_1}^{x_2}(G_1-G_2)u^2dx+\int_{x_1}^{x_2}(K_1-K_2)u'^2dx$$
$$+\int_{x_1}^{x_2}K_2\frac{(u'v-uv')^2}{v^2}dx,$$

which is known as *the Picone formula.*

Let x_1 and x_2 be consecutive zeros of u, and suppose that v is not zero at any point of the *closed* interval $x_1 \leqslant x \leqslant x_2$. Then the right-hand member of the Picone formula is positive (apart from the exceptional case mentioned below) and the left-hand member is zero. This contradiction proves that v has at least one zero in the interval (x_1, x_2).

The theorem also holds if v is zero at one or both of x_1 and x_2; a slight modification of the form of the left-hand member of the Picone formula is all that is necessary. Suppose, for instance, that v vanishes at x_1, then the indeterminate quantity u/v must be replaced by its limiting value u'/v', which is determinate since u' and v' are not zero at points where u and v respectively vanish. Consequently,

$$\lim_{x\to x_1}\left[\frac{u}{v}(K_1u'v-K_2uv')\right]=[(K_1-K_2)uu']_{x=x_1}$$
$$=0.$$

Thus, whether v is zero at x_1 and x_2 or not, the left-hand side of the Picone formula is zero, and the right-hand side positive, a contradiction which leads to the conclusion that v has at least one zero in the open interval $x_1 < x < x_2$.

If in any finite part of the interval (x_1, x_2) $G_1 > G_2$, then the first term of the right-hand member of the Picone formula is positive and not zero. The only conceivable case in which the right-hand member could become zero is when $G_1 = G_2$ throughout the interval (x_1, x_2), and $K_1 = K_2$ in part of the interval, whilst in the remainder of the interval $u' = 0$ (which implies $G_1 = 0$ in that range). The first and second integrals are then zero, the third is zero if v is proportional to u. The essence of the exception lies in the fact that if, in any part of (x_1, x_2), G is identically zero, then, within that range, K can be changed in any continuous way without increasing the oscillation of solution which is constant in that range. This exceptional case may be met by imposing the condition that G_1 and G_2 are not both identically zero in any finite part of (a, b).

10·32. Conditions that the Solutions of an Equation may be Oscillatory or Non-oscillatory.—The coefficients K and G in the equation

$$\text{(A)} \qquad \frac{d}{dx}\left\{K\frac{dy}{dx}\right\}-Gy=0$$

* *Ann. Scuola Norm. Pisa*, 11 (1909), p. 1.

being supposed to be continuous and bounded in the interval $a \leqslant x \leqslant b$, let the upper bounds of K and G in this interval be $\mathbf{K}$ and $\mathbf{G}$ and their lower bounds $\mathbf{k}$ and $\mathbf{g}$ respectively. Thus, throughout (a, b),

$$\mathbf{K} \geqslant K \geqslant \mathbf{k} > 0,$$
$$\mathbf{G} \geqslant G \geqslant \mathbf{g}.$$

As a first comparison equation, consider

$$\text{(B)} \qquad \frac{d}{dx}\left\{\mathbf{k}\frac{dy}{dx}\right\} - \mathbf{g}y = 0,$$

which may be written

$$\frac{d^2y}{dx^2} - \frac{\mathbf{g}}{\mathbf{k}}y = 0.$$

Then the solutions of equation (A) do not oscillate more rapidly in (a, b) than the solutions of (B). The latter equation is (as its alternative form shows) immediately integrable; its solutions are as follows:

1°. If $\mathbf{g} > 0$, there is the exponential solution $\exp\{\sqrt{(\mathbf{g}/\mathbf{k})}x\}$, which has no zero in (a, b). Similarly, if $\mathbf{g} = 0$, the comparison solution may be taken as unity. Hence, if $\mathbf{g} \geqslant 0$ the solutions of (B) are non-oscillatory. This leads to the conclusion that *if $G \geqslant 0$ throughout the interval (a, b), the solutions of the given equation* (A) *are non-oscillatory.*

2°. If $\mathbf{g} < 0$, there is the oscillatory solution $\sin\{\sqrt{(-\mathbf{g}/\mathbf{k})}x\}$; the interval between its consecutive zeros, or between consecutive zeros of any other solution of the comparison equation, is $\pi\sqrt{(-\mathbf{k}/\mathbf{g})}$. If, therefore,

$$\pi\sqrt{(-\mathbf{k}/\mathbf{g})} > b - a,$$

no solution of the given equation can have more than one zero in the interval (a, b). Consequently, the solutions of (A) are non-oscillatory provided that

$$-\frac{\mathbf{g}}{\mathbf{k}} < \frac{\pi^2}{(b-a)^2}.$$

Now consider, as a second comparison equation,

$$\text{(C)} \qquad \frac{d}{dx}\left\{\mathbf{K}\frac{dy}{dx}\right\} - \mathbf{G}y = 0,$$

or

$$\frac{d^2y}{dx^2} - \frac{\mathbf{G}}{\mathbf{K}}y = 0;$$

then the solutions of (A) oscillate at least as rapidly as those of (C). Let $\mathbf{G}$ be negative; then the solutions of (C) are oscillatory, and the interval between consecutive zeros of any solution is $\pi\sqrt{(-\mathbf{K}/\mathbf{G})}$. It follows that *a sufficient condition that the solutions of the given equation* (A) *should have at least m zeros in (a, b) is that*

$$m\pi\sqrt{(-\mathbf{K}/\mathbf{G})} \leqslant b - a,$$

or

$$-\frac{\mathbf{G}}{\mathbf{K}} \geqslant \frac{m^2\pi^2}{(b-a)^2}.$$

In particular, a sufficient condition that the equation (A) should possess a solution which oscillates in (a, b) is that

$$-\frac{\mathbf{G}}{\mathbf{K}} \geqslant \frac{\pi^2}{(b-a)^2}.$$

10·33. Application to the Sturm-Liouville Equation.—The equation

$$\frac{d}{dx}\left\{k\frac{dy}{dx}\right\} + (\lambda g - l)y = 0$$

is typical of a large class of equations which arise in problems of mathematical physics.* The oscillatory or non-oscillatory character of its solutions, and, in the oscillatory case, the number of zeros in an interval (a, b), are questions of considerable interest to the physicist.

If $k>0$ and $g>0$, which is the case in many physical problems, the equation can be regarded as a particular case of

$$\frac{d}{dx}\left\{K\frac{dy}{dx}\right\}-Gy=0,$$

with

$$K=k, \quad G=l-\lambda g.$$

In this case an increment in λ leaves k unaltered, but diminishes G and therefore increases the rapidity of the oscillation.

Another, and apparently distinct, case is that in which $k>0$, $l\geqslant 0$ and g changes sign within the interval (a, b). This case may, however, be brought under the general type by writing

$$K=\frac{k}{|\lambda|}, \quad G=\frac{l-\lambda g}{|\lambda|}.$$

If $|\lambda|$ increases whilst λ remains continually of one sign, both K and G diminish in general. If l is identically zero, K diminishes but G is unchanged. In either case an increment in $|\lambda|$ produces a more rapid oscillation of the solution.

10·4. The First Comparison Theorem.—This theorem aims at comparing the distribution of the zeros of the solution $u(x)$ of the equation

$$\frac{d}{dx}\left\{K_1\frac{du}{dx}\right\}-G_1u=0$$

which satisfies the initial conditions

$$u(a)=\alpha_1, \quad u'(a)=\alpha_1',$$

with the distribution of the zeros of the solution $v(x)$ of

$$\frac{d}{dx}\left\{K_2\frac{dv}{dx}\right\}-G_2v=0$$

which satisfies the conditions

$$v(a)=\alpha_2, \quad v'(a)=\alpha_2',$$

when, throughout the interval (a, b),

$$K_1\geqslant K_2>0, \quad G_1\geqslant G_2.$$

The following assumptions are made :

1°. α_1 and α_1' are not both zero, nor are α_2 and α_2 .

2°. If $\alpha_1\neq 0$, then

$$\frac{K_1(a)\alpha_1'}{\alpha_1}\geqslant\frac{K_2(a)\alpha_2'}{\alpha_2},$$

which implies that $\alpha_2\neq 0$.

3°. The identity $G_1\equiv G_2\equiv 0$ is not satisfied in any finite part of (a, b).

Then Sturm's first comparison theorem states that *if $u(x)$ has m zeros in the interval $a<x\leqslant b$, then $v(x)$ has at least m zeros in the same interval, and the i th. zero of $v(x)$ is less than the i th. zero of $u(x)$.*

Let $x_1, x_2, \ldots, x_m$ be the zeros of $u(x)$ which lie in (a, b) ; if these zeros are so enumerated that

$$a<x_1<x_2<\ldots<x_m\leqslant b,$$

then Sturm's fundamental theorem shows that between each pair of consecu-

* See § 9·41.

tive zeros x_i and x_{i+1} there lies at least one zero of $v(x)$. The comparison theorem follows at once if it can be proved that at least one zero of $v(x)$ lies between a and x_1.

If $u(x)$ has also a zero at the end-point a, that is to say, if $\alpha_1=0$, then $v(x)$ certainly has a zero between a and x_1; it will therefore be supposed that $\alpha_1 \neq 0$. Then, since $v(a)=\alpha_2 \neq 0$, the Picone formula

$$\left[u^2\left(K_1\frac{u'}{u}-K_2\frac{v'}{v}\right)\right]_a^{x_1}=\int_a^{x_1}(G_1-G_2)u^2dx+\int_a^{x_1}(K_1-K_2)u'^2dx$$
$$+\int_a^{x_1}K_2\frac{(u'v-uv')^2}{v^2}dx$$

may be applied. The right-hand member is positive; if the left-hand member is evaluated, on the supposition that v has no zero in (a, x_1), it is found to reduce to

$$-u^2(a)\left(\frac{K_1(a)\alpha_1'}{\alpha_1}-\frac{K_2(a)\alpha_2'}{\alpha_2}\right),$$

which is negative or zero in virtue of the second assumption. This contradiction proves that there is at least one zero of $v(x)$ between a and x_1. The theorem is therefore true.

If the zeros of the solution of the differential system

$$\begin{cases}\dfrac{d}{dx}\left\{K\dfrac{dy}{dx}\right\}-Gy=0,\\ y(a)=\alpha, \quad y'(a)=\alpha'\end{cases}$$

are marked in order on the line AB, where A is the point $x=a$, and B is $x=b$ $(a<b)$, then the effect of diminishing K and G, but leaving α and α' invariant, is to cause all the roots to move in the direction from B towards A. When K and G diminish continuously,* a stage may arrive when a new zero enters the segment AB. This new zero will first appear in the segment† at B; a further diminution of K and G will cause the zero to enter into the segment and to travel towards A.

10·41. The Second Comparison Theorem.—Let c be any interior point of the interval (a, b) which is not a zero of $u(x)$ or of $v(x)$, then in the open interval (a, c), $v(x)$ has by the first comparison theorem at least as many zeros as $u(x)$. The second comparison theorem states that *if c is such that $u(x)$ and $v(x)$ have the same number of zeros in the interval $a<x<c$, then*

$$\frac{K_1(c)u'(c)}{u(c)}>\frac{K_2(c)v'(c)}{v(c)}.$$

Let x_i be the zero next before c; it is necessarily a zero of $u(x)$ and not of $v(x)$, for between a and x_i there lie not less than i (and by supposition exactly i) zeros of $v(x)$. Then the Picone formula, taken between the limits x_i and c shows that

$$\left[u^2\left(K_1\frac{u'}{u}-K_2\frac{v'}{v}\right)\right]_{x_i}^{c}>0.$$

This gives at once the desired inequality. If $u(x)$ and $v(x)$ had no zero in

* This process may most easily be affected by supposing K and G to depend upon an auxiliary parameter λ, as in the Sturm-Liouville equation.

† The boundary conditions preclude the possibility of a new zero entering at A; since the solution is continuous and varies continuously with K and G, any new zero appearing at an interior point of (a, b) would appear as a double zero, which is contrary to the supposition that K does not vanish in (a, b). Any new zero which appears, therefore, enters the segment at B.

(a, c), the theorem would be proved in a similar manner by considering the Picone formula taken between the limits a and c.

Thus, in the system

$$\text{(A)} \qquad \begin{cases} L(y) \equiv \dfrac{d}{dx}\left\{K\dfrac{dy}{dx}\right\} - Gy = 0, \\ y(a) = \alpha, \quad y'(a) = \alpha', \end{cases}$$

the effect of continuously diminishing K and G is to cause the value of $K(x)y'(x)/y(x)$ at any point of (a, b), which was not originally a zero of $y(x)$, to diminish until that point becomes a zero of $y(x)$.

It may be noted that the comparison theorems which have been proved for the system (A) hold equally well in the case of the system

$$\text{(B)} \qquad \begin{cases} L(y) = 0, \\ y(a) = \rho\alpha, \quad y'(a) = \rho\alpha', \end{cases}$$

where ρ is any constant. For if $y(x)$ is the solution of (A), then $\rho y(x)$ will be the solution of (B). The truth of the remark is now obvious. But if ρ is regarded as arbitrary, then (B) is equivalent to the system

$$\text{(C)} \qquad \begin{cases} L(y) = 0, \\ \alpha' y(a) - \alpha y'(a) = 0, \end{cases}$$

in which the two non-homogeneous boundary conditions have been replaced by one homogeneous condition. Since the solution of (C) is $\rho y(x)$, the two comparison theorems hold in the case of the completely homogeneous system (C).

10·5. Boundary Problems in One Dimension.—By a boundary problem in its general sense is meant the question as to whether a given dfferential equation possesses or does not possess solutions which satisfy certain boundary, or end-point, conditions, and assuming that such solutions exist, to determine their functional nature and to investigate those modifications which arise through variations either in the differential equation itself, or in the assigned boundary conditions.

A boundary problem in one dimension is that aspect of the general problem which arises when the equation is an ordinary differential equation, in particular an ordinary linear equation, and the boundary conditions are relations which hold between the values of the solution and its successive derivatives for particular values of the independent variable x. The fundamental existence theorems of Chapter III. are in reality solutions of one-point boundary problems, for the initial conditions are such as refer to a single point x_0. In the following pages a wider aspect of the problem will be taken up, namely the two-point boundary problem, in which the boundary conditions relate to the two-end points of the interval $a \leqslant x \leqslant b$.

It will be supposed that the coefficients in the differential equation, and possibly also those which enter into the boundary conditions, depend upon a parameter λ. Thus it will be supposed that in

$$\frac{d}{dx}\left\{K\frac{dy}{dx}\right\} - Gy = 0,$$

K and G are continuous functions of (x, λ) when $a \leqslant x \leqslant b$, $\Lambda_1 < \lambda < \Lambda_2$, that K is positive and is uniformly differentiable with respect to x, its derived function being continuous in (a, b).* The coefficients in the boundary conditions are also assumed to be continuous functions of λ when $\Lambda_1 < \lambda < \Lambda_2$.

* It may happen that K has only an R-derivative at a and an L-derivative at b.

The questions which arise are now of two categories:

1°. *Questions of Existence.*—For what values of λ does a solution exist which satisfies all the conditions of the problem?

2°. *Questions of Oscillation.*—When a solution exists, how many zeros does it possess in the interval (a, b)?

For the one-point boundary problem, the first question is answered by the fundamental existence theorem, which states that for every value of λ in (Λ_1, Λ_2) a solution exists, and is a continuous function of (x, λ). The second question is then answered, in part at least, by the theorems which have been developed in this chapter. These theorems will now be developed and expanded in such a way that they become applicable to the more delicate two-point problem.*

10·6. Sturm's Oscillation Theorems.—The differential system which furnishes the simplest type of two-point boundary problem is the following, known as a Sturmian system:

$$\text{(A)}\qquad \begin{cases} L(y) \equiv \dfrac{d}{dx}\left\{K\dfrac{dy}{dx}\right\} - Gy = 0, \\ \alpha' y(a) - \alpha y'(a) = 0, \\ \beta' y(b) + \beta y'(b) = 0. \end{cases}$$

The particular boundary conditions which are here imposed are of a very special type, for each is, in itself, a one-point boundary condition. The equation, taken together with the first condition, has one and only one distinct solution, say $y = Y(x, \lambda)$. The association of this solution with the second boundary condition furnishes the characteristic equation

$$F(\lambda) \equiv \beta' Y(b, \lambda) + \beta Y'(b, \lambda) = 0,$$

whose roots are the characteristic numbers.

It will be supposed that K and G are real monotonic decreasing functions of λ, and, in accordance with the provisions of § 10·31, that G is not identically zero in any finite sub-interval of (a, b). The upper bounds $\mathbf{G}$ and $\mathbf{K}$, and the lower bounds $\mathbf{g}$ and $\mathbf{k}$ are continuous monotonic decreasing functions of λ in the interval (Λ_1, Λ_2).

It was seen in § 10·32 that if, for any particular value of λ, the equation

$$L(y) = 0$$

is such that

$$-\frac{\mathbf{G}}{\mathbf{K}} \geqslant \frac{m^2\pi^2}{(b-a)^2},$$

then, for that value of λ, the equation admits of a real solution, satisfying the boundary condition

$$\alpha' y(a) - \alpha y'(a) = 0,$$

and having at least m zeros in the interval (a, b). Now suppose that the further condition that

$$-\mathbf{G}/\mathbf{K} \rightarrow +\infty \quad \text{as} \quad \lambda \rightarrow \Lambda_2$$

is imposed; it will be proved that the solution in question can be caused to have any number of zeros, however great, in (a, b) by taking λ sufficiently near to Λ_2. The coefficients α and α' may be functions of λ, in which case it will be supposed that $K(a)\alpha'/\alpha$ is a monotonic decreasing function of λ.

* The oscillation theorem which immediately follows occurs in the famous paper by Sturm, already quoted, *J. de Math*, 1 (1836), p. 106. The boundary conditions are there, however, of a very special type. The investigation was brought to successive degrees of completion by Mason, *Trans. Am. Math. Soc.* 7 (1906), p. 337; Bôcher, *C. R. Acad. Sc. Paris*, 140 (1905), p. 928; Birkhoff, *Trans. Am. Math. Soc.* 10 (1909), p. 259.

Let λ be caused to increase from a number arbitrarily close to Λ_1, and suppose that the solution considered has initially i zeros in the open interval $a<x<b$. As λ increases, the number of zeros increases, and each zero tends to move in the direction of the end-point a. Consequently, for a certain value of λ, say $\lambda=\mu_i$, the solution will acquire an additional zero, which appears at the end-point b and then travels, as λ increases, towards a. For the value $\lambda=\mu_{i+1}$ another zero appears, and so on. Thus, there exists a sequence of numbers

$$\mu_i, \quad \mu_{i+1}, \quad \mu_{i+2}, \ldots$$

which have the limit-point Λ_2, and which are such that when

$$\mu_m<\lambda<\mu_{m+1},$$

the equation admits of a unique solution which has exactly $m+1$ zeros in (a, b), and which satisfies the first boundary condition.

Moreover, it was seen, by the second comparison theorem, that when λ varies from μ_m to μ_{m+1}, the expression

$$K(b)y'(b)/y(b)$$

is a monotonic decreasing function of λ. It must necessarily decrease from $+\infty$ to $-\infty$ because when $\lambda=\mu_m$ and $\lambda=\mu_{m+1}$, $y(b)=0$, but $y'(b)\neq 0$.

The effect of imposing the second boundary condition

$$\beta' y(b)+\beta y'(b)=0$$

in addition to the first will now be considered. The coefficients β and β' may be functions of λ; it will be supposed that β is not identically zero,* and that

$$K(b)\beta'/\beta$$

is a monotonic decreasing function of λ.

Since $K(b)y'(b)/y(b)$ is a function which, as λ increases from μ_m to μ_{m+1}, steadily decreases from $+\infty$ to $-\infty$, and since $-K(b)\beta'/\beta$ steadily increases in the same interval, there must be a single value of λ between μ_m and μ_{m+1} for which these two expressions become equal, that is to say, for which the second boundary condition is satisfied as well as the first. For this value of λ, say λ_{m+1}, the system is compatible; it admits of a solution which has precisely $m+1$ zeros in the interval $a<x<b$. The results which have been obtained so far may in part be summed up as:

THEOREM I. *The system* (A) *has an infinite number of real characteristic numbers which have no limit point but* Λ_2. *For each integer* $m\geqslant i$ *there exists one and only one characteristic number* λ_{m+1}, *to which corresponds a solution having* $m+1$ *zeros in the open interval* (a, b).

In order to obtain a degree of precision which is lacking in this theorem as it stands, a further assumption is made, namely that

$$-\mathbf{g}/\mathbf{k}\to-\infty \quad \text{as} \quad \lambda\to\Lambda_1.$$

Since $\mathbf{k}$ is positive for all relevant values of x and λ, this implies that, in the neighbourhood of Λ_1, $\mathbf{g}$ is positive.

Consider, then, as a comparison equation

$$\text{(B)} \qquad \frac{d}{dx}\left\{\mathbf{k}\frac{du}{dx}\right\}-\mathbf{g}u=0,$$

which may be written

$$u''-s^2u=0,$$

where

$$s^2=\mathbf{g}/\mathbf{k}>0$$

for values of λ sufficiently near to Λ_1.

* The case $\beta\equiv 0$ may be dismissed at once; the second boundary condition reduces to $y(b)=0$, the characteristic numbers are therefore $\mu_i, \mu_{i+1}, \ldots$

Let $u(x)$ be that solution of (B) which satisfies the initial conditions

(C) $$u(a)=\alpha,\ u'(a)=\alpha',$$

then

$$u(x)=\tfrac{1}{2}\left(\alpha+\frac{\alpha'}{s}\right)e^{s(x-a)}+\tfrac{1}{2}\left(\alpha-\frac{\alpha'}{s}\right)e^{-s(x-a)}.$$

For sufficiently large values of s, that is to say for values of λ sufficiently near to Λ_1, $u(x)$ approximates to $\alpha \cosh s(x-a)$ and therefore has no zeros.

Now let $y(x)$ be the solution of the original equation

$$\frac{d}{dx}\left\{K\frac{dy}{dx}\right\}-Gy=0$$

which satisfies the conditions (C).

Then the conditions of the first comparison theorem, viz.

$$K\geqslant \mathbf{k},\quad G\geqslant \mathbf{g},$$

$$K\alpha'/\alpha\geqslant \mathbf{k}\alpha'/\alpha,$$

are satisfied. Consequently $y(x)$ has no more zeros for $a<x<b$ than $u(x)$, and therefore, for values of λ sufficiently near to Λ_1, $y(x)$ has no zeros in (a, b). It follows that $i=0$.

It may now be proved that there exists one and only one characteristic number λ_0 in the interval (Λ_1, μ_0). Since, for values of λ in that interval, $y(x)$ and $u(x)$ have no zeros for $a<x<b$, it follows from the second comparison theorem that

$$\frac{K(b)y'(b)}{y(b)}\geqslant\frac{\mathbf{k}u'(b)}{u(b)}.$$

But as $\lambda\rightarrow\Lambda_1$, $s\rightarrow+\infty$ and therefore

$$u'(b)/u(b)\rightarrow+\infty,$$

and since $\mathbf{k}>0$,

$$K(b)y'(b)/y(b)\rightarrow+\infty.$$

Consequently, as λ increases from Λ_1 to μ_0, $K(b)y'(b)/y(b)$ steadily decreases from $+\infty$ to $-\infty$. The system has therefore one characteristic number, and one only, in the interval (Λ_1, μ_0). The sum total of these results is contained in the main theorem of oscillation:

THEOREM II.—*The real characteristic numbers of the system* (A) *may be arranged in increasing order of magnitude and may be denoted by*

$$\lambda_0,\quad \lambda_1,\quad \lambda_2,\quad \ldots,\quad \lambda_m,\quad \ldots.$$

if the corresponding characteristic functions are

$$y_0,\quad y_1,\quad y_2,\quad \ldots,\quad y_m,\quad \ldots,$$

then y_m will have exactly m zeros in the interval $a<x<b$.

The supposition that

$$\lim\{-\mathbf{g}/\mathbf{k}\}=-\infty,$$

upon which Theorem II. depends, was made for the express purpose of ensuring that the characteristic number λ_0 should exist. This condition, though sufficient, is very far from being necessary for the existence of λ_0; its chief importance lies in its practical applicability. Another set of conditions, sufficient to ensure the existence of λ_0 and of some utility in later work is as follows.

Up to the present it has been supposed that K, G, α, α', β and β' are defined in the open interval $\Lambda_1<\lambda<\Lambda_2$; it will now be supposed that the interval is closed at its left-hand end-point, that is to say that Λ_1 belongs to the interval. Let

$$K_1,\quad G_1,\quad \alpha_1,\quad \alpha_1',\quad \beta_1,\quad \beta_1'$$

be the values of the corresponding quantities when $\lambda=\Lambda_1$, and suppose that

$$\mathbf{g}_1\geqslant 0, \quad \alpha_1\alpha_1'\geqslant 0, \quad \beta_1\beta_1'\geqslant 0,$$

but that α_1 and α_1' are not both zero, nor are β_1 and β_1'.

Now consider the comparison system

$$\text{(C)} \qquad \begin{cases} \dfrac{d}{dx}\left\{\mathbf{k}_1\dfrac{du}{dx}\right\}-\mathbf{g}_1u=0, \\ \alpha_1'u(a)-\alpha_1u'(a)=0\,; \end{cases}$$

the differential equation may be written as

$$u''-s^2u=0,$$

in which

$$s^2=\mathbf{g}_1/\mathbf{k}_1\geqslant 0.$$

Suppose for the moment that $s>0$, then the solution of the comparison system may be taken as

$$u(x)=\alpha_1\cosh s(x-a)+\frac{\alpha_1'}{s}\sinh s(x-a),$$

so that $u(x)$ is definitely positive or definitely negative for $x>a$.

Now if $v(x)$ is the solution of the system

$$\text{(D)} \qquad \begin{cases} \dfrac{d}{dx}\left\{K_1\dfrac{dv}{dx}\right\}-G_1v=0, \\ \alpha_1'v(a)-\alpha_1v'(a)=0, \end{cases}$$

in which K_1 and G_1 represent K and G when $\lambda=\Lambda_1$, the first comparison theorem states that $v(x)$ can have no more zeros in (a, b) than $u(x)$ has ; it therefore has no zeros in (a, b), in other words $i=0$. For a certain value of λ greater than Λ_1, namely $\lambda=\mu_0$, the solution $y(x)$ of

$$\begin{cases} \dfrac{d}{dx}\left\{K\dfrac{dy}{dx}\right\}-Gy=0, \\ \alpha'y(a)-\alpha y'(a)=0 \end{cases}$$

(which reduces to $v(x)$ when $\lambda=\Lambda_1$) will have a zero at $x=b$. Since neither $u(x)$ nor $y(x)$ has a zero in (a, b) when $\Lambda_1\leqslant\lambda<\mu_0$, the second comparison theorem may be applied ; it shows that

$$\frac{K_1(b)y'(b)}{y(b)}>\frac{\mathbf{k}_1u'(b)}{u(b)},$$

when $\Lambda_1\leqslant\lambda<\mu_0$. The right-hand member of the inequality may be calculated directly ; it is readily found to be positive, from which it follows that the left-hand member is also positive. Thus the expression

$$K(b)y'(b)/y(b),$$

which assumes the value $K_1(b)v'(b)/v(b)$ when $\lambda=\Lambda_1$, steadily diminishes from a value greater than zero to negative infinity as λ increases from Λ_1 to μ_0. Since

$$-K(b)\beta'/\beta$$

steadily increases from a negative value when $\lambda=\Lambda_1$, a point must come at which the two expressions become equal, and for that value of λ, say λ_0, $y(x)$ satisfies also the second boundary condition

$$\beta y'(b)+\beta'y(b)=0.$$

There is, therefore, a characteristic number λ_0 in the interval (Λ_1, μ_0) distinct from Λ_1 and μ_0 (except when $\beta=0$, in which case $\lambda_0=\mu_0$) such that the system (A) has a solution which has no zeros in the interval $a<x<b$.

The special case $s=0$ may now be considered very briefly. The solution

$u(x)$ is here a linear function of the argument $x-a$. Furthermore $u(x)$ is definitely positive or negative in (a, b) and $\mathbf{k}u'(b)/u(b)$ is in general positive but may be zero. Thus, as before, the characteristic number λ_0 exists but may, in a special case, coincide with Λ_1. This case arises when

$$\alpha_1'=\beta_1'=0, \quad G_1 \equiv 0,$$

but in no other circumstances. Hence follows :

THEOREM III.—*Under the assumption that*

$$\mathbf{g}_1 \geqslant 0, \quad \alpha_1\alpha_1' \geqslant 0, \quad \beta_1\beta_1' \geqslant 0,$$

the system (A) *has an infinite set of real characteristic numbers*

$$\lambda_0, \quad \lambda_1, \quad \lambda_2, \quad \ldots, \quad \lambda_m, \ldots,$$

to which correspond the characteristic functions

$$y_0, \quad y_1, \quad y_2, \quad \ldots, \quad y_m, \ldots,$$

such that y_m has exactly m zeros in the interval $a<x<b$. The least characteristic number λ_0 is distinct from Λ_1 except in the case

$$G_1 \equiv 0, \quad \alpha_1'=0, \quad \beta_1'=0.$$

10·61. Application to the Sturm-Liouville System.—The group of theorems now known as the oscillation theorems were first proved by Sturm * in the case of the system

$$\begin{cases} \dfrac{d}{dx}\left\{k\dfrac{dy}{dx}\right\}+(\lambda g-l)y=0, \\ \alpha' y(a)-\alpha y'(a)=0, \\ \beta' y(a)+\beta y'(a)=0, \end{cases}$$

which has already been met with.†

In this case it will be supposed that k, g and l are real continuous functions of x when $a \leqslant x \leqslant b$, are independent of λ and are such that $k>0$, $g>0$. The coefficients α, α', β and β' are also independent of λ. Since $G \equiv l-\lambda g$ steadily decreases, or at most remains constant for any value of x in (a, b) as λ increases from $\Lambda_1=-\infty$ to $\Lambda_2=+\infty$, the conditions which were imposed in the course of the proof of Theorem II. (§ 10·6) are satisfied. In particular

$$-\mathbf{G}=\min(\lambda g-l) \to +\infty,$$

as $\lambda \to +\infty$. Consequently *there exists an infinite set of real characteristic numbers $\lambda_0, \lambda_1, \lambda_2, \ldots$, which have no limit-point except $\lambda=+\infty$; if the corresponding characteristic functions are $y_0, y_1, y_2, \ldots$, then y_m has exactly m zeros in the interval $a<x<b$.*

If the additional conditions

$$l \geqslant 0, \quad \alpha\alpha' \geqslant 0, \quad \beta\beta' \geqslant 0$$

for $\lambda=0$ are imposed, then Λ_1 may be taken to be zero. In this case, when $\lambda=0$,

$$\mathbf{g}=\min l \geqslant 0.$$

and the characteristic numbers are all positive. This case is important from the physical point of view.

Now consider the case in which $k>0$, $l \geqslant 0$ and g changes sign in the interval (a, b). The problem may be attacked by precisely the same device as that which was adopted in § 10·33. Rewrite the equation as

$$\frac{d}{dx}\left\{\frac{k}{|\lambda|}\frac{dy}{dx}\right\}-\frac{l-\lambda g}{|\lambda|}y=0 ;$$

* *J. de Math.* 1 (1836), pp. 139, 143. † §§ 9·41, 10·33.

it is now of the general type considered in § 10·6 if

$$K=\frac{k}{|\lambda|},\quad G=\frac{l}{|\lambda|}-g \qquad \text{when } \lambda>0,$$

$$K=\frac{k}{|\lambda|},\quad G=\frac{l}{|\lambda|}+g \qquad \text{when } \lambda<0.$$

In either case, K and G steadily diminish as $|\lambda|$ increases; if the conditions $\alpha\alpha'\geqslant 0$, $\beta\beta'\geqslant 0$ are also satisfied,

$$\frac{\alpha' k(a)}{\alpha\,|\lambda|} \quad\text{and}\quad \frac{\beta' k(b)}{\beta\,|\lambda|}$$

steadily diminish as $|\lambda|$ increases. Up to the present point the required conditions are satisfied, but if it is noted that, since g changes sign in (a, b),

$$\mathbf{G}>0 \quad\text{and}\quad \mathbf{K}>0,$$

it is seen that

$$-\mathbf{G}/\mathbf{K}\rightarrow-\infty$$

as $|\lambda|\rightarrow\infty$.

Thus the conditions of Theorem I. (§ 10·6) are not satisfied; it does not, however, follow that the theorem is false in the case now considered. On the contrary, since g changes sign in (a, b) a sub-interval (a', b') can be found in which

$$g>0 \text{ in the case } \lambda>0,$$
$$g<0 \text{ in the case } \lambda<0.$$

In either case, values of λ may be taken sufficiently large in absolute value to make it certain that $G<0$ in (a', b'). Consequently the required condition that

$$-\mathbf{G}/\mathbf{K}\rightarrow+\infty$$

as $|\lambda|\rightarrow\infty$ is fulfilled in the interval (a', b'). Thus λ may be taken sufficiently great to ensure that the solution of the system

$$\begin{cases}\dfrac{d}{dx}\left\{k\dfrac{dy}{dx}\right\}+(\lambda g-l)y=0,\\ \alpha' y(a)-\alpha y'(a)=0\end{cases}$$

oscillates in (a', b') and *a fortiori* in (a, b). The number of zeros in (a, b) may be increased indefinitely by taking λ sufficiently large.

But on the other hand the solution of the system

$$\begin{cases}\dfrac{d}{dx}\left\{k\dfrac{du}{dx}\right\}-lu=0,\\ \alpha' y(a)-\alpha y'(a)=0\end{cases}$$

(which is the case $\lambda=0$) has no zero in (a, b) if $l\geqslant 0$ except possibly in the case $l\equiv 0$, when one zero may exist.

Let it be supposed that

$$l\geqslant 0,\quad \alpha\alpha'\geqslant 0,\quad \beta\beta'\geqslant 0,$$

and let the special case

$$l\equiv 0,\quad \alpha'=\beta'=0,$$

which requires special treatment,* be excluded. Then the methods by which Theorem III. (§ 10·6) was proved may be utilised here to demonstrate the existence of characteristic numbers to which correspond characteristic functions having $0, 1, 2, \ldots, m, \ldots$ zeros in (a, b). The only real difference is that the case $\lambda<0$ separates itself from the case $\lambda>0$ so that

* Such a treatment is given by Picone, *Ann. Scuola Norm. Pisa*, 11 (1909), p. 39; Bôcher, *Bull. Am. Math. Soc.* 21 (1914), p. 6.

there is an infinite set of negative characteristic numbers with the limit-point $\lambda = -\infty$ as well as an infinite set of positive characteristic numbers with the limit-point $\lambda = +\infty$. The oscillation theorem now reads as follows: *

If g changes sign in (a, b), and

$$l \geqslant 0, \quad \alpha\alpha' \geqslant 0, \quad \beta\beta' \geqslant 0,$$

there exists an infinite set of real characteristic numbers which have the limit-points $+\infty$ and $-\infty$. If the positive and negative characteristic numbers are arranged each in order of increasing numerical value, and are denoted by

$$\lambda_0^+, \ \lambda_1^+, \ \lambda_2^+, \ \ldots, \ \lambda_m^+, \ \ldots,$$
$$\lambda_0^-, \ \lambda_1^-, \ \lambda_2^-, \ \ldots, \ \lambda_m^-, \ \ldots;$$

and the corresponding characteristic functions by

$$y_0^+, \ y_1^+, \ y_2^+, \ \ldots, \ y_m^+, \ \ldots,$$
$$y_0^-, \ y_1^-, \ y_2^-, \ \ldots, \ y_m^-, \ \ldots,$$

then y_m^+ and y_m^- have exactly m zeros in the interval $a<x<b$.

10·7. The Orthogonal Property of Characteristic Functions and its Consequences.—Consider the differential system

$$\text{(A)} \quad \begin{cases} L(u)+\lambda gu \equiv p_0\dfrac{d^n u}{dx^n}+p_1\dfrac{d^{n-1}u}{dx^{n-1}}+\ \ldots\ +p_{n-1}\dfrac{du}{dx}+(p_n+\lambda g)u=0, \\ U_i(u)=0 \qquad\qquad (i=1, 2, \ldots, n), \end{cases}$$

in which the coefficients $p_0, p_1, \ldots, p_{n-1}, p_n, g$ in the differential equation and the coefficients which enter into the expressions $U_i(u)$ are independent of the parameter λ. The adjoint system is

$$\text{(B)} \quad \begin{cases} \bar{L}(v)+\lambda gv \equiv (-1)^n\dfrac{d^n(p_0v)}{dx^n}+(-1)^{n-1}\dfrac{d^{n-1}(p_1v)}{dx^{n-1}}+\ \ldots\ -\dfrac{d(p_{n-1}x)}{dx} \\ \qquad\qquad\qquad\qquad +(p_n+\lambda g)v=0, \\ V_i(v)=0 \qquad\qquad (i=1, 2, \ldots, n). \end{cases}$$

Let the system (A) admit of at least two characteristic numbers, say λ_i and λ_j, and let the corresponding characteristic functions be u_i and u_j. Then the system (B) is compatible for λ_i and λ_j; let the characteristic functions be v_i and v_j.

Now Green's formula

$$\int_a^b \{vL(u)-u\bar{L}(v)\}=U_1V_{2n}+U_2V_{2n-1}+\ \ldots\ +U_{2n}V_1$$

(§ 9·31) holds whatever u and v may be. Let $u=u_i$ and $v=v_j$, then the right-hand member vanishes since

$$U_1(u_i)=U_2(u_i)=\ \ldots\ =U_n(u_i)=0,$$
$$V_1(v_j)=V_2(v_j)=\ \ldots\ =V_n(v_j)=0.$$

Consequently

$$\int_a^b \{v_jL(u_i)-u_i\bar{L}(v_j)\}dx=0,$$

which reduces to

$$(\lambda_j-\lambda_i)\int_a^b gu_iv_jdx=0,$$

* Sanlievici, *Ann. Éc. Norm.* (3) 26 (1909), p. 19; Picone, *loc. cit.*; Richardson, *Math. Ann.* 68 (1910), p. 279.

and since λ_i and λ_j are distinct

$$\int_a^b g u_i v_j dx = 0.$$

In particular when the system (A) is self-adjoint,

$$\int_a^b g u_i u_j dx = 0 \qquad (i \neq j).$$

A set of functions

$$u_1, u_2, \ldots, u_i, \ldots, u_j, \ldots$$

which are such that, the function g being assigned,

$$\int_a^b g u_i u_j dx = 0 \qquad (i \neq j),$$

are said to be *orthogonal* with respect to the function g; if, in addition,

$$\int_a^b g u_i^2 dx > 0,$$

then each function u_i may be multiplied by a constant so that

$$\int_a^b g u_i^2 dx = 1,$$

when so adjusted the functions are said to be *normal.* The characteristic functions of the system (A), when the latter is self-adjoint, therefore form an orthogonal set. In certain cases, and in particular when $g > 0$, they can also be normalised.

From this orthogonal property follows the important theorem that *if $g > 0$ throughout the interval (a, b), the characteristic numbers are all real.* For suppose that $\lambda_i = \sigma + i\tau$ is a complex characteristic number, then since the coefficients of the system are all real, among the remaining characteristic numbers is the number conjugate to λ_i, say $\lambda_j = \sigma - i\tau$. If the characteristic function u_i is $s + it$, then u_j will be its conjugate $s - it$. Then

$$\int_a^b g u_i u_j dx = \int_a^b g(s^2 + t^2) dx,$$

which cannot be zero unless $s \equiv t \equiv 0$. Thus when $g > 0$, the assumption of the existence of complex characteristic numbers leads to a contradiction, which proves the theorem. The condition $g > 0$ may be replaced by the less stringent condition $g \geqslant 0$ provided that the equality does not hold at all points of any finite sub-interval of (a, b).

10·71. Application to Sturm-Liouville Systems.—The preceding investigation is immediately applicable to the Sturm-Liouville system,

$$\text{(A)} \qquad \begin{cases} \dfrac{d}{dx}\left\{k\dfrac{dy}{dx}\right\} + (\lambda g - l)y = 0, \\ \alpha' y(a) - \alpha y'(a) = 0, \\ \beta' y(b) + \beta y'(b) = 0\,; \end{cases}$$

if g is of one sign throughout the interval (a, b), every characteristic number is real.*

If, on the other hand, g changes sign in (a, b), then all the characteristic numbers may be proved to be real provided that the conditions

$$k > 0, \quad l \geqslant 0, \quad \alpha\alpha' \geqslant 0, \quad \beta\beta' \geqslant 0$$

hold (cf. § 10·61). Let it be supposed, for the moment, that λ_i is a complex

* This theorem can be traced back to Poisson, *Bull. Soc. Philomath. Paris,* 1826, p. 145.

characteristic number, say $\sigma+i\tau$; the corresponding characteristic function y_i will be complex, say $s+it$. Then the equation

$$\frac{d}{dx}\left\{k\left(\frac{ds}{dx}+i\frac{dt}{dx}\right)\right\}+\{(\sigma+i\tau)g-l\}(s+it)=0$$

is satisfied identically. The real and imaginary parts, equated separately to zero, give respectively

$$S\equiv\frac{d}{dx}\left\{k\frac{ds}{dx}\right\}+(\sigma g-l)s-\tau gt=0,$$

$$T\equiv\frac{d}{dx}\left\{k\frac{dt}{dx}\right\}+\tau gs+(\sigma g-l)t=0.$$

From these equations it follows that

$$\int_a^b(sS+tT)dx=\Big[k(ss'+tt')\Big]_a^b-\int_a^b k(s'^2+t'^2)dx$$

$$+\sigma\int_a^b g(s^2+t^2)dx-\int_a^b l(s^2+t^2)dx=0.$$

Now

$$\Big[k(ss'+tt')\Big]_a^b\leqslant 0,$$

by virtue of the restrictions

$$k>0,\ \alpha\alpha'\geqslant 0,\ \beta\beta'\geqslant 0\,;$$

also

$$-\int_a^b k(s'^2+t'^2)dx<0,$$

since $k>0$ in (a, b), and s' and t' are not identically zero; * and finally

$$\int_a^b g(s^2+t^2)dx=0,$$

$$-\int_a^b l(s^2+t^2)dx\leqslant 0.$$

The contradiction which evidently follows proves that no complex or imaginary characteristic number can exist in the case under consideration. In this case also it may be proved that, if y_i be any characteristic function,

$$\int_a^b gy_i^2dx\neq 0.$$

Let λ_i be the characteristic number to which y_i corresponds, then

$$\lambda_i gy_i=ly_i-\frac{d}{dx}\left\{k\frac{dy_i}{dx}\right\}.$$

If this identity is multiplied through by y_i and integrated between the limits a and b, it gives rise to the relation

$$\lambda_i\int_a^b gy_i^2dx=\int_a^b ly_i^2dx+\int_a^b ky_i'^2dx-\Big[ky_iy_i'\Big]_a^b.$$

The first term in the right-hand member is positive or zero, the second is definitely positive and the third is positive or zero. Hence

$$\int_a^b gy_i^2dx>0 \quad\text{if}\quad \lambda_i>0,$$
$$<0 \quad\text{if}\quad \lambda_i<0.$$

* $s'=t'=0$ would imply $(\sigma+i\tau)g-l\equiv 0$, and therefore $\tau g\equiv 0$; since $\tau\neq 0$, $g\equiv 0$ contrary to the supposition that g changes sign in (a, b).

In the notation of § 10·51, the characteristic functions y_i^+ and y_i may be multiplied by appropriate *real* constants so that

$$\int_a^b g(y_i^+)^2 dx = +1,$$

$$\int_a^b g(y_i^-)^2 dx = -1.$$

Now consider the more general system : *

$$\text{(B)} \qquad \begin{cases} \dfrac{d}{dx}\left\{k\dfrac{dy}{dx}\right\} + (\lambda g - l)y = 0, \\ \alpha_1 y(a) + \alpha_2 y(b) + \alpha_3 y'(a) + \alpha_4 y'(b) = 0, \\ \beta_1 y(a) + \beta_2 y(b) + \beta_3 y'(a) + \beta_4 y'(b) = 0, \end{cases}$$

(cf. § 9·41). It is supposed that at least two of the ratios

$$\frac{\alpha_1}{\beta_1}, \quad \frac{\alpha_2}{\beta_2}, \quad \frac{\alpha_3}{\beta_3}, \quad \frac{\alpha_4}{\beta_4}$$

are unequal. If

$$\frac{\alpha_1}{\beta_1} = \frac{\alpha_3}{\beta_3}, \quad \frac{\alpha_2}{\beta_2} = \frac{\alpha_4}{\beta_4},$$

the system reduces to (A). This particular case is rejected as having been dealt with ; in any other case the boundary conditions are reducible to

$$\text{(C)} \qquad \begin{aligned} y(a) &= \gamma_1 y(b) + \gamma_1' y'(b) \\ y'(a) &= \gamma_2 y(b) + \gamma_2' y'(b). \end{aligned}$$

It will be supposed that the condition

$$\text{(D)} \qquad k(b) = (\gamma_1\gamma_2' - \gamma_1'\gamma_2)k(a),$$

that the system may be self-adjoint, is satisfied.

Now the relation

$$\Big[k(ss' + tt')\Big]_a^b - \int_a^b k(s'^2 + t'^2)dx - \int_a^b l(s^2 + t^2)dx = 0,$$

which is a necessary consequence of the supposition that the system (B) admits of a complex characteristic number, is violated when $k > 0$, $l \geqslant 0$, if

$$\Big[k(ss' + tt')\Big]_a^b \leqslant 0,$$

that is to say, if

$$k(a)s(a)s'(a) - k(b)s(b)s'(b) \geqslant 0,$$
$$k(a)t(a)t'(a) - k(b)t(b)t'(b) \geqslant 0.$$

It follows from (C) that these two inequalities are satisfied if

$$k(a)\{\gamma_1\xi + \gamma_1'\eta\}\{\gamma_2\xi + \gamma_2'\eta\} - k(b)\xi\eta \geqslant 0,$$

where $\xi = s(b)$, $\eta = s'(b)$, or $\xi = t(b)$, $\eta = t'(b)$. By means of (D) this inequality reduces to

$$\gamma_1\gamma_2\xi^2 + 2\gamma_1'\gamma_2\xi\eta + \gamma_1'\gamma_2'\eta^2 \geqslant 0,$$

which may also be written

$$\frac{(\gamma_1\gamma_2\xi + \gamma_1'\gamma_2\eta)^2 + \gamma_1'\gamma_2(\gamma_1\gamma_2' - \gamma_1'\gamma_2)\eta^2}{\gamma_1\gamma_2} \geqslant 0.$$

The condition (C) implies that $\gamma_1\gamma_2' - \gamma_1'\gamma_2 > 0$; it follows that the above inequalities are satisfied when both $\gamma_1'\gamma_2 \geqslant 0$ and $\gamma_1\gamma_2 \geqslant 0$.

The system (B) then admits of none but real characteristic numbers.

* Mason, *Trans. Am. Math. Soc.* 7 (1906), p. 337.

These conditions are satisfied in a very important case, namely that of the periodic boundary conditions,

$$y(a)=y(b), \quad y'(a)=y'(b).$$

Thus if $k>0$, $l\geqslant 0$ and $k(a)=k(b)$, the characteristic numbers of the system are all real.

10·72. The Index and Multiplicity of the Characteristic Numbers.—Consider again the simple Sturm-Liouville system:

$$\begin{cases} \dfrac{d}{dx}\left\{k\dfrac{dy}{dx}\right\}+(\lambda g-l)y=0, \\ \alpha' y(a)-\alpha y'(a)=0, \\ \beta' y(b)+\beta y'(b)=0. \end{cases}$$

If, for any particular value of λ, the index of the system were 2, then the most general solution of the equation would satisfy the first boundary-condition, which is clearly impossible. The index of the system, for each characteristic number, is therefore unity.

Let $y(x, \lambda)$ be the solution of the differential equation which satisfies the first boundary-condition. Then the second boundary-condition imposed upon $y(x, \lambda)$ gives the characteristic equation, viz.

$$F(\lambda)\equiv\beta' y(b, \lambda)+\beta y'(b, \lambda)=0.$$

Let λ_i be a characteristic number, and $y(x, \lambda_i)$ the corresponding characteristic function, then

$$\frac{d}{dx}\left\{k\frac{d}{dx}y(x, \lambda)\right\}+(\lambda g-l)y(x, \lambda)=0,$$

$$\frac{d}{dx}\left\{k\frac{d}{dx}y(x, \lambda_i)\right\}+(\lambda_i g-l)y(x, \lambda_i)=0.$$

By eliminating l between these equations and then integrating the eliminant between the limits a and b, there is obtained the relation

$$\Big[k\{y(x, \lambda)y'(x, \lambda_i)-y'(x, \lambda)y(x, \lambda_i)\Big]_a^b+(\lambda_i-\lambda)\int_a^b gy(x, \lambda_i)y(x, \lambda)dx=0,$$

which, in view of the fact that $y(x, \lambda)$ and $y(x, \lambda_i)$ both satisfy the first boundary-condition, while $y(x, \lambda_i)$ satisfies also the second boundary-condition, reduces to

$$\beta'\int_a^b gy(x, \lambda_i)y(x, \lambda)dx=k(b)y'(b, \lambda_i)\{\beta' y(b, \lambda)+\beta y'(b, \lambda)\}/(\lambda-\lambda_i)$$
$$=k(b)y'(b, \lambda_i)F(\lambda)/(\lambda-\lambda_i).$$

Now as $\lambda\to\lambda_i$,

$$F(\lambda)/(\lambda-\lambda_i)\to F'(\lambda_i) \quad \text{since} \quad F(\lambda_i)=0,$$
$$y(x, \lambda)\to y(x, \lambda_i)$$

uniformly because $y(x, \lambda)$ is an integral function of λ. Consequently in the limit,

$$\beta'\int_a^b g\{y(x, \lambda_i)\}^2dx=k(b)y'(b, \lambda_i)F'(\lambda_i).$$

If $\beta'\neq 0$, the left-hand member of the equation is not zero. It follows that $F'(\lambda_i)\neq 0$, that is to say, λ_i is a *simple* root of the characteristic equation.

If $\beta'=0$, a modification of the method leads to the same result with the possible exception of the case in which g changes sign in (a, b), l is identically zero, and $\alpha'=\beta'=0$. In that case the characteristic numbers may occur as double roots of the characteristic equation.

10·8. Periodic Boundary Conditions.—The system which will now be considered is the following: *

$$\text{(A)} \qquad \begin{cases} L(y) \equiv \dfrac{d}{dx}\left\{K\dfrac{dy}{dx}\right\} - Gy = 0, \\ y(a) = y(b), \\ y'(a) = y'(b), \end{cases}$$

in which the condition that the system be self-adjoint, viz. $K(a)=K(b)$, is satisfied. It includes, as a most important particular case, that in which K and G are periodic functions, with period $(b-a)$, but in reality it goes far beyond this case.

It is, as before, assumed that K and G are continuous functions of (x, λ) when $a \leqslant x \leqslant b$, $\Lambda_1 < \lambda < \Lambda_2$, and that both decrease as λ increases. The slightly more stringent restriction that

$$\frac{\partial G}{\partial \lambda} < 0$$

is also made; this does not exclude the most important of all cases, the Sturm-Liouville case where $G = l - \lambda g$, $g > 0$. It is also assumed that

$$\lim_{\lambda=\Lambda_1} \frac{-\mathbf{g}}{\mathbf{k}} = -\infty, \quad \lim_{\lambda=\Lambda_2} \frac{-\mathbf{G}}{\mathbf{K}} = +\infty.$$

Let $y_1(x, \lambda)$ and $y_2(x, \lambda)$ be two fundamental solutions of the differential equation chosen so as to satisfy the initial conditions

$$y_1(a, \lambda) = 1, \quad y_2(a, \lambda) = 0,$$
$$y_1'(a, \lambda) = 0, \quad y_2'(a, \lambda) = 1,$$

then, by Abel's formula (§ 9·4),

$$\text{(B)} \qquad y_1(b, \lambda) y_2'(b, \lambda) - y_2(b, \lambda) y_1'(b, \lambda) = K(a)/K(b) = 1,$$

a relation satisfied identically for all values of λ.

The characteristic equation is

$$\begin{vmatrix} y_1(a, \lambda) - y_1(b, \lambda), & y_2(a, \lambda) - y_2(b, \lambda) \\ y_1'(a, \lambda) - y_1'(b, \lambda), & y_2'(a, \lambda) - y_2'(b, \lambda) \end{vmatrix} = 0,$$

or

$$\begin{vmatrix} 1 - y_1(b, \lambda), & -y_2(b, \lambda) \\ -y_1'(b, \lambda), & 1 - y_2'(b, \lambda) \end{vmatrix} = 0,$$

which, by virtue of the above identity (B), reduces to

$$\text{(C)} \qquad F(\lambda) \equiv y_1(b, \lambda) + y_2'(b, \lambda) - 2 = 0.$$

A number λ such that $F(\lambda)=0$, but not all the elements of the characteristic determinant are zero, is said to be a *simple characteristic number*. If all these elements are zero, then there will exist two linearly independent solutions of the system (A). Such a value of λ, for which

$$y_1(b, \lambda) = 1, \quad y_2(b, \lambda) = 0,$$
$$y_1(b, \lambda) = 0, \quad y_2'(b, \lambda) = 1,$$

is said to be a *double characteristic number*.

The immediate problem is to prove that, under the conditions stated, the characteristic equation admits, as its roots, of an infinite set of real character-

* Tzitzéica, *C. R. Acad. Sc. Paris*, 140 (1905), p. 223; Bôcher, *ibid.* p. 928; Mason, *ibid.* p. 1086; *Math. Ann.* 58 (1904), p. 528; *Trans. Am. Math. Soc.* 7 (1906), p. 337. See also Picard, *Traité d'Analyse*, 3 (1st ed.), p. 140; (2nd ed.), p. 138. Extensions to the general self-adjoint linear system of the second order have been made by Birkhoff, *Trans. Am. Math. Soc.* 10 (1909), p. 259; and Ettlinger, *ibid.* 19 (1918), p. 79; 22 (1921), p. 136.

istic numbers.* This problem is attacked, in an indirect manner, by studying the sign of $F(\lambda)$ for certain values of λ corresponding to which the solutions of $L(y)=0$ have certain ascertainable properties.

In the first place, let $\lambda=\mu_i$ be a characteristic number of the system

$$\text{(D)} \qquad \begin{cases} L(u)=0, \\ u(a)=u(b)=0. \end{cases}$$

This system is of the Sturmian type, in fact it is the particular case of the Sturmian system (§ 10·6, A) in which $\alpha=\beta=0$. It has therefore an infinite number of characteristic numbers μ_i $(i \geqslant 1)$ such that each of the corresponding characteristic functions $u_i(x)$ has, in the interval † $a \leqslant x < b$, a number of zeros equal to the suffix i.

The characteristic numbers μ_i of (D) are not in general, but in particular cases may be, roots of the characteristic equation (C). Now $u_i(x)$ may be identified with $y_2(x, \mu_i)$. Since in this case

$$y_2(b, \mu_i)=0,$$

the identity (B) reduces to

$$y_1(b, \mu_i)y_2'(b, \mu_i)=1,$$

and hence

$$\begin{aligned} F(\mu_i) &= y_1(b, \mu_i)-2+\frac{1}{y_1(b, \mu_i)} \\ &= \frac{\{y_1(b, \mu_i)-1\}^2}{y_1(b, \mu_i)} = \frac{\{y_2'(b, \mu_i)-1\}^2}{y_2'(b, \mu_i)}. \end{aligned}$$

Consequently

$$\begin{aligned} F(\mu_i)>0 \quad &\text{when} \quad y_1(b, \mu_i)>0 \quad \text{but} \quad \neq 1, \\ &\text{or when} \quad y_2'(b, \mu_i)>0 \quad \text{but} \quad \neq 1, \\ F(\mu_i)=0 \quad &\text{when} \quad y_1(b, \mu_i)=y_2'(b, \mu_i)=1, \\ F(\mu_i)<0 \quad &\text{when} \quad y_1(b, \mu_i) \quad \text{or} \quad y_2'(b, \mu_i)<0. \end{aligned}$$

Now since $y_2'(a, \mu_i)=1$ and $y_2(b, \mu_i)=y_2(a, \mu_i)=0$, $y_2'(b, \mu_i)$ is positive or negative according to whether $y_2(x, \mu_i)$ has an even or an odd number of zeros in the interval $a \leqslant x < b$. Therefore, when i is even, $F(\mu_i) \geqslant 0$ and consequently μ_i may be a root of the characteristic equation (C), and when i is odd, $F(\mu_i)<0$ ‡ and μ_i is not a root of (C).

The sign of $F(\lambda)$ at the points $\mu_1, \mu_2, \mu_3, \ldots$ may be exhibited graphically as follows :

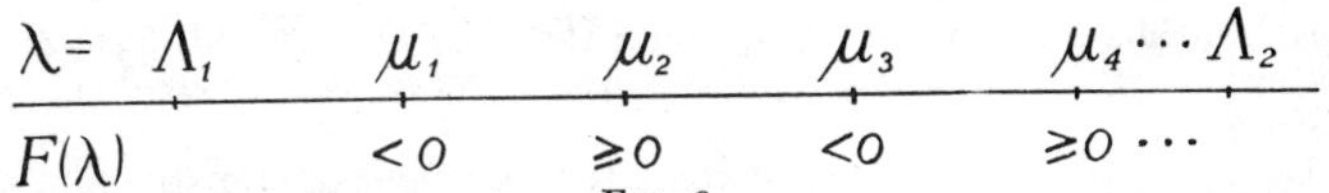

FIG. 6.

The characteristic equation $F(\lambda)=0$ has therefore an even number of roots § in each interval (μ_1, μ_3), (μ_3, μ_5), . . ., thus it is seen that there exists an infinite set of real characteristic numbers of the system (A).

In the second place consider the system

$$\text{(E)} \qquad \begin{cases} L(v)=0, \\ v'(a)=v'(b)=0 \, ; \end{cases}$$

* The methods of the preceding section may be employed to prove that in a very large class of cases, the system has no complex characteristic numbers.

† The first end-point a is included, but the second end-point b is excluded because $u(b)=u(a)$; there is no characteristic number μ_0 since each $u_i(x)$ has a zero when $x=a$.

‡ A very slight modification of the argument shows that $F(\mu_i) \leqslant -4$ when i is odd.

§ A possible double root is counted twice.

it admits of an infinite set of characteristic numbers $\nu_i (i \geqslant 0)$ such that each characteristic function $v_i(x)$ has i zeros in the interval $a \leqslant x < b$. By identifying $v_i(x)$ with $y_1(x, \nu_i)$ it is found, as before, that

$$y_1(b, \nu_i) y_2'(b, \nu_i) = 1,$$

$$F(\nu_i) = \frac{\{y_1(b, \nu_i) - 1\}^2}{y_1(b, \nu_i)}.$$

Consequently

$$\begin{aligned} F(\nu_i) &> 0 \quad \text{when} \quad y_1(b, \nu_i) > 0 \quad \text{but} \neq 1, \\ F(\nu_i) &= 0 \quad \text{when} \quad y_1(b, \nu_i) = 1, \\ F(\nu_i) &< 0 \quad \text{when} \quad y_1(b, \nu_i) < 0. \end{aligned}$$

Now $y_1(x, \nu_i)$ has an even or an odd number of zeros in $a \leqslant x < b$ according as i is even or odd. Since $y_1(a, \nu_i) = 1$ it follows that $y_1(b, \nu_i)$ is positive or negative according to whether i is even or odd. Therefore, when i is even, $F(\nu_i) \geqslant 0$ and ν_i may be a root of the characteristic equation (C), and when i is odd, $F(\nu_i) < 0$ and ν_i is not a root of (C).

The sign of $F(\lambda)$ therefore runs as follows :

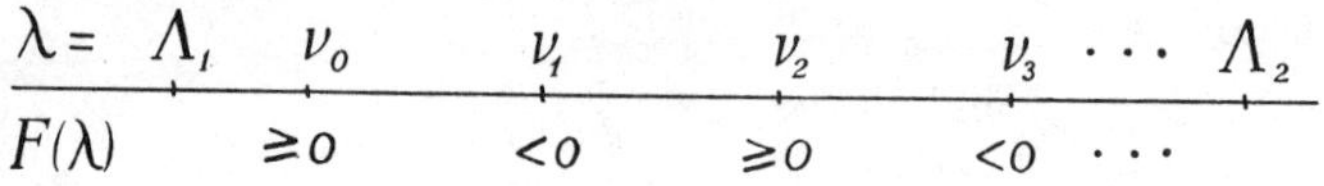

FIG. 7.

$F(\lambda)$ has thus an even number of roots in each interval (Λ_1, ν_1), (ν_1, ν_3), (ν_3, ν_5), ... Now it is clear that

$$\nu_i < \mu_{i+1} < \nu_{i+2}, \quad \mu_i < \nu_{i+1} < \mu_{i+2},$$

because an increase in the number of zeros in $a \leqslant x < b$ implies an increase in the value of λ. But, on the other hand, nothing can be said as to the relative magnitudes of μ_i and ν_i. Supposing, merely for purposes of illustration, that $\mu_i < \nu_i$, the change in the sign of $F(\lambda)$ may be exhibited thus :

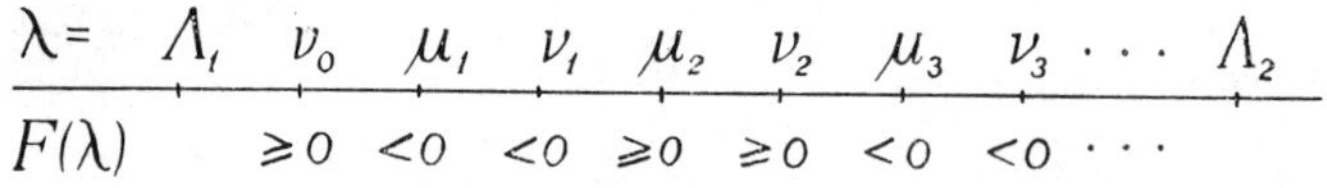

FIG. 8.

It has thus been proved that, under the conditions stated, *there exists at least one characteristic number for the system* (A) *in each interval* (μ_i, μ_{i+1}), (ν_i, ν_{i+1}).

The next step is to show that there is only one characteristic number for the system (A) in each interval (μ_i, μ_{i+1}) or (ν_i, ν_{i+1}). In order to do so it will be sufficient to prove that $F'(\lambda)$ has the same sign at every root of $F(x) = 0$ which occurs in any such interval. Since ascending and descending nodes must succeed one another in the graph of a continuous function, the result will then follow immediately. To simplify the working it will now be assumed that $K(x)$ is independent of λ. Now

$$F(\lambda) = y_1(b, \lambda) + y_2'(b, \lambda) - 2,$$

and therefore

$$F'(\lambda) = \frac{\partial y_1(b, \lambda)}{\partial \lambda} + \frac{\partial y_2'(b, \lambda)}{\partial \lambda}.$$

Let $u(x, \lambda)$ be the unique solution of the system

$$\begin{cases} L(u) = 0, \\ u(a) = \alpha, \ u'(a) = \alpha', \end{cases}$$

in which a and a' are real numbers, independent of λ. Then clearly $\frac{\partial u}{\partial \lambda}$ satisfies the non-homogeneous equation

$$\frac{d}{dx}\left\{K\frac{d}{dx}\left(\frac{\partial u}{\partial \lambda}\right)\right\}-G\frac{\partial u}{\partial \lambda}=\frac{\partial G}{\partial \lambda}u.$$

But the corresponding homogeneous equation

$$\frac{d}{dx}\left\{K\frac{d}{dx}\left(\frac{\partial v}{\partial \lambda}\right)\right\}-G\frac{\partial v}{\partial \lambda}=0$$

is known to possess the fundamental pair of solutions

$$\frac{\partial v}{\partial \lambda}=y_1(x, \lambda), \quad \frac{\partial v}{\partial \lambda}=y_2(x, \lambda),$$

from which $\frac{\partial u}{\partial \lambda}$ and $\frac{\partial}{\partial \lambda}\left(\frac{du}{dx}\right)$ are to be derived by the method of variation of parameters (§ 5·23), thus *

$$\frac{\partial u}{\partial \lambda}=\frac{1}{K(a)}\int_a^x \frac{\partial G(t, \lambda)}{\partial \lambda}u(t, \lambda)\left\{y_1(t, \lambda)y_2(x, \lambda)-y_2(t, \lambda)y_1(x, \lambda)\right\}dt$$

and

$$\frac{\partial}{\partial \lambda}\left(\frac{du}{dx}\right)=\frac{1}{K(a)}\int_a^x \frac{\partial G(t, \lambda)}{\partial \lambda}u(t, \lambda)\left\{y_1(t, \lambda)y_2'(x, \lambda)-y_2(t, \lambda)y_1'(x, \lambda)\right\}dt.$$

Therefore, taking $x=b$ and $u=y_1$ in the expression for $\frac{\partial u}{\partial \lambda}$ it is found that

$$\frac{\partial y_1(b, \lambda)}{\partial \lambda}=\frac{1}{K(a)}\int_a^b \frac{\partial G(t, \lambda)}{\partial \lambda}y_1(t, \lambda)\left\{y_1(t, \lambda)y_2(b, \lambda)-y_2(t, \lambda)y_1(b, \lambda)\right\}dt,$$

and taking $x=b$ and $u=y_2$ in the expression for $\frac{\partial}{\partial \lambda}\left(\frac{du}{dx}\right)$ it is similarly found that

$$\frac{\partial y_2'(b, \lambda)}{\partial \lambda}=\frac{1}{K(a)}\int_a^b \frac{\partial G(t, \lambda)}{\partial \lambda}y_2(t, \lambda)\left\{y_1(t, \lambda)y_2'(b, \lambda)-y_2(t, \lambda)y_1'(b, \lambda)\right\}dt.$$

It follows that

$$F'(\lambda)=\frac{1}{K(a)}\int_a^b \frac{\partial G(t, \lambda)}{\partial \lambda}\left\{y_2(b, \lambda)y_1^2(t, \lambda)+[y_2'(b, \lambda)-y_1(b, \lambda)]y_1(t, \lambda)y_2(t, \lambda) -y_1'(b, \lambda)y_2^2(t, \lambda)\right\}dt.$$

Since $K(a)>0$ and $\frac{\partial G}{\partial \lambda}<0$, the sign of $F'(\lambda)$ is opposite to that of the quadratic form

$$\Phi(\xi, \eta)=y_2(b, \lambda)\xi^2+[y_2'(b, \lambda)-y_1(b, \lambda)]\xi\eta-y_1'(b, \lambda)\eta^2,$$

in which $\xi=y_1(t, \lambda)$, $\eta=y_2(t, \lambda)$. The discriminant of this form is

$$[y_2'(b, \lambda)-y_1(b, \lambda)]^2+4y_2(b, \lambda)y_1'(b, \lambda)$$

which, by virtue of Abel's formula,

$$y_1(b, \lambda)y_2'(b, \lambda)-y_2(b, \lambda)y_1'(b, \lambda)=1,$$

reduces to

$$[y_2'(b, \lambda)+y_1(b, \lambda)]^2-4,$$

and therefore, for those values of λ for which the characteristic equation

$$y_1(b, \lambda)+y_2'(b, \lambda)=2$$

* It is to be remembered that $u(a, \lambda)=a$, $u'(a, \lambda)=a'$ for all values of λ, and therefore

$$\frac{\partial}{\partial \lambda}u(a, \lambda)=0, \quad \frac{\partial}{\partial \lambda}u'(a, \lambda)=0.$$

is satisfied, the discriminant is zero. For such values of λ, the quadratic form may be written as

$$\Phi(\xi, \eta)=\frac{\{y_2\xi+\frac{1}{2}(y_2'-y_1)\eta\}^2}{y_2(b, \lambda)}$$
$$=-\frac{\{y_1'\eta-(1-y_1)\xi\}^2}{y_1'(b, \lambda)}.$$

Now at a simple characteristic value of λ

$$y_2\xi+\tfrac{1}{2}(y_2'-y_1)\eta, \quad y_1'\eta-(1-y_1)\xi$$

cannot both be zero. It follows therefore that $F'(\lambda)$ is not zero and its sign is that of $y_1'(b, \lambda)$ or $-y_2(b, \lambda)$. Consequently $F(\lambda)$ *changes sign at a simple characteristic value of* λ.

When, for any particular value of λ,

$$y_2(b, \lambda)=y_1'(b, \lambda)=0,$$

Abel's formula reduces to

$$y_1(b, \lambda)+y_2'(b, \lambda)=0,$$

and it then follows from the characteristic equation

$$y_1(b, \lambda)+y_2'(b, \lambda)=2$$

that

$$y_1(b, \lambda)=y_2'(b, \lambda)=1.$$

The value of λ in question is therefore a double characteristic number and for such a value

$$F(\lambda)=0, \quad F'(\lambda)=0.$$

Now it may be proved, by a method similar to that adopted in finding $F'(\lambda)$, that

$$F''(\lambda)=-\frac{2}{\{K(a)\}^2}\int_a^b\int_a^s \frac{\partial G(s, \lambda)}{\partial\lambda}\cdot\frac{\partial G(t, \lambda)}{\partial\lambda}\{y_1(s, \lambda)y_2(t, \lambda)-y_2(s, \lambda)y_1(t, \lambda)\}^2 dt ds.$$

Since y_1 and y_2 are independent solutions of the differential equation, and s and t are independent variables,

$$y_1(s, \lambda)y_2(t, \lambda)-y_2(s, \lambda)y_1(t, \lambda)$$

is not identically zero. Consequently $F''(\lambda)$ is negative for a double characteristic value of λ, and therefore, *in the neighbourhood of a double characteristic number,* $F(\lambda)$ *preserves a constant negative sign.*

Now since $F(\lambda)$ is negative at μ_{2m-1} and at μ_{2m+1} and is positive or zero at μ_{2m}, there must exist at least two simple characteristic numbers λ_p and λ_q, or one double characteristic number $\lambda_p=\lambda_q$, in the double interval (μ_{2m-1}, μ_{2m+1}) such that

$$\mu_{2m-1}<\lambda_p\leqslant\mu_{2m}\leqslant\lambda_q<\mu_{2m+1}.$$

No double characteristic number can lie in this interval except at μ_{2m}. If then there are additional characteristic numbers in (μ_{2m-1}, μ_{2m}) they must be simple, and even in number. But for these values of λ, $F'(\lambda)$ is of opposite sign to $y_2(b, \lambda)$ which is impossible since $y_2(b, \lambda)$ does not change sign at any interior point of the interval. Thus there are no characteristic numbers other than λ_p and λ_q in the double interval (μ_{2m-1}, μ_{2m+1}). In the same way it may be proved that there are only two characteristic numbers in the double interval (ν_{2m-1}, ν_{2m+1}); obviously these characteristic numbers are λ_p and λ_q, and therefore

$$\nu_{2m-1}<\lambda_p\leqslant\nu_{2m}\leqslant\lambda_q<\nu_{2m+1}.$$

It follows immediately that no characteristic number can lie in the open interval (μ_{2m}, ν_{2m}) or in the closed interval (μ_{2m+1}, ν_{2m+1}). In the same way it may be proved that $F(\lambda)>0$ in the interval $\Lambda_1<\lambda<\nu_0$, and therefore no characteristic number lies in that interval.

Since λ_p and λ_q are interior points of the double interval (μ_{2m-1}, μ_{2m+1}), the corresponding characteristic functions y_p and y_q cannot have less than $2m-1$ nor more than $2m+1$ zeros in the interval $a\leqslant x<b$. But, on account of the periodic boundary conditions, the number of zeros in that interval must be even. Consequently y_p and y_q both have precisely $2m$ zeros in the interval $a\leqslant x<b$.

Let the interval (Λ_1, ν_0) be denoted by (κ_0), and the intervals (μ_1, ν_1), (μ_2, ν_2), . . . by (κ_1), (κ_2), . . . (Fig. 9). Then no characteristic number can be an interior point of any interval (κ_i). On the other hand, between any two consecutive intervals (κ_i) and (κ_{i+1}) there lies one and only one characteristic number; * let it be denoted by λ_i and let $y_i(x)$ be the corresponding characteristic function. Then $y_0(x)$ does not vanish in the interval $a\leqslant x<b$, $y_1(x)$ and $y_2(x)$ vanish twice, $y_3(x)$ and $y_4(x)$ vanish four times, and so on. This leads to the following *Oscillation Theorem:*

There exists for the system (A) *an infinite set of characteristic numbers* $\lambda_0, \lambda_1, \lambda_2, \ldots, \lambda_i, \ldots$ *such that, if the corresponding characteristic functions are denoted by* $y_0, y_1, y_2, \ldots, y_i, \ldots$, *then* y_i *has an even number of zeros in the interval* $a\leqslant x<b$, *namely* i *or* $i+1$ *zeros.*

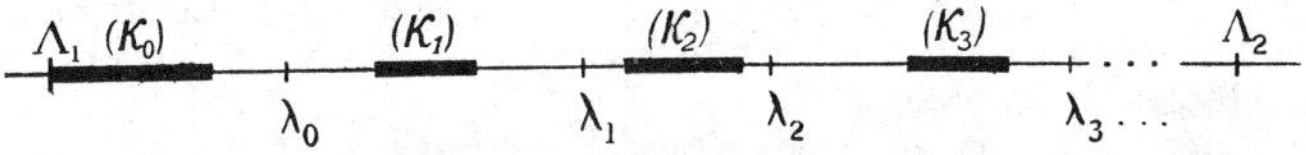

FIG. 9.

10·81. Equations with Periodic Coefficients.—The most important application of the theory of systems with periodic boundary conditions is to the case in which the coefficients of the differential equation are periodic functions of x with a period commensurable with $(b-a)$. In particular, let K and G be even periodic functions, with period π, and let the boundary conditions be

$$y(-\pi)=y(\pi), \quad y'(-\pi)=y'(\pi),$$

then it will follow from the differential equation that if y_i is any characteristic function, $y_i^{(r)}(-\pi)=y_i^{(r)}(\pi)$, and therefore every characteristic function will be purely periodic and of period 2π.

It is convenient to define the fundamental solutions $y_1(x, \lambda)$ and $y_2(x, \lambda)$ thus:

$$y_1(0, \lambda)=1, \quad y_2(0, \lambda)=0,$$
$$y_1'(0, \lambda)=0, \quad y_2'(0, \lambda)=1,$$

then $y_1(x, \lambda)$ will be an even, and $y_2(x, \lambda)$ an odd, function of x. For if $y_1(x, \lambda)$, for instance, were not even, then $y_1(x, \lambda)-y_1(-x, \lambda)$ would be a solution of the equation, vanishing, together with its first derivative for $x=0$, which is impossible.

If, for any value of λ, $y_1(-\pi, \lambda)=0$, then $y_1(x, \lambda)$ would have an even number of zeros in the interval $-\pi\leqslant x\leqslant\pi$, which would violate the condition $y'(-\pi)=y'(\pi)$, and consequently that value of λ would not be characteristic. For any other value of λ, $y_1(x, \lambda)$ satisfies the condition

$$y(-\pi)=y(\pi)\neq 0.$$

The further condition

$$y'(-\pi)=y'(\pi)=0$$

* The modification of this statement when double characteristic numbers occur is obvious.

is satisfied when $\lambda=\nu_{2m}$. Similarly, for all relevant values of λ, $y_2(x, \lambda)$ satisfies the condition

$$y'(-\pi)=y'(\pi)\neq 0,$$

and also satisfies the condition

$$y(-\pi)=y(\pi)=0$$

when $\lambda=\mu_{2m}$. In this case, therefore, λ_i is to be identified with ν_i when i is even, and with μ_{i+1} when i is odd.

An interesting and important extension of this case is to periodic solutions of the second kind; that is to say $y(\pi)$ and $y'(\pi)$ are not equal to, but are merely proportional to $y(-\pi)$ and $y'(-\pi)$. The two linear boundary conditions are now replaced by a single quadratie boundary condition, viz.

$$y(-\pi)y'(\pi)-y'(-\pi)y(\pi)=0.$$

The problem is essentially that dealt with in a later chapter under the name of the Floquet Theory. The system will there be seen always to have one solution, and in general, for all values of λ, to have two linearly independent solutions.

10·9. Klein's Oscillation Theorem.—An example of an oscillation theorem will now be given, whose scope far outreaches that of the theorems due to Sturm. It gives an indication of the lines upon which further generalisations of the problem have proceeded.

Consider the equation known as the Lamé equation,*

$$\frac{d^2y}{dx^2}+\tfrac{1}{2}\left\{\frac{1}{x-e_1}+\frac{1}{x-e_2}+\frac{1}{x-e_3}\right\}\frac{dy}{dx}-\frac{Ax+B}{(4x-e_1)(x-e_2)(x-e_3)}y=0,$$

in which $e_1<e_2<e_3$. Let two closed intervals (a_1, b_1), (a_2, b_2) be taken, such that each lies wholly within one or other of the open intervals (e_1, e_2), (e_2, e_3), (e_3, ∞), but not both within the same interval. In this way the continuity of the coefficients of the differential equation is ensured in each of the intervals (a_1, b_1), (a_2, b_2). The constants A and B are to be regarded as parameters; the problem which is suggested by physical considerations is, if possible, so to determine A and B that the equation possesses, at the same time, a solution y_1 which satisfies certain boundary conditions relative to (a_1, b_1), and a solution y_2 which satisfies other boundary conditions relative to (a_2, b_2). Or, more particularly, it may be required to determine A and B such that the equation admits of a solution y_1 which vanishes at a_1 and b_1 and has m_1 zeros between a_1 and b_1, and also admits of a solution y_2 which vanishes at a_2 and b_2 and has m_2 zeros between m_1 and m_2. This was the problem actually discussed by Klein; † his method of attack forms the basis of the rather more general theory which will now be discussed.

In the differential equation

$$\text{(A)} \qquad \frac{d}{dx}\left\{K\frac{dy}{dx}\right\}-Gy=0,$$

let G be of the form

$$G=l(x)-\{\lambda_0+\lambda_1 x+\ .\ .\ .\ +\lambda_n x^n\}g(x),$$

being thus dependent upon $n+1$ parameters. Further, let there be $n+1$ closed intervals

$$(a_0, b_0),\quad (a_1, b_1),\quad .\ .\ .,\quad (a_n, b_n),$$

* See Whittaker and Watson, *Modern Analysis* (3rd ed.), Chap. XXIII.

† *Math. Ann.* 18 (1881), p. 410; *Gött. Nach.* (1890), p. 91; [*Ges. Math. Abh.*, 2, pp. 512, 540]; Bôcher, *Bull. Am. Math. Soc.* 4 (1898), p. 295; 5 (1899), p. 365. The case of a pair of equations of the second order with two parameters is treated by Richardson, *Trans. Am. Math. Soc.* 13 (1912), p. 22; *Math. Ann.* 73 (1912), p. 289.

where

$$a_0 < b_0 < a_1 < b_1 < \ldots < a_n < b_n,$$

such that K, l and g are continuous and $g > 0$ for values of x lying in any of these intervals.*

The problem now set is to investigate the possibility of determining $\lambda_0, \lambda_1, \ldots \lambda_n$ in such a manner that $n+1$ particular solutions of the equation can be found, say $y_0, y_1, \ldots, y_n$, where y_r satisfies the pair of boundary conditions

$$\text{(B)} \qquad \begin{cases} \alpha_r' y_r(a_r) - \alpha_r y_r'(a_r) = 0, \\ \beta_r' y_r(b_r) - \beta_r y_r'(b_r) = 0 \qquad (r=0, 1, \ldots, n), \end{cases}$$

and has an assigned number of zeros, say m_r, in (a_r, b_r).

The oscillation theorem which provides a complete solution of the problem stated is as follows: *There exists an infinite set of simultaneous characteristic numbers* $(\lambda_0, \lambda_1, \ldots, \lambda_n)$, *such that to each particular set there corresponds a set of characteristic functions. If* $(n+1)$ *positive integers or zeros* $(m_0, m_1, \ldots, m_n)$ *are assigned, then the characteristic numbers* $(\lambda_0, \lambda_1, \ldots, \lambda_n)$ *can be chosen, in one way only, so that in each interval* $a_r < x < b_r$, *the corresponding characteristic function* y_r *has precisely* m_r *zeros.*

The theorem is proved by induction; it is certainly true when $n=0$, for then it reduces to the older oscillation theorem of § 10·6. Let it be supposed that the theorem is true up to and including the case of n parameters; it will then be proved to be true for the case of $n+1$ parameters. Now if G is rewritten in the form

$$G = \{l(x) - \lambda_n x^n g(x)\} - \{\lambda_0 + \lambda_1 x + \ldots + \lambda_{n-1} x^{n-1}\} g(x),$$

and the parameter λ_n is, for the moment, fixed, then G may be regarded as dependent upon the n parameters $\lambda_0, \lambda_1, \ldots, \lambda_{n-1}$. Now the hypothesis is that these n constants may be chosen in one way, and in one way only, so that the characteristic functions $y_0, y_1, \ldots, y_{n-1}$ exist such that each satisfies its peculiar boundary conditions, and each has an assigned number of zeros in the corresponding interval. The n characteristic numbers $\lambda_0, \lambda_1, \ldots, \lambda_{n-1}$ so determined naturally depend upon λ_n, and therefore, if $\lambda_0, \lambda_1, \ldots, \lambda_{n-1}$ are expressed in terms of λ_n, G may now be regarded as a function of x and of the single parameter λ_n. If Sturm's oscillation theorem can be applied to the equation

$$\frac{d}{dx}\left\{K \frac{dy}{dx}\right\} - G(x, \lambda_n) y = 0$$

so as to demonstrate the existence of a solution y_n having m_n zeros in the interval $a_n < x < b_n$, the theorem is proved. It is therefore imperative to make certain that $G(x, \lambda_n)$ is such that the conditions requisite for the validity of the oscillation theorem are satisfied.

In the first place, it will be proved that $G(x, \lambda_n)$ is a continuous function of (x, λ_n) for values of x which lie in the interval (a_n, b_n). Now if λ_n' is any fixed value of the parameter λ_n, the difference

$$G(x, \lambda_n) - G(x, \lambda_n')$$

must vanish for at least one value of x in each interval $a_r \leqslant x \leqslant b_r$ $(r \leqslant n-1)$, for if this difference were constantly of one sign in any interval (a_r, b_r) then $y_r(x, \lambda_n)$ would, by the comparison theorem, oscillate more (or less) rapidly than $y_r(x, \lambda_n')$, which contradicts the fact that y_r has exactly m_r zeros in (a_r, b_r). Hence there is at least one point x_r in each interval (a_r, b_r) such that

$$G_r(x_r, \lambda_n) = G(x_r, \lambda_n') \qquad (r=0, 1, \ldots, n-1).$$

* Nothing is assumed as to the nature of K, l and g, for values of x which do not lie in one or other of these intervals; in fact, in the case of the Lamé equation, the coefficients become infinite for certain values of x (viz. e_1, e_2, e_3) outside the intervals chosen.

But

$$G(x, \lambda_n)-G(x, \lambda_n')=\{(\lambda_0'-\lambda_0)+(\lambda_1'-\lambda_1)x+ \dots +(\lambda_n'-\lambda_n)x^n\}g(x)$$
$$=(\lambda_n'-\lambda_n)(x-x_0)(x-x_1) \dots (x-x_{n-1})g(x).$$

Thus when x lies in (a_n, b_n)

$$|G(x, \lambda_n)-G(x, \lambda_n')|<|\lambda_n'-\lambda_n||b_n-a_0||b_n-a_1| \dots |b_n-a_{n-1}||g(x)|,$$

from which the continuity of $G(x, \lambda_n)$ follows. Also

$$x-x_r>0 \qquad (r=0, 1, \dots, n-1),$$

when x lies in (a_n, b_n), and consequently

$$\frac{G(x, \lambda_n)-G(x, \lambda_n')}{\lambda_n-\lambda_n'}<0$$

for $a_n \leqslant x \leqslant b_n$. More precisely,

$$G(x, \lambda_n) \to -\infty \quad \text{as} \quad \lambda_n \to +\infty,$$
$$G(x, \lambda_n) \to +\infty \quad \text{as} \quad \lambda_n \to -\infty.$$

The conditions requisite for Sturm's oscillation theorem are therefore satisfied. Consequently there exists one and only one characteristic number λ_n such that y_n admits of exactly m_n zeros in the interval $a_n<x<b_n$. The induction is now complete, and the theorem proved.

The characteristic numbers which have been under consideration are real. As in Sturm's case, the question arises as to whether or not there may also exist complex characteristic numbers, and as before the assumption of the existence of complex characteristic numbers leads to a contradiction.

Let $\lambda_0, \lambda_1, \dots, \lambda_n$ be a set of simultaneous characteristic numbers, to which corresponds the set of characteristic functions $u_0, u_1 \dots, u_n$. If, as is supposed, at least one of $\lambda_0, \lambda_1, \dots, \lambda_n$ is a complex number, while all other coefficients in the differential equation and in the boundary conditions are real, then the differential system admits as a set of characteristic numbers the set $\mu_0, \mu_1, \dots, \mu_n$, conjugate to $\lambda_0, \lambda_1, \dots, \lambda_n$, together with the set of characteristic functions $v_0, v_1, \dots, v_n$ conjugate to $u_0, u_1, \dots, u_n$. Then

$$\frac{d}{dx}\left\{K\frac{du_r}{dx}\right\}+\{(\lambda_0+\lambda_1 x+ \dots +\lambda_n x^n)g-l\}u_r=0,$$
$$\frac{d}{dx}\left\{K\frac{dv_r}{dx}\right\}+\{(\mu_0+\mu_1 x+ \dots +\mu_n x^n)g-l\}v_r=0$$
$$(r=0, 1, \dots, n).$$

On eliminating l between the two equations and integrating the eliminant between the limits a_r and b_r the following set of equations is obtained:

$$\int_{a_r}^{b_r}\{(\lambda_0-\mu_0)+(\lambda_1-\mu_1)x+ \dots +(\lambda_n-\mu_n)x^n\}gu_r v_r dx=0$$
$$(r=0, 1, \dots, n).$$

The $(n+1)$ numbers $\lambda_r-\mu_r$ are not all zero; let it be supposed, in the first place, that no one of them is zero. Then there are $n+1$ equations between the $(n+1)$ quantities $(\lambda_r-\mu_r)$; the condition that these equations should be consistent is that (C)

$$\int_{a_0}^{b_0}\dots\int_{a_n}^{b_n}\Delta(x_0, \dots, x_n)g(x_0)\dots g(x_n)u_0(x_0)v_0(x_0)\dots u_n(x_n)v_n(x_n)dx_0\dots dx_n=0,$$

where

$$\Delta(x_0, \ldots, x_n) = \begin{vmatrix} 1, & x_0, & \ldots, & x_0^n \\ 1, & x_1, & \ldots, & x_1^n \\ \ldots & \ldots & \ldots & \ldots \\ 1, & x_n, & \ldots, & x_n^n \end{vmatrix}$$

$$= \Pi(x_r - x_s) \qquad (r>s).$$

If p of the quantities $\lambda_r - \mu_r$ vanish (which implies that the corresponding numbers λ_r are real) there will be $n+1$ equations between the $n-p+1$ remaining quantities. The condition for their consistence is expressible as a number of equations of the form (C), in each of which the order of the multiple integral is $n-p+1$. The remainder of the argument is essentially the same in all cases.

When $n=0$ the formula (C) reduces to

$$\int_a^b guvdx = 0.$$

Now in (C),

$\Delta(x_0, \ldots, x_n) > 0$ since $x_0 < x_1 < \ldots < x_n$,

$g_r(x_r) > 0$,

$u_r(x_r)v_r(x_r) > 0$ since u_r and v_r are conjugate quantities.

The integral therefore cannot be zero, a contradiction which proves the non-existence of complex or imaginary characteristic numbers.

The theory can be extended, without any real difficulty, to the case of an equation in which

$$G = l - \lambda_0 g_0 - \lambda_1 g_1 - \ldots - \lambda_n g_n.$$

In the multiple integral, the product

$$\Delta(x_0, \ldots, x_n) g(x_0) \ldots g(x_n)$$

is replaced by the determinant

$$\begin{vmatrix} g_0(x_0), & g_1(x_0), & \ldots, & g_n(x_0) \\ g_0(x_1), & g_1(x_1), & \ldots, & g_n(x_1) \\ \ldots & \ldots & \ldots & \ldots \\ g_0(x_n), & g_1(x_n), & \ldots, & g_n(x_n) \end{vmatrix}$$

The non-existence of complex characteristic numbers is assured if $g_0, g_1, \ldots, g_n$ are such that the determinant maintains a fixed sign when $a_0 \leqslant x_0 \leqslant b_0, \ldots, a_n \leqslant x_n \leqslant b_n$.

MISCELLANEOUS EXAMPLES.

1. Prove that the Wronskian of k linearly independent solutions of a linear differential equation of order $n > k$ cannot have an infinite number of zeros in any interval (a, b) in which the coefficients are continuous.

[Bôcher, *Bull. Am. Math. Soc.* 8 (1901), p. 53.]

2. Let y be any solution of

$$\frac{d}{dx}\left\{K\frac{dy}{dx}\right\} - Gy = 0,$$

and ϕ_1 and ϕ_2 be functions of x which, with their first derivatives, are continuous in the interval (a, b). Let

$$\Phi = \phi_1 y - \phi_2 K y',$$

$$\{\phi_1, \phi_2\} = \phi_1'\phi_2 - \phi_1\phi_2' + \frac{\phi_1^2}{K} - G\phi_2^2,$$

then if $\{\phi_1, \phi_2\}$ does not vanish in (a, b), Φ cannot vanish more than a finite number of times there, and Φ and Φ' do not both vanish at any point of (a, b).

[Bôcher, *Trans. Am. Math. Soc.* 2 (1901), p. 430.]

3. If y_1 and y_2 are distinct solutions of the equation of (2), and if

$$\Phi_1=\phi_1 y_1-\phi_2 K y_1', \quad \Phi_2=\phi_1 y_2-\phi_2 K y_2',$$

then between any two consecutive zeros of Φ_1, there lies one and only one zero of Φ_2.

[Bôcher, *ibid.* 3 (1902), p. 214.]

4. Let ψ_1 and ψ_2 be functions of the same nature as ϕ_1 and ϕ_2, and let

$$\Phi=\phi_1 y-\phi_2 K y', \quad \Psi=\psi_1 y-\psi_2 K y',$$

then if neither of

$$\phi_1\psi_2-\phi_2\psi_1, \quad \{\phi_1, \phi_2\}$$

vanishes in (a, b), then in any portion of (a, b) in which Ψ does not vanish, Φ cannot vanish more than once.

[Bôcher, *ibid.*, 2 (1901), p. 430.]

5. If none of the functions

$$\phi_1\psi_2-\phi_2\psi_1, \quad \{\phi_1, \phi_2\}, \quad \{\psi_1, \psi_2\}$$

vanish in (a, b), then between two consecutive zeros of Φ lies one and only one root of Ψ and *vice versa.*

[Bôcher, *ibid.*, p. 431.]

6. If to the conditions of (5) is added the condition that $\{\phi_1, \phi_2\}$ and $\{\psi_1, \psi_2\}$ are of opposite sign, then neither Φ nor Ψ vanishes more than once in (a, b), and if one of these functions vanishes, the other does not. Consider the special case $\psi_1=1$, $\psi_2=0$.

[Bôcher, *ibid.* p. 431.]

7. Let χ_1 and χ_2 be similar to ϕ_1 and ϕ_2, and let

$$\Phi=\phi_1 y-\phi_2 K y', \quad \Psi=\psi_1 y-\psi_2 K y', \quad X=\chi_1 y-\chi_2 K y',$$

then if none of the six functions

$$\phi_1\psi_2-\phi_2\psi_1, \quad \psi_1\chi_2-\psi_2\chi_1, \quad \chi_1\phi_2-\chi_2\phi_1, \quad \{\phi_1, \phi_2\}, \quad \{\psi_1, \psi_2\}, \quad \{\chi_1, \chi_2\}$$

vanish in (a, b), if the last three have the same sign, and if the product of all six is negative, then between any root of Φ and a larger root of X lies a root of Ψ, between any root of Ψ and a larger root of Φ lies a root of X, and between any root of X and a larger root of Ψ lies a root of Φ.

[Bôcher, *ibid.* p. 432 ; in a special case, Sturm, *J. de Math.* 1 (1836), p. 165.]

8. If, throughout the interval (a, b)

$$K>0, \quad K'\neq 0, \quad G<0, \quad \frac{d}{dx}\left(\frac{K'}{G}\right)<1$$

then the zeros of y, y', y'' follow one another cyclically in that order if $K'>0$, and in the reverse order if $K'<0$.

9. The positive zeros of the Bessel functions $J_n(x)$, $J_{n+1}(x)$, $J_{n+2}(x)$ follow one another cyclically in that order if $n>-1$, and in the reverse order if $n<-1$.

[Bôcher. *Bull. Am. Math. Soc.* (1897), p. 207 ; *loc. cit. ante*, p. 434.]

10. For a system

$$(1)\quad \begin{cases} \dfrac{d}{dx}\left\{K\dfrac{dy}{dx}\right\}-Gy=0, \\ L_0[y(a)]=M_0[y(b)], \quad L_1[y(a)]=M_1[y(b)], \end{cases}$$

where

$$L_i[y(x)]=\alpha_i y(x)-\beta_i K y'(x),$$
$$M_i[y(x)]=\gamma_i y(x)+\delta_i K y'(x) \qquad (i=1, 2),$$

and K, G, α_i, β_i, γ_i, δ_i depend upon λ, let the following conditions for $(a\leqslant x\leqslant b)$, $(\Lambda_1<\lambda<\Lambda_2)$ be imposed namely:

(A_{I}) K and G are continuous and $K>0$ for all values of (x, λ) considered ;

(A_{II}) K and G do not increase as λ increases, and for any λ there exists a value of x for which K or G actually decreases ;

(A_{III}) the eight coefficients $\alpha_i, \ldots, \delta_i$ are continuous real functions of λ in the interval considered and

$$|\alpha_i| + |\beta_i|>0, \quad |\gamma_i| + |\delta_i|>0 ;.$$

(A_{IV}) either β_i is identically zero or α_i/β_i does not increase as λ increases, and either δ_i is identically zero or γ_i/δ_i does not increase with λ ;

(B) the conditions which will ensure the correctness of Sturm's oscillation theorem for the system

$$(2)\quad \begin{cases} \dfrac{d}{dx}\left\{K\dfrac{dy}{dx}\right\}-Gy=0, \\ L_0[y(a)]=0, \quad M_0[y(b)]=0 ; \end{cases}$$

(C)
$$\begin{vmatrix} \alpha_0, & \beta_0, & \gamma_0, & \delta_0 \\ \alpha_1, & \beta_1, & \gamma_1, & \delta_1 \\ \alpha_0', & \beta_0', & \gamma_0', & \delta_0' \\ \alpha_1', & \beta_1', & \gamma_1', & \delta_1' \end{vmatrix} \leqslant 0.$$

Let $y_0(x, \lambda)$ and $y_1(x, \lambda)$ denote the two linearly independent solutions of the differential equation satisfying the conditions

$$L_0[y_0(a)]=0, \quad L_1[y_0(a)]=1,$$
$$L_0[y_1(a)]=1, \quad L_1[y_1(a)]=0,$$

then the characteristic equation for the system (1) is

$$F(\lambda) \equiv M_1[y_0(b, \lambda)]+M_0[y_1(b, \lambda)]-2=0,$$

and there exists one and only one characteristic number between every pair of characteristic numbers of the Sturmian system (2). If $\mu_0, \mu_1, \ldots$ are the ordered characteristic numbers of the system (2) and $\lambda_0, \lambda_1, \ldots$ those of the system (1), account being taken of their multiplicity, then the following cases are possible:

I_a. $\quad \Lambda_1 < \mu_0 \leqslant \lambda_0 < \mu_1 < \lambda_1 \leqslant \mu_2 \leqslant \lambda_2 < \mu_3 < \lambda_3 < \ldots < \Lambda_2,$

I_b. $\quad \Lambda_1 < \lambda_0 \leqslant \mu_0 \leqslant \lambda_1 < \mu_1 < \lambda_2 \leqslant \mu_2 \leqslant \lambda_3 < \mu_3 < \ldots < \Lambda_2,$

II_a. $\quad \Lambda_1 < \lambda_0 < \mu_0 < \lambda_1 \leqslant \mu_1 \leqslant \lambda_2 < \mu_2 < \lambda_3 \leqslant \mu_3 \leqslant \ldots < \Lambda_2,$

II_b. $\quad \Lambda_1 < \mu_0 < \lambda_0 \leqslant \mu_1 \leqslant \lambda_1 < \mu_2 < \lambda_2 \leqslant \mu_3 \leqslant \lambda_3 < \ldots < \Lambda_2.$

The conditions for these cases are respectively

I_a. $\quad M_1[y_0(b, \lambda_0)]>0, \quad F(\Lambda_1+\epsilon)>0,$

I_b. $\quad M_1[y_0(b, \lambda_0)]>0, \quad F(\Lambda_1+\epsilon)<0,$

II_a. $\quad M_1[y_0(b, \lambda_0)]<0, \quad F(\Lambda_1+\epsilon)>0,$

II_b. $\quad M_1[y_0(b, \lambda_0)]<0, \quad F(\Lambda_1+\epsilon)<0.$

The characteristic function corresponding to the characteristic number λ_p will have $p-2$, $p-1$, p, $p+1$ or $p+2$ zeros in the interval $a<x<b$.

[Ettlinger, *Trans. Am. Math. Soc.* 19 (1918), p. 79 ; 22 (1921), p. 136.]

CHAPTER XI

FURTHER DEVELOPMENTS IN THE THEORY OF BOUNDARY PROBLEMS

11·1. Green's Functions in One Dimension.—The most powerful instrument for carrying the theory of boundary problems beyond the stage to which it was brought in the previous chapter is the so-called Green's function, which will now be defined.* Consider the completely-homogeneous linear differential system:

$$\text{(A)} \quad \begin{cases} L(u) \equiv p_0 \dfrac{d^n u}{dx^n} + p_1 \dfrac{d^{n-1}u}{dx^{n-1}} + \cdots + p_{n-1}\dfrac{du}{dx} + p_n u = 0, \\ U_i(u) = 0 \qquad\qquad\qquad\qquad\qquad\qquad (i = 1, 2, \ldots, n). \end{cases}$$

It will be supposed that this system is incompatible, that is to say, it admits of no solution, not identically zero, which together with its first $n-1$ derivatives, is continuous throughout the interval (a, b). But though (A) possesses no solution in this strict sense, there possibly exists a function which formally satisfies the system but violates, at least in part, the conditions of continuity.

Such is a *Green's Function* $G(x, \xi)$ which

(1°) is continuous and possesses continuous derivatives of orders up to and including $(n-2)$ when $a \leqslant x \leqslant b$,

(2°) is such that its derivative of order $(n-1)$ is discontinuous at a point ξ within (a, b), the discontinuity being an upward jump of amount $1/p_0(\xi)$,

(3°) formally satisfies the system at all points of (a, b) except ξ.

It will first of all be proved that such a function $G(x, \xi)$ actually does exist, and, moreover, is unique. Let

$$u_1(x), \quad u_2(x), \quad \ldots, \quad u_n(x)$$

be a fundamental set of solutions of the equation

$$L(u) = 0,$$

then, since $G(x, \xi)$ satisfies the equation in the interval $a \leqslant x < \xi$, it must be expressible in the form

$$G(x, \xi) = a_1 u_1(x) + a_2 u_2(x) + \cdots + a_n u_n(x)$$

in that interval; similarly it must be expressible as

$$G(x, \xi) = b_1 u_1(x) + b_2 u_2(x) + \cdots + b_n u_n(x)$$

in the interval $\xi < x \leqslant b$. But $G(x, \xi)$ and its first $(n-2)$ derivatives are continuous at ξ, and therefore

$$\{a_1u_1(\xi)+a_2u_2(\xi)+\cdots+a_nu_n(\xi)\}-\{b_1u_1(\xi)+b_2u_2(\xi)+\cdots+b_nu_n(\xi)\}=0,$$
$$\{a_1u_1'(\xi)+a_2u_2'(\xi)+\cdots+a_nu_n'(\xi)\}-\{b_1u_1'(\xi)+b_2u_2'(\xi)+\cdots+b_nu_n'(\xi)\}=0,$$
$$\cdots\cdots\cdots\cdots\cdots\cdots\cdots\cdots$$
$$\{a_1u_1^{(n-2)}(\xi)+a_2u_2^{(n-2)}(\xi)+\cdots+a_nu_n^{(n-2)}(\xi)\}$$
$$-\{b_1u_1^{(n-2)}(\xi)+b_2u_2^{(n-2)}(\xi)+\cdots+b_nu_n^{(n-2)}(\xi)\}=0.$$

* Bôcher, *Bull. Am. Math. Soc.* 7 (1901), p. 297; Hilbert, *Grundzüge einer allgemeinen Theorie der linearen Integralgleichungen*, vii–ix.

The discontinuity in $G^{(n-1)}(x, \xi)$, when $x=\xi$, gives rise to the equation

$$\{a_1u_1^{(n-1)}(\xi)+a_2u_2^{(n-1)}(\xi)+\ . \ . \ . \ +a_nu^{n(n-1)}(\xi)\}$$
$$-\{b_1u_1^{(n-1)}(\xi)+b_2u_2^{(n-1)}(\xi)+\ . \ . \ . \ +b_nu_n^{(n-1)}(\xi)\}=-\frac{1}{p_0(\xi)}.$$

These equations may be written

$$\begin{aligned} c_1u_1(\xi)+c_2u_2(\xi)+\ . \ . \ . \ +c_nu_n(\xi) &=0, \\ c_1u_1'(\xi)+c_2u_2'(\xi)+\ . \ . \ . \ +c_nu_n'(\xi) &=0, \\ . \ . \ . \ . \ . \ . \ . \ . \ . \ \\ c_1u_1^{(n-2)}(\xi)+c_2u_2^{(n-2)}(\xi)+\ . \ . \ . \ +c_nu_n^{(n-2)}(\xi) &=0, \\ c_1u_1^{(n-1)}(\xi)+c_2u_2^{(n-1)}(\xi)+\ . \ . \ . \ +c_nu_n^{(n-1)}(\xi) &=\frac{1}{p_0(\xi)}, \end{aligned}$$

where

$$c_i=b_i-a_i \qquad (i=1, 2, \ . \ . \ ., n).$$

The discriminant of these n equations is the value of the Wronskian of $u_1(x), u_2(x), \ . \ . \ ., u_n(x)$ when $x=\xi$; it is not zero since the n solutions chosen form a fundamental set. Consequently the numbers $c_1, c_2, \ . \ . \ ., c_n$ may be determined uniquely.

Thus far the boundary conditions in (A) have not been utilised. Let

$$U_i(u)=A_i(u)+B_i(u),$$

where the terms relative to the end-point a are grouped under A_i, and those relative to b under B_i. Then, taking into consideration the fact that the representation of G in (a, ξ) differs from that in (ξ, b), it is seen that

$$\begin{aligned} U_i(G)=a_1A_i(u_1)+a_2A_i(u_2)+\ . \ . \ . \ +a_nA_i(u_n) \\ +b_1B_i(u_1)+b_2B_i(u_2)+\ . \ . \ . \ +b_nB_i(u_n)=0, \end{aligned}$$

which may be rewritten as

$$b_1U_i(u_1)+b_2U_i(u_2)+\ . \ . \ . \ +b_nU_i(u_n)=c_1A_i(u_1)+c_2A_i(u_2)+\ . \ . \ . \ +c_nA_i(u_n)$$
$$(i=1, 2, \ . \ . \ ., n).$$

The determinant $|U_i(u_j)|$ is not zero since the n boundary conditions are linearly independent and the system is incompatible. The equations are therefore sufficient to determine $b_1, b_2, \ . \ . \ ., b_n$ uniquely in terms of the known quantities $c_1, c_2, \ . \ . \ ., c_n$ and the coefficients of U_i.

Thus the coefficients a_i and b_i are determined uniquely; $G(x, \xi)$ is therefore unique. Also $G(x, \xi)$ and its first $(n-2)$ derivatives are continuous in (a, b), whilst the next derivative has the discontinuity postulated, viz.

$$\lim_{\epsilon\to 0}\left\{\frac{\partial^{n-1}G(\xi+\epsilon, \xi)}{\partial x^{n-1}}-\frac{\partial^{n-1}G(\xi-\epsilon, \xi)}{\partial x^{n-1}}\right\}=\frac{1}{p_0(\xi)}.$$

Now let $H(x, \xi)$ denote the corresponding Green's function for the adjoint system

$$\left\{\begin{aligned} \bar{L}(v)&=0, \\ V_i(v)&=0 \end{aligned}\right. \qquad (i=1, 2, \ . \ . \ ., n).$$

Let the interval (a, b) be divided up into three parts (a, ξ_1), (ξ_1, ξ_2), (ξ_2, b), and consider the two Green's functions

$$u=G(x, \xi_1), \quad v=H(x, \xi_2).$$

Then Green's formula

$$\int_a^b \{vL(u)+u\bar{L}(v)\}dx=\Big[P(u, v)\Big]_a^b$$

may be applied, with the proviso that the range of integration is regarded as the limiting case of the aggregate of the three ranges $(a, \xi_1-\epsilon)$, $(\xi_1+\epsilon, \xi_2-\epsilon)$, $(\xi_2+\epsilon, b)$ when ϵ tends to zero. In each of these ranges,

$$L(G)=0, \quad \bar{L}(H)=0,$$

and therefore

$$\lim \Big[P(G, H)\Big]_a^{\xi_1-\epsilon} + \lim \Big[P(G, H)\Big]_{\xi_1+\epsilon}^{\xi_2-\epsilon} + \lim \Big[P(G, H)\Big]_{\xi_2+\epsilon}^{b} = 0.$$

Since, by virtue of the boundary conditions,

$$P(G, H)=0$$

when $x=a$ and when $x=b$, this relation reduces to

$$\lim \Big[P(G, H)\Big]_{\xi_1-\epsilon}^{\xi_1+\epsilon} + \lim \Big[P(G, H)\Big]_{\xi_2-\epsilon}^{\xi_2+\epsilon} = 0.$$

On referring back to § 9·31, it is seen that the only discontinuous term in $P(G, H)$ is

$$p_0\left[H\frac{d^{n-1}G}{dx^{n-1}} - G\frac{d^{n-1}H}{dx^{n-1}}\right],$$

and therefore

$$p_0(\xi_1)H(\xi_1, \xi_2) \lim \left[\frac{d^{n-1}G}{dx^{n-1}}\right]_{\xi_1-\epsilon}^{\xi_1+\epsilon} - p_0(\xi_2)G(\xi_2, \xi_1) \lim \left[\frac{d^{n-1}H}{dx^{n-1}}\right]_{\xi_2-\epsilon}^{\xi_2+\epsilon} = 0,$$

and since

$$p_0(\xi_1) \lim \left[\frac{d^{n-1}G}{dx^{n-1}}\right]_{\xi_1-\epsilon}^{\xi_1+\epsilon} = p_0(\xi_2) \lim \left[\frac{d^{n-1}H}{dx^{n-1}}\right]_{\xi_2-\epsilon}^{\xi_2+\epsilon} = 1,$$

it follows that

$$H(\xi_1, \xi_2)=G(\xi_2, \xi_1).$$

This formula has been proved when $\xi_2>\xi_1$, it may equally well be proved when $\xi_2<\xi_1$. Consequently, if x and ξ are any two points in (a, b),

$$H(x, \xi)=G(\xi, x),$$

or in other words, *the Green's function of the adjoint system* (B) *is* $G(\xi, x)$. Furthermore, if the given system is self-adjoint, the Green's function is *symmetrical*, that is to say,

$$G(\xi, x)=G(x, \xi).$$

Since the Green's function of a given system is unique, the converse follows, namely that *if the Green's function of a given system is symmetrical, the system is self-adjoint.*

11·11. Solution of the Non-Homogeneous System.—It is known that, since the homogeneous system

$$\text{(A)} \qquad \begin{cases} L(u) = 0, \\ U_i(u)=0 \qquad (i=1, 2, \ldots, n) \end{cases}$$

is incompatible, any non-homogeneous system corresponding to it, and in particular the system

$$\text{(B)} \qquad \begin{cases} L(y) = r(x), \\ U_i(y)=0 \qquad (i=1, 2, \ldots, n) \end{cases}$$

admits of one, and only one, solution. When the Green's function $G(x, \xi)$ of (A) is known, an explicit solution of (B) can immediately be obtained, namely,

$$\text{(C)} \qquad y(x)=\int_a^b G(x, \xi)r(\xi)d\xi.$$

For

$$y^{(\nu)}(x)=\int_a^b \frac{\partial^\nu G(x, \xi)}{\partial x^\nu} r(\xi)d\xi \quad (\nu=1, 2, \ldots, n-2),$$

and since $\dfrac{\partial^{n-2}G(x, \xi)}{\partial x^{n-2}}$ is uniformly continuous in (a, b) it follows that

$$y^{(n-1)}(x)=\int_a^b \frac{\partial^{n-1}G(x, \xi)}{\partial x^{n-1}} r(\xi)d\xi.$$

But the integrand is now discontinuous at $\xi=x$, and therefore

$$\begin{aligned} y^{(n)}(x) &= \frac{\partial}{\partial x}\int_a^x \frac{\partial^{n-1}G(x, \xi)}{\partial x^{n-1}} r(\xi)d\xi+\frac{\partial}{\partial x}\int_x^b \frac{\partial^{n-1}G(x, \xi)}{\partial x^{n-1}} r(\xi)d\xi \\ &=\int_a^b \frac{\partial^n G(x, \xi)}{\partial x^n} r(\xi)d\xi + \lim \left[\frac{\partial^{n-1}G(x, \xi)}{\partial x^{n-1}} r(\xi)\right]_{\xi=x-\epsilon}^{\xi=x+\epsilon} \\ &=\int_a^b \frac{\partial^n G(x, \xi)}{\partial x^n} r(\xi)d\xi+\frac{r(x)}{p_0(x)}, \end{aligned}$$

and therefore

$$\begin{aligned} L(y)&=\int_a^b L(G)r(\xi)d\xi+r(x) \\ &=r(x), \end{aligned}$$

since $L(G)=0$. The differential equation of (B) is therefore satisfied.

Since $U_i(y)$ involves no derivatives of y of higher order than $(n-1)$ it follows that

$$\begin{aligned} U_i(y)&=\int_a^b U_i(G)r(\xi)d\xi \\ &=0 \qquad\qquad (i=1, 2, \ldots, n), \end{aligned}$$

since $U_i(G)=0$. Thus the boundary conditions are also satisfied. The expression (C) is therefore the solution of the system (B).

The solution of the more general non-homogeneous system

$$\text{(D)} \qquad \begin{cases} L(y) =r(x), \\ U_i(y)=\gamma_i \end{cases} \qquad (i=1, 2, \ldots, n)$$

may now be obtained in a very simple way. Let $G_i(x)$ be the unique solution of the system

$$\begin{cases} L(G_i) =0, \\ U_1(G_i)= \ldots =U_{i-1}(G_i)=U_{i+1}(G_i)= \ldots =U_n(G_i)=0 \\ U_i(G_i) =1\,; \end{cases}$$

then it may immediately be verified that the solution of (D) is

$$y(x)=\int_a^b G(x, \xi)r(\xi)d\xi+\gamma_1 G_1(x)+\gamma_2 G_2(x)+ \ldots +\gamma_n G_n(x).$$

Let $u_1(x)$ and $u_2(x)$ be linearly distinct solutions of the equation

$$p_0(x)\frac{d^2u}{dx^2}+p_1(x)\frac{du}{dx}+p_2(x)u=0,$$

and consider the function

$$F(x, \xi)=Au_1(x)+Bu_2(x)\pm \frac{u_1(x)u_2(\xi)-u_2(x)u_1(\xi)}{2p_0(\xi)\{u_1(\xi)u_2'(\xi)-u_2(\xi)u_1'(\xi)\}},$$

where the positive sign is taken when $a\leqslant x\leqslant \xi$, and the negative sign when $\xi\leqslant x\leqslant b$. $F(x, \xi)$ is continuous in (a, b); its differential coefficient has the finite discontinuity $1/p_0(\xi)$ when $x=\xi$ but is elsewhere continuous. The third term is independent of the solutions $u_1(x)$ and $u_2(x)$ chosen. $F(x, \xi)$ is therefore of the nature of a Green's function, and by a choice of the constants A and B so that $F(x, \xi)$ satisfies assigned boundary conditions, becomes the Green's function of that system.

Examples.—

$$(1^\circ)\quad \begin{cases} \dfrac{d^2u}{dx^2}=0, \\ u(0)=u(1)=0. \end{cases}$$

$$F(x, \xi)=A+Bx\pm\tfrac{1}{2}(\xi-x)$$

$$G(x, \xi)=x(\xi-1) \qquad (x\leqslant\xi),$$
$$=\xi(x-1) \qquad (x\geqslant\xi).$$

$$(2^\circ)\quad \begin{cases} \dfrac{d^2u}{dx^2}-n^2u=0, \\ u(0)=u(1)=0. \end{cases}$$

$$F(x, \xi)=A\cosh nx+B\sinh nx\pm\frac{1}{2n}\sinh n(\xi-x),$$

$$G(x, \xi)=\frac{\sinh nx\sinh n(\xi-1)}{n\sinh n} \qquad (x\leqslant\xi),$$
$$=\frac{\sinh n\xi\sinh n(x-1)}{n\sinh n} \qquad (x\geqslant\xi).$$

$$(3^\circ)\quad \begin{cases} \dfrac{d^2u}{dx}+n^2u=0, \\ u(0)=u(1), \\ u'(0)=u'(1). \end{cases}$$

$$F(x, \xi)=A\cos nx+B\sin nx\pm\frac{1}{2n}\sin n(\xi-x),$$

$$G(x, \xi)=\frac{1}{2n}\left\{\cot\frac{n}{2}\cos n(\xi-x)+\sin n\,|\,\xi-x\,|\right\}.$$

This last example shows that, when the system becomes compatible, *i.e.* when $n=2k\pi$, where k is an integer, the Green's function becomes infinite.

11·12. The Green's Function of a System involving a Parameter.—The preceding investigation shows that when λ is not a characteristic number of the system

$$\begin{cases} L(u)+\lambda u=r(x), \\ U_i(u)=0 \qquad (i=1, 2, \ldots, n), \end{cases}$$

a unique Green's function $G(x, \xi;\ \lambda)$ exists, and the solution of the system is

$$u(x)=\int_a^b G(x, \xi;\ \lambda)r(\xi)d\xi.$$

Similarly the solution of the adjoint system

$$\begin{cases} \overline{L}(v)+\lambda v=r(x), \\ V_i(v)=0 \qquad (i=1, 2, \ldots, n) \end{cases}$$

is

$$v(x)=\int_a^b G(\xi, x;\ \lambda)r(\xi)d\xi.$$

As an important corollary it follows that if λ_i is a characteristic number which renders the homogeneous system

$$\begin{cases} L(u)+\lambda u=0, \\ U_i(u)=0 \qquad (i=1, 2, \ldots, n) \end{cases}$$

singly-compatible, and if $u_i(x)$ is the corresponding characteristic function, then

$$u_i(x)=(\lambda-\lambda_i)\int_a^b G(x, \xi;\ \lambda)u_i(\xi)d\xi.$$

This result follows immediately from the fact that the differential equation

$$L(u)+\lambda u=(\lambda-\lambda_i)u_i(x)$$

admits of the solution $u_i(x)$.

If $y_1(x), y_2(x), \ldots, y_n(x)$ form a linearly independent set of solutions of the homogeneous equation

$$L(u)+\lambda u=0,$$

the explicit form of $G(x, \xi; \lambda)$ may be written down,* namely

$$G(x, \xi; \lambda)=\frac{N(x, \xi; \lambda)}{\Delta(\lambda)},$$

where

$$N(x, \xi, \lambda)=\begin{vmatrix} y_1(x), & y_2(x), & \ldots, & y_n(x), & g(x, \xi; \lambda) \\ U_1(y_1), & U_1(y_2), & \ldots, & U_1(y_n), & U_1(g) \\ \cdot & \cdot & \cdot & \cdot & \cdot \\ U_n(y_1), & U_n(y_2), & \ldots, & U_n(y_n), & U_n(g) \end{vmatrix},$$

$$\Delta(\lambda)=(-1)^n\begin{vmatrix} U_1(y_1), & U_1(y_2), & \ldots, & U_1(y_n) \\ U_2(y_1), & U_2(y_2), & \ldots, & U_2(y_n) \\ \cdot & \cdot & \cdot & \cdot \\ U_n(y_1), & U_n(y_2), & \ldots, & U_n(y_n) \end{vmatrix}$$

and

$$g(x, \xi; \lambda)=\pm\tfrac{1}{2}\frac{\begin{vmatrix} y_1(x), & y_2(x), & \ldots, & y_n(x) \\ y_1^{(n-2)}(\xi), & y_2^{(n-2)}(\xi), & \ldots, & y_n^{(n-2)}(\xi) \\ \cdot & \cdot & \cdot & \cdot \\ y_1(\xi), & y_2(\xi), & \ldots, & y_n(\xi) \end{vmatrix}}{\begin{vmatrix} y_1^{(n-1)}(\xi), & y_2^{(n-1)}(\xi), & \ldots, & y_n^{(n-1)}(\xi) \\ y_1^{(n-2)}(\xi), & y_2^{(n-2)}(\xi), & \ldots, & y_n^{(n-2)}(\xi) \\ \cdot & \cdot & \cdot & \cdot \\ y_1(\xi), & y_2(\xi), & \ldots, & y_n(\xi) \end{vmatrix}},$$

the positive or negative sign being taken according as $x<$ or $>\xi$.

The existence theorem of § 3·31 shows that if $L(u)$, $r(x)$ and $U_i(u)$ are independent of λ, the solutions $y_1(x), y_2(x), \ldots, y_n(x)$ are integral functions of λ. It follows that $G(x, \xi; \lambda)$ is an analytic function of λ for all values of λ except the zeros of $\Delta(\lambda)$, that is, for all values of λ except the characteristic numbers.† The form which $G(x, \xi; \lambda)$ assumes in the neighbourhood of a simple characteristic number λ_i which occurs as a simple zero of $\Delta(\lambda)$ will now be determined.

If $\Delta(\lambda)$ has the simple zero λ_i, the Green's function may be written

$$G(x, \xi; \lambda)=\frac{R(x, \xi)}{\lambda-\lambda_i}+G_i(x, \xi; \lambda),$$

where $G_i(x, \xi; \lambda)$ is analytic at $\lambda=\lambda_i$.

Now

$$R(x, \xi)=\lim_{\lambda\to\lambda_i}\frac{(\lambda-\lambda_i)N(x, \xi; \lambda)}{\Delta(\lambda)}$$

$$=\frac{N(x, \xi; \lambda_i)}{\Delta'(\lambda_i)}.$$

* Birkhoff, *Trans. Am. Math. Soc.* 9 (1908), p. 377. It is assumed that the coefficient of $u^{(n)}$ in $L(u)$ is unity.

† In fact, $G(x, \xi; \lambda)$ is a *meromorphic* function of λ.

In the expansion of the determinant for $N(x, \xi; \lambda_i)$, the coefficient of $g(x, \xi; \lambda_i)$ is zero. Consequently $N(x, \xi; \lambda_i)$ and its first n derivatives with respect to x and ξ are continuous functions of (x, ξ) for $a \leqslant x \leqslant b$, $a \leqslant \xi \leqslant b$. Moreover, $N(x, \xi; \lambda)$ satisfies the system

$$\begin{cases} L_x(u)+\lambda u=0, \\ U_i(u)=0 \qquad (i=1, 2, \ldots, n) \end{cases}$$

for all values of λ, and therefore $R(x, \xi)$, regarded as a function of x, satisfies this system for the characteristic number λ_i. This characteristic number is simple, and therefore $R(x, \xi)$ is of the form

$$C_i u_i(x),$$

where $u_i(x)$ is the characteristic function corresponding to λ_i, and C_i depends upon ξ only. But regarded as a function of ξ, $R(x, \xi)$ satisfies the system

$$\begin{cases} \overline{L}_\xi(v)+\lambda v=0, \\ V_i(v)=0, \qquad (i=1, 2, \ldots, n) \end{cases}$$

for the characteristic number λ_i; C_i is therefore of the form

$$c_i v_i(\xi),$$

where c_i is a constant. Hence

$$R(x, \xi)=c_i u_i(x) v_i(\xi),$$

and it remains to determine the constant c_i.

Now

$$(\lambda-\lambda_i)G(x, \xi; \lambda)-R(x, \xi)$$

is analytic in λ if λ is sufficiently near to λ_i, and is continuous in x and ξ, since both G and R are continuous in x and ξ; also

$$\lim_{\lambda\to\lambda_i} \{(\lambda-\lambda_i)G(x, \xi; \lambda)-c_i u_r(x) v_i(\xi)\}=0.$$

It follows that

$$\lim_{\lambda\to\lambda_i} (\lambda-\lambda_i)\int_a^b G(x, \xi; \lambda)u_i(\xi)d\xi-c_i u_i(x)\int_a^b u_i(\xi)v_i(\xi)d\xi=0.$$

But

$$(\lambda-\lambda_i)\int_a^b G(x, \xi; \lambda)u_i(\xi)d\xi=u_i(x),$$

which is not identically zero, and therefore

$$c_i\int_a^b u_i(\xi)v_i(\xi)d\xi=1.$$

The following theorem has thus been established: *If $\lambda=\lambda_i$ is a simple root of the characteristic equation, the Green's function has the form*

$$\frac{u_i(x)v_i(\xi)}{(\lambda-\lambda_i)\int_a^b u_i(\xi)v_i(\xi)d\xi}+R(x, \xi; \lambda),$$

where

$$\int_a^b u_i(\xi)v_i(\xi)d\xi\neq 0,$$

and $R(x, \xi; \lambda)$ is regular in the neighbourhood of λ_i.

If all the characteristic numbers λ_i whose moduli are less than a number Λ are simple roots of the characteristic equation, then

$$G(x, \xi; \lambda)=\sum_i \frac{u_i(x)v_i(\xi)}{(\lambda-\lambda_i)\int_a^b u_i(\xi)v_i(\xi)d\xi}+E(x, \xi; \lambda),$$

where $E(x, \xi; \lambda)$ does not become infinite for any value of λ such that $|\lambda| < \Lambda$.

Since $u_i(x)$ and $v_i(x)$ satisfy homogeneous systems, they may be normalised so that

$$\int_a^b u_i(\xi)v_i(\xi)d\xi = 1,$$

and then

$$G(x, \xi; \lambda) = \sum_i \frac{u_i(x)v_i(\xi)}{\lambda - \lambda_i} + E(x, \xi; \lambda).$$

11·2. The Relationship between a Linear Differential System and an Integral Equation.—Any non-homogeneous linear differential system with boundary conditions equal in number to n, the order of the equation, may be written in the form

$$\text{(A)} \qquad \begin{cases} L(y) = g(x)y + r(x), \\ U_i(y) = \gamma_i \qquad (i = 1, 2, \ldots, n), \end{cases}$$

and, moreover, the main theorem of § 9·6 shows that when the system is given, $g(x)$ may be so chosen that the homogeneous system

$$\text{(B)} \qquad \begin{cases} L(u) = 0, \\ U_i(u) = 0 \end{cases}$$

is incompatible. It does not follow that (A) has a unique solution, or in fact any solution at all. Let it be assumed, however, for the moment, that (A) has a solution $y_1(x)$. Then the system

$$\begin{cases} L(y) = g(x)y_1(x) + r(x), \\ U_i(y) = \gamma_i \end{cases}$$

has a unique solution, and this solution is $y_1(x)$. As in § 11·11, $y_1(x)$ satisfies the relation

$$y(x) = \int_a^b G(x, \xi)\{g(\xi)y(\xi) + r(\xi)\}d\xi + \gamma_1 G_1(x) + \gamma_2 G_2(x) + \ldots + \gamma_n G_n(x),$$

where $G(x, \xi)$ is the Green's function of the system (B).

But now $y(x)$ occurs under the integral sign; the relation has therefore taken the form of an integral equation, of which $G(x, \xi)$ is the nucleus. Write

$$K(x, \xi) = G(x, \xi)g(\xi),$$
$$f(x) = \gamma_1 G_1(x) + \gamma_2 G_2(x) + \ldots + \gamma_n G_n(x) + \int_a^b G(x, \xi)r(\xi)d\xi,$$

expressions which, theoretically at least, are regarded as known. Then the integral equation which would be satisfied by a solution of (A) is

$$\text{(C)} \qquad y(x) = f(x) + \int_a^b K(x, \xi)y(\xi)d\xi,$$

which is known as a Fredholm equation of the second kind.*

It has thus been proved that *any solution of the differential system* (A), *supposed compatible, satisfies the integral equation* (C).

Conversely if $y_2(x)$ is a solution of (C), then

$$y(x) = f(x) + \int_a^b K(x, \xi)y_2(\xi)d\xi$$

satisfies the system

$$\begin{cases} L(y) = g(x)y_2(x) + r(x), \\ U_i(y) = 0. \end{cases}$$

* Whittaker and Watson, *Modern Analysis*, § 11·2.

But, in the integral equation, $y(x)=y_2(x)$; the differential system therefore admits of the solution $y_2(x)$, that is to say, *any solution of the integral equation* (C) *satisfies the differential system* (A).

These two theorems are included in the general statement that *the differential system and the integral equation are equivalent to one another.*

In particular, if λ is not a characteristic number of the system

$$\text{(D)}\qquad \begin{cases} L(u)+\lambda u=0, & \\ U_i(u)=0 & (i=1, 2, \ldots, n), \end{cases}$$

in which $L(u)$ and $U_i(u)$ are independent of λ, then the system

$$\text{(E)}\qquad \begin{cases} L(y)+\lambda y=r(x), & \\ U_i\,(y)=0 & (i=1, 2, \ldots, n) \end{cases}$$

is equivalent to the integral equation

$$\text{(F)}\qquad y(x)+\lambda\int_a^b G(x, \xi)y(\xi)d\xi=f(x),$$

where

$$f(x)=\int_a^b G(x, \xi)r(\xi)d\xi.$$

$G(x, \xi)$ is, as before, the Green's function of the system (B); let $\Gamma(x, \xi; \lambda)$ be the Green's function of the system

$$\text{(G)}\qquad \begin{cases} \bar{L}(v)+\lambda v=0, & \\ V_i(v)=0 & (i=1, 2, \ldots, n), \end{cases}$$

adjoint to (D). Then by applying Green's formula

$$\int_a^b \{vL(u)-u\bar{L}(v)\}dx=\Big[P(u, v)\Big]_a^b,$$

it is found, as in § 11·1, that

$$\text{(H)}\quad \lambda\int_a^b G(x, \xi_1)\Gamma(x, \xi_2; \lambda)dx$$
$$=p_0(\xi_1)\Gamma(\xi_1, \xi_2; \lambda)\lim\left[\frac{d^{n-1}G}{dx^{n-1}}\right]_{\xi_1-\epsilon}^{\xi_1+\epsilon}-p_0(\xi_2)G(\xi_2, \xi_1)\lim\left[\frac{d^{n-1}\Gamma}{dx^{n-1}}\right]_{\xi_2-\epsilon}^{\xi_2+\epsilon}$$
$$=\Gamma(\xi_1, \xi_2; \lambda)-G(\xi_2, \xi_1).$$

The function $\Gamma(x, \xi; \lambda)$ which enters into this relation is known as the *resolvent function* of the nucleus $G(x, \xi)$, for now the integral equation (F) and therefore the differential system (E) have solutions explicitly given by

$$\text{(I)}\qquad y(x)=f(x)-\lambda\int_a^b \Gamma(x, \xi; \lambda)f(\xi)d\xi,$$

as is seen by substituting this expression in (F) and making use of (H).

But since the characteristic numbers of the system (G) are the poles of its Green's function $\Gamma(x, \xi; \lambda)$, and since the poles of $\Gamma(x, \xi; \lambda)$ are precisely the characteristic numbers of the homogeneous integral equation

$$\text{(J)}\qquad u(x)+\lambda\int_a^b G(x, \xi)u(\xi)d\xi=0,$$

it follows that this integral equation is equivalent to the system (D), and in the same way the adjoint integral equation

$$\text{(K)}\qquad v(x)+\lambda\int_a^b G(\xi, x)v(\xi)d\xi=0$$

is equivalent to the adjoint system (G).

If the solutions of the system (D) are denoted by $u_i(x)$, and those of (G) by $v_i(x)$, then it is known from the theory of adjoint integral equations that the systems $u_i(x)$, $v_i(x)$ are biorthogonal, that is to say,

$$\int_a^b u_i(x)v_j(x)dx=0 \qquad (i \neq j).$$

The systems may also be normalised, so that

$$\int_a^b u_i(x)v_i(x)dx=1.$$

Then $G(x, \xi)$, regarded as the nucleus of the homogeneous integral equation (J) may be developed thus :

$$G(x, \xi)=\sum_{i=1}^{n} \frac{u_i(x)v_i(\xi)}{\lambda-\lambda_i}+E(x, \xi),$$

where $\lambda_1, \lambda_2, \ldots, \lambda_n$ are arranged in order of increasing modulus and $E(x, \xi)$ is a nucleus which has no characteristic numbers of modulus less than $|\lambda_n|$. This agrees with the development in the preceding section.

When the given differential system is self-adjoint, and therefore the Green's function is symmetrical, the results of the well-developed theory of integral equations with symmetrical nuclei can be taken over bodily. Thus, for instance, the theorems that at least one characteristic number exists, and that there can be no imaginary characteristic numbers are true for self-adjoint differential systems.

Moreover, it may be shown that when the given system is of the form (D) the Green's function is closed, that is to say, there exists no continuous function $\phi(x)$ such that

$$\int_a^b G(x, \xi)\phi(\xi)d\xi=0$$

identically. In such a case there always exists an infinite set of characteristic numbers.

11·3. Application of the Method of Successive Approximations.—The demonstration of the existence theorems of Chapter III. by means of the method of successive approximations is equivalent to the theoretical solution of a one-point boundary problem. By a modification of the method the two-point problem may also be approached.* This new aspect of the problem is valuable because it brings out very clearly the part played by the characteristic numbers.

The differential system may be written in a variety of ways as

$$\text{(A)} \qquad \begin{cases} L(y) =M(y)+r(x), \\ U_i(y)=V_i(y)+\gamma_i \end{cases} \qquad (i=1, 2, \ldots, n),$$

in which parts of the differential expression and of the boundary expressions have been transferred to the right-hand members of the equations. Thus $L(y)$ is a differential expression of order n, and $M(y)$ a differential expression of order lower than n; $U_i(y)$ and $V_i(y)$ are linear forms in

$$y(a),\quad y'(a),\quad \ldots,\quad y^{(n-1)}(a),\quad y(b),\quad y'(b),\quad \ldots,\quad y^{(n-1)}(b).$$

The coefficient of $y^{(n)}(x)$ in $L(y)$ will be taken to be unity, the remaining coefficients in $L(y)$ and those of $M(y)$ will be supposed to be continuous in (a, b).

Now the given system may, by § 9·6, be so written in the form (A) that the system

$$\text{(B)} \qquad \begin{cases} L(u) =0, \\ U_i(u)=0 \end{cases} \qquad (i=1, 2, \ldots, n)$$

is incompatible.

* Liouville, *J. de Math.* 5 (1840), p. 356.

Now let y_0 be a function of x such that $M(y_0)$ is continuous in (a, b) and the expressions $V_i(y_0)$ are finite. Then since (B) is incompatible, a system of functions

$$y_1(x), \quad y_2(x), \quad \ldots, \quad y_r(x), \ldots$$

is determined uniquely by the recurrence-relations

$$\begin{cases} L(y_r) = M(y_{r-1}) + r(x), \\ U_i(y_r) = V_i(y_{r-1}) + \gamma_i \end{cases} \qquad (i=1, 2, \ldots, n).$$

In fact, if $G(x, \xi)$ is the Green's function of the system (B),

$$y_r(x) = \int_a^b G(x, \xi)\Big[M\{y_{r-1}(\xi)\} + r(\xi)\Big]d\xi + \sum_{i=1}^{n}\Big[V_i\{y_{r-1}(x)\} + \gamma_i\Big]G_i(x),$$

so that if

$$v_1 = y_1, \quad v_2 = y_2 - y_1, \quad \ldots, \quad v_r = y_r - y_{r-1}, \ldots$$

then

$$\text{(C)} \qquad v_r(x) = \int_a^b G(x, \xi)M\{v_{r-1}(\xi)\}d\xi + \sum_{i=1}^{n} V_i\{v_{r-1}(x)\}G_i(x),$$

where the functions $G_i(x)$ are as defined in § 11·11.

The question now at issue is whether or not the process converges, that is to say whether or not the series

$$v_1 + v_2 + \ldots + v_r + \ldots$$

and the first $n-1$ derived series obtained by term-by-term differentiation converge uniformly in the interval (a, b). It will be seen that the question is now by no means as simple as it was in the case of the one-point boundary problem.

Let A be a number at least equal to the greatest of the upper bounds of

$$|G(x, \xi)|, \quad \left|\frac{\partial G}{\partial x}\right|, \quad \ldots, \quad \left|\frac{\partial^{n-1} G}{\partial x^{n-1}}\right|,$$

$$|G_i(x)|, \quad |G_i'(x)|, \quad \ldots, \quad |G_i^{(n-1)}(x)|, \quad (i=1, 2, \ldots, n)$$

in the interval (a, b).

Let $F(x)$ be the sum of the moduli of the coefficients of $M(v)$ and Ω the sum of the moduli of the coefficients of all the n expressions $V_i(v)$. Also let ω_r be the greatest of the upper bounds of

$$|v_r|, \quad |v_r'|, \quad \ldots, \quad |v_r^{(n-1)}|$$

in (a, b). Then

$$|v_r^{(\nu)}(x)| < \int_a^b A\omega_{r-1}F(t)dt + A\Omega\omega_{r-1} \qquad (\nu=0, 1, \ldots, n-1)$$

for all values of x in (a, b), or

$$\omega_r < AB\omega_{r-1},$$

where

$$B = \int_a^b F(t)dt + \Omega.$$

The process therefore converges if $AB<1$. Now it will be seen that A depends only on the coefficients of $L(v)$ and $U_i(v)$ and on $r(x)$ and γ_i, and B depends only upon the coefficients of $M(v)$ and $V_i(v)$. If, therefore, $M(v)$ and $V_i(v)$ can be chosen so that AB is sufficiently small, the process will converge.

The most satisfactory way of attacking the problem is to consider the auxiliary system

$$\text{(D)} \qquad \begin{cases} L(y) = \lambda\{M(y) + r_1(x)\} + r_2(x), \\ U_i(y) = \lambda\{V_i(y) + \eta_i\} + \theta_i \end{cases} \qquad (i=1, 2, \ldots, n),$$

where

$$r_1(x)+r_2(x)=r(x),$$
$$\eta_i+\theta_i=\gamma_i.$$

This system reduces to the original system (A) when $\lambda=1$.

Let $y_1(x)$ be chosen so as to satisfy the system

$$\begin{cases} L(y) = r_2(x), \\ U_i(y)=\theta_i \end{cases} \qquad (i=1, 2, \ldots, n),$$

and let

$$y_2(x), \quad \ldots, \quad y_r(x), \ldots$$

be defined by successive approximation in (D). Then $y_1(x)$ will be independent of λ and $y_r(x)$ will be a polynomial in λ of degree $r-1$. In the limit this polynomial becomes a power series in λ which will presumably converge for sufficiently small values of $|\lambda|$. The point at issue is whether or not it converges for $\lambda=1$.

To settle this question, consider for the moment a system of a more general character than (D), namely

$$\text{(E)} \qquad \begin{cases} \mathbf{L}(w) = \mathbf{r}(x), \\ \mathbf{U}_i(w)=\beta_i \end{cases} \qquad (i=1, 2, \ldots, n),$$

in which $\mathbf{r}(x)$ and the coefficients of $\mathbf{L}(w)$ are analytic functions of λ throughout a given domain, and are uniformly continuous functions of x in (a, b). Similarly β_i and the coefficients of $\mathbf{U}_i(w)$ are analytic functions of λ in the given domain.

The formal expression of the solution of this system is

$$w(x)=\begin{vmatrix} w_0, & y_1, & \ldots, & y_n \\ \mathbf{U}_1(w_0)-\beta_1, & \mathbf{U}_1(y_1), & \ldots, & \mathbf{U}_1(y_n) \\ \cdot & \cdot & \cdot & \cdot \\ \mathbf{U}_n(w_0)-\beta_n, & \mathbf{U}_n(y_1), & \ldots, & \mathbf{U}_n(y_n) \end{vmatrix} \div \begin{vmatrix} \mathbf{U}_1(y_1), & \ldots, & \mathbf{U}_1(y_n) \\ \cdot & \cdot & \cdot \\ \mathbf{U}_n(y_1), & \ldots, & \mathbf{U}_n(y_n) \end{vmatrix},$$

in which w_0 is a solution of the equation

$$\mathbf{L}(w)=\mathbf{r}(x),$$

and $y_1, \ldots, y_n$ are linearly-independent solutions of

$$\mathbf{L}(y)=0.$$

Now since $w_0, y_1, \ldots, y_n$ are solutions of equations whose coefficients are analytic in λ and uniformly continuous in x, the two determinants which figure in the expression for $w(x)$ are themselves analytic in λ and uniformly continuous in x. Hence $w(x)$ is also analytic in λ and uniformly continuous in x except for those values of λ for which the determinant in the denominator vanishes, that is to say, except for characteristic values of λ.

This result may now be applied to the system (D) to the effect that the power series in λ which represents the limiting value of $y_r(x)$ converges in any circle whose centre lies at the point $\lambda=0$ and which does not contain any characteristic number of the system

$$\text{(F)} \qquad \begin{cases} L(u) = \lambda M(u), \\ U_i(u)=\lambda V_i(u) \end{cases} \qquad (i=1, 2, \ldots, n).$$

It follows that the method of successive approximations as applied to the system (A) will converge if the system (F) has no characteristic number of modulus less than or equal to unity.

A much more precise result can now be obtained. Let $\lambda=\lambda_1$, be a characteristic number of the homogeneous system corresponding to (E). Then $(\lambda-\lambda_1)$ will be a factor of the denominator of $w(x)$ and the multiplicity of this factor will be at least equal to the index of λ_1. If it so happens that $(\lambda-\lambda_1)$ is also a factor of the numerator of the same multiplicity as in the denominator, then the solution $w(x)$ will exist even for the characteristic number λ_1. This will occur, for instance, when the multiplicity of λ_1 is equal to its index k, and the non-homogeneous system (E) has a solution when $\lambda=\lambda_1$. For then every minor of order $n-k$ which can be extracted from the numerator of $w(x)$ will be zero when $\lambda=\lambda_1$, and therefore the numerator, as well as the denominator, will contain the factor $(\lambda-\lambda_1)$ repeated exactly k times. Thus $w(x)$ will remain analytic when $\lambda=\lambda_1$.

Applied to the system (D) this result proves that *the process will converge when* $|\lambda|\leqslant 1$ *provided that if any characteristic numbers of* (F) *lie within or on the circumference of the circle* $|\lambda|=1$, *the index of each such characteristic number is equal to its multiplicity and that for each such characteristic number the system* (D) *is compatible.*

11·31. Conditions for the Compatibility of a Non-Homogeneous System for Characteristic Values of the Parameter.—When, as in the case of any consistent one-point boundary problem, there exist no characteristic numbers, the method of successive approximations certainly converges for all values of the parameter for which the coefficients of the equation remain continuous. On the other hand, the system corresponding to a two-point boundary problem has, in general, characteristic numbers, and in order that the method of successive approximation may be applicable, it is necessary that the system should remain compatible at least for those characteristic numbers whose moduli do not exceed a certain magnitude. Necessary and sufficient conditions for the existence of solutions of a non-homogeneous system for a characteristic value of the parameter are known.* In the present section such conditions will be given in the case of the self-adjoint system of the second order

$$\text{(A)}\qquad \begin{cases} L(y)\equiv \dfrac{d}{dx}\left\{K\dfrac{dy}{dx}\right\}-Gy=R, \\ U_1(y)=\alpha_1 y(a)+\alpha_2 y(b)+\alpha_3 y'(a)+\alpha_4 y'(b)=A, \\ U_2(y)=\beta_1 y(a)+\beta_2 y(b)+\beta_3 y'(a)+\beta_4 y'(b)=B. \end{cases}$$

All coefficients which occur in the system are supposed to be analytic functions of the parameter λ in a given domain, K, G and R are further supposed to be uniformly continuous functions of x in (a, b). The condition that the system may be self-adjoint is that

$$\text{(B)}\qquad \delta_{24}K(a)=\delta_{13}K(b),$$

where

$$\delta_{ij}=\alpha_i\beta_j-\alpha_j\beta_i.$$

Let $u_1(x)$ and $u_2(x)$ be solutions of

$$L(u)=0$$

such that

$$\text{(C)}\qquad u_1'u_2-u_2'u_1=1/K,$$

then the general solution of the equation

$$L(y)=R$$

is

$$y=c_1u_1+c_2u_2+u_1\int_a^x KRu_2dx+u_2\int_x^b KRu_1dx.$$

* Such conditions are given in the case of equations of the second order by Mason, *Trans. Am. Math. Soc.* 7 (1906), p. 337; and in the case of equations of higher order by Dini, *Ann. di Mat.* (3), 12 (1906), p. 243.

The constants c_1 and c_2 are determined by imposing the boundary conditions thus:

$$c_1U_1(u_1)+c_2U_1(u_2)=A-\{\alpha_1u_2(a)+\alpha_3u_2'(a)\}\int_a^b KRu_1dx$$
$$-\{\alpha_2u_1(b)+\alpha_4u_1'(b)\}\int_a^b KRu_2dx,$$

$$c_1U_2(u_1)+c_2U_2(u_2)=B-\{\beta_1u_2(a)+\beta_3u_2'(a)\}\int_a^b KRu_1dx$$
$$-\{\beta_2u_1(b)+\beta_4u_1'(b)\}\int_a^b KRu_2dx.$$

Everything depends upon the determinant

$$\Delta(u_1, u_2)=\begin{vmatrix} U_1(u_1), & U_1(u_2) \\ U_2(u_1), & U_2(u_2) \end{vmatrix}.$$

Values of λ for which Δ is not zero are not characteristic numbers; the given system is then compatible. The purpose of the present investigation is to discover what conditions must be imposed upon R, A and B in order that when Δ is zero, the system may admit of a solution. Two cases arise:

(1°) *The minors of Δ are not all zero.*

The reduced system is now singly-compatible; it admits of one and only one independent solution. Let $u_1(x)$ be this solution, then

$$U_1(u_1)=U_2(u_1)=0,$$

but $U_1(u_2)$ and $U_2(u_2)$ are not both zero.

A necessary and sufficient condition that the system (A) should be compatible is that

$$U_1(u_2)\left[B-\{\beta_1u_2(a)+\beta_3u_2'(a)\}\int_a^b KRu_1dx-\{\beta_2u_1(b)+\beta_4u_1'(b)\}\int_a^b KRu_2dx\right]$$
$$-U_2(u_2)\left[A-\{\alpha_1u_2(a)+\alpha_3u_2'(a)\}\int_a^b KRu_1dx-\{\alpha_2u_1(b)+\alpha_4u_1'(b)\}\int_a^b KRu_2dx\right]=0.$$

In the left-hand member, the coefficient of $\int_a^b KRu_2dx$ may be written

$$\{\alpha_1u_2(a)+\alpha_3u_2'(a)\}\{\beta_1u_1(a)+\beta_3u_1'(a)\}-\{\alpha_2u_2(b)+\alpha_4u_2'(b)\}\{\beta_2u_1(b)+\beta_4u_1'(b)\}$$
$$-\{\beta_1u_2(a)+\beta_3u_2'(a)\}\{\alpha_1u_1(a)+\alpha_3u_1'(a)\}+\{\beta_2u_2(b)+\beta_4u_2'(b)\}\{\alpha_2u_1(b)+\alpha_4u_1'(b)\}$$
$$=\delta_{13}\{u_1'(a)u_2(a)-u_2'(a)u_1(a)\}-\delta_{24}\{u_1'(b)u_2(b)-u_2'(b)u_1(b)\}$$
$$=\frac{\delta_{13}}{K(a)}-\frac{\delta_{24}}{K(b)}=0,$$

so that the condition becomes

$$\text{(D)}\qquad U_1(u_2)\left[B-\{\beta_1u_2(a)+\beta_3u_2'(a)\}\int_a^b KRu_1dx\right]$$
$$-U_2(u_2)\left[A-\{\alpha_1u_2(a)+\alpha_3u_2'(a)\}\int_a^b KRu_1dx\right]=0.$$

Now

$$\text{(E)}\qquad \begin{cases} \delta_{21}u_1(b)+\delta_{31}u_1'(a)+\delta_{41}u_1'(b)=0, \\ \delta_{12}u_1(a)\qquad\qquad +\delta_{32}u_1'(a)+\delta_{42}u_1'(b)=0, \\ \delta_{13}u_1(a)+\delta_{23}u_1(b)\qquad\qquad +\delta_{43}u_1'(b)=0, \\ \delta_{14}u_1(a)+\delta_{24}u_1(b)+\delta_{34}u_1'(a)\qquad\qquad =0, \end{cases}$$

for the left-hand members of these equations are of the form

$$\beta_iU_1(u_1)-\alpha_iU_2(u_1),$$

where $i=1, 2, 3, 4$ respectively.

By means of the relations (B), (C) and (E) it may be verified that

$$
\text{(F)}\begin{cases}
U_1(u_2)\{\beta_2K(a)u_1'(a)+\beta_1K(b)u_1'(b)\}-U_2(u_2)\{\alpha_2K(a)u_1'(a)+\alpha_1K(b)u_1'(b)\}=0,\\
U_1(u_2)\{\beta_4K(a)u_1(a)\ +\beta_3K(b)u_1(b)\}\ -U_2(u_2)\{\alpha_4K(a)u_1(a)\ +\alpha_3K(b)u_1(b)\}\ =0,\\
U_1(u_2)\{\beta_2K(a)u_1(a)\ -\beta_3K(b)u_1'(b)\}-U_2(u_2)\{\alpha_2K(a)u_1(a)\ -\alpha_3K(b)u_1'(b)\}=0,\\
U_1(u_2)\{\beta_4K(a)u_1'(a)-\beta_1K(b)u_1(b)\}\ -U_2(u_2)\{\alpha_4K(a)u_1'(a)\ -\alpha_1K(b)u_1(b)\}\ =0.
\end{cases}
$$

Now $U_1(u_2)$ and $U_2(u_2)$ are not both zero; they may therefore be eliminated between (D) and any one of the four equations of (F). Thus the eliminant of (D) and the first equation of (F) is

$$
\{\alpha_2K(a)u_1'(a)+\alpha_1K(b)u_1'(b)\}\left[B-\{\beta_1u_2(a)+\beta_3u_2'(a)\}\int_a^b KRu_1dx\right]
$$
$$
-\ \beta_2K(a)u_1'(a)+\beta_1K(b)u_1'(b)\}\left[A-\{\alpha_1u_2(a)+\alpha_3u_2'(a)\}\int_a^b KRu_1dx\right]=0,
$$

which reduces to

$$
(\alpha_2B-\beta_2A)K(a)u_1'(a)+(\alpha_1B-\beta_1A)K(b)u_1'(b)+\delta_{12}\int_a^b KRu_1dx=0.
$$

Moreover, the process may be reversed, that is to say, the eliminant so obtained and the first equation of (F) lead back to (D), except when

$$
\alpha_2K(a)u_1'(a)+\alpha_1K(b)u_1'(b)=0,
$$
$$
\beta_2K(a)u_1'(a)+\beta_1K(b)u_1'(b)=0,
$$

that is to say, except when $\delta_{12}=0$ or when $u_1'(a)=u_1'(b)=0$. In the latter case, both $u_1(a)$ and $u_1(b)$ must be distinct from zero, and therefore the first and second equations of (E) show that $\delta_{12}=0$, which is thus the only exceptional case. The equations obtained by eliminating $U_1(u_2)$ and $U_2(u_2)$ between (D) and the four equations (F) are respectively

$$
\text{(G)}\quad\begin{cases}
(\alpha_2B-\beta_2A)K(a)u_1'(a)+(\alpha_1B-\beta_1A)K(b)u_1'(b)+\delta_{12}\int_a^b KRu_1dx=0,\\
(\alpha_4B-\beta_4A)K(a)u_1(a)+(\alpha_3B-\beta_3A)K(b)u_1(b)\ \ +\delta_{43}\int_a^b KRu_1dx=0,\\
(\alpha_2B-\beta_2A)K(a)u_1(a)-(\alpha_3B-\beta_3A)K(b)u_1'(b)\ \ +\delta_{23}\int_a^b KRu_1dx=0,\\
(\alpha_4B-\beta_4A)K(a)u_1'(a)-(\alpha_1B-\beta_1A)K(b)u_1(b)\ \ +\delta_{14}\int_a^b KRu_1dx=0.
\end{cases}
$$

Any one of these is equivalent to the condition (D) provided that the corresponding determinant δ_{12}, δ_{43}, δ_{23} or δ_{14} is not zero Now if any three of these determinants are zero, then all determinants δ_{ij} are zero, which is impossible, since the expressions $U_1(u)$ and $U_2(u)$ are independent. Hence at least two of the equations (G) are significant.

Hence *a necessary and sufficient condition that the system* (A) *be compatible when the corresponding reduced system has only one distinct solution* u_1 *is that* A, B *and* R *should satisfy one or other of the relations* (G) *with non-zero determinant* δ_{ij}.

When A and B are both zero, the condition is that

$$
\int_a^b KRu_1dx=0.
$$

(2°) *The minors of* Δ *are all zero.*

The reduced system is now doubly-compatible and admits of the two solutions $u_1(x)$ and $u_2(x)$. The equations (E) still hold, but there is now also a precisely similar set of equations in u_2. Suppose for the moment that $\delta_{13}=0$, then, by (B), $\delta_{24}=0$. The first equation in (E) becomes

$$\delta_{21}u_1(b)+\delta_{41}u_1'(b)=0,$$

and similarly

$$\delta_{21}u_2(b)+\delta_{41}u_2'(b)=0.$$

But, by (C),

$$K(b)\{u_1'(b)u_2(b)-u_2'(b)u_1(b)\}=1,$$

and therefore

$$\delta_{21}=\delta_{41}=0.$$

Similarly

$$\delta_{23}=\delta_{43}=0.$$

All determinants are thus zero, which is impossible; it follows that δ_{13} and δ_{24} are not zero.

Since

$$U_1(u_1)=U_1(u_2)=U_2(u_1)=U_2(u_2)=0,$$

necessary and sufficient conditions for the existence of a solution of thc system (A) are that

$$A-\{\alpha_1u_2(a)+\alpha_3u_2'(a)\}\int_a^b KRu_1dx-\{\alpha_2u_1(b)+\alpha_4u_1'(b)\}\int_a^b KRu_2dx=0,$$

$$B-\{\beta_1u_2(a)+\beta_3u_2'(a)\}\int_a^b KRu_1dx-\{\beta_2u_1(b)+\beta_4u_1'(b)\}\int_a^b KRu_2dx=0.$$

These equations are equivalent to

$$\text{(H)}\quad\begin{cases}(\alpha_1B-\beta_1A)K(a)u_2(a)+(\alpha_3B-\beta_3A)K(a)u_2'(a)+\delta_{13}\displaystyle\int_a^b KRu_2dx=0,\\[2ex](\beta_2A-\alpha_2B)K(b)u_1(b)+(\beta_4A-\alpha_4B)K(b)u_1'(b)+\delta_{24}\displaystyle\int_a^b KRu_1dx=0.\end{cases}$$

Other equations of the same type may be found, but only two are independent.

A necessary and sufficient condition that the system (A) *may be compatible when the corresponding reduced system has two linearly distinct solutions* u_1 *and* u_2 *is that* A, B *and* R *should satisfy one or other of the relations* (H).

When A and B are zero, R must satisfy the relations

$$\int_a^b KRu_1dx=0, \qquad \int_a^b KRu_2dx=0.$$

11·32. Development of the Solution of a Non-Homogeneous System.—Consider the particular system *

$$\text{(A)}\quad\begin{cases}\dfrac{d}{dx}\left\{k\dfrac{dy}{dx}\right\}+(\lambda g-l)y+p(x)=0,\\ y'(a)-hy(a)=0,\\ y'(b)+Hy(b)=0,\end{cases}$$

where k, g, l and $p(x)$ are continuous, and k does not vanish when $a\leqslant x\leqslant b$. Let u_1 and u_2 be a fundamental pair of solutions of the homogeneous equation

$$\frac{d}{dx}\left\{k\frac{du}{dx}\right\}+(\lambda g-l)u=0,$$

such that

$$u_1(a)=1, \quad u_1'(a)=0,$$
$$u_2(a)=0, \quad u_2'(a)=1.$$

* Kneser, *Math. Ann.* 58 (1904), p. 109.

Then the general solution of the differential equation in (A) is

$$y=C_1u_1(x)+C_2u_2(x)+u_1(x)\int_a^x u_2(t)p(t)dt-u_2(x)\int_a^x u_1(t)p(t)dt,$$

where C_1 and C_2 are arbitrary constants. Each of the four terms which enter into this expression is an integral function of λ when $a\leqslant x\leqslant b$.

The boundary conditions of (A) lead to the relations

$$C_2-hC_1=0,$$

$$C_1\{u_1'(b)+Hu_1(b)\}+C_2\{u_2'(b)+Hu_2(b)\}$$
$$+\{u_1'(b)+Hu_1(b)\}\int_a^b u_2(t)p(t)dt-\{u_2'(b)+Hu_2(b)\}\int_a^b u_1(t)p(t)dt=0,$$

which determine C_1 and C_2. Thus

$$y=w(x, \lambda)/\Delta(\lambda),$$

where $w(x, \lambda)$ is, for all values of x in (a, b), an integral function of λ, and $\Delta(\lambda)$, the characteristic determinant, is an integral function of λ alone.

Let λ_i be a characteristic number of the homogeneous system

$$\text{(B)}\qquad \begin{cases} \dfrac{d}{dx}\left\{k\dfrac{dv}{dx}\right\}+(\lambda g-l)v=0, \\ v'(a)-hv(a)=v'(b)+Hv(b)=0, \end{cases}$$

and let $v_i(x)$ be the corresponding characteristic function. Then since this system is simply-compatible, a necessary and sufficient condition that the non-homogeneous system may have a solution when $\lambda=\lambda_i$ is that

$$\text{(C)}\qquad \int_a^b k(x)p(x)v_i(x)dx=0.$$

If this condition is satisfied, the function $w(x, \lambda)/\Delta(\lambda)$ will be finite when $\lambda=\lambda_i$. Let it be supposed that the condition is satisfied by all characteristic functions

$$v_1(x),\ v_2(x),\ \ldots,\ v_n(x),\ \ldots$$

then $w(x, \lambda)/\Delta(\lambda)$ will be finite when λ assumes any of the values

$$\lambda_1,\ \lambda_2,\ \ldots,\ \lambda_n,\ \ldots,$$

that is to say, it is finite for all values of λ for which Δ vanishes. Consequently, when (C) is satisfied for all integral values of i, $y(x)$ is an integral function of λ and may be developed, when $a\leqslant x\leqslant b$, in the convergent series

$$y(x)=a_0+a_1\lambda+\ \ldots\ +a_n\lambda^n+\ \ldots$$

in which the coefficients $a_0, a_1, \ldots, a_n, \ldots$ may be determined by the method of successive approximations.

11·4. The Asymptotic Development of Characteristic Numbers and Functions.—In the Sturm-Liouville equation

$$\frac{d}{dx}\left\{k\frac{dy}{dx}\right\}+(\lambda g-l)y=0,$$

it will be supposed that, throughout the interval $a\leqslant x\leqslant b$, the functions k, g and l are continuous and k and g do not vanish, that k possesses a continuous derivative, and that gk has a continuous second derivative. Then if the following transformations are made

$$z=\frac{1}{K}\int_a^x\left(\frac{g}{k}\right)^{\frac{1}{2}}dx, \quad u=(gk)^{\frac{1}{4}}y, \quad \rho^2=K^2\lambda,$$

where K is the constant

$$\frac{1}{\pi}\int_a^b\left(\frac{g}{k}\right)^{\frac{1}{2}}dx,$$

the equation assumes the normal form

$$\frac{d^2u}{dz^2}+\{\rho^2-q(z)\}u=0,$$

where

$$q(z)=\left\{\frac{\theta''(z)}{\theta(z)}-K^2\phi(z)\right\},$$

and $\theta(z)$ and $\phi(z)$ are respectively $(gk)^{\frac{1}{4}}$ and l/g, expressed as functions of z. The interval $a\leqslant x\leqslant b$ becomes $0\leqslant z\leqslant \pi$. Throughout this interval $q(z)$ is continuous; for the present no further restrictions are necessary,* but later work requires also the existence and continuity of the first two derivatives of $q(z)$.

The boundary conditions are not altered in form by the transformation; they will be supposed to be

$$\begin{cases} u'(0)-hu(0)=0, \\ u'(\pi)+Hu(\pi)=0, \end{cases}$$

where the constants h and H are real.

If now the equation is written as

$$\frac{d^2u}{dz^2}+\rho^2u=q(z)u,$$

its general solution may be expressed symbolically as

$$u(z)=A\cos\rho z+B\sin\rho z+(D^2+\rho^2)^{-1}q(z)u(z)$$

$$=A\cos\rho z+B\sin\rho z+\frac{1}{\rho}\int_0^z \sin\rho(z-t)q(t)u(t)dt.$$

The differential system as it stands is homogeneous; to make its solution quite definite, the first boundary condition will be replaced by the non-homogeneous conditions

$$u(0)=1;\quad u'(0)=h.$$

The constants A and B are then uniquely determined,† and

$$u(z)=\cos\rho z+\frac{h}{\rho}\sin\rho z+\frac{1}{\rho}\int_0^z \sin\rho(z-t)q(t)u(t)dt.$$

The fundamental existence theorem affirms that $|u(z)|$ is bounded in $(0, \pi)$. Let M be its upper bound, then

$$|u(z)|\leqslant\left(1+\frac{h^2}{\rho^2}\right)^{\frac{1}{2}}+\frac{M}{\rho}\int_0^\pi |q(t)|dt.$$

Since $|u(z)|$ is continuous in the closed interval $0\leqslant x\leqslant\pi$, it attains its upper bound, and therefore

$$M\leqslant\left(1+\frac{h^2}{\rho^2}\right)^{\frac{1}{2}}+\frac{M}{\rho}\int_0^\pi |q(t)|dt,$$

whence

$$M\leqslant\left(1+\frac{h^2}{\rho^2}\right)^{\frac{1}{2}}\Big/\left\{1-\frac{1}{\rho}\int_0^\pi |q(t)|dt\right\}$$

for all values of ρ greater than a fixed positive number.

If now the second boundary condition is applied, it is found that ρ is determined by the equation

$$\tan\pi\rho=\frac{P}{\rho-P'},$$

* These restrictions may be considerably lightened by adopting the methods of Dixon, *Phil. Trans. R. S.* (A) 211 (1911), p. 411.

† The relation thus obtained is interesting historically as being the first recorded instance of an integral equation of the first kind, Liouville, *J. de Math.* 2 (1837), p. 24.

where

$$P = h + H + \int_0^\pi \left\{\cos \rho t - \frac{H}{\rho} \sin \rho t\right\} q(t)u(t)dt,$$

$$P' = \frac{Hh}{\rho} + \int_0^\pi \left\{\sin \rho t + \frac{H}{\rho} \cos \rho t\right\} q(t)u(t)dt.$$

Since

$$M \leqslant 1 + O(\rho^{-1}),$$

in $(0, \pi)$, it follows that $|P|$ and $|P'|$ are both less than finite numbers independent of ρ.

The development will now be carried a step further.* Since $u(t)$ is of the form

$$\cos \rho t + \frac{\alpha(\rho, t)}{\rho},$$

where $\alpha(\rho, t)$ is bounded,

$$u(z) = \cos \rho z \left\{1 - \frac{1}{\rho}\int_0^z \sin \rho t\left(\cos \rho t + \frac{\alpha(\rho, t)}{\rho}\right) q(t)dt\right\}$$
$$+ \sin \rho z \left\{\frac{h}{\rho} + \frac{1}{\rho}\int_0^z \cos \rho t\left(\cos \rho t + \frac{\alpha(\rho, t)}{\rho}\right) q(t)dt\right\},$$

and therefore

$$u(z) = \cos \rho z \{1 + O(\rho^{-2})\} + \sin \rho z \{Q(z)\rho^{-1} + O(\rho^{-2})\},$$

where

$$Q(z) = h + \tfrac{1}{2}\int_0^z q(t)dt.$$

It is now easy to verify that

$$P = h + H + h_1 + O(\rho^{-1}), \quad P' = O(\rho^{-1}),$$

where

$$h_1 = \tfrac{1}{2}\int_0^\pi q(t)dt.$$

The characteristic equation now becomes

$$\tan \pi\rho = \frac{h + H + h_1 + O(\rho^{-1})}{\rho + O(\rho^{-1})},$$

and therefore, for sufficiently large values of ρ

$$\rho = n + \frac{h + H + h_1}{\pi\rho} + O(\rho^{-2})$$
$$= n + \frac{h + H + h_1}{n\pi} + O(n^{-2}),$$

or

$$\rho_n = n + cn^{-1} + O(n^{-2}),$$

where c is independent of n. This expression incidentally furnishes a new proof of the theorem that there exists an infinite set of characteristic numbers.

Now

$$\cos \rho_n z = \cos nz \{1 + O(n^{-2})\} - \sin nz \{czn^{-1} + O(n^{-2})\},$$
$$\sin \rho_n z = \sin nz \{1 + O(n^{-2})\} + \cos nz \{czn^{-1} + O(n^{-2})\},$$

and therefore the characteristic function corresponding to ρ_n is

$$u_n(z) = \cos nz \{1 + O(n^{-2})\} + \sin nz \{a(z)n^{-1} + O(n^{-2})\},$$

* Hobson, *Proc. London Math. Soc.* (2) 6 (1908), p. 374.

where

$$\alpha(z)=Q(z)-cz.$$

Let the characteristic function be normalised, and then denoted by $v_n(z)$, thus

$$v_n(z)=\left(\frac{2}{\pi}\right)^{\frac{1}{2}}\cos nz\,\{1+O(n^{-2})\}+\sin nz\,\{\beta(z)n^{-1}+O(n^{-2})\}.$$

This is the asymptotic expression for the characteristic functions; it is of particular utility in computing the characteristic functions for large values of n. The expression may be carried to any desired degree of approximation.*

Two exceptional cases deserve mention, namely (1°) when either h or H is infinite, (2°) when both h and H are infinite.† In the first case, at one of the end-points, but not at the other, $u(x)$ is zero; then

$$\rho_n=n+\tfrac{1}{2}+O(n^{-1}).$$

In the second case, $u(x)$ vanishes at both end-points, and

$$\rho_n=n+1+O(n^{-1}).$$

11·5. The Sturm-Liouville Development of an Arbitrary Function.—Let

$$u_0(x),\quad u_1(x),\quad \ldots,\quad u_n(x),\quad \ldots$$

be the set of normalised characteristic functions of the system

$$\text{(A)}\qquad \begin{cases} \dfrac{d^2u}{dx^2}+\{\rho^2-q(x)\}u=0, \\ u'(0)-hu(0)=0, \\ u'(\pi)+Hu(\pi)=0, \end{cases}$$

corresponding respectively to the characteristic numbers

$$\rho_0,\quad \rho_1,\quad \ldots,\quad \rho_n,\quad \ldots$$

where, as in § 11·4,

$$\rho_n=n+cn^{-1}+O(n^{-2}).$$

It will first be shown that this set of characteristic functions is *closed*, that is to say, if $p(x)$ is any function continuous in $(0, \pi)$ and if

$$\text{(B)}\qquad \int_0^\pi p(x)u_n(x)dx=0$$

for all values of n, then

$$p(x)=0$$

identically.‡

Consider the system

$$\text{(C)}\qquad \begin{cases} \dfrac{d^2v}{dx^2}+\{\rho^2-q(x)\}v+p(x)=0, \\ v'(0)-hv(0)=v'(\pi)+Hv(\pi)=0. \end{cases}$$

When ρ is not a characteristic number, this system has a unique solution which may be expressed in the form of the infinite series

$$\text{(D)}\qquad v(x)=v_0+\rho^2v_1+\ \ldots\ +\rho^{2n}v_n+\ \ldots,$$

* Horn, *Math. Ann.* 52 (1899), pp. 271, 340; Schlesinger, *ibid.* 63 (1907), p. 277; Birkhoff, *Trans. Am. Math. Soc.* 9 (1908), pp. 219, 373; Blumenthal, *Archiv d. Math. u. Phys.* (3), 19 (1912), p. 136.

† Kneser, *Math. Ann.* 58 (1904), p. 136.

‡ *Ibid.*, p. 113.

where $v_0, v_1, \ldots, v_n, \ldots$ satisfy the equations

$$\frac{d^2v_0}{dx^2} - qv_0 + p(x) = 0,$$

$$\frac{d^2v_1}{dx_2} - qv_1 + v_0 = 0,$$

$$\cdot \quad \cdot \quad \cdot$$

$$\frac{d^2v_n}{dx^2} - qv_n + v_{n-1} = 0,$$

$$\cdot \quad \cdot \quad \cdot$$

From these equations it is easily verified that

$$\int_0^\pi \left\{ v_{m+1}\frac{d^2v_n}{dx^2} - v_n\frac{d^2v_{m+1}}{dx^2} \right\} dx = \int_0^\pi \{v_m v_n - v_{m+1}v_{n-1}\} dx.$$

Now the left-hand member of this relation reduces to

$$\left[v_{m+1}v_n' - v_n v'_{m+1} \right]_0^\pi$$

which is zero on account of the boundary conditions. Hence

$$\int_0^\pi v_{m+1}v_{n-1}dx = \int_0^\pi v_m v_n dx.$$

The common value of these integrals therefore depends only upon the sum of the suffixes; it will be denoted by W_{m+n}. Now

$$\int_0^\pi (\alpha v_{m-1} + \beta v_{m+1})^2 dx = W_{2m-2}\alpha^2 + 2W_{2m}\alpha\beta + W_{2m+2}\beta^2,$$

which cannot be negative for any real values of α and β, and therefore considering $\alpha=0$, $\beta=0$ in turn,

$$W_{2m+2} \geqslant 0, \quad W_{2m-2} \geqslant 0.$$

Moreover, since the quadratic form in α, β is positive,

$$W_{2m}^2 - W_{2m-2}W_{2m+2} \leqslant 0,$$

and therefore W_{2m} is either zero for all values of m or always positive. Suppose that $W_0 > 0$, then

(E) $$\frac{W_2}{W_0} \leqslant \frac{W_4}{W_2} \leqslant \cdots \leqslant \frac{W_{2m+2}}{W_{2m}} \leqslant \cdots$$

Now it follows from § 11·31 that if the system (C) has a solution $V_n(x)$ when $\rho = \rho_n$, then

$$\int_0^\pi p(x)V_n(x)dx = 0,$$

and conversely. Moreover it was proved in § 11·32 that if this relation holds for all integral values of n, the system (C) has a solution $v(x)$ for all values of ρ, and this solution, by the fundamental existence theorem, is represented by the series (D) which then converges for all values of ρ and for all values of x in $(0, \pi)$. Consequently the development

$$\int_0^\pi v_0 v(x) dx = W_0 + \rho^2 W_1 + \cdots + \rho^{2n}W_n + \cdots$$

is finite for all values of ρ, which is impossible in view of the inequalities (E). It therefore follows that

$$W_0 = W_2 = \cdots = W_{2m} = \cdots = 0.$$

Consequently

$$v_0=0 \quad \text{and} \quad p(x)=0$$

identically in (a, b).

Now let $f(x)$ be an arbitrary function of the real variable x. The theory of Fourier series suggests that it may be possible to develop $f(x)$ as an infinite series of normal functions, thus

$$f(x)=c_0u_0(x)+c_1u_1(x)+\ .\ .\ .\ +c_nu_n(x)+\ .\ .\ .$$

If this development is possible, then on account of the orthogonal properties of the functions $u_n(x)$, it is easily found that

$$c_n=\int_0^\pi f(t)u_n(t)dt,$$

so that the coefficients c_n are determined uniquely.

The two main questions which arise are

(1°) whether the series

$$\sum_{r=0}^{\infty} u_r(x)\int_0^\pi f(t)u_r(t)dt$$

converges uniformly in $(0, \pi)$ or not,

(2°) when the series converges, whether it converges to the value $f(x)$ or to some other limit. These questions will be dealt with in the succeeding sections.

11·51. The Convergence of the Development.—In the first place, a very special function $\phi(x)$ will be dealt with, which is continuous and has continuous first and second derivatives in $(0, \pi)$. Consider the series *

$$\sum_{r=1}^{\infty} u_r(x)\int_0^\pi \phi(t)u_r(t)dt.$$

Now

$$\begin{aligned}\int_0^\pi \phi(t)u_r(t)dt &= \int_0^\pi \frac{\phi(t)}{\rho_r^2-q(t)}\{\rho_r^2-q(t)\}u_r(t)dt\\ &= -\int_0^\pi \frac{\phi(t)}{\rho_r^2-q(t)}\cdot\frac{d^2u_r(t)}{dt^2}dt\\ &= \left[-\frac{\phi(t)u_r'(t)}{\rho_r^2-q(t)}+u_r(t)\frac{d}{dt}\left\{\frac{\phi(t)}{\rho_r^2-q(t)}\right\}\right]_0^\pi-\int_0^\pi u_r(t)\frac{d^2}{dt^2}\left\{\frac{\phi(t)}{\rho_r^2-q(t)}\right\}dt,\end{aligned}$$

on integrating by parts; in view of the boundary conditions this reduces to

$$\frac{H\phi(\pi)u_r(\pi)}{\rho_r^2-q(\pi)}+\frac{h\phi(0)u_r(0)}{\rho_r^2-q(0)}+\left[u_r(t)\frac{d}{dt}\left\{\frac{\phi(t)}{\rho_r^2-q(t)}\right\}\right]_0^\pi-\int_0^\pi u_r(t)\frac{d^2}{dt^2}\left\{\frac{\phi(t)}{\rho_r^2-q(t)}\right\}dt.$$

Now since $\phi(t)$ is continuous and has continuous first and second derivatives, and the same hypothesis has been made with regard to $q(t)$, it is clear that

$$\rho^2\frac{d}{dt}\left\{\frac{\phi(t)}{\rho^2-q(t)}\right\} \quad \text{and} \quad \rho^2\frac{d^2}{dt^2}\left\{\frac{\phi(t)}{\rho^2-q(t)}\right\}$$

are bounded for sufficiently large values of ρ, say $\rho \geqslant \rho_\nu$, and for all values of t in $(0, \pi)$. Hence

$$\sum_{r=\nu}^{\infty}u_r(x)\int_0^\pi \phi(t)u_r(t)dt=\sum_{r=\nu}^{\infty}u_r(x)\left\{\frac{H\phi(\pi)u_r(\pi)}{\rho_r^2-q(\pi)}+\frac{h\phi(0)u_r(0)}{\rho_r^2-q(0)}+\frac{A_r}{\rho_r^2}\right\},$$

where the constants A_r are finite for all values of r. The series is therefore

* Kneser, *Math. Ann.* 58 (1904), p. 121.

absolutely and uniformly convergent in the interval $0 \leqslant x \leqslant \pi$. The sum of the series

$$\text{(A)} \qquad \sum_{r=0}^{\infty} u_r(x) \int_0^{\pi} u_r(t)\phi(t)dt$$

is therefore a continuous function of x in $(0, \pi)$; let it be denoted by $\psi(x)$. Then since term-by-term integration of the series for $\psi(x)u_n(x)$ is justified by its uniform convergence,

$$\int_0^{\pi} \psi(x)u_n(x)dx = \sum_{r=0}^{\infty} \int_0^{\pi} u_r(x)u_n(x)dx \int_0^{\pi} u_r(t)\phi(t)dt$$
$$= \int_0^{\pi} u_n(t)\phi(t)dt,$$

on account of the orthogonality of the functions $u_n(x)$. Thus it is seen that

$$\int_0^{\pi} \{\psi(x) - \phi(x)\}u_n(x)dx = 0,$$

for all values of n, and therefore

$$\psi(x) = \phi(x)$$

identically in $(0, \pi)$. *The series* (A) *therefore converges absolutely and uniformly in the interval* $0 \leqslant x \leqslant \pi$, *and in that interval its value is* $f(x)$.

11·52. Comparison of the Sturm-Liouville Development with the Fourier Cosine Development.—It will now be supposed that $f(x)$ is a continuous function of the real variable x in $(0, \pi)$; no further restrictions will be put upon it. Let $s_n(x)$ be the sum of the first $(n+1)$ terms of the Sturm-Liouville development, thus

$$s_n(x) = \int_0^{\pi} f(t) \sum_{r=0}^{n} u_r(x)u_r(t)dt.$$

The behaviour of $s_n(x)$ as n tends to infinity will now be investigated.*

The Fourier cosine development is a particular case of the above; the differential system to which the normal set of orthogonal functions

$$\left(\frac{1}{\pi}\right)^{\frac{1}{2}}, \quad \left(\frac{2}{\pi}\right)^{\frac{1}{2}} \cos x, \; \ldots, \quad \left(\frac{2}{\pi}\right)^{\frac{1}{2}} \cos nx, \; \ldots$$

corresponds is

$$\begin{cases} \dfrac{d^2v}{dx^2} + \rho^2 v = 0, \\ v'(0) = v'(\pi) = 0. \end{cases}$$

It will now be shown that the Sturm-Liouville development of $f(x)$ behaves in all respects exactly like the Fourier cosine development. Let

$$\sigma_n(x) = \int_0^{\pi} f(t) \left\{ \frac{1}{\pi} + \frac{2}{\pi} \sum_{r=1}^{\infty} \cos rx \cos rt \right\} dt,$$

then if

$$\Phi_n(x, t) = \sum_{r=0}^{n} u_r(x)u_r(t) - \left\{ \frac{1}{\pi} + \frac{2}{\pi} \sum_{r=1}^{n} \cos rx \cos rt \right\},$$

it follows that

$$s_n(x) - \sigma_n(x) = \int_0^{\pi} \Phi_n(x, t) f(t) dt.$$

* Haar, *Math. Ann.* 69 (1910), p. 339; Mercer, *Phil. Trans. R. S.* (A) 211 (1910), p. 111.

By means of this relation will be proved the remarkable theorem that

$$s_n(x)-\sigma_n(x)\to 0$$

uniformly as $n\to\infty$. The proof depends upon two lemmas.

LEMMA I.—*There exists an absolute constant M such that*

$$|\Phi_n(x, t)|<M.$$

for all values of n.

On account of the asymptotic form of $u_r(x)$ it is easily seen that

$$u_r(x)u_r(t)-\frac{2}{\pi}\cos rx\cos rt$$

$$=\left(\frac{2}{\pi}\right)^{\frac{1}{2}}\{\beta(t)\cos rx\sin rt+\beta(x)\cos rt\sin rx\}\frac{1}{r}+O\left(\frac{1}{r^2}\right)$$

$$=\frac{1}{(2\pi)^{\frac{1}{2}}}\{\beta(x)+\beta(t)\}\frac{\sin r(x+t)}{r}+\frac{1}{(2\pi)^{\frac{1}{2}}}\{\beta(x)-\beta(t)\}\frac{\sin r(x-t)}{r}+O\left(\frac{1}{r^2}\right).$$

Since the sums of the series

$$\sum_{r=1}^{\infty}\frac{\sin r(x+t)}{r}\quad\text{and}\quad\sum_{r=1}^{\infty}\frac{\sin r(x-t)}{r}$$

are bounded, and $\beta(x)$ is bounded in $(0, \pi)$, the lemma follows.

LEMMA II.—*If $\phi(x)$ is continuous in $(0, \pi)$ and has continuous first and second derivatives in that interval, then*

$$\int_0^{\pi}\Phi_n(x, t)\phi(t)dt\to 0$$

uniformly in $(0, \pi)$ as $n\to\infty$.

For if $g_n(x)$ and $h_n(x)$ represent the first $n+1$ terms of the Sturm-Liouville and the cosine developments of $\phi(x)$ respectively, then

$$g_n(x)-h_n(x)=\int_0^{\pi}\Phi_n(x, t)\phi(t)dt.$$

But $g_n(x)$ and $h_n(x)$ both approach $\phi(x)$ uniformly, which proves the lemma. The main theorem may now be attacked.

Since $f(x)$ is continuous in $(0, \pi)$, a sequence of continuous functions

$$\phi_1(x),\quad \phi_2(x),\ \ldots,\ \phi_n(x),\ \ldots$$

having continuous first and second derivatives can be formed which tends to $f(x)$ uniformly in $(0, \pi)$. These functions may, for example, be polynomials of degree equal to the suffix.* Then

$$s_n(x)-\sigma_n(x)=\int_0^{\pi}\Phi_n(x, t)\{f(t)-\phi_m(t)\}dt+\int_0^{\pi}\Phi_n(x, t)\phi_m(t)dt.$$

Since ϕ_m approaches f uniformly, m may be chosen such that for all values of t in $(0, \pi)$

$$|f(t)-\phi_m(t)|<\epsilon/2\pi M,$$

where M is the absolute constant of Lemma I. Then m having been so chosen, n may by Lemma II. be taken sufficiently large to make the absolute value of the second integral less than $\frac{1}{2}\epsilon$. Consequently

$$|s_n(x)-\sigma_n(x)|<\epsilon$$

uniformly for sufficiently large values of n. This proves the theorem :

The Sturm-Liouville development of any continuous function $f(x)$ converges or diverges at any point of the interval $(0, \pi)$ according as the cosine development converges or diverges at that point. It converges uniformly in any sub-interval

* Weierstrass. *Math. Werke*, 3, p. 1.

of $(0, \pi)$ *when and only when the cosine series converges uniformly in that sub-interval.*

This result is of far-reaching importance because it implies that the enormous volume of work which has been done concerning convergence or divergence of the Fourier development of an arbitrary continuous function applies with merely verbal changes to any Sturm-Liouville development of that function, when the conditions of continuity and differentiability which have been imposed upon the coefficients k, g and l are satisfied.*

But more lies in the theorem than appears on the surface. Thus let $S_n(x)$ be the arithmetic mean of

$$s_0(x), \quad s_1(x), \quad . \ . \ ., \ s_n(x),$$

and let $\Sigma_n(x)$ be the arithmetic mean of

$$\sigma_0(x), \quad \sigma_1(x), \quad . \ . \ ., \ \sigma_n(x).$$

Then from the fact that

$$s_n(x) - \sigma_n(x) \rightarrow 0$$

uniformly as $n \rightarrow \infty$, it follows immediately that

$$S_n(x) - \Sigma_n(x) \rightarrow 0$$

uniformly. Now the cosine development of a continuous function is always uniformly summable by the method of arithmetic means.† Consequently the Sturm-Liouville development is summable (C.1).

* It is also supposed that the constants h and H in the boundary conditions are real and finite.

† Fejér, *Math. Ann.* 58 (1904), p. 59.

PART II

DIFFERENTIAL EQUATIONS IN THE COMPLEX DOMAIN

CHAPTER XII

EXISTENCE THEOREMS IN THE COMPLEX DOMAIN

12·1. General Statement.—The purpose of the present chapter is to extend the work of Chapter III. concerning the existence and nature of solutions of differential equations with one real independent variable to equations with a complex independent variable. In the first place a single equation of the first order

$$\frac{dw}{dz}=f(z, w)$$

will be considered.

In order that the equation may have a meaning, $\frac{dw}{dz}$ must exist, that is to say w is to be an analytic function of z. Let $f(z, w)$ be an analytic function * of the two variables z and w. With this assumption, the Method of Successive Approximations (§ 3·2) can be applied with merely verbal alterations. The main theorem may be stated as follows : †

The differential equation admits of a unique solution $w=w(z)$, which is analytic within the circle $|z-z_0|=h$, and which reduces to w_0 when $z=z_0$.

The Cauchy-Lipschitz method can also be extended so as to be applicable to the complex domain.‡ But perhaps the method most appropriate to the complex domain is that known as the Method of Limits,§ to which the following section is devoted.

* By Cauchy's definition, $f(z, w)$ is an analytic function of z and w in a domain D if (i) $f(z, w)$ is a continuous function of z and w in D; and (ii) $\frac{\partial f}{\partial z}$, $\frac{\partial f}{\partial w}$ both exist at every point of D. This definition implies the Riemann conditions that if $z=x+iy$, $w=u+iv$, $f(z, w)$ $=P(x, y, u, v)+iQ(x, y, u, v)$, then P and Q are differentiable, in D, with respect to their four real arguments and their first partial differential coefficients are continuous and satisfy the equations

$$\frac{\partial P}{\partial x}=\frac{\partial Q}{\partial y}, \quad \frac{\partial P}{\partial y}=-\frac{\partial Q}{\partial x}, \quad \frac{\partial P}{\partial u}=\frac{\partial Q}{\partial v}, \quad \frac{\partial P}{\partial v}=-\frac{\partial Q}{\partial u}.$$

(See Picard, *Traité d'Analyse*, 2, Chap. IX.)

The condition of analyticity when the variables are complex, replaces the condition that, when the variables are real, f is continuous and satisfies a Lipschitz condition. The fact that, when $f(z, w)$ is analytic, $\frac{\partial f}{\partial w}$ is bounded takes the place of the Lipschitz condition in the proof of the existence theorems.

† The number h is here defined precisely as in § 3·1. Painlevé, *Bull. Soc. Math. France*. 27 (1899), p. 152, has shown that, in certain cases, the radius of convergence may exceed h,

‡ Painlevé, *C. R. Acad. Sc. Paris*, 128 (1899), p. 1505, and Picard, *ibid.* p. 1363; *Ann. Éc. Norm.* (3), 21 (1904), p. 56, have shown that the method leads to convergent developments representing the solution throughout the domain in which it is analytic.

§ Cauchy, *C. R. Acad. Sc. Paris*, 9–11, 14, 15, 23 (1839–46) *passim*, *Œuvres* (1), 4–7, 10; simplified by Briot and Bouquet, *C. R.* 36, 39, 40 (1853–55), *passim* ; *J. Éc. Polyt.* (1) cah. 36 (1856), pp. 85, 131. The method was apparently independently discovered by Weierstrass, *Math. Werke*, 1, pp. 67, 75 (dated 1842); *J. für Math.* 51 (1856), p. 1, [*Math. Werke*, 1, p. 153]. Weierstrass' treatment was simplified by Kœnigsberger, *J. für Math.* 104 (1889), p. 174; *Lehrbuch*, p. 25. See also Briot and Bouquet, *Théorie des Fonctions Elliptiques*, p. 325.

12·2. The Method of Limits.—In the equation

$$\frac{dw}{dz}=f(z, w),$$

the function $f(z, w)$ is supposed to be analytic in the neighbourhood of (z_0, w_0). There is, however, no loss in generality in supposing z and w to be written in place of $z-z_0$ and $w-w_0$ respectively; which amounts to assuming that $z_0=w_0=0$. The conditions of the problem may therefore be re-stated as follows:

Let $f(z, w)$ be analytic when z and w remain respectively within circles C and Γ, of radii a and b, drawn about the origin of the z- and w-planes. Further let $f(z, w)$ be continuous on the circumferences C and Γ. In these conditions $|f(z, w)|$ is bounded within this domain; let M be its upper bound. Thus

$$|f|\leqslant M \text{ when } |z|\leqslant a,\ |w|\leqslant b.$$

By repeated differentiation in the equation, the successive differential coefficients

$$\frac{d^2w}{dz^2},\quad \frac{d^3w}{dz^3},\ \ldots,\ \frac{d^rw}{dz^r},\ \ldots$$

are found, thus

$$\frac{d^2w}{dz^2}=\frac{\partial f}{\partial z}+\frac{\partial f}{\partial w}\cdot\frac{dw}{dz},$$

$$\frac{d^3w}{dz^3}=\frac{\partial^2 f}{\partial z^2}+2\frac{\partial^2 f}{\partial z\partial w}\cdot\frac{dw}{dz}+\frac{\partial^2 f}{\partial w^2}\left(\frac{dw}{dz}\right)^2+\frac{\partial f}{\partial w}\cdot\frac{d^2w}{dz^2},$$

$$\cdot\quad\cdot\quad\cdot\quad\cdot$$

and it is to be noted that these expressions are formed by the operations of addition and multiplication only. With the relation

$$\left(\frac{dw}{dz}\right)_0=f(0, 0)$$

as the starting point, these relations determine in succession the values of the coefficients in the Maclaurin series

$$w=\left(\frac{dw}{dz}\right)_0\frac{z}{1}+\left(\frac{d^2w}{dz^2}\right)_0\frac{z^2}{2!}+\ldots+\left(\frac{d^rw}{dz^r}\right)_0\frac{z^r}{r!}+\ldots$$

It is clear that the series for w, so defined, formally satisfies the differential equation; the essential point is to prove that it converges for sufficiently small values of z.

To this end let the Maclaurin development of $f(z, w)$ in the neighbourhood of $z=w=0$ be

$$f(z, w)=\sum A_{pq}z^pw^q,$$

where

$$A_{pq}=\frac{1}{p!\,q!}\left(\frac{\partial^{p+q}f}{\partial z^p\partial w^q}\right)_0.$$

But *

$$\left|\frac{\partial^{p+q}f}{\partial z^p\partial w^q}\right|_0<\frac{p!\,q!}{a^pb^q}M,$$

and hence

$$|A_{pq}|<\frac{M}{a^pb^q},$$

* Picard, *Traité d'Analyse*, 2 (1st ed.), p. 239; (2nd ed.), p. 259.

from which it follows that, if

$$F(z, w)=\sum \frac{M}{a^p b^q} z^p w^q,$$

$$\left(\frac{\partial^{p+q}F}{\partial z^p \partial w^q}\right)_0 \geqslant \left|\frac{\partial^{p+q}f}{\partial z^p \partial w^q}\right|_0$$

for all positive integral or zero values of p and q. But

$$F(z, w)=\frac{M}{\left(1-\frac{z}{a}\right)\left(1-\frac{w}{b}\right)},$$

and therefore, if

$$W=\left(\frac{dW}{dz}\right)_0 \frac{z}{1}+\left(\frac{d^2W}{dz^2}\right)_0 \frac{z^2}{2!}+\ldots+\left(\frac{d^rW}{dz^r}\right)_0 \frac{z^r}{r!}+\ldots$$

is the solution of

$$\frac{dW}{dz}=F(z, W)$$

which reduces to zero when $z=0$, then

$$\left(\frac{d^rW}{dz^r}\right)_0 \geqslant \left(\frac{d^rw}{dz^r}\right)_0,$$

for the successive terms $\left(\frac{d^rW}{dz^r}\right)_0$ are formed from the coefficients $\left(\frac{\partial^{p+q}F}{\partial z^p \partial z^q}\right)_0$ by precisely the same law of addition and multiplication as that by which the terms $\left(\frac{d^rw}{dz^r}\right)_0$ were derived from the coefficients $\left(\frac{\partial^{p+q}f}{\partial z^p \partial z^q}\right)_0$.

The series for W is therefore a *dominant series* for the function w, that is to say the Maclaurin series for w converges absolutely and uniformly within any circle concentric with and interior to the circle of convergence of the series for W. But an explicit expression for the radius of convergence of the series for W can easily be found, for if the differential equation

$$\frac{dW}{dz}=F(z, W)$$

is written in the form

$$\left(1-\frac{W}{b}\right)\frac{dW}{dz}=\frac{M}{1-\frac{z}{a}},$$

the variables are separate, and the solution which reduces to zero when $z=0$ is readily found to be *

$$W=b-b\sqrt{\left\{1+\frac{2Ma}{b}\log\left(1-\frac{z}{a}\right)\right\}}.$$

The radius of convergence ρ is therefore determined by the equation

$$1+\frac{2Ma}{b}\log\left(1-\frac{\rho}{a}\right)=0,$$

or

$$\rho=a\left(1-e^{-\frac{b}{2Ma}}\right),$$

and therefore the series formally obtained converges absolutely and uniformly within any circle $|z|=\rho-\epsilon$, where $0<\epsilon<\rho$, and is in consequence a solution

* The principal value of the radical is taken, *i.e.* that which becomes $+1$ when $z=0$.

of the differential equation.* Since the coefficients in the Maclaurin series for w are obtained in a definite manner by operations of addition and multiplication, and since the Maclaurin development of an analytic function is unique, the equation admits of one and only one solution which satisfies the assigned conditions.

12·21. Extension to Systems of Equations.—The method of limits can be extended so as to apply to the system of m equations of the first order,

$$\frac{dw_1}{dz}=f_1(z, w_1, w_2, \ldots, w_m),$$

$$\frac{dw_2}{dz}=f_2(z, w_1, w_2, \ldots, w_m),$$

$$\cdots$$

$$\frac{dw_m}{dz}=f_m(z, w_1, w_2, \ldots, w_m).$$

Again, without loss of generality, the initial conditions may be taken to be such that $w_1=w_2=\ldots=w_m=0$ as $z=0$. Let the functions $f_1, f_2, \ldots, f_m$ be analytic in the domain $|z|\leqslant a$, $|w_1|\leqslant b$, $|w_2|\leqslant b, \ldots, |w_m|\leqslant b$, and let M be the upper bound of the set $f_1, f_2, \ldots, f_m$ in this domain. Then the dominant functions may be taken as the appropriate solutions of the equations

$$\frac{dW_1}{dz}=\frac{dW_2}{dz}=\ldots=\frac{dW_m}{dz}=\frac{M}{\left(1-\frac{z}{a}\right)\left(1-\frac{W_1}{b}\right)\ldots\left(1-\frac{W_m}{b}\right)}.$$

The functions $W_1, W_2, \ldots, W_m$ are all zero, when $z=0$, and are therefore all equal. The set may therefore be replaced by a single dominant function W which satisfies the equation

$$\frac{dW}{dz}=\frac{M}{\left(1-\frac{z}{a}\right)\left(1-\frac{W}{b}\right)^m},$$

or, taking into account the initial conditions,

$$Y=b-b\left\{1+\frac{(m+1)Ma}{b}\log\left(1-\frac{z}{a}\right)\right\}^{\frac{1}{m+1}}$$

and therefore the radius of convergence is

$$\rho=a\left(1-e^{-\frac{b}{(m+1)Ma}}\right).$$

12·22. An Existence Theorem for the Linear Differential Equation of Order n.—In view of the very great importance, theoretical and practical, of ordinary linear equations, an independent proof of the existence of solutions satisfying assigned initial conditions for $z=z_0$ will now be given.† The analogy with the theory as it is in the case of an equation, or system of equations, of the first order will be clear.

Let

$$\frac{d^nw}{dz^n}+p_1(z)\frac{d^{n-1}w}{dz^{n-1}}+\ldots+p_{n-1}(z)\frac{dw}{dz}+p_n(z)w=0$$

* It may be noted that the radius of convergence of the series obtained by the Method of Limits is less than that obtained by the Method of Successive Approximations. Note also that, within the circle $|z|=\rho$, $|w|<b$; the original hypotheses are therefore not violated by the solution.

† Fuchs, *J. für Math.* 66, (1866) p. 121; [*Math. Werke*, 1, p. 159].

be a homogeneous linear differential equation of order n, in which the coefficients $p_1(z), \ldots, p_n(z)$ are analytic throughout a domain D in the z-plane. In the Taylor series

$$w(z)=\sum_{r=0}^{\infty}\frac{1}{r!}w^{(r)}(z_0)(z-z_0)^r$$
$$=\sum c_r(z-z_0)^r,$$

in which z_0 and z are in D, let the coefficients $w^{(r)}(z_0)$, or the corresponding coefficients c_r, be so determined that the series formally satisfies the differential equation. The n initial values

$$w(z_0), \quad w'(z_0), \ldots, w^{(n-1)}(z_0)$$

are to be assigned arbitrarily; the succeeding values

$$w^{(n)}(z_0), \quad w^{(n+1)}(z_0), \ldots$$

may be determined from the differential equation as it stands and from the equations obtained by its successive differentiation with respect to z. Thus the constants $w^{(r)}(z_0)$ or c_r may be determined uniquely; since they are determined from the initial values by processes of addition and multiplication only, they remain finite so long as the initial values are themselves finite. Let the recurrence-relations which determine $w^{(r)}(z_0)$ be

$$w^{(r)}(z_0)=\sum_{s=1}^{r} A_{rs}w^{(r-s)}(z_0) \qquad (r \geqslant n).$$

The coefficients $p_\nu(z)$ are bounded throughout the circular domain $|z-z_0| \leqslant a$ which is supposed to lie entirely within D; let the upper bound of $|p_\nu(z)|$ on the circle Γ or $|z-z_0|=a$ be M_ν. Then since

$$p_\nu(z)=p_\nu^{0}+p_\nu^{1}(z-z_0)+ \ldots +p_\nu^{(r)}(z-z_0)^r+ \ldots,$$

where $p_\nu^{(r)}$ is the value of $\frac{1}{r!}\cdot\frac{d^rp_\nu(z)}{dz^r}$ when $z=z_0$, it follows by the Cauchy integral theorem that

$$p_\nu^{(r)}=\frac{1}{2\pi i}\int_\Gamma \frac{p_\nu(z)dz}{(z-z_0)^{r+1}}$$
$$\leqslant \frac{M_\nu}{a^r}.$$

Hence if $P_\nu(z)$ is defined by the equation

$$P_\nu(z)=\frac{M_\nu}{1-\dfrac{z-z_0}{a}},$$

then $|p_\nu(z)| \leqslant |P_\nu(z)|$ within the circle Γ and on its circumference.

Now consider the differential equation

$$\frac{d^nW}{dz^n}=P_1(z)\frac{d^{n-1}W}{dz^{n-1}}+ \ldots +P_{n-1}(z)\frac{dW}{dz}+P_n(z)W;$$

let it be satisfied by the Taylor series

$$W(z)=\sum_{r=0}^{\infty}\frac{1}{r!}W^{(r)}(z_0)(z-z_0)^r$$
$$=\sum C_r(z-z_0)^r,$$

which is such that $C_0=|c_0|, C_1=|c_1|, \ldots, C_{n-1}=|c_{n-1}|$. Let the recurrence relation determining $W^{(r)}(z_0)$ be

$$W^{(r)}(z_0)=\sum_{s=1}^{r} B_{rs}W^{(r-s)}(z_0).$$

Since the coefficients of the expansion of $P_\nu(z)$ are positive real numbers, and since B_{rs} is derived from those coefficients and from $C_0, C_1, \ldots, C_{n-1}$ by addition and multiplication, B_{rs} is a positive real number, and

$$|A_{rs}| \leqslant B_{rs},$$

whence it follows by induction that

$$|w^{(r)}(z_0)| \leqslant W^{(r)}(z_0),$$

and hence

$$\sum |c_r(z-z_0)^r| \leqslant \sum C_r(z-z_0)^r.$$

The circle of convergence of the dominant series $\sum C_r(z-z_0)^r$ may be found without difficulty; in fact it will be shown to be $|z-z_0|=a$. Write $z-z_0=a\zeta$, then the differential equation which determines $W(z)$ becomes

$$(1-\zeta)\frac{d^nW}{d\zeta^n}=M_1a\frac{d^{n-1}W}{d\zeta^{n-1}}+\ldots+M_{n-1}a^{n-1}\frac{dW}{d\zeta}+M_na^nW;$$

if it is satisfied by the power series $\sum \gamma_r\zeta^r$, the following recurrence relation must hold

$$(n+r)!\,\gamma_{n+r}-r(n+r-1)!\,\gamma_{n+r-1}=\sum_{s=1}^{n}(n+r-s)!\,M_sa^s\gamma_{n+r-s}.$$

But in order that $\sum \gamma_r\zeta^r$ may be formally identical with $\sum C_r(z-z_0)^r$, $\gamma_r=a^rC_r$ (when $r=0, 1, 2, \ldots, n-1$). It follows by induction that $\gamma_r>0$ for all r.

Hence

$$\gamma_{n+r}=\frac{r+M_1a}{r+n}\gamma_{n+r-1}+\theta_{n+r-2},$$

where $\theta_{n+r-2}\geqslant 0$. Now M_1 is not restricted except by the condition that $|p_1(z)|\leqslant M_1$ on the circle Γ; let M_1 be chosen so large that $M_1a>n$, then

$$\gamma_{n+r}>\gamma_{n+r-1}$$

for all values of r, and consequently

$$\gamma_{n+r-1}>\gamma_{n+r-s},$$

when $s\geqslant 2$. Now

$$\frac{\gamma_{n+r}}{\gamma_{n+r-1}}=\frac{r+M_1a}{r+n}+\sum_{s=2}^{n}\frac{(n+r-s)!\,M_sa^s\gamma_{n+r-s}}{(n+r)!\,\gamma_{n+r-1}},$$

whence

$$\lim_{r\to\infty}\frac{\gamma_{n+r}}{\gamma_{n+r-1}}=1.$$

Hence the series $\sum \gamma_r\zeta^r$ converges when $|\zeta|<1$, and therefore the dominant series is convergent when $|z-z_0|<a$. Consequently the differential equation admits of a solution which satisfies the specified initial conditions when $z=z_0$, and which is expressible as a power series which is absolutely and uniformly convergent within any circle with z_0 as centre in which the coefficients $p_1(z), \ldots, p_n(z)$ are analytic.

12·3. Analytical Continuation of the Solution; Singular Points.—The method of limits shows that there exists a solution

$$W(z-z_0)=w_0+\sum a_r(z-z_0)^r$$

of the differential equation

$$\frac{dw}{dx}=f(z, w),$$

which is analytic throughout the domain $|z-z_0|<\rho$, where

$$\rho=a\left(1-e^{-\frac{b}{2Ma}}\right).$$

Since M is the upper bound of $|f(z, w)|$ in the domain $|z-z_0|\leqslant a$, $|w-w_0|\leqslant b$, it is clear that M in general depends upon the choice of z_0 and w_0.

Now the solution obtained is the only analytic solution which corresponds to the initial value-pair (z_0, w_0). But there still remains the question as to whether or not there may exist non-analytic solutions which satisfy the initial conditions. This question has been completely answered in the negative ; * for the purposes of the work which follows it will be sufficient to show that there can be no solution, satisfying the initial conditions, which proceeds as a series of other than positive integral powers of $z-z_0$.† In this case the conclusion is obvious, for if the series involved negative or fractional powers of the variable, then on and after a certain order the differential coefficients would become infinite when $z=z_0$. But the values of $y_0^{(r)}$ as obtained from the differential equation and its successive derivatives are necessarily finite, which leads to a contradiction.

In the statement that only one solution corresponds to the initial value-pair (z_0, w_0), the supposition is implied that these values are actually attained. Let it now be supposed merely that $w\to w_0$ as $z\to z_0$ along a definite simple curve C in the z-plane. Since the path described is a simple curve, given $\epsilon>0$, it is possible to find a point z_1 on the curve such that

$$|z_1-z_0|<\epsilon,$$

and it is also supposed that there exists $\delta>0$ such that

$$|w-w_0|<\delta \quad \text{when} \quad |z-z_0|<\epsilon.$$

Let W be the analytic solution, and let $W+\overline{W}$ be supposed to be a distinct solution satisfying the modified initial conditions. Then

$$\overline{W}\to 0 \quad \text{as} \quad z\to z_0 \text{ along } C.$$

Now

$$\begin{aligned}\frac{d\overline{W}}{dz}&=f(z, W+\overline{W})-f(z, W)\\&=\overline{W}F(z, W, \overline{W}),\end{aligned}$$

where F represents a series which converges when z is a point on C such that $|z-z_0|\leqslant a$, and when

$$|W-w_0|\leqslant b, \quad |W+\overline{W}-w_0|\leqslant b.$$

Assuming that $\overline{W}\neq 0$,

$$\log \overline{W}=\int\frac{d\overline{W}}{\overline{W}}=\int_C F dz,$$

and if $|z-z_0|\leqslant a$, $|F|$ has an upper bound M so that

$$\left|\int_C F dz\right|\leqslant M\int_C |dz|\leqslant Ml,$$

where l is the length of the path considered. On the other hand, since $\overline{W}\to 0$ as $z\to z_0$, the value of

$$\left|\int_C F dz\right|$$

may be made indefinitely great by carrying the corresponding integration

* Briot and Bouquet, *J. Éc. Polyt.* (1), cah. 36 (1856), p. 133 ; Picard, *Traité d'Analyse*, 2, p. 314 ; (2nd ed.), 2, p. 357 ; Painlevé, *Leçons sur la théorie analytique des équations différentielles* (Stockholm, 1895), p. 394.

† Hamburger, *J. für Math.* 112 (1893), p. 211.

along C sufficiently close to z_0. This leads to a contradiction provided that l is *finite*,* and consequently there does not exist a solution of the kind postulated, other than the original analytic solution.

Let z_1 be a point within the circle $|z-z_0|=\rho$, then all the coefficients in the series

$$W_1(z-z_1)=W(z_1-z_0)+W'(z_1-z_0)(z-z_1)+\ldots+W^{(r)}(z_1-z_0)\frac{(z-z_1)^r}{r!}+\ldots$$

can be determined, and are finite, and the series $W_1(z-z_1)$ has a radius of convergence at least equal to $\rho-|z_1-z_0|$. In the sector common to their circles of convergence $W_1(z-z_1)$ and $W(z-z_0)$ are formally identical. $W_1(z-z_1)$ is therefore, at all points at which it is analytic, a solution of the differential equation, and is the only solution which reduces to the value $W(z_1-z_0)$ when $z=z_1$. If $f(z, w)$ is analytic within and continuous on the boundary of the domain $|z-z_1|=a_1$, $|w-w_1|=b_1$, and if M_1 is the upper bound of $|f(z, w)|$ within this domain, the function $W_1(z-z_1)$ is analytic throughout the domain $|z-z_1|\leqslant\rho_1$ where

$$\rho_1=a_1\left(1-e^{-\frac{b_1}{2M_1a_1}}\right).$$

If $\rho_1>\rho-|z_1-z_0|$, the circle of convergence of $W_1(z-z_1)$ will extend beyond the circle of convergence of $W(z-z_0)$; this in general will be the case.† Let z_2 be a point within the circle of convergence of $W_1(z-z_1)$, though not necessarily within the circle of convergence of $W(z-z_0)$, then the series

$$W_2(z-z_2)=W_1(z_2-z_1)+W_1'(z_2-z_1)(z-z_2)+\ldots+W_1^{(r)}(z_2-z_1)\frac{(z-z_2)^r}{r!}+\ldots$$

is formally identical with $W_1(z-z_1)$ in the region common to their circles of convergence and therefore satisfies the differential equation. It is therefore an analytic continuation of the solution $W(z-z_0)$.

The process may be repeated a finite number of times, giving in succession the solutions

$$W_1(z-z_1),\quad W_2(z-z_2),\ \ldots,\ W_k(z-z_k),$$

which are analytic continuations of the solution $W(z-z_0)$.

The series-solution $W(z-z_0)$ together with all the series obtained by analytical continuation defines a function $F(z\,;\ z_0, w_0)$ in which the initial values z_0, w_0 appear as parameters. This function is analytic at all points of the domain ‡ D defined by the aggregate of the circles of convergence of $W, W_1, W_2, \ldots, W_k$.

If $z=\zeta$ is a point such that for the value-pair $z=\zeta$, $w=F(\zeta\,;\ z_0, w_0)$ the function $f(z, w)$ is not analytic, then this point ζ is not an internal point of the domain D. Such points, together with the points for which $F(\zeta\,;\ z_0, w_0)$ becomes infinite, and possibly the point at infinity are the *singular points* of the differential equation. These singular points will now be studied more closely.

12·4. Initial Values for which $f(z, w)$ is Infinite.—It has been seen that if $f(z, w)$ is uniform and continuous in the neighbourhood of (z_0, w_0), then w

* For discussions of the case when l is infinite (for example, when C is a curve encircling z spirally) see Painlevé, *Leçons*, p. 19; Young, *Proc. London Math. Soc.* (1), 34 (1902), p. 234.

† Picard, *Bull. Sc. Math.* (2), 12 (1888), p. 148; *Traité d'Analyse*, 2, p. 311; (2nd ed.), 2, p. 351. A representation valid throughout the whole of the domain in which an analytic solution exists can be obtained by replacing the Taylor series by series of the polynomials of Mittag-Leffler, *C. R. Acad. Sc. Paris*, 128 (1899), p. 1212.

‡ But not (unless the domain is simply connected) necessarily analytic throughout the domain D. For instance, the function $\log z$ is analytic at every point of the domain $0<\delta\leqslant|z|\leqslant\Delta$, but is not analytic throughout this domain.

may be expressed as a convergent power series in $(z-z_0)$. In other words, if (z_0, w_0) is an ordinary point of the function $f(z, w)$ it is also an ordinary point of the solution $w=F(z; z_0, w_0)$. On the other hand, there will in general be points for which the conditions of uniformity and continuity imposed upon $f(z, w)$ are not fulfilled; in the first place it will be supposed that $f(z, w)$ becomes infinite at (z_0, w_0), but in such manner that the reciprocal $1/f(z, w)$ is analytic in the neighbourhood of this value-pair. In this case

$$\frac{1}{f(z, w)}=A_0(z)+A_1(z)(w-w_0)+A_2(z)(w-w_0)^2+\ldots,$$

in which the coefficients $A_0(z), A_1(z), A_2(z), \ldots$ are themselves developable in series of ascending powers of $z-z_0$, and $A_0(z_0)=0$.

It will be assumed * that not all of the coefficients $A(z)$ are zero when $z=z_0$; for definiteness it will be supposed that

$$A_0(z_0)=A_1(z_0)=\ldots=A_{k-1}(z_0)=0, \quad A_k(z_0)\neq 0.$$

The differential equation may now be written in the form

$$\frac{dz}{dw}=\frac{1}{f(z, w)},$$

in which z is regarded as the dependent, and w as the independent variable. The method of limits may be applied to it; since the successive differential coefficients

$$\frac{dz}{dw}, \quad \frac{d^2z}{dw^2}, \ldots, \frac{d^kz}{dw^k}$$

are zero for $z=z_0$, $w=w_0$, whereas $\dfrac{d^{k+1}z}{dw^{k+1}}$ is not zero, the equation admits of a unique solution whose development is

$$z-z_0=(w-w_0)^{k+1}\{c_0+c_1(w-w_0)+c_2(w-w_0)^2+\ldots\},$$

in which $c_0\neq 0$. It follows that $w-w_0$ can be expressed as a series of powers of the $(k+1)$*ple* valued function $(z-z_0)^{\frac{1}{k+1}}$, *i.e.* $w-w_0=\mathbf{P}_1\left\{(z-z_0)^{\frac{1}{k+1}}\right\}$, where $\mathbf{P}_1$ denotes a power-series whose leading term is of the first degree in the argument. There are therefore $k+1$ solutions which satisfy the initial conditions; and the point z_0 is a *branch point* around which these solutions are permutable.

In particular, let the differential equation be

$$\frac{dw}{dz}=\frac{g(z, w)}{h(z, w)},$$

in which $g(z, w)$ and $h(z, w)$ are polynomials in w whose coefficients are analytic functions of z; let the degree of $h(z, w)$ be n. Let z_0 be such that the equations $g(z_0, w)=0$ and $h(z_0, w)=0$ have no common root, then to z_0 correspond n values of w_0 such that to each of these initial value-pairs (z_0, w_0) there corresponds a set of solutions having a branch point at z_0. If the point z_0 is supposed to describe a curve in the z-plane, such that for no point z_0 on this curve do the equations $g(z_0, w)=0$, $h(z_0, w)=0$ have a common root, then every point of such a curve is a branch point for one or more sets of solutions. The branch points may therefore be regarded as movable singularities. On the other hand, any other singularities which may appear,

* If all these coefficients vanish when $z=z_0$, it is possible to write

$$\frac{1}{f(z, w)}=G(z)g(z, w),$$

where $g(z, w)$ is analytic near (z_0, w_0), and $G(z)$ is a function of z only which vanishes when $z=z_0$. The point z_0 is then a singular point of the equation. See §§ 12·6, 12·61.

and in particular any essential singularities, arise through the coefficients of the polynomials $g(z, w)$ and $h(z, w)$ ceasing to be analytic. Since this occurs quite independently of w, such singularities are fixed * as to their position in the z-plane.

12·41. Values of z for which the Function $F(z; z_0, w_0)$ becomes Infinite.—Let z_1 be a value of z for which the solution

$$w=F(z; z_0, w_0)$$

becomes infinite; the mode in which $F(z; z_0, w_0)$ becomes infinite will now be investigated, certain assumptions being made as to the behaviour of $f(z, w)$ when $z=z_1$, $w=\infty$.

Write $w=W^{-1}$, so that the differential equation appears as

$$\frac{dW}{dz}=-W^2f(z, W^{-1})$$
$$=\phi(z, W), \text{ say.}$$

In the first place assume $\phi(z, W)$ to be analytic in the neighbourhood of $z=z_1$, $W=0$. The initial value-pair $(z_1, 0)$ thus relates to an ordinary point of W, and the corresponding development of W is

$$W=(z-z_1)^k\{c_0+c_1(z-z_1)+c_2(z-z_1)^2+\ldots\},$$

in which k is a positive integer (not zero). Consequently

$$w=W^{-1}=(z-z_1)^{-k}\{\gamma_0+\gamma_1(z-z_1)+\gamma_2(z-z_1)^2+\ldots\},$$

that is to say the solution $w=F(z; z_0, w_0)$ has a pole of order k at $z=z_1$ or

$$w=\mathbf{P}_{-k}(z-z_1).$$

Now let $\phi(z, W)$ become infinite at $(z_1, 0)$, but in such a way that $1/\phi(z, W)$ is analytic in the neighbourhood of that value-pair. Then (as in § 12·4) there exists a set of solutions which permute among themselves around the branch point $z=z_1$ thus

$$W=(z-z_1)^{\frac{1}{k+1}}\{c_0+c_1(z-z_1)+c_2(z-z_1)^2+\ldots\},$$

and consequently

$$w=W^{-1}=(z-z_1)^{-\frac{1}{k+1}}\{\gamma_0+\gamma_1(z-z_1)+\gamma_2(z-z_1)^2+\ldots\},$$

or

$$w=\mathbf{P}_{-1}\left\{(z-z_1)^{\frac{1}{k+1}}\right\},$$

and therefore z_1 is both an infinity and a branch point for the solution $w=F(z; z_0, w_0)$. The fixed singular points of the two types met with in this section are collectively termed *regular*.

12·5. Fixed and Movable Singular Points.—In this section $f(z, w)$ will be restricted to be a rational function of w, say

$$f(z, w)=\frac{g(z, w)}{h(z, w)},$$

where

$$g(z, w)=p_0(z)+p_1(z)w+\ldots+p_m(z)w^m,$$
$$h(z, w)=q_0(z)+q_1(z)w+\ldots+q_n(z)w^n.$$

Then any singularities of solutions of the differential equation, which do not fall into one or other of the classes discussed in the two preceding sections,

* What is here said concerning the fixity of essential singularities refers only to an equation of the first order; it is not true in the case of equations of higher order than the first.

can only arise for discrete values of z; in other words they are independent of the initial value of the dependent variable w. Such singularities may arise at the point $z=z_1$, where

(a) z_1 is a singular point for any of the coefficients * p and q,

$$e.g. \quad \frac{dw}{dz}=\frac{w}{\sqrt{(z-z_1)}}.$$

$$\text{Solution:} \quad w=Ce^{2\sqrt{(z-z_1)}}$$

(b) z_1 is such that $h(z_1, w)$ is identically zero,

$$e.g. \quad \frac{dw}{dz}=\frac{w}{(z-z_1)^2}.$$

$$\text{Solution:} \quad w=Ce^{-\frac{1}{z-z_1}}.$$

The preceding example also illustrates this case.

(c) z_1 is such that the equations

$$g(z_1, w)=0, \quad h(z_1, w)=0$$

are satisfied simultaneously by particular values † of w.

$$e.g. \quad \frac{dw}{dz}=\frac{w+\sin(z-z_1)}{z-z_1}.$$

Here $g=h=0$ when $z=z_1$, $w=0$.

$$\text{Solution:} \quad w=(z-z_1)\int_a^z \frac{\sin(t-z_1)}{(t-z_1)^2}\,dt.$$

Now let $W=w^{-1}$ and

$$\begin{aligned}\phi(z, W)&=-W^2 f(z, W^{-1})\\ &=\frac{g_1(z, W)}{h_1(z, W)},\end{aligned}$$

this fraction being reduced to its lowest terms in W. Singularities may then arise at $z=z_1$ in the cases

(d) z_1 is such that $h_1(z, w)$ is identically zero,

(e) z_1 is such that the equations

$$g_1(z_1, W)=0, \quad h(z_1, W)=0$$

are satisfied simultaneously by particular values of W.

The point at infinity is also examined for singularity by transforming the differential equation by the substitution $z=\zeta^{-1}$ and testing the point $\zeta=0$ in the light of the above investigation.

The singular points which then arise are known as the *Fixed or Intrinsic Singular Points* of the Differential Equation; they can be determined *a priori* by inspection of the function $f(z, w)$. Let each of those fixed singular points which lie in the finite part of the plane be surrounded by a small circle, such that no two circles intersect and let each circle be joined to the point at infinity by a rectilinear cut, in such a way that no two cuts intersect. In this way a simply-connected region R is defined in the z-plane, such that at every point of the surface, every solution $F(z; z_0, w_0)$ is regular.

Now let z_1 be an interior point of the region R, and let it be supposed that

* It is assumed that if any such singular point can be eliminated by multiplying $g(z, w)$ and $h(z, w)$ by an appropriate function of z, it has been so removed.

† The equations cannot be satisfied simultaneously for a continuous sequence of values of w without g and h having a common factor in z, such a factor is supposed to have been removed. The singular values z_1 are obtained by eliminating w between $g(z, w)=0$ and $h(z, w)=0$.

as the variable z tends to the value z_1, $F(z\,;\ z_0, w_0)$ tends to a limiting value, finite or infinite ; let

$$F(z\,;\ z_0, w_0) \to w_1 \quad \text{as } z \to z_1.$$

The following cases may then arise:

1°. If $f(z, w)$ is analytic in the neighbourhood of (z_1, w_1) then $F(z\,;\ z_0, w_0)$ is analytic in the neighbourhood of the point z_1, which is an *ordinary* point of the equation.

2°. If w_1 is infinite, but $\phi(z, w)$ analytic in the neighbourhood of $(z_1, 0)$, then, as was seen in § 12·41, $F(z\,;\ z_0, w_0)$ has a pole at the point z_1, which is a *regular singularity* of the equation.

3°. If w_1 is finite, but $f(z_1, w_1)$ is infinite, then, since the coefficients p or q are analytic in the neighbourhood of z_1, $h(z_1, w_1)$ must be zero. But the possibility of $g(z_1, w_1)$ being also zero cannot arise in the region R. Hence $1/f(z, w)$ is analytic in the neighbourhood of (z_1, w_1), and therefore z_1 is a branch point of $F(z\,;\ z_0, w_0)$ and a *regular singular point* of the equation.

4°. If, w_1 is infinite, and $\phi(z, w)$ also infinite in the neighbourhood of $(z_1, 0)$, $1/\phi(z, w)$ is analytic near $(z_1, 0)$. Then z_1 is an infinity and branch point of $F(z\,;\ z_0, w_0)$ and a *regular singular* point of the equation.

The only possibility which remains is that as z tends to z_1, $F(z\,;\ z_0, w_0)$ may not tend to any definite limit. It will be shown that this is impossible.*

Let a moving point start from z_0 and describe a simple curve C in the region R. Suppose z_1 to be the first point encountered at which there is any doubt as to the existence of a limiting value of $F(z;\ z_0, w_0)$. Then let the roots of the equation $h(z_1, w)=0$ be $\omega_1, \ldots, \omega_n$, multiple roots being enumerated once only.

Let δ be an arbitrarily small positive number, and define a region Δ in the w-plane as the aggregate of the points which satisfy the inequalities

$$|w-\omega_1|<\delta, \ \ldots, \ |w-\omega_n|<\delta, \quad |w|>1/\delta.$$

Now suppose that as z approaches z_1, $F(z;\ z_0, w_0)$ assumes values corresponding to points lying ultimately within Δ. Then a positive number ϵ exists such that for every point z of C, for which $|z-z_1|<\epsilon$, one or other of the inequalities

$$|w-\omega_1|<\delta, \ \ldots, \ |w-\omega_n|<\delta, \quad |w|>1/\delta$$

is satisfied. But $F(z\,;\ z_0, w_0)$ varies continuously as z moves on C from z_0 to z_1 and therefore only one of these inequalities can be satisfied. Hence w assumes one or other of the definite values $\omega_1, \ldots, \omega_n, \infty$, when $z=z_1$.

The alternative supposition is that as z approaches z_1, the values of $w=F(z\,;\ z_0, w_0)$ ultimately correspond to points lying outside the circles Δ. Then a number γ exists such that when $|z-z_1|<\gamma$, $|h(z, w)|$ has a positive lower bound, and therefore $|f(z, w)|$ is bounded. Now let $\bar{z}$ be a point of C such that $|z_1-\bar{z}|<\tfrac{1}{2}\gamma$ then whatever number $\bar{w}$ is associated with $\bar{z}$, the series-solution of the differential equation corresponding to the initial values $z=\bar{z}$, $w=\bar{w}$ has a radius of convergence not less than some definite number μ, provided only that $\bar{w}$ lies outside the circles Δ. Choose then a point $\bar{z}_1$ on C whose distance from z_1 is less than the smaller of μ and $\tfrac{1}{2}\delta$, and let the value $\bar{w}_1$ associated with it be $\bar{w}_1=F(\bar{z}_1\,;\ z_0, w_0)$. Then the circle of convergence of development of z as a power series in $z-\bar{z}_1$ includes the point z_1 and therefore the function $F(z\,;\ z_0, w_0)$ is analytic at z_1.

Thus in all cases $w=F(z\,;\ z_0, w_0)$ tends to a definite limit, finite or infinite at every interior point of the region R. A singularity arises at z_1 only for particular values of w, which depend in their turn upon (z_0, w_0). A change in (z_0, w_0) will in general move the singularity from z_1 to another point of the z-plane. Any point of the z-plane may be a singularity of one or more

* Painlevé, *Ann. Fac. Sc. Toulouse* (1888), p. 38 ; *Leçons*, p. 32 ; Picard, *Traité d'Analyse*, 2 (2nd ed.), p. 370.

solutions of the equation. Take the point z_k for instance, and let w_k be any root of the equation $h(z_k, w)=0$. Then if $g(z_k, w_k)\neq 0$, a singularity arises for $z=z_k$, $w=w_k$. Such singularities, which move in the z-plane as the initial values are varied are known as *Movable* or *Parametric Singularities.** The theorem proved in the preceding paragraph is equivalent to the statement that there cannot be, when the equation is of the first order and first degree, any movable essential singularities.

As an example consider the equation

$$\frac{dw}{dz}+\frac{z}{w}=0,$$

in which case $g(z, w)=-z$, $h(z, w)=w$. The solution which corresponds to the initial pair of values (z_0, w_0) is

$$z^2+w^2=z_0^2+w_0^2$$

or

$$w=\sqrt{(z_0^2+w_0^2-z^2)}.$$

The singularity, in this case a branch point, arises when $h(z, w)=w=0$. Any point z_k can be made a singular point by choosing z_0 and w_0 such that

$$z_k^2=z_0^2+w_0^2.$$

In conclusion it is to be noted that whereas the movable singularities of an equation of the first order are regular and not essential singularities, this is not generally true of equations of higher order than the first.

12·51. The Generalised Riccati Equation.—It was seen in the previous section that singularities of solutions of the equation

$$\frac{dw}{dz}=f(z, w)=\frac{g(z, w)}{h(z, w)}$$

fall into two categories :

(*a*) The *fixed* singular points, which are points in the z-plane whose positions are independent of the initial values.

(*b*) The *movable* singular points, which depend upon the initial values, and move over the z-plane as the initial conditions are varied. The movable singularities may be either poles or branch points.

The question now arises as to what restrictions must be imposed upon $f(z, w)$ if no solutions with movable branch points are to be possible. Let z_0 be any point of the z-plane which is not one of the fixed singular points. Then it is *necessary* that there should be no value of w for which the equation

$$h(z_0, w)=0$$

is satisfied. But this equation always has roots unless $h(z_0, w)$ is independent of w. Since z_0 is any non-singular point of the z-plane it follows that $h(z, w)$ is a function of z only. In other words, $f(z, w)$ is a polynomial in w, say

$$f(z, w)=p_0(z)+p_1(z)w+\ .\ .\ .\ +p_n(z)w^n.$$

For a similar reason,

$$\begin{aligned}\phi(z, W)&=-W^2f(z, W^{-1})\\&=-p_0(z)W^2-p_1(z)W-p_2(z)-p_3(z)W^{-1}-\ .\ .\ .\ -p_n(z)W^{-n+2}\end{aligned}$$

must be a polynomial in W, and consequently it is necessary that

$$p_3(z)=p_4(z)=\ .\ .\ .\ =p_n(z)=0$$

identically.

* Hamburger, *J. für Math.* 83 (1877), p. 185; Fuchs, *Sitz. Akad. Wiss. Berlin,* 32 (1884), p. 699 [*Math. Werke*, 2, p. 355].

The differential equation is therefore necessarily of the form

$$\frac{dw}{dz}=p_0(z)+p_1(z)w+p_2(z)w^2,$$

and the condition that it be of this form is easily seen to be sufficient for the non-appearance of movable branch points. The equation thus obtained is the *generalised Riccati Equation*; * when $p_2(z)$ is identically zero, it reduces to the linear equation.

The condition that the equation should be completely dissociated from movable branch points leads to an important conclusion as to the form of the general solution. Let z_0 and z be two points in the region R (§ 12·5); they can be joined by a simple curve which does not pass through any branch point. Let w_0 be the initial value of the dependent variable chosen to correspond to z_0, and let w be the value at z obtained by analytical continuation through the medium of a finite number of circles which in the aggregate completely enclose the path z_0z. Through all the steps of this continuation the solution or its reciprocal remains an analytic function of w_0, and the final value of w is an analytic function of w_0. Whatever value, finite or infinite, w_0 may have, w is uniquely determined, for the region R is completely free from branch points. Hence w_1 regarded as a function of w_0 is one-valued, analytic, and devoid of singularities other than poles, and is therefore a rational function of w_0.

But the process may be reversed, w being considered as an arbitrary initial value and w_0 the value derived from it by analytical continuation; w_0 is thus a rational function of w. This rational one-to-one correspondence between w and w_0 can only be if w is a linear fractional function of w_0, *i.e.*

$$w=\frac{Aw_0+B}{Cw_0+D},$$

where A, B, C, and D are functions of z.

It follows from the properties of the anharmonic ratio that if w_1, w_2 and w_3 are any three particular solutions of the Riccati equation, then the general solution is expressible in the form

$$\frac{w-w_1}{w-w_2}=A\frac{w_3-w_1}{w_3-w_2},$$

where A is a constant.

An alternative method of finding the general solution is as follows. The equations

$$\begin{aligned}w' &=p_0+p_1w+p_2w^2,\\ w_1' &=p_0+p_1w_1+p_2w_1^2,\\ w_2' &=p_0+p_1w_2+p_2w_2^2,\\ w_3' &=p_0+p_1w_3+p_3w_3^2\end{aligned}$$

are consistent if

$$\begin{vmatrix} w', & 1, & w, & w^2 \\ w_1', & 1, & w_1, & w_1^2 \\ w_2', & 1, & w_2, & w_2^2 \\ w_3', & 1, & w_3, & w_3^2 \end{vmatrix}=0.$$

This condition is equivalent to

$$\frac{d}{dz}\left[\frac{w-w_1}{w-w_2}\cdot\frac{w_3-w_2}{w_3-w_1}\right]=0,$$

whence the result follows.

* d'Alembert, *Hist. Acad. Berlin*, 19 (1763), p. 242; Liouville, *J. Éc. Polyt.* cah. 22 (1833), p. 1; *J. de Math.* 6 (1841), p. 1. The particular case to which the name of Riccati is more commonly attached has been studied in § 2·15.

12·52. Reduction to a Linear Equation of the Second Order.—If $p_2(z)$ is identically zero, the Riccati equation degenerates into a linear equation of the first order. Set this case aside, and write

$$w=-\frac{u'}{p_2(z)u},$$

then the Riccati equation becomes

$$-\frac{u''}{p_2u}+\frac{1}{p_2}\left(\frac{u'}{u}\right)^2+\frac{p_2'}{(p_2)^2}\cdot\frac{u'}{u}=p_0-\frac{p_1}{p_2}\cdot\frac{u'}{u}+p_2\left(\frac{u'}{p_2u}\right)^2,$$

and reduces to the homogeneous linear equation of the second order

$$p_2(z)\frac{d^2u}{dz^2}-\{p_2'(z)+p_1(z)p_2(z)\}\frac{du}{dz}+p_0(z)p_2^2(z)u=0.$$

Conversely the equation of the second order

$$\frac{d^2u}{dz^2}+P(z)\frac{du}{dz}+Q(z)u=0$$

is transformed by the substitution

$$w=u'/u$$

into the Riccati equation

$$\frac{dw}{dz}=-Q(z)-P(z)w-w^2.$$

The theory of the Riccati equation is therefore equivalent to the theory of the homogeneous linear equation of the second order.

The general solution of the linear equation is of the form

$$u=C_1u_1(z)+C_2u_2(z),$$

and therefore the general solution of the Riccati equation is

$$w=-\frac{C_1u_1'(z)+C_2u_2'(z)}{p_2(z)\{C_1u_1(z)+C_2u_2(z)\}}.$$

Example.—Deduce that the movable singularities of the Riccati equation are all poles.

12·6. Initial Values for which $f(z, w)$ is indeterminate.—The equation of Briot and Bouquet,*

$$z\frac{dw}{dz}-\lambda w=a_{10}z+a_{20}z^2+a_{11}zw+a_{02}w^2+\ldots,$$

is characterised by the fact that for the pair of initial values $z=w=0$ the differential coefficient, being of the form $0/0$, is indeterminate. The question of interest is whether or not there may exist one or more solutions, analytic in the neighbourhood of $z=0$, and reducing to zero when $z=0$.

Let the series

$$w(z)=c_1z+c_2z^2+\ldots+c_nz^n+\ldots$$

be supposed formally to satisfy the differential equation, then its successive coefficients are determined by the relations

$$\begin{aligned}&(1-\lambda)c_1=a_{10},\\&(2-\lambda)c_2=a_{20}+a_{11}c_1+a_{02}c_1^2,\\&\quad\cdot\qquad\cdot\qquad\cdot\qquad\cdot\\&(n-\lambda)c_n=P_n(a_{n0},\ldots,a_{0n};\ c_1,\ldots,c_{n-1}),\\&\quad\cdot\qquad\cdot\qquad\cdot\qquad\cdot\qquad\cdot\qquad\cdot\qquad\cdot\end{aligned}$$

* *J. Éc. Polyt.* cah. 36 (1856), p. 161. See Picard, *Traité d'Analyse*, 3, Chap. II.

in which P_n is a polynomial in its arguments, whose coefficients are positive integers. Thus the successive coefficients $c_1, c_2, \ldots, c_n \ldots$ may be calculated, provided that λ is not a positive integer. The series $w(z)$ then represents a solution of the differential equation if it converges for sufficiently small values of $|z|$. That the series does actually converge may be proved by an adaptation of the method of limits, as follows.

Since $n-\lambda$ is supposed not to be zero, a number B can be found such that $|n-\lambda| \geqslant B$ for all values of n. Let the series

$$a_{10}z+a_{20}z^2+a_{11}zw+a_{02}w^2+\ldots$$

converge within the domain $|z|=r$, $|w|=R$, and let it be bounded on the frontier of the domain; let M be the upper bound of its modulus on the frontier. Then the function

$$\Phi(z, W)=\frac{M}{\left(1-\frac{z}{r}\right)\left(1-\frac{W}{R}\right)}-M\left(1+\frac{W}{R}\right)$$
$$=A_{10}z+A_{20}z^2+A_{11}zW+A_{02}W^2+\ldots$$

is a dominant function for this series.

Consider that root of the quadratic equation

$$BW=\Phi(z, W),$$

which is zero when z is zero; it may be developed as a Maclaurin series

$$W=C_1z+C_2z^2+\ldots+C_nz^n+\ldots,$$

which has a finite, non-zero radius of convergence.

The coefficients of this series are successively determined by the relations

$$BC_1=A_{10},$$
$$BC_2=A_{20}+A_{11}C_1+A_{02}C_2,$$
$$\cdot\quad\cdot\quad\cdot\quad\cdot\quad\cdot$$
$$BC_n=P_n(A_{n0}, \ldots, A_{0n};\ C_1, \ldots, C_{n-1}),$$
$$\cdot\quad\cdot\quad\cdot\quad\cdot\quad\cdot\quad\cdot\quad\cdot\quad\cdot$$

where the polynomial P_n is formally the same as that which determines $(n-\lambda)c_n$. Consequently, since

$$B \leqslant |n-\lambda|;\ A_{10} \geqslant |a_{10}|, \ldots, A_{0n} \geqslant |a_{0n}|,$$

it follows, by induction, that

$$C_n \geqslant |c_n|.$$

Thus when λ is not a positive integer, the equation admits of a solution, analytic in the neighbourhood of $z=0$, which vanishes when z is zero. This analytic solution may easily be proved to be unique.*

In the case when $\lambda=1$, no analytic solution can exist unless $a_{10}=0$. When this is the case, c_1 may be chosen arbitrarily, and the remaining coefficients determined. In the same way, if $\lambda=n>1$, there exists an analytic solution if, and only if, there is a certain algebraic relation between the coefficients a_{rs}, where $r+s \leqslant n$. Thus, when $\lambda=2$, this relation is

$$a_{20}-a_{11}a_{10}+a_{02}a_{10}^2=0.$$

In the case $\lambda=n$, the coefficient c_n is arbitrary.

* Briot and Bouquet proved that when the real part of λ is negative, there exists none but the analytic solution so long as z tends to zero along a path of finite length which does not wind an infinite number of times around the origin. On the other hand, when the real part of λ is positive, the equation admits of an infinite number of *non-analytic* solutions which reduce to zero when z is zero. Representations of these non-analytic solutions have been given by Picard, *C. R. Acad. Sc. Paris*, 87 (1878), pp. 430, 743; *Bull. Soc. Math. France*, 12 (1883), p. 48, and Poincaré, *J. Éc. Polyt.* cah. 45 (1878), p. 13; *J. de Math.* (3), 7 (1881), p. 375; 8 (1882), p. 251; (4), 1 (1885), p. 167.

As an example consider the simple case,

$$z\frac{dw}{dz}-\lambda w=az.$$

The general solution is

$$w=\frac{a}{1-\lambda}z+Cz^{\lambda} \quad \text{if } \lambda\neq 1,$$

$$w=az\log z+Cz \quad \text{if } \lambda=1,$$

where C is an arbitrary constant.

12·61. The Generalised Problem of Briot and Bouquet ; the First Reduced Type.—The problem of the previous section will now be generalised and restated as follows. It is required to investigate the existence of solutions of the equations

$$\text{(A)} \qquad \frac{dw}{dz}=\frac{g(z, w)}{h(z, w)},$$

which vanish when $z=0$, where *

$$g(0, 0)=h(0, 0)=0.$$

It is assumed that $g(z, w)$ and $h(z, w)$ may be expanded as convergent ascending double series in z and w near the origin, and also that neither g nor h is divisible by any power of z or w.

In $g(z, w)$ let the term involving w to the lowest power and not multiplied by a power of z be that in w^m. Then let

z^{r_1} be the lowest power of z which multiplies w^{m-1},
z^{r_2} . . . . w^{m-2},
. . .
and z^{r_m} the lowest power of z which has a constant coefficient.

Both w^m and z^{r_m} must exist, for g is not divisible by any power of z or w, but any of the other terms mentioned may be absent.

The numbers $r_1, r_2, \ldots, r_m$ are positive integers, not zero. If all the terms of higher order than those corresponding to these indices are omitted, $g(z, w)$ is reduced to a polynomial in z and w.

Similarly $h(z, w)$ involves terms such as

$$w^n, \quad z^{s_1}w^{n-1}, \quad z^{s_2}w^{n-2}, \ldots, z^{s_n},$$

the first and last of which must exist, together with terms of higher order.

The problem in hand is that of investigating the possibility of a solution which is $O(z^{\mu})$ at the origin. The equation (A) itself may be written in the form

$$h(z, w)z\frac{dw}{dz}=zg(z, w).$$

Now construct a diagram similar to the classical Newton's diagram, representing any term $z^{\xi}w^{\eta}$ be the point whose Cartesian co-ordinates are (ξ, η). Let the points P_i represent the various terms of $zg(z, w)$ and the points Q_i represent the terms of $h(z, w)z\frac{dw}{dz}$ which, for the purposes of the diagram is regarded as equivalent to $w\,h(z, w)$. †

Among these points there is one point $Q_0(0, n+1)$ on the η-axis and no

* Briot and Bouquet, *C. R. Acad. Sc. Paris*, 39 (1854), p. 368 ; *J. Éc. Polyt.* cah. 36 (1856), p. 133 ; Poincaré, *loc. cit. ante* ; *J. de Math.* (4), 2 (1886), p. 151.

† Note that since $w=O(z^{\mu})$ it follows that

$$z\frac{dw}{dz}=O(z^{\mu})=O(w).$$

point P on that axis. Also there is one point $P_m(r_m+1, 0)$, and no point Q on the ξ-axis. Nor are there any points in the segments OQ_0 and OP_m.

The figure below illustrates the case

$$\frac{dw}{dz}=\frac{[w^6, zw^5, z^5w^4, zw^3, zw^2, z^3w, z^6]}{[w^4, z^3w^3, zw^2, z^2w, z^3]}$$

in which the terms of lowest order in $w^6, w^5, \ldots, w^0$ are given without numerical coefficients.

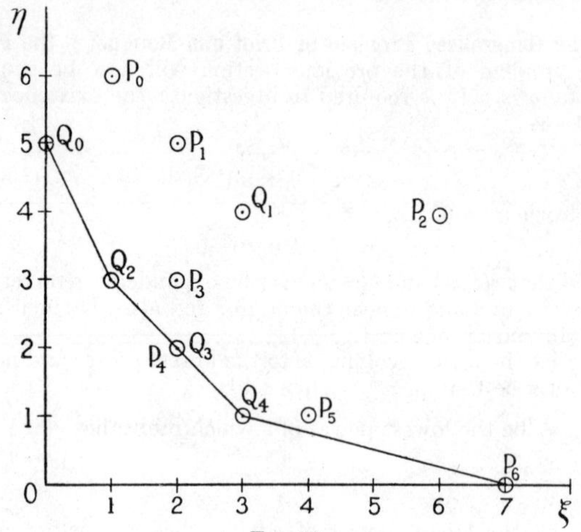

FIG. 10.

Construct the polygon Q_0P_m, which is known as the *Puiseux diagram.** It is the broken line everywhere convex to the origin such that all points P_i and Q_i either lie upon the line or on the side remote from the origin. Since the line begins at Q_0 and ends at P_m there must be at least one side which contains a point P and a point Q. Put aside the case where these two points coincide, and let these points correspond respectively to terms

$$z^{\alpha+a}w^{\beta}, \quad z^{\alpha}w^{\beta+b}.$$

These terms are associated with one another as terms of equal order; if any other points occur on the side of the polygon considered, the corresponding terms are of the same order, all points not on this side relate to terms of higher order. Now since

$$O(z^{\alpha+a}w^{\beta})=O(z^{\alpha}w^{\beta+b}),$$

it follows that

$$w=O(z^{a/b})=O(z^{h/k}),$$

where h/k is the fraction a/b in its lowest terms. This association of terms may therefore be expected to lead to a solution which is $O(z^{\mu})$ at the origin, where

$$\mu=h/k,$$

and therefore $-\mu$ is the slope of the side of the polygon considered.

To investigate a possible solution, let

$$z=t^k, \quad w=t^h+\text{higher terms},$$

* The application of the Puiseux diagram to the theory of differential equations is discussed in detail by Fine, *Amer. J. Math.* 11 (1889), p. 317.

then

$$z\frac{dw}{dz}=O(t^h),$$

and if

$$zg(z, w)=O(t^N),$$

then

$$h(z, w)=O(t^{N-h}).$$

Thus if

$$w=t^h u,$$

where u is $O(1)$ at the origin, then

$$zg(z, w)=t^N U_0+t^{N+1}U_1+\text{higher terms},$$
$$h(z, w)=t^{N-h}V_0+t^{N-h+1}V_1+\text{higher terms},$$

where $U_0, U_1, \ldots, V_0, V_1, \ldots$ are polynomials in u.

The equation (A) then reduces to

$$\text{(B)} \qquad \left(t\frac{du}{dt}+hu\right)\{V_0+V_1 t+\ldots\}=k\{U_0+U_1 t+\ldots\},$$

and if

$$u=u_0+O(t),$$

where $n_0\neq 0$, the roots of

$$F(u)\equiv huV_0-kU_0=0$$

give the initial values u_0.

Equation (B) may be written

$$(V_0+V_1 t+\ldots)t\frac{du}{dt}=F(u)+(kU_1-huV_1)t+\ldots$$

To avoid complications, assume in the first place that $u=u_0$ is a simple root of the equation $F(u)=0$; then

$$F(u)=(u-u_0)F'(u_0)+\ldots,$$

where

$$F'(u_0)\neq 0.$$

Assume also that

$$V_0\neq 0 \quad \text{when } u=u_0,$$

and write

$$V_0=a_0+a_1(u-u_0)+\ldots \qquad (a_0\neq 0).$$

Then

$$t\frac{du}{dt}=\frac{(u-u_0)F'(u_0)+(kU_1-huV_1)t+\ldots}{a_0+a_1(u-u_0)+V_1 t+\ldots}$$

or

$$t\frac{dv}{dt}=\lambda v+at+\text{higher terms},$$

where $v=u-u_0$ and $\lambda\neq 0$. The equation is now said to be of the *First Reduced Type*, and is of the form studied in the preceding section. Thus, apart from the exceptional case where λ is a positive integer, the original equation has a solution

$$w=t^h\left(u_0+\sum_{r=1}^{\infty}c_r t^r\right)$$
$$=\mathbf{P}_h(z^{1/k}).$$

where $\mathbf{P}_h(z^{1/k})$ denotes a power-series in $z^{1/k}$ whose leading term is of degree h.

Suppose now that $u=u_0$ is a multiple root of $F(u)=0$, so that $F'(u_0)=0$, then if, as before, $u=u_0$ is not a root of $V_0=0$,

$$t\frac{du}{dt}=\frac{(kU_1-huV_1)t+\ldots}{a_0+a_1(u-u_0)+V_1 t+\ldots},$$

or if $v=u-u_0$,

$$t\frac{dv}{dt}=at+\text{terms of the second and higher orders.}$$

This is merely the particular case of the First Reduced Type where $\lambda=0$, and calls for no special remark.

12·62. The Second Reduced Type.—Consider now the case in which u_0 is a common zero of $F(u)$ and of V_0, so that both

$$F'(u_0)=0 \quad \text{and} \quad a_0=0.$$

If, as before, $v=u-u_0$, the equation assumes the form

$$t\frac{dv}{dt}=\frac{\alpha'v+\beta't+\ldots}{\alpha v+\beta t+\ldots},$$

where α, β, α', β' are constants, any or all of which may be zero.* The right-hand member has still an indeterminate form at the origin. An examination of the polygon corresponding to this case leads to the tentative assumption that the first approximation to a solution at the origin is

$$\alpha'v+\beta't=0.$$

Write therefore (assuming that $\alpha'\neq 0$)

$$v=\left(-\frac{\beta'}{\alpha'}+v_1\right)t,$$

and the equation becomes

$$t^2\frac{dv_1}{dt}=\frac{\alpha'v_1+\ldots}{\dfrac{\alpha'\beta-\alpha\beta'}{\alpha'}+\alpha v_1+\ldots}.$$

Then if $\alpha'\beta-\alpha\beta'\neq 0$, the equation is reduced to the new form

$$t^2\frac{dv_1}{dt}=\lambda v_1+at+\text{higher terms.}$$

If, on the other hand, $\alpha'\beta-\alpha\beta'=0$, the right-hand member is still indeterminate at the origin. The process is then repeated and either leads to an equation of the form

$$t^3\frac{dv_2}{dt}=\lambda v_2+at+\text{higher terms}$$

or to one in which the right-hand member is of the form $0/0$ at the origin. In the latter case, the reduction is continued. It can be proved that after a finite number of reductions, the right-hand member ceases to be indeterminate at the origin, and thus an equation of the form

$$t^{m+1}\frac{dv_m}{dt}=\lambda v_m+at+\text{higher terms,}$$

where m is a positive integer $\geqslant 1$ is arrived at. This is the *Second Reduced Type.*†

The origin is, in general, an essential singular point of the equation of the Second Reduced Type, for if $\lambda\neq 0$, $m>1$, the equation cannot be satisfied by an ascending series of powers of t in which the leading term is t^p. If

* Necessary but not sufficient conditions for the existence of this case are that the side of the polygon considered contains (a) at least two points P and two points Q, or (b) no points P, or (c) no points Q.

† For a study of the behaviour of solutions this equation in the neighbourhood of the origin, see Bendixson, *Öfv. Vet.-Akad. Stockholm*, 55 (1898), pp. 69, 139, 171; Horn, *J. für Math.* 118 (1897), p. 257; 119 (1898), pp. 196, 267; *Math. Ann.* 51 (1898), pp. 346, 360. A further generalisation is due to Perron, *Math. Ann.* 75 (1914), p. 256.

$\lambda \neq 0$, $m=1$, the equation can *formally* be satisfied by a Maclaurin series, which, however, diverges for all values of t.

For instance, the equation

$$z^2 \frac{dw}{dz} = \lambda w - z$$

has the formal solution

$$w = \sum_{n=1}^{\infty} \frac{(n-1)!}{\lambda^n} z^n,$$

which obviously converges only if $z=0$.

12·63. Special Cases of the Reduced Forms of the Equation.—(i) The equation

$$z^{m+1} \frac{dw}{dz} = \lambda w + az$$

is of the Second Reduced Type. But it is also a linear equation and can therefore be integrated by quadratures. Its solution is

$$w = e^{-\lambda/mz^m} \left\{ a \int e^{\lambda/mz^m} z^{-m} dz + C \right\}.$$

If $\lambda=0$ the integral is algebraic, but in the general case, $\lambda \neq 0$, there is an essential singularity at the origin.

(ii) $$z^2 \frac{dw}{dz} = \alpha z^3 + \beta w^2 \qquad (\alpha \neq 0,\ \beta \neq 0)$$

is a case of the Riccati equation. The polygon corresponding to this equation (Fig. 11) has two sides, P_1Q_0 and Q_0P_0.

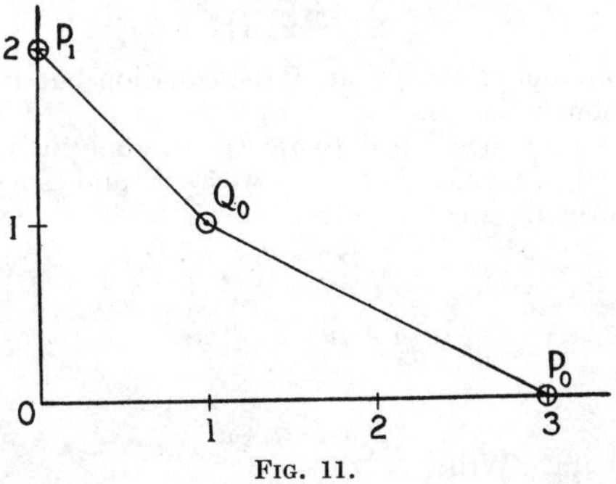

FIG. 11.

In the side P_1Q_0, w^2 is associated with zw, which suggests a solution $w=O(z)$ at the origin. Let

$$w = zu,$$

then the equation becomes

$$z \frac{du}{dz} + u = \alpha z + \beta u^2.$$

The equation which determines u_0 is

$$F(u) \equiv \beta u^2 - u = 0,$$

and has the non-zero root $u_0 = 1/\beta$.

Then if $u=v+1/\beta$,

$$z\frac{dv}{dz}=az+v+\beta v^2,$$

which is that case of the Briot and Bouquet equation where $\lambda=1$. It has no analytic solution unless $a=0$. To find the nature of the solution, if any, near $z=0$, write the equation in the form

$$\frac{d}{dz}\left(\frac{v}{z}\right)=\frac{a}{z}+\beta\left(\frac{v}{z}\right)^2,$$

which is a Riccati equation in v/z. Now transform it into a linear equation of the second order by writing

$$\frac{v}{z}=-\frac{W'}{\beta W}.$$

It then assumes the very simple form

$$zW''=-a\beta W,$$

and this equation has the two distinct solutions

$$W_1=z-\frac{a\beta}{2\,!}z^2+\frac{a^2\beta^2}{2\,.\,3\,!}z^3-\;.\;.\;.$$

and

$$W_2=W_1\log z+W_0,$$

where W_0 is a power series in z.

Thus W_1 alone leads to a solution of the Riccati equation, and this solution is

$$v=-\frac{1-a\beta z+\dfrac{a^2\beta^2}{2\,.\,2\,!}z^2-\;.\;.\;.}{\beta\left\{1-\dfrac{a\beta}{2\,!}z+\dfrac{a^2\beta^2}{2\,.\,3\,!}z^2-\;.\;.\;.\right\}},$$

which is indeed an analytic solution of the equation but it does not satisfy the initial conditions $z=w=0$.

Since the side P_1Q_0 has failed to reveal an analytic solution, the side Q_0P_0 is now tried. It associates zw with z^3 and suggests a solution $w=O(z^2)$ at the origin. Let

$$w=z^2u,$$

then

$$z\frac{du}{dz}+2u=a+\beta zu^2.$$

In this case

$$F(u)=2u-a$$

and therefore $u_0=\frac{1}{2}a$. Write

$$u=\tfrac{1}{2}a+v,$$

then the equation becomes

$$z\frac{dv}{dz}+2v=\beta z(\tfrac{1}{2}a+v)^2$$

and is a Briot and Bouquet equation of the first type, with $\lambda=-2$. There is here no complication; there exists an analytic solution

$$v=\tfrac{1}{12}a^2\beta z+\;.\;.\;.,$$

and therefore there is one solution of the original equation which is analytic in the neighbourhood of the origin, and assumes the value zero there, namely,

$$w=z^2(\tfrac{1}{2}a+\tfrac{1}{12}a^2\beta z+\;.\;.\;.).$$

MISCELLANEOUS EXAMPLES.

1. In the equation

$$\frac{dw}{dz}=\frac{P(z, w)}{Q(z, w)},$$

let

$$P(z, w)=az+bw+\ .\ .\ .\ ,\quad Q(z, w)=\alpha z+\beta w+\ .\ .\ .,$$

and let $\lambda=\lambda_1/\lambda_2$, where λ_1 and λ_2 are the roots of the equation

$$\begin{vmatrix} \alpha-\lambda, & \beta \\ a\quad, & b-\lambda \end{vmatrix}=0.$$

Prove that if neither λ nor $1/\lambda$ is a positive integer, or if λ is not a negative real number, two particular solutions exist analytic near $z=0$, $w=0$ and are of the forms

$$U(z, w)\equiv gz+hw+\ .\ .\ .\ =0,\quad V(z, w)\equiv \gamma z+\kappa w+\ .\ .\ .\ =0,$$

where $g\kappa-h\gamma\neq 0$, and that the general solution is

$$U(z, w)=c[V(z, w)]^{\lambda},$$

where c is an arbitrary constant.

[Poincaré.]

2. When, in the notation of the preceding question, λ or $1/\lambda$ is a positive integer, prove that there exists in general one and only one analytic solution such that $w=0$ when $z=0$. Let this solution be $V(z, w)$, then the general solution is of the form

$$\frac{S(z, w)}{V(z, w)}+h\log V(z, w)=\text{const.},$$

where $S(z, w)$ is analytic in the neighbourhood of $z=0$, $w=0$. The number h depends upon the earlier coefficients $a, b, \ldots, \alpha, \beta, \ldots$ in P and Q. Discuss the particular case $h=0$.

[Poincaré, Bendixson, Horn.]

3. If λ is a negative real number, two particular analytic solutions exist such that $w=0$ when $z=0$, but the general solution is not of the form specified in Ex. 1. Transform the equation into one of a similar type in which $\alpha=1$, $\beta=0$, $a=0$, $b=\lambda$, and writing

$$zw^{-\lambda}=\rho^{1-\lambda},\quad w=ze^{u},$$

prove that the general solution admits of the development

$$\rho+\rho^2A_2(u)+\rho^3A_3(u)+\ .\ .\ .\ =\text{const.},$$

where $A_2, A_3, \ldots$ are analytic near $u=0$, and the series converges when $|\rho|<\delta$, $|u|<G$, where G is arbitrary and δ depends upon G and tends to zero as G tends to infinity.

[Bendixson.]

CHAPTER XIII

EQUATIONS OF THE FIRST ORDER BUT NOT OF THE FIRST DEGREE

13·1. Specification of the Equations Considered.—In the differential equations which are now to be dealt with, the differential coefficient is not defined explicitly in terms of z and w, but is related implicitly * to z and w, thus

$$F\left(z, w, \frac{dw}{dz}\right)=0.$$

Of this general class of equations only those equations in which the left-hand member is a polynomial in w and $\frac{dw}{dz}$ will be considered. Writing

$$p \equiv \frac{dw}{dz},$$

it is then possible to express $F(z, w, p)$ in the form

$$A_0(z, w)p^m + A_1(z, w)p^{m-1} + \ldots + A_{m-1}(z, w)p + A_m(z, w),$$

where the functions $A(z, w)$ are assumed to be polynomials in w, whose coefficients are analytic functions of z. It is now further supposed that the above expression is *irreducible*, that is to say not decomposable into factors of the same analytical character as itself.

The main problem is to determine necessary and sufficient conditions for the absence of movable branch points,† and thus to obtain generalisations of the Riccati equation.

Let $D(z, w)$ be the p-discriminant of the equation

$$F(z, w, p)=0;$$

it is a polynomial in w, whose coefficients are analytic functions of z.

A number of values of z are excluded from the following discussion, namely those for which

(*a*) $D(z, w)=0$ independently of w,
(*b*) $A_0(z, w)=0$ independently of w,
(*c*) the coefficients A possess singular points for general values of w,
(*d*) the roots of $D(z, w)=0$, regarded as an equation in w, have singular points.

All these values of z are fixed, and depend only upon the coefficients A. They correspond to singular points fixed in the z-plane. Henceforward z_0 will be considered as an initial value of z distinct from one of the singular

* A knowledge of the elementary properties of implicit algebraic functions will be assumed.

† Fuchs, *Sitz. Akad. Wiss. Berlin*, 32 (1884), p. 699 [*Math. Werke*, 2, p. 355].

values enumerated; let w_0 be the corresponding initial value of w. Then there are four distinct cases to consider, according as

$$\begin{array}{lll} \text{(i)} & D(z_0, w_0) \neq 0, & A_0(z_0, w_0) \neq 0, \\ \text{(ii)} & D(z_0, w_0) \neq 0, & A_0(z_0, w_0) = 0, \\ \text{(iii)} & D(z_0, w_0) = 0, & A_0(z_0, w_0) \neq 0, \\ \text{(iv)} & D(z_0, w_0) = 0, & A_0(z_0, w_0) = 0. \end{array}$$

These four cases will now be considered in detail.

13·2. Case (i).—When neither $D(z, w)$ nor $A_0(z, w)$ is zero for $z=z_0$, $w=w_0$, it follows from the theory of algebraic functions that the equation

$$\text{(A)} \qquad F(z, w, p)=0$$

determines, in the neighbourhood of (z_0, w_0), m distinct finite values of p. Let w assume the fixed value w_0, then the equation

$$F(z, w_0, p)=0$$

will have p distinct roots which are analytic in the neighbourhood of z_0. Let these roots be

$$\varpi_1, \quad \varpi_2, \quad . \; . \; . \; \varpi_m,$$

then in the neighbourhood of (z_0, w_0) m expressions of the form

$$p=\varpi_i+C_i^{(1)}(w-w_0)+C_i^{(2)}(w-w_0)^2+ \; . \; . \; . \qquad (i=1, 2, \; . \; . \; ., m)$$

exist. Since ϖ_i and the coefficients C_i are analytic in the neighbourhood of z_0, these expressions may be written as

$$\text{(B)} \qquad p=\varpi_i^{(0)}+P_i(z-z_0, w-w_0) \qquad (i=1, 2, \; . \; . \; ., m),$$

where $\varpi_i^{(0)}$ is the value of ϖ_i when $z=z_0$, and P_i denotes a double series which converges for sufficiently small values of $|z-z_0|$ and $|w-w_0|$, and vanishes when $z=z_0$, $w=w_0$. Thus the original equation (A) is replaced by the set of m distinct equations (B), each of which is known to possess one and only one analytic solution which reduces to w_0 when $z=z_0$. The equation (A) has therefore m distinct analytic solutions which satisfy the initial conditions. Nor has it any other solution.

13·3. Case (ii).—When $A_0(z_0, w_0)=0$ but $D(z_0, w_0)\neq 0$, the equation (A) determines m values of p, one of which becomes infinite at (z_0, w_0). There cannot be two values of p which thus become infinite, for this would necessitate $A_1(z_0, w_0)=0$ and $D(z_0, w_0)=0$. So there are $m-1$ distinct expressions for p, analytic in the neighbourhood of (z_0, w_0), and these lead to a set of $m-1$ solutions of the equation which satisfy the initial conditions.

To investigate that root which becomes infinite at (z_0, w_0), let

$$P=\frac{dz}{dw}=\frac{1}{p},$$

then the equation

$$F(z, w, P^{-1})=0$$

has a root

$$P=P(z-z_0, w-w_0)$$

vanishing at, and analytic in the neighbourhood of (z_0, w_0). This equation has a solution

$$z=z_0+\mathbf{P}_r(w-w_0),$$

where $r \leqslant 2$ since

$$\frac{dz}{dw}=0, \quad \text{when } w=w_0.$$

The solution of (A) corresponding to the value of p which becomes infinite for (z_0, w_0) is therefore

$$w-w_0=\mathbf{P}_1\left\{(z-z_0)^{\frac{1}{r}}\right\}.$$

Thus Case (ii) always leads to a solution which has a branch point at z_0, *i.e.* a *movable* branch point. This leads to the first necessary condition for the absence of movable branch points, namely :

The equation $A_0(z, w)=0$ *has no solution* $w=\zeta(z)$ *such that* $D(z, w)\neq 0$.

13·4. Case (iii).—The left-hand member of the algebraic equation

$$D(z, w)=0$$

is a polynomial in w with coefficients which are analytic in z. Let $w=\eta(z)$ satisfy this algebraic equation, then $\eta(z)$ ceases to be analytic only at the singular points of $D(z, w)$ and possibly at a limited number of other points. Let these points, which are fixed, be excluded in what follows.

The equation

$$F(z, \eta, p)=0$$

has at least one multiple root in p, say $p=\varpi$; let it be of multiplicity λ. On the other hand, for general values of w, the equation

$$F(z, w, p)=0$$

has m distinct roots. Let those roots which become equal to one another and to ϖ when $w=\eta$ be

$$p_1, \ p_2, \ \ldots, \ p_\lambda.$$

Let z be fixed for the moment and let w describe a small circuit around the point η corresponding to the value of z chosen. On the completion of this circuit, p_1 returns either to its initial value or to one of the values $p_2, \ldots, p_\lambda$. After $\alpha(\leqslant\lambda)$ complete circuits have been described, p_1 returns to its initial value. Let the sequence of values assumed by p_1 during this process be

$$p_1, \ p_2, \ \ldots, \ p_\alpha, \ p_1 ;$$

this sequence is said to form a *cycle* of order α.

Thus p_1, regarded as a function of w, has a branch point of order $\alpha-1$ at $w=\eta$; write

$$w-\eta=W^\alpha,$$

then p_1 becomes a uniform function of W. But $p_1=\varpi$ when $w=\eta$, and is bounded when w is in the neighbourhood of η. Therefore p_1 is developable in the Maclaurin series

$$p_1=\varpi+\sum_{r=1}^{\infty} c_r W^r,$$

whose coefficients depend upon z, and which converges when z takes non-singular values and W is sufficiently small. Let c_k be the first of the coefficients which does not vanish identically, then

$$p_1=\varpi+c_k\{w-\eta(z)\}^{\frac{k}{\alpha}}+c_{k+1}\{w-\eta(z)\}^{\frac{k+1}{\alpha}}+\ldots,$$

and thus $w-\eta(z)$ satisfies the differential equation

$$\frac{d\{w-\eta(z)\}}{dz}=\varpi-\frac{d\eta}{dz}+c_k\{w-\eta(z)\}^{\frac{k}{\alpha}}+c_{k+1}\{w-\eta(z)\}^{\frac{k+1}{\alpha}}+\ldots.$$

13·41. Condition for the Absence of Branch Points in Case (ii).—In the particular case $\alpha=1$, the right-hand member of this equation is analytic (except for isolated points) in z and in $w-\eta(z)$; the equation then has an analytic solution. If, however, $\alpha > 1$, the right-hand member is non-uniform and then p is said to have a *branched* value. Consider first the case in which

$$\varpi \neq \frac{d\eta}{dz}$$

identically. The *isolated* values of z for which ϖ and $\frac{d\eta}{dz}$ are equal are excluded. Let

$$\varpi - \frac{d\eta}{dz} = a_0 + a_1(z-z_0) + a_2(z-z_0)^2 + \dots \qquad (a_0 \neq 0),$$

$$c_r = c_r^{(0)} + c_r^{(1)}(z-z_0) + c_r^{(2)}(z-z_0)^2 + \dots \qquad (r \geqslant k),$$

then if, as before,

$$w - \eta(z) = W^\alpha, \qquad (\alpha \geqslant 2)$$

$$\alpha W^{\alpha-1}\frac{dW}{dz} = a_0 + a_1(z-z_0) + a_2(z-x_0)^2 + \dots$$
$$+\{c_k^{(0)} + c_k^{(1)}(z-z_0) + c_k^{(2)}(z-z_0)^2 + \dots\}W^k$$
$$+\{c_{k+1}^{(0)} + c_{k+1}^{(1)}(z-z_0) + c_{k+1}^{(2)}(z-z_0)^2 + \dots\}W^{k+1} + \dots,$$

and the right-hand member of this equation is analytic for sufficiently small values of $z-z_0$ and W. Consequently

$$\frac{dz}{dW} = \frac{\alpha}{a_0} W^{\alpha-1} + \text{higher terms},$$

and this equation has a unique analytic solution of the form

$$z - z_0 = \mathbf{P}_\alpha(W).$$

On inverting, this becomes

$$W = \mathbf{P}_1\left\{(z-z_0)^{\frac{1}{\alpha}}\right\},$$

and the original equation has a solution

$$w = \eta(z) + \mathbf{P}_1\left\{(z-z_0)^{\frac{1}{\alpha}}\right\}^\alpha$$
$$= \eta(z) + \mathbf{P}_\alpha\left\{(z-z_0)^{\frac{1}{\alpha}}\right\}$$
$$= \eta(z) + (z-z_0)\left\{\gamma_0 + \gamma_1(z-z_0)^{\frac{1}{\alpha}} + \gamma_2(z-z_0)^{\frac{2}{\alpha}} + \dots\right\}.$$

Thus there is a parametric branch point whenever the equation

$$\varpi = \frac{d\eta}{dz}$$

is not satisfied identically. A necessary condition for the absence of parametric branch points is therefore:

If $p = \varpi_1, \varpi_2, \dots$ are multiple roots of $F(z, \eta, p) = 0$, and correspond to branched values of p, then

$$\frac{d\eta}{dz} = \varpi_1 = \varpi_2 = \dots$$

identically.

Consider further the condition that

$$\varpi = \frac{d\eta}{dz}$$

identically. The equation now becomes

$$aW^{a-1}\frac{dW}{dz}=\{c_k^{(0)}+c_k^{(1)}(z-z_0)+c_k^{(2)}(z-z_0)^2+\ .\ .\ .\ \}W^k$$
$$+\{c_{k+1}^{(0)}+c_{k+1}^{(1)}(z-z_0)+c_{k+1}^{(2)}(z-z_0)^2+\ .\ .\ .\ \}W^{k+1}+\ .\ .\ ..$$

One solution is obvious, namely $W=0$ or

$$w=\eta(z).$$

It is the *Singular Solution* of the equation, which has arisen as a root of the p-discriminant.

There may possibly be other solutions ; this possibility will be considered (*a*) when $a-1>k$, (*b*) $a-1\leqslant k$.

When $a-1>k$, let $a-1=k+r$; the equation may be divided out by W^k and becomes

$$aW^r\frac{dW}{dz}=c_k^{(0)}+\text{terms in } W \text{ and } (z-z_0),$$

where $r\geqslant 1$ and $c_k^{(0)}\neq 0$ when z_0 is not one of the fixed singular points of the equation. Thus

$$\frac{dz}{dW}=\frac{a}{c_k^{(0)}}W^r+\text{higher terms},$$

an equation having the analytic solution

$$z=z_0+\mathbf{P}_{r+1}(W),$$

which in turn leads to

$$W=\mathbf{P}_1\left\{(z-z_0)^{\frac{1}{r+1}}\right\},$$

and thus the solution of the original equation is

$$w=\eta(z)+\mathbf{P}_a\left\{(z-z_0)^{\frac{1}{r+1}}\right\},$$

and since $r\geqslant 1$ this solution always has a movable branch point.

Alternatively, when $a-1\leqslant k$, let

$$k=a+s-1 \qquad (s\geqslant 0).$$

After division by W^{a-1} the equation becomes

$$a\frac{dW}{dz}=c_k^{(0)}W^s+\text{higher terms}.$$

In this case $\frac{dW}{dz}$ is an analytic function of W and $z-z_0$, and therefore there is an analytic solution

$$W=\mathbf{P}_1(z-z_0).$$

If $s>0$, an obvious solution is $W=0$, and by the fundamental existence theorem it is the only solution reducing to zero when $z=z_0$. Thus the singular solution

$$w=\eta(z)$$

is the only solution when $s>0$.

If $s=0$, there exists the analytic solution

$$W=\frac{c_k^{(0)}}{a}(z-z_0)+\mathbf{P}_2(z-z_0),$$

and therefore the solution of the original equation is

$$w=\eta(z)+\mathbf{P}_a(z-z_0)$$

which has not a branch point at $z=z_0$.

Thus *the condition $k\geqslant a-1$ is necessary for the absence of movable branch points.*

13·5. Case (iv).—In this case $w=\eta(z)$ is a solution common to the two equations

$$D(z, w)=0, \quad A_0(z, w)=0,$$

and the equation

$$F(z, \eta, p)=0$$

has a multiple infinite root. Of the roots $p_1, p_2, \ldots, p_\lambda$ of

$$F(z, w, p)=0,$$

which become infinite when $w=\eta(z)$, let $p_1, p_2, \ldots, p_a$ form a cycle of order $a(\geqslant 1)$; then p_1, for instance, will be expressible in the form

$$p_1=\{w-\eta(z)\}^{-\frac{k}{a}}[c_0+c_1\{w-\eta(z)\}^{\frac{1}{a}}+c_2\{w-\eta(z)\}^{\frac{2}{a}}+\ldots],$$

where the coefficients c depend upon z and k is a positive integer which has been so chosen that c_0 is not identically zero. As before, it is supposed that z_0 is such that

$$c_0^{(0)}=c_0(z_0)\neq 0.$$

Let

$$\{w-\eta(z)\}=W^a,$$

so that the equation becomes

$$aW^{a-1}\frac{dW}{dz}=-\frac{d\eta}{dz}+W^{-k}\{c_0+c_1W+c_2W^2+\ldots\}$$

or

$$\frac{dz}{dW}=aW^{k+a-1}\left\{-W^k\frac{d\eta}{dz}+c_0+c_1W+\ldots\right\}^{-1}$$

$$=\frac{a}{c_0^{(0)}}W^{k+a-1}+\text{higher terms.}$$

Since $k+a-1>0$ this equation has a unique analytic solution

$$z-z_0=\mathbf{P}_{k+a}(W),$$

whence, by inversion,

$$W=\mathbf{P}_1\left\{(z-z_0)^{\frac{1}{k+a}}\right\},$$

and therefore

$$w=\eta(z)+\mathbf{P}_a\left\{(z-z_0)^{\frac{1}{k+a}}\right\}.$$

Since $k>0$, this solution has a movable branch point, and this is true even when $a=1$, and the expression for p_1 is one-valued.

Hence *a further necessary condition for the absence of movable branch points is that $A_0(z, w)$ and $D(z, w)$ should have no common factor of the form $w-\eta(z)$.*

The conditions thus obtained may be summed up as follows: Necessary conditions for the non-appearance of movable branch points are:

(A) *The coefficient $A_0(z, w)$ is independent of w and therefore reduces to a function of z alone or to a constant* (§§ 13·3, 13·5). The equation may then be divided throughout by A_0 and takes the form

$$p^m+\psi_1(z, w)p^{m-1}+\ldots+\psi_{m-1}(z, w)p+\psi_m(z, w)=0,$$

in which the coefficients ψ are polynomials in w, and analytic, except for isolated singular points, in z.

(B) *If $w=\eta(z)$ is a root of $D(z, w)=0$, and $p=\varpi(z)$ is a multiple root of $F(z, \eta, p)=0$, such that the corresponding root of $F(z, w, p)=0$, regarded as a function of $w-\eta(z)$ is branched, then* (§ 13·41)

$$\varpi(z)=\frac{d\eta}{dz}.$$

(C) *If the order of any branch is α, so that the equation is effectively of the form*

$$\frac{d}{dz}\{w-\eta(z)\}=c_k\{w-\eta(z)\}^{\frac{k}{\alpha}}$$

then (§ 13·41) $k \geqslant \alpha-1$.

13·6. The Dependent Variable initially Infinite.—To consider the possibility of the dependent variable becoming infinite at a branch point, let it be assumed that

$$w \to \infty \quad \text{as} \quad z \to z_0.$$

Make the substitution

$$w=W^{-1},$$

so that

$$W \to 0 \quad \text{as} \quad z \to z_0,$$

and write

$$P=\frac{dW}{dz}=-\frac{p}{W^2},$$

then the equation becomes

$$P^m-\psi_1(z, W^{-1})W^2P^{m-1}+\ldots+(-1)^m\psi_m(z, W^{-1})W^{2m}=0.$$

In order that the coefficient of each power of P may be rational in W, the coefficient of P^m being unity, it is necessary (and sufficient) that $\psi_1(z, w)$, $\psi_2(z, w), \ldots, \psi_m(z, w)$ should be of degrees $2, 4, \ldots, 2m$ at most in w. When $m=1$, and this condition is satisfied, the equation simply reduces to the Riccati equation.

Thus condition (A) of the previous section must be supplemented by

(A′) $\psi_r(z, w)$ *is at most of degree* $2r$ *in* w.

Now let $D'(z, W)$ be the P-discriminant of the transformed equation. If the discriminant $D(z, w)$ of the original equation has a factor $w-\eta(z)$, then $D'(z, W)$ will have a corresponding factor $W-1/\eta(z)$, and therefore, if conditions (B) and (C) are satisfied for the original equation, they are satisfied for the transformed equation. But, in addition to such factors, the discriminant $D'(z, W)$ may also contain W as a factor. More exactly, when condition (A′) is satisfied, $D(z, w)$ is at most of degree $2m(m-1)$ in w, but may be of a lower degree, say $2m(m-1)-s$. $D'(z, W)$ will then contain the factor W^s.

This last case has to be considered apart, and gives rise to special conditions for the absence of movable branch points.* If P, as deduced from the transformed equation and regarded as a function of W, has a branch point corresponding to $W=0$, then (condition B) $P=0$ when $W=0$. It follows that W must be a factor of the term $W^{2m}\psi_m(z, W^{-1})$. But since W is also a factor of the discriminant, it must also be a factor of the preceding coefficient $W^{2m-2}\psi_{m-1}(z, W^{-1})$.

It then follows, as in § 13·41, that the equation, when solved for P, gives

$$P=c_k^{(0)}W^{\frac{k}{\alpha}}+\text{higher terms},$$

* The necessity for the special treatment of this case was first pointed out by Hill and Berry, *Proc. London Math. Soc.* (2), 9 (1910), p. 231. These writers give the equation

$$\left[\frac{d\{w-f(z)\}}{dz}\right]^m=\{w-f(z)\}^{m+r},$$

where m and r are positive integers prime to one another and $r<m$ as an instance of the necessity of special conditions. The equation satisfies conditions (A), (A′), (B), (C), but, as the solution

$$w=f(z)+\left\{\frac{r}{m}(z_0-z)\right\}^{-m/r}$$

shows, it has a movable branch point, for which w is infinite.

$c_k^{(0)}$ being a constant, not zero. In order that this expression for P may give rise to a solution which has not a movable branch point it is necessary that

$$k \geqslant a - 1.$$

The two new conditions which have been obtained may be formulated as follows :

(B′) *If the equation is transformed by the substitution* $w = W^{-1}$, *and* W *is a factor of the discriminant of the transformed equation, then if* P *is a many-valued function of* W, W *must be a factor of the last two coefficients in the transformed equation.*

(C′) *If the order of a branch is* a *so that the equation is effectively of the form*

$$\frac{dW}{dz} = c_k^{(0)} W^{\frac{k}{a}},$$

then $k \geqslant a - 1$.

The conditions (A), (B), (C) and the supplementary conditions (A′), (B′), (C′) are necessary, and are clearly also sufficient for the non-appearance of movable branch points.

By adopting a line of argument not essentially different from that applied in § 12·5 to the case of the equation of the first degree, it is not difficult to prove that solutions of the equation

$$p^m + \psi_1(z, w)p^{m-1} + \ldots + \psi_{m-1}(z, w)p + \psi_m(z, w) = 0$$

have no movable essential singularities.* This is true whether the equation has movable branch points or not.

13·7. Equations into which z does not enter explicitly.—Consider the case in which the equation is of the form †

$$p^m + A_1(w)p^{m-1} + \ldots + A_{m-1}(w)p + A_m(w) = 0,$$

in which the coefficients A are polynomials in w with constant coefficients and the polynomial A_r is of degree not exceeding $2r$. Further, let the equation be such that its solutions have no movable branch points.

Now, except possibly for the point at infinity, the equation admits of no fixed singular points, for such singular points are singularities of the coefficients A, and these coefficients are independent of z.

Let $w = \phi(z)$ be any solution of the equation, then since the equation is unaltered by writing $z + c$ for z, where c is an arbitrary constant,

$$w = \phi(z + c)$$

is a solution. Since it contains an arbitrary constant it is the general solution.

Since, therefore, all solutions of the equation are free from branch points and essential singularities in the finite part of the z-plane, such solutions partake of the nature of rational functions. Consequently any solution, continued analytically from a point z_0 along any closed simple curve in the z-plane, returns to its initial value at z_0, and therefore the point at infinity cannot be a branch point. It may, however, be an essential singular point.

In the case of the Riccati equation, when z does not appear explicitly, the equation may be integrated by elementary methods. Let the equation be

$$\frac{dw}{dz} = a_0 + a_1 w + a_2 w^2,$$

where a_0, a_1, a_2 are constants ; the variables may be separated, thus

$$\frac{dw}{a_2 w^2 + a_1 w + a_0} = dz.$$

* Painlevé, *Leçons*, p. 56.

† Briot and Bouquet, *J. Éc. Polyt.* (1) cah. 36 (1856), p. 199.

Let ρ_1 and ρ_2 be the zeros of $a_2w^2+a_1w+a_0$, then if $\rho_2 \neq \rho_1$,

$$\frac{dw}{a_2(w-\rho_1)(w-\rho_2)}=dz,$$

whence

$$\frac{w-\rho_1}{w-\rho_2}=Ce^{a_2(\rho_1-\rho_2)z},$$

or if $\rho_2=\rho_1$

$$w-\rho_1=\frac{1}{C-a_2z},$$

C being an arbitrary constant in each case.

13·8. Binomial Equations of Degree m.—Consider now the class of equations included in the type *

$$\text{(A)} \qquad p^m+A(z, w)=0,$$

which is assumed to be irreducible. It is to be supposed that the conditions for the absence of movable branch points are all fulfilled. In particular, $A(z, w)$ must be a polynomial of degree $2m$ at most; suppose for the moment that its degree is less than $2m$, and that it is not exactly divisible by w. Write $w=W^{-1}$, then the equation becomes

$$\left(\frac{dW}{dz}\right)^m+(-1)^m W^{2m}A(z, W^{-1})=0;$$

but here the term $W^{2m}A(z, W^{-1})$ is of degree $2m$ in W. On the other hand, if $A(z, w)$ is of lower degree than $2m$ in w and contains the factor w, let a be such that $w-a$ is not a factor. Then, by writing $w-a=W^{-1}$ and proceeding as before, an equation is obtained which does contain W^{2m}. There is thus no loss of generality in supposing that $A(z, w)$ is exactly of degree $2m$ in w.

Since (A) has equal roots if, and only if, $A(z, w)=0$, the p-discriminant is effectively $A(z, w)$. Let $w-\eta(z)$ be a factor of $A(z, w)$, then $p=0$ is a root of

$$p^m+A(z, \eta)=0.$$

First let the corresponding root of (A) be branched when $w=\eta(z)$. Then by condition B (§ 13·5), which here reduces to

$$\frac{d\eta}{dz}=0,$$

$\eta(z)$ is a mere constant.

Secondly, suppose that the corresponding root of $A(z, w)$ is not branched. Then $A(z, w)$ contains either $\{w-\eta(z)\}^{2m}$ or $\{w-\eta(z)\}^m$ as a factor.

If $\{w-\eta(z)\}^{2m}$ is a factor, the equation becomes

$$p^m+K(z)\{w-\eta(z)\}^{2m}=0$$

and is reducible, contrary to supposition. If $\{w-\eta(z)\}^m$ is a factor and the remaining factor can be written as $k(z)\{w-\eta_1(z)\}^m$, the equation is again reducible. Hence if $\{w-\eta(z)\}^m$ is a factor, any other factor $w-\eta_1(z)$ can only occur to a degree less than m, from which it follows that the value of p corresponding to $w-\eta_1(z)$ is branched and therefore $\eta_1(z)$ is a mere constant.

Consider first of all the case in which $A(z, w)$ does not contain a factor $\{w-\eta(z)\}^m$. The equation may then be written in the form

$$p^m+K(z)\prod(w-a_i)^{\mu_i}=0,$$

where a_i is a constant, and

$$\sum \mu_i=2m,$$

* Briot and Bouquet, *Fonctions Elliptiques*, p. 388; for the simpler type $p^m=f(w)$, see Briot and Bouquet, *C. R. Acad. Sc. Paris*, 40 (1855), p. 342.

and p may be developed in a series in which the leading term is

$$c(z)(w-a_i)^{\mu_i/m}.$$

Let μ_i/m be reduced to its lowest terms and written k_i/α_i, then, by condition C (§ 13·5),

$$k_i \geqslant \alpha_i - 1$$

or

$$\frac{k_i}{\alpha_i} \geqslant 1 - \frac{1}{\alpha_i} \geqslant \tfrac{1}{2},$$

since $\alpha_i \geqslant 2$. Hence

$$\mu_i \geqslant \tfrac{1}{2}m.$$

Thus the problem of finding all possible types of binomial equations of the form

$$p^m + K(z)F(w) = 0$$

is that of finding sets of rational numbers $\frac{\mu_i}{m}$ such that

$$\frac{\mu_i}{m} \geqslant \tfrac{1}{2}, \quad \sum \frac{\mu_i}{m} = 2.$$

But since

$$\frac{\mu_i}{m} - \frac{k_i}{\alpha_i} \geqslant 1 - \frac{1}{\alpha_i},$$

any fraction $\frac{\mu_i}{m}$ which is less than unity is of the form $\frac{\alpha-1}{\alpha}$ where $\alpha \geqslant 2$. There are six cases to consider, in which the equation is of a degree higher than the first, and irreducible.

Type I.—There is one factor whose exponent μ_1 exceeds m. Let the remaining exponents (none of which can exceed m) be

$$m\left(1-\frac{1}{\alpha_1}\right), \quad m\left(1-\frac{1}{\alpha_2}\right), \quad \ldots, \quad m\left(1-\frac{1}{\alpha_r}\right),$$

then

$$m\left(1-\frac{1}{\alpha_1}\right) + \ldots + m\left(1-\frac{1}{\alpha_r}\right) < m,$$

whence

$$r-1 < \frac{1}{\alpha_1} + \frac{1}{\alpha_2} + \ldots + \frac{1}{\alpha_r}$$
$$< \tfrac{1}{2}r,$$

since $\alpha_1, \alpha_2, \ldots, \alpha_r$ are integers greater than unity. Hence $r=1$, and thus the only possibility is that of two factors whose exponents are $m+1$ and $m-1$ respectively. The equation then is

I. $$p^m + K(z)(w-a_1)^{m+1}(w-a_2)^{m-1} = 0,$$

where m is any positive integer.

Type II.—Let $\mu_1 = m$, then if the remaining exponents are as before,

$$m\left(1-\frac{1}{\alpha_1}\right) + \ldots + m\left(1-\frac{1}{\alpha_r}\right) = m,$$

whence

$$r-1 = \frac{1}{\alpha_1} + \frac{1}{\alpha_2} + \ldots + \frac{1}{\alpha_r}$$
$$\leqslant \tfrac{1}{2}r.$$

Here arise two possibilities $r=1$ and $r=2$. If $r=1$ the equation reduces to one of the first degree, viz.

$$p + K(z)(w-a_1)(w-a_2) = 0.$$

But if $r=2$,

$$1=\frac{1}{a_1}+\frac{1}{a_2},$$

whence $a_1=a_2=2$. The exponents are therefore m, $\frac{1}{2}m$, $\frac{1}{2}m$, and the equation is reducible unless $m=2$. Thus the only irreducible equation of this type is

II. $$p^2+K(z)(w-a_1)^2(w-a_2)(w-a_3)=0.$$

Types III.–VI.—All the exponents are now less than m. The only sets of numbers of the form $\dfrac{a-1}{a}$ whose sum is 2 are:

$$\tfrac{1}{2},\ \tfrac{1}{2},\ \tfrac{1}{2},\ \tfrac{1}{2};\qquad \tfrac{2}{3},\ \tfrac{2}{3},\ \tfrac{2}{3};\qquad \tfrac{3}{4},\ \tfrac{3}{4},\ \tfrac{1}{2};\qquad \tfrac{5}{6},\ \tfrac{2}{3},\ \tfrac{1}{2}.$$

These give rise respectively to the four types of equation:

III. $$p^2+K(z)(w-a_1)(w-a_2)(w-a_3)(w-a_4)=0,$$

IV. $$p^3+K(z)(w-a_1)^2(w-a_2)^2(w-a_3)^2=0,$$

V. $$p^4+K(z)(w-a_1)^3(w-a_2)^3(w-a_3)^2=0,$$

VI. $$p^6+K(z)(w-a_1)^5(w-a_2)^4(w-a_3)^3=0,$$

where, in all cases, a_1, a_2, a_3, a_4 are *distinct* constants.

Now return to the case in which the factor $\{w-\eta(z)\}^m$ occurs. The equation can be written

$$p^m+K(z)\{w-\eta(z)\}^m\prod(w-a_i)^{\mu_i}=0,$$

and as before the condition

$$\mu_i\geqslant\tfrac{1}{2}m$$

must hold. But since, in this case, $\mu_i<m$, $\sum\mu_i=m$, the only possibility is that of two exponents μ_1 and μ_2 such that

$$\mu_1=\mu_2=\tfrac{1}{2}m.$$

But the equation is now reducible unless $m=2$. The only possible equation of this type is therefore

$$p^2+K(z)\{w-\eta(z)\}^2(w-a_2)(w-a_3)=0,$$

where $a_2\neq a_3$. This is a generalisation of Type II., to which it degenerates when $\eta(z)$ becomes a constant a_1, distinct from a_2 and a_3.

These six types (including under Type II. its generalised form) exhaust all those cases in which the binomial equation

$$p^m+A(z, w)=0,$$

where $m>1$ and $A(z, w)$ is exactly of degree $2m$ in w, have solutions free from movable branch points.

Corresponding to each of the six main types of equation are equations in which w occurs to a lower degree than $2m$. Such equations are obtained by the substitution $W=(w-a_i)^{-1}$, where $w-a_i$ occurs as a factor in $A(z, w)$. It may be verified that the following list of equations so derived is exhaustive. The type given is the main type from which the new equations are derived.

Type I. $$p^m+K(z)(w-a)^{m-1}=0,$$
$$p^m+K(z)(w-a)^{m+1}=0.$$

Type II. $$p^2+K(z)(w-a_1)^2(w-a_2)=0,$$
$$p^2+K(z)(w-a_1)(w-a_2)=0.$$

Type III. $$p^2+K(z)(w-a_1)(w-a_2)(w-a_3)=0.$$

Type IV. $$p^3+K(z)(w-a_1)^2(w-a_2)^2=0.$$

Type V. $$p^4+K(z)(w-a_1)^3(w-a_2)^3=0,$$
$$p^4+K(z)(w-a_1)^3(w-a_2)^2=0.$$

Type VI. $$p^6+K(z)(w-a_1)^5(w-a_2)^4=0,$$
$$p^6+K(z)(w-a_1)^5(w-a_2)^3=0,$$
$$p^6+K(z)(w-a_1)^4(w-a_2)^3=0.$$

13·81. Integration of the six Types of Binomial Equations.—The equation of Type I. may be written in the form

$$p^m=\{A(z)\}^m(w-a_1)^{m+1}(w-a_2)^{m-1}.$$

Let

$$t^m=\frac{w-a_1}{w-a_2},$$

then the equation is transformed into

$$\frac{dt}{dz}=\frac{(a_1-a_2)}{m}A(z),$$

and therefore its general integral is

$$\left(\frac{w-a_1}{w-a_2}\right)^{\frac{1}{m}}=C+\frac{(a_1-a_2)}{m}\int A(z)dz.$$

Consider now the most general equation of Type II., which may be written as

$$p^2=\{A(z)\}^2\{w-\eta(z)\}^2(w-a_1)(w-a_2).$$

Let

$$t^2=\frac{w-a_1}{w-a_2},$$

then the equation becomes

$$\frac{dt}{dz}=\pm\tfrac{1}{2}A(z)[a_1-\eta(z)-\{a_2-\eta(z)\}t^2],$$

which is a case of the Riccati equation.

In the case where $\eta(z)$ is a mere constant, say η, let

$$a_1-\eta=b, \quad a_2-\eta=b_2,$$

the equation is then reduced to

$$\frac{dt}{dz}=\pm\tfrac{1}{2}A(z)(b_1-b_2t^2)$$

and, so far as t is concerned, its integration involves only elementary quadratures.

The integration of the four remaining types involves the introduction of elliptic functions. In these types there is no loss of generality in replacing $K(z)$ by -1, for since each equation is of the form

$$\left(\frac{dw}{dz}\right)^m=-K(z)\prod(w-a_i),$$

the substitution

$$(dz')^m=-K(z)(dz)^m$$

leaves terms in w unchanged. Nor is there any loss in considering, not the main equation of any type, but any equivalent equation.

Thus the equation

$$\left(\frac{dw}{dz}\right)^2=(w-a_1)(w-a_2)(w-a_3)$$

can be taken to illustrate Type III.

Type IV. may be represented by

$$\left(\frac{dw}{dz}\right)^3=(w-a_1)^2(w-a_2)^2.$$

Let

$$(w-a_1)(w-a_2)=t^3, \quad \frac{dw}{dz}=t^2,$$

then, by differentiation,

$$3t^2\frac{dt}{dz}=(2w-a_1-a_2)t^2$$

or

$$\frac{dt}{dz}=\frac{2}{3}\sqrt{\left\{t^3+\left(\frac{a_1-a_2}{2}\right)^2\right\}}.$$

Type V. may be represented by

$$\left(\frac{dw}{dz}\right)^4=(w-a_1)^3(w-a_2)^2.$$

Let $w=t^2+a_1$, then this equation reduces to

$$\frac{dt}{dz}=\tfrac{1}{2}\sqrt{\{t(t^2+a_1-a_2)\}}.$$

Type VI. may be represented by

$$\left(\frac{dw}{dz}\right)^6=(w-a_1)^4(w-a_2)^3.$$

Let $w=t^3+a_1$, then this equation reduces to

$$\frac{dt}{dz}=\tfrac{1}{3}\sqrt{(t^3+a_1-a_2)}.$$

Thus in every case the equation is reducible to one of the form

$$\frac{dt}{dz}=\sqrt{P_3(t)}$$

where $P_3(t)$ represents a cubic function of t. This differential equation is integrable only by means of elliptic functions.*

By a linear transformation, this equation may be brought into the form

$$\frac{dt}{dz}=\sqrt{\{4t^3-g_2t-g_3\}},$$

whence $t=\wp(z+\alpha, g_2, g_3)$, where α is an arbitrary constant.

* Whittaker and Watson, *Modern Analysis*, § 20·22.

CHAPTER XIV

NON-LINEAR EQUATIONS OF HIGHER ORDER

14·1. Statement of the Problem.—The study of the uniform functions defined by differential equations of the first order, certain aspects of which were treated in the preceding two chapters, may be regarded as fairly complete, and does not present any very serious analytical difficulties. The comparative simplicity of this investigation is accounted for, at least in the case of equations which involve p and w rationally, by the absence of movable essential singularities. In the case of equations of the second and higher orders, even of a very simple form, movable essential singularities may arise, and add greatly to the difficulty of the problem.

To take a simple case, the equation

$$\frac{d^2w}{dz^2}=\left(\frac{dw}{dz}\right)^2 \cdot \frac{2w-1}{w^2+1}$$

has the general solution

$$w=\tan\{\log(Az-B)\},$$

where A and B are arbitrary constants.

As z tends to B/A, either in an arbitrary manner, or along any special path, w tends to no limit, finite or infinite. In fact an infinite number of distinct branches of the function spring from the point B/A, which is both a branch point and an essential singularity. As this point depends upon the constants of integration, it is a movable singular point.

The problem thus arises of determining whether or not equations of the form

$$\text{(A)} \qquad \frac{d^2w}{dz^2}=F(z, w, p)$$

exist, where F is rational in p, algebraic in w, and analytic in z, which have all their critical points (that is their branch points and essential singularities) fixed.*

An obvious extension is to the more general equation of the second order

$$F(z, w, p, q)=0 \qquad \left(q\equiv\frac{d^2w}{dz^2}\right),$$

but this generalisation is not at present of any great interest.

* Picard, *C. R. Acad. Sc. Paris*, 104 (1887), p. 41 ; 110 (1890), p. 877 ; *J. de Math.* (4), 5 (1889), p. 263 ; *Acta Math.* 17 (1893), p. 297 ; Painlevé, *C. R.* 116 (1893), pp. 88, 173, 362, 566 ; 117 (1893), pp. 211, 611, 686 ; 126 (1898), pp. 1185, 1329, 1697 ; 127 (1898), pp. 541, 945 ; 129 (1899), pp. 750, 949 ; 133 (1901), p. 910 ; *Bull. Soc. Math. France* 28 (1900), p. 201 ; *Acta Math.* 25 (1902), p. 1 ; Gambier, *C. R.* 142 (1906), pp. 266, 1403, 1497 ; 143 (1906), p. 741 ; 144 (1907), pp. 827, 962 ; *Acta Math.* 33 (1910), p. 1.

14·11. The General Solution as a Function of the Constants of Integration.—In the case of equations of the second order, it is important to distinguish between the various modes in which the constants of integration may enter into the solution. The fundamental existence theorems show that when the critical points are fixed, and z is a non-critical point, the solution is completely and uniquely specified by the knowledge of the values w_0 and w_0' which the dependent variable w and its first derivative w' assume at the point z_0. The solution may thus be regarded as a function of w_0 and w_0' whose coefficients are functions of $z-z_0$.

Three cases may arise, as follows :

(i) The solution may be an algebraic, or in particular a rational, function of w_0 and w_0' or of an equivalent pair of constants of integration ; thus, for instance, the equation

$$w''+3ww'+w^3=q(z)$$

has the general solution $w=u'/u$, where u is a general solution of the linear equation of the third order

$$u'''=q(z)u.$$

Since u is of the form

$$u=A_1u_1+A_2u_2+A_3u_3,$$

where u_1, u_2, u_3 form a fundamental system for the linear equation, and A_1, A_2, A_3 are arbitrary constants, the general solution of the equation in w is

$$w=\frac{u_1'+Bu_2'+Cu_3'}{u_1+Bu_2+Cu_3},$$

and is a rational function of the two constants of integration, B and C.

(ii) The general solution is not an algebraic function of two constants of integration, but nevertheless the equation admits of a first integral which involves a constant of integration algebraically. In this case the general solution is said to be a *semi-transcendental* function of the constants of integration. Thus the first integral of

$$w''+2ww'=q(z)$$

is

$$w'+w^2=\int q(z)dz+A,$$

and depends linearly upon the constant A. The general solution is therefore a semi-transcendental function of A and the second constant of integration.

(iii) Neither (i) nor (ii) is true. The general solution is in this case said to be an *essentially-transcendental* function of the two constants of integration.

Only those equations which come into the last category can be regarded as sources of new transcendental functions, that is to say of functions distinct from the transcendental functions defined by equations of the first order with algebraic coefficients.

14·12. Outline of the Method of Procedure.—The equation (A) may be replaced by the system

$$\begin{cases} \dfrac{dw}{dz}=p, \\ \dfrac{dp}{dz}=F(z, w, p), \end{cases}$$

or, more generally, by a system of the form

$$\text{(B)} \qquad \begin{cases} \dfrac{dw}{dz}=H(z, w, u), \\ \dfrac{du}{dz}=K(z, w, u). \end{cases}$$

It is convenient to suppose, for the moment, that H and K are functions of a parameter α, which are analytic in α throughout a domain D of which $\alpha=0$ is an interior point. The following lemma will be found to be of importance in all that follows. *If the general solution of the differential system is uniform in z for all values of α in D except (possibly) $\alpha=0$, then it will be uniform also for $\alpha=0$.*

For let $w(z, \alpha)$, $u(z, \alpha)$ be that pair of solutions of the system which corresponds to the initial conditions

$$z=z_0, \quad w=w_0, \quad u=u_0.$$

Let C be a closed contour in the z-plane, beginning and ending at z_0, on which $w_0(z)$ and $u_0(z)$ are analytic, where

$$w_0(z)=w(z, 0), \quad u_0(z)=u(z, 0).$$

Then if the functions $w(z, \alpha)$ and $u(z, \alpha)$ are developed as series of ascending powers of the parameter α, thus

$$w(z, \alpha)=w_0(z)+\alpha w_1(z)+\alpha^2 w_2(z)+ \ldots,$$
$$u(z, \alpha)=u_0(z)+\alpha u_1(z)+\alpha^2 u_2(z)+ \ldots,$$

these series will converge for values of z on C and for sufficiently small values of $|\alpha|$. Let $w_\nu(z)$ increase by k_ν as z describes the circuit C, then

$$\sum_{\nu=0}^{\infty} k_\nu \alpha^\nu=0$$

for $0<|\alpha|\leqslant\alpha_0$, and consequently

$$k_\nu=0$$

for all ν. It follows that $w_0(z)$, $w_1(z)$, ..., and in a similar manner $u_0(z)$, $u_1(z)$, ... are uniform.*

The method to be adopted breaks up into two distinct stages. First of all a set of necessary conditions for the absence of movable critical points is obtained. Then a comprehensive set of equations which satisfy these necessary conditions is derived, and it is shown, by direct integration or otherwise, that the general solutions of these equations are free from movable critical points, thus proving the sufficiency of the conditions. In order to obtain the set of necessary conditions, a parameter α is introduced into the system (B) in such a way that the new system has the same fixed critical points as (B) and, in addition, is integrable when $\alpha=0$. The functions $w_0(z)$, $w_1(z)$, ..., $u_0(z)$, $u_1(z)$, ... are determined by quadratures; the conditions that their critical points are all fixed are necessary conditions for the absence of movable critical points in the given system.

Let

$$u=g(z_0, w_0)$$

be a pole of one or both of the functions $H(z_0, w_0, u)$, $K(z_0, w_0, u)$. There is no loss in generality in supposing $g(z, w)$ to be identically zero, since $u-g(z, w)$ could be replaced by a new variable U. This being the case, the system (B) may be written as

$$u^m \frac{dw}{dz}=H_0(z, w)+uH_1(z, w)+ \ldots,$$

$$u^n \frac{du}{dz}=K_0(z, w)+uK_1(z, w)+ \ldots,$$

where one at least of the numbers m, n is greater than zero.

* It would be sufficient to know that $w(z, \alpha)$, $u(z, \alpha)$ are uniform in z for an infinite sequence $\alpha_1, \alpha_2, \ldots$ of values of α, having $\alpha=0$ as a limit point.

First suppose that $m<n+1$, where n is greater than zero, and introduce the parameter α by writing

$$z=z_0+\alpha^{n+1}Z, \quad w=w_0+\alpha^{n-m+1}W, \quad u=\alpha U,$$

then the system becomes

$$\begin{cases} U^m \dfrac{dW}{dZ}=H_0(z_0, w_0)+O(\alpha), \\ U^n \dfrac{dU}{dZ}=K_0(z_0, w_0)+O(\alpha). \end{cases}$$

Except possibly when $\alpha=0$, this new system has fixed critical points when those of (B) are fixed. When $\alpha=0$, it becomes

$$U^m \frac{dW}{dZ}=H_0(z_0, w_0), \quad U^n \frac{dU}{dZ}=K_0(z_0, w_0),$$

and this system has a solution of the form

$$U=\{(n+1)K_0Z+A\}^{\frac{1}{n+1}}.$$

where A is an arbitrary constant. The solution of the system when $\alpha=0$ has therefore a movable branch point, and in consequence of the lemma cannot have fixed critical points when $\alpha \neq 0$. The system (B) therefore has solutions which have movable critical points when $m<n+1$.

Suppose, on the other hand, that $m \geqslant n+1$. It is sufficient to suppose that $m=n+1$, allowing at the same time the hypothesis that $K_0(z, w)$ and possibly others of the functions $K(z, w)$ may be identically zero. Write

$$z=z_0+\alpha^m Z, \quad u=\alpha U,$$

then the system becomes

$$\begin{cases} U^m \dfrac{dw}{dZ}=H_0(z_0, w)+O(\alpha), \\ U^{m-1} \dfrac{dU}{dZ}=K_0(z_0, w)+O(\alpha)\,; \end{cases}$$

if

$$H_0(z_0, w)=\eta(w), \quad K_0(z_0, w)=\kappa(w),$$

the system reduces, when $\alpha=0$, to

$$\begin{cases} U^m \dfrac{dw}{dZ}=\eta(w), \\ U^{m-1} \dfrac{dU}{dZ}=\kappa(w). \end{cases}$$

This new system may be integrated by quadratures; in order that the original system may have no movable critical points, it is necessary that the branch points of the solutions of this reduced system should be fixed.

This condition, when applied to all the poles of $H(z, w, u)$ and $K(z, w, u)$ of the form $u=g(z, w)$ or $w=h(z)$, and to the values $u=\infty$ and $w=\infty$, gives a set of conditions which are necessary for the non-appearance of movable branch points in the general solution. The same process must also be applied to any values $u=g(z)$, $w=h(z)$ which render H or K indeterminate, as well as to the singular points of H and K, should any occur.

14·2. Application of the Method.—Consider the equation

$$\text{(C)} \qquad \frac{d^2w}{dz^2}=R(z, w, p),$$

where R is a rational function of w and p, with coefficients analytic in z. It is to be supposed that R is irreducible and therefore expressible in the form

$$R(z, w, p)=\frac{P(z, w, p)}{Q(z, w, p)},$$

where P and Q are polynomials in p with no common factor.

The equation is equivalent to the system

$$\begin{cases} \dfrac{dw}{dz}=p, \\ \dfrac{dp}{dz}=R(z, w, p). \end{cases}$$

In this case m is zero; if R were to have a pole $p=g(z, w)$, the condition $m \geqslant n+1$ could not be satisfied. Hence if no movable critical point is to appear, R can have no such pole, and must be a polynomial in p; let it be of degree q.

But (C) is also equivalent to the system

$$\begin{cases} \dfrac{dw}{dz}=\dfrac{1}{u}, \\ u^{q-2}\dfrac{du}{dz}=-u^q R(z, w, u^{-1})=R_0(z, w)+uR_1(z, w)+\ldots \end{cases}$$

In this case $m=1$, $n=q-2$, and thus the inequality $m \geqslant n+1$ leads to the condition that q is at most 2. Consequently, if the general solution of (C) has no movable critical points, it is necessary that it should be of the form

$$\text{(D)} \qquad \frac{d^2w}{dz^2}=L(z, w)p^2+M(z, w)p+N(z, w),$$

in which L, M and N are rational functions of w, with coefficients analytic in z.

Now let

$$z=z_0+\alpha Z, \quad w=\alpha W,$$

then (D) is equivalent to a system which reduces, when $\alpha=0$, to

$$\begin{cases} \dfrac{dW}{dZ}=\dfrac{1}{u}, \\ \dfrac{du}{dZ}=-L(z_0, W), \end{cases}$$

and this system is equivalent, in turn, to the equation

$$\frac{d^2W}{dZ^2}=\left(\frac{dW}{dZ}\right)^2 L(z_0, W).$$

It is thus necessary, in the first place, to determine explicitly those equations of the form

$$\text{(E)} \qquad \frac{d^2w}{dz^2}=p^2 l(w),$$

whose solutions have only fixed branch points.

14·21. The First Necessary Condition for the Absence of Movable Critical Points.—The first step is to show that the function $l(w)$ has only simple poles; let $w=w_1$ be a pole of order r. Since this pole may be made to coincide with the origin by a translation, which does not alter the form of the equation, w_1 may be taken to be zero. The equation is then equivalent to the system

$$\begin{cases} \dfrac{dw}{dz}=p, \\ \dfrac{dp}{dz}=p^2w^{-r}\{k+O(w)\}, \end{cases}$$

where k is a constant. Write

$$w=aW, \quad p=a^rP,$$

then the system becomes

$$\begin{cases} \dfrac{dW}{dz}=a^{r-1}P, \\ \dfrac{dP}{dz}=\dfrac{kP^2}{W^r}+O(a). \end{cases}$$

When arranged in ascending powers of a, the solution of this system is

$$W=W_0-\frac{a^{r-1}}{k}W_0{}^r\log\left(\frac{z+C}{z_0+C}\right)+O(a^r),$$

$$P=-\frac{W_0{}^r}{k(z+C)}+O(a),$$

where

$$C=-z_0-\frac{W_0{}^r}{P_0k},$$

and $r>1$. Thus when $r>1$ the critical points are certainly not fixed.

When $r=1$, the system becomes

$$\begin{cases} \dfrac{dW}{dz}=P, \\ \dfrac{dP}{dz}=\dfrac{kP^2}{W}+O(a); \end{cases}$$

when $a=0$, this system has the solution

$$W=(Az+B)^{\frac{1}{1-k}} \qquad \text{when } k\neq 1,$$

or

$$W=e^{Az+B} \qquad \text{when } k=1,$$

and this solution has a movable branch point except when

$$k=1+\frac{1}{n} \quad \text{or} \quad k=1,$$

where n is an integer, positive or negative.

Now the equation

$$\frac{d^2w}{dz^2}=p^2l(w)$$

may be integrated once; the first integral is

$$p=Ce^{\int l(w)dw}.$$

At any pole, $w=w_1$, $l(w)$ is to have a principal part of the form

$$\frac{1+\dfrac{1}{n_1}}{w-w_1},$$

and therefore contributes a factor $(w-w_1)^{1+\frac{1}{n_1}}$ to the expression $e^{\int l(w)dw}$. Let ν be the least common denominator of the exponents of all such factors, then if

$$\text{(F)} \qquad p^\nu=\phi(w),$$

$\phi(w)$ will be a function having no singularities other than poles in the finite part of the plane. The transformation $w=W^{-1}$ shows that it has, at most, a pole at infinity and is therefore a rational function. Thus the problem is made to depend upon the question of determining those equations of type (F) whose solutions are free from movable branch points; this question was disposed of in § 13·8. Now (F), on differentiation, becomes identified with (E) if

$$l(w)=\frac{\phi'(w)}{\nu\phi(w)},$$

and thus the knowledge of the types of equation (F) which have no movable branch points leads to the conclusion that $l(w)$ must be either identically zero, or of one of the following types:

$$\text{Type I.}\quad \nu=m,\quad l(w)=\frac{m+1}{m(w-a_1)}+\frac{m-1}{m(w-a_2)}\qquad (m\geqslant 1),$$

$$\text{,, III.}\quad \nu=2,\quad l(w)=\frac{\frac{1}{2}}{w-a_1}+\frac{\frac{1}{2}}{w-a_2}+\frac{\frac{1}{2}}{w-a_3}+\frac{\frac{1}{2}}{w-a_4},$$

$$\text{,, IV.}\quad \nu=3,\quad l(w)=\frac{\frac{2}{3}}{w-a_1}+\frac{\frac{2}{3}}{w-a_2}+\frac{\frac{2}{3}}{w-a_3},$$

$$\text{,, V.}\quad \nu=4,\quad l(w)=\frac{\frac{3}{4}}{w-a_1}+\frac{\frac{3}{4}}{w-a_2}+\frac{\frac{1}{2}}{w-a_3},$$

$$\text{,, VI.}\quad \nu=6,\quad l(w)=\frac{\frac{5}{6}}{w-a_1}+\frac{\frac{2}{3}}{w-a_2}+\frac{\frac{1}{2}}{w-a_3}.$$

The constants a_1, a_2, a_3, a_4 may have any values, any one of which may be infinite, and are not necessarily all unequal. Type II. is omitted, as for the present purpose it may be regarded as a degenerate case of Type III.

The manner in which $l(w)$ arose from $L(z, w)$ leads to the conclusion that *a necessary condition that solutions of the equation*

$$\frac{d^2w}{dz^2}=L(z, w)p^2+M(z, w)p+N(z, w)$$

may be free from movable critical points is that **L** (z, w) *should be identically zero or else belong to one or other of the five main types enumerated above, where* a_1, a_2, a_3, a_4 *are now to be regarded as functions of* z.

14·22. The Second Necessary Condition for the Absence of Movable Critical Points.—The next step taken is to show that the poles of $M(z, w)$ and $N(z, w)$, regarded as functions of w, are simple, and are included among the poles of $L(z, w)$. Let $w=h(z)$ be a pole of order j of $M(z, w)$ and a pole of order k of $N(z, w)$. Since the substitution $W=w-h(z)$, while not essentially altering the form of the equation, changes the pole in question into $W=0$, it may be assumed that $h(z)$ is identically zero. The equation may then be written in the expanded form

$$\frac{d^2w}{dz^2}=\frac{p^2}{w}\left\{\left(1-\frac{1}{n}\right)+O(w)\right\}+\frac{p}{w^j}\{M(z)+O(w)\}+\frac{k}{w^k}\{N(z)+O(w)\},$$

where, if $w=0$ is a pole of $L(z, w)$, n is a positive or negative integer distinct from 0 or 1; if $w=0$ is not a pole of $L(z, w)$, then $n=1$.

Make the following transformation:

$$w=aW,\quad z=z_0+a^jZ,\qquad \text{if}\quad k\leqslant 2j-1,$$

or

$$w=aW,\quad z=z_0+a^{\frac{1}{2}(k+1)}Z,\qquad \text{if}\quad k\geqslant 2j-1,$$

and write

$$P=\frac{dW}{dZ},\quad M_0=M(z_0),\quad N_0=N(z_0).$$

Then

$$\frac{d^2W}{dZ^2}=\frac{P^2}{W}\left(1-\frac{1}{n}\right)+\frac{M_0P}{W^j}+O(\alpha), \qquad \text{if } k<2j-1,$$

$$\frac{d^2W}{dZ^2}=\frac{P^2}{W}\left(1-\frac{1}{n}\right)+\frac{N_0}{W^k}+O(\alpha), \qquad \text{if } k>2j-1,$$

$$\frac{d^2W}{dZ^2}=\frac{P^2}{W}\left(1-\frac{1}{n}\right)+\frac{M_0P}{W^j}+\frac{N_0}{W^k}+O(\alpha), \quad \text{if } k=2j-1.$$

The third of these equations effectively contains the other two when $\alpha=0$, on the supposition that M_0 or N_0 may be zero. Now the equation

$$\frac{d^2W}{dZ^2}=\frac{P^2}{W}\left(1-\frac{1}{n}\right)+\frac{M_0P}{W^j}+\frac{N_0}{W^{2j-1}},$$

in which, when $M_0=0$, $2j-1$ is to be regarded as a symbol standing for the positive integer k, and therefore $2j$ is an integer not less than 2, may be replaced by the system

$$\begin{cases} \dfrac{dW}{du}=-\dfrac{W}{u\left(j-\dfrac{1}{n}+M_0u+N_0u^2\right)}, \\[2ex] \dfrac{dZ}{dW}=uW^{j-1}. \end{cases}$$

Now assume, in the first place, that $j>1$, then since n is an integer, $j\neq\frac{1}{n}$. Moreover, M_0 and N_0 are not both zero, from which it follows that the equation

$$N_0u^2+M_0u+j-\frac{1}{n}=0$$

has at least one non-zero root, say $u=u_1$, a constant. Then $u=u_1$ is a particular solution of the first equation of the above system. But the second equation of the system, which becomes

$$\frac{dZ}{dW}=u_1W^{j-1},$$

has then (since $j>1$) a solution with a movable branch point, and consequently the general solution is not free from movable branch points. Thus the possibility $j>1$, and similarly the possibility $k>1$, must be ruled out. Thus if $M(z, w)$ and $N(z, w)$ have poles $w=h(z)$, these poles are simple.

Now assume $j=k=n=1$, that is to say, suppose that $W=0$ is a simple pole of $M(z, w)$, or of $N(z, w)$, or of both, but not a pole of $L(z, w)$. Then the reduced equation is

$$\frac{d^2W}{dZ^2}=\frac{M_0P+N_0}{W},$$

and is equivalent to the system

$$\begin{cases} \dfrac{dW}{dZ}=\dfrac{1}{u}, \\[2ex] \dfrac{du}{dZ}=-\dfrac{u(M_0+N_0u)}{W}. \end{cases}$$

This system in turn becomes, on replacing u by αu, Z by αZ,

$$\begin{cases} \dfrac{dW}{dZ}=\dfrac{1}{u}, \\[2ex] \dfrac{du}{dZ}=-\dfrac{\alpha u(M_0+\alpha N_0u)}{W}; \end{cases}$$

when solved for W and u in series of ascending powers of a, with coefficients which are functions of Z, it has the solution

$$W=W_0+\frac{Z-Z_0}{u_0}+O(a),$$

$$u=u_0-aM_0u_0^2\log\left(W_0+\frac{Z-Z_0}{u_0}\right)+O(a^2),\quad \text{if } M_0\neq 0,$$

or

$$u=u_0-a^2N_0u_0^3\log\left(W_0+\frac{Z-Z_0}{u_0}\right)+O(a^3),\quad \text{if } M_0=0.$$

Here the solution has a movable critical point, so that the only possibility which remains is that expressed by

$$n\neq 1,\quad j=k=1,$$

that is to say $w=h(z)$ can only be a pole of $M(z, w)$ or of $N(z, w)$ if it is also a pole of $L(z, w)$. Thus the poles of $M(z, w)$ and $N(z, w)$ are simple, and are included among the poles of $L(z, w)$.

Now, on referring back to the Types I.–VI., it will be seen that $L(z, w)$ can be written in the form $\dfrac{\lambda(z, w)}{D(z, w)}$, where $D(z, w)$ is at most of the fourth degree in w, and $\lambda(z, w)$ is at least one degree lower than D. M and N can therefore be expressed in the forms

$$M(z, w)=\frac{\mu(z, w)}{D(z, w)},\quad N(z, w)=\frac{\nu(z, w)}{D(z, w)},$$

where μ and ν are polynomials in w whose maximum degree is to be determined. Let $D(z, w)$ be of degree δ in w.

In equation (D) write $w=W^{-1}$, then that equation becomes

$$\frac{d^2W}{dz^2}=\{2W-L(z, W^{-1})\}W^{-2}\left(\frac{dW}{dz}\right)^2+M(z, W^{-1})\frac{dW}{dz}-W^2N(z, W^{-1}).$$

If the numbers a which occur in the expressions for $L(z, w)$ in Types I.–VI. are all finite, then $\{2W-L(z, W^{-1})\}W^{-2}$ will be finite (or zero) at $W=0$. Consequently $W=0$ cannot be a pole of $M(z, W^{-1})$ or of $W^2N(z, W^{-1})$, from which it follows that the degree of $\mu(z, w)$ in w is at most δ, and that of $\nu(z, w)$ at most $\delta+2$. If, on the other hand, $L(z, w)$ is either identically zero or of a degenerate type, in which one of the numbers a_1, a_2, a_3, a_4 has been made infinite, then $W=0$ is a simple pole of $\{2W-L(z, W^{-1})\}W^{-2}$, and therefore may be a simple pole of $M(z, W^{-1})$ and of $W^2N(z, W^{-1})$. In this case $\mu(z, w)$ and $\nu(z, w)$ are of degrees not exceeding $\delta+1$ and $\delta+3$ respectively.

Thus, in general terms, *the second necessary condition for the absence of movable critical points is that if $D(z, w)$ is the least common denominator of the partial fractions in $L(z, w)$ and is of degree δ in w, then $M(z, w)$ and $N(z, w)$ are respectively expressible in the forms*

$$\frac{\mu(z, w)}{D(z, w)},\quad \frac{\nu(z, w)}{D(z, w)},$$

where μ and ν are polynomials in w of degrees not exceeding $\delta+1$ and $\delta+3$.

14·3. Reduction to Standard Form.—It has been seen that if the solutions of an equation of the second order have no movable critical points, the equation is necessarily of the form

$$\text{(D)}\qquad \frac{d^2w}{dz^2}=L(z, w)\left(\frac{dw}{dz}\right)^2+M(z, w)\frac{dw}{dz}+N(z, w),$$

where $L(z, w)$ is either identically zero, or of one of the five main types enumerated in § 14·21. To simplify the form of $L(z, w)$ make one or other of the following transformations:

(i) If $L(z, w)$ has only one pole, $w=a_1$, write $W=\dfrac{1}{w-a_1}$,

(ii) If $L(z, w)$ has two poles, $w=a_1, a_2$, write $W=\dfrac{w-a_2}{w-a_1}$.

(iii) If $L(z, w)$ has three poles, $w=a_1, a_2, a_3$, or four poles, $w=a_1, a_2, a_3, a_4$, write *

$$W=\frac{a_1-a_3}{a_2-a_3}\cdot\frac{w-a_2}{w-a_1}.$$

The equation is then transformed into one of the type

$$\frac{d^2W}{dz^2}=A(z, W)\left(\frac{dW}{dz}\right)^2+B(z, w)\frac{dW}{dz}+C(z, W),$$

in which $A(z, W)$ has one of the following eight distinct forms:

i. 0, ii. $\dfrac{1}{W}$,

iii. $\dfrac{m-1}{mW}$ (m an integer greater than unity),

iv. $\dfrac{1}{2W}+\dfrac{1}{W-1}$, v. $\dfrac{2}{3}\left\{\dfrac{1}{W}+\dfrac{1}{W-1}\right\}$,

vi. $\dfrac{3}{4}\left\{\dfrac{1}{W}+\dfrac{1}{W-1}\right\}$, vii. $\dfrac{2}{3W}+\dfrac{1}{2(W-1)}$,

viii. $\dfrac{1}{2}\left\{\dfrac{1}{W}+\dfrac{1}{W-1}+\dfrac{1}{W-\eta}\right\}$.

Of these forms iii. arises from Type I.; ii., iv. and viii. from Type III.; v. from Type IV.; vi. from Type V.; and vii. from Type VI.

In viii.

$$\eta=\frac{a_1-a_3}{a_2-a_3}\cdot\frac{a_2-a_4}{a_1-a_4}.$$

This quantity may be a constant, or it may depend upon z. In the latter case it is taken as a new independent variable Z.

14·31. Case i.—In Case i. $L(z, w)$ is identically zero; the second set of necessary conditions (§ 14·22) shows that the equation is of the form

$$\text{(G)}\qquad \frac{d^2w}{dz^2}=\{A(z)w+B(z)\}\frac{dw}{dz}+C(z)w^3+D(z)w^2+E(z)w+F(z).$$

But nothing has been found which would immediately settle the question as to whether solutions of this equation are, or are not, free from movable critical points. In fact, the conditions which have been found are necessary, but by no means sufficient. The investigation has thus to be continued to a further stage, though without any essential alteration in the method.

Let

$$w=\frac{W}{a},\quad z=z_0+aZ,$$

then the equation becomes

$$\frac{d^2W}{dZ^2}=A(z_0)W\frac{dW}{dZ}+C(z_0)W^3+O(a),$$

* In the case of Type V. it is more convenient to write

$$W=\frac{a_2-a_3}{a_2-a_1}\cdot\frac{w-a_1}{w-a_3}.$$

and, when $\alpha=0$, this equation is equivalent to the system

$$\begin{cases} \dfrac{dW}{dZ}=\dfrac{W^2}{u}, \\ \dfrac{du}{dZ}=(2-a_0u-c_0u^2)W, \end{cases}$$

where $a_0=A(z_0)$, $c_0=C(z_0)$. When $c_0=0$, the reduced equation

$$\frac{d^2W}{dZ^2}=a_0W\frac{dW}{dZ}$$

has the first integral

$$\frac{dW}{dZ}=\tfrac{1}{2}a_0W^2+\gamma,$$

where γ is the constant of integration; its general solution is uniform. Let $c_0\neq0$, then if W is replaced by $W\{-c(z)\}^{-\frac{1}{2}}$, $c(z)$ is replaced by -1; it may therefore be supposed, with no loss of generality, that $c_0=-1$. The system may now be written

$$\begin{cases} \dfrac{dW}{dZ}=\dfrac{W^2}{u}, \\ \dfrac{du}{dZ}=(u-h)(u-k)W, \end{cases}$$

where

$$h+k=a_0, \quad hk=2.$$

Write

$$u=h+\alpha v,$$

then if $h=k$ the system becomes

$$\begin{cases} \dfrac{dW}{dZ}=\dfrac{W^2}{h+\alpha v}, \\ \dfrac{dv}{dZ}=\alpha v^2W, \end{cases}$$

and this system has the solution

$$W=-\frac{h}{Z-c_1}+O(\alpha),$$
$$v=c_2-\alpha c_2{}^2h\log(Z-c_1)+O(\alpha),$$

where c_1 and c_2 are constants of integration. But this solution has a movable critical point; the supposition $h=k$ must be rejected.

On the other hand, if $h\neq k$, the system becomes

$$\begin{cases} \dfrac{dW}{dZ}=\dfrac{W^2}{h+\alpha v}, \\ \dfrac{dv}{dZ}=(h-k+\alpha v)vW, \end{cases}$$

and now the solution is

$$W=-\frac{h}{Z-c_1}+O(\alpha),$$
$$v=c_2(Z-c_1)^{2-h^2}+O(\alpha).$$

The movable singular point $Z=c_1$ will be a branch point unless $2-h^2$ is an integer n, positive or negative (but not zero since $h^2=2$ implies the rejected possibility $h=k$). Thus

$$h^2=2-n,$$

and similarly

$$k^2=2-n',$$

and therefore

$$(2-n)(2-n')=h^2k^2$$
$$=4.$$

The only three distinct possibilities are

$$\text{i. } 2-n=1, \quad 2-n'=4,$$
$$\text{ii. } 2-n=-1, \quad 2-n'=-4,$$
$$\text{iii. } 2-n=-2, \quad 2-n'=-2,$$

and these correspond respectively to

$$\text{i. } h=\pm 1, \quad k=\pm 2, \quad a_0=\pm 3,$$
$$\text{ii. } h=\pm i, \quad k=\mp 2i, \quad a_0=\mp i,$$
$$\text{iii. } h=\pm i\sqrt{2}, \quad k=\mp i\sqrt{2}, \quad a_0=0\,;$$

in each case either all the upper signs, or all the lower signs are to be taken.

The case $a_0=+3$ is deducible from the case $a_0=-3$ by reversing the sign of w_1 and is therefore not distinct from the latter case; in ii. the transformation $w=\pm iw_1$ results in changing $C(z)$ from -1 to $+1$ and in changing a_0 into -1. Now since z_0 is arbitrary, any relation such as

$$a_0=A(z_0)=-3$$

holds for all values of z_0, and thus $A(z)$ is constant.

When $A=0$, $C\neq 0$, if W is replaced by $W\sqrt{(2/C)}$, C is replaced by 2; if $A\neq 0$, $C=0$, and W is replaced by $-2W/A$, A is replaced by -2.

To sum up, if in Case i. the general solution of the equation is free from movable critical points, it is necessary that the equation should be reducible. by a substitution of the form

$$w=\lambda(z)W,$$

to an equation in which $A(z)$ and $C(z)$ have the pairs of constant values given in the table:

(a) $A=0, \quad C=0.$ (b) $A=-2, C=0.$
(c) $A=-3, C=-1.$ (d) $A=-1, C=1.$
(e) $A=0, \quad C=2.$

The more general transformation

$$w=\lambda(z)W+\mu(z), \quad Z=\phi(z)$$

does not alter the main features of equation (G), which becomes

$$\frac{d^2W}{dZ^2}=\frac{dW}{dZ}\cdot\frac{1}{\phi'}\left\{A\lambda W+A\mu+B-\frac{2\lambda'}{\lambda}-\frac{\phi''}{\phi'}\right\}+\frac{C\lambda^2W^3}{\phi'^2}$$
$$+\{A\lambda'+3C\lambda\mu+D\lambda\}\frac{W^2}{\phi'^2}+\left\{A\frac{\lambda'\mu}{\lambda}+\frac{B\lambda'}{\lambda}+A\mu'+3C\mu^2+2D\mu+E-\frac{\lambda''}{\lambda}\right\}\frac{W}{\phi'^2}$$
$$+\{A\mu\mu'+B\mu'+C\mu^3+D\mu^2+E\mu+F-\mu''\}\frac{1}{\lambda\phi'^2},$$

where dashes denote differentiation with respect to z. Particular forms of this transformation will be of use in the sections which follow.

14·311. Sub-case i(a).—When $A=C=0$, let λ, μ and ν be chosen to satisfy the relations

$$\frac{2\lambda'}{\lambda}+\frac{\phi''}{\phi'}=B, \quad D\lambda=6\phi'^2,$$
$$2D\mu=\frac{\lambda''}{\lambda}-\frac{B\lambda'}{\lambda}-E.$$

When D is identically zero, the equation is linear; this simple case will be put aside and D supposed not to be identically zero. Then λ, ϕ and μ are determined by quadratures, and the transformation

$$w=\lambda(z)W+\mu(z), \quad Z=\phi(z)$$

brings the equation into the form

$$\frac{d^2W}{dZ^2}=6W^2+S(z),$$

where $S(z)$ is expressible in terms of B, D, E and F.

To determine whether the solutions of this equation are free from movable critical points or not, let

$$W=a^{-2}V, \quad Z=a+au,$$

where a is an arbitrary constant. Then

$$\frac{d^2V}{du^2}=6V^2+a^4S(a)+a^5uS'(a)+\tfrac{1}{2}a^6u^2S''(a)+O(a^7).$$

This equation in V has a solution which may be developed in ascending powers of a, thus

$$V=v+a^4v_0+a^5v_1+a^6v_2+\ldots,$$

where

$$v''=6v^2,$$

$$v_r''-12vv_r=\frac{u^r}{r!}S^{(r)}(a) \qquad (r=0, 1, 2, 3).$$

When $r\geqslant 4$ the recurrence equation is more complicated; fortunately it will be sufficient to proceed only as far as $r=2$.

The first integral of

$$v''=6v^2$$

is

$$v'^2=4v^3-h,$$

where h is the constant of integration. The general solution is therefore

$$v=\wp(u-k, 0, h),$$

where k is the second constant of integration.

Now consider the homogeneous equation

$$v_r''-12\wp(u-k, 0, h)v_r=0;$$

its general solution is

$$v_r=C_1\{u\wp'+2\wp\}+C_2\wp',$$

where C_1 and C_2 are constants of integration. The non-homogeneous equation

$$v_r''-12\wp(u-k, 0, h)v_r=\frac{u^r}{r!}S^{(r)}(a) \qquad (r=0, 1, 2, 3)$$

can now be integrated by the method of Variation of Parameters; its general solution is

$$v_r=U_1(u)\{u\wp'+2\wp\}+U_2(u)\wp',$$

where

$$U_1'(u)=\frac{1}{24}\cdot\frac{S^{(r)}(a)}{r!}u^r\wp'(u-k),$$

$$U_2'(u)=\frac{1}{24}\cdot\frac{S^{(r)}(a)}{r!}u^r\{u\wp'(u-k)+2\wp(u-k)\}.$$

Now

$$\wp(u-k)=\frac{1}{(u-k)^2}+O\{(u-k)^2\},$$

$$\wp'(u-k)=\frac{-2}{(u-k)^3}+O(u-k),$$

$$u\wp'(u-k)+2\wp(u-k)=\frac{-2k}{(u-k)^3}+O(u-k),$$

and therefore, on integrating to obtain U_1 and U_2, a term in $\log (u-k)$ will appear when $r=2$. It follows that if the solution is to be free from movable critical points,

$$S''(a)=0.$$

But a is arbitrary, and therefore

$$S''(Z)=0,$$

from which it follows that $S(Z)$ is of the linear form $pZ+q$.

Thus if solutions of the equation of sub-case (a) are free from movable critical points, the equation is reducible to the form

$$\frac{d^2w}{dz^2}=6w^2+pz+q.$$

By trivial changes in the variables, this equation may be brought into one or other of the three standard forms

$$\text{(i)} \quad \frac{d^2w}{dz^2}=6w^2 \qquad \text{(when } p=q=0),$$

$$\text{(ii)} \quad \frac{d^2w}{dz^2}=6w^2+\tfrac{1}{2} \qquad \text{(when } p=0,\ q\neq0),$$

$$\text{(iii)} \quad \frac{d^2w}{dz^2}=6w^2+z \qquad \text{(when } p\neq0).$$

Of these forms, the first two may be integrated by elliptic functions, giving respectively the *uniform* solutions

$$w=\wp(z-k,\ 0,\ h), \quad w=\wp(z-k,\ 1,\ h),$$

where h and k are arbitrary constants. The solutions are thus semi-transcendental functions of the constants of integration; they have no movable critical points, but do have movable poles. The third equation is not integrable in terms of elementary functions, algebraic or transcendental; * its general solution is in fact an essentially transcendental function of two constants. It is therefore to be regarded as defining a new type of transcendent, which is, in fact, free from movable critical points. The study of this equation will be taken up, in greater detail, in § 14·41.

14·312. Sub-case i(b).—When $A=-2$ and $C=0$, let $\lambda=1$, $\varphi=z$ and

$$2\mu'=2D\mu+E.$$

The equation then takes the form

$$\frac{d^2W}{dz^2}=-2W\frac{dW}{dz}+P(Z)\frac{dW}{dz}+Q(Z)W^2+S(Z).$$

Let

$$W=a^{-1}w, \quad Z=z_0+az, \quad P(z_0)=P_0, \quad Q(z_0)=Q_0,$$

then the equation becomes

$$\frac{d^2w}{dz^2}=-2w\frac{dw}{dz}+a\left\{P_0\frac{dw}{dz}+Q_0w^2\right\}+O(a^2),$$

and in this form may be satisfied by

$$w=w_0+aw_1+a^2w_2+\ldots,$$

where

$$w_0''=-2w_0w_0',$$

$$w_1''=-2(w_0w_1'+w_0'w_1)+P_0w_0'+Q_0w_0^2,$$

$$\cdot \quad \cdot \quad \cdot \quad \cdot \quad \cdot \quad \cdot \quad \cdot$$

From these relations it follows that

$$w_0=c\frac{e^{2c(z-a)}-1}{e^{2c(z-a)}+1},$$

* That is, the exponential, circular and elliptic functions. In future the term *classical transcendents* will be used to signify the class of elementary transcendents and transcendents defined by linear differential equations.

where a and c are arbitrary constants, and

$$w_1 = e^{-2\int w_0 dz} + \int\{\int(P_0 w_0' + Q_0 w_0^2)dz\}e^{2\int w_0 dz}dz.$$

But

$$P_0 w_0' + Q_0 w_0^2 = P_0 c^2 + (Q_0 - P_0)w_0^2,$$

and since w_0^2 possesses a double pole, the twofold integration which the expression for w_1 involves will lead to a logarithmic term, depending upon a unless $P(z_0) = Q(z_0)$ for all values of z_0, that is $P(z)$ and $Q(z)$ must be identically equal.

The equation is thus reduced to

$$\frac{d^2w}{dz^2} = -2w\frac{dw}{dz} + P(z)\left\{\frac{dw}{dz} + w^2\right\} + S(z).$$

It is now integrable, its first integral is

$$\frac{dw}{dz} + w^2 = u,$$

where

$$\frac{du}{dz} = P(z)u + S(z).$$

This first integral is of the Riccati type, the singular points of the function u are fixed, and therefore the general solution has fixed critical points.

An equivalent form of the equation is

$$\frac{d^2W}{dz^2} = -2W\frac{dW}{dz} + q(z)\frac{dW}{dz} + q'(z)W,$$

for this equation has also the first integral

$$\frac{dw}{dz} + w^2 = u,$$

where

$$w = W - \tfrac{1}{2}q, \quad u = \tfrac{1}{4}q^2 - \tfrac{1}{2}q'.$$

The general solution is a semi-transcendental function of the constants of integration.

14·313. Sub-case i(c).—When $A = -3$, $C = -1$, the typical equation whose general solution has only fixed critical points is

$$\frac{d^2W}{dZ^2} = -3W\frac{dW}{dZ} - W^3 + q(Z)\left\{\frac{dW}{dZ} - W^2\right\}.$$

The general solution is

$$W = -\frac{1}{u} \cdot \frac{du}{dZ},$$

where $u\ (Z)$ is the general solution of the linear equation of the third order

$$u''' = q(Z)u'';$$

it is therefore a rational function of the constants of integration.

14·314. Sub-case i(d).—When $A = -1$, $C = 1$, let

$$\lambda = 1, \quad \phi = z,$$
$$3\mu + D = -3\mu + 3B = 3P(z),$$

then the equation takes the form

$$\frac{d^2W}{dZ^2} = -W\frac{dW}{dZ} + W^3 + P(Z)\left\{3\frac{dW}{dZ} + W^2\right\} + R(Z)W + S(Z).$$

Solutions free from movable critical points arise in five distinct instances, as follows :

1° $\qquad R(z) = P'(z) - 2P^2(z), \quad S(z) = 0.$

The equation may be written

$$\frac{d^2w}{dz^2} = -w\frac{dw}{dz} + w^3 + P(z)\left\{3\frac{dw}{dz} + w^2\right\} + \{P'(z) - 2P^2(z)\}w\ ;$$

its solution is arrived at by the following steps, let

$$\frac{u'}{u}=P(z), \qquad \frac{v'}{\sqrt{(4v^2-1)}}=u,$$

then

$$w=v'/v,$$

where dashes denote differentiation with respect to z. By writing

$$w=\phi'(z)W, \quad Z=\phi(z),$$

where

$$\phi''=P(z)\phi,$$

the equation is brought into the standard form

$$\frac{d^2W}{dZ^2}=-W\frac{dW}{dZ}+W^3.$$

2° $$P(z)=\frac{q'(z)}{2q(z)}, \quad R(z)=\frac{q''(z)}{2q(z)}-\frac{q'^2(z)}{q^2(z)}-q(z), \quad S(z)=0.$$

The equation may be written

$$\frac{d^2w}{dz^2}=-w\frac{dw}{dz}+w^3+\frac{q'(z)}{2q(z)}\left\{3\frac{dw}{dz}+w^2\right\}+\left\{\frac{q''(z)}{2q(z)}-\frac{q'^2(z)}{q^2(z)}-q(z)\right\}w\,;$$

its solution is obtained as follows, let

$$\frac{u'}{\sqrt{(4u^3-12u+K)}}=\frac{1}{2\sqrt{3}}\sqrt{\{q(z)\}} \qquad (K \text{ an arbitrary constant}),$$

then

$$w=\frac{u'}{u-1}.$$

If

$$w=\phi'(z)W, \quad Z=\phi(z),$$

where

$$\phi'^2(z)=\tfrac{1}{12}q(z),$$

the equation is brought into the standard form

$$\frac{d^2W}{dZ^2}=-W\frac{dW}{dZ}+W^3-12W.$$

3° $$P(z)=\frac{q'(z)}{q(z)}+q(z), \quad R(z)=P'(z)-2P^2(z)-12q^2(z), \quad S(z)=-24q^3(z).$$

The equation may be written

$$\frac{d^2w}{dz^2}=-w\frac{dw}{dz}+w^3+\left\{\frac{q'(z)}{q(z)}+q(z)\right\}\left\{3\frac{dw}{dz}+w^2\right\}$$
$$+\left\{\frac{q''(z)}{q(z)}-3\frac{q'^2(z)}{q^2(z)}-3q'(z)-14q^2(z)\right\}w-24q^3(z)\,;$$

to integrate it, let

$$u=e^{\int P dz}, \quad v=\frac{1}{q(z)u}, \quad V=\wp(v+K,\,0,\,1),$$

where K is an arbitrary constant, then

$$w=\frac{v^3\sqrt{(4V^3-1)}+2}{uv(v^2V-1)}.$$

By the substitution

$$w=\phi'(z)W, \quad Z=\phi(z),$$

where

$$3\phi''=P(z)\phi,$$

the equation is brought into the standard form

$$\frac{d^2W}{dZ^2}=-W\frac{dW}{dZ}+W^3-12\wp(Z,\,0,\,1)W+12\wp'(Z,\,0,\,1).$$

4° $$P(z)=-\frac{2q(z)}{q'(z)}, \quad R(z)=-\frac{24z}{q(z)}, \quad S(z)=\frac{12}{q(z)},$$

where

$$q(z)=4z^3-\varepsilon z-\kappa,$$

and either $\varepsilon=0$ and $\kappa=1$ or else $\varepsilon=1$ and κ is arbitrary. The equation

$$\frac{d^2w}{dz^2}=-w\frac{dw}{dz}+w^3-\frac{2q(z)}{q'(z)}\left\{3\frac{dw}{dz}+w^2\right\}-\frac{24z}{q(z)}w+\frac{12}{q(z)}$$

is integrated as follows, let

$$\frac{u'}{\sqrt{\{4u^3-\varepsilon u+K\}}}=\frac{1}{\sqrt{\{4z^3-\varepsilon z-\kappa\}}} \qquad (K \text{ arbitrary}),$$

then

$$w=\frac{u'-1}{u-z}.$$

By means of the transformation

$$W=w\sqrt{\{4z^3-\varepsilon z-a\}}, \quad Z=\wp(Z, \varepsilon, a)$$

the equation is brought into the standard form

$$\frac{d^2W}{dz^2}=-W\frac{dW}{dZ}+W^3-12\wp(Z, \varepsilon, a)W+12\wp'(Z, \varepsilon, a).$$

5° $\quad P(z)=0, \quad R(z)=-12q(z), \quad S(z)=12q'(z),$

where $q(z)$ is the new transcendental function which satisfies the differential equation

$$q''=6q^2+z.$$

The solution of the equation

$$\frac{d^2w}{dz^2}=-w\frac{dw}{dz}+w^3-12q(z)w+12q'(z)$$

is

$$w=\frac{u'(z)-q'(z)}{u(z)-q(z)},$$

where $u(z)$ is any solution of

$$u''=6u^2+z$$

distinct from $q(z)$.

Thus every equation which comes under sub-class i(d) and has its general solution free from movable critical points is reducible to the standard form

$$\frac{d^2W}{dZ^2}=-W\frac{dW}{dZ}+W^3-12q(Z)W+12q'(Z),$$

where either

(α) q is zero,
or (β) q is a constant, not zero,
or (γ and δ) $q(z)$ satisfies the equation
$q''=6q^2+\eta \qquad (\eta=0 \text{ or } 1),$
or (ε) $q(z)$ satisfies the equation
$q''=6q^2+z.$

In (α)—(δ) the solution is a semi-transcendental, and in (ε) an essentially transcendental function of the constants of integration.

14·315. Sub-case i(e).—In this case $A=0$, $C=2$; suppose, in the first place, that B is also zero, and let

$$\lambda=1, \quad 2\mu=-D,$$

then the equation becomes

$$\frac{d^2W}{dz^2}=2W^3+R(z)W+S(z).$$

If $R(z)$ and $S(z)$ are constants, say β and γ, the equation is at once integrable in terms of elliptic functions. If $R(z)$ is not a constant, then for the absence of movable critical points it is necessary that

$$R(z)=z+\beta, \quad S(z)=\gamma,$$

where β and γ are again constants. The transformation

$$Z=z+\beta$$

then brings the equation into the standard form

$$\frac{d^2W}{dZ^2}=2W^3+ZW+\gamma.$$

This equation is not integrable in terms of the elementary transcendental functions; its general solution, nevertheless, can be shown to be free from movable critical points.

If B is not zero, then the only admissible case is found to be

$$\frac{d^2w}{dz^2}=-3q(z)\frac{dw}{dz}+2w^3-\{q'(z)+2q^2(z)\}w.$$

The transformation

$$W=we^{\int q\,dz},\quad Z=\int e^{-\int q\,dz}dz$$

reduces the equation to the standard form

$$\frac{d^2W}{dZ^2}=2W^3.$$

14·316. Canonical Equations of Type I.—To sum up, the following set of ten equations may be regarded as canonical equations of the type characterised by $L(z, w)\equiv 0$.

I. $\dfrac{d^2W}{dZ^2}=0.$ II. $\dfrac{d^2W}{dZ^2}=6W^2.$ III. $\dfrac{d^2W}{dZ^2}=6W^2+\tfrac{1}{2}.$ IV. $\dfrac{d^2W}{dZ^2}=6W^2+Z.$

V. $\dfrac{d^2W}{dZ^2}=-2W\dfrac{dW}{dZ}+q(Z)\dfrac{dW}{dZ}+q'(Z)W.$

VI. $\dfrac{d^2W}{dZ^2}=-3W\dfrac{dW}{dZ}-W^3+q(Z)\left\{\dfrac{dW}{dZ}+W^2\right\}.$

VII. $\dfrac{d^2W}{dZ^2}=2W^3.$ VIII. $\dfrac{d^2W}{dZ^2}=2W^3+\beta W+\gamma.$ IX. $\dfrac{d^2W}{dZ^2}=2W^3+ZW+\gamma.$

X. $\dfrac{d^2W}{dZ^2}=-W\dfrac{dW}{dZ}+W^3-12q(Z)W+12q'(Z).$

In V. and VI. $q(Z)$ is arbitrary, in X. $q(Z)$ is as defined in § 14·314.

14·32. Case ii.—The equation is, in the present case, necessarily of the form

$$\frac{d^2w}{dz^2}=\frac{1}{w}\left(\frac{dw}{dz}\right)^2+\left\{A(z)w+B(z)+\frac{C(z)}{w}\right\}\frac{dw}{dz}$$
$$+D(z)w^3+E(z)w^2+F(z)w+G(z)+\frac{H(z)}{w}.$$

Let

$$w=\alpha^{-1}W,\quad z=z_0+\alpha Z,\quad A(z_0)=a_0,\quad D(z_0)=d_0,$$

then the equation becomes

$$\frac{d^2W}{dZ^2}=\frac{1}{W}\left(\frac{dW}{dZ}\right)^2+a_0W\frac{dW}{dZ}+d_0W^3+O(\alpha),$$

and this is equivalent, when $\alpha=0$, to the system

$$\begin{cases}\dfrac{dW}{dZ}=\dfrac{W^2}{u},\\[2ex] \dfrac{du}{dZ}=(1-a_0u+d_0u^2)W.\end{cases}$$

When $d_0=0$, the solutions of this system are uniform; if $d_0\neq 0$, it may be proved, as in § 14·31 that the only possibility is $a_0=0$. It follows that either $A(z)$ or $D(z)$ is identically zero. Similarly it may be proved, by writing

$$w=\frac{1}{aW}, \quad z=z_0+aZ$$

and proceeding as before, that either $C(z)$ or $H(z)$ is identically zero.

14·321. Canonical Equations of Type II.—1°. When $A=C=0$ there are three canonical equations.

XI. $$\frac{d^2W}{dZ^2}=\frac{1}{W}\left(\frac{dW}{dZ}\right)^2.$$

XII. $$\frac{d^2W}{dZ^2}=\frac{1}{W}\left(\frac{dW}{dZ}\right)^2+\alpha W^3+\beta W^2+\gamma+\frac{\delta}{W}.$$

First integral :

$$\left(\frac{dW}{dZ}\right)^2=\alpha W^4+2\beta W^3-2\gamma W-\delta+KW^2,$$

where K is an arbitrary constant. The integration may be completed by the use of elliptic functions.

XIII. $$\frac{d^2W}{dZ^2}=\frac{1}{W}\left(\frac{dW}{dZ}\right)^2-\frac{1}{Z}\cdot\frac{dW}{dZ}+\frac{1}{Z}(\alpha W^2+\beta)+\gamma W^3+\frac{\delta}{W},$$

or, if $Z=e^z$,

XIII[1]. $$\frac{d^2W}{dz^2}=\frac{1}{W}\left(\frac{dW}{dz}\right)^2+e^z(\alpha W^2+\beta)+e^{2z}\left(\gamma W^3+\frac{\delta}{W}\right).$$

This equation is not integrable in terms of the classical transcendents.

2°. When $A\neq 0$, $C\neq 0$, there is one canonical equation.

XIV. $$\frac{d^2W}{dZ^2}=\frac{1}{W}\left(\frac{dW}{dZ}\right)^2+\left\{q(Z)W+\frac{r(Z)}{W}\right\}+q'(Z)W^2-r'(Z).$$

The first integral is of the Riccati type :

$$\frac{dW}{dZ}=q(Z)W^2+KW-r(Z),$$

where K is an arbitrary constant.

3°. Where $A=0$, $C\neq 0$ there are two canonical equations.

XV. $$\frac{d^2W}{dZ^2}=\frac{1}{W}\left(\frac{dW}{dZ}\right)^2+\frac{1}{W}\cdot\frac{dW}{dZ}+r(Z)W^2-W\frac{d}{dZ}\left\{\frac{r'(Z)}{r(Z)}\right\}.$$

The first integral is

$$\left\{\frac{dW}{dZ}+\frac{r'(Z)}{r(Z)}W+1\right\}^2=2W^2\{r(Z)W+\int r(Z)dZ+K\},$$

where K is an arbitrary constant.

XVI. $$\frac{d^2W}{dZ^2}=\frac{1}{W}\left(\frac{dW}{dZ}\right)^2-q'(Z)\frac{1}{W}\cdot\frac{dW}{dZ}+W^3-q(Z)W^2+q''(Z).$$

The first integral is

$$\left\{\frac{dW}{dZ}-q'(Z)\right\}^2=W^2\{[W-q(Z)]^2+K\}.$$

The case $A\neq 0$, $C=0$ is deducible from the preceding by writing $1/W$ for W. The general solution of each canonical equation is a semi-transcendental function of the constants of integration, with the exception of equations XIII. and XIII[1]. which are irreducible.

14·33. Case iii.—The equation is of the form

$$\frac{d^2w}{dz^2}=\frac{m-1}{mw}\left(\frac{dw}{dz}\right)^2+\left\{A(z)w+B(z)+\frac{C(z)}{w}\right\}\frac{dw}{dz}$$
$$+D(z)w^3+E(z)w^2+F(z)w+G(z)+\frac{H(z)}{w}.$$

Let

$$w=\alpha^{-1}W,\quad z=z_0+\alpha Z,\quad A(z_0)=a_0,\quad D(z_0)=d_0,$$

then the equation becomes

$$\frac{d^2W}{dZ^2}=\frac{m-1}{mW}\left(\frac{dW}{dZ}\right)^2+a_0W\frac{dW}{dZ}+d_0W^3+O(\alpha).$$

The treatment of § 14·31 may be applied here, but an alternative procedure is as follows. Let

$$\frac{dW}{dZ}=uW^2,$$

then, when $\alpha=0$, the equation reduces to

$$\frac{u''}{u'^2}=\frac{-2\frac{m+1}{m}u+a_0}{-\frac{m+1}{m}u^2+a_0u+d_0}+\frac{u}{-\frac{m+1}{m}u^2+a_0u+d_0}.$$

But if the critical points are to be fixed, the right-hand member of this equation must, when decomposed into partial fractions, be of one or other of the eight forms enumerated in § 14·3, W being, of course, replaced by u. This leads to several distinct possibilities, namely,

(*a*) if m is unrestricted, then either
 (i) both $A(z)$ and $D(z)$ are identically zero, or
 (ii) $D(z)=-\frac{m}{(m+2)^2}A^2(z)$,
(*b*) when $m=2$, either
 (i) $A(z)=0$, identically, $D(z)\neq 0$, or
 (ii) $D(z)=\frac{1}{2}A^2(z)$,
(*c*) when $m=3$, $D(z)=\frac{3}{2}A^2(z)$,
(*d*) when $m=5$, $D(z)=5A^2(z)$.

By writing $w=1/v$, the original equation is transformed into

$$\frac{d^2v}{dz^2}=\frac{m+1}{mv}\left(\frac{dv}{dz}\right)^2+\left\{C(z)v+B(z)+\frac{A(z)}{v}\right\}\frac{dv}{dz}$$
$$-H(z)v^3-G(z)v^2-F(z)v-E(z)-\frac{D(z)}{v^2}.$$

It follows that
(*a*) if m is unrestricted, then either
 (i) both $C(z)$ and $H(z)$ are identically zero, or
 (ii) $H(z)=-\frac{m}{(m-2)^2}C^2(z)$,
(*b*) when $m=2$, either
 (i) both $C(z)$ and $H(z)$ are identically zero, or
 (ii) $C(z)=0$, identically, $H(z)\neq 0$.

Consider, in particular, the case in which $A(z)$ and $D(z)$ are both identically zero, then if the equation is first transformed by writing

$$w=\alpha^{-2}W,\quad z=z_0+\alpha Z,\quad e_0=E(z_0),$$

and a is then made zero, the equation becomes

$$\frac{d^2W}{dZ^2}=\frac{m-1}{mW}\left(\frac{dW}{dZ}\right)^2+e_0W^2.$$

An evident possibility is that $E(z)$ is also identically zero. When $E(z)\neq 0$ it may be proved, as in §14·22, that the only possibilities are $m=2$, $m=4$, and $m=-4$. In the same way, if $C(z)$ and $H(z)$ are both identically zero, then either $G(z)$ is identically zero or $m=4$.

This discussion limits the number of cases to be considered.* By continuing the investigation it is found that the equations whose solutions are free from movable critical points are of the canonical forms enumerated in the following sub-section.

14·331. Canonical Equations of Type III.—

1°. When A, C, D and H are identically zero, there are seven canonical equations.

XVII.

$$\frac{d^2W}{dZ^2}=\frac{m-1}{mW}\left(\frac{dW}{dZ}\right)^2.$$

The general solution $W=(K_1Z+K_2)^m$ is rational in the constants of integration K_1 and K_2.

XVIII.

$$\frac{d^2W}{dZ^2}=\frac{1}{2W}\left(\frac{dW}{dZ}\right)^2+4W^2.$$

First integral :

$$\left(\frac{dW}{dZ}\right)^2=4W(K+W^2).$$

XIX.

$$\frac{d^2W}{dZ^2}=\frac{1}{2W}\left(\frac{dW}{dZ}\right)^2+4W^2+2W.$$

First integral :

$$\left(\frac{dW}{dZ}\right)^2=4W(K+W+W^2).$$

XX.

$$\frac{d^2W}{dZ^2}=\frac{1}{2W}\left(\frac{dW}{dZ}\right)^2+4W^2+2ZW.$$

Equivalent to

$$\frac{d^2u}{dZ^2}=2u^3+Zu \qquad (u^2=W),$$

a particular case of equation IX.

XXI.

$$\frac{d^2W}{dZ^2}=\frac{3}{4W}\left(\frac{dW}{dZ}\right)^2+3W^2$$

Equivalent to

(XXIX.)

$$\frac{d^2u}{dZ^2}=\frac{1}{2u}\left(\frac{du}{dZ}\right)^2+\frac{3u^3}{2} \qquad (u^2=W).$$

XXII.

$$\frac{d^2W}{dZ^2}=\frac{3}{4W}\left(\frac{dW}{dZ}\right)^2-1.$$

Equivalent to

(XXXII.)

$$\frac{d^2u}{dZ^2}=\frac{1}{2u}\left(\frac{du}{dZ}\right)^2-\frac{1}{2u} \qquad (u^2=W).$$

XXIII.

$$\frac{d^2W}{dZ^2}=\frac{3}{4W}\left(\frac{dW}{dZ}\right)^2+3W^2+aW+\beta.$$

* There are fourteen cases to be discussed of which nine are essentially distinct. The discussion is, in its complete form, due to Gambier, *C. R. Acad. Sc. Paris*, 142 (1906), pp. 1403, 1497 ; the previous discussion by Painlevé was not exhaustive.

Equivalent to

$$\frac{d^2u}{dZ^2}=\frac{1}{2u}\left(\frac{du}{dZ}\right)^2+\frac{3u^3}{2}+\frac{\alpha u}{2}+\frac{\beta}{2u} \qquad (u^2=W),$$

a particular case of equation XXX.

2°. When C and H are identically zero, and $(m+2)^2D+mA^2=0$, there are two canonical equations.

$$\text{XXIV.} \quad \frac{d^2W}{dZ^2}=\frac{m-1}{mW}\left(\frac{dW}{dZ}\right)^2+qW\frac{dW}{dZ}-\frac{mq^2}{(m+2)^2}W^3+\frac{mq'}{m+2}W^2.$$

Solution :

$$W=-\frac{(m+2)(K_1Z+K_2)}{m\int(K_1Z+K_2)^m q(Z)dZ}.$$

$$\text{XXV.} \quad \frac{d^2W}{dZ^2}=\frac{3}{4W}\left(\frac{dW}{dZ}\right)^2-\frac{3W}{2}\frac{dW}{dZ}-\frac{W^3}{4}+\frac{q'}{2q}\left(W^2+\frac{dW}{dZ}\right)+rW+q.$$

Solution :

$$W=\frac{q}{2u'+u^2-\dfrac{q'}{q}u-r},$$

where $u=t'/t$, t being the general solution of the linear equation

$$t'''=\frac{3q'}{2q}t''+\left(r+\frac{q''}{q}-\frac{q'^2}{q^2}\right)t'+\tfrac{1}{2}\left(r'+q-\frac{q'r}{q}\right)t.$$

3°. When A and D are identically zero, and $(m-2)^2H+mC^2=0$, there is one canonical equation.

$$\text{XXVI.} \quad \frac{d^2W}{dZ^2}=\frac{3}{4W}\left(\frac{dW}{dZ}\right)^2+\frac{6q'}{W}\cdot\frac{dW}{dZ}+3W^2+12qW-12q''-\frac{36q'^2}{W},$$

where

$$q''=6q^2, \quad \text{or} \quad q''=6q^2+\tfrac{1}{2}, \quad \text{or} \quad q''=6q^2+Z.$$

Solution :

$$3W=2V'+V^2-12q,$$

where

$$V=\frac{Q'-q'}{Q-q}$$

and

$$Q''=6Q^2, \quad \text{or} \quad Q''=6Q^2+\tfrac{1}{2}, \quad \text{or} \quad Q''=6Q^2+Z,$$

as the case may be, but $Q\neq q$.

Other equations in which A and D are identically zero are particular cases of the following :

4°. When m is unrestricted, and $(m-2)^2H+mC^2=0$ there is one equation.* Its general form is

$$\text{XXVII.} \quad \frac{d^2W}{dZ^2}=\frac{m-1}{mW}\left(\frac{dW}{dZ}\right)^2+\left(fW+\phi-\frac{m-2}{mW}\right)\frac{dW}{dZ}-\frac{mf^2}{(m+2)^2}W^3$$
$$+\frac{m(f'-f\phi)}{m+2}W^2+\psi W-\phi-\frac{1}{mW},$$

where f, ϕ and ψ are definite rational functions of two arbitrary analytic functions $q(Z)$ and $r(Z)$ and of their derivatives. In the particular case $m=2$, the canonical equation is

$$\frac{d^2W}{dZ^3}=\frac{1}{2W}\left(\frac{dW}{dZ}\right)^2-2W\frac{dW}{dZ}-\frac{W^3}{2}+F(Z)W-\frac{1}{2W},$$

and its solution is

$$W=u'/u,$$

where

$$2u'''u'=u''^2+2Fu'^2-u^2.$$

* This difficult case was studied in special detail and solved by Gambier, *Acta Math.* 33 (1910), p. 51.

On differentiation, this last equation becomes linear, and of the fourth order

$$u^{\mathrm{IV}}=2Fu''+F'u'-u.$$

Thus, when $m=2$, the general solution is a rational function of the constants of integration.

5°. When $m=2$, $D=\frac{1}{2}A$ and C identically zero, there is one equation.*

XXVIII.
$$\frac{d^2W}{dZ^2}=\frac{1}{2W}\left(\frac{dW}{dZ}\right)^2-(W-q)\frac{dW}{dZ}+\frac{W^3}{2}-2qW^2+3(q'+\tfrac{1}{2}q^2)W-\frac{72r^2}{W},$$

in which q and r are determined as follows. Let V_1 and V_2 be any two solutions of either

$$v''=6v^2,\quad \text{or}\quad v''=6v^2+\tfrac{1}{2},\quad \text{or}\quad v''=6v^2+Z,$$

then

$$q=\frac{V_2'-V_1'}{V_2-V_1},\quad r=\tfrac{1}{2}(V_2-V_1).$$

Solution :

$$W=\frac{6(V-V_1)(V-V_2)}{V'-\frac{1}{2}(V_1'+V_2')-q\{V-\frac{1}{2}(V_1+V_2)\}},$$

where V satisfies the same equation as V_1 and V_2. When V_1 and V_2 are made equal,†

$$q=\lim\frac{V_2'-V_1'}{V_2-V_1},\quad r=0.$$

6°. When $m=2$, A and C identically zero, D not identically zero, there are three canonical equations.

XXIX.
$$\frac{d^2W}{dZ^2}=\frac{1}{2W}\left(\frac{dW}{dZ}\right)^2+\frac{3W^3}{2}.$$

First integral :

$$\left(\frac{dW}{dZ}\right)^2=W^4+KW.$$

XXX.
$$\frac{d^2W}{dZ^2}=\frac{1}{2W}\left(\frac{dW}{dZ}\right)^2+\frac{3W^3}{2}+4\alpha W^2+2\beta W-\frac{\gamma^2}{2W}.$$

First integral :

$$\left(\frac{dW}{dZ}\right)^2=W^4+4\alpha W^3+4\beta W^2+4KW+\gamma^2.$$

XXXI.
$$\frac{d^2W}{dZ^2}=\frac{1}{2W}\left(\frac{dW}{dZ}\right)^2+\frac{3W^3}{2}+4ZW^2+2(Z^2-\alpha)W-\frac{\beta^2}{2W}.$$

Not integrable in terms of classical transcendents.

7°. When $m=2$, A, C and D identically zero, H not identically zero, there are three canonical equations.

XXXII.
$$\frac{d^2W}{dZ^2}=\frac{1}{2W}\left(\frac{dW}{dZ}\right)^2-\frac{1}{2W}.$$

First integral :

$$\left(\frac{du}{dZ}\right)^2=K+\frac{1}{u^2}\qquad (u^2=W).$$

XXXIII.
$$\frac{d^2W}{dZ^2}=\frac{1}{2W}\left(\frac{dW}{dZ}\right)^2+4W^2+\alpha W-\frac{1}{2W}.$$

* This case also was given special notice by Gambier, *ibid.* p. 49.

† V depends in general upon two parameters, say α and β, and may be written $V(Z, \alpha, \beta)$. V_1 is obtained by giving α and β the special values α_1 and β_1 and

$$\lim\frac{V_2'-V_1'}{V_2-V_1}=\left[\left(\lambda\frac{\partial^2V}{\partial\alpha\partial Z}+\mu\frac{\partial^2V}{\partial\beta\partial Z}\right)\Big/\left(\lambda\frac{\partial V}{\partial\alpha}+\mu\frac{\partial V}{\partial\beta}\right)\right]_{\alpha_1,\beta_1},$$

where λ and μ are constants whose ratio is arbitrary.

First integral :

$$\left(\frac{dW}{dZ}\right)^2=4W^3+2\alpha W^2+4KW+1.$$

XXXIV. $$\frac{d^2W}{dZ^2}=\frac{1}{2W}\left(\frac{dW}{dZ}\right)^2+4\alpha W^2-ZW-\frac{1}{2W} \qquad (\alpha\neq 0).$$

Solution :

$$2\alpha W=V'+V^2+\tfrac{1}{2}Z,$$

where

(IX.) $$V''=2V^3+ZV-2\alpha-\tfrac{1}{2}.$$

8°. When $n=3$, $D=\frac{3}{2}A^2$, $H=-3C^2$ there is one canonical equation.*

XXXV. $$\frac{d^2W}{dZ^2}=\frac{2}{3W}\left(\frac{dW}{dZ}\right)^2-\left(\tfrac{2}{3}W-\tfrac{2}{3}q-\frac{r}{W}\right)\frac{dW}{dZ}+\tfrac{2}{3}W^3-\tfrac{10}{3}qW^2$$
$$+(4q'+r+\tfrac{8}{3}q^2)W+2qr-3r'-\frac{3r^2}{W},$$

where, if $2u^3+Su+T$ represents either $2u^3$ or $2u^3+\alpha u+\beta$ or $2u^3+Zu+\alpha$,

$$q''=2q^3+Su+T, \quad r=-\tfrac{1}{3}S-\tfrac{2}{3}(q'+q^2).$$

Solution :

$$W=\frac{V'-q'+V^2-q^2}{V-q},$$

where V is any solution of

$$V''=2V^3+SV+T.$$

9°. When $n=5$, $D=5A^2$, $H=-\frac{5}{9}C^2$, there is one canonical equation.

$$\frac{d^2W}{dZ^2}=\frac{4}{5W}\left(\frac{dW}{dZ}\right)^2-\left(\tfrac{2}{5}W+\tfrac{4}{5}q-\frac{r}{W}\right)\frac{dW}{dZ}+\tfrac{4}{5}W^3+\tfrac{14}{5}qW^2$$
$$+(r-3q'+\tfrac{6}{5}q^2)W-\tfrac{1}{3}(qr+5r')-\tfrac{5}{9}\frac{r^2}{W},$$

where

$$q=\frac{V_2'-V_1'}{V_2-V_1}, \quad r=\tfrac{72}{5}V_1+\tfrac{36}{5}V_2-\tfrac{9}{5}\left(\frac{V_2'-V_1'}{V_2-V_1}\right)^2,$$

V_1 and V_2 being solutions of

$$V''=6V+S \qquad (S=0,\ \tfrac{1}{2} \text{ or } Z).$$

Solution :

$$W=\frac{V'-q'}{V-q}-\tfrac{1}{2}(q+Z),$$

V being the general solution of $V''=6V+S$.

14·34. Canonical Equations of Type IV.—In Case iv. there are four canonical equations.†

XXXVII. $$\frac{d^2W}{dZ^2}=\left\{\frac{1}{2W}+\frac{1}{W-1}\right\}\left(\frac{dW}{dZ}\right)^2.$$

First integral :

$$\left(\frac{dW}{dZ}\right)^2=4K_1{}^2W(W-1)^2.$$

Solution :

$$W=\tanh^2(K_1Z+K_2).$$

XXXVIII. $$\frac{d^2W}{dZ^2}=\left\{\frac{1}{2W}+\frac{1}{W-1}\right\}\left(\frac{dW}{dZ}\right)^2+W(W-1)\left\{\alpha(W-1)+\beta\frac{W-1}{W^2}\right.$$
$$\left.+\frac{\gamma}{W-1}+\frac{\delta}{(W-1)^2}\right\}.$$

* For details of this case, see Gambier, *ibid.* p. 32.

† Gambier, *C. R. Acad. Sc. Paris*, 143 (1906), p. 741.

First integral:

$$\left(\frac{dW}{dZ}\right)^2 = W(W-1)^2\left\{2\alpha W - \frac{2\beta}{W} - \frac{2\gamma}{W-1} - \frac{\delta}{(W-1)^2} + K\right\}.$$

XXXIX. $$\frac{d^2W}{dZ^2} = \left\{\frac{1}{2W} + \frac{1}{W-1}\right\}\left(\frac{dW}{dZ}\right)^2 - \frac{1}{Z}\cdot\frac{dW}{dZ} + \frac{(W-1)^2}{Z^2}\left(\alpha W + \frac{\beta}{W}\right) + \frac{\gamma W}{Z} + \frac{\delta W(W+1)}{W-1}.$$

Not integrable in terms of the classical transcendents.

XL. $$\frac{d^2W}{dZ^2} = \left\{\frac{1}{2W} + \frac{1}{W-1}\right\}\left(\frac{dW}{dZ}\right)^2 + 2\frac{qW+r}{W-1}\cdot\frac{dW}{dZ} + \tfrac{1}{2}(W-1)^2\left\{S^2W - \frac{t^2}{W}\right\} + 2\{q^2 - r^2 - (q'+r')\}W,$$

where

$$s' = 2qs, \quad t' = -2rt.$$

Method of integration: let

$$\frac{W'+2(q+r)}{W-1} - sW = -2u,$$

then u is given by the Riccati equation

$$u' + u^2 + 2ru - \tfrac{1}{4}t^2 = \tfrac{1}{2}v + su,$$

where

$$v' = 2(q-r)v.$$

It may be noted that if s and t are not both zero,

$$\frac{v'}{v} = \frac{s'}{s} + \frac{t'}{t},$$

and therefore $v = Kst$.

14·35. Canonical Equations of Type V.—In Case v. there are two canonical equations.*

XLI. $$\frac{d^2W}{dZ^2} = \frac{2}{3}\left\{\frac{1}{W} + \frac{1}{W-1}\right\}\left(\frac{dW}{dZ}\right)^2.$$

First integral:

$$\left(\frac{dW}{dZ}\right)^3 = 27K_1^3W^2(W-1)^2.$$

Solution:

$$2W = 1 + \wp'(K_1Z + K_2, \ 0, \ -1).$$

XLII. $$\frac{d^2W}{dZ^2} = \frac{2}{3}\left\{\frac{1}{W} + \frac{1}{W-1}\right\}\left(\frac{dW}{dZ}\right)^2 + \left\{qW + \frac{r}{W} - \frac{s}{W-1} - \tfrac{1}{3}(q+r+s)\right\}\frac{dW}{dZ} + W(W-1)\left\{3q^2W + \frac{3r^2}{W^2} - \frac{3s^2}{(W-1)^2} + 3q' + \tfrac{3}{2}q(r+s-q) + \frac{3r' - \frac{3}{2}r(r+s-q)}{W} + \frac{3s' - \frac{3}{2}s(q+r+s)}{W-1}\right\},$$

where

$$3q = \frac{V_1'}{V_1} - V_1 + \frac{E}{V_1} - 2C,$$

$$3r = \frac{V_1'}{V_1} + V_1 + \frac{E}{V_1} + 2C,$$

$$3s = 2V_1,$$

* Gambier, *C. R.* 144 (1907), p. 827.

and V_1 is any solution of the equation

$$V''=\frac{V'^2}{2V}+\tfrac{3}{2}V^3+4CV^2+2DV-\frac{E^2}{2V},$$

in which C, D, and E are all zero (Equation XXIX.), or are all constants (Equation XXX.) or $C=Z$, $D=Z^2-\alpha$, $E=\beta$ (Equation XXXI.). If V is the general solution of this equation, then *

$$W=1+\frac{2V(V-V_1)}{V'-V_1'-V^2+V_1^2-\tfrac{3}{2}(q+r-s)(V-V_1)}.$$

14·36. Canonical Equations of Type VI.—In Case vi. there are five canonical equations.†

XLIII. $$\frac{d^2W}{dZ^2}=\frac{3}{4}\left\{\frac{1}{W}+\frac{1}{W-1}\right\}\left(\frac{dW}{dZ}\right)^2.$$

First integral :

$$\left(\frac{dW}{dZ}\right)^4=256K_1^4W^3(W-1)^3.$$

Solution :

$$\frac{W-1}{W}=\wp^2(K_1Z+K_2,\ 4,\ 0).$$

XLIV. $$\frac{d^2W}{dZ^2}=\frac{3}{4}\left\{\frac{1}{W}+\frac{1}{W-1}\right\}\left(\frac{dW}{dZ}\right)^2+W(W-1)\left\{\frac{\alpha}{W}+\frac{\beta}{W-1}+2\gamma(W-1)\right\}.$$

Solution :

$$\frac{W-1}{W}=u^2,$$

where

$$u'^2=\left\{\alpha u-\frac{\beta}{u}-\frac{\gamma}{u-1}+K\right\}u(1-u^2).$$

XLV. $$\frac{d^2W}{dZ^2}=\frac{3}{4}\left\{\frac{1}{W}+\frac{1}{W-1}\right\}\left(\frac{dW}{dZ}\right)^2+\left\{A+\frac{B}{W}-\frac{C}{W-1}\right\}\frac{dW}{dZ}$$
$$+W(W-1)\left\{4D^2(2W-1)+\frac{B^2}{W^2}-\frac{C^2}{(W-1)^2}+\frac{H}{W}+\frac{K}{W-1}\right\},$$

where

$$A=\frac{V_2'-V_1'}{V_2-V_1},\quad B-C=-\tfrac{3}{2}(V_1+V_2),\quad B+C=-\tfrac{3}{2}\frac{V_2'-V_1'}{V_2-V_1},\quad D=\tfrac{1}{2}(V_2-V_1),$$
$$H=2B'+AB,\quad K=2C'+AC,$$

in which V_1 and V_2 are any two solutions of the equation

$$V''=2V^3+SV+T$$

(Equation VII., VIII., or IX.). If V is the general solution of this equation,

$$2W-1=\frac{2V'-V_1'-V_2'-A(2V-V_1-V_2)}{2(V-V_1)(V-V_2)}.$$

XLVI. $$\frac{d^2W}{dZ^2}=\frac{3}{4}\left\{\frac{1}{W}+\frac{1}{W-1}\right\}\left(\frac{dW}{dZ}\right)^2-\frac{H'}{H}\left\{1+\frac{3}{2(W-1)}\right\}\frac{dW}{dZ}$$
$$+W(W-1)\left\{\frac{4\beta^2}{H^2}(2W-1)-\left(\frac{3H'}{2H}\right)^2\frac{1}{(W-1)^2}+\frac{H}{W}+\left(\frac{3H''}{H}-\frac{9H'^2}{2H^2}\right)\frac{1}{W-1}\right\},$$

where

$$H=2(V_1'+V_1^2)+\alpha,$$

* For the complete discussion of this case see Gambier, *Acta Math.* 33 (1910), p. 38.
† Gambier, *C. R.* 144 (1907), p. 962.

in which V_1 is any solution of

(VIII.) $$V''=2V^3+\alpha V+\beta.$$

If V is the general solution of this equation, and

$$T=\frac{V'-V_1'}{V-V_1}+V+V_1,$$

then

$$W=\frac{3T^2}{2T'+2T^2-(4C+4D+3A)T+H}.$$

XLVII. $$\frac{d^2W}{dZ^2}=\frac{3}{4}\left\{\frac{1}{W}+\frac{1}{W-1}\right\}\left(\frac{dW}{dZ}\right)^2-\frac{H'}{H}\left\{1+\frac{3}{2(W-1)}\right\}\frac{dW}{dZ}$$
$$+W(W-1)\left\{\frac{(2\alpha+1)^2}{H^2}(2W-1)-\left(\frac{3H'}{2H}\right)^2\frac{1}{(W-1)^2}+\frac{H}{W}+\left(\frac{3H''}{H}-\frac{9H'^2}{2H^2}\right)\frac{1}{W-1}\right\},$$

where

$$H=2(V_1'+V_1^2)+Z,$$

in which V_1 is any solution of

(IX.) $$V''=2V^3+ZV+\alpha.$$

The integration proceeds on the lines indicated under XLVI.

14·37. Canonical Equations of Type VII.—In Case vii. the equation is: *

XLVIII. $$\frac{d^2W}{dZ^2}=\left\{\frac{2}{3W}+\frac{1}{2(W-1)}\right\}\left(\frac{dW}{dZ}\right)^2+\left\{AW+B+\frac{C}{W}\right\}\frac{dW}{dZ}$$
$$+W(W-1)\left\{\frac{3A^2W}{8}+F+\frac{3C^2}{W^2}+\frac{H}{(W-1)^2}+\frac{K}{W}+\frac{H}{3(W-1)}\right\},$$

where

$$A=-\tfrac{10}{9}(t+u),\quad B=\tfrac{1}{9}(2t+5u),\quad C=-\tfrac{4}{9}(u-2t),\quad F=\tfrac{3}{2}(a'-ab)-\tfrac{3}{4}a^2,$$
$$H=-\tfrac{9}{2}v^2,\quad K=3(c'-bc)-\tfrac{3}{2}c^2,$$

in which

$$t=\tfrac{1}{2}\left\{\frac{V_2'-V_1'}{V_2-V_1}+\frac{V_3'-V_1'}{V_3-V_1}\right\},\quad v=\tfrac{1}{2}\left\{\frac{V_2'-V_1'}{V_2-V_1}-\frac{V_3'-V_1'}{V_3-V_1}\right\},\quad u=-\frac{v'}{v},$$

and V_1, V_2, V_3 are any three particular solutions of

$$V''=6V^2+S \qquad (S=0,\ \tfrac{1}{2},\text{ or } Z).$$

Solution:

$$W=1+\tfrac{3}{2}\frac{(Y-t)^2-v^2}{\{Y'-t'+(Y-t)u\}-\{(Y-t)^2-v^2\}},$$

where

(X.) $$Y''=-YY'+Y^3-12V_1Y+12V_1'.$$

14·38. Canonical Equations of Type VIII.—There are two typical equations in Case viii., in the first of which η is a constant, say α, and in the second of which $\eta=Z$.†

XLIX. $$\frac{d^2W}{dZ^2}=\frac{1}{2}\left\{\frac{1}{W}+\frac{1}{W-1}+\frac{1}{W-\alpha}\right\}\left(\frac{dW}{dZ}\right)^2$$
$$+W(W-1)(W-\alpha)\left\{\beta+\frac{\gamma}{W^2}+\frac{\delta}{(W-1)^2}+\frac{\epsilon}{(W-\alpha)^2}\right\}.$$

* Gambier, *C. R.* 144 (1907), p. 962; *Acta Math.* 33 (1910), p. 45.
† *Ibid.* 143 (1906), p. 741.

First integral:

$$\left(\frac{dW}{dZ}\right)^2 = W(W-1)(W-a)\left\{2\beta W - \frac{2\gamma}{W} - \frac{2\delta}{W-1} - \frac{2\epsilon}{W-a} + K\right\}.$$

The general solution is expressible in terms of elliptic functions.

L. $$\frac{d^2W}{dZ^2} = \frac{1}{2}\left\{\frac{1}{W} + \frac{1}{W-1} + \frac{1}{W-Z}\right\}\left(\frac{dW}{dZ}\right)^2 - \left\{\frac{1}{Z} + \frac{1}{Z-1} + \frac{1}{W-Z}\right\}\frac{dW}{dZ}$$
$$+ \frac{W(W-1)(W-Z)}{2Z^2(Z-1)^2}\left\{\alpha - \frac{\beta Z}{W^2} + \frac{\gamma(Z-1)}{(W-1)^2} - \frac{(\delta-1)Z(Z-1)}{(W-Z)^2}\right\}.$$

This equation is not, in general, integrable in terms of the classical transcendents. When $\alpha=\beta=\gamma=\delta=0$ it may be integrated as follows. Let $\Lambda(u, Z)$ be the elliptic function defined by

$$u = \int_0^{\Lambda} \frac{dw}{\sqrt{\{w(w-1)(w-Z)\}}},$$

and let $2\omega_1$ and $2\omega_2$ be its periods, which are functions of Z. Then the general solution of the equation is

$$W = \Lambda(K_1\omega_1 + K_2\omega_2, Z),$$

where K_1 and K_2 are arbitrary constants.*

14·39. General Conclusion.—The repeated application of the conditions necessary for the absence of movable critical points has thus led, by a process of exhaustion, to fifty types of the equation

$$\frac{d^2W}{dZ^2} = F\left(\frac{dW}{dZ}, W, Z\right),$$

in which F is rational in W and in W', and analytic in Z. Of these fifty types all but six are integrable in terms of known functions and the general solution is found in each of these cases to be free from movable critical points. This latter fact is true in the remaining six cases; the lines upon which the demonstration proceeds will be indicated in later sections (§§ 14·41 *et seq.*). Thus when the restrictions stated are imposed upon F, the aggregate of conditions is sufficient as well as necessary. The fifty canonical types which have been enumerated may be generalised by the transformation

$$W = \frac{l(z)w + m(z)}{p(z)w + q(z)}, \quad Z = \phi(z),$$

where l, m, p, q and ϕ are analytic functions of z, and the new types obtained contain all the equations of the second order, rational in w and w', whose general solutions have fixed critical points.

But when the equation is algebraic in w, and is not reducible to an equivalent equation in which w appears rationally, the state of affairs is altogether different. This is clearly shown by the following example: †

$$\frac{d^2w}{dz^2} = \left\{\frac{w[2k^2w^2-(1+k^2)]}{(1-w^2)(1-k^2w^2)} - \frac{1}{\lambda(1-w^2)^{\frac{1}{2}}(1-k^2w^2)^{\frac{1}{2}}}\right\}\left(\frac{dw}{dz}\right)^2.$$

It is not difficult to prove that the general solution of this equation has no algebraic singularities other than poles; with rather greater difficulty it can be proved that any solution, which tends to a determinate value when z tends to z_0 along any path, is analytic or has a pole at z_0. But it

* In its general form Equation L. was first discovered by R. Fuchs, *C. R. Acad. Sc. Paris*, 141 (1905), p. 555. The integration, when α, β, γ and δ are zero, is due to Painlevé.

† Painlevé, *Bull. Soc. Math. France*, 28 (1900), p. 230.

does not follow that the solution is meromorphic throughout the z-plane. In fact the general solution is

$$w = \text{sn}\{\lambda \log (Az - B)\} \qquad (\text{mod } k),$$

where A and B are arbitrary constants. The point $z = B/A$ is an essential singularity of the solution: as z tends to B/A along any definite path, w tends to no limit whatsoever.

This example shows clearly why it is that the necessary conditions may not be sufficient, and consequently why each of the fifty canonical types obtained in the foregoing sections has to be examined separately in order that the absence of movable critical points may be confirmed.

14·4. The Painlevé Transcendents.—The most interesting of the fifty types enumerated are those which are *irreducible* * and serve to define new transcendents. These irreducible equations are those numbered IV., IX., XIII., XXXI., XXXIX. and L., six types in all. It is convenient to tabulate and renumber them, thus :

$$\text{(i)} \quad \frac{d^2w}{dz^2} = 6w^2 + z, \qquad\qquad \text{(ii)} \quad \frac{d^2w}{dz^2} = 2w^3 + zw + a,$$

$$\text{(iii)} \quad \frac{d^2w}{dz^2} = \frac{1}{w}\left(\frac{dw}{dz}\right)^2 - \frac{1}{z}\cdot\frac{dw}{dz} + \frac{1}{z}(\alpha w^2 + \beta) + \gamma w^3 + \frac{\delta}{w},$$

$$\text{(iv)} \quad \frac{d^2w}{dz^2} = \frac{1}{2w}\left(\frac{dw}{dz}\right)^2 + \frac{3w^3}{2} + 4zw^2 + 2(z^2 - \alpha)w + \frac{\beta}{w},$$

$$\text{(v)} \quad \frac{d^2w}{dz^2} = \left\{\frac{1}{2w} + \frac{1}{w-1}\right\}\left(\frac{dw}{dz}\right)^2 - \frac{1}{z}\cdot\frac{dw}{dz} + \frac{(w-1)^2}{z^2}\left\{\alpha w + \frac{\beta}{w}\right\} + \frac{\gamma w}{z} + \frac{\delta w(w+1)}{w-1},$$

$$\text{(vi)} \quad \frac{d^2w}{dz^2} = \frac{1}{2}\left\{\frac{1}{w} + \frac{1}{w-1} + \frac{1}{w-z}\right\}\left(\frac{dw}{dz}\right)^2 - \left\{\frac{1}{z} + \frac{1}{z-1} + \frac{1}{z-x}\right\}\frac{dw}{dz}$$
$$+ \frac{w(w-1)(w-z)}{z^2(z-1)^2}\left\{\alpha + \frac{\beta z}{w^2} + \frac{\gamma(z-1)}{(w-1)^2} + \frac{\delta z(z-1)}{(w-z)^2}\right\}.$$

The new transcendental functions defined by these equations are known as the *Painlevé Transcendents*.† The solutions of (i), (ii), and (iii) have no branch points, and are therefore uniform functions of z. If, in (iv) and (v), the independent variable is changed by the transformation $z = e^z$, the solutions are uniform functions of z. But in equation (vi) the points $z = 0$, $z = 1$ and $z = \infty$ are critical points.

Equation (vi) contains, in reality, the first five equations, which may be derived from it by a process of coalescence.‡ As it can be proved that the solutions of (i) are indeed new transcendents, it follows that the solutions of the remaining five equations cannot (except possibly for special values of α, β, γ and δ) be expressible in terms of the classical transcendental functions alone.

This process of step-by-step degeneration may be carried out as follows :

In (vi) replace z by $1 + \varepsilon z$, δ by $\frac{\delta}{\varepsilon^2}$, γ by $\frac{\gamma}{\varepsilon} - \frac{\delta}{\varepsilon^2}$, and let $\varepsilon \to 0$. The limiting form of the equation is (v).

* By irreducible is meant not replaceable by a simpler equation or combination of simpler equations.

† Only the first three types were discovered by Painlevé, the last three were subsequently added by Gambier.

‡ Painlevé, *C. R. Acad. Sc. Paris*, 143 (1906), p. 1111. The solutions of (vi.) in the neighbourhood of a singular point were studied by Garnier, *C. R.* 162 (1916), p. 939 ; 163 (1916), pp. 8, 118.

In (v) replace w by $1+\varepsilon w$, β by $-\dfrac{\beta}{\varepsilon^2}$, α by $\dfrac{\beta}{\varepsilon^2}+\dfrac{\alpha}{\varepsilon}$, γ by $\gamma\varepsilon$ and δ by $\delta\varepsilon$. In the limit, when $\varepsilon \to 0$, the equation becomes (iii).

Similarly in (v) replace w by $\varepsilon w\sqrt{2}$, z by $1+\varepsilon z\sqrt{2}$, α by $\dfrac{1}{2\varepsilon^4}$, γ by $-\dfrac{1}{\varepsilon^4}$ and δ by $-\left(\dfrac{1}{2\varepsilon^4}+\dfrac{\delta}{\varepsilon^2}\right)$. In the limit equation (iv) arises.

In (iii) replace z by $1+\varepsilon^2 z$, w by $1+2\varepsilon w$, γ by $\dfrac{1}{4\varepsilon^6}$, δ by $-\dfrac{1}{4\varepsilon^6}$, α by $-\dfrac{1}{2\varepsilon^6}$, β by $\dfrac{1}{2\varepsilon^6}+\dfrac{2\beta}{\varepsilon^3}$. In the limit the equation becomes (ii).

Similarly (ii) may be obtained from (iv) by replacing z by $\dfrac{\varepsilon z}{2^{\frac{1}{3}}}-\dfrac{1}{\varepsilon^3}$, w by $2^{\frac{2}{3}}\varepsilon w+\dfrac{1}{\varepsilon^3}$, α by $-\dfrac{1}{2\varepsilon^6}-\alpha$, β by $-\dfrac{1}{2\varepsilon^{12}}$ and taking the limit.

Finally, in (ii) replace z by $\varepsilon^2 z-\dfrac{6}{\varepsilon^{10}}$, w by $\varepsilon w+\dfrac{1}{\varepsilon^5}$, α by $\dfrac{4}{\varepsilon^{15}}$, and in the limit the equation degenerates into (i).

14·41. The First Painlevé Transcendent: Freedom from Movable Branch Points.—The equation

$$\text{(i)}\qquad \frac{d^2w}{dz^2}=6w^2+z,$$

satisfied by the first Painlevé transcendent, will now be studied in greater detail. It will first of all be proved that its general solution is free from movable critical points.* The principle of the method is applicable to the five equations which define the remaining transcendents.

The first step is to show that the equation admits of solutions possessing movable poles, but not movable branch points. In the neighbourhood of any arbitrary point z_0, the equation is satisfied by the series

$$w=\frac{1}{(z-z_0)^2}-\tfrac{1}{10}z_0(z-z_0)^2-\tfrac{1}{6}(z-z_0)^3+h(z-z_0)^4+\tfrac{1}{300}z_0^2(z-z_0)^6+\ldots,$$

where h is the second arbitrary parameter; this series may also be written in the form

$$w=\frac{1}{(z-z_0)^2}-\tfrac{1}{10}z(z-z_0)^2-\tfrac{1}{15}(z-z_0)^3+h(z-z_0)^4+\tfrac{1}{300}z_0^2(z-z_0)^6+\ldots$$

On eliminating $z-z_0$ between the latter series, and that for w', namely,

$$w'=-\frac{2}{(z-z_0)^3}-\tfrac{1}{5}z(z-z_0)-\tfrac{3}{10}(z-z_0)^2+4h(z-z_0)^3+\tfrac{1}{50}z_0^2(z-z_0)^6+\ldots,$$

and writing $w=v^{-2}$, it is found that

$$w'=-2\epsilon v^{-3}-\tfrac{1}{2}\epsilon zv-\tfrac{1}{2}v^2+7\epsilon hv^3+\ldots,$$

where $\epsilon=\pm 1$. Transform equation (i) by writing †

$$w=v^{-2},\quad w'=-2v^{-3}-\tfrac{1}{2}zv-\tfrac{1}{2}v^2+uv^3,$$

* Painlevé, *Bull. Soc. Math. France*, 28 (1900), p. 227; *C. R. Acad. Sc. Paris*, 135 (1902), pp. 411, 641, 757, 1020.

† Alternatively, the transformation

$$w=v^{-2},\quad w'=2v^{-3}+\tfrac{1}{2}zv-\tfrac{1}{2}v^2-uv^3$$

may be made

the equation then becomes the system

$$\text{(i}a\text{)}\qquad \begin{cases} \dfrac{dv}{dz}=1+\frac{1}{4}zv^4+\frac{1}{4}v^5-\frac{1}{2}uv^6, \\ \dfrac{du}{dz}=\frac{1}{8}z^2v+\frac{3}{8}zv^2+(\frac{1}{4}-zu)v^3-\frac{5}{4}uv^4+\frac{3}{2}u^2v^5. \end{cases}$$

This system has a unique solution which is analytic in the neighbourhood of z_0 and satisfies the initial conditions $u=u_0$, $v=0$ when $z=z_0$. The corresponding solution $w(z)$ has a pole at z_0 and the constant h is equal to $\frac{1}{7}u_0$.

Thus the general solution has a movable pole at any arbitrary point z_0. No solution can have an algebraic branch point at any point z_1, for if $A(z-z_1)^r$ is the dominant term of a solution having an algebraic singularity at z_1, r is necessarily -2, and then the solution is analytic in the neighbourhood of z_1.

14·42. Freedom from Movable Essential Singularities.—It has now to be shown that no solution of equation (i) can have a movable essential singularity in the finite part of the plane.* With this end in view, a number of preliminary theorems, relating to special solutions of (i) will first be proved. Let $w(z)$ be the particular solution which assumes the finite value w_0, while $w'(z)$ assumes the finite value w_0', when $z=z_0$. This solution is analytic in the neighbourhood of z_0; let Γ be the greatest circle whose centre is at z_0, within which $w(z)$ has no singularities other than poles. If the radius of Γ is infinite, the solution has no essential singularity except possibly at infinity, so that the theorem is proved. If the radius of Γ were finite, then on the circumference of Γ there would be an essential singularity of $w(z)$. It will be shown that this hypothesis is untenable.

Let the supposed essential singularity occur at $z=a$, and let M be the upper bound of $|w(z)|$ and $|w'(z)|$ as z tends to a along the radius z_0a. Assume first of all that the solution $w(z)$ is such that M is finite. Then if z_1 is a point on the radius, and $w(z_1)=w_1$, $w'(z_1)=w_1'$, and ϵ is arbitrary,

$$|w-w_1|\leqslant A, \quad |w'-w_1'|\leqslant A, \quad \text{when } |z-z_1|\leqslant\epsilon, \quad |z_1-a|\leqslant\epsilon,$$

where A is finite. Now (i) can be written as the system

$$\begin{cases} \dfrac{dw}{dz}=w', \\ \dfrac{dw'}{dz}=6w^2+z, \end{cases}$$

and the right-hand member of each equation of this system is finite for all finite values of z, w and w'. By the fundamental existence theorem (§ 12·2), there will exist a solution $w(z)$, satisfying the assigned initial conditions with respect to z_1, which will be analytic throughout the circle $|z-z_1|=\epsilon$. The solution will thus be analytic at a, contrary to hypothesis. It must, therefore, be supposed that if a is an essential singularity, $|w(z)|$ is not bounded on z_0a.

It will now be shown that, if $w(z)$ is any particular solution of (i) such that $|w(z)|$ is not bounded on z_0a, the point a is a pole of $w(z)$ provided that there exists a set of points z_1 on the radius, having a as their limit-point, such that $|w(z)|$ is unbounded, but, for a particular sign of $\pm w^{\frac{1}{2}}$, $|u(z_1)|<C$, where C is a fixed number.

Returning to the transformation

$$u(z)=\pm w^{\frac{1}{2}}\left\{\frac{dw}{dz}+\frac{1}{2w}\right\}+2w^3+\tfrac{1}{2}zw,$$

* The necessity for this discussion is illustrated by the examples in §§ 14·1, 14·39.

which is equivalent either to

$$u=v^{-3}(w'+2v^{-3}+\tfrac{1}{2}zv+\tfrac{1}{2}v^2)$$

or to

$$u=-v^{-3}(w'-2v^{-3}-\tfrac{1}{2}zv+\tfrac{1}{2}v^2),$$

where $w=v^{-2}$, it is seen that, if w is a solution having a pole at the point a, then, in the neighbourhood of a, one determination of u is such that

$$u(z)=7h+O\{(z-a)^2\}.$$

Now, from the assumption that, for one of the determinations of u, $|u(z_1)|<C$, it follows that one or other of the expressions

$$v^{-3}(w'+2v^{-3}+\tfrac{1}{2}zv+\tfrac{1}{2}v^2), \quad v^{-3}(w'-2v^{-3}-\tfrac{1}{2}zv+\tfrac{1}{2}v^2)$$

will, when $z=z_1$, be of modulus less than C. Suppose, for definiteness, that the first of these expressions satisfies this condition. As before, let (i) be transformed by the substitution

$$w=v^{-2}, \quad w'=-2v^{-3}-\tfrac{1}{2}zv-\tfrac{1}{2}v^2+uv^3;$$

the resulting system (i*a*) will have a solution $u(z)$, $v(z)$ such that u, v assume assigned initial values u_1, v_1 when $z=z_1$. Then, if ϵ is arbitrary,

$$|u-u_1|\leqslant K, \quad |v-v_1|\leqslant K, \text{ when } |z-z_1|\leqslant\epsilon, \quad |z_1-a|\leqslant\epsilon,$$

where K is finite, from which it follows, by the fundamental existence theorem that $u(z)$ and $v(z)$ are analytic throughout the circle $|z-a|=\epsilon$. Consequently $w(z)$ has a pole at a.

It is possible to find any number of functions $U(z)$ having the same property as $u(z)$, namely that if, for points z_1 on z_0a, having a as their limit-point, $|U(z_1)|$ is bounded whenever $|w(z_1)|$ is unbounded, then $w(z)$ has a pole at a. One such function may be constructed as follows, and has the advantage of being a rational expression in z, w, w'.

The two-valued function

$$u=w^{\frac{3}{2}}(w'+\tfrac{1}{2}w^{-1})+2w^3+\tfrac{1}{2}zw$$

is such that if w has a pole at z_0, one of the two determinations of u assumes the arbitrary value $7h$ when $z=z_0$. Whichever determination is the correct one, u satisfies the equation

$$\{w'+\tfrac{1}{2}w^{-1}-w^{\frac{3}{2}}(2+\tfrac{1}{2}zw^{-2}-uw^{-3})\}\{w'+\tfrac{1}{2}w^{-1}+w^{\frac{3}{2}}(2+\tfrac{1}{2}zw^{-2}-uw^{-3})\}=0.$$

The left-hand member of this equation, when expanded, is free from fractional powers of w and may be written

$$w'^2+\frac{w'}{w}-4w^3-2zw+4u+\ldots,$$

where the omitted terms involve w^{-1}, w^{-2} and w^{-3} but not w'. Let

$$U=w'^2+\frac{w'}{w}-4w^3-2zw,$$

then on substituting for w the series

$$w=(z-z_0)^{-2}-\tfrac{1}{10}z_0(z-z_0)^2-\tfrac{1}{6}(z-z_0)^3+h(z-z_0)^4+\ldots,$$

it is found that

$$U(z)=-28h+O\{(z-z_0)^2\}.$$

The fact that $U'(z_0)=0$ would introduce apparent complications into the later work. To avoid this difficulty, let

$$V(z)=U(z)+z,$$

then in the neighbourhood of z_0,

$$\begin{aligned}V(z)&=-28h+z+O\{(z-z_0)^2\}\\&=-4u(z)+z+O(w^{-1}),\end{aligned}$$

where $u(z)$ is that determination of u which is finite at z_0. Since $u'(z_0)=0$,

$$V'(z)=1+O(z-z_0).$$

Let z_1 be any value of z for which $|w|$ is unbounded, but $|V|$ bounded, the corresponding value of w' is either root of the equation

$$w'^2+w'w^{-1}-4w^3-2zw+z=V.$$

But since the corresponding value of u is determined by

$$w'^2+w'w^{-1}-4w^3-2zw+O(w^{-1})=-4u,$$

$$u=\tfrac{1}{4}(z-V)+O(w^{-1}),$$

and therefore $|u(z_1)|$ is bounded.

It follows that if there exists a set of points z_1 on the radius z_0a, having a as their limit-point, and such that $|w(z_1)|$ is unbounded, but $|V(z_1)|$ is bounded, then, for one determination of $u(z)$, $|u(z_1)|$ will be bounded, and consequently $w(z)$ will have a pole at a.

14·421. The Main Proof in the Case when $|w|$ has a positive Lower Bound.—An important restriction will now be imposed, and removed at a later stage, namely that if $w(z)$ is a solution having an essential singularity at a, then for all points on the radius z_0a, $|w(z)|\geqslant\rho$, a positive number. Then there must be a set of points z_1 on the radius, such that $|V(z_1)|$ is unbounded. For if $|w(z_1)|$ and $|V(z_1)|$ were both bounded, then, by the definition of V, $|w'(z_1)|$ would be bounded and $w(z)$ would be analytic at a. If, on the other hand, $|V(z_1)|$ were bounded, but $|w(z_1)|$ unbounded, then, by the concluding theorem of the preceding section, $w(z)$ would have a pole at a. Thus if a is an essential singular point, a set of points z_1 for which $|V(z_1)|$ is unbounded, certainly exists.

It will now be proved that, as a consequence of this result, another set of points z_2, having a as a limit-point, exists such that $|V(z_2)|$ is arbitrarily small. For consider the expression

$$W=\frac{V'}{V}=\frac{2w'w''+w^{-1}w''-w^{-2}w'^2-12w^2w'-2zw'-2w+1}{w'^2+w^{-1}w'-4w^3-2zw+z}$$

$$=\frac{4w^3-w'^2+zw+w^2}{w(ww'^2+w'-4w''-2zw^2+zw)}.$$

If $|W|$ were bounded on the radius z_0a, $|V|$ would be bounded, even for the set of points z_1, which is not true. Thus a set of points z_2 arbitrarily close to a, must exist such that $|W(z_2)|$ is unbounded. Moreover $|w(z_2)|$ is also unbounded. For if $|w(z_2)|$ and $|w'(z_2)|$ are bounded then $|V(z_2)|$ is arbitrarily small and $w(z)$ is analytic at a; if $|w(z_2)|$ is bounded, and $|w'(z_2)|$ unbounded, then $|W(z_2)|$ would be bounded, contrary to hypothesis.

Now if w' is eliminated between the expressions for V and W, it is found that

$$W=V^{-1}+O(w^{-\frac{1}{2}}),$$

and since, for the set of points z_2, having the limit-point a, $|w(z_2)|$ and $|W(z_2)|$ are unbounded, $|V(z_2)|$ is arbitrarily small. It follows from the conclusion of the preceding section that $w(z)$ has a pole for $z=a$.

The case in which $w(z)$ tends to a unique limit g as z approaches a along the radius can be dismissed at once, for the preceding investigation is not altered except in the non-essential point that in the expression for V, the term w'/w is replaced by $w'/(w-g)$. In particular, the proof holds good if $|w(z)|$, instead of having a positive lower bound, had the limit zero when $z=a$.

The choice of the radius z_0a as the line of approach to a is not an essential part of the proof; any curve of finite length, ending at a, no point of which, with the assumed exception of a, is an essential singularity of $w(z)$, would serve equally well.

14·422. Discussion of the Case in which the Lower Bound of $|w(z)|$ is zero.—All possible hypotheses have now been disposed of except one, namely that there exists, on the radius z_0a, a set of points z_1 having the limit-point a, such that $|w(z_1)|<\rho$, and another set of points z_2, also having the limit-point a, such that $|w(z_2)|>\rho$. It will be shown that, even in this case, a is a pole of $w(z)$.

Let $\lambda_1, \lambda_2, \ldots$ be a sequence of non-overlapping segments of the radius z_0a, at the end points of which $|w(z)|=\rho$, and within which $|w(z)|<\rho$; let $l_1, l_2, \ldots$ be the lengths of these segments. The existence of the set of points z_2 implies that the number of intervals λ is infinite. It will be shown that every segment λ_ν can be replaced by a curved segment Λ_ν, of length L_ν, where $1<L_\nu/l_\nu<3\pi$ along which $|w(z)|=\rho$ and such that, in the region between λ_ν and Λ_ν, $w(z)$ is analytic.

When z is regarded as dependent, and w as independent, variable, equation (i) becomes

$$(ib) \qquad \frac{d^2z}{dw^2}=-\left\{\frac{dw}{dz}\right\}^3(6w^2+z).$$

Let Z_ν be an end-point of λ_ν and let W_ν be the corresponding value of $w(z)$, so that $|W_\nu|=\rho$. Let $z(w)$ be the solution of (ib) such that

$$z(W_\nu)=Z_\nu, \quad z'(W_\nu)=Z_\nu'.$$

If $Z_\nu'=0$ this solution is merely $z=Z_\nu$; it does not involve w and therefore corresponds to no solution $w(z)$ of (i). It may therefore be supposed that $z'(W_\nu)\neq 0$. But if ε is a positive number less than $\frac{1}{2}$, a number τ can be found such that, when

$$|w|\leqslant\rho, \quad |Z_\nu'|\leqslant\tau,$$

then

$$z'=Z_\nu'(1+\delta),$$

where δ is analytic in w and Z_ν' and

$$|\delta|<\varepsilon.$$

As z describes the segment λ_ν, w will describe a curve C_ν in the w-plane; this curve C_ν will lie within a certain circle Γ_ν described about the point $w=0$ with radius ρ; the initial and final values of w will correspond to points on the circumference of Γ_ν. Let S_ν denote the length of C_ν. On the radius z_0a, let

$$z=a+re^{i\alpha},$$

where α is constant. Then

$$l_\nu=\int_0^{S_\nu}\left|z'(w)\frac{dw}{ds}\right|ds=\int_0^{S_\nu}\left|\frac{dr}{ds}\right|ds,$$

where the path of integration is the curve C_ν. Since

$$\left|\frac{dr}{ds}\right|=|z'|=|Z_\nu'|\,|1+\delta|>\tfrac{1}{2}|Z_\nu'|,$$

it follows that

$$l_\nu>\tfrac{1}{2}|Z_\nu'|S_\nu.$$

Now let w describe the smaller arc of Γ_ν between the end points of C_ν, let σ_ν be the length of this arc and k_ν the length of its chord. Then

$$\sigma_\nu\leqslant\pi k_\nu\leqslant\pi S_\nu.$$

But

$$L_\nu=\int_0^{\sigma_\nu}\left|\frac{dr}{d\sigma}\right|d\sigma\leqslant|Z_\nu'|\int_0^{\sigma_\nu}|1+\delta|\,d\sigma<\tfrac{3}{2}|Z_\nu'|\,\sigma_\nu,$$

that is

$$L_\nu<\tfrac{3}{2}\pi|Z_\nu'|S_\nu,$$

and consequently

$$1<\frac{L_\nu}{l_\nu}<3\pi.$$

Since $z'(w)$ is analytic and not zero within the circle Γ_ν and on its circumference, $w(z)$ will be free from poles in the region between the curve Λ_ν and the segment λ_ν. But $w(z)$ can have no singularities but poles in this region, and therefore λ_ν can be deformed into Λ_ν without meeting any singular point of $w(z)$. Thus, if each segment λ_ν is replaced by the corresponding arc Λ_ν, there is formed a path Λ, leading from z_0 to a, composed of an infinite number of arcs, whose total length does not exceed $3\pi R$, where R is the length of the radius z_0a. For all points of the path Λ,

$$|w(z)|\geqslant\rho,$$

and at its end-point a $w(z)$ is supposed to have an essential singularity. But the discussion of the previous section shows that this is impossible, and therefore, finally, $w(z)$ *has no essential singularity at any finite point of the z-plane.*

14·43. Representation of the Transcendent as the Quotient of two Integral Functions.—Let $w(z)$ be the first Painlevé transcendent, then since

$$\frac{d^2w}{dz^2}=6w^2+z,$$

if

$$\eta(z)=\tfrac{1}{2}\left(\frac{dw}{dz}\right)^2-2w^3-zw,$$

it follows that

$$\frac{d\eta}{dz}=-w$$

and $\eta(z)$ satisfies the equation

$$\frac{d^3\eta}{dz^3}+6\left(\frac{d\eta}{dz}\right)^2+z=0.$$

Since the only singular points of $w(z)$ are poles at which the development takes the form

$$w=(z-z_0)^{-2}+O\{(z-z_0)^2\},$$

the only singularities of $\eta(z)$ are simple poles. Let

$$\zeta(z)=e^{\int\eta(z)dz},$$

then $\zeta(z)$ is uniform, for although $\int\eta dz$ is infinitely many-valued, its values differ by additive multiples of $2\pi i$. But $\zeta(z)$ has no poles, it is therefore an integral function of z.

Thus $w(z)$ can be expressed in the form

$$w=\frac{\zeta'^2-\zeta\zeta''}{\zeta^2},$$

and both numerator and denominator of this expression are integral functions of z.

14·44. The Arbitrary Constants which enter into the Transcendent.—It will be shown that the transcendent is an essentially transcendental function of the two constants of integration. In the first place, it cannot be a rational function of two parameters, for, if it were, the solution of the equation

$$\frac{d^2w}{dz^2}=6w^2+a^5z,$$

obtained from (i) by replacing z by az and w by $a^{-2}w$, would also be rational in the constants of integration. But, when $a=0$, the solution

$$w=\wp(z+\beta, 0, \gamma)$$

is not rational in β and γ; it is therefore not rational in its parameters when $a\neq0$.

Suppose then that $w(z)$ were a semi-transcendental function of the constants of integration. Then (i) would admit of a first integral, polynomial in w and w', say

$$P(z, w, w')\equiv w'^m+Q_1(z, w)w'^{m-1}+\ \ldots\ +Q_{m-1}(z, w)w'+Q_m(z, w,)=0.$$

Since the solution of this first integral, that is the transcendent itself, is free from movable branch-points, Q_i is a polynomial in w of degree not exceeding $2i$. Replace z by z_0+az, w by $a^{-2}w$ and w' by $a^{-3}w'$, then

$$P(z, w, w')=a^{-k}P_0(w, w')+O(a^{-k+1}) \qquad (k\geqslant 3m),$$

where $P_0(w, w')$ is a homogeneous polynomial in $\sqrt[3]{w'}$, $\sqrt{w}$. But $P_0=0$ is a first integral of the equation

$$w''=6w^2,$$

and therefore P_0 is of the form

$$P_0=K(w'^2-4w^3)^j,$$

where K and j are constants. It is easily verified that

$$k=3m=6j$$

and that, in consequence, $Q_m(z, w)$ is of degree $\frac{3}{2}m$ in w.

Now $w(z)$ admits of movable poles, and in the neighbourhood of such a pole there is a relation of the form (§ 14·41)

$$w'+\frac{1}{2w}+\ldots=\pm(2w^{\frac{3}{2}}+\tfrac{1}{2}zw^{-\frac{1}{2}}-7hw^{-\frac{3}{2}}+\ldots)$$

(where h is a constant), in which the integral and fractional powers of w have been disposed on opposite sides of the equation. For large values of w, every root w' of the equation

$$P(z, w, w')=0$$

must be expressible in this form, and therefore

$$\begin{aligned}P(z, w, w')&=\prod_{i=1}^{j}\{(w'+\tfrac{1}{2}w^{-1}+\ldots)^2-w^3(2+\tfrac{1}{2}zw^{-2}-7h_i w^{-3}+\ldots)^3\}\\&=w'^{2j}+j\frac{w'^{2j-1}}{w}+\ldots,\end{aligned}$$

which is impossible, since the right-hand member is not a polynomial in w. Consequently *the first Painlevé transcendent is an essentially-transcendental function of two parameters.*

Yet it might be supposed that equation (i) could possess particular solutions which are either algebraic or expressible in terms of the classical transcendents. If the solution $w(z)$ were algebraic, it would be developable, for large values of $|z|$ as a series

$$w=a_\nu z^\nu+a_{\nu-1}z^{\nu-1}+a_{\nu-2}z^{\nu-2}+\ldots.$$

If ν were negative or zero, w and w' would be finite for $z=\infty$, and therefore the equation would not be satisfied. If $\nu>0$, ν must be an integer on account of the term z in the equation, but when ν is a positive integer, the term $z^{2\nu}$ introduced by the term w^2 in the equation, is uncompensated, and the equation cannot be satisfied. Consequently $w(z)$ is transcendental.

Suppose that $w(z)$ is a classical transcendent, then it must satisfy an algebraic differential equation distinct from (i). By eliminating the higher differential coefficients between (i) and equations derived from it, on the one hand, and the new equation, on the other, an equation of the form

$$P(z, w, w')=0$$

is arrived at, in which P is a polynomial in w and w'. But it has just been shown that this is impossible, and therefore *no particular solution exists which reduces to a known function.*

14·45. The Asymptotic Relationship between the First Painlevé Transcendent and the Weierstrassian Elliptic Function.—Although the first Painlevé transcendent is an essentially new function, yet it is, in a certain sense, asymptotic to the elliptic $\wp$-function.* This property is somewhat analogous to the property of the Bessel function $J_n(z)$ that, when $|z|$ is large,†

$$J_n(z) \sim \left(\frac{2}{\pi z}\right)^{\frac{1}{2}} \cos\left(z-\tfrac{1}{2}n\pi-\tfrac{1}{4}\pi\right).$$

The equation

$$\frac{d^2w}{dz^2}=6w^2-6z^\mu$$

is not essentially different, when $\mu=1$, from the equation satisfied by the transcendent. Make the transformation

$$w=z^{\frac{1}{2}\mu}W, \quad Z=\frac{4}{\mu+4}z^{\frac{1}{4}\mu+1},$$

then the equation becomes

$$\frac{d^2W}{dZ^2}=6W^2-6-\frac{5\mu}{(\mu+4)Z}\cdot\frac{dW}{dZ}+\frac{4\mu(2-\mu)}{(\mu+4)^2Z^2}W.$$

This last equation may be compared with

$$\frac{d^2V}{dZ^2}=6V^2-6,$$

an equation whose general solution is

$$V=\wp(Z-\beta, 12, \gamma),$$

where β and γ are constants of integration. This comparison suggests that, for large values of $|Z|$,

$$W \sim \wp(Z-\beta, 12, \gamma).$$

and that, if $w(z)$ is the Painlevé transcendent,

$$w(z) \sim \wp(\tfrac{4}{5}z^{\frac{5}{4}}-\beta, 12, \gamma).$$

This question was thoroughly investigated by Boutroux, who determined the region wherein the asymptotic relation, for determinate values of β and γ, was valid.

[For details of the proof, the reader is referred to the papers quoted.]

In conclusion, a theorem due to Painlevé may be stated: the equation

$$w(z)=A$$

has an infinite number of roots for any value of the constant A.

14·5.—Equations of the Second Order, algebraic in w.—The general problem of finding necessary and sufficient conditions that the general solution of

$$\frac{d^2w}{dz^2}=F(z, w, p) \qquad \left(p\equiv\frac{dw}{dz}\right)$$

should be free from movable critical points, when F is rational in p, algebraic in w, and analytic in z, demands a knowledge of the theory of algebraic functions.‡

* Boutroux, *Ann. Éc. Norm.* (3), 30 (1913), p. 255; 31 (1914), p. 99. The second Painlevé transcendent (Equation ii, § 14·4) is asymptotically related to the Jacobian elliptic function sn(z).

† Whittaker and Watson, *Modern Analysis*, § 17·5; Watson, *Bessel Functions*, § 7·1.

‡ The essential point is that when the equation is expressed, as is always possible, in the form

$$\frac{d^2w}{dz^2}=\Phi(z, w, u, p),$$

where Φ is rational in w, u and p, and w and u are connected by the relation

$$H(z, w, u)=0,$$

in which H is a polynomial in w and u whose coefficients are analytic functions of z, the *genus* of the relation $H=0$ is 0 or 1. When the genus is 0, the equation is reducible to one or other of the fifty types already enumerated; when the genus is 1, the equation belongs to one of the three new classes.

Apart from the types already enumerated, there are three, and only three types of equation whose critical points are fixed. They are as follows:

$$\text{(i)} \qquad \frac{d^2w}{dz^2}=\frac{6w^2-\frac{1}{2}g_2}{4w^3-g_2w-g_3}\left(\frac{dw}{dz}\right)^2+q(z)\frac{dw}{dz}+r(z)\sqrt{\{4w^3-g_2w-g_3\}}.$$

This equation is equivalent to the system

$$\begin{cases} \dfrac{dw}{dz}=u\sqrt{\{4w^3-g_2w-g_3\}}, \\ \dfrac{du}{dz}=q(z)u+r(z)\,; \end{cases}$$

its solution is therefore a semi-transcendental function of the constants of integration. By a change of variables the system may be reduced to

$$\begin{cases} \dfrac{dW}{dZ}=U\sqrt{\{4W^3-g_2W-g_3\}}, \\ \dfrac{dU}{dZ}=0, \end{cases}$$

and is therefore equivalent to

$$\frac{d^2W}{dZ^2}=\frac{6W^3-\frac{1}{2}g_2}{4W^3-g_2W-g_3}\left(\frac{dW}{dZ}\right)^2$$

(§ 14·38, equation XLIX.).

$$\text{(ii)} \qquad \frac{d^2w}{dz^2}=\frac{1}{2}\left\{\frac{1}{w}+\frac{1}{w-1}+\frac{1}{w-z}\right\}\left(\frac{dw}{dz}\right)^2-\left\{\frac{1}{z}+\frac{1}{z-1}+\frac{1}{w-z}\right\}\frac{dw}{dz}$$
$$+\frac{w(w-1)}{2z(z-1)(w-z)}+q(z)\sqrt{\{w(w-1)(w-z)\}}.$$

The general solution is an essentially-transcendental function of two constants; it may be arrived at as follows: Let $u_1(z)$ be any solution of

$$u''-\frac{2z-1}{z(z-1)}u'+\frac{u}{4z(z-1)}=q(z)\,;$$

let $\Lambda(u, z)$ be defined by the inversion of

$$\int_0^w \frac{dw}{\sqrt{\{w(w-1)(w-z)\}}}=u,$$

and let $2\omega_1$, $2\omega_2$ be its periods. Then the general solution of the equation considered is

$$u=\Lambda(u_1+K_1\omega_1+K_2\omega_2, z),$$

where K_1 and K_2 are the constants of integration. Thus the equation does not lead to any new type of transcendental function.

$$\text{(iii)} \qquad \frac{d^2w}{dz^2}=\left\{\frac{6w^2-\frac{1}{2}g_2}{4w^3-g_2w-g_3}+\frac{i\pi}{\omega\sqrt{(4w^3-g_2w-g_3)}}\right\}\left(\frac{dw}{dz}\right)^2+q(z)\frac{dw}{dz}$$
$$+r(z)\sqrt{(4w^3-g_2w-g_3)},$$

in which 2ω is any period of $\wp(u, g_2, g_3)$. The equation is equivalent to the system

$$\begin{cases} \dfrac{dw}{dz}=u\sqrt{(4w^3-g_2w-g_3)}, \\ \dfrac{du}{dz}=\dfrac{i\pi}{\omega}u^2+q(z)u+r(z)\,; \end{cases}$$

its solution is thus a semi-transcendental function of the constants of integration. The system may be transformed into

$$\begin{cases} \dfrac{dW}{dZ} = U\sqrt{(4W^3 - g_2W - g_3)}, \\ \dfrac{dU}{dZ} = \dfrac{i\pi}{\omega}U^2, \end{cases}$$

and therefore the original equation is equivalent to

$$\frac{d^2W}{dZ^2} = \left\{ \frac{6W^2 - \frac{1}{2}g_2}{4W^3 - g_2W - g_3} + \frac{i\pi}{\omega\sqrt{(4W^3 - g_2W - g_3)}} \right\} \left(\frac{dW}{dZ} \right)^2,$$

which is the simplest equation of this particular type.

Another question now arises, but cannot be dealt with in full here, namely whether or not it is possible, when the general solution of an equation is free from movable critical points, to have a singular solution whose critical points are not fixed.*

The following example shows that this may actually happen :

The general solution of the equation

$$w'' = -w^3w' + ww'\sqrt{(4w' + w^4)}$$

is

$$w = A \tan (A^3z + B),$$

a singular solution is

$$w = \sqrt[3]{\left(\frac{4}{3(z-C)} \right)},$$

where A, B and C are arbitrary constants.

14·6. Equations of the Third and Higher Orders.—The principle of Painlevé's α-method, which enabled a complete discussion of equations of the second order to be carried out, may be applied to the discussion of equations of the third and higher orders.†

As before the method naturally divides itself into two stages, the determination of conditions which are necessary for the absence of movable critical points, and the subsequent proof of the sufficiency of these conditions. There is no difficulty whatever in extending the method for the determination of the necessary conditions, but the difficulty of proving that these conditions are sufficient increases with the order of the equations discussed.

* Chazy, *C. R. Acad. Sc. Paris*, 148 (1909), p. 157.

† Painlevé, *Bull. Soc. Math. France*, 28 (1900), p. 252; Chazy, *C. R. Acad. Sc. Paris*, 145 (1907), p. 305, 1263; 149 (1909), p. 563; 150 (1910), p. 456; 151 (1910), p. 203; 155 (1912), p. 132; *Acta Math.* 34 (1911), p. 317. Garnier, *C. R.* 145 (1907), p. 308; 147 (1908), p. 915; *Ann. Éc. Norm.* (3), 29 (1912), p. 1.

CHAPTER XV

LINEAR EQUATIONS IN THE COMPLEX DOMAIN

15·1. The a priori Knowledge of the Singular Points.—It will be convenient to begin this present chapter by recalling a number of established theorems relating to the homogeneous linear equation of order n

$$(A)\qquad \frac{d^nw}{dz^n}+p_1(z)\frac{d^{n-1}w}{dz^{n-1}}+\ldots+p_{n-1}(z)\frac{dw}{dz}+p_n(z)w=0.$$

Let z_0 be any point in the neighbourhood of which the n coefficients are analytic. Then, by the existence theorem of § 12·22, there exists a unique solution, such that this solution and its first $n-1$ derivatives assume any arbitrarily-assigned values when $z=z_0$. This solution is expressible as a power series in $z-z_0$, which converges at least within the circle whose centre is z_0 and whose circumference passes through that singular point of the coefficients which lies nearest to z_0. In other words, the singularities of the solutions can be none other than the singularities of the equation, and therefore movable singularities, even movable poles, cannot arise when the equation is linear.

Again, the general theory of the linear equation with real coefficients, as expounded in Chapter V., may be transferred to the complex domain when obvious verbal changes in the investigation have been made. In particular, if

$$w_1,\ w_2,\ \ldots,\ w_n$$

are n distinct solutions, forming a fundamental set, the Wronskian

$$\Delta(w_1,\ w_2,\ \ldots,\ w_n)$$

cannot vanish when $z=z_0$. Since

$$\Delta=\Delta_0\exp\{-\int_{z_0}^{z}p_1(z)dz\},$$

where Δ_0 is the value of Δ when $z=z_0$, and the path of integration is restricted to lie within the region containing z_0 within which $p_1(z)$ is analytic, it is clear that Δ cannot vanish at any point except possibly a singular point of $p_1(z)$.

The point at infinity is or is not a singular point, according as the coefficients of the equation obtained by the substitution

$$z=\zeta^{-1}$$

followed by a reduction to the form (A) have or have not singularities at the origin.

Thus the singular points can immediately be found by mere inspection of the equation. For any non-singular point a fundamental set of n distinct solutions can be found; the question now at issue is to determine whether there also exists a fundamental set of solutions relative to any given singular point, and having demonstrated the existence of these solutions, to investigate their behaviour in the neighbourhood of the singular point. This investigation leads to what is known as the *Fuchsian Theory* of linear differential equations.*

15·2. Closed Circuits enclosing Singular Points.—Let the coefficients of the equation (A) be one-valued and have only isolated singular points. Let

$$w_1, w_2, \ldots, w_n$$

be a fundamental set of solutions and let z_0 be any ordinary (*i e.* non-singular) point of the equation. A simple closed circuit γ is drawn, beginning and ending at z_0, not passing through any singular point, but possibly enclosing one or more singular points in its interior. Let $W_1, W_2, \ldots W_n$ be what $w_1, w_2, \ldots, w_n$ respectively become after the variable z has described the circuit γ in the positive direction. The determination of $W_1, W_2, \ldots, W_n$ may be carried out by the process of analytical continuation in a finite number of steps.†

Since the coefficients $p_1(z), p_2(z), \ldots, p_n(z)$ are unaltered by the description of this circuit, the equation as a whole is unchanged, that is to say, the functions

$$W_1, W_2, \ldots, W_n$$

are solutions of (A); they may therefore be expressed linearly in terms of the fundamental system $w_1, w_2, \ldots, w_n$, thus

$$\text{(B)} \qquad \begin{cases} W_1=a_{11}w_1+a_{12}w_2+\ldots+a_{1n}w_n, \\ W_2=a_{21}w_1+a_{22}w_2+\ldots+a_{2n}w_n, \\ \cdot \quad \cdot \quad \cdot \quad \cdot \\ W_n=a_{n1}w_1+a_{n2}w_2+\ldots+a_{nn}w_n, \end{cases}$$

where the coefficients a are numerical constants.

At any point z on the contour,

$$\Delta(w_1, w_2, \ldots, w_n)=\Delta_0 \exp\{-\int_{z_0}^{z} p_1(z)dz\},$$

the integral being described from z_0 to z along that branch of the contour which has the interior of the contour on its left-hand side. Let Δ_1 be the value of the Wronskian after a complete description of the circuit γ, then

$$\begin{aligned}\Delta_1&=\Delta_0 \exp\{-\int_{\gamma} p_1(z)dz\}\\ &=e^{-2\pi iR}\Delta_0,\end{aligned}$$

where R denotes the sum of the residues of $p_1(z)$ at the poles which lie within the contour. Thus

$$\Delta(W_1, W_2, \ldots, W_n)$$

is not zero at $z=z_0$, and since, at any ordinary point z,

$$\Delta(W_1, W_2, \ldots, W_n)=\Delta_1 \exp\{-\int_{z_0}^{z} p_1(z)dz\}\neq 0,$$

$W_1, W_2, \ldots, W_n$ form a fundamental set of solutions.

It may be remarked in passing that

$$\begin{aligned}|a_{rs}| &= \frac{\Delta(W_1, W_2, \ldots, W_n)}{\Delta(w_1, w_2, \ldots, w_n)}\\ &=\Delta_1/\Delta_0\neq 0.\end{aligned}$$

* Riemann (Posthumous Fragment dated 1857), *Ges. Werke* (2nd ed.), p. 379; Fuchs, *J. für Math.* 66 (1866), p. 121; 68 (1868), p. 354 [*Ges. Werke*, 1, pp. 159, 205].

† If the length of the circuit is l, and the distance of any singular point from any point of the circuit is greater than d, the number of steps required will not be greater than N where N is the integer next above $l/2d$.

Now that these preliminary results are established, it is possible to determine constants $\lambda_1, \lambda_2, \ldots, \lambda_n$ such that the particular solution

$$u=\lambda_1 w_1+\lambda_2 w_2+\ldots+\lambda_n w_n$$

becomes su after the circuit has been completely described once, where s is a numerical constant. For let u become U after description of the circuit, then

$$U=\lambda_1 W_1+\lambda_2 W_2+\ldots+\lambda_n W_n,$$

so that, if $U=su$,

$$s(\lambda_1 w_1+\lambda_2 w_2+\ldots+\lambda_n w_n)=\sum_{r=i}^{n}\lambda_r(a_{r1}w_1+a_{r2}w_2+\ldots+a_{rn}w_n).$$

This relation is to hold identically, and therefore

(C) $$s\lambda_r=\lambda_1 a_{1r}+\lambda_2 a_{2r}+\ldots+\lambda_r a_{rr}+\ldots+\lambda_n a_{nr} \qquad (r=1, 2, \ldots, n).$$

When the undetermined constants λ_r are eliminated from this set of simultaneous equations, the equation to be satisfied by s is found, namely,

$$\begin{vmatrix} a_{11}-s, & a_{21}, & \ldots, & a_{n1} \\ a_{12}, & a_{22}-s, & \ldots, & a_{n2} \\ \cdot & \cdot & \cdot & \cdot \\ a_{1n}, & a_{2n}, & \ldots, & a_{nn}-s \end{vmatrix}=0.$$

This determinantal equation is known as the *characteristic equation* of the system chosen. It cannot have a zero root as otherwise $|a_{rs}|$ would be zero, contrary to the hypothesis that the system chosen is fundamental. To any value of s which satisfies the characteristic equation corresponds a set of constants $\lambda_1, \lambda_2, \ldots, \lambda_n$ whose ratios may be evaluated from equations (C). These lead to a solution u determinate apart from a constant factor, which becomes su after the point z has completely described the circuit γ.

The characteristic equation is *invariant*, that is to say, it is independent of the initial choice of a fundamental system. For let

$$v_1, v_2, \ldots, v_n$$

be a fundamental system distinct from that originally chosen; it must be linearly related to the former one, thus

$$\begin{aligned} v_1&=c_{11}w_1+c_{12}w_2+\ldots+c_{1n}w_n, \\ v_2&=c_{21}w_1+c_{22}w_2+\ldots+c_{2n}w_n, \\ &\cdots\cdots \\ v_n&=c_{n1}w_1+c_{n2}w_2+\ldots+c_{nn}w_n, \end{aligned}$$

where the coefficients c_{rs} are constants such that $|c_{rs}|\neq 0$. Suppose that, after the circuit has been described, the solutions $v_1, v_2, \ldots, v_n$ become respectively $V_1, V_2, \ldots, V_n$, then

$$\begin{aligned} V_1&=A_{11}v_1+A_{12}v_2+\ldots+A_{1n}v_n, \\ V_2&=A_{21}v_1+A_{22}v_2+\ldots+A_{2n}v_n, \\ &\cdots\cdots \\ V_n&=A_{n1}v_1+A_{n2}v_2+\ldots+A_{nn}v_n, \end{aligned}$$

where $|A_{rs}|\neq 0$. Hence

$$V_r=\sum_{s=1}^{n}A_{rs}(c_{s1}w_1+c_{s2}w_2+\ldots+c_{sn}w_n).$$

But also

$$V_r = c_{r1}W_1 + c_{r2}W_2 + \dots + c_{rn}W_n$$
$$= \sum_{s=1}^{n} c_{rs}(a_{s1}w_1 + a_{s2}w_2 + \dots + a_{sn}w_n).$$

Thus, by comparison of the coefficient of w_t,

$$\sum_{s=1}^{n} A_{rs}c_{st} = \sum_{s=1}^{n} c_{rs}a_{st} \qquad \begin{pmatrix} r=1, 2, \dots, n \\ t=1, 2, \dots, n \end{pmatrix}.$$

Now, by virtue of these relations, the product *

$$\begin{vmatrix} c_{11}, & c_{12}, & \dots, & c_{1n} \\ c_{21}, & c_{22}, & \dots, & c_{2n} \\ \cdot & \cdot & \cdot & \cdot \\ c_{n1}, & c_{n2}, & \dots, & c_{nn} \end{vmatrix} \begin{vmatrix} a_{11}-s, & a_{21}, & \dots, & a_{n1} \\ a_{12}, & a_{22}-s, & \dots, & a_{n2} \\ \cdot & \cdot & \cdot & \cdot \\ a_{1n}, & a_{2n}, & \dots, & a_{nn}-s \end{vmatrix}$$

and the product

$$\begin{vmatrix} c_{11}, & c_{21}, & \dots, & c_{n1} \\ c_{12}, & c_{22}, & \dots, & c_{n2} \\ \cdot & \cdot & \cdot & \cdot \\ c_{1n}, & c_{2n}, & \dots, & c_{nn} \end{vmatrix} \begin{vmatrix} A_{11}-s, & A_{12}, & \dots, & A_{1n} \\ A_{21}, & A_{22}-s, & \dots, & A_{2n} \\ \cdot & \cdot & \cdot & \cdot \\ A_{n1}, & A_{n2}, & \dots, & A_{nn}-s \end{vmatrix}$$

are exactly equal. It follows that

$$\begin{vmatrix} a_{11}-s, & a_{21}, & \dots, & a_{n1} \\ a_{12}, & a_{22}-s, & \dots, & a_{n2} \\ \cdot & \cdot & \cdot & \cdot \\ a_{1n}, & a_{2n}, & \dots, & a_{nn}-s \end{vmatrix} = \begin{vmatrix} A_{11}-s, & A_{21}, & \dots, & A_{n1} \\ A_{12}, & A_{22}-s, & \dots, & A_{n2} \\ \cdot & \cdot & \cdot & \cdot \\ A_{1n}, & A_{2n}, & \dots, & A_{nn}-s \end{vmatrix}$$

identically with respect to s.

15·21. Non-Repeated Roots of the Characteristic Equation.—In the first place, let the characteristic equation have n unequal roots $s_1, s_2, \dots, s_n$. Then there exist n solutions $u_1, u_2, \dots, u_n$ which, after the circuit has been once described, become $U_1, U_2, \dots, U_n$ respectively, where

$$U_1 = s_1u_1, \quad U_2 = s_2u_2, \quad \dots, \quad U_n = s_nu_n.$$

The solutions $u_1, u_2, \dots, u_n$ are fully equivalent to the original set, and form a fundamental system.

Consider in particular the case where the contour encloses one singular point only,† say $z=\zeta$, and consider the multiform function $(z-\zeta)^\rho$. After one complete circuit has been described, this function becomes $e^{2\pi i\rho}(z-\zeta)^\rho$. Let ρ_k be chosen so that

$$s_k = e^{2\pi i\rho_k},$$

then the function

$$\phi(z-\zeta) = (z-\zeta)^{-\rho_k}u_k$$

will return to its initial value after the description of a complete circuit about ζ; in other words $\phi(z-\zeta)$ is a uniform function of z in the domain of the point ζ.

Moreover ρ_k is undetermined, in the sense that it may be replaced by $\rho_k \pm m$ where m is any positive integer. If ρ_k can be so determined that

* For the rule for multiplying together two determinants of the same order, see Scott and Mathews, *Theory of Determinants*, Chap. V.

† The contour might now conveniently be taken to be the circle $|z-\zeta|=R$ where, if z_1 is the nearest singular point to ζ, R is any number less than $|z_1-\zeta|$.

$\phi_k(0)$ is finite, but not zero, the solution is said to be regular. A regular solution is therefore one which is expressible in the form

$$u_k=(z-\zeta)^{\rho_k}\phi_k(z-\zeta),$$

where

$$\phi(z-\zeta)=O(1) \text{ as } z\to\zeta.$$

The index ρ_k is known as the kth *exponent* relative to the *regular singular point* $z=\zeta$.

If ρ_k cannot be determined in this way, $\phi_k(z-\zeta)$ (and therefore u_k) has an essential singularity * at $z=\zeta$; the solution is then said to be irregular.

This occurs, for instance, when

$$\phi(z-\zeta)=e^{1/(z-\zeta)}.$$

15·22. The Case of Repeated Roots.—Suppose now that the characteristic equation has repeated roots, for instance let the root s_1 be repeated m times, s_2 repeated m_2 times and so on until the enumeration of the roots is complete. Then

$$m_1+m_2+\ \ldots\ =n.$$

It will now be proved † that, corresponding to any root s of multiplicity m, there exists a sub-set of $\mu(\leqslant m)$ linearly distinct solutions

$$v_1,\ v_2,\ \ldots,\ v_\mu,$$

which become respectively, after the circuit has been described,

$$sv_1,\quad s(v_2+v_1),\quad \ldots,\quad s(v_\mu+v_{\mu-1}).$$

The remaining solutions $v_{\mu+1}, v_{\mu+2}, \ldots, v_m$ give rise to other sub-sets with the same multiplier s. In other words, what has to be proved is that the set of n linear transformations (§ 15·2, B) may be replaced by the aggregate of a number of sub-sets of which

$$V_1=sv_1,\quad V_2=s(v_2+v_1),\quad \ldots,\quad V_\mu=s(v_\mu+v_{\mu-1})$$

is typical, $v_1, v_2, \ldots, v_\mu$ being linear combinations of $w_1, w_2, \ldots, w_n$. This will be proved by induction, the first step being to assume it true with regard to an $(n-1)$-fold system, and to deduce from this assumption its truth in the case of an n-fold system.

Let σ be any root of the characteristic equation; then there exists a solution v such that

$$V=\sigma v.$$

Of the solutions $w_1, w_2, \ldots, w_n$, at least $n-1$ are linearly independent of v; let them be $w_2, \ldots, w_n$. After the circuit has been described they become $W_2, \ldots, W_n$ respectively, where

$$\text{(C)}\qquad \begin{cases} W_2=b_2v+b_{22}w_2+\ \ldots\ +b_{2n}w_n, \\ \cdot\quad\cdot\quad\cdot\quad\cdot\quad\cdot\quad\cdot \\ W_n=b_nv+b_{n2}w_2+\ \ldots\ +b_{nn}w_n. \end{cases}$$

But

$$\begin{vmatrix} s, & 0, & \ldots, & 0 \\ b_2, & b_{22}, & \ldots, & b_{2n} \\ \cdot & \cdot & \cdot & \cdot \\ b_n, & b_{n2}, & \ldots, & b_{nn} \end{vmatrix} \neq 0,$$

from which it follows that

$$|b_{ij}|\neq 0 \qquad \begin{pmatrix} i=2, 3, \ldots, n \\ j=2, 3, \ldots, n \end{pmatrix}.$$

* ζ is also said to be a *point of indetermination*.

† Fuchs, *J. für Math.* 66 (1866), p. 136 [*Ges. Werke*, 1, p. 174]; Hamburger, *J. für Math.* 76 (1873), p. 121.

Write

$$W_2 - b_2 v = W_2', \quad \ldots, \quad W_n - b_n v = W_n',$$

then

$$(\mathrm{C}') \quad \begin{cases} W_2' = b_{22}w_2 + \ldots + b_{2n}w_n, \\ \cdot \quad \cdot \quad \cdot \quad \cdot \quad \cdot \quad \cdot \\ W_n' = b_{n2}w_2 + \ldots + b_{nn}w_n \end{cases}$$

is a set of linear transformations on $n-1$ symbols, with non-zero determinant. It follows from the assumption made, that $w_2, \ldots, w_n$ may be replaced by linear combinations of these symbols, say

$$u_1, \quad u_2, \quad \ldots, \quad u_{n-1},$$

which become $U_1', U_2', \ldots, U_{n-1}'$ after description of the circuit.

Then the system (C′) is transformed into

$$U_1' = su_1, \quad U_2' = s(u_2 + u_1), \quad \ldots, \quad U_\mu' = s(u_\mu + u_{\mu-1}),$$

together with other similar sub-sets giving in all $n-1$ equations. But if the transformation which changes $w_2, \ldots, w_n$ into $u_1, \ldots, u_{n-1}$ is applied to the system (C) instead of to the system (C′), the former system will become

$$U_1 = su_1 + k_1 v, \quad U_2 = s(u_2 + u_1) + k_2 v, \quad \ldots, \quad U_\mu = s(u_\mu + u_{\mu-1}) + k_\mu v,$$

where $k_1, k_2, \ldots, k_\mu$ are definite constants depending upon certain of the coefficients b_r. Now write

$$u_1 = v_1 + \lambda_1 v, \quad u_2 = v_2 + \lambda_2 v, \quad \ldots, \quad u_\mu = v_\mu + \lambda_\mu v,$$

where $\lambda_1, \lambda_2, \ldots, \lambda_\mu$ are arbitrary constants. Let the quantities $v_1, v_2, \ldots, v_\mu$ thus defined become $V_1, V_2, \ldots, V_\mu$ when $u_1, u_2, \ldots, u_\mu$ become $U_1, U_2, \ldots, U_\mu$, so that

$$U_1 = V_1 + \lambda_1 \sigma v, \quad U_2 = V_2 + \lambda_2 \sigma v, \quad \ldots, \quad U_\mu = V_\mu + \lambda_\mu \sigma v.$$

Then

$$(\mathrm{C}'') \quad \begin{cases} V_1 = sv_1 + \{k_1 - (\sigma - s)\lambda_1\}v, \\ V_r = s(v_r + v_{r-1}) + \{k_r - (\sigma - s)\lambda_r + s\lambda_{r-1}\}v \qquad (r = 2, 3, \ldots, \mu). \end{cases}$$

In the first place, let $\sigma \neq s$, then $\lambda_1, \lambda_2, \ldots, \lambda_\mu$ may be chosen so that the coefficient of v is zero in each case. Then the set of substitutions assumes the *canonical form*

$$V_1 = sv_1, \quad V_2 = s(v_2 + v_1), \quad \ldots, \quad V_\mu = s(v_\mu + v_{\mu-1}).$$

In the second place, let $\sigma = s$, then if $k_1 = 0$, $\lambda_1, \lambda_2, \ldots \lambda_{\mu-1}$ may be chosen to make the coefficient of v disappear, and the set of substitutions again assumes the canonical form as above. On the other hand, if $k_1 \neq 0$, v may be replaced by sv/k_1 throughout and $\lambda_1, \lambda_2, \ldots, \lambda_{\mu-1}$ chosen so as to make the coefficients of v, in all equations but the first, vanish. The canonical set of substitutions then becomes

$$V = sv, \quad V_1 = s(v_1 + v), \quad V_2 = s(v_2 + v_1), \quad \ldots, \quad V_\mu = s(v_\mu + v_{\mu-1}).$$

There may also arise two or more sets of substitutions (C″) with the same factor* $s = \sigma$. They may be reduced, by proper choice of the constant, λ, to

$$\begin{aligned} &V_1 = sv_1 + k_1 v, && V_2 = s(v_2 + v_1), \quad \ldots, && V_\mu = s(v_\mu + v_{\mu-1}), \\ &V_{\mu+1} = sv_{\mu+1} + k_1' v, && V_{\mu+2} = s(v_{\mu+2} + v_{\mu+1}), \quad \ldots, && V_\nu = s(v_\nu + v_{\nu-1}), \end{aligned}$$

etc., and it is assumed that $k_1 \neq 0$, $k_1' \neq 0, \ldots$. As before, by replacing v

* No special treatment is required when there are several sets of substitutions with a factor $s \neq \sigma$, as the reduction of each set to canonical form is immediate. The only case which calls for special mention is the one treated, where $s = \sigma$, $k_1 \neq 0$, $k_1' \neq 0$, etc.

by sv/k_1, k_1 is replaced by s, and the first set, taken together with the substitution $V=sv$ becomes canonical. In the second set, write

$$v_{\mu+r}=v'_{\mu+r}+k_1'v_r/k_1,$$
$$V_{\mu+r}=V'_{\mu+r}+k_1'V_r/k_1,$$

then

$$V'_{\mu+1}=sv'_{\mu+1},\quad V'_{\mu+2},=s(v'_{\mu+2}+v'_{\mu+1})\quad \ldots \quad V'_\nu=s(v'_\nu+v'_{\nu-1}),$$

which is of canonical form. The remaining sub-sets, if there are any, are dealt with in the same way. Thus the first part of the theorem is proved, namely that if a set of $n-1$ substitutions can be reduced to canonical form, a set of n substitutions can similarly be reduced. But when $n=1$ the theorem is obviously true, in fact trivial; it is therefore true generally.

15·23. Solutions of a Canonical Sub-Set.—It has thus been proved that corresponding to an m-ple root s of the characteristic equation there exists a set of m solutions,

$$v_1,\ v_2,\ \ldots,\ v_m$$

which may be arranged in sub-sets so that, if the solutions become

$$V_1,\ V_2,\ \ldots,\ V_m$$

when the circuit has been described,

$$V_1=sv_1,\qquad V_2=s(v_2+v_1),\ \ldots,\qquad V_\mu=s(v_\mu+v_{\mu-1}),$$
$$V_{\mu+1}=sv_{\mu+1},\ V_{\mu+2}=s(v_{\mu+2}+v_{\mu+1}),\ \ldots,\ V_\nu=s(v_\nu+v_{\nu-1}),$$
$$\cdot\quad\cdot\quad\cdot$$

Consider the first sub-set, supposing as before that the contour encloses only one singular point $z=\zeta$. The nature of the μ solutions which compose this sub-set will now be examined.

As before

$$v_1=(z-\zeta)^\rho\phi_1(z-\zeta),$$

where

$$s=e^{2\pi i\rho}$$

and $\phi_1(z-\zeta)$ is uniform in the domain of the point ζ.

Now

$$\frac{V_2}{V_1}=\frac{v_2}{v_1}+1,$$

that is to say, v_2/v_1 is a quasi-periodic function of $z-\zeta$. But the function $\frac{1}{2\pi i}\log(z-\zeta)$ has the same quasi-periodicity, for after a circuit described in the positive sense around the point ζ, $\frac{1}{2\pi i}\log(z-\zeta)$ becomes $\frac{1}{2\pi i}\log(z-\zeta)+1$.

Consequently the difference

$$\frac{v_2}{v_1}-\frac{1}{2\pi i}\log(z-\zeta)$$

returns to its initial value after the circuit has been described, and therefore

$$\frac{v_2}{v_1}-\frac{1}{2\pi i}\log(z-\zeta)=\psi_1(z-\zeta),$$

where $\psi_1(z-\zeta)$ is uniform in the domain of ζ. Hence

$$v_2=(z-\zeta)^\rho\left\{\frac{1}{2\pi i}\phi_1(z-\zeta)\log(z-\zeta)+\phi_2(z-\zeta)\right\},$$

where

$$\phi_2(z-\zeta)=\phi_1(z-\zeta)\psi_1(z-\zeta).$$

Now make the substitution

$$t=\frac{1}{2\pi i}\log(z-\zeta)$$

and let

$$v_r=(z-\zeta)^\rho u_r.$$

Then as the variable z describes a simple circuit, in the positive direction around the point ζ, t increases to $t+1$, and thus the functions u_r, regarded as functions of t, satisfy the quasi-periodic relations

$$u_r(t+1)=u_r(t)+u_{r-1}(t) \qquad (r\geqslant 2).$$

These relations can be satisfied by taking $u_1(t)=1$, $u_2(t)=t$, and in general by taking $u_r(t)$ to be the polynomial

$$C_r t(t-1)\ .\ .\ .\ (t-r+2).$$

The constant C_r has to satisfy the relation

$$(r-1)C_r=C_{r-1} \qquad (C_1=1),$$

and thus

$$C_r=1/(r-1)! \qquad (r\geqslant 2).$$

Thus a particular solution of the functional equation satisfied by $u_r(t)$ has been found. Denote this solution by $\theta_r(t)$, so that

$$\theta_r(t)=\frac{t(t-1)\ .\ .\ .\ (t-r+2)}{(r-1)!},$$

and consider the function

$$\Theta_r(t)=\theta_r(t)\chi_1(t)+\theta_{r-1}(t)\chi_2(t)+\ .\ .\ .\ +\theta_1(t)\chi_r(t),$$

where each function $\chi(t)$ is such that

$$\chi_s(t+1)=\chi_s(t).$$

Then

$$\Theta_r(t+1)-\Theta_r(t)=\sum_{s=1}^{r}\chi_{r-s+1}(t)\{\theta_s(t+1)-\theta_s(t)\}$$

$$=\sum_{s=2}^{r}\chi_{r-s+1}(t)\theta_{s-1}(t)$$

$$=\Theta_{r-1}(t),$$

and therefore,

$$u_r(t)=\Theta_r(t) \qquad (r=2,3,\ .\ .,\ \mu)$$

is a general solution of the system of relations

$$u_r(t+1)=u_r(t)+u_{r-1}(t).$$

Now referring back to the variable z, it will be seen that the functions $v_1, v_2, \ .\ .\ ., v_\mu$ are of the following forms:

$$v_1=(z-\zeta)^\rho\phi_1(z-\zeta),$$
$$v_2=(z-\zeta)^\rho\{\theta_2\phi_1(z-\zeta)+\phi_2(z-\zeta)\},$$
$$.\qquad .\qquad .$$
$$v_\mu=(z-\zeta)^\rho\{\theta_\mu\phi_1(z-\zeta)+\theta_{\mu-1}\phi_2(z-\zeta)+\ .\ .\ .\ +\phi_\mu(z-\zeta)\},$$

in which θ_r is written in short for

$$\theta_r\left\{\frac{1}{2\pi i}\log(z-\zeta)\right\},$$

where the same determination of the logarithm is taken throughout, and the functions $\phi_r(z-\zeta)$ are uniform in the neighbourhood of the point ζ.

The remaining sub-sets having the same multiplier s may be treated in precisely the same way. Thus in general, when s is a repeated root of the characteristic equation, terms having logarithmic factors enter into the general solution. This case is frequently spoken of as the *logarithmic case* (see § 6·3).

Example.—The equation

$$z^2(z+1)\frac{d^2w}{dz^2}-z^2\frac{dw}{dz}+\tfrac{1}{4}(3z+1)w=0$$

has the two linearly-independent solutions

$$w_1=z^{\frac{1}{2}}, \quad w_2=z^{\frac{1}{2}}\log z+z^{\frac{3}{2}}.$$

If z describes a circuit in the positive direction around the origin, these solutions become respectively,

$$W_1=-z^{\frac{1}{2}}=-w_1, \quad W_2=-z^{\frac{1}{2}}(\log z+2\pi i)-z^{\frac{3}{2}}=-w_2-2\pi i w_1.$$

The characteristic equation is therefore

$$\begin{vmatrix} -1-s, & 0 \\ -2\pi i, & -1-s \end{vmatrix}=0,$$

or

$$(s+1)^2=0.$$

Any solution of the form

$$w=(z-\zeta)^\rho\{t^r\phi_1(z-\zeta)+t^{r-1}\phi_{r1}(z-\zeta)+\ \dots\ +\phi_{rr}(z-\zeta)\}$$

is said to be *regular*,* when the point ζ is an ordinary point or pole of the functions ϕ. If all the n solutions relative to the point ζ are regular, ζ is said to be a *regular singular point* of the equation. If any one of the functions ϕ has an essential singularity at ζ, the point ζ is said to be an *irregular singular point* of the equation.

15·24. Alternative Method of Obtaining the Solutions of a Canonical Set.—Starting from the solution

$$v_1=(z-\zeta)^\rho\phi_1(z-\zeta),$$

write

$$w=v_1\int v_{12}dz$$

then v_{12} satisfies a homogeneous linear equation of order $n-1$, which has at least one uniform solution; let this uniform solution be v_{12}. The corresponding characteristic equation is of degree $n-1$, for one root s has dropped out and the canonical sub-set

$$V_1=sv_1, \quad V_2=s(v_2+v_1), \ \dots,\ V_\mu=s(v_\mu+v_{\mu-1})$$

is now replaced by

$$V_{12}=sv_{12}, \quad V_{13}=s(v_{13}+v_{12}), \ \dots,\ V_{1\mu}=s(v_{1\mu}+v_{1,\mu-1}).$$

Now write

$$w=v_1\int v_{12}\int v_{23}(dz)^2$$

and repeat the process. In this way there arises a set of μ solutions corresponding to the canonical sub-set, namely (*cf.* § 5·21),

$$v_1=(z-\zeta)\phi_1(z-\zeta),$$
$$v_2=v_1\int v_{12}dz,$$
$$\cdot \quad \cdot \quad \cdot \quad \cdot \quad \cdot$$
$$v_r=v_1\int v_{12}\int v_{23}\ \dots\ \int v_{r-1,r}(dz)^{r-1} \qquad (r=2, 3, \dots, \mu),$$

in which $v_{12}, v_{23}, \dots, v_{r-1,r}$ are all one-valued in the domain of ζ. Since these functions are one-valued, v_r must necessarily be of the form

$$v_r=(z-\zeta)^\rho\{t^r\phi_{r0}(z-\zeta)+t^{r-1}\phi_{r1}(z-\zeta)+\ \dots\ +\phi_{rr}(z-\zeta)\},$$

where $t^r=\log(z-\zeta)$ and ϕ_{r0} is a constant multiple of ϕ_1.

* Thomé, *J. für Math.* 75 (1873), p. 266.

15·3. A Necessary Condition for a Regular Singularity.—The preceding theory is of great theoretical importance in that it reveals the character of the general solution of an equation relative to any of its singular points, but it contributes little towards the more difficult problem of determining the explicit form of the general solution. In fact a point has now been reached where it is practically impossible to proceed further without imposing some convenient restrictions upon the equation or upon its solutions. The path to take is pointed out very clearly by the following theorem.*

A necessary and sufficient condition that the point $z=\zeta$ *should be a regular singular point of the equation*

$$\frac{d^n w}{dz^n}+p_1(z)\frac{d^{n-1}w}{dz^{n-1}}+\ldots+p_{n-1}(z)\frac{dw}{dz}+p_n(z)w=0$$

is that

$$p_r(z)=(z-\zeta)^{-r}P(z) \qquad (r=1,\ 2,\ \ldots,\ n),$$

where $P(z)$ *is analytic in the neighbourhood of* ζ.

There is no loss in generality in supposing the point ζ to be the origin. The necessity of the condition relative to $z=0$ will first be proved. It has been seen that there always exists a solution

$$w_1=z^\rho\phi(z),$$

where $\phi(z)$ is uniform in the domain of the origin, and assuming this solution to be regular, $\phi(0)\neq 0$. Now let

$$w=w_1\int v\,dz$$

be a solution of the equation, then v will satisfy a differential equation of the form

$$\frac{d^{n-1}v}{dz^{n-1}}+q_1(z)\frac{d^{n-2}v}{dz^{n-2}}+\ldots+q_{n-1}(z)v=0,$$

and if w is to be a regular solution, v must be regular. But the coefficients q are expressible in terms of w_1 and the coefficients p, thus

$$q_1=\frac{1}{w_1}\left\{n\frac{dw_1}{dz}+p_1w_1\right\},$$

$$q_r=\frac{1}{w_1}\left\{{}_nC_r\frac{d^rw_1}{dz^r}+{}_{n-1}C_{r-1}p_1\frac{d^{r-1}w_1}{dz^{r-1}}+\ldots+(n-r+1)p_{r-1}\frac{dw_1}{dz}+p_rw_1\right\}.$$

Take first of all the simple case $n=1$; the equation

$$\frac{dw}{dz}+p_1w=0$$

has the solution

$$w=Ce^{-\int p_1 dx},$$

and if this solution is to be regular it will be necessary for p_1 to have the form $z^{-1}f_1(z)$, where $f_1(z)$ is analytic near the origin. Next proceed to the case $n=2$. The equation in v will be of the first order and consequently near the origin,

$$q_1(z)=O(z^{-1}).$$

Also

$$\frac{1}{w_1}\cdot\frac{dw_1}{dz}=O(z^{-1}).$$

Hence, as before, $p_1(z)$ is of the form

$$p_1(z)=z^{-1}f_1(z),$$

* Fuchs, *J. für Math.* 66 (1866), p. 143; 68 (1868), p. 358; Tannery, *Ann. Éc. Norm.* (2), 4 (1875), p. 135.

where $f_1(z)$ is analytic in the neighbourhood of the origin. But

$$p_2 = -\frac{1}{w_1}\left\{\frac{d^2w_1}{dz^2} + p_1\frac{dw_1}{dz}\right\},$$

and since, near the origin,

$$\frac{1}{w_1}\cdot\frac{d^2w_1}{dz^2} = O(z^{-2}),$$

$$\frac{1}{w}\cdot\frac{dw}{dz} = O(z^{-1}),$$

$$p_1 = O(z^{-1}),$$

p_2 is of the form $z^{-2}f_2(z)$, where $f_2(z)$ is analytic in the neighbourhood of $z=0$.

The proof is now completed by induction. The theorem is supposed true for an equation of order $n-1$, thus in the equation for v, it is assumed that

$$q_r(z) = z^{-r}g_r(z) \qquad (r=1, 2, \ldots, n-1),$$

where $g_r(z)$ is analytic at the origin. Then it follows immediately from the expressions for the coefficients p that

$$p_r(z) = z^{-r}f_r(z) \qquad (r=1, 2, \ldots, n-1),$$

$f_r(z)$ being analytic at the origin. It therefore remains only to prove that $p_r(z)$ is of this form when $r=n$. But this follows at once from the equation

$$p_n(z) = -\frac{1}{w_1}\left\{\frac{d^nw_1}{dz^n} + p_1\frac{d^{n-1}w_1}{dz^{n-1}} + \ldots + p_{n-1}\frac{dw_1}{dz}\right\}.$$

The condition stated is therefore necessary.

A proof of the sufficiency of this condition could be supplied by proving that when the condition is satisfied, convergent expressions for the n solutions of the equation can be obtained explicitly. This proof will be given at the beginning of the next chapter; in the meanwhile an independent and somewhat more general proof of sufficiency will be outlined.

15·31. Sufficiency of the Condition for a Regular Singular Point.—It has now to be proved that if, in the equation

$$\frac{d^nw}{dz^n} + z^{-1}P_1(z)\frac{d^{n-1}w}{dz^{n-1}} + \ldots + z^{-n+1}P_{n-1}(z)\frac{dw}{dz} + z^{-n}P_n(z)w = 0,$$

all the functions $P(z)$ are analytic in the neighbourhood of the origin, the equation possesses a fundamental set of n solutions regular at the origin.

Now the equation may be replaced by the system

$$\begin{cases} w = w_1, \quad z\dfrac{dw_1}{dz} = w_2, \quad \ldots, \quad z\dfrac{dw_{n-1}}{dz} = w_n, \\[2ex] z\dfrac{dw_n}{dz} = A_1(z)w_1 + A_2(z)w_2 + \ldots + A_n(z)w_n, \end{cases}$$

where $A_1(z), \ldots, A_n(z)$ are linear combinations of $P_1(z), \ldots, P_n(z)$ with constant coefficients, and are therefore analytic near $z=0$.

It is convenient to consider, in place of the above system, the more general system

$$\begin{cases} z\dfrac{dw_1}{dz} = A_{11}w_1 + A_{12}w_2 + \ldots + A_{1n}w_n, \\[2ex] z\dfrac{dw_2}{dz} = A_{21}w_1 + A_{22}w_2 + \ldots + A_{2n}w_n, \\[1ex] \quad \cdot \quad \cdot \quad \cdot \quad \cdot \quad \cdot \quad \cdot \quad \cdot \quad \cdot \\[1ex] z\dfrac{dw_n}{dz} = A_{n1}w_1 + A_{n2}w_n + \ldots + A_{nn}w_n, \end{cases}$$

wherein all the coefficients A are analytic in the neighbourhood of the origin. It will first of all be proved that, when a certain restriction (to be removed later) is imposed, there exists a set of solutions of this system, regular at the origin *and also free from logarithmic terms*, namely,

$$w_1=z^r u_1, \quad w_2=z^r u_2, \quad . \; . \; ., \quad w_n=z^r u_n,$$

where r is a certain constant, and $u_1, u_2, \ldots, u_n$ are all analytic at the origin. The constant r may be so chosen that if $c_1, c_2, \ldots, c_n$ are the values of $u_1, u_2, \ldots, u_n$ when $z=0$, at least one of the numbers c is not zero. Let a_{rs} be the value of A_{rs} when $z=0$, then by substituting $w_1, w_2, \ldots, w_n$ in the system and equating to zero the coefficient of z^r in each equation, the following set of relations is found :

$$\begin{aligned}
&(a_{11}-r)c_1+a_{12}c_2+\ldots+a_{1n}c_n=0,\\
&a_{21}c_1+(a_{22}-r)c_2+\ldots+a_{2n}c_n=0,\\
&\cdot \quad \cdot \quad \cdot \quad \cdot \quad \cdot \quad \cdot \quad \cdot \quad \cdot \quad \cdot\\
&a_{n1}c_1+a_{n2}c_2+\ldots+(a_{nn}-r)c_n=0.
\end{aligned}$$

By eliminating the unknown coefficients c_r from this system the *indicial equation* or equation to determine r is found, namely,

$$\begin{vmatrix}
a_{11}-r, & a_{12}, & \ldots, & a_{1n}\\
a_{21}, & a_{22}-r, & \ldots, & a_{2n}\\
\cdot & \cdot & \cdot & \cdot\\
a_{n1}, & a_{n2}, & \ldots, & a_{nn}-r
\end{vmatrix}=0;$$

let its roots, which may not all be distinct, be denoted by

$$r_1, \quad r_2, \quad \ldots, \quad r_n.$$

Now if $W_1, W_2, \ldots, W_n$ are written for $z\dfrac{dw_1}{dz}, z\dfrac{dw_2}{dz}, \ldots, z\dfrac{dw_n}{dz}$ respectively, the system under consideration is

$$\begin{aligned}
W_1&=a_{11}w_1+a_{12}w_2+\ldots+a_{1n}w_n+O(z, w),\\
W_2&=a_{21}w_1+a_{22}w_2+\ldots+a_{2n}w_n+O(z, w),\\
&\cdot \quad \cdot \quad \cdot \quad \cdot \quad \cdot \quad \cdot \quad \cdot \quad \cdot \quad \cdot\\
W_n&=a_{n1}w_1+a_{n2}w_2+\ldots+a_{nn}w_n+O(z, w),
\end{aligned}$$

where $O(z, w)$ is written in brief for linear expressions in $w_1, w_2, \ldots w_n$ whose coefficients are analytic functions of z which vanish at the origin. Apart from the terms $O(z, w)$, this set of linear substitutions is quite analogous to that which arose in § 15·2 although its source is completely different. Let the terms $O(z, w)$ be ignored for the moment, then $w_1, w_2, \ldots, w_n$ may be replaced by linear combinations of these quantities, namely $v_1, v_2, \ldots, v_n$, such that the system becomes, when the roots of the indicial equation are all unequal,

$$V_1=r_1v_1, \quad V_2=r_2v_2, \quad \ldots, \quad V_n=r_nv_n.$$

By performing exactly the same reduction on the system when the terms $O(z, w)$ are present, the system considered may be replaced by

$$\begin{aligned}
V_1&=r_1v_1+O(z, v),\\
V_2&=r_2v_2+O(z, v),\\
&\cdot \quad \cdot \quad \cdot \quad \cdot\\
V_n&=r_nv_n+O(z, v).
\end{aligned}$$

If, on the other hand, the roots of the indicial equation are not all distinct, the system may be replaced by the aggregate of a number of sub-systems such as

$$V_1=r_1v_1+O(z, v), \qquad V_{\mu+1}=r_2v_{\mu+1}+O(z, v),$$
$$V_2=r_1(v_2+v_1)+O(z, v), \qquad V_{\mu+2}=r_2(v_{\mu+2}+v_{\mu+1})+O(z, v),$$
$$\cdot \quad \cdot \quad \cdot \quad \cdot \quad \cdot \quad \cdot \quad \cdot \quad \cdot \quad \cdot \quad \cdot$$
$$V_\mu=r_1(v_\mu+v_{\mu-1})+O(z, v), \quad V_\nu=r_2(v_\nu+v_{\nu-1})+O(z, v),$$

and so forth. As the latter case includes the former, only the latter will be considered. Transform the system by writing

$$v_1=z^{r_1}\phi_1(z), \quad v_2=z^{r_1}\phi_2(z), \quad . \ . \ ., \quad v_n=z^{r_1}\phi_n(z),$$

then since $V_1, V_2, \ldots, V_n$ are the same linear combinations of $W_1, W_2, \ldots, W_n$ as $v_1, v_2, \ldots, v_n$ are of $w_1, w_2, \ldots, w_n$, it follows that

$$V_1 = z\frac{dv_1}{dz}, \quad . \ . \ ., \quad V_n = z\frac{dv_n}{dz}.$$

The system therefore becomes

$$z\frac{d\phi_1}{dz} = O_1(z, \phi),$$
$$z\frac{d\phi_2}{dz} = r_1\phi_1+O_2(z, \phi),$$
$$\cdot \quad \cdot \quad \cdot \quad \cdot \quad \cdot$$
$$z\frac{d\phi_\mu}{dz} = r_1\phi_{\mu-1}+O_\mu(z, \phi),$$
$$z\frac{d\phi_{\mu+1}}{dz} = (r_2-r_1)\phi_{\mu+1}+O_{\mu+1}(z, \phi),$$
$$\cdot \quad \cdot \quad \cdot \quad \cdot \quad \cdot \quad \cdot \quad \cdot$$

Since the terms $O(z, \phi)$ can be found explicitly, and are linear in $\phi_1, \phi_2, \ldots, \phi_n$ with coefficients analytic in z and vanishing at the origin, the functions ϕ can be determined from the equations, as power series in z, by a method of successive approximation. It can be seen almost immediately that $\phi_1(z), \ldots, \phi_{\mu-1}(z)$ must be zero when $z=0$, whereas $\phi_\mu(0)$ may have any arbitrary value a. Thus, for instance, if $\phi_{\mu-1}(0)$ were not zero, $\phi_\mu(z)$ would involve a logarithmic term, contrary to hypothesis. If r_2-r_1 is a positive integer, say m, then in general the process of determining successive coefficients in the expansion of $\phi_{\mu+1}(z)$ breaks down at the term in z^m, for then there is nothing to balance the term in z^m proceeding from the term $O(z, \phi)$. Thus, for the development of all the functions ϕ as power series in z to be possible, it is necessary to restrict r_k-r_1 to be not a positive integer (though it may be zero) for any value of k. This is the restriction mentioned earlier in this section. When this restrictive condition is satisfied, it is possible to determine all the coefficients in the series developments of the functions ϕ. It only remains to prove that these developments converge for sufficiently small values of $|z|$. An outline of one possible method of proving this convergence is as follows.

Let ϵ be the numerical difference between r_2-r_1 and the nearest positive integer, and consider the system of ordinary linear equations.

$$\psi_1=Q_1(z, \psi),$$
$$\psi_2=r_1\psi_1+Q_2(z, \psi),$$
$$\cdot \quad \cdot \quad \cdot \quad \cdot \quad \cdot$$
$$\psi_\mu-|a|=r_1\psi_{\mu-1}+Q_\mu(z, \psi),$$
$$\epsilon\psi_{\mu+1}=Q_{\mu+1}(z, \psi),$$
$$\cdot \quad \cdot \quad \cdot \quad \cdot \quad \cdot \quad \cdot$$

in which $Q_1, Q_2, \ldots, Q_n$ are linear expressions in $\psi_1, \psi_2, \ldots, \psi_n$ whose coefficients, vanishing at the origin, are dominant functions for the corresponding coefficients in the terms $O(z, \phi)$ of the system in $\phi_1, \phi_2, \ldots, \phi_n$. But this present system may be solved for the functions ψ in series of ascending powers of z with *positive* coefficients, and these series converge for sufficiently small values of $|z|$. If the coefficient of the leading term in the series for each of the functions ψ is the modulus of the leading term in the series for the corresponding function ϕ, the moduli of the remaining coefficients in the series for the functions ϕ will be at most equal to the corresponding coefficients in the series for the functions ψ. The series for the functions ϕ therefore converge absolutely and uniformly within a definite circle whose centre is at the origin.

It follows that *the system of n linear differential equations of the first order possesses the set of regular solutions*

$$w_1 = z^{r_1}u_1, \quad w_2 = z^{r_1}u_2, \quad \ldots, \quad w_n = z^{r_1}u_n,$$

where $u_1, u_2, \ldots, u_n$ are analytic in the neighbourhood of $z=0$, and r_1 is a root of the indicial equation such that the difference

$$r_k - r_1,$$

where r_k is any other root of the indicial equation, is not a positive integer.

When no two of the roots of the indicial equations differ by an integer, the system possesses n distinct sets of solutions of the above type.

In the case of the single equation of order n, to which the system is equivalent, the indicial equation is

$$[r]_n + P_1(0)[r]_{n-1} + \ldots + P_{n-1}(0)r + P_n(0) = 0,$$

where $[r]_n = r(r-1) \ldots (r-n+1)$. If the roots of this equation are

$$r_1, \quad r_2, \quad \ldots, \quad r_n,$$

the differential equation will possess a solution

$$w = z^{r_k}u_k(z)$$

corresponding to each root r_k, where $u_k(z)$ is analytic near $z=0$ and $u_k(0) \neq 0$, provided that none of the differences

$$r_1 - r_k, \quad r_2 - r_k, \quad \ldots, \quad r_n - r_k$$

are positive integers, though one or more of these differences may possibly be zero.

15·311. The Logarithmic Case.—To complete the proof of the sufficiency of Fuchs' conditions, it is now necessary to admit the possibility of the roots of the indicial equation differing by an integer. Let the roots

$$r_1, \quad r_2, \quad \ldots, \quad r_\mu$$

differ from one another by integers, and from all other roots by numbers other than integers. Let

$$r_1 \geqslant r_2 \geqslant \ldots \geqslant r_\mu.$$

The solution

$$w_1 = z^{r_1}u_1(z)$$

corresponding to r_1 exists in consequence of the work of the previous section. Let

$$w = w_1 \int v\,dz$$

be a solution, then (§ 15·3) v satisfies an equation of order $n-1$ satisfying Fuchs' conditions with respect to $z=0$. But since

$$v = \frac{d}{dz}\left(\frac{w}{w_1}\right),$$

the roots of the characteristic equation relative to the equation in v are

$$r_2 - r_1 - 1, \quad r_3 - r_1 - 1, \quad \ldots, \quad r_\mu - r_1 - 1,$$

and of these the first $\mu - 1$ are negative integers.

Since $r_2 \geqslant r_3$, there will be a solution

$$v = z^{r_2 - r_1 - 1}\psi(z),$$

where $\psi(z)$ is analytic near the origin and $\psi(0) \neq 0$. Consequently there exists the solution

$$w_2 = w_1 \int z^{r_2 - r_1 - 1}\psi(z)dz$$

which, multiplied if necessary by a constant factor, reduces in general * to

$$w_2 = z^{r_1}\{u_1(z) \log z + u_{22}(z)\}.$$

The process may be repeated, giving in general

$$w_\nu = z^{r_1}\{u_1(z)t^\nu + u_{\nu 1}(z)t^{\nu-1} + \ldots + u_{\nu\nu}(z)\} \qquad (\nu = 2, 3, \ldots, \mu),$$

where the functions $u(z)$ are all analytic in the neighbourhood of $z=0$. The remaining groups of indices are treated in the same way and the proof of the sufficiency of the condition is complete.

15·4. Equations of Fuchsian Type.—An equation of Fuchsian type is one in which every singular point, including the point at infinity, is a regular singularity. Let there be ν regular singular points

$$a_1, \; a_2, \; \ldots, \; a_\nu$$

in the finite part of the plane. It is an immediate consequence of the theorem of Fuchs that the coefficient $p_m(z)$ will be of the form

$$p_m(z) = (z-a_1)^{-m}(z-a_2)^{-m} \ldots (z-a_\nu)^{-m}P_m(z),$$

where, since there are no other singular points in the finite part of the z-plane, $P_m(z)$ is an integral function of z.

Now consider the behaviour of these coefficients at infinity; if the equation is to have a regular singularity at infinity, the point at infinity must be at most a pole of the function $p_m(z)$. Consequently, $P_m(z)$ is a polynomial in z, and $p_m(z)$ is expressible in the form

$$p_m(z) = \sum_{s=1}^{\nu} \frac{P_{ms}}{(z-a_s)^m} + \frac{Q_m(z)}{(z-a_1)^{m-1}(z-a_2)^{m-1} \ldots (z-a_\nu)^{m-1}},$$

where P_{ms} is a constant † and Q_m is a polynomial whose maximum degree is to be determined. On the other hand $p_m(z)$ admits of the development

$$p_m(z) = z^{\sigma_m}(b_{m0} + b_{m1}z^{-1} + b_{m2}z^{-2} + \ldots),$$

convergent for sufficiently large values of $|z|$; let

$$w = z^r(c_0 + c_1 z^{-1} + c_2 z^{-2} + \ldots)$$

be assumed to be a solution of the equation, regular at infinity. The exponent r is determined by the indicial equation relative to the point at infinity; if there are to be n distinct regular solutions this indicial equation must not degenerate to an order lower than n. Since, therefore, the indicial equation arises by equating to zero the terms of highest order in z, it must involve the term of highest order in $w^{(n)}$ which is $O(z^{r-n})$, and no other term can be of an order greater than this. But the dominant term arising out of $p_m(z)w^{(m)}$ is $O(z^{\sigma_m + r - n + m})$ and therefore

$$\sigma_m \leqslant -m.$$

It follows that

$$Q_m(z) = O(z^{m\nu - m - \nu})$$

at most, when $m > 1$, and that Q_1 is identically zero.

There remains the question as to what degree of definiteness is introduced

* In the very particular case in which the series development of $\psi(z)$ does not involve the term $z^{r_1 - r_2}$ no logarithmic term appears in w_2.

† $P_{ms} = (a_s - a_1)^{-m} \ldots (a_s - a_{s-1})^{-m}(a_s - a_{s+1})^{-m} \ldots (a_s - a_\nu)^{-m}P_m(a_s)$.

into the equation by the knowledge of the n exponents which correspond to each singular point. Consider the singularity $z=a_s$; if the regular solution

$$w=(z-a_s)^r\sum_{\kappa=0}^{\infty} C_\kappa(z-a_s)^\kappa$$

is assumed, the indicial equation is found to be

$$[r]_n+\sum_{m=1}^{n} P_{ms}[r]_{n-m}=0.$$

Consequently, if the exponents

$$\alpha_{s1}, \alpha_{s2}, \ldots, \alpha_{sn}$$

relative to a_s, are pre-assigned, the constants P_{ms} are uniquely determined, thus

$$P_{1s}-\tfrac{1}{2}n(n-1)=-\sum_{k=1}^{n}\alpha_{sk},$$

$$P_{2s}-\tfrac{1}{2}(n-1)(n-2)P_{1s}+\tfrac{1}{24}n(n-1)(n-2)(3n-1)=\sum_{k=1}^{n}\sum_{l=1}^{n}\alpha_{sk}\alpha_{sl} \quad (k\neq l),$$

and so on.

Now suppose that the leading term in $Q_m(z)$ is $A_m z^{m\nu-m-\nu}$, so that for large values of z,

$$p_m(z)=z^{-m}\left\{\sum_{s=1}^{\nu} P_{ms}+A_m\right\}+O(z^{-m-1}).$$

If a solution of the type

$$w=z^\sigma(b_0+b_1z^{-1}+b_2z^{-2}+\ldots)$$

is assumed, the corresponding indicial equation is found to be

$$[\sigma]_n+\sum_{m=1}^{n}\left\{\sum_{s=1}^{\nu} P_{ms}+A_m\right\}[\sigma]_{n-m}=0.$$

The exponents relative to the point at infinity are defined as the roots of this equation in σ with their signs changed.

If the exponents are $\delta_1, \delta_2, \ldots, \delta_n$ then, since $A_1=0$,

$$\sum_{s=1}^{\nu} P_{1s}-\tfrac{1}{2}n(n-1)=\sum_{k=1}^{n}\delta_k.$$

But

$$\sum_{s=1}^{\nu}P_{1s}-\tfrac{1}{2}\nu n(n-1)=-\sum\alpha_{sk} \qquad \begin{pmatrix}s=1, 2, \ldots, \nu\\ k=1, 2, \ldots, n\end{pmatrix},$$

and therefore

$$\sum\alpha_{sk}+\sum\delta_k=\tfrac{1}{2}n(n-1)(\nu-1),$$

that is, the sum of all the exponents is constant. Thus if there are $\nu+1$ singular points (including the point at infinity) there are $n(\nu+1)$ exponents with one relation between them. The coefficient $p_m(z)$ contains $m(\nu-1)+1$ constants, namely the ν constants P_{ms} and the $m\nu-m-\nu+1$ coefficients of the polynomial $Q_m(z)$. Thus the equation contains, in all,

$$\tfrac{1}{2}n(n+1)(\nu-1)+n$$

distinct constants, of which $n(\nu+1)-1$ are accounted for by the exponents. There remain

$$\tfrac{1}{2}(n-1)(n\nu-n-2)$$

arbitrary constants.

The n solutions corresponding to each of the $\nu+1$ singular points are grouped together under one symbol known as the Riemann P-function * :

$$P\left\{\begin{array}{cccccc} a_1 & . & . & . & a_\nu & \infty \\ a_{11} & . & . & . & a_{\nu 1} & \delta_1 \\ . & & & & . & . \quad z \\ a_{1n} & . & . & . & a_{\nu n} & \delta_n \end{array}\right\}$$

which indicates the location of the singular points, and the exponents relative to each singularity.

15·5. A Class of Equations whose general Solution is Uniform.—Consider the equation

$$p_0(z)\frac{d^n w}{dz^n}+p_1(z)\frac{d^{n-1}w}{dz^{n-1}}+\ .\ .\ .\ +p_{n-1}(z)\frac{dw}{dz}+p_n(z)w=0\ ;$$

it will be assumed

(a) that the coefficients are polynomials in z and that the degree of $p_0(z)$ is not less than that of any other coefficient ;

(b) that the singular points which lie in the finite part of the z-plane are regular ; the point at infinity may or may not be regular ;

(c) that the general solution of the equation is uniform.

In order that (c) may be true, it is necessary, in the first place, that the exponents relative to every singular point be integral, and in the second place that no logarithmic terms appear in the solution.

It will now be proved that, when these conditions are fulfilled, the general solution of the equation is of the form

$$w=C_1e^{\lambda_1 z}R_1(z)+C_2e^{\lambda_2 z}R_2(z)+\ .\ .\ .\ +C_ne^{\lambda_n z}R_n(z),$$

where $C_1, C_2, \ldots, C_n$ are the constants of integration, $\lambda_1, \lambda_2, \ldots, \lambda_n$ are definite constants which need not be all unequal, and the functions $R(z)$ are rational.†

Let the finite singular points be $a_1, a_2, \ldots, a_\nu$, and let the least negative exponent relative to a_s be α_s ; if the exponents corresponding to a_s are all positive, let α_s be zero. Then the change of dependent variable

$$w_1=(z-a_1)^{\alpha_1}\ .\ .\ .\ (z-a_\nu)^{\alpha_\nu}w$$

transforms the equation into one in which all the exponents relative to the finite singularities are positive integers or zero ; let the transformed equation be

$$q_0(z)\frac{d^n w_1}{dz^n}+q_1(z)\frac{dw_1{}^{n-1}}{dz^{n-1}}+\ .\ .\ .\ +q_{n-1}(z)\frac{dw_1}{dz}+q_n(z)w_1=0.$$

This equation has the properties (a), (b) and (c) specified for the original equation.

Now let

$$w_1=w_2e^{\lambda z},$$

then there arises an equation in which the coefficient of w_2 is

$$\lambda^n q_0(z)+\lambda^{n-1}q_1(z)+\ .\ .\ .\ +\lambda q_{n-1}(z)+q_n(z),$$

and λ can be so chosen as to make the coefficient of the highest power of z zero. The equation may then be written

$$Q_0(z)\frac{d^n w_2}{dz^n}+Q_1(z)\frac{d^{n-1}w_2}{dz^{n-1}}+\ .\ .\ .\ +Q_{n-1}(z)\frac{dw_2}{dz}+Q_n(z)w_2=0,$$

* Riemann, *Abh. Ges. Wiss. Gött.*, 7 (1857), p. 3 [*Math. Werke* (2nd ed.), p. 67]. Cf. § 7·23.

† Halphen, *C. R. Acad. Sc. Paris*, 101 (1885), p. 1238.

in which, if $Q_0(z)$ is of degree m in z, $Q_n(z)$ is at most of degree $m-1$ and the remaining coefficients are of degrees not exceeding m.

Now

$$\frac{Q_1(z)}{Q_0(z)}=\gamma_0+\sum_{s=1}^{\nu}\frac{\gamma_s}{z-a_s},$$

where γ_0 is a constant. The sum of the exponents relative to a_s is

$$\tfrac{1}{2}n(n-1)-\gamma_s.$$

Now these exponents are unequal positive integers, their sum is therefore not less than

$$0+1+2+\ . \ . \ . \ +(n-1)=\tfrac{1}{2}n(n-1).$$

Consequently γ_s is zero or a negative integer, and therefore

$$S=\sum_{s=1}^{\nu}\gamma_s\leqslant 0.$$

Suppose, for the moment, that $Q_n(z)$ is not identically zero; it will be shown that a finite chain of transformations can be set up which leads to an equation in which the term corresponding to $Q_n(z)$ is identically zero. Let

$$W_1=\frac{dw_2}{dz},$$

then

$$Q_0(z)\frac{d^{n-1}W_1}{dz^{n-1}}+Q_1(z)\frac{d^{n-2}W_1}{dz^{n-2}}+\ . \ . \ . \ +Q_{n-1}(z)W_1+Q_n(z)w_2=0.$$

Differentiate with respect to z, obtaining the equation

$$Q_0(z)\frac{d^nW_1}{dz^n}+\{Q_0'(z)+Q_1(z)\}\frac{d^{n-1}W_1}{dz^{n-1}}+\ . \ . \ . \ +Q_n'(z)w_2=0,$$

and then eliminate w_2 between the last two equations. The eliminant is

$$Q_0(z)Q_n(z)\frac{d^nW_1}{dz^n}+[\{Q_0'(z)+Q_1(z)\}Q_n(z)-Q_0(z)Q_n'(z)]\frac{d^{n-1}W_1}{dz^{n-1}}+\ . \ . \ . \ =0,$$

and is an equation of the same type as that in w_2. Let S' be the number which, in this equation, replaces the number S in the equation in w_2; S' is the coefficient of z^{-1} in the expansion in descending powers of z of

$$\frac{Q_0'(z)}{Q_0(z)}+\frac{Q_1(z)}{Q_0(z)}-\frac{Q_n'(z)}{Q_n(z)},$$

and this coefficient is m in $Q_0'(z)/Q_0(z)$, S in $Q_1(z)/Q_0(z)$ and is not greater than $m-1$ in $Q_n'(z)/Q_n(z)$. Consequently,

$$S'\geqslant S+1.$$

The process may be repeated, provided that the coefficient of W_1 in the above equation is not zero, by finding the equation in W_2 where

$$W_2=\frac{dW_1}{dz};$$

a number S'' is obtained such that

$$S''\geqslant S+2,$$

and so on. The process must, however, terminate, because the numbers S', S'', . . . are negative integers. Thus there will come a stage at which the coefficient of the dependent variable

$$W_\rho=\frac{d^\rho w_2}{dz^\rho}$$

is zero. The equation then has the solution

$$W_\rho=\text{constant},$$

and therefore w_2 is a polynomial in z of degree ρ. Thus, since

$$w=\frac{w_2 e^{\lambda z}}{(z-a_1)^{\alpha_1}\ldots(z-a_\nu)^{\alpha_\nu}},$$

there exists a solution

$$w=e^{\lambda z}R(z)$$

of the given equation, where $R(z)$ is a rational function of z.

To complete the proof it is necessary to show that there are distinct solutions of this type equal in number to the order of the equation. This will be assumed when the order is $n-1$ and then proved for an equation of order n.

The given equation possesses one solution of the type considered, let it be

$$w_1=e^{\lambda_1 z}R_1(z)$$

and write

$$w=w_1\int u\,dz.$$

The new dependent variable satisfies an equation of order $n-1$ and this equation will be of precisely the same type as that in w. It therefore has a solution

$$u=e^{\lambda z}R(z),$$

where $R(z)$ is a rational function of z; let

$$w=w_1\int e^{\lambda z}R(z)dz.$$

Now, since w is to be uniform, the integral

$$\int e^{\lambda z}R(z)dz$$

can introduce no logarithmic terms; it must therefore be of the form

$$e^{\lambda z}\mathfrak{R}(z),$$

where $\mathfrak{R}(z)$ is rational in z. The $n-1$ independent solutions $u(z)$ therefore lead to $n-1$ solutions

$$w_r=e^{\lambda_r z}R_r(z) \qquad (r=2, 3, \ldots, n),$$

which together with w_1 form a set of n independent solutions of the given equation. Since the theorem is true when $n=1$, it is true always.

The converse of this theorem is also true, namely, that if $e^{\lambda_1 z}R_1(z)$, $e^{\lambda_2 z}R_2(z), \ldots, e^{\lambda_n z}R_n(z)$ are linearly distinct, these n functions satisfy a differential equation of order n, with polynomial coefficients, such that the degree of the coefficient of $\dfrac{d^n w}{dz^n}$ is not less than the degree of any other coefficient in the equation. Consider, in the first place, the single function

$$w=e^{\lambda_1 z}R_1(z)$$
$$=e^{\lambda_1 z}\frac{P(z)}{Q(z)},$$

where P and Q are polynomials in z. Then

$$PQ\frac{dw}{dz}=(\lambda_1 PQ+P'Q-PQ')w,$$

and therefore the coefficient of w is a polynomial of degree not exceeding that of the coefficient of $\dfrac{dw}{dz}$.

Now suppose that for an equation of degree $n-1$ the coefficient of $\dfrac{d^{n-1}w}{dz^{n-1}}$ is a polynomial of degree not less than that of the remaining coefficients.

The n functions

$$1,\quad e^{(\lambda_2-\lambda_1)z}\frac{R_2(z)}{R_1(z)},\ \ldots,\ e^{(\lambda_n-\lambda_1)z}\frac{R_n(z)}{R_1(z)}$$

satisfy a differential equation

$$Q_0(z)\frac{d^nW}{dz^n}+Q_1(z)\frac{d^{n-1}W}{dz^{n-1}}+\ldots+Q_{n-1}(z)\frac{dW}{dz}=0,$$

whose coefficients are polynomials in z multiplied by exponentials. If

$$u=\frac{dW}{dz},$$

there arises an equation of order $n-1$ in u whose solutions are

$$\frac{d}{dz}\left\{e^{(\lambda_2-\lambda_1)z}\frac{R_2(z)}{R_1(z)}\right\},\ \ldots,\ \frac{d}{dz}\left\{e^{(\lambda_n-\lambda_1)z}\frac{R_n(z)}{R_1(z)}\right\},$$

each of which is of the type $e^{\lambda z}R(z)$. By reason of the assumption made, the exponential factors in $Q_0(z), \ldots, Q_{n-1}(z)$ cancel out, and the degree of $Q_0(z)$ is at most equal to that of the remaining coefficients. Now make the substitution

$$w=e^{\lambda_1 z}R_1(z)W,$$

then the equation satisfied by w is of order n and is of the type specified. In particular the degree of the coefficient of $\dfrac{d^nw}{dz^n}$ is at least as great as that of the other coefficients. The converse theorem is therefore proved.

This investigation gives a clue to the nature of the solutions when the point at infinity is a point of indetermination of a simple character.

15·6. Equations whose Coefficients are Doubly-Periodic Functions.—Another class of equations whose general solution, when uniform, is expressible in terms of known functions is revealed by the following theorem.* *When the coefficients of a homogeneous linear differential equation are doubly-periodic functions of the independent variable, the equation possesses a fundamental set of solutions which, if uniform, are in general doubly-periodic functions of the second kind.*

Let the differential equation be

$$\frac{d^nw}{dz^n}+p_1(z)\frac{d^{n-1}w}{dz^{n-1}}+\ldots+p_{n-1}(z)\frac{dw}{dz}+p_n(z)w=0,$$

and let the coefficients $p(z)$ be doubly-periodic functions with the periods 2ω and $2\omega'$. It will also be assumed that the number of singular points in a period-parallelogram is finite, and that the general solution of the equation is uniform, for which it is necessary that the exponents relative to every singular point should be unequal integers.

Let $w_1(z), w_2(z), \ldots, w_n(z)$ be a fundamental set of solutions of the equation. Then

$$w_1(z+2\omega),\quad w_2(z+2\omega),\ \ldots,\ w_n(z+2\omega)$$

will also be solutions forming a fundamental set, and there arises a set of n linear relations

$$w_r(z+2\omega)=a_{r1}w_1(z)+\ldots+a_{rn}w_n(z)\qquad (r=1, 2, \ldots, n).$$

* Hermite, *C. R. Acad. Sc. Paris*, 85–94 (1877–82) *passim* [*Œuvres*, 3, p. 266]; Picard, *C. R.* 89 (1879), p. 140; 90 (1880), p. 128; *J. für Math.* 90 (1881), p. 281. Mittag-Leffler, *C. R.* 90 (1880), p. 299; Floquet, *C. R.* 98 (1884), pp. 38, 82; *Ann. Éc. Norm.* (3), 1 (1884), pp. 181, 405.

By following a line of reasoning very similar to that used in § 15·2 it can be proved that there is at least one solution $u_1(z)$ such that

$$u_1(z+2\omega)=su_1(z),$$

where s is a numerical constant. Now consider the other period; the functions

$$u_1(z), \quad u_1(z+2\omega'), \quad u_1(z+4\omega'), \ldots$$

are all solutions of the equation. Since the equation has only n distinct solutions, there will be a number m ($\leqslant n$), such that $u_1(z+2m\omega')$ is expressible as the linear combination

$$b_1u_1(z)+b_2u_1(z+2\omega')+\ldots+b_mu_1\{z+2(m-1)\omega'\},$$

and supposing m to be the least integer for which this is true, the constant b_1 is not zero.

Let

$$\begin{aligned} u_1(z+2\omega')&=u_2(z), \\ u_2(z+2\omega')&=u_3(z), \\ &\cdots \\ u_{m-1}(z+2\omega')&=u_m(z), \end{aligned}$$

then

$$u_m(z+2\omega')=b_1u_1(z)+b_2u_2(z)+\ldots+b_mu_m(z),$$

and $u_1(z), u_2(z), \ldots, u_m(z)$ are linearly distinct, and

$$u_r(z+2\omega)=su_r(z) \qquad (r=1, 2, \ldots, m).$$

The existence of the above set of transformations shows that there is at least one function $v(z)$ which is a linear combination of $u_1(z), \ldots, u_m(z)$ such that

$$v(z+2\omega')=s'v(z),$$

where s' is a constant.

Consequently the equation has a solution $w=v(z)$ such that

$$v(z+2\omega)=sv(z), \quad v(z+2\omega')=s'v(z),$$

in other words $v(z)$ is a doubly-periodic function of the second kind, or a quasi-doubly-periodic function.

In the general case, when the characteristic equation corresponding to the substitution of $z+2\omega$ (or $z+2\omega'$) for z has n distinct roots, the equation will have a set of n fundamental solutions each of which has a quasi-periodicity of this nature.

In any case, an analytic expression of the general solution can be arrived at. Let

$$w_1=\phi_1(z)$$

be any quasi-periodic solution of the given equation, and write

$$w=\phi_1(z)\int W\,dz.$$

Then W will be a uniform solution of an equation of order $n-1$. On account of the fact that $\phi_1'(z)/\phi(z)$ and its successive derivatives are purely periodic, the coefficients of this equation, after division throughout by $\{\phi(z)\}^n$, will be purely periodic. This equation in turn has a quasi-periodic solution $\phi_2(z)$ and therefore

$$w_2=\phi_1(z)\int\phi_2(z)dz$$

is a solution of the original equation. This process may be continued, and the n distinct solutions

$$\begin{aligned} w_1&=\phi_1(z), \\ w_2&=\phi_1(z)\int\phi_2(z)dz, \\ &\cdots \\ w_n&=\phi_1(z)\int\phi_2(z)\ldots\int\phi_n(z)(dz)^{n-1} \end{aligned}$$

are obtained.

15·61. The Explicit Form of the Solution.—Let

$$w=\phi(z)$$

be a solution of the equation, such that

$$\phi(z+2\omega)=s\phi(z), \quad \phi(z+2\omega')=s'\phi(z).$$

Consider the function

$$\psi(z)=e^{\lambda z}\frac{\sigma(z-a)}{\sigma(z)},$$

where λ and a are constants, and $\sigma(z)$ is the Weierstrassian σ-function.* Then

$$\psi(z+2\omega)=e^{2\lambda\omega-2\eta a}\psi(z), \quad \psi(z+2\omega')=e^{2\lambda\omega'-2\eta' a}\psi(z),$$

and therefore the quotient $\phi(z)/\psi(z)$ will be doubly-periodic if

$$2\lambda\omega-2\eta a=\log s,$$
$$2\lambda\omega'-2\eta' a=\log s'.$$

Since it is known that †

$$\eta\omega'-\omega\eta'=\tfrac{1}{2}\pi i\neq 0,$$

these equations determine λ and a in terms of ω, ω', η, η', $\log s$ and $\log s'$. Thus

$$\phi(z)=e^{\lambda z}\frac{\sigma(z-a)}{\sigma(z)}\mathfrak{P}(z),$$

where $\mathfrak{P}(z)$ is an elliptic function.

Now restrict the equation to the second order; when the roots of both characteristic equations are unequal, both solutions are doubly-periodic functions of the second kind. Let the two solutions be $\phi_1(z)$ and $\phi_2(z)$ and consider first of all the case in which both characteristic equations have double roots; suppose

$$\phi_1(z+2\omega)=s\phi_1(z), \quad \phi_2(z+2\omega)=s\phi_2(z),$$
$$\phi_1(z+2\omega')=s'\phi_1(z), \quad \phi_2(z+2\omega')=s'\phi_2(z)+t'\phi_1(z).$$

If $t'=0$, $\phi_1(z)$ and $\phi_2(z)$ are doubly-periodic functions of the second kind; let $t'\neq 0$, then $\phi_1(z)$ is expressible in the form obtained above. Also if

$$\chi(z)=\phi_2(z)/\phi_1(z),$$

then

$$\chi(z+2\omega)=\chi(z),$$
$$\chi(z+2\omega')=\chi(z)+\frac{t'}{s'}.$$

Compare this with the function

$$A\zeta(z)+Bz,$$

* Whittaker and Watson, *Modern Analysis*, § 20·42. It may here be noted that

$$\frac{\sigma(z+2\omega)}{\sigma(z)}=-e^{2\eta(z+\omega)}, \quad \frac{\sigma(z+2\omega')}{\sigma(z)}=-e^{2\eta'(z+\omega')},$$

where η and η' are constants. Also

$$\frac{d}{dz}\log\sigma(z)=\zeta(z),$$

so that

$$\zeta(z+2\omega)=\zeta(z)+2\eta, \quad \zeta(z+2\omega')=\zeta(z)+2\eta',$$
$$\frac{d}{dz}\zeta(z)=-\wp(z).$$

† Whittaker and Watson, *loc. cit.* § 20·411.

which increases by $2A\eta+2B\omega$ when z increases by 2ω, and by $2A\eta'+2B\omega'$ when z increases by $2\omega'$. Thus if

$$A\eta+B\omega = 0, \quad A\eta'+B\omega' = \frac{t'}{2s'},$$

the function

$$\chi(z)-A\zeta(z)-Bz$$

will be a doubly-periodic function of z. In this case therefore

$$\phi_2(z)=\{\mathfrak{P}_1(z)+A\zeta(z)+Bz\}\phi_1(z),$$

where $\mathfrak{P}_1(z)$ is an elliptic function and A and B are definite constants. On the other hand, let

$$\phi_1(z+2\omega)=s\phi_1(z), \quad \phi_2(z+2\omega)=s\phi_2(z)+t\phi_1(z),$$
$$\phi_1(z+2\omega')=s'\phi_1(z), \quad \phi_2(z+2\omega')=s'\phi_2(z)+t'\phi_1(z).$$

These are consistent since $\phi\{(z+2\omega)+2\omega'\}=\phi\{(z+2\omega')+2\omega\}$. Now $\phi_1(z)$ is a doubly-periodic function of the second kind as before, but in this case

$$\chi(z+2\omega)=\chi(z)+\frac{t}{s}, \quad \chi(z+2\omega')=\chi(z)+\frac{t'}{s'},$$

and the constants A and B have to be determined by the equations

$$A\eta+B\omega=\frac{t}{2s}, \quad A\eta'+B\omega'=\frac{t'}{2s'}.$$

The form of $\phi_2(z)$ is, however, as before.

The equation of the third order may be treated similarly; the only case needing special discussion is that in which the characteristic equation has a triple root. In that case $\phi_1(z)$ and $\phi_2(z)$ are of the forms given; the third solution $\phi(z)$ will be found to involve terms in

$$z^2, \quad z\zeta(z) \quad \text{and} \quad \zeta^2(z).$$

In general, if the characteristic equation has an m-ple root, there will be solutions involving z and $\zeta(z)$ up to the $(m-1)^{\text{th}}$ power. This corresponds to the logarithmic case in an equation of Fuchsian type.

15·62. The Lamé Equation.—In the equation of Lamé,*

$$\frac{d^2w}{dz^2}-\{h+n(n+1)\wp(z)\}w=0,$$

where n is a positive integer and h a constant, the singular points are the origin and its congruent points $2m\omega+2m'\omega'$. The exponents relative to any singular point are $-n$ and $n+1$. The Fuchsian theory makes clear the existence of one uniform solution, namely

$$w_1(z)=(z-2m\omega-2m'\omega')^{n+1}W(z),$$

where $W(z)$ is analytic in the domain of the point $2m\omega+2m'\omega'$ and not zero at that point. The difference of the exponents is $2n+1$, and since this is a positive integer, the possibility of the second solution $w_2(z)$ containing a logarithmic term has to be considered. But since

$$w_2(z)w_1'(z)-w_1(z)w_2'(z)=0,$$

* Whittaker and Watson, *Modern Analysis* (3rd ed.), Chap. XXIII. The Jacobian form of the equation, namely

$$\frac{d^2w}{dx^2}=\{n(n+1)k^2 \operatorname{sn}^2x-\eta\}w$$

is obtained by the transformations

$$\wp(z)-e_3=\frac{e_1-e_3}{\operatorname{sn}^2t}, \quad t=z(e_1-e_3)^{\frac{1}{2}}, \quad k^2=\frac{e_2-e_3}{e_1-e_3}, \quad x=t-iK'.$$

the second solution is

$$w_2(z)=Cw_1(z)\int\frac{dz}{\{w_1(z)\}^2},$$

where C is a constant. But $1/\{w_1(z)\}^2$ is easily seen to be an even function of z; its residues relative to the origin and congruent points are zero, and therefore a logarithmic term cannot arise. In particular, let $n=1$, so that the equation is

$$\frac{d^2w}{dz^2}-\{h+2\wp(z)\}w=0.$$

Introduce a parameter a, connected with h by means of the transcendental equation

$$\wp(a)=h.$$

Then the equation has the solutions

$$w_1=e^{-z\zeta(a)}\frac{\sigma(z+a)}{\sigma(z)},\quad w_2=e^{z\zeta(a)}\frac{\sigma(z-a)}{\sigma(z)},$$

and these solutions are in general distinct. If, however, h is equal to e_1, e_2 or e_3, the solutions are not distinct. For example, if h is equal to e_1, a becomes equal to ω_1 and the two solutions which in general are distinct now both reduce in effect to

$$w_1=e^{-\eta_1 z}\frac{\sigma(z+\omega_1)}{\sigma(z)}.$$

When $h=e_1$ the second solution may be obtained by means of a quadrature, but it is more convenient to arrive at it by a limiting process, supposing, in the first place, that h is not equal to e_1, but differs only infinitesimally from it. Then the equation

$$\wp(a)=h$$

has the roots $a=\omega_1\pm\epsilon$, where ϵ is infinitesimal. Consider the function

$$\frac{1}{2\epsilon}(W_1-W_2),$$

where

$$W_1=e^{-z\zeta(\omega_1+\epsilon)}\frac{\sigma(z+\omega_1+\epsilon)}{\sigma(z)},$$

$$W_2=e^{-z\zeta(\omega_1-\epsilon)}\frac{\sigma(z+\omega_1-\epsilon)}{\sigma(z)}.$$

This function is a solution of the equation; its limit will be the second solution w_2 required.

Now

$$\begin{aligned}\zeta(\omega_1+\epsilon)&=\zeta(\omega_1)+\epsilon\zeta'(\omega_1)+\dots\\&=\eta_1-\epsilon e_1+\dots,\end{aligned}$$

and therefore

$$e^{-z\zeta(\omega_1+\epsilon)}=e^{-\eta_1 z}(1+\epsilon e_1 z+\dots).$$

Also

$$\begin{aligned}\sigma(z+\omega_1+\epsilon)&=\sigma(z+\omega_1)+\epsilon\sigma'(z+\omega_1)+\dots\\&=\sigma(z+\omega_1)\{1+\epsilon\zeta(z+\omega_1)+\dots\},\end{aligned}$$

and thus

$$W_1=e^{-\eta_1 z}\frac{\sigma(z+\omega_1)}{\sigma(z)}[1+\epsilon\{\zeta(z+\omega_1)+e_1z\}+\dots].$$

W_2 differs from W_1 only in the sign of ϵ. Finally

$$\begin{aligned}w_2&=\lim\frac{1}{2\epsilon}(W_1-W_2)\\&=e^{-\eta_1 z}\frac{\sigma(z+\omega_1)}{\sigma(z)}\{\zeta(z+\omega_1)+e_1z\}.\end{aligned}$$

It will be noted that the solution in general is not doubly-periodic, but consists of a doubly-periodic function multiplied by an exponential factor. Thus when a has one of the *characteristic* values ω_1, ω_2 or ω_3, the first solution w_1 is periodic, but the second solution is not periodic.

The two independent solutions of the Lamé equation

$$\frac{d^2w}{dz^2}-\left\{e_1+2\wp(z)\right\}w=0$$

may also be expressed in the forms

$$\{\wp(z)-e_1\}^{\frac{1}{2}}, \quad \{\wp(z)-e_1\}^{\frac{1}{2}}\{\zeta(z+\omega_1)+e_1z\}.$$

15·63. Equations with Doubly-Periodic Coefficients such that the Ratio of any two Solutions is Uniform.—As before, let the equation be

$$\frac{d^nw}{dz^n}+p_1(z)\frac{d^{n-1}w}{dz^{n-1}}+\ldots+p_{n-1}(z)\frac{dw}{dz}+p_n(z)w=0,$$

and let the coefficients be doubly-periodic functions with periods 2ω and $2\omega'$. It will now be supposed that although the general solution is not uniform, nevertheless the ratio of any two particular solutions is a uniform function of z. It will be shown that this case can be reduced to that in which the general solution is uniform.*

Let a_1 be a singular point; the exponents relative to this singularity must differ by integers. Let

$$\nu_1, \quad \nu_1+e_{11}, \quad \nu_1+e_{12}, \quad \ldots$$

be the exponents, arranged in increasing order of magnitude so that $e_{11}, e_{12}, \ldots$ are positive integers. Let α_1 be the residue of $p_1(z)$ relative to the pole $z=a_1$. Then the sum of the roots of the indicial equation relative to a_1 is

$$\tfrac{1}{2}n(n-1)-\alpha_1,$$

and this is equal to the sum of the exponents, that is to

$$n\nu_1+\sum_{s=1}^{n-1}e_{1s}.$$

Now let there be k singular points

$$a_1, \quad a_2, \quad \ldots, \quad a_k$$

in one and the same period-parallelogram, then

$$n(\nu_1+\nu_2+\ldots+\nu_k)+\sum e_{rs}=\tfrac{1}{2}kn(n-1)-\sum\alpha_r.$$

But $\sum\alpha_r$, the sum of the residues relative to the poles within a period-parallelogram, is zero, and consequently

$$n(\nu_1+\nu_2+\ldots+\nu_k)$$

is an integer. Let m be the least integer for which

$$m^2(\nu_1+\nu_2+\ldots+\nu_k)=\lambda$$

is an integer and consider the function

$$\mathbb{P}(z)=\left\{\sigma\left(\frac{z}{m}\right)\right\}^{-\lambda}\{\sigma(z-a_1)\}^{\nu_1}\ldots\{\sigma(z-a_k)\}^{\nu_k}.$$

Since

$$\sigma\left(\frac{z+2m\omega}{m}\right)=e^{2\eta\left(\frac{z}{m}+\omega\right)+\pi i}\sigma\left(\frac{z}{m}\right),$$

$$\sigma(z-a+2m\omega)=e^{2m\eta(z-a+m\omega)+m\pi i}\sigma(z-a),$$

$$\mathbb{P}(z+2m\omega)=e^{-\lambda\left\{2\eta\left(\frac{z}{m}+\omega\right)+\pi i\right\}+\Sigma\nu\{2m\eta(z-a+m\omega)+m\pi i\}}\mathbb{P}(z).$$

* Halphen, *Mém. Acad. Sc. Paris* (2) 28 (1884) [*Œuvres*, 3, p. 55].

It follows that the logarithmic derivative of $\wp(z+2m\omega)$ exceeds the logarithmic derivative of $\wp(z)$ by

$$2m\eta\sum\nu-2\eta\lambda/m,$$

which is zero. The same is true with regard to the period $2m\omega'$. Thus the function $\wp'(z)/\wp(z)$ is a doubly-periodic function with periods $2m\omega$, $2m\omega'$.

Now make the substitution

$$w=\wp(z)W,$$

then the equation in W has coefficients which are doubly-periodic with the periods $2m\omega$, $2m\omega'$. But in this equation the exponents relative to each singular point are positive integers. The equation therefore has one uniform solution. But since the ratio of any two solutions of the equation in w is uniform, the same is true of the solutions of the equation in W. Consequently the general solution of the equation in W is uniform, which was the theorem to be proved.

15·7. Equations with Simply-Periodic Coefficients.—In the equation

$$\frac{d^nw}{dz^n}+p_1(z)\frac{d^{n-1}w}{dz^{n-1}}+\ldots+p_{n-1}(z)\frac{dw}{dz}+p_n(z)w=0,$$

let the coefficients be uniform purely-periodic functions of z with period 2ω, devoid of any singularities but poles in the finite part of the z-plane. There is no loss in generality in supposing ω to be a positive real number. The theory of equations of this type is very similar to that of equations with doubly-periodic coefficients, by which it appears to have been suggested, and is generally known as the *Floquet Theory.**

Let $w_1(z)$, $w_2(z)$, . . ., $w_n(z)$ be a fundamental set of solutions of the equation. Then $w_1(z+2\omega)$, $w_2(z+2\omega)$, . . ., $w_n(z+2\omega)$ likewise satisfy the equations, and therefore there exists a set of linear relations

$$w_r(z+2\omega)=a_{r1}w_1(z)+a_{r2}w_2(z)+\ldots+a_{rn}w_n(z) \quad (r=1,\ 2,\ \ldots,\ n)$$

and, as in § 15·2, the determinant $|a_{rs}|$ is not zero.

The problem of determining a solution $u(z)$ such that

$$u(z+2\omega)=su(z)$$

is equivalent to that of reducing the above set of linear relations to its canonical form, which in turn depends upon the characteristic equation

$$\begin{vmatrix} a_{11}-s, & a_{12}, & \ldots, & a_{1n} \\ a_{21}, & a_{22}-s, & \ldots, & a_{2n} \\ \cdot & \cdot & \cdot & \cdot \\ a_{n1}, & a_{n2}, & \ldots, & a_{nn}-s \end{vmatrix}=0.$$

If this equation has n distinct roots s_1, s_2, . . ., s_n, then a fundamental set of n solutions $u_1(z)$, $u_2(z)$, . . ., $u_n(z)$ can be found such that

$$u_1(z+2\omega)=s_1u_1(z),\ \ldots,\ \ u_n(z+2\omega)=s_nu_n(z).$$

If, on the other hand, s_1 is a repeated root, there will be a sub-set of solutions $u_1(z)$, . . ., $u_\mu(z)$ such that

$$\begin{aligned} u_1(z+2\omega)&=s_1u_1(z), \\ u_2(z+2\omega)&=s_1\{u_2(z)+u_1(z)\}, \\ &\cdots\cdots \\ u_\mu(z+2\omega)&=s_1\{u_\mu(z)+u_{\mu-1}(z)\}, \end{aligned}$$

and possibly other sub-sets of a similar nature.

* Floquet, *Ann. Éc. Norm.* (2) 13 (1883), p. 47.

Consider the analytic expression of the solutions in these two cases. In either case there is at least one solution $u_1(z)$ such that

$$u_1(z+2\omega)=s_1u_1(z).$$

Now

$$e^{-a(z+2\omega)}u_1(z+2\omega)=s_1e^{-2a\omega}e^{-az}u_1(z),$$

and therefore

$$e^{-az}u(z)$$

will be a purely periodic function, with period 2ω, provided a is so chosen that

$$e^{2a\omega}=s_1.$$

A number a satisfying the equation

$$e^{2a\omega}=s_r,$$

for any particular value of r is called a *characteristic exponent*; its imaginary part is ambiguous in that any integral multiple of $\pi i/\omega$ may be added to it. The real part of a, on the other hand, is perfectly definite, and plays an important part in the theory.

Thus, when the n roots of the characteristic equation are distinct, there exists a linearly independent set of u solutions $u_1(z), u_2(z), \ldots, u_n(z)$ such that

$$u_r(z)=e^{a_rz}\phi_r(z),$$

where a_r is a characteristic exponent corresponding to s_r and $\phi_r(z)$ is a purely periodic function with period 2ω.

Now consider the case where s_1 is a repeated root. By writing

$$u_\nu(z)=e^{a_1z}v_\nu(z),$$

the canonical sub-set is reduced to

$$\begin{aligned} v_1(z+2\omega)&=v_1(z),\\ v_2(z+2\omega)&=v_2(z)+v_1(z),\\ &\cdots\cdots\\ v_\mu(z+2\omega)&=v_\mu(z)+v_{\mu-1}(z). \end{aligned}$$

Thus

$$\frac{v_2(z+2\omega)}{v_1(z+2\omega)}=\frac{v_2(z)}{v_1(z)}+1,$$

and therefore

$$\frac{v_2(z)}{v_1(z)}-\frac{z}{2\omega}$$

is a purely periodic function of z, with period 2ω. In general, it may be proved, precisely as in § 15·23, that if

$$P_\nu(z)=\frac{z(z-2\omega)\ldots\{z-(2\nu-2)\omega\}}{(2\omega)^\nu\nu!},$$

then

$$\begin{aligned} u_1(z)&=e^{a_1z}\phi_1(z),\\ u_2(z)&=e^{a_1z}\{P_1(z)\phi_1(z)+\phi_2(z)\},\\ &\cdots\cdots\\ u_\nu(z)&=e^{a_1z}\{P_{\nu-1}(z)\phi_1(z)+P_{\nu-2}(z)\phi_2(z)+\ldots+P_1(z)\phi_{\nu-1}(z)+\phi_\nu(z)\} \end{aligned}$$

$$(\nu=2, 3, \ldots, \mu),$$

where $\phi_1(z), \phi_2(z) \ldots, \phi_\mu(z)$ are purely periodic, with period 2ω.

15·71. The Characteristic Exponents.—When the characteristic exponent a is a pure imaginary, the corresponding solution remains finite as z tends to infinity along the real axis. On the other hand, if the real part of a is not zero, the modulus of the term e^{az} becomes infinite either for $z=+\infty$ or for

$z=-\infty$. In the former case the solution is said to be *stable*, in the latter, *unstable*.

The problem of determining the characteristic exponents is in general a very difficult one.* The theory which has been outlined, and which reveals the functional character of the general solution, does not provide a practical method for obtaining the solution explicitly. The problem has therefore to be attacked indirectly, as follows.

Consider the equation of the second order

$$\frac{d^2w}{dz^2}=p(z)w,$$

where $p(z)$ is a function having the real period 2ω, which is analytic throughout a strip $-\eta<y<\eta$, including the real axis in its interior. The characteristic equation is, in this case, of the form

$$s^2-As+1=0,$$

where A is a constant depending only upon the function $p(z)$. Let $f(z)$ and $g(z)$ be two solutions of the equation such that

$$f(0)=1, \quad f'(0)=0,$$
$$g(0)=0, \quad g'(0)=1,$$

and let

$$f(z+2\omega)=a_{11}f(z)+a_{12}g(z),$$
$$g(z+2\omega)=a_{21}f(z)+a_{22}g(z),$$

so that

$$g'(z+2\omega)=a_{21}f'(z)+a_{22}g'(z).$$

By writing $z=0$, it is seen that

$$f(2\omega)=a_{11}, \quad g'(2\omega)=a_{22},$$

and since the characteristic equation is

$$\begin{vmatrix} a_{11}-s, & a_{12} \\ a_{21}, & a_{22}-s \end{vmatrix}=0,$$

it follows that

$$\begin{aligned} A&=a_{11}+a_{22} \\ &=f(2\omega)+g'(2\omega). \end{aligned}$$

Now consider, instead of the original equation, the equation

$$\frac{d^2w}{dz^2}=\lambda p(z)w\,;$$

it possesses solutions

$$f(z, \lambda)=1+\lambda f_1(z)+\ldots+\lambda^n f_n(z)+\ldots,$$
$$g(z, \lambda)=z+\lambda g_1(z)+\ldots+\lambda^n g_n(z)+\ldots,$$

such that the functions $f_n(z)$ and $g_n(z)$ are zero at $z=0$, and the series are convergent for all values of λ when z lies within the parallel strip enclosing the axis of reals.

Now the functions $f_n(z)$ and $g_n(z)$ satisfy the relations

$$\frac{d^2f_n(z)}{dz^2}=p(z)f_{n-1}(z),$$

$$\frac{d^2g_n(z)}{dz^2}=p(z)g_{n-1}(z),$$

* Liapounov, *Ann. Fac. Sc. Toul.* (2), 9 (1907), pp. 203-469 [originally published in Russian, Kharkov, 1892]. Poincaré, *Les Méthodes nouvelles de la Mécanique céleste*, 1, Chap. IV.; Horn, *Z. Math. Phys.* 48 (1903), p. 400.

and therefore $f_n(z)$ and $g_n(z)$ may be evaluated from the equations

$$f_n(z)=\int_0^z\int_0^z p(z)f_{n-1}(z)(dz)^2,$$
$$g_n(z)=\int_0^z\int_0^z p(z)g_{n-1}(z)(dz)^2,$$

with the initial conditions

$$f_0(z)=1, \quad g_0(z)=z.$$

When the functions $f_n(z)$, $g_n(z)$ have been found, λ may be made equal to unity; it then follows that

$$A=2+\sum_{r=1}^{\infty}\{f_n(2\omega)+g_n'(2\omega)\}.$$

In the first place, suppose that $p(z)$ is positive for all real values of z, then the functions $f_n(z)$, $g_n(z)$ and $g_n'(z)$ are all positive when $z>0$. It follows that $A>2$, and consequently that the roots of the characteristic equation are real. The characteristic exponents may be taken to be real, and therefore any solution is unstable. For a stable solution it is therefore necessary that $p(z)$ be negative for some real values of z.*

15·72. Hill's Equation.—Suppose now that $p(z)$ is an even periodic function of period π. The equation may be written in the form

$$\frac{d^2w}{dz^2}+\{\theta_0+2\theta_1\cos 2z+2\theta_2\cos 4z+\ldots\}w=0,$$

$p(z)$ being replaced by the equivalent Fourier-cosine series. It will be assumed that this series converges absolutely and uniformly throughout a parallel strip enclosing the real axis.

Assume a solution

$$w=e^{az}\sum_{r=-\infty}^{\infty}b_re^{riz},$$

then, on substituting in the equation it is found that the coefficients b_r satisfy the recurrence-relations

$$(a+2ri)^2b_r+\sum_{\nu=-\infty}^{\infty}\theta_\nu b_{r-\nu}=0$$

for all integral values of r. By dividing this relation throughout by $(a+2ri)^2$ and then eliminating the coefficients b, the characteristic exponent a is found to satisfy the convergent determinantal equation

$$\begin{vmatrix}
 & \cdot & \cdot & \cdot & \cdot & \cdot & \\
\cdots & \frac{(ia+4)^2-\theta_0}{4^2-\theta_0}, & \frac{-\theta_1}{4^2-\theta_0}, & \frac{-\theta_2}{4^2-\theta_0}, & \frac{-\theta_3}{4^2-\theta_0}, & \frac{-\theta_4}{4^2-\theta_0}, & \cdots \\
\cdots & \frac{-\theta_1}{2^2-\theta_0}, & \frac{(ia+2)^2-\theta_0}{2^2-\theta_0}, & \frac{-\theta_1}{2^2-\theta_0}, & \frac{-\theta_2}{2^2-\theta_0}, & \frac{-\theta_3}{2^2-\theta_0}, & \cdots \\
\cdots & \frac{-\theta_2}{-\theta_0}, & \frac{-\theta_1}{-\theta_0}, & \frac{(ia)^2-\theta_0}{-\theta_0}, & \frac{-\theta_1}{-\theta_0}, & \frac{-\theta_2}{-\theta_0}, & \cdots \\
\cdots & \frac{-\theta_3}{2^2-\theta_0}, & \frac{-\theta_2}{2^2-\theta_0}, & \frac{-\theta_1}{2^2-\theta_0}, & \frac{(ia-2)^2-\theta_0}{2^2-\theta_0}, & \frac{-\theta_1}{2^2-\theta_0}, & \cdots \\
\cdots & \frac{-\theta_4}{4^2-\theta_0}, & \frac{-\theta_3}{4^2-\theta_0}, & \frac{-\theta_2}{4^2-\theta_0}, & \frac{-\theta_1}{4^2-\theta_0}, & \frac{(ia-4)^2-\theta_0}{4^2-\theta_0}, & \cdots \\
 & \cdot & \cdot & \cdot & \cdot & \cdot &
\end{vmatrix}=0.$$

* It was shown by Liapounov (*loc. cit.*) that if $p(z)$ is negative for all real values of z and $2\omega\int_0^{2\omega}p(z)dz$ is in absolute magnitude not greater than 4, $|A|<2$ and the roots of the characteristic equation are conjugate complex numbers, of modulus unity.

The problem now takes on a two-fold aspect: either the constants θ may all be given explicitly, and it is required to determine a to correspond, or else the problem may be to find what relation must exist between the constants θ in order that a may be zero, and the solution purely periodic with period π.

The first aspect of the problem is at once soluble, for writing Hill's determinantal equation in the form

$$\Delta(ia)=0,$$

it is found that *

$$\Delta(ia)=\Delta(0)-\frac{\sin^2(\frac{1}{2}\pi ai)}{\sin^2(\frac{1}{2}\pi\sqrt{\theta_0})},$$

and therefore a is a root of the transcendental equation

$$\sin^2(\tfrac{1}{2}\pi ai)=\Delta(0)\sin^2(\tfrac{1}{2}\pi\sqrt{\theta_0}).$$

The second aspect of the problem reduces to determining a relation between the constants θ so that

$$\Delta(0)=0.$$

15·8. Analogies with the Fuchsian Theory.—An equation, such as that of Hill, may be brought into the form

$$(1-t^2)\frac{d^2w}{dt^2}-t\frac{dw}{dt}+\left(\sum_{\nu=0}^{\infty}c_\nu t^{2\nu}\right)w=0$$

by writing $t=\cos z$. This is an equation with regular singularities at $t=\pm 1$, the exponents being, in each case, 0 and $\frac{1}{2}$, and an irregular singular point at infinity. By considering the equation in this algebraic form, from the point of view of the Fuchsian theory, certain interesting properties are brought into view.†

The fundamental solutions relative to $t=+1$ may be written

$$F_1(1-t)=1+\sum_{\nu=1}^{\infty}a_\nu(1-t),$$

$$F_2(1-t)=\sqrt{(1-t)}\left\{1+\sum_{\nu=1}^{\infty}b_\nu(1-t)\right\};$$

in each case the series converges within the circle $|1-t|=2$; in the second case $\sqrt{(1-t)}$ is initially positive when $-1<t\leq+1$. Since the equation is unchanged when t is replaced by $-t$, the solutions relative to the singular point $t=-1$ are

$$F_1(1+t) \quad\text{and}\quad F_2(1+t),$$

the series now being convergent within the circle $|1+t|=2$. Within the region common to both circles of convergence,

$$F_1(1-t)=\alpha F_1(1+t)+\beta F_2(1+t),$$
$$F_2(1-t)=\gamma F_1(1+t)+\delta F_2(1+t),$$

where α, β, γ and δ are constants. Also

$$\begin{aligned}F_1(1+t)&=\alpha F_1(1-t)+\beta F_2(1-t)\\&=(\alpha^2+\beta\gamma)F_1(1+t)+\beta(\alpha+\delta)F_2(1+t),\\F_2(1+t)&=\gamma F_1(1-t)+\delta F_2(1-t)\\&=\gamma(\alpha+\delta)F_1(1+t)+(\beta\gamma+\delta^2)F_2(1+t),\end{aligned}$$

* Hill, *Acta Math.* 8 (1886); *see* Whittaker and Watson, *Modern Analysis*, § 19·42.

† Poole, *Proc. London Math. Soc.* (2), 20 (1922), p. 374.

and these relations must be satisfied identically, for otherwise $F_1(1+t)$ and $F_2(1+t)$ would be linearly related. Hence

$$\alpha^2+\beta\gamma=\beta\gamma+\delta^2=1, \quad \beta(\alpha+\delta)=\gamma(\alpha+\delta)=0,$$

and there are only two possibilities, namely, either

$$\text{(i)} \quad \alpha=\delta=\pm 1, \quad \beta=0, \quad \gamma=0,$$

$$\text{or} \quad \text{(ii)} \quad \alpha=-\delta, \quad \beta\gamma=1-\alpha^2.$$

Consider first the possibility $\alpha=\delta=+1$, $\beta=\gamma=0$. Then

$$F_1(1-t)=F_1(1+t), \quad F_2(1-t)=F_2(1+t),$$

and this relation holds in the common region of convergence of the series, and therefore it certainly holds near the origin. But the origin is an ordinary point of the equation, so that there cannot be two distinct even solutions both valid near $t=0$. This first hypothesis must therefore be rejected. The hypothesis $\alpha=\delta=-1$, $\beta=\gamma=0$ similarly implies the existence of two distinct odd solutions, valid at the origin, and must likewise be rejected. Thus there only remains the hypothesis $\alpha=-\delta$, $\beta\gamma=1-\alpha^2$, which, however, admits of a multitude of particular cases. The following are the more important:

(*a*) Let $\alpha=-\delta=\pm 1$, $\beta=0$, so that

$$F_1(1-t)=\pm F_1(1+t)$$

when $|1\pm t|<2$. This is a solution, even if $\alpha=+1$, odd if $\alpha=-1$, having no singularity in the finite part of the plane. The substitution $t=\cos z$ expresses it when $\alpha=+1$ as a series of cosines of even multiples of z, and when $\alpha=-1$ as a series of odd multiples of z.

(*b*) Let $\alpha=-\delta=\pm 1$, $\gamma=0$, so that

$$F_2(1-t)=\pm F_2(1+t)$$

when $|1\pm t|<2$. The solution is the product of $\sqrt{(1-t^2)}$ and an integral function of t, for it changes sign when t describes a small circuit about $t=+1$ or about $t=-1$. This integral function is even if $\delta=+1$, and odd if $\delta=-1$. By writing $t=\cos z$ the solution becomes, when $\delta=+1$ a series of even multiples of z, and when $\delta=-1$ a series of sines of odd multiples of z.

(*c*) Let $\alpha=\delta=0$, $\beta\gamma=1$. Then

$$F_1(1-t)=\beta F_2(1+t), \quad F_2(1-t)=\frac{1}{\beta}F_1(1+t),$$

when $|1\pm t|<2$. The solutions may be written

$$F_1(1-t)=\sqrt{(1+t)}\phi(t), \quad F_2(1-t)=\sqrt{(1-t)}\phi(-t),$$

where $\phi(t)$ is an integral function of t. By writing $t=\cos z$, they are transformed into

$$F_1=\cos\tfrac{1}{2}z f(z), \quad F_2=\sin\tfrac{1}{2}z f(\pi-z),$$

where $f(z)$ is a series of cosines of integral multiples of z, converging throughout the finite part of the z-plane. Thus the equation admits of two independent solutions having the period 4π.

15·81. The Existence of Periodic Solutions in General.—The existence of solutions of period 4π which has just been proved raises the question of the possibility of the existence of solutions of period $2m\pi$ where m is any positive integer.

Consider the circuit illustrated in the figure, which is in the form of a loop enclosing the two singular points $t=\pm 1$. Start at the point A with the two solutions

$$u_A=F_1(1-t),\quad v_A=F_2(1-t),$$

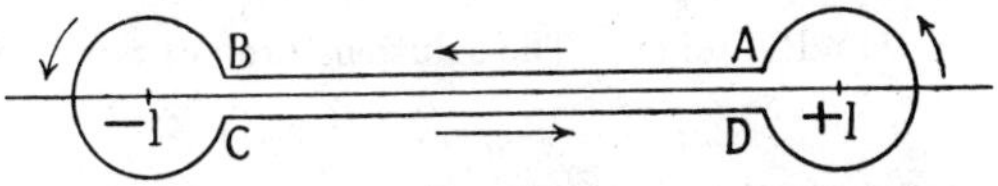

FIG. 12.

and proceed along the cut AB. At B the solutions become

$$u_B=\alpha F_1(1+t)+\beta F_2(1+t),\quad v_B=\gamma F_1(1+t)+\delta F_2(1+t).$$

The effect of describing the circle BC is to change the sign of F_2, F_1 remaining unchanged in sign, so that

$$u_C=\alpha F_1(1+t)-\beta F_2(1+t),\quad v_C=\gamma F_1(1+t)-\delta F_2(1+t).$$

Next describe the cut CD; at D the solutions become

$$u_D=(\alpha^2-\beta\gamma)F_1(1-t)+\beta(\alpha+\delta)F_2(1-t),$$
$$v_D=\gamma(\alpha-\delta)F_1(1-t)+(\beta\gamma-\delta^2)F_2(1-t).$$

Lastly, after describing the circuit DA the solutions become $\bar{u}_A$, $\bar{v}_A$ where

$$\bar{u}_A=(\alpha^2-\beta\gamma)F_1(1-t)-\beta(\alpha-\delta)F_2(1-t),$$
$$\bar{v}_A=\gamma(\alpha-\delta)F_1(1-t)-(\beta\gamma-\delta^2)F_2(1-t).$$

But, as before

$$\alpha=-\delta,\quad \beta\gamma=1-\alpha^2,$$

and therefore

$$\bar{u}=(2\alpha^2-1)u-2\alpha\beta v,$$
$$\bar{v}=2\alpha\gamma u+(2\alpha^2-1)v.$$

Now let

$$W=au+bv$$

be a solution such that

$$a\bar{u}+b\bar{v}=s(au+bv),$$

then the equation which determines s is

$$\begin{vmatrix} 2\alpha^2-1-s, & 2\alpha\gamma \\ -2\alpha\beta, & 2\alpha^2-1-s \end{vmatrix}=0$$

or

$$(2\alpha^2-1-s)^2+4\alpha^2(1-\alpha^2)=0.$$

This equation reduces to

$$s^2+2s(1-2\alpha^2)+1=0.$$

If $\alpha^2>1$ this equation gives rise to two real and distinct values of s; leading to two solutions W_1 and W_2, which become respectively s^nW_1 and $s^{-n}W_2$ after n circuits have been described. These solutions are not periodic. On the other hand, if $\alpha^2<1$ the roots of the equation in s are conjugate complex numbers of modulus unity. Suppose, in the first place, that $s^m=1$ where m is a positive integer. Then

$$s=e^{\pm 2r\pi i/m}$$

and

$$2\alpha^2-1=\cos\frac{2r\pi}{m}.$$

Solutions which return to their initial values after m circuits thus arise; in terms of the variable z they are

$$W_1=e^{rzi/m}f(z),\quad W_2=e^{-rzi/m}f(-z),$$

where $f(z)$ is a function of period 2π, finite for all finite values of z. These solutions are of period $2m\pi$. If, on the other hand, s is not a complex root of unity the solutions will be of the form

$$W_1=e^{\theta zi}f(z), \quad W_2=e^{-\theta zi}f(-z),$$

where θ is an irrational number. The solutions are not now periodic; they are, however, stable.

15·9. Linear Substitutions.—Consider a simple closed contour in the z-plane, defined in terms of the vectorial angle θ by the equation

$$z=\phi(\theta),$$

where $\phi(\theta)$ is a one-valued periodic function of θ. It will be supposed that the contour does not pass through any singular point. Now any solution of the differential equation

$$\frac{d^n w}{dz^n}+p_1(z)\frac{d^{n-1}w}{dz^{n-1}}+\ldots+p_{n-1}(z)\frac{dw}{dz}+p_n(z)w=0$$

may be developed, by the method of successive approximations, as a series which converges for all values of the real variable θ. Let

$$w_1(\theta), \quad w_2(\theta), \quad \ldots, \quad w_n(\theta)$$

be a fundamental set of solutions, and assuming that the coefficients of the equation are one-valued,

$$w_1(\theta+2\pi), \quad w_2(\theta+2\pi), \quad \ldots, \quad w_n(\theta+2\pi)$$

is also a fundamental set. Consequently

$$\begin{aligned} w_1(\theta+2\pi)&=a_{11}w_1(\theta)+a_{12}w_2(\theta)+\ldots+a_{1n}w_n(\theta),\\ w_2(\theta+2\pi)&=a_{21}w_1(\theta)+a_{22}w_2(\theta)+\ldots+a_{2n}w_n(\theta),\\ &\cdots\cdots\cdots\cdots\cdots\\ w_n(\theta+2\pi)&=a_{n1}w_1(\theta)+a_{n2}w_2(\theta)+\ldots+a_{nn}w_n(\theta), \end{aligned}$$

where the coefficients a are constants, with of course a non-vanishing determinant, which can be evaluated from n sets of n equations of which the following is typical:

$$\begin{aligned} w_r(\theta+2\pi)&=a_{r1}w_1(\theta)+a_{r2}w_2(\theta)+\ldots+a_{rn}w_n(\theta),\\ w_r'(\theta+2\pi)&=a_{r1}w_1'(\theta)+a_{r2}w_2'(\theta)+\ldots+a_{rn}w_n'(\theta),\\ &\cdots\cdots\cdots\cdots\cdots\\ w_r^{(n-1)}(\theta+2\pi)&=a_{r1}w_1^{(n-1)}(\theta)+a_{r2}w_2^{(n-1)}(\theta)+\ldots+a_{rn}w_n^{(n-1)}(\theta). \end{aligned}$$

Thus the linear substitutions undergone by a set of fundamental solutions when z describes a simple closed circuit may be considered as known.

In particular, suppose that the coefficients of the equation are rational functions of z, which when decomposed into partial fractions are of the form

$$\sum\frac{A_{ik}}{(z-a_i)^k}.$$

Then it follows from the general existence theorems that if the coefficients A_{ik} are regarded as parameters in the equation the solutions $w_1(z)$, $w_2(z)$, . . ., $w_n(z)$ are integral functions of these parameters and therefore, *the coefficients a_{rs} in the set of linear substitutions are meromorphic functions of these parameters.**

* Further developments depend to a great extent upon the theory of the invariants of the general linear differential equation. See Hamburger, *J. für Math.* 83 (1877), p. 193; Poincaré, *Acta Math.* 4 (1883), p. 212; Mittag-Leffler, *Acta Math.* 15 (1890), p. 1; von Koch, *ibid.* 16 (1892), p. 217.

15·91. The Group of a Linear Differential Equation.—It will be assumed that the coefficients of the equation are uniform in z and that there are only a finite number of singular points. Then the effect of causing z to describe a closed circuit not passing through any singular point is that of a linear substitution S which transforms $w_1, w_2, \ldots, w_n$ respectively into

$$\begin{array}{l} a_{11}w_1+a_{12}w_2+\;\ldots\;+a_{1n}w_n, \\ a_{21}w_1-a_{22}w_2+\;\ldots\;+a_{2n}w_n, \\ \cdot\quad\cdot\quad\cdot\quad\cdot\quad\cdot\quad\cdot \\ a_{n1}w_1+a_{n2}w_2+\;\ldots\;+a_{nn}w_n, \end{array}$$

where the determinant of the constants a_{rs} is not zero.

Let S' be the linear substitution corresponding to a circuit distinct from the first. Then the result of performing the second circuit followed by the first is a substitution of the same general form, namely the product $S'S$. This is in general distinct from the substitution SS'.

Now any circuit in the z-plane enclosing a number of singularities is equivalent to a succession of closed circuits or loops described in a definite order and such that each loop encircles one and only one singular point.

Let there be m singular points $a_1, a_2, \ldots, a_m$ and let S_r be the *simple* substitution which arises from a circulation around the point a_r in the positive direction. Then S_r^{-1} is the inverse substitution due to the same circulation made in the negative direction. Any arbitrary substitution can thus be decomposed into a succession of simple substitutions of the form

$$S_l^\lambda S_m^\mu \ldots S_p^\varpi S_r^\rho,$$

where $\lambda, \mu, \ldots, \varpi, \rho$ are positive or negative integers, and S_r^ρ denotes S_r described $|\rho|$ times in the positive or negative direction according as ρ is positive or negative.

The aggregate of these substitutions is known as the *group* of the equation.*

The group has been defined with reference to a particular fundamental set of solutions. Now consider a second fundamental set; it is derived from the first set by a definite substitution Σ. Then if S is any substitution carried out on the first set, $\Sigma^{-1}S\Sigma$ is a substitution carried out on the second set. Clearly if the substitutions S form a group the substitutions $\Sigma^{-1}S\Sigma$ will also form a group and these groups will be intimately related to one another.

15·92. The Riemann Problem.—The following classical problem † will serve as an illustration of the general theory of linear differential equations. It is proposed to determine a function

$$P\left\{\begin{array}{cccc} a & b & c & \\ \alpha & \beta & \gamma & z \\ \alpha' & \beta' & \gamma' & \end{array}\right\}$$

which satisfies the following conditions:

(i) It is uniform and continuous throughout the whole plane except at the singular points a, b, c.

* More specifically it is known as the *monodromic group* of the equation to distinguish it from a more extensive group known as the *rationality group*. It may be noted that a set of linear substitutions forms a group if the set contains (a) the identical substitution, (b) the inverse of each substitution, (c) the product of any two substitutions.

† Riemann, *Abh. Ges. Wiss. Gött.* 7 (1857), p. 3; [*Math. Werke* (2nd ed.), p. 67].

(ii) Between any three determinations P_1, P_2, P_3 of this function there exists a linear relation

$$c_1P_1+c_2P_2+c_3P_3=0,$$

where c_1, c_2 and c_3 are constants.

(iii) In the neighbourhood of the point a there are two distinct determinations:

$$(z-a)^{\alpha}f_1(z), \quad (z-a)^{\alpha'}f_2(z),$$

where $f_1(z)$ and $f_2(z)$ are analytic in the neighbourhood of $z=a$ and not zero at a. Similarly in the neighbourhood of $x=b$ there are two determinations:

$$(z-b)^{\beta}g_1(z), \quad (z-b)^{\beta'}g_2(z),$$

and in the neighbourhood of $z=c$ there are also two determinations:

$$(z-c)^{\gamma}h_1(z), \quad (z-c)^{\gamma'}h_2(z).$$

Let P_1 and P_2 be any two linearly distinct determinations of the required function. Then since any other determination is linearly dependent upon P_1 and P_2, the required function will satisfy the differential equation of the second order

$$\begin{vmatrix} \dfrac{d^2w}{dz^2}, & \dfrac{dw}{dz}, & w \\ P_1'', & P_1', & P_1 \\ P_2'', & P_2', & P_2 \end{vmatrix} =0,$$

which may be written

$$\frac{d^2w}{dz^2}+p\frac{dw}{dz}+qw=0,$$

where

$$p=-\frac{P_2P_1''-P_1P_2''}{P_2P_1'-P_1P_2'}, \quad q=\frac{P_2'P_1''-P_1'P_2''}{P_2P_1'-P_1P_2'}.$$

Consider the behaviour of the function p in the neighbourhood of the singular point $z=a$. Let

$$P_1=(z-a)^{\alpha}f_1(z), \quad P_2=(z-a)^{\alpha'}f_2(z),$$

then it is found that

$$p=\frac{1-\alpha-\alpha'}{z-a}+\varpi(z),$$

where $\varpi(z)$ is analytic in the neighbourhood of $z=a$. It follows that

$$p=\frac{1-\alpha-\alpha'}{z-a}+\frac{1-\beta-\beta'}{z-b}+\frac{1-\gamma-\gamma'}{z-c}+u(z),$$

where $u(z)$ is analytic everywhere. Now since the P-function is analytic at infinity it is necessary that, for large values of $|z|$,

$$p=\frac{2}{z}+O(z^{-2}).$$

But

$$p=\frac{3-\alpha-\alpha'-\beta-\beta'-\gamma-\gamma'}{z}+O(z^{-2})+u(z),$$

and since $u(z)=O(1)$ it is necessary that $u(z)=0$ and therefore

$$p=\frac{1-\alpha-\alpha'}{z-a}+\frac{1-\beta-\beta'}{z-b}+\frac{1-\gamma-\gamma'}{z-c},$$

where

$$\alpha+\alpha'+\beta+\beta'+\gamma+\gamma'=1.$$

In the same way it is found that in the neighbourhood of a

$$q = \frac{\alpha\alpha'}{(z-a)^2} + O\left\{\frac{1}{z-a}\right\},$$

and therefore q may be written

$$q = \frac{\alpha\alpha'}{(z-a)^2} + \frac{\beta\beta'}{(z-b)^2} + \frac{\gamma\gamma'}{(z-c)^2} + \frac{A}{z-a} + \frac{B}{z-b} + \frac{C}{z-c},$$

where A, B, C are finite for all finite values of z. It is more convenient, however, to adopt the equivalent expression

$$q = \frac{1}{(z-a)(z-b)(z-c)}\left\{\frac{L}{z-a} + \frac{M}{z-b} + \frac{N}{z-c}\right\},$$

where L, M and N are finite for all finite values of z.

But since the point at infinity is an ordinary point, for large values of $|z|$,

$$q(z) = O(z^{-4}),$$

and therefore L, M, N are constants, and it may easily be verified that

$$L = \alpha\alpha'(a-b)(a-c),$$
$$M = \beta\beta'(b-c)(b-a),$$
$$N = \gamma\gamma'(c-a)(c-b).$$

Thus Riemann's P-function satisfies the differential equation,*

$$\frac{d^2w}{dz^2} + \sum \frac{1-\alpha-\alpha'}{z-a} \cdot \frac{dw}{dz} + \sum \frac{\alpha\alpha'(a-b)(a-c)}{z-a} \cdot \frac{w}{(z-a)(z-b)(z-c)} = 0.$$

This equation is known as the *generalised hypergeometric equation*; when $a=0$, $b=1$, $c=\infty$,† $\alpha'=\beta'=0$, it becomes the ordinary hypergeometric equation

$$z(1-z)\frac{d^2w}{dz^2} + \{(\alpha+\beta-2)z+1-\alpha\}\frac{dw}{dz} - \gamma\gamma' w = 0.$$

The solutions of the generalised hypergeometric equation thus furnish the required functions. In order that they may be of the form postulated it is only necessary that no one of the exponent differences

$$\alpha-\alpha', \quad \beta-\beta', \quad \gamma-\gamma'$$

should be an integer; otherwise logarithmic terms would enter into one or other of the solutions.

15·93. The Group of the Hypergeometric Equation.—Let P_α and $P_{\alpha'}$ be the two solutions appropriate respectively to the exponents α and α' at the singularity a, P_β and $P_{\beta'}$ those relative to the singularity b, and P_γ and $P_{\gamma'}$ those relative to the singularity c. Let Γ be any closed simple curve, for example the circle which passes through the points a, b and c. Then within Γ the six solutions are analytic and there exist between them relations such as

$$P_\alpha = A_\beta P_\beta + A_{\beta'} P_{\beta'},$$
$$P_{\alpha'} = A'_\beta P_\beta + A'_{\beta'} P_{\beta'},$$
$$P_\alpha = A_\gamma P_\gamma + A_{\gamma'} P_{\gamma'},$$
$$P_{\alpha'} = A'_\gamma P_\gamma + A'_{\gamma'} P_{\gamma'},$$

wherein the coefficients A are constants. These constants are not all independent; there exist relations between them which will now be determined.

Since the point at infinity is an ordinary point, a circuit in the positive

* First obtained by Papperitz, *Math. Ann.* 25 (1885), p. 213. In Riemann's exposition, simplifications were introduced which led to the ordinary hypergeometric equation.

† $z-c$ is replaced by $1/z$.

direction around the point c is equivalent to a circuit in the negative direction around the two points a and b. The third of the above relations shows that the effect of the first circuit is to change P_a into

$$A_\gamma e^{2\pi i\gamma}P_\gamma+A_{\gamma'}e^{2\pi i\gamma'}P_{\gamma'},$$

whilst the first relation shows that the second circuit changes P_a into

$$e^{-2\pi i a}(A_\beta e^{-2\pi i\beta}P_\beta+A_{\beta'}e^{-2\pi i\beta'}P_{\beta'}).$$

Consequently

$$A_\gamma e^{2\pi i\gamma}P_\gamma+A_{\gamma'}e^{2\pi i\gamma'}P_{\gamma'}=e^{-2\pi i a}\{A_\beta e^{-2\pi i\beta}P_\beta+A_{\beta'}e^{-2\pi i\beta'}P_{\beta'}\},$$

and similarly

$$A'_\gamma e^{2\pi i\gamma}P_\gamma+A'_{\gamma'}e^{2\pi i\gamma'}P_{\gamma'}=e^{-2\pi i a'}\{A'_\beta e^{-2\pi i\beta}P_\beta+A'_{\beta'}e^{-2\pi i\beta'}P_{\beta'}\}.$$

But

$$A_\gamma P_\gamma+A_{\gamma'}P_{\gamma'}=A_\beta P_\beta+A_{\beta'}P_{\beta'},$$
$$A'_\gamma P_\gamma+A'_{\gamma'}P_{\gamma'}=A'_\beta P_\beta+A'_{\beta'}P_{\beta'}.$$

On eliminating P_γ, $P_{\gamma'}$, P_β, $P_{\beta'}$ between these four relations it is found that

$$\frac{A_\gamma}{A'_\gamma}=\frac{A_\beta}{A'_\beta}\cdot\frac{e^{-\pi i a}\sin(a+\beta+\gamma')\pi}{e^{-\pi i a'}\sin(a'+\beta+\gamma')\pi}=\frac{A_{\beta'}}{A'_{\beta'}}\cdot\frac{e^{-\pi i a}\sin(a+\beta'+\gamma')\pi}{e^{-\pi i a'}\sin(a'+\beta'+\gamma')\pi},$$
$$\frac{A_{\gamma'}}{A'_{\gamma'}}=\frac{A_\beta}{A'_\beta}\cdot\frac{e^{-\pi i a}\sin(a+\beta+\gamma)\pi}{e^{-\pi i a'}\sin(a'+\beta+\gamma)\pi}=\frac{A_{\beta'}}{A'_{\beta'}}\cdot\frac{e^{-\pi i a}\sin(a+\beta'+\gamma)\pi}{e^{-\pi i a'}\sin(a'+\beta'+\gamma)\pi}.$$

Thus any one of the ratios

$$\frac{A_\beta}{A'_\beta},\quad \frac{A_{\beta'}}{A'_{\beta'}},\quad \frac{A_\gamma}{A'_\gamma},\quad \frac{A_{\gamma'}}{A'_{\gamma'}}$$

is a known multiple of the others. The four relations given are consistent if

$$\frac{\sin(a+\beta'+\gamma')\pi\,.\sin(a'+\beta+\gamma')\pi}{\sin(a+\beta+\gamma')\pi\,.\sin(a'+\beta'+\gamma')\pi}=\frac{\sin(a+\beta'+\gamma)\pi\,.\sin(a'+\beta+\gamma)\pi}{\sin(a'+\beta'+\gamma)\pi\,.\sin(a+\beta+\gamma)\pi},$$

which is satisfied in virtue of the relation

$$a+a'+\beta+\beta'+\gamma+\gamma'=1.$$

In order to determine the group of the equation it is sufficient to consider the substitutions which any pair of fundamental solutions, for example P_a and $P_{a'}$, undergoes when the point z describes a circuit around each of two singular points, a and b for example. The description of a circuit round a in the positive sense transforms P_a and $P_{a'}$ respectively into

$$e^{2\pi i a}P_a,\quad e^{2\pi i a'}P_{a'},$$

and similarly when a positive circuit round b is completed, P_a and $P_{a'}$ respectively become

$$A_\beta e^{2\pi i\beta}P_\beta+A_{\beta'}e^{2\pi i\beta'}P_{\beta'},\quad A'_\beta e^{2\pi i\beta}P_\beta+A'_{\beta'}e^{2\pi i\beta'}P_{\beta'}.$$

But since

$$P_a=A_\beta P_\beta+A_{\beta'}P_{\beta'},$$
$$P_{a'}=\lambda A_\beta P_\beta+\lambda' A_{\beta'}P_{\beta'},$$

where

$$\lambda=\frac{A'_\beta}{A_\beta},\quad \lambda'=\frac{A'_{\beta'}}{A_{\beta'}},$$

the final forms which P_a and $P_{a'}$ take after description of the circuit round b may be expressed in terms of P_a and $P_{a'}$ as follows:

$$\frac{\lambda' e^{2\pi i\beta}-\lambda e^{2\pi i\beta'}}{\lambda'-\lambda}P_a+\frac{e^{2\pi i\beta'}-e^{2\pi i\beta}}{\lambda'-\lambda}P_{a'},$$
$$\frac{\lambda\lambda'(e^{2\pi i\beta}-e^{2\pi i\beta'})}{\lambda'-\lambda}P_a+\frac{\lambda' e^{2\pi i\beta'}-\lambda e^{2\pi i\beta}}{\lambda'-\lambda}P_{a'}.$$

To obtain a more symmetrical expression, let

$$u=(\lambda'-\lambda)P_a, \quad v=P_{a'},$$

then if S_a is the operation of describing a positive circuit around a,

$$S_a u=e^{2\pi i a}u, \quad S_a v=e^{2\pi i a'}v,$$

and if S_b is the similar operation with regard to b

$$S_b u=\frac{\mu e^{2\pi i\beta}-e^{2\pi i\beta'}}{\mu-1}u+(e^{2\pi i\beta'}-e^{2\pi i\beta})v,$$

$$S_b v=\frac{\mu(e^{2\pi i\beta}-e^{2\pi i\beta'})}{(\mu-1)^2}u+\frac{\mu e^{2\pi i\beta'}-e^{2\pi i\beta}}{\mu-1}v,$$

where

$$\mu=\frac{\lambda'}{\lambda}=\frac{\sin(\alpha+\beta'+\gamma')\pi\,.\,\sin(\alpha'+\beta+\gamma')\pi}{\sin(\alpha'+\beta'+\gamma')\pi\,.\,\sin(\alpha+\beta+\gamma')\pi}.$$

The two substitutions S_a and S_b may be regarded as the fundamental substitutions of the group; any other substitution is compounded of integral powers of S_a and S_b.

If it is postulated that all the solutions of the equation are *algebraic* functions of z, and are therefore the roots of an algebraic equation, then each solution can have but a finite number of values at each singular point. Consequently the number of distinct substitutions is finite and the group is a finite group. It is evident that a necessary condition for the finiteness of the group is that

$$\alpha, \quad \alpha', \quad \beta, \quad \beta', \quad \gamma, \quad \gamma'$$

are all rational numbers.

When the equation is reduced to its normal form by removing the term in $\frac{dw}{dz}$ by means of the substitution

$$w=(z-a)^{\frac{1}{2}(\alpha+\alpha'-1)}(z-b)^{\frac{1}{2}(\beta+\beta'-1)}(z-c)^{\frac{1}{2}(\gamma+\gamma'-1)}v,$$

it becomes

$$\frac{d^2v}{dz^2}+\sum_{r=1}^{3}\frac{1-\lambda_r^2}{(z-a_r)^2}\cdot\frac{\psi'(a_r)}{4\psi}v=0,$$

where

$$a_1=a, \quad a_2=b, \quad a_3=c,$$

$$\lambda_1=\tfrac{1}{2}(\alpha-\alpha'), \quad \lambda_2=\tfrac{1}{2}(\beta-\beta'), \quad \lambda_3=\tfrac{1}{2}(\gamma-\gamma').$$

There are fifteen different cases in which an algebraic solution is possible; the values which $\lambda_1, \lambda_2, \lambda_3$ may assume are as follows: *

I.	1/2	1/2	1/n	II.	1/2	1/3	1/3
III.	2/3	1/3	1/3	IV.	1/2	1/3	1/4
V.	2/3	1/4	1/4	VI.	1/2	1/3	1/5
VII.	2/5	1/3	1/3	VIII.	2/3	1/5	1/5
IX.	1/2	2/5	1/5	X.	3/5	1/3	1/5
XI.	2/5	2/5	2/5	XII.	2/3	1/3	1/5
XIII.	4/5	1/5	1/5	XIV.	1/2	2/5	1/3
XV.	3/5	2/5	1/3				

[For a detailed discussion of linear equations of the second order whose general solutions are algebraic, and for practical methods of constructing such solutions, see Forsyth, *Theory of Differential Equations*, Vol. 4, pp. 176–190.]

* Schwarz, *J. für Math.* 75 (1872), p. 293; Cayley, *Trans. Camb. Phil. Soc.* 13 (1881), p. 5 [*Coll. Math. Papers*, 11, p. 148]; Klein, *Math. Ann.* 11 (1877), p. 115; 12, p. 167 [*Ges. Math. Abhand.* 2, pp. 302, 307]; *Vorlesungen über das Ikosaeder*, p. 115.

Miscellaneous Examples.

1. Prove that if w satisfies the algebraic equation

$$w^n+a_2w^{n-2}+\ .\ .\ .\ +a_n=0,$$

whose coefficients are polynomials in z, then w satisfies a linear differential equation of order $n-1$, whose coefficients are rational functions of z.

2. If u is any function of z, and

$$w^3+3w=u,$$

prove that

$$\frac{d^2w}{dz^2}+\left(\frac{uu'}{u^2+4}-\frac{u''}{u'}\right)\frac{dw}{dz}-\frac{1}{9}\frac{u'^2}{u^2+4}w=0.$$

3. Prove that the differential equation of the scheme

$$P\left\{\begin{matrix} 0 & 1 & a & \infty & \\ 0 & 0 & 0 & \sigma & z \\ 1-\lambda & 1-\mu & \nu & \tau & \end{matrix}\right\},$$

in which

$$\lambda+\mu-\nu-\sigma-\tau=0,$$

is

$$\frac{d^2w}{dz^2}+\left(\frac{\lambda}{z}+\frac{\mu}{z-1}+\frac{1-\nu}{z-a}\right)\frac{dw}{dz}+\frac{\sigma\tau(z-q)}{z(z-1)(z-a)}w=0,$$

where q is an arbitrary constant. If the solution relative to the singular point $z=0$ with exponent 0 is denoted by

$$W(a, q\ ;\ \sigma, \tau, \lambda, \mu\ ;\ z)$$

show that there are in general eight possible solutions of the form

$$w=z^\alpha(z-1)^\beta(z-a)^\gamma W(a, q\ ;\ \sigma', \tau', \lambda', \mu'\ ;\ z).$$

[When $a=1$, $q=1$, or when $a=0$, $q=0$, the equation degenerates into the hypergeometric equation. A set of 64 solutions can be constructed analogous to the set of 24 solutions of the hypergeometric equation. See Heun, *Math. Ann.*, 33 (1889), pp. 161, 180.]

4. The equation

$$\frac{d^2w}{dz^2}+p\frac{dw}{dz}+qw=0$$

is transformed by the substitution

$$w=W.\exp\left(-\tfrac{1}{2}\int pdz\right)$$

into

$$\frac{d^2W}{dz^2}+IW=0,$$

where

$$I=q-\tfrac{1}{4}p^2-\tfrac{1}{2}\frac{dp}{dz}.$$

[This is known as the *normal form* of the equation. Equations which have the same normal form are *equivalent*, and I is their *invariant*.]

If z is a function of s, the expression

$$\{s, z\}\equiv-\left[\frac{z'''}{z''}-\frac{3}{2}\left(\frac{z''}{z'}\right)^2\right]\Big/z'^2,$$

where dashes denote differentiation with respect to s, is known as the *Schwarzian derivative*. Let w_1 and w_2 be two distinct solutions of the above equation in w, and let $s=w_1/w_2$. Then

$$\{s, z\}=2q-\tfrac{1}{2}p^2-\frac{dp}{dz}=2I.$$

Prove that, for a change of independent variable from z to Z,

$$\{s, z\}=\{s, Z\}\left(\frac{dZ}{dz}\right)^2+\{Z, z\}.$$

5. Prove that, for the hypergeometric equation

$$z(1-z)\frac{d^2w}{dz^2}+\{\gamma-(\alpha+\beta+1)z\}\frac{dw}{dz}-\alpha\beta w=0,$$

$$\{s, z\}=\frac{\frac{1}{2}\left(1-\frac{1}{\lambda^2}\right)}{z^2}+\frac{\frac{1}{2}\left(1-\frac{1}{\mu^2}\right)}{(z-1)^2}+\frac{\frac{1}{2}\left(\frac{1}{\lambda^2}+\frac{1}{\mu^2}-\frac{1}{\nu^2}\right)}{z(z-1)},$$

where γ, μ, ν depend upon α, β, γ.

[For the connection of this result with the construction of algebraic solutions, see Forsyth, *Theory of Differential Equations*, Vol. 4, pp. 182–184.]

6. When the Lamé equation with $n=1$ is expressed in the Jacobian form

$$\frac{d^2w}{dz^2}=\{2k^2\,\mathrm{sn}^2\,z-\eta\}w,$$

its general solution is

$$w=A\frac{H(z+a)}{\Theta(z)}e^{-zZ(a)}+B\frac{H(z-a)}{\Theta(z)}e^{zZ(a)},$$

where $\mathrm{dn}^2 a=\eta-k^2$.

Discuss the particular cases

$$h=1+k^2,\ 1,\ k^2.$$

[Hermite.]

7. Show that, when n is a positive integer, the Lamé equation

$$\frac{d^2w}{dz^2}-\{h+n(n+1)\wp(z)\}w=0$$

has, for appropriate values of h, solutions of the forms

(i) $w=P_m$ $(n=2m)$,

(ii) $w=[\{\wp(z)-e_\lambda\}\{\wp(z)-e_\mu\}]^{\frac{1}{2}}P_{m-1}$ $(n=2m)$,

(iii) $w=[\wp(z)-e_\lambda]P_{m-1}$ $(n=2m-1)$,

(iv) $w=\wp'(z)P_{m-2}$ $(n=2m-1)$,

where P_r denotes a polynomial of degree r in $\wp(z)$, and e_λ, e_μ are any two of the constants e_1, e_2, e_3.

Investigate the corresponding solutions of the Jacobian form of the Lamé equation.

8. Integrate the equation

$$\frac{d^2w}{dz^2}=\left\{\frac{\lambda(\lambda+1)}{\mathrm{sn}^2 z}+\frac{\mu(\mu+1)\mathrm{dn}^2 z}{\mathrm{cn}^2 z}+\frac{\nu(\nu+1)k^2\mathrm{cn}^2 z}{\mathrm{dn}^2 z}+n(n+1)k^2\mathrm{sn}^2 z+\eta\right\}w.$$

[Darboux.]

9. Find the linear differential equation whose solutions are the products of solutions of the equation

$$\frac{d^2w}{dz^2}+Iw=0,$$

and explain why it is of the third order. [Lindemann.]

10. Show that the equation

$$z(1-z)\frac{d^2w}{dz^2}+\tfrac{1}{2}(1-2z)\frac{dw}{dz}+(az+b)w=0$$

has two particular solutions the product of which is a single-valued transcendental function $F(z)$, and show that these solutions are

$$w_1=\{F(z)\}^{\frac{1}{2}}.\exp\left[c\int\frac{dz}{\{z(1-z)\}^{\frac{1}{2}}F(z)}\right],$$

$$w_2=\{F(z)\}^{\frac{1}{2}}.\exp\left[-c\int\frac{dz}{\{z(1-z)\}^{\frac{1}{2}}F(z)}\right],$$

where c is a determinate constant. In what circumstances are these two particular solutions coincident?

[Math. Tripos, II. 1898.]

CHAPTER XVI

SOLUTION OF LINEAR DIFFERENTIAL EQUATIONS IN SERIES

16·1. The Method of Frobenius.—It was shown in the preceding chapter (§ 15·3) that if all the solutions of a linear differential equation are regular in the neighbourhood of a singular point, the coefficients of the equation are subject to certain definite restrictions. Thus, if the singular point in question is the origin, the equation may be written in the form

$$z^n \frac{d^n w}{dz^n} + z^{n-1} P_1(z) \frac{d^{n-1} w}{dz^{n-1}} + \ldots + z P_{n-1}(z) \frac{dw}{dz} + P_n(z) w = 0,$$

in which $P_1(z), \ldots, P_n(z)$ are analytic throughout the neighbourhood of $z=0$. In this case it is possible to obtain an explicit development of the n fundamental solutions relative to the singularity at the origin, and incidentally to prove that these developments are convergent for sufficiently small values of $|z|$.*

16·11. The Formal Solution.—Set up a series

$$W(z, \rho) = \sum_{\nu=0}^{\infty} c_\nu z^{\rho+\nu} \qquad (c_0 \neq 0),$$

in which the number ρ and the coefficients c_ν are so to be determined that W is a solution of the differential equation. Let the differential equation be represented symbolically as

$$Lw = 0,$$

then

$$\begin{aligned} LW(z, \rho) &= \sum c_\nu L z^{\rho+\nu} \\ &= \sum c_\nu z^{\rho+\nu} f(z, \rho+\nu), \end{aligned}$$

where $f(z, \rho+\nu)$ represents the expression

$$[\rho+\nu]_n + [\rho+\nu]_{n-1} P_1(z) + \ldots + [\rho+\nu]_1 P_{n-1}(z) + P_n(z),$$

in which $[\rho+\nu]_n$ is written for $(\rho+\nu)(\rho+\nu-1) \ldots (\rho+\nu-n+1)$. Now let $f(z, \rho+\nu)$, which is an analytic function of z in the neighbourhood of $z=0$, be developed as a power series in z, thus

$$f(z, \rho+\nu) = \sum_{\lambda=0}^{\infty} f_\lambda(\rho+\nu) z^\lambda,$$

then

$$LW(z, \rho) = \sum \{c_\nu f_0(\rho+\nu) + c_{\nu-1} f_1(\rho+\nu-1) + \ldots + c_0 f_\nu(\rho)\} z^{\rho+\nu}.$$

Now if

$$LW(z, \rho) = 0,$$

* Frobenius, *J. für Math.* 76 (1873), p. 214. Modifications of the original exposition are due to Forsyth, *Differential Equations*, Vol. 4, pp. 78–97.

the coefficient of each separate power of z must be zero. There thus arises the set of recurrence-relations :

$$\begin{aligned} &c_0 f_0(\rho)=0, \\ &c_1 f_0(\rho+1)+c_0 f_1(\rho)=0, \\ &\cdot \quad \cdot \quad \cdot \quad \cdot \quad \cdot \\ &c_\nu f_0(\rho+\nu)+c_{\nu-1} f_1(\rho+\nu-1)+\ldots+c_0 f_\nu(\rho)=0, \end{aligned}$$

and so on.

Since c_0 is not zero, the first equation of the set, viz.

$$f_0(\rho) \equiv [\rho]_n+[\rho]_{n-1} P_1(0)+\ldots+\rho P_{n-1}(0)+P_n(0)=0,$$

that is to say the indicial equation, determines n values of ρ which may, or may not, be distinct. If one of these values is so chosen that $f_0(\rho+\nu) \neq 0$ for any positive integral value of ν, then the recurrence-relations determine the constants c_ν uniquely, thus

$$c_\nu=\frac{(-1)^\nu c_0 F_\nu(\rho)}{f_0(\rho+1) f_0(\rho+2) \ldots f_0(\rho+\nu)},$$

where

$$F_\nu(\rho)=\left|\begin{array}{cccccc} f_1(\rho+\nu-1), & f_2(\rho+\nu-2), & \ldots, & f_{\nu-1}(\rho+1), & f_\nu(\rho) \\ f_0(\rho+\nu-1), & f_1(\rho+\nu-2), & \ldots, & f_{\nu-2}(\rho+1), & f_{\nu-1}(\rho) \\ 0, & f_0(\rho+\nu-2), & \ldots, & f_{\nu-3}(\rho+1), & f_{\nu-2}(\rho) \\ \cdot & \cdot & \cdot & \cdot & \cdot \\ 0, & 0, & \ldots, & f_0(\rho+1), & f_1(\rho) \end{array}\right|$$

Assuming for the moment the convergence of the series $W(z, \rho)$ for each particular value of ρ chosen, it is seen that, if the n roots of the indicial equation are distinct and no two of them differ by an integer, to each ρ corresponds a determinate sequence of coefficients c_ν, and altogether n distinct solutions, forming a fundamental system, are obtained.

If the n indices are not such that no two of them differ by an integer, they may be arranged in order in distinct sets,

$$\begin{aligned} &\rho_0, \quad \rho_1, \quad \ldots, \quad \rho_{\alpha-1}, \\ &\rho_\alpha, \quad \rho_{\alpha+1}, \quad \ldots, \quad \rho_{\beta-1}, \\ &\cdot \quad \cdot \quad \cdot \quad \cdot \quad \cdot \end{aligned}$$

in such manner that the numbers in each set differ only by integers, and are so arranged that their real parts form a non-increasing sequence. The first member only of each set gives rise to a solution of the type just discussed, since, for instance, any member $\rho_{\alpha+k}$ of the set $\rho_\alpha \ldots \rho_{\beta-1}$ is either equal to ρ_α or is less than ρ_α by a positive integer. In the first case, the solution corresponding to $\rho_{\alpha+k}$ is formally identical with that proceeding from ρ_α; in the second case, the solution corresponding to $\rho_{\alpha+k}$ is nugatory owing to the violation of the condition $f_0(\rho+\nu) \neq 0$, when $\nu=\rho_\alpha-\rho_{\alpha+k}$.

The difficulty in the second case could be removed by replacing the initial constant c_0 by $c_0 f_0(\rho_{\alpha+k}+\nu)$; a series is then obtained in which all the coefficients c_ν are finite, but it will be seen that the first ν terms vanish and the series differs from that corresponding to ρ_α only by a constant multiplier, and is therefore not a distinct solution.

16·12. Modification of the Formal Method of Solution.—In order to obtain the material from which all the solutions corresponding to each set may be deduced it is necessary to modify the preceding method as follows.

Let σ be a parameter whose variation is restricted to a circle drawn round a root of $f_0(\rho)=0$ with radius sufficiently small to exclude all other roots.* Assume the series

$$W(z, \sigma)=\sum_{\nu=0}^{\infty} c_\nu z^{\sigma+\nu},$$

in which c_0 is arbitrary, and c_ν is in general determined as a function of σ by the recurrence-relations

$$c_1 f_0(\sigma+1)+c_0 f_1(\sigma)=0,$$
$$\cdot \quad \cdot \quad \cdot \quad \cdot$$
$$c_\nu f_0(\sigma+\nu)+c_{\nu-1} f_1(\sigma+\nu-1)+ \ldots +c_0 f_\nu(\sigma)=0,$$
$$\cdot \quad \cdot \quad \cdot \quad \cdot$$

in which the functional operators $f_0, f_1, \ldots f_\nu, \ldots$ are as previously defined. Then

$$\begin{aligned} LW(z, \sigma) &= \sum\{c_\nu f_0(\sigma+\nu)+c_{\nu-1} f_1(\sigma+\nu-1)+ \ldots +c_0 f_\nu(\sigma)\} z^{\sigma+\nu} \\ &= c_0 f_0(\sigma) z^\sigma \end{aligned}$$

in virtue of the recurrence relations. The previous solution is now obtained by taking $\sigma=\rho$, where ρ is an appropriate solution of $f_0(\sigma)=0$.

16·2. The Convergence of the Development.—Let Γ be the radius of the largest circle, with its centre at the origin, within which all of the functions $P_1(z), P_2(z), \ldots, P_n(z)$ are analytic. Then the series

$$f(z, \sigma+\nu)=\sum_\lambda f_\lambda(\sigma+\nu) z^\lambda,$$

and the series

$$f'(z, \sigma+\nu)=\sum_\lambda (\lambda+1) f_{\lambda+1}(\sigma+\nu) z^\lambda,$$

obtained by differentiating the former term-by-term with respect to z, are convergent for $|z|<\Gamma$. Let $M(\sigma+\nu)$ be the upper bound of $|f'(z, \sigma+\nu)|$ on the circle $|z|=R=\Gamma-\epsilon$, where ϵ is an arbitrarily small positive number. Then, by Cauchy's integral theorem,

$$(\lambda+1) f_{\lambda+1}(\sigma+\nu)=\frac{1}{2\pi i}\int_R \frac{f'(z, \sigma+\nu)}{z^{\lambda+1}}\, dz,$$

whence

$$|(\lambda+1) f_{\lambda+1}(\sigma+\nu)|<\frac{1}{2\pi}\cdot\frac{M(\sigma+\nu)}{R^{\lambda+1}} 2\pi R=\frac{M(\sigma+\nu)}{R^\lambda},$$

and

$$|f_{\lambda+1}(\sigma+\nu)|<M(\sigma+\nu) R^{-\lambda} \qquad (\lambda=0, 1, 2, \ldots).$$

Since σ is restricted to vary in the neighbourhood of the roots of $f_0(\sigma)=0$, and since the number of such roots is finite, a positive integer N may be so chosen that $f(\sigma+\nu+1)\neq 0$ when $\nu \geqslant N$. This being the case,

$$c_{\nu+1}=-\frac{1}{f_0(\sigma+\nu+1)}\{c_\nu f_1(\sigma+\nu)+c_{\nu-1} f_2(\sigma+\nu-1)+ \ldots +c_0 f_{\nu+1}(\sigma)\},$$

and if each term is replaced by its modulus,

$$\begin{aligned} |c_{\nu+1}| &\leqslant \frac{1}{|f_0(\sigma+\nu+1)|}\{|c_\nu|\,|f_1(\sigma+\nu)|+|c_{\nu-1}|\,|f_2(\sigma+\nu-1)|+ \ldots +|c_0|\,|f_{\nu+1}(\sigma)|\} \\ &< \frac{1}{|f_0(\sigma+\nu+1)|}\{|c_\nu| M(\sigma+\nu)+|c_{\nu-1}| M(\sigma+\nu-1) R^{-1}+ \ldots +|c_0| M(\sigma) R^{-\nu}\} \\ &= C_{\nu+1} \quad \text{say.} \end{aligned}$$

* $f_0(\rho)=0$ is an algebraic equation in ρ of degree n; its roots are therefore isolated, and each root, a multiple root being reckoned once only, can be surrounded by a circle of non-zero radius which excludes all other roots.

Then as a consequence of this definition of $C_{\nu+1}$,

$$C_{\nu+1}=\frac{|c_\nu|M(\sigma+\nu)}{|f_0(\sigma+\nu+1)|}+\frac{|f_0(\sigma+\nu)|}{|f_0(\sigma+\nu+1)|}R^{-1}C_\nu,$$

and since $|c_\nu|<C_\nu$, it follows that

$$\frac{C_{\nu+1}}{C_\nu}<\frac{M(\sigma+\nu)}{|f_0(\sigma+\nu+1)|}+\frac{|f_0(\sigma+\nu)|}{|f_0(\sigma+\nu+1)|}R^{-1}.$$

Let positive numbers A_ν be chosen to satisfy the recurrence relation

$$\frac{A_{\nu+1}}{A_\nu}=\frac{M(\sigma+\nu)}{|f_0(\sigma+\nu+1)|}+\frac{|f_0(\sigma+\nu)|}{|f_0(\sigma+\nu+1)|}R^{-1},$$

and such that $A_N=C_N$, then

$$|c_{\nu+1}|<C_{\nu+1}<A_{\nu+1} \qquad (\nu\geqslant N).$$

Now

$$f(z,\ \sigma+\nu)=[\sigma+\nu]_n+[\sigma+\nu]_{n-1}P_1(z)+\ .\ .\ .\ +P_n(z),$$

whence

$$f'(z,\ \sigma+\nu)=[\sigma+\nu]_{n-1}P_1'(z)+[\sigma+\nu]_{n-2}P_2'(z)+\ .\ .\ .\ +P_n'(z),$$

i.e. $f'(z,\ \sigma+\nu)$ is a polynomial in $\sigma+\nu$ of degree $n-1$ whose coefficients depend upon z only. Consequently

$$M(\sigma+\nu)=\text{Max}\,|f'(z,\ \sigma+\nu)| \qquad (|z|\leqslant R)$$
$$\leqslant M_1|[\sigma+\nu]_{n-1}|+M_2|[\sigma+\nu]_{n-2}|+\ .\ .\ .\ +M_n,$$

where

$$M_r=\text{Max}\,|P_r'(z)| \qquad (|z|\leqslant R),$$

and therefore, given ν_0, a number K, independent of σ, exists, such that

$$M(\sigma+\nu)<K\nu^{n-1},$$

when $\nu\geqslant\nu_0$. Similarly, since $f_0(\sigma+\nu)$ is a polynomial in $\sigma+\nu$ of degree n, a number K_1, independent of σ, exists, such that

$$|f_0(\sigma+\nu)|<K_1\nu^n.$$

Hence, as $\nu\rightarrow\infty$,

$$\frac{M(\sigma+\nu)}{|f_0(\sigma+\nu+1)|}\rightarrow 0$$

and

$$\left|\frac{f_0(\sigma+\nu)}{f_0(\sigma+\nu+1)}\right|\rightarrow 1,$$

both uniformly with respect to σ, from which it follows that

$$\frac{A_{\nu+1}}{A_\nu}\rightarrow R^{-1},$$

uniformly with respect to σ.

Hence * the power-series

$$\sum A_n z^n$$

has R as its radius of convergence, and therefore, since

$$|c_n|\leqslant A_n,$$

the radius of convergence of the series

$$\sum c_n z^n$$

is not less than R. Since A_n is independent of σ, the convergence is uniform in σ.

* Bromwich, *Theory of Infinite Series*, § 84.

16·3. The Solutions corresponding to a Set of Indices.—Consider one of the sets of indices, for instance the set *

$$\rho_0, \quad \rho_1, \ldots, \quad \rho_{a-1}$$

which is so arranged that, if $\kappa<\lambda$, $\rho_\kappa-\rho_\lambda$ is a positive integer or zero. Since these indices are not necessarily equal to one another, they may be divided into sub-sets such that the members of each sub-set are equal to one another. Thus suppose $\rho_0=\rho_1= \ldots =\rho_{i-1}$ to correspond to a root of $f_0(\sigma)=0$ of multiplicity i; $\rho_i=\rho_{i+1}= \ldots =\rho_{j-1}$ to correspond to a root of multiplicity $j-i$; $\rho_j=\rho_{j+1}= \ldots =\rho_{k-1}$ to correspond to a root of multiplicity $k-j$, and so on until the set is exhausted.

In order to avoid any of the coefficients c_ν, as determined by the recurrence relations of § 16·12, becoming infinite, c_0 is replaced by

$$c_0 f_0(\sigma+1) f_0(\sigma+2) \ldots f_0(\sigma+\omega)=c_0 f(\sigma),$$

where $\omega=\rho_0-\rho_{a-1}$, which amounts to multiplying the series for $W(z, \sigma)$ throughout by $f(\sigma)$. Then

$$\begin{aligned}\overline{W}(z, \sigma)&=f(\sigma)W(z, \sigma)\\ &=\sum_{\nu=0}^{\infty} c_\nu f(\sigma) z^{\sigma+\nu},\end{aligned}$$

and is finite when σ is restricted to vary in the neighbourhood of any one of $\rho_0, \rho_1, \ldots, \rho_{a-1}$. Also

$$\begin{aligned}L\overline{W}(z, \sigma)&=c_0 f_0(\sigma) f(\sigma) z^\sigma\\ &=c_0 F(\sigma) z^\sigma,\end{aligned}$$

where $F(\sigma)$ is written for the product $f_0(\sigma)f_0(\sigma+1) \ldots f_0(\sigma+\omega)$.

Now in $F(\sigma)$, the factor $f_0(\sigma)$ is of degree i in $(\sigma-\rho_0)$, of degree $j-i$ in $(\sigma-\rho_i)$, of degree $k-j$ in $(\sigma-\rho_j)$ and so on. No other factor contains $(\sigma-\rho_0)$, but $f_0(\sigma+\rho_0-\rho_i)$ is of degree i in $(\sigma-\rho_i)$. Similarly $(\sigma-\rho_j)$ appears as a factor of degree $j-i$ in $f_0(\sigma+\rho_i-\rho_j)$ and as a factor of degree i in $f_0(\sigma+\rho_0-\rho_i)$. Thus $F(\sigma)$ is of degree i in $(\sigma-\rho_0)$, of degree j in $(\sigma-\rho_i)$, of degree k in $(\sigma-\rho_j)$ and so on.

When σ lies in a certain domain in the σ-plane containing the point ρ_μ, where ρ_μ is an index of the set under consideration, the coefficients c_ν are analytic (in fact rational) functions of σ. When also $|z|<R$, the series $\sum c_\nu z^\nu$ is a uniformly convergent series of analytic functions of σ and can therefore be differentiated any number of times with respect to σ. Furthermore the operators L and $\dfrac{\partial^s}{\partial\sigma^s}$ are permutable. Hence

$$\left[L\left\{\frac{\partial^s}{\partial\sigma^s}\overline{W}(z, \sigma)\right\}\right]_{\sigma=\rho_\mu}=\left[\frac{\partial^s}{\partial\sigma^s}\left\{c_0 F(\sigma) z^\sigma\right\}\right]_{\sigma=\rho_\mu}$$

for $s=0, 1, 2, \ldots, m-1$, where m is the degree of $F(\sigma)$ in $(\sigma-\rho_\mu)$, and consequently for any one of these values of s

$$\left[\frac{\partial^s}{\partial\sigma^s}\overline{W}(z, \sigma)\right]_{\sigma=\rho_\mu}$$

is a solution of the differential equation.

Now

$$\overline{W}(z, \sigma) = z^\sigma \sum_{\nu=0}^{\infty} g_\nu(\sigma) z^\nu,$$

* Each index is written a number of times equal to the multiplicity of the corresponding root of $f_0(\sigma)=0$.

where $g_\nu(\sigma)=c_\nu f_1(\sigma)$, and therefore

$$\frac{\partial^s \overline{W}(z,\sigma)}{\partial\sigma^s}=z^\sigma\left\{\sum_{\nu=0}^{\infty}g_\nu^{(s)}(\sigma)z^\nu+s\log z\sum_{\nu=0}^{\infty}g_\nu^{(s-1)}(\sigma)z^\nu+\ldots+(\log z)^s\sum_{\nu=0}^{\infty}g_\nu(\sigma)z^\nu\right\}$$
$$=w_s(z,\sigma)+s\log z\, w_{s-1}(z,\sigma)+\ldots+(\log z)^s w_1(z,\sigma),$$

where $w_r(z,\sigma)$ is written for

$$\sum_{\nu=0}^{\infty}g_\nu^{(r)}(\sigma)z^{\sigma+\nu}.$$

Consider the index ρ_0 of the first sub-set. In this case $g_\nu(\rho_0)=c_\nu f(\rho_0)$ is finite or zero for all values of ν and $g_0(\rho_0)\neq 0$. Thus there arises the sub-set of i solutions

$$W_0=w_0(z,\rho_0),$$
$$W_1=w_0(z,\rho_0)\log z+w_1(z,\rho_0),$$
$$W_2=w_0(z,\rho_0)(\log z)^2+2w_1(z,\rho_0)\log z+w_2(z,\rho_0),$$
$$\cdots\cdots\cdots\cdots\cdots$$
$$W_{i-1}=w_0(z,\rho_0)(\log z)^{i-1}+(i-1)w_1(z,\rho_0)(\log z)^{i-2}+\ldots+w_{i-1}(z,\rho_0).$$

The presence of the term $w_0(z,\rho_0)(\log z)^{r-1}$ in W_r shows that the i solutions are linearly distinct.

Next consider the index ρ_i of the second sub-set. Here $g_\nu(\rho_i)$ is zero to the order i when $\nu=0, 1, 2, \ldots, \rho_0-\rho_i-1$, and finite or zero when $\nu\geqslant\rho_0-\rho_i$. Hence

$$\left[\frac{\partial^s}{\partial\sigma^s}\left\{z^\sigma\sum_{\nu=0}^{\rho_0-\rho_i-1}g_\nu(\sigma)z^\nu\right\}\right]_{\sigma=\rho_i}=0,$$

when $s=0, 1, 2, \ldots, i-1$. The leading significant term in $W(z,\sigma)$ is therefore of degree $\sigma+\rho_0-\rho_i$ in z, that is to say of degree ρ_1 when $\sigma=\rho_i$.

The solutions corresponding to the sub-set of index i have been completely enumerated; they are $W_0, W_1, \ldots, W_{i-1}$. Since the solution

$$\left[z^\sigma\sum_{\nu=\rho_0-\rho_i}^{\infty}g_\nu(\sigma)z^\nu\right]_{\sigma=\rho_i}$$

is free from logarithmic terms, it is a constant multiple of W_0, and in general, when $s\leqslant i-1$,

$$\left[\frac{\partial^s}{\partial\sigma^s}\left\{z^\sigma\sum_{\nu=\rho_0-\rho_i}^{\infty}g_\nu(\sigma)z^\nu\right\}\right]_{\sigma=\rho_i}$$

is a linear combination of the solutions $W_0, W_1, \ldots W_s$.

There remain the $j-i$ solutions

$$\left[\frac{\partial^s}{\partial\sigma^s}\left\{z^\sigma\sum_{\nu=0}^{\infty}g_\nu(\sigma)z^\nu\right\}\right]_{\sigma=\rho_i}$$

where $s=i, i+1, \ldots, j-1$. These solutions form the sub-set

$$W_i=w_0(z,\rho_i)(\log z)^i+iw_1(z,\rho_i)(\log z)^{i-1}+\ldots+w_i(z,\rho_i),$$
$$\cdots\cdots\cdots\cdots\cdots$$
$$W_{j-1}=w_0(z,\rho_i)(\log z)^{j-1}+(j-1)w_1(z,\rho_i)(\log z)^{j-2}+\ldots+w_{j-1}(z,\rho_i),$$

in which $w_r(z,\rho_i)$, when $r\leqslant i-1$, is a linear combination of $w_0(z,\rho_0)$, $w_1(z,\rho_0)$, $\ldots, w_r(z,\rho_0)$. The term $w_i(z,\rho_i)$ is not identically zero for

$$\left[\frac{\partial^i}{\partial\sigma^i}g_0(\sigma)\right]_{\sigma=\rho_i}\neq 0.$$

The remaining members of the sub-set of index i involve $w_i(z,\rho_i)$ multiplied by a logarithmic factor, thus W_{i+r} involves the term $w_i(z, \rho_i)(\log z)^r$. The members of the sub-set are therefore linearly independent of one another; it will be proved in the next section that they are also linearly independent of the members of the first sub-set.

In the same way it may be proved that the sub-set of index j furnishes $k-j$ solutions which are given by

$$\left[\frac{\partial^s}{\partial\sigma^s}\left\{z^\sigma \sum_{\nu=0}^{\infty} g_\nu(\sigma)z^\nu\right\}\right]_{\sigma=\rho_i}$$

where $s=j, j+1, \ldots, k-1$, and so on until the complete set of indices $\rho_0, \rho_1, \ldots, \rho_{a-1}$ has been exhausted.

Similarly the set of indices

$$\rho_a, \quad \rho_{a+1}, \quad \ldots, \quad \rho_{\beta-1}$$

is divided into sub-sets of equal indices and dealt with in the same way. Thus finally an aggregate of n solutions of the equation is obtained; it remains to prove that they form a fundamental system.

16·31. Proof of the Linear Independence of the Solutions.—Consider the solutions which correspond to a particular set of indices, for example the set $\rho_0, \rho_1, \ldots, \rho_{a-1}$, and suppose that these solutions are connected by the linear relation

$$A_0W_0+A_1W_1+ \ldots +A_{a-1}W_{a-1}=0.$$

Arrange the left-hand member in descending powers of $\log z$, then the aggregate of terms which are of the highest degree k in $\log z$ must vanish identically, thus

$$A_rW_r+ \ldots +A_sW_s=0.$$

But each of $W_r, \ldots, W_s$ proceeds from a distinct sub-set; they therefore correspond to different indices. The coefficient of the term of highest index must therefore vanish, likewise the coefficient of the term of second highest index and so on. Thus finally

$$A_r= \ldots =A_s=0.$$

The expression $A_0W_0+A_1W_1+ \ldots +A_{a-1}W_{a-1}$ is now of degree $k-1$ in $\log z$, the aggregate of terms involving $(\log z)^{k-1}$ are now equated to zero; each coefficient which enters into these terms is then proved to be zero. The process is continued until finally it is proved that

$$A_0=A_1= \ldots =A_{a-1}=0.$$

The solutions of any particular set are therefore linearly independent of one another.

Now consider the aggregate of the solutions

$$W_1, \quad W_2, \quad \ldots, \quad W_n$$

and suppose that a linear relationship of the form

$$A_1W_1+A_2W_2+ \ldots +A_nW_n=0$$

exists. The aggregate of the terms of highest degree k in $\log z$ must vanish identically thus

$$(A_rW_r+ \ldots +A_sW_s)+(A_tW_t+ \ldots +A_uW_u)+ \ldots =0,$$

where the terms bracketed together are of the same set. Let the multipliers of these sets, corresponding to a circuit of the point z around the origin, be $\theta_1, \theta_2, \ldots$. Then after λ circuits

$$\theta_1^\lambda(A_rW_r+ \ldots +A_sW_s)+\theta_2^\lambda(A_tW_t+ \ldots +A_uW_u)+ \ldots =0.$$

Since $\theta_1 \neq \theta_2 \neq \ldots$, these equations, for $\lambda=0, 1, 2, \ldots$, are inconsistent unless

$$A_r W_r + \ldots + A_s W_s = 0, \quad A_t W_t + \ldots + A_u W_u = 0, \ldots$$

which has been proved impossible unless

$$A_r = \ldots = A_s = A_t = \ldots = A_u = \ldots = 0.$$

Now deal with the terms of degree $k-1$ in $\log z$; the coefficients they involve are likewise proved to be zero. The process is continued until finally it is proved that

$$A_1 = A_2 = \ldots = A_n = 0.$$

The n solutions are therefore linearly independent and form a fundamental system.

16·32. Application to the Bessel Equation.—Take the Bessel equation in the form *

$$z^2 \frac{d^2 w}{dz^2} + z \frac{dw}{dz} + (z^2 - n^2) w = 0,$$

or symbolically, $Lw=0$. Then if

$$W(z, \sigma) = \sum c_\nu z^{\sigma+\nu},$$

it is found that

$$LW = c_0(\sigma^2 - n^2) z^\sigma,$$

provided that

$$c_1\{(\sigma+1)^2 - n^2\} = 0,$$

$$c_\nu\{(\sigma+\nu)^2 - n^2\} + c_{\nu-2} = 0 \qquad (\nu \geqslant 2).$$

The roots of the indicial equation

$$\sigma^2 - n^2 = 0$$

are $\pm n$; when n is not an integer the corresponding solutions are distinct. The solutions are, in fact, $J_n(z)$ and $J_{-n}(z)$, where

$$J_n(z) = \sum_{r=0}^{\infty} \frac{(-1)^r z^{n+2r}}{2^{n+2r} r! \, \Gamma(n+r+1)}.$$

The first exceptional case to consider is that in which n is zero. In this case $J_n(z)$ and $J_{-n}(z)$ coincide in the one function

$$J_0(z) = 1 - \frac{z^2}{2^2} + \frac{z^4}{2^2 . 4^2} - \frac{z^6}{2^2 . 4^2 . 6^2} + \ldots.$$

Since $\sigma=0$ is a double root of the indicial equation the second solution is

$$K_0(z) = \lim_{\sigma=0} \frac{\partial}{\partial \sigma}\left[z^\sigma \left\{ 1 - \frac{z^2}{(\sigma+2)^2} + \frac{z^4}{(\sigma+2)^2(\sigma+4)^2} - \ldots \right\} \right]$$

$$= J_0(z) \log z + \sum_{r=1}^{\infty} \frac{(-1)^{r-1}}{\{\Gamma(r+1)\}^2} \cdot \left(\frac{z}{2}\right)^{2r} \psi(r),$$

where

$$\psi(r) = \left[\frac{d}{dt} \log \Gamma(t+1) \right]_{t=r}.$$

Now let n be a positive integer; the solution

$$w = J_n(z)$$

is the one and only solution free from logarithms. The function $J_{-n}(z)$ has now no meaning, because the coefficients, on and after the coefficient of z^{2n},

* This application is due to Forsyth, *Differential Equations*, Vol. 4, p. 101.

become infinite through the occurrence of the factor $(\sigma+2n)^2-n^2$ in the denominator. Write

$$c_0=C\{(\sigma+2n)^2-n^2\},\quad (-1)^n C=E\prod_{r=1}^{n-1}\{(\sigma+2r)^2-n^2\},$$

so that

$$w=C\{(\sigma+2n)^2-n^2\}z^\sigma\left[1-\frac{z^2}{(\sigma+2)^2-n^2}+\ldots+(-1)^{n-1}\frac{z^{2n-2}}{\prod_{r=1}^{n-1}\{(\sigma+2r)^2-n^2\}}\right]$$

$$+Ez^{\sigma+2n}\left[1-\frac{z^2}{(\sigma+2n+2)^2-n^2}+\frac{z^4}{\{(\sigma+2n+2)^2-n^2\}\{(\sigma+2n+4)^2-n^2\}}-\ldots\right]$$

$=w_1+w_2$ say.

When $\sigma=-n$, w_1 becomes zero, and w_2 reduces to a multiple of $J_n(z)$. The second solution is obtained from

$$\lim_{\sigma=-n}\frac{\partial w}{\partial\sigma}.$$

Let

$$\lim_{\sigma=-n}\frac{\partial w_1}{\partial\sigma}=W_1,\quad \lim_{\sigma=-n}\frac{\partial w_2}{\partial\sigma}=W_2,$$

then

$$W_1=z^{-n}\frac{2Cn}{\Gamma(n)}\sum_{r=0}^{n-1}\frac{\Gamma(n-r)}{\Gamma(r+1)}\left(\frac{z}{2}\right)^{2r},$$

$$W_2=Ez^n\log z\left[1-\frac{z^2}{2^2(n+1)}+\frac{z^4}{2^4 2!\,(n+1)(n+2)}-\ldots\right]$$

$$+\tfrac{1}{2}Ez^n\sum_{r=0}^{\infty}\frac{(-1)^{r-1}\Gamma(n+1)}{\Gamma(r+1)\Gamma(n+r+1)}\{\psi(r)+\psi(n+r)-\psi(n)\}\left(\frac{z}{2}\right)^{2r}.$$

The term

$$\tfrac{1}{2}Ez^n\sum_{r=0}^{\infty}\frac{(-1)^r\psi(n)\Gamma(n+1)}{\Gamma(r+1)\Gamma(n+r+1)}\left(\frac{z}{2}\right)^{2r},$$

which occurs in W_2, is a constant multiple of $J_n(z)$ and can be discarded altogether. Let

$$C=-\frac{2^{n-1}\Gamma(n)}{n},$$

so that

$$E=\frac{1}{2^{n-1}\Gamma(n+1)},$$

then that part of the solution $w=W_1+W_2$ which remains is

$$w=\sum_{r=0}^{\infty}\frac{(-1)^r}{\Gamma(r+1)\Gamma(n+r+1)}\{2\log z-\psi(r)-\psi(n+r)\}\left(\frac{z}{2}\right)^{n+2r}-\sum_{r=0}^{n-1}\frac{\Gamma(n-r)}{\Gamma(r+1)}\left(\frac{z}{2}\right)^{2r-n}$$

and this may be taken as the second solution of the Bessel equation. It differs only by a constant multiple of $J_n(z)$ from Hankel's function * $Y_n(z)$.

16·33. Conditions that all Solutions relative to a particular Index may be free from Logarithms.—The first solution corresponding to a set of indices, such as the solution W_0 of § 16·3, is free from logarithms ; the subsequent solutions of the first sub-set certainly involves logarithmic terms. In general the leading solution of the second sub-set, the solution W_i, for instance, also involves logarithms, but in particular cases may not do so, whereas the remaining solutions of the second sub-set must involve logarithms. It is

* Whittaker and Watson, *Modern Analysis*, § 17·61 ; Watson, *Bessel Functions*, § 3·52.

likewise true that for every sub-set after the first, the only solution which may not involve logarithms is the leading solution of that sub-set.

Consider any set of indices

$$\rho_0, \rho_1, \rho_2, \ldots, \rho_\mu, \ldots$$

so arranged that

$$\rho_\kappa - \rho_\mu$$

is a positive integer for $\mu > \kappa$. A set of conditions which are necessary and sufficient for the absence of logarithmic terms from every solution W_μ corresponding to the index ρ_μ will now be investigated.*

In the first place, ρ_μ must be a simple root of the indicial equation, for a multiple root always introduces logarithmic terms. Moreover, since every index ρ_κ whose suffix κ is less than μ exceeds ρ_μ by a positive integer, any solution of the form

$$W_\mu + b_1 W_{\mu-1} + \ldots + b_{\mu-1} W_1 + b_\mu W_0,$$

where $b_1, \ldots, b_\mu$ are arbitrary constants, is a solution of index ρ_μ. Consequently the solutions $W_0, W_1, \ldots, W_{\mu-1}$ must be free from logarithms. It is therefore necessary that the indices $\rho_1, \rho_2, \ldots, \rho_\mu$ should be distinct.

Now

$$W_\mu = \left[\frac{\partial^\mu}{\partial\sigma^\mu}\left\{z^\sigma \sum_{\nu=0}^{\infty} g_\nu(\sigma) z^\nu\right\}\right]_{\sigma=\rho_\mu}$$

$$= z^{\rho_\mu}\left[\sum_{\nu=0}^{\infty} \frac{\partial^\mu g_\nu(\sigma)}{\partial\sigma^\mu} z^\nu + \mu \log z \sum_{\nu=0}^{\infty} \frac{\partial^{\mu-1} g_\nu(\sigma)}{\partial\sigma^{\mu-1}} z^\nu + \ldots + (\log z)^\mu \sum_{\nu=0}^{\infty} g_\nu(\sigma) z^\nu\right]_{\sigma=\rho_\mu}.$$

In order therefore that W_μ may be free from logarithms it is necessary and sufficient that

$$\left[\frac{\partial^s g_\nu(\sigma)}{\partial\sigma^s}\right]_{\sigma=\rho_\mu} = 0$$

for $s=0, 1, 2, \ldots, \mu-1$ and for all values of ν. Consequently $g_\nu(\sigma)$ must contain the factor $(\sigma-\rho_\mu)^\mu$ for all values of ν.

But

$$\frac{g_\nu(\sigma)}{g_0(\sigma)} = (-1)^\nu \frac{F_\nu(\sigma)}{f_0(\sigma+1) f_0(\sigma+2) \ldots f_0(\sigma+\nu)} = H_\nu(\sigma),$$

and since $g_0(\sigma) = c_0 f(\sigma)$, $g_0(\sigma)$ contains the factor $(\sigma-\rho_\mu)^\mu$. A necessary and sufficient condition is therefore that $H_\nu(\rho_\mu)$ should be finite or zero for all values of ν. Now the recurrence relations for $g_\nu(\sigma)$ and therefore those for $H_\nu(\sigma)$ are the same as those for c_ν, namely,

$$H_\nu(\sigma) f_0(\sigma+\nu) + H_{\nu-1}(\sigma) f_1(\sigma+\nu-1) + \ldots + H_0(\sigma) f_\nu(\sigma) = 0,$$

where $H_0(\sigma)=1$. If therefore $H_1(\rho_\mu), H_2(\rho_\mu), \ldots, H_{\nu-1}(\rho_\mu)$ are finite, $H_\nu(\rho_\mu)$ will be finite unless ν is such that $\rho_\mu+\nu$ is a root of the indicial equation

$$f_0(\sigma) = 0,$$

which occurs when ν assumes one or other of the increasing positive integers

$$\rho_{\mu-1} - \rho_\mu, \quad \rho_{\mu-2} - \rho_\mu, \quad \ldots, \quad \rho_0 - \rho_\mu.$$

When $\nu = \rho_{\mu-1} - \rho_\mu$, the factor $f_0(\sigma+\nu)$ in the denominator of $H_\nu(\sigma)$ has a simple zero $\sigma=\rho_\mu$ and no other factor vanishes. Consequently it is necessary that

$$F_\nu(\rho_\mu) = 0,$$

when $\nu = \rho_{\mu-1} - \rho_\mu$, and sufficient that $F_\nu(\sigma)$ should vanish to the first order when $\sigma = \rho_\mu$.

* Frobenius, *loc. cit.*, p. 224.

When $\nu=\rho_{\mu-2}-\rho_\mu$, two factors in the denominator of $H_\nu(\sigma)$ have simple zeros for $\sigma=\rho_\mu$, namely

$$f_0(\sigma+\nu-\rho_{\mu-2}+\rho_{\mu-1}) \quad\text{and}\quad f_0(\sigma+\nu).$$

It is therefore necessary and sufficient that for this particular value of ν $F_\nu(\sigma)$ should vanish to the second order when $\sigma=\rho_\mu$, or

$$F_\nu(\rho_\mu)=0, \quad \left[\frac{\partial F_\nu(\sigma)}{\partial\sigma}\right]_{\sigma=\rho_\mu}=0 \quad\text{when}\quad \nu=\rho_{\mu-2}-\rho_\mu.$$

When $\nu=\rho_{\mu-3}-\rho_\mu$ three factors in the denominator of $H_\nu(\sigma)$ have simple zeros for $\sigma=\rho_\mu$, namely,

$$f_0(\sigma+\nu-\rho_{\mu-3}+\rho_{\mu-1}), \quad f_0(\sigma+\nu-\rho_{\mu-3}+\rho_{\mu-2}), \quad\text{and}\quad f_0(\sigma+\nu),$$

and therefore for this value of ν, $F_\nu(\sigma)$ must vanish to the third order when $\sigma=\rho_\mu$. Therefore it is necessary and sufficient that

$$F_\nu(\rho_\mu)=0, \quad \left[\frac{\partial F_\nu(\sigma)}{\partial\sigma}\right]_{\sigma=\rho_\mu}=0, \quad \left[\frac{\partial^2 F_\nu(\sigma)}{\partial\sigma^2}\right]_{\sigma=\rho_\mu}=0,$$

where $\nu=\rho_{\mu-3}-\rho_\mu$.

In the same way, when $\nu=\rho_{\mu-r}-\rho_\mu$, r factors in the denominator of $H_\nu(\sigma)$ have simple zeros for $\sigma=\rho_\mu$, and therefore $F_\nu(\sigma)$ must vanish to order r when $\sigma=\rho_\mu$. The last condition is that, when $\nu=\rho_0-\rho_\mu$, $F_\nu(\sigma)$ must vanish to order μ for $\sigma=\rho_\mu$.

But it has been assumed that the solutions relative to $\rho_1, \rho_2, \ldots, \rho_{\mu-1}$ are free from logarithms. The number of conditions to be satisfied is respectively 1, 2, . . ., $\mu-1$ which together with the μ conditions relating particularly to ρ_μ itself make up an aggregate of $\frac{1}{2}\mu(\mu+1)$ conditions which are necessary and sufficient for all solutions relative to the index ρ_μ to be free from logarithms.

16·4. Real and Apparent Singularities.—The singularities of solutions of a linear differential equation are necessarily singularities of the equation, but the converse is not always true. In general when the point $z=a$ satisfies the conditions for a regular singularity, some, if not all, of the solutions involve negative or fractional powers of $(z-a)$ and possibly also powers of $\log(z-a)$. In these cases the singularity is said to be *real*. But it may happen in special circumstances that every solution is analytic at $z=a$, in which case the singularity is said to be *apparent*. A set of conditions, sufficient to ensure that the singularity is only apparent will now be derived.*

Let the equation be written in the form

$$\frac{d^n w}{dz^n}+\frac{P_1(z)}{z-a}\cdot\frac{d^{n-1}w}{dz^{n-1}}+\ldots+\frac{P_{n-1}(z)}{(z-a)^{n-1}}\cdot\frac{dw}{dz}+\frac{P_n(z)}{(z-a)^n}w=0,$$

where $P_1(z), \ldots, P_n(z)$ are analytic at $z=a$. Let the point $z=a$ be an apparent singularity, so that each solution of the fundamental set

$$w_1, \quad w_2, \quad \ldots, \quad w_n$$

is an analytic function of $z-a$ in the neighbourhood of the singularity.

Let

$$\Delta(z)=\begin{vmatrix} w_1^{(n-1)}, & w_1^{(n-2)}, & \ldots, & w_1', & w_1 \\ w_2^{(n-1)}, & w_2^{(n-2)}, & \ldots, & w_2', & w_2 \\ \cdot & \cdot & \cdot & \cdot & \cdot \\ w_n^{(n-1)}, & w_n^{(n-2)}, & \ldots, & w_n', & w_n \end{vmatrix}$$

* Fuchs, *J. für Math.* 68 (1868), p. 378.

and let $\Delta_r(z)$ be the determinant derived from Δ by replacing $w_1^{(n-r)}$, . . ., $w_n^{(n-r)}$ respectively by $w_1^{(n)}$, . . ., $w_n^{(n)}$. Then

$$\frac{P_r(z)}{(z-a)^r}=-\frac{\Delta_r(z)}{\Delta(z)},$$

but for at least one value of r, $P_r(z)$ does not contain the factor $(z-a)^r$ and therefore, for that value of r, $\Delta_r(a)/\Delta(a)$ is infinite. But $\Delta_r(z)$ is analytic for $z=a$ and therefore

$$\Delta(a)=0.$$

Now

$$\frac{1}{\Delta(z)}\cdot\frac{d\Delta(z)}{dz}=-\frac{P_1(z)}{z-a}$$
$$=-\frac{P_1(a)}{z-a}+\frac{dG(z-a)}{dz},$$

where $G(z-a)$ is analytic near $z=a$ and therefore

$$\Delta(z)=A(z-a)^{-P_1(a)}e^{G(z-a)},$$

where A is a constant. But $\Delta(z)$ is analytic at $z=a$ and therefore $P_1(a)$ must be a negative integer.

The indicial equation relative to $z=a$ is

$$[\rho]_n+[\rho]_{n-1}P_1(a)+\ .\ .\ .\ +\rho P_{n-1}(a)+P_n(a)=0.$$

The roots of this equation must be positive integers, and must be unequal, for equal roots necessarily lead to logarithmic terms. The least root may of course be zero. The condition that the exponents are positive integers includes the condition that $P_1(a)$ is a negative integer; the latter may be regarded as a preliminary test, when it is not satisfied the singularity is undoubtedly real.

Finally, a set of conditions sufficient to ensure that no logarithmic terms appear must be imposed. Let the roots of the indicial equation, arranged in descending order of magnitude be $\rho_0, \rho_1, \ldots, \rho_{n-1}$. The solution with the exponent ρ_0 certainly does not involve logarithms. One condition suffices to ensure that every solution with the exponent ρ_1 is free from logarithms, two further conditions are sufficient for the exponent ρ_2, and so on until finally $n-1$ further conditions suffice for the exponent ρ_{n-1}. Thus in all

$$1+2+\ .\ .\ .\ +(n-1)=\tfrac{1}{2}n(n-1)$$

conditions suffice to ensure the absence of logarithmic terms from the general solution.

The conditions that the exponents are positive integers or zero and that no logarithmic terms appear ensure that the singularity is apparent.

16·401. An Example illustrating the Conditions for an Apparent Singularity.—The equation

$$L(w)\equiv z^2\frac{d^2w}{dz^2}-(4z+\lambda z^2)\frac{dw}{dz}+(4-\kappa z)w=0$$

contains two parameters λ, κ. It will be shown that when certain relations exist between these parameters the singularity $z=0$ is only apparent.*

Assuming, as in the general method, that

$$w=\sum_{\nu=0}^{\infty}c_\nu z^{\sigma+\nu},$$

it is found that

$$L(w)=c_0(\sigma-4)(\sigma-1)z^\sigma,$$

* Forsyth, *Differential Equations*, Vol. 4, p. 119. Note that $P_1(a)=-4$, a negative integer, and therefore the singularity may be apparent.

provided that the coefficients c_ν satisfy the recurrence relations

$$(\sigma+\nu-4)(\sigma+\nu-1)c_\nu=\{\lambda(\sigma+\nu-1)+\kappa\}c_{\nu-1}.$$

The exponents $\rho_0=4$ and $\rho_1=1$ are positive integers ; corresponding to the greater exponent there is a solution analytic at $z=0$, namely $w=c_0u$, where

$$u=z^4\{1+\gamma_1z+\gamma_2z^2+ . . . +\gamma_\nu z^\nu + . . .\}$$

and

$$\gamma_\nu=\frac{4\lambda+\kappa}{1.4}\cdot\frac{5\lambda+\kappa}{2.5} . . . \frac{(\nu+3)\lambda+\kappa}{\nu.(\nu+3)}.$$

The solution corresponding to the smaller exponent $\rho_1=1$ will, in general, involve logarithms. In order that it may be free from logarithms one condition must be imposed. Since $\rho_0-\rho_1=3$, the necessary and sufficient condition is that $F_3(1)=0$. Now

$$f_0(\sigma)=(\sigma-4)(\sigma-1),$$

$$F_3(\sigma)=f_0(\sigma+1)f_0(\sigma+2)f_0(\sigma+3)\frac{c_3}{c_0}$$

$$=\{\lambda(\sigma+2)+\kappa\}\{\lambda(\sigma+1)+\kappa\}\{\lambda\sigma+\kappa\},$$

and therefore the necessary and sufficient conditions reduce to

$$(3\lambda+\kappa)(2\lambda+\kappa)(\lambda+\kappa)=0.$$

Thus there are three possibilities:

(i) $\kappa=-\lambda$ when the relevant solution is $w=z$,

(ii) $\kappa=-2\lambda$,, ,, ,, $w=z+\frac{1}{2}\lambda z^2$,

(iii) $\kappa=-3\lambda$,, ,, ,, $w=z+\lambda z^2+\frac{1}{2}\lambda^2z^3$.

In these cases, and these only, is the origin an apparent singularity.

16·5. The Peano-Baker Method of Solution.—The solution of a linear differential equation obtained in the form of an infinite series by the Frobenius or a similar method, is, from the practical point of view, quite satisfactory. But from the theoretical point of view it suffers from the disadvantage of being valid only within the circle of convergence which, in general, covers but an insignificant part of the plane of the independent variable. The method * which will now be expounded is of great theoretical interest in that it leads to an analytic expression for the general solution, which is valid almost throughout the whole plane. As an offset against this extended region of validity, it would appear that the convergence of the development is slow,† and that therefore the method is not adaptable to computation.

Consider the system of n simultaneous linear equations

$$\frac{dw_i}{dz}=u_{i1}w_1+u_{i2}w_2+ . . . +u_{in}w_n \qquad (i=1, 2, . . ., n),$$

where the coefficients u_{ij} are functions of z. The point z_0 will be supposed not to be a singular point of any of the coefficients. Consider the Mittag-Leffler star ‡ bounded by non-intersecting straight lines drawn from every singular point of the coefficients to infinity. For definiteness these barriers may be taken to be the continuations of the radii vectores drawn from the point z_0 to the singular points. It will be supposed that the coefficients u_{ij} are analytic throughout the star.

Now the system of n linear equations may be represented symbolically as

$$\frac{dw}{dz}=uw,$$

* Peano, *Math. Ann.* 32 (1888), p. 455 ; Baker, *Proc. London Math. Soc.* 34 (1902), p. 354 ; 35 (1902), p. 334 ; (2), 2 (1904), p. 293 (giving a historical summary) ; *Phil. Trans. R. S.* (A), 216 (1915), p. 155. See also Bôcher, *Am. J. Math.* 24 (1902), p. 311.

† Milne, *Proc. Edin. Math. Soc.* 34 (1915), p. 41.

‡ Mittag-Leffler, *C. R. Acad. Sc. Paris,* 128 (1889), p. 1212.

where u represents, not a single function of z, but the square matrix

$$\begin{pmatrix} u_{11}, & \cdot\cdot\cdot, & u_{1n} \\ \cdot & \cdot \quad \cdot \quad \cdot & \cdot \\ u_{n1}, & \cdot\cdot\cdot, & u_{nn} \end{pmatrix}$$

and w represents the aggregate $(w_1, w_2, \ldots, w_n)$.

The symbol Qu will be defined as representing the matrix obtained by integrating every element of the matrix u from z_0 to z along a path which does not encounter any of the barriers of the star. The symbol uQu denotes the matrix obtained by multiplying the matrix u into the integrated matrix Qu.* $Q(uQu)$ is written $QuQu$, and so on.

Now form the series of matrices

$$\Omega(u) = 1 + Qu + QuQu + QuQuQu + \ldots;$$

its sum is a matrix. It will be proved that the elements of the matrix $\Omega(u)$ converge absolutely and uniformly throughout any finite domain D containing z_0 and lying wholly within the Mittag-Leffler star. In the domain D the functions u_{ij} are bounded; let M_{ij} be such that

$$|u_{ij}| \leqslant M_{ij}$$

for all points of D, and let M be such that

$$M_{ij} \leqslant M$$

for all values of i and j. Let

$$u_{ij}^{(1)}(z) = \int_{z_0}^{z} u_{ij}(z)dz,$$

$$u_{ij}^{(2)}(z) = \int_{z_0}^{z} \{u_{i1}(z)u_{1j}^{(1)}(z) + \ldots + u_{in}(z)u_{nj}^{(1)}(z)\}dz,$$

$$\cdot \quad \cdot \quad \cdot \quad \cdot$$

the path of integration being a simple curve lying wholly within D. Let z_1 be any particular point on the path (z_0, z), s_1 the length of the path (z_0, z_1) and s that of the whole path (z_0, z). Then

$$\begin{aligned} |u_{ij}^{(1)}(z_1)| &\leqslant s_1 M_{ij} \\ &\leqslant s_1 M, \\ |u_{ij}^{(2)}(z)| &\leqslant \int_0^s M s_1 (M_{i1} + M_{i2} + \ldots + M_{in}) ds_1 \\ &\leqslant nM^2 \int_0^s s_1 ds_1 = \tfrac{1}{2} n s^2 M^2, \end{aligned}$$

and, in particular,

$$|u_{ij}^{(2)}(z_1)| \leqslant \tfrac{1}{2} n s_1^2 M^2.$$

Similarly,

$$\begin{aligned} |u_{ij}^{(3)}(z)| &\leqslant \int_0^s \tfrac{1}{2} n s_1^2 M^2 (M_{i1} + M_{i2} + \ldots + M_{in}) ds_1 \\ &\leqslant \frac{1}{3!} n^2 s^3 M^3, \end{aligned}$$

* The product of two square matrices $u = (u_{ij})$ and $v = (v_{ij})$ of the same order n is formed according to the law $uv = (u_{i1}v_{1j} + \ldots + u_{in}v_{nj})$, and is in general distinct from vu.

The sum of the two matrices u and v is the matrix $(u_{ij} + v_{ij})$.

The symbol 1, regarded as a matrix, represents

$$\begin{pmatrix} 1, & 0, & \ldots, & 0 \\ 0, & 1, & \ldots, & 0 \\ 0, & 0, & \ldots, & 1 \end{pmatrix}$$

and so on indefinitely. But $u_{ij}^{(1)}(z)$ is the $(i, j)^{th}$ element of the matrix Qu, $u_{ij}^{(2)}(z)$ that of the matrix $QuQu$, etc. Consequently the series

$$1+sM+\frac{1}{2!}ns^2M^2+\frac{1}{3!}n^2s^3M^3+\ldots$$

is a dominant series for every element of the matrix $\Omega(u)$, and therefore the elements of $\Omega(u)$ are series which are absolutely and uniformly convergent throughout the domain D. Hence if

$$w=(1+Qu+QuQu+\ldots)w_0,$$

where w_0 denotes the aggregate of arbitrary initial values $(w_1{}^0, w_2{}^0, \ldots, w_n{}^0)$, then, by term-by-term differentiation,

$$\begin{aligned}\frac{dw}{dz}&=u(1+Qu+QuQu+\ldots)w_0\\&=uw,\end{aligned}$$

and therefore

$$w=\Omega(u)w_0$$

is a solution of the system of linear equations which converges throughout any region lying wholly within the star, and which is such that $(w_1, w_2, \ldots, w_n)$ reduces to $(w_1{}^0, w_2{}^0, \ldots, w_n{}^0)$ when $z=z_0$.

16·51. Properties of $\Omega(u)$.—Let Ω_{ij} be the typical element of $\Omega(u)$; if

$$w=\Omega(u)W,$$

where W denotes the aggregate

$$(W_1, W_2, \ldots W_n),$$

then

$$w_i=\Omega_{i1}W_1+\ldots+\Omega_{in}W_n,$$

and

$$\frac{dw_i}{dz}=\frac{d\Omega_{i1}}{dz}W_1+\ldots+\frac{d\Omega_{in}}{dz}W_n+\Omega_{i1}\frac{dW_1}{dz}+\ldots+\Omega_{in}\frac{dW_n}{dz}.$$

When translated back into matrix symbolism, this result becomes

$$\begin{aligned}\frac{dw}{dz}&=\left\{\frac{d}{dz}\Omega(u)\right\}W+\Omega(u)\frac{dw}{dz}\\&=u\Omega(u)W+\Omega(u)\frac{dW}{dz}\\&=uw+\Omega(u)\frac{dW}{dz}.\end{aligned}$$

Now let $\Omega^{-1}(u)$ be the matrix inverse to $\Omega(u)$, that is to say, the matrix which is such that

$$\Omega^{-1}(u)\Omega(u)=\Omega(u)\Omega^{-1}(u)=1.$$

It will now be proved that, if u and v are square matrices each of n^2 elements,

$$\Omega(u+v)=\Omega(u)\Omega\{\Omega^{-1}(u)v\Omega(u)\},$$

provided that the determinant of the matrix $\Omega(u)$ is not zero.

For consider the system of linear differential equations

$$\frac{dw}{dz}=(u+v)w,$$

and in it make the change of dependent variables

$$w=\Omega(u)W,$$

or, what is the same thing,

$$W=\Omega^{-1}(u)w.$$

Then

$$\frac{dw}{dz}=uw+\Omega(u)\frac{dW}{dz},$$

and thus

$$(u+v)w=uw+\Omega(u)\frac{dW}{dz},$$

that is

$$\Omega(u)\frac{dW}{dz}=vw$$
$$=v\Omega(u)W,$$

or

$$\frac{dW}{dz}=\Omega^{-1}(u)v\Omega(u)W.$$

Consequently

$$w=\Omega(u)W$$
$$=\Omega(u)\Omega\{\Omega^{-1}(u)v\Omega(u)\}w_0.$$

But on the other hand

$$w=\Omega(u+v)w_0,$$

which, in view of the known uniqueness of the solutions of a system with given initial values, proves the theorem.

It is not difficult to calculate the determinant Δ of $\Omega(u)$; in fact

$$\Delta=\exp\int_{z_0}^{z}(u_{11}+u_{22}+\ .\ .\ .\ +u_{nn})dz.$$

For since Ω_{ij} represents the typical element of $\Omega(u)$, the equations

$$\frac{d}{dz}\Omega(u)=u\Omega(u),$$

when written out in full, are of the form

$$\frac{d\Omega_{ij}}{dz}=u_{i1}\Omega_{1j}+\ .\ .\ .\ +u_{in}\Omega_{nj}.$$

Now $\dfrac{d\Delta}{dz}$ can be written as the sum of n determinants, each of which is obtained by differentiating all the elements of one particular column of Δ. By using the above expression for the derivative of Ω_{ij} it is easily seen that

$$\frac{d\Delta}{dz}=(u_{11}+u_{22}+\ .\ .\ .\ +u_{nn})\Delta,$$

from which the result follows at once.

In particular, if

$$u_{11}+u_{22}+\ .\ .\ .\ +u_{nn}=0,$$

Δ is independent of z, and in fact $\Delta=1$.

16·52. Conversion of a Linear Equation of Order n into a Linear System.—A linear differential equation of order n may be expressed as a system of n simultaneous equations of the first order in an infinity of ways. There is, however, one method which is particularly adapted to the matrix notation, as follows:

Let the given equation be written in the form

$$\frac{d^nw}{dz^n}=\frac{P_{n-1}}{\phi_n}\cdot\frac{d^{n-1}w}{dz^{n-1}}+\frac{P_{n-2}}{\phi_{n-1}\phi_n}\cdot\frac{d^{n-2}w}{dz^{n-2}}+\ .\ .\ .\ +\frac{P_0}{\phi_1\phi_2\ .\ .\ .\ \phi_n}w,$$

and write

$$w_1 = w, \quad w_2 = \phi_1 \frac{dw}{dz}, \quad w_2 = \phi_1 \phi_2 \frac{d^2w}{dz^2}, \ldots,$$

then

$$\frac{dw_1}{dz} = \frac{w_2}{\phi_1}, \quad \frac{dw_2}{dz} = \frac{\phi_1' w_2}{\phi_1} + \frac{w_3}{\phi_2}, \quad \frac{dw_3}{dz} = \left(\frac{\phi_1'}{\phi_1} + \frac{\phi_2'}{\phi_2}\right) w_3 + \frac{w_4}{\phi_3},$$

$$\cdot \quad \cdot \quad \cdot \quad \cdot \quad \cdot$$

$$\frac{dw_{n-1}}{dz} = \left(\sum_{r=1}^{n-2} \frac{\phi_r'}{\phi_r}\right) w_{n-1} + \frac{w_n}{\phi_{n-1}},$$

$$\frac{dw_n}{dz} = \left(\sum_{r=1}^{n-1} \frac{\phi_r'}{\phi_r}\right) w_n + \frac{P_0}{\phi_n} w_1 + \frac{P_1}{\phi_n} w_2 + \ldots + \frac{P_{n-1}}{\phi_n} w_n.$$

Thus if $H_m = \sum_{r=1}^{m} \frac{\phi_r'}{\phi_r}$, the equation is equivalent to the system

$$\frac{dw}{dz} = uw,$$

where u represents the matrix

$$\begin{pmatrix} 0, & \frac{1}{\phi_1}, & 0, & 0, & \ldots, & 0, & 0 \\ 0, & H_1, & \frac{1}{\phi_2}, & 0, & \ldots, & 0, & 0 \\ 0, & 0, & H_2, & \frac{1}{\phi_3}, & \ldots, & 0, & 0 \\ \cdot & \cdot & \cdot & \cdot & \cdot & & \\ 0, & 0, & 0, & 0, & \ldots, & H_{n-2}, & \frac{1}{\phi_{n-1}} \\ \frac{P_0}{\phi_n}, & \frac{P_1}{\phi_n}, & \frac{P_2}{\phi_n}, & \frac{P_3}{\phi_n}, & \ldots, & \frac{P_{n-2}}{\phi_n}, & \frac{P_{n-1}}{\phi_n} + H_{n-1} \end{pmatrix}$$

The following cases are those of greatest interest :

(a) The functions P and ϕ are polynomials, and no one of the functions ϕ has a multiple factor. The linear system then has the form

$$\frac{dw}{dz} = \left(V + \sum_s \frac{A_s}{z - a_s}\right) w,$$

where V is a matrix each of whose elements is a polynomial in z, $z - a_s$ is a factor of one or more of the functions ϕ and A_s is a matrix of constants.

For example,

$$(z+1)z^3 w''' - \{(a_2+b_2)z + a_2\} z^2 w'' - \{(a_1+b_1)z + a_1\} z w' - \{(a+b)z + a\} w = 0$$

leads to the equivalent system

$$\frac{dw}{dz} = \left\{ \begin{pmatrix} 0, & 1, & 0 \\ 0, & 1, & 1 \\ a, & a_1, & a_2+2 \end{pmatrix} \frac{1}{z} + \begin{pmatrix} 0, & 0, & 0 \\ 0, & 0, & 0 \\ b, & b_1, & b_2 \end{pmatrix} \frac{1}{z+1} \right\} w.$$

(b) The functions P and ϕ are as above, and $\phi_1 = \phi_2 = \ldots = \phi_n = \phi$. For example,

$$w'' = \frac{Az+B}{z(z-1)} w' + \frac{Cz^2+Dz+E}{z^2(z-1)^2} w$$

leads to

$$\frac{dw}{dz} = \left\{ \begin{pmatrix} 0, & 0 \\ C, & 0 \end{pmatrix} + \begin{pmatrix} 0, & -1 \\ -E, & 1-B \end{pmatrix} \frac{1}{z} + \begin{pmatrix} 0, & 1 \\ C+D+E, & A+B+1 \end{pmatrix} \frac{1}{z-1} \right\} w.$$

(c) The functions P are polynomials, $\phi_1=\phi_2= \ldots =\phi_{n-1}=1$ and ϕ_n is a polynomial without multiple roots. The functions H are then all zero. Thus

$$w''=\left\{\lambda+\sum_{r=1}^{N}\frac{\lambda_r}{z-a_r}\right\}w'+\left\{\mu+\sum_{r=1}^{N}\frac{\mu_r}{z-a_r}\right\}w$$

leads to

$$\frac{dw}{dz}=\left\{\begin{pmatrix}0, & 1\\ \mu, & \lambda\end{pmatrix}+\sum_{r=1}^{N}\begin{pmatrix}0, & 0\\ \mu_r, & \lambda_r\end{pmatrix}\frac{1}{z-a_r}\right\}w.$$

(d) The functions P are analytic functions of z, and each of the functions ϕ is either unity or $z-a$, where a is not a singular point of the functions P. For instance let the equation be

$$w^{(n)}=\frac{p_{n-1}+zQ_{n-1}}{z}w^{(n-1)}+\frac{p_{n-2}+zQ_{n-2}}{z^2}w^{(n-2)}+\ldots+\frac{p_0+zQ_0}{z^n}w,$$

where $p_0, p_1, \ldots, p_{n-1}$ are constants, and $Q_0, Q_1, \ldots, Q_{n-1}$ are functions developable about the origin in positive integral powers of z. The equivalent system is

$$\frac{dw}{dz}=\left(V+\frac{A}{z}\right)w,$$

where

$$V=\begin{pmatrix}0, & 0, & \ldots, & 0\\ 0, & 0, & \ldots, & 0\\ & & & \\ Q_0, & Q_1, & \ldots, & Q_{n-1}\end{pmatrix}\quad A=\begin{pmatrix}0, & 1, & 0, & 0, & \ldots, & 0\\ 0, & 1, & 1, & 0, & \ldots, & 0\\ 0, & 0, & 2, & 1, & \ldots, & 0\\ \cdot & \cdot & \cdot & \cdot & \cdot & \cdot & \cdot\\ p_0, & p_1, & p_2, & p_3, & \ldots, & p_{n-1}+n-1\end{pmatrix}$$

16·53. Particular Examples.—In the first place, consider the single equation of the first order

$$\frac{dw}{dz}=uw.$$

In this case it may easily be verified that

$$QuQu=\frac{1}{2!}(Qu)^2,\quad QuQuQu=\frac{1}{3!}(Qu)^3,$$

and so on. Thus the solution is

$$w=w_0\exp Qu,$$

which is identical with the solution obtained by elementary methods.

Now consider the linear equation of the second order

$$\frac{d^2w}{dz^2}=vw;$$

this equation is equivalent to the system

$$\left(\frac{dw}{dz}, \frac{dw'}{dz}\right)=\begin{pmatrix}0, & 1\\ v, & 0\end{pmatrix}(w, w'),$$

where $w'=\dfrac{dw}{dz}$. It may be verified that, if the initial value of z is taken to be zero,

$$u=\begin{pmatrix}0, & 1\\ v, & 0\end{pmatrix},\qquad Qu=\begin{pmatrix}0, & z\\ Qv, & 0\end{pmatrix},$$

$$uQu=\begin{pmatrix}Qv, & 0\\ 0, & vz\end{pmatrix},\qquad QuQu=\begin{pmatrix}Q^2v, & 0\\ 0, & Qvz\end{pmatrix},$$

$$uQuQu = \begin{pmatrix} 0, & Qvz \\ vQ^2v, & 0 \end{pmatrix}, \qquad QuQuQu = \begin{pmatrix} 0, & Q^2vz \\ QvQ^2v, & 0 \end{pmatrix},$$

$$uQuQuQu = \begin{pmatrix} QvQ^2v, & 0 \\ 0, & vQ^2vz \end{pmatrix}, \quad QuQuQuQu = \begin{pmatrix} Q^2vQ^2v, & 0 \\ 0, & QvQ^2vz \end{pmatrix},$$

and so on. Thus the general solution is

$$w = w_0 W_1 + w_0' W_1,$$

where W_1 and W_2 are given by the series

$$W_1 = 1 + Q^2v + Q^2vQ^2v + Q^2vQ^2vQ^2v + \ldots,$$
$$W_2 = z + Q^2vz + Q^2vQ^2vz + Q^2vQ^2vQ^2vz + \ldots,$$

and (w_0, w_0') are the values of (w, w') when $z=0$. It will be observed that Q^2v, Q^2vz, Q^2vQ^2v, Q^2vQ^2vz vanish to orders 2, 3, 4, 5 when $z=0$.

As a particular instance, consider the Bessel equation

$$z^2\frac{d^2w}{dz^2} + z\frac{dw}{dz} + (z^2 - n^2)w = 0.$$

Write

$$z = 4ce^{\frac{1}{2}t}, \quad m = \tfrac{1}{4}n^2,$$

where c is arbitrary ; then the equation becomes

$$\frac{d^2v}{dt^2} = (m - ce^t)w.$$

In this case,

$$W_1 = 1 + \left\{m\frac{t^2}{2\,!} - c(e^t - 1 - t)\right\} + \left\{m^2\frac{t^4}{4\,!} + mc\left[4 + 2t + \frac{t^2}{2\,!} + \frac{t^3}{3\,!} - e^t(4 - 2t + \tfrac{1}{2}t^2)\right]\right.$$
$$\left. + c^2[-\tfrac{5}{4} - \tfrac{1}{2}t + e^t(1-t) + \tfrac{1}{4}e^{2t}]\right\} + \ldots,$$

$$W_2 = t + \left\{m\frac{t^5}{3\,!} - c[(t-2)e^t + t + 2]\right\}$$
$$+ \left\{m^2\frac{t^5}{5\,!} + mc\left[-8 - 4t - t^2 - \frac{t^3}{3\,!} + e^t\left(8 - 4t - t^2 - \frac{t^3}{3\,!}\right)\right]\right.$$
$$\left. + c^2[\tfrac{3}{4} + \tfrac{1}{4}t + te^t + \tfrac{1}{4}(t-3)e^{2t}]\right\} + \ldots.$$

These series are convergent for all values of t ; when rearranged in powers of t, they agree with the expressions for the solutions obtained by direct calculation of the coefficients.

Lastly, consider the linear system

$$\frac{dw}{dz} = \left[A_0 + A_1 z + \ldots + A_\mu z^\mu + \sum_{r=1}^{\sigma} C_r(z - c_r)^{-1}\right]w,$$

and suppose that a new variable s can be found so that $\log(z - c_r)$ is a uniform analytic function of s, for a certain range of values of s and for $r = 1, 2, \ldots \sigma$. Then every solution of the linear system is a one-valued function of s.

Let

$$\log(z - c_r) = \psi_r(s),$$

so that

$$z = c_r + \exp\psi_r(s)$$
$$= \Psi(s), \text{ say.}$$

Thus the system is

$$\frac{dw}{ds} = \left\{\Psi'(s)[A_0 + A_1\Psi + \ldots + A_\mu\Psi^\mu] + \sum_{r=1}^{\sigma} C_r\psi_r'(s)\right\}w$$
$$= uw.$$

The terms

$$Qu=\int_{s_0}^{s} u\,ds, \quad QuQu=\int_{s_0}^{s} u\,ds\int_{s_0}^{s} u\,ds, \ldots$$

are all uniform analytic functions of s; the solution

$$w=\Omega(u)w_0=(1+Qu+QuQu+\ldots)w_0$$

is also analytic in the neighbourhood of s.

For instance, the Bessel equation may be written as the system

$$\frac{dw}{dz}=\left[-z\begin{pmatrix}0, & 0\\ 1, & 0\end{pmatrix}+\frac{1}{z}\begin{pmatrix}0\ , & 1\\ n^2, & 0\end{pmatrix}\right]w;$$

its solution is expressible as a one-valued function of the new variable $s=\log z$.

The scope of the matrix method is very wide, but its successful application demands a knowledge of theorems in the calculus of matrices which cannot be given here. There is, however, a simple application of some theoretical importance which will be outlined in the following section.

16·54. Application to an Equation with Periodic Coefficients.—Consider the equation

$$\frac{d^2w}{dz^2}+(n^2+\Psi)w=0,$$

where n is an integer, and Ψ a periodic function of z.

Write

$$X=\tfrac{1}{2}e^{inz}\left(\frac{dw}{dz}-inw\right), \quad Y=\tfrac{1}{2}e^{-inz}\left(\frac{dw}{dz}+inw\right),$$

then

$$\frac{dX}{dz}=-\frac{i\Psi}{2n}e^{inz}(Xe^{-inz}-Ye^{inz}),$$

$$\frac{dY}{dz}=-\frac{i\Psi}{2n}e^{-inz}(Xe^{-inz}-Ye^{inz}).$$

If $\tau=2iz$, $\zeta=e^{\tau}$, the system can be written

$$\frac{d}{d\tau}(X,\ Y)=-\frac{\Psi}{4n}\begin{pmatrix}1\ , & -\zeta^{n}\\ \zeta^{-n}, & -1\end{pmatrix}(X,\ Y).$$

In particular, let $n=1$, $\Psi=4a\cos hz+4b\cos kz$, then

$$\frac{d}{d\tau}(X,\ Y)=(ap+bq)(X,\ Y),$$

where p, q denote the matrices

$$p=\tfrac{1}{2}(\zeta^{\frac{1}{2}h}+\zeta^{-\frac{1}{2}h})\begin{pmatrix}-1\ , & \zeta\\ -\zeta^{-1}, & 1\end{pmatrix}, \quad q=\tfrac{1}{2}(\zeta^{\frac{1}{2}k}+\zeta^{-\frac{1}{2}k})\begin{pmatrix}-1\ , & \zeta\\ -\zeta^{-1}, & 1\end{pmatrix}.$$

The solution

$$(X,\ Y)=\Omega(ap+bq)(X_0,\ Y_0),$$

where

$$\Omega(ap+bq)=1+aQp+bQq+a^2QpQp+ab(QpQq+QqQp)+b^2QqQq+\ldots,$$

is absolutely and uniformly convergent for all values of z.

[For further developments of this application of the method, and in particular for a discussion of the stability of solutions of the linear differential equation of the second order with periodic coefficients, the reader is referred to Baker's *Phil. Trans.* memoir already quoted.]

Miscellaneous Examples.

1. Solve in series of ascending powers of z

$$\text{(i) } 4z\frac{d^2w}{dz^2}+2\frac{dw}{dz}+w=0.$$

$$\text{(ii) } (2z+z^3)\frac{d^2w}{dz^2}-\frac{dw}{dz}-6zw=0.$$

$$\text{(iii) } z\frac{d^2w}{dz^2}+\frac{dw}{dz}-w=0.$$

2. Find the complete solution of the hypergeometric equation

$$z(1-z)\frac{d^2w}{dz^2}+\{\gamma-(\alpha+\beta+1)z\}\frac{dw}{dz}-\alpha\beta w=0$$

(i) when $\gamma=1$; (ii) when γ is a negative integer.

3. Show that the equation

$$z^3\frac{d^4w}{dz^4}+(\rho+\sigma+\tau+3)z^2\frac{d^3w}{dz^3}+(1+\rho+\sigma+\tau+\rho\sigma+\sigma\tau+\tau\rho)z\frac{d^2w}{dz^2}-(z-\rho\sigma\tau)\frac{dw}{dz}-\alpha w=0$$

is satisfied by the function

$${}_1F_3(\alpha\,;\ \rho,\sigma,\tau\,;\ z)=1+\frac{\alpha}{\rho\sigma\tau}z+\frac{\alpha(\alpha+1)}{2\,!\,\rho(\rho+1)\sigma(\sigma+1)\tau(\tau+1)}z^2+\ .\ .\ .$$

and find the remaining solutions relative to the singularity $z=0$.

When $\alpha=\tau$, ${}_1F_3(\alpha\,;\ \rho\,;\ \sigma,\tau\,;\ z)$ reduces to ${}_0F_2(\rho,\sigma\,;\ z)$; prove that this function satisfies the equation

$$z^2\frac{d^3w}{dz^3}+(\rho+\sigma+1)z\frac{d^2w}{dz^2}+\rho\sigma\frac{dw}{dz}-w=0.$$

Establish the relationship between the two equations. [Pochhammer.]

4. Show that every solution of the following equations, relative to the singularity at the origin is free from logarithms:

$$\text{(i) } \frac{d^2w}{dz^2}+\frac{dw}{dz}-\frac{2}{z^2}w=0,$$

$$\text{(ii) } z(2-z^2)\frac{d^2w}{dz^2}-(z^2-4z+2)\left\{(1-z)\frac{dw}{dz}+w\right\}=0.$$

5. Prove that the origin is an apparent singularity of the equation

$$z\frac{d^2w}{dz^2}-(1+z)\frac{dw}{dz}+2(1-z)w=0.$$

CHAPTER XVII

EQUATIONS WITH IRREGULAR SINGULAR POINTS

17·1. The Possible Existence of Regular Solutions.—The theorems which were established in the two preceding chapters show that, when the point z_0 is a regular singularity, the functional nature of the fundamental set of solutions appropriate to z_0 is known. Moreover, each solution of the set can be developed in series of ascending powers of $z-z_0$, whose coefficients are determined in succession by a system of recurrence-relations.

Let it now be supposed that, in the neighbourhood of z_0, every coefficient of

$$\text{(A)}\quad L(w)\equiv\frac{d^n w}{dz^n}+p_1(z)\frac{d^{n-1}w}{dz^{n-1}}+p_2(z)\frac{d^{n-2}w}{dz^{n-2}}+\ldots+p_{n-1}(z)\frac{dw}{dz}+p_n(z)w=0$$

is analytic, but that one at least of the coefficients $p_r(z)$ has a pole at z_0 of order exceeding the suffix r. Then since the condition for a regular singularity is violated, not all of the n solutions appropriate to the point z_0 will be regular. The problem which now arises is whether any of these solutions can be regular, and if so to obtain analytic expressions for them.*

Let $\varpi_1, \varpi_2, \ldots, \varpi_n$ be the orders of the poles which $p_1, p_2, \ldots, p_n$ respectively have at z_0, and consider the numbers

$$\varpi_1+n-1,\quad \varpi_2+n-2,\quad \ldots,\quad \varpi_{n-1}+1,\quad \varpi_n,$$

of which, by hypothesis, at least one exceeds n. Let g_{n-1} be the greatest of these numbers, excluding ϖ_n, and suppose that the equation has a regular solution

$$w(z)=(z-z_0)^\rho\phi(z-z_0),$$

where $\phi(0)\neq 0$, then by substituting this solution in the equation

$$p_n(z)=-\frac{1}{w}\left\{\frac{d^n w}{dz^n}+p_1(z)\frac{d^{n-1}w}{dz^{n-1}}+\ldots+p_{n-1}(z)\frac{dw}{dz}\right\},$$

it will be seen that $p_n(z)$ will have a pole at z_0 of order not exceeding g_{n-1}. Thus a necessary condition for the existence of a regular solution is that $\varpi_n\leqslant g_{n-1}$.

17·11. A necessary Condition for the Existence of $n-r$ Regular Solutions.—The previous theorem will now be generalised. Let g_r be the greatest of the r numbers

$$\varpi_1+n-1,\quad \varpi_2+n-2,\quad \ldots,\quad \varpi_r+n-r;$$

then if there are $n-r$ regular solutions, the remaining numbers

$$\varpi_{r+1}+n-r-1,\quad \ldots,\quad \varpi_{n-1}+1,\quad \varpi_n$$

will all be less than g_r. A proof by induction will be adopted; let the theorem be supposed to be true when the order of the equation is $n-1$, it will then be proved true when the order of the equation is n.

* Thomé, *J. für Math.* 74 (1872), p. 193; 75 (1873), p. 265; 76 (1873), p. 273.

Let the equation be subjected to the transformation

$$w=w_1(z)\int v\,dz,$$

where

$$w_1(z)=(z-z_0)^\rho\phi(z-z_0)$$

is a regular solution of (A), then v will satisfy an equation of the form

$$\text{(B)}\qquad \frac{d^{n-1}v}{dz^{n-1}}+q_1(z)\frac{d^{n-2}v}{dz^{n-2}}+\ldots+q_{n-2}(z)\frac{dv}{dz}+q_{n-1}(z)v=0,$$

and, on the supposition that (A) has $n-r$ regular solutions, (B) will have $n-r-1$ regular solutions. Let

$$\sigma_1,\quad \sigma_2,\ \ldots,\quad \sigma_{n-1}$$

be the order of the pole at z_0 in $q_1, q_2, \ldots, q_{n-1}$ respectively.

Now $q_s(z)$ may be expressed explicitly as follows :

$$q_s(z)=\frac{1}{w_1}\left\{{}_nC_s\frac{d^sw_1}{dz^s}+\ldots+(n-s+1)p_{s-1}(z)\frac{dw_1}{dz}+p_s(z)w_1\right\},$$

and therefore

$$\sigma_s\leqslant g_r+s-n \qquad (s=1, 2, \ldots, r).$$

Thus each of the numbers

$$\sigma_1+(n-1)-1,\quad \sigma_2+(n-1)-2,\quad \ldots,\quad \sigma_r+(n-1)-r.$$

is at most equal to g_r-1.

The assumption made is that when equation (B) has $n-r-1$ regular solutions, the remaining numbers

$$\sigma_{r+1}+(n-1)-r-1,\ \ldots,\ \sigma_{n-1}$$

are also at most equal to g_r-1. Then, referring back to the expression for q_s, it will be seen in succession that

$$\varpi_{r+1}\leqslant\sigma_{r+1},\ \ldots,\ \varpi_{n-1}\leqslant\sigma_{n-1},$$

and consequently that each of the numbers

$$\varpi_{r+1}+n-r-1,\ \ldots,\ \varpi_{n-1}+1$$

is at most equal to g_r. It then follows, as in the last section, that ϖ_n is also at most equal to g_r.

Now the theorem is true in the case of an equation of order $r+1$ which has one solution regular at z_0 ; it is therefore true for an equation of order $r+2$ having two regular solutions, and therefore, finally, for an equation of order n having $n-r$ solutions regular at z_0.

From this theorem a very important corollary can be deduced, as follows. Let g be the greatest of the numbers

$$\varpi_1+n-1,\quad \varpi_2+n-2,\quad \ldots,\quad \varpi_{n-1}+1,\quad \varpi_n,$$

and let r be the least integer for which

$$\varpi_r+n-r=g,$$

then *the equation will have at most $n-r$ distinct solutions regular at z_0*. For if it had a greater number, $n-s$, of independent solutions, regular at z_0, then, since $s<r$, each of the s numbers

$$\varpi_1+n-1,\ \ldots,\ \varpi_s+n-s$$

is at most equal to a number h, itself less than g. But as there are now supposed to be $n-s$ regular solutions, each of the remaining numbers

$$\varpi_{s+1}+n-s-1,\ \ldots,\ \varpi_n$$

is at most equal to h. In particular

$$\varpi_r+n-r\leqslant h<g,$$

contrary to hypothesis. The theorem is therefore true.

The number r is known as the *Class* * of the singular point z_0. When all the solutions appropriate to z_0 are regular, each of the numbers

$$\varpi_0+n, \quad \varpi_1+n-1, \quad \ldots, \quad \varpi_{n-1}+1, \quad \varpi_n,$$

$(\varpi_0=0)$ is less than or equal to n. In this case, therefore, r is zero.

When $r \geqslant 1$, the number of distinct solutions regular at z_0 has been proved at most equal to $n-r$, but may fall short of this upper bound.

Thus, in the equation

$$z^2\frac{d^2w}{dz^2}+a\frac{dw}{dz}+bw=0,$$

where a and b are constants $(a \neq 0)$, $\varpi_1=\varpi_2=2$, and, considering the singularity at the origin,

$$\varpi_1+n-1=3, \quad \varpi_2+n-2=2.$$

Consequently

$$g=3, \quad r=1,$$

and therefore there is at most one solution, regular at the origin. It is easily seen that if this regular solution exists, its development is

$$w=A_0+A_1z+A_2z^2+\ldots+A_mz^m+\ldots,$$

where

$$(m+1)aA_{m+1}+\{b+m(m-1)\}A_m=0.$$

But $\lim|A_{m+1}/A_m|=\infty$, and therefore the series does not converge for any value of z except $z=0$. Thus in this case no solution, regular at the origin, exists.

17·2. The Indicial Equation.—For simplicity, let the singular point z_0 be the origin. In place of (A) consider the equivalent equation

$$L_1(w)=0,$$

where $L_1=z^gL$. Now L_1 may be written in the form

$$z^nQ_0(z)D^n+z^{n-1}Q_1(z)D^{n-1}+\ldots+zQ_{n-1}(z)D+Q_n(z),$$

where

$$Q_0(z)=z^{g-n},$$
$$Q_\nu(z)=z^{g-n+\nu}p_\nu(z) \qquad (\nu=1, 2, \ldots, n).$$

The functions Q are analytic in the neighbourhood of the origin; from the definitions of g and r it follows that, when $z=0$ is not a regular singularity,

$$Q_0(0)=Q_1(0)=\ldots=Q_{r-1}(0)=0,$$
$$Q_r(0)\neq 0,$$

and the remaining coefficients Q are finite or zero at the origin.

Let it be assumed that there exists a solution regular at the origin, say

$$w=z^\rho\{1+O(z)\},$$

then the coefficient of the term in z^ρ proceeding from

$$z^{n-\nu}Q_\nu(z)D^{n-\nu}w$$

will be

$$Q_\nu(0)[\rho]_{n-\nu}$$

and will vanish when $\nu<r$ but not when $\nu=r$. Since z^ρ is the lowest power of z which occurs in $L_1(w)$, ρ must satisfy the *indicial equation*

$$Q_r(0)[\rho]_{n-r}+\ldots=0,$$

where the omitted terms are of lower degree in ρ than the term written. The degree of the indicial equation is therefore $n-r$, which confirms the

* The accepted term is *Characteristic Index*, but the terms "characteristic" and "index" are already sufficiently overworked. The excess of the number $n-r$ over the actual number of regular solutions could conveniently be called the *Deficiency*.

theorem already proved, that there cannot be more than $n-r$ distinct regular solutions.

In particular, when $n=r$ the indicial equation becomes

$$Q_n(0)=0.$$

Thus when the left-hand member of the indicial equation is independent of ρ there can be no regular solution.

17·21. Reducibility of an Equation which has Regular Solutions.—Let it be supposed that the equation (A) has k distinct solutions which are regular at the singular point $z=0$. These solutions form a fundamental set for an equation of order k whose coefficients satisfy Fuchs' conditions with respect to the origin. Let this equation be

$$M(w)=0,$$

where the coefficient of D^k in M is unity, and write $M_1=z^kM$. Then

$$L_1=R_1M_1,$$

where R_1 is a differential operator of order $n-k$ formed as indicated in § 5·4. Since the coefficients of both L_1 and M_1 are finite or zero at the origin, and are analytic in the neighbourhood of the origin, the same is true of the coefficients of R_1. Consequently the equation $L_1(w)=0$, and the equivalent equation $L(w)=0$, are reducible if one or more regular solutions exist.

Now the equation

$$[z^{-\rho}L_1(z^\rho)]_{z=0}=0,$$

which, from the definition of g, is not an identity, is the indicial equation of $L_1(w)=0$ or of $L(w)=0$ with respect to the singularity $z=0$. Let

$$L_1(z^\rho)=f(z,\ \rho)z^\rho=\sum f_\lambda(\rho)z^{\lambda+\rho},$$

$$M_1(z^\rho)=g(z,\ \rho)z^\rho=\sum g_\lambda(\rho)z^{\lambda+\rho},$$

$$R_1(z^\rho)=h(z,\ \rho)z^\rho=\sum h_\lambda(\rho)z^{\lambda+\rho},$$

where the summation begins, in each case, with $\lambda=0$, then since

$$f_0(\rho)=0,\quad g_0(\rho)=0$$

are the indicial equations of $L(w)=0$ and $M(w)=0$, neither $f_0(\rho)$ nor $g_0(\rho)$ is identically zero. Now

$$\begin{aligned}L_1(z^\rho)&=R_1M_1(z^\rho)\\&=R_1\{\sum g_\lambda(\rho)z^{\lambda+\rho}\}\\&=\sum g_\lambda(\rho)R_1z^{\lambda+\rho}\\&=\sum_\lambda g_\lambda(\rho)\sum_\mu h_\mu(\lambda+\rho)z^{\lambda+\mu+\rho}.\end{aligned}$$

Thus

$$\sum f_\lambda(\rho)z^\lambda=\sum_\kappa g_\kappa(\rho)\sum_\mu h_\mu(\kappa+\rho)z^{\kappa+\mu}$$

and therefore

$$f_\lambda(\rho)=\sum_{\kappa=0}^{\lambda}g_\kappa(\rho)h_{\lambda-\kappa}(\kappa+\rho),$$

a set of relations which determine, in turn, the polynomials $h_0(\rho)$, $h_1(\rho)$, $h_2(\rho)$, ... In particular

$$f_0(\rho)=g_0(\rho)h_0(\rho),$$

which proves that $h_0(\rho)$ is not identically zero.

When the polynomials $h_\lambda(\rho)$ have been evaluated, R_1 can be determined explicitly, as follows: The degrees of the polynomials $f_\lambda(\rho)$ have the upper bound n, which is attained; those of $g_\lambda(\rho)$ have the upper bound k, which is also attained. The upper bound of the degrees of $h_\lambda(\rho)$ will therefore be $n-k$, and will be attained. Let $n-k=m$, then since

$$h(z, \rho)=\sum h_\lambda(\rho)z^\lambda,$$

$h(z, \rho)$ is expressible in the form

$$h(z, \rho)=\sum_{r=0}^{m-1}\rho(\rho-1)\ .\ .\ .\ (\rho-m+r+1)u_{m-r}(z)+u_0(z),$$

where the coefficients $u(z)$ are determined by the formula

$$s\,!\,u_s(z)=[\Delta^r h(z, \rho)]_{\rho=0},$$

where

$$\Delta h(z, \rho)=h(z, \rho+1)-h(z, \rho),$$
$$\Delta^2 h(z, \rho)=\Delta h(z, \rho+1)-\Delta h(z, \rho),$$
$$.\quad .\quad .$$

Hence

$$\begin{aligned}R_1(z^\rho)&=h(z, \rho)z^\rho\\&=\{[\rho]_m u_m(z)+[\rho]_{m-1}u_{m-1}(z)+\ .\ .\ .\ +\rho u_1(z)+u_0(z)\}z^\rho,\end{aligned}$$

and therefore R_1 is the operator

$$z^m u_m(z)D^m+\ .\ .\ .\ +zu_1(z)D+u_0(z).$$

Now $f_0(\rho)$ is of degree $n-r$ in ρ and $g_0(\rho)$ of degree k. Hence $h_0(\rho)$ is of degree $n-r-k$. Thus since $h_0(\rho)=0$ is the indicial equation of $R_1(w)=0$, the degree of this indicial equation is the number by which the degree of the indicial equation of $L(w)=0$ exceeds the number of regular solutions. In particular, if $R_1(w)=0$ has no indicial equation, $L(w)=0$ has precisely $n-r$ regular solutions.*

17·3. Proof of the general Non-Existence of Regular Solutions.—In § 17·11 it was shown by an example that even when the equation $L(w)=0$ possesses an indicial equation for the singularity $z=0$, the corresponding formal development of the solution may diverge for all values of $|z|$. This phenomenon is in no way exceptional, in fact the exceptional case is for a regular solution to exist at all.

Consider, as before, the modified equation

$$L_1(w)=0,$$

then, if there exists a regular solution

$$w=z^\rho(c_0+c_1z+c_2z^2+\ .\ .\ .),$$

ρ will be determined by the indicial equation

$$f_0(\rho)=0.$$

By equating to zero the coefficients of successive powers of z in

$$L_1(w)=z^\rho\Big\{\sum c_0 f_\lambda(\rho)z^\lambda+\sum c_1 f_\lambda(\rho+1)z^{\lambda+1}+\sum c_2 f_\lambda(\rho+2)z^{\lambda+2}+\ .\ .\ .\Big\},$$

the following set of recurrence-relations is obtained:

$$c_1f_0(\rho+1)+c_0f_1(\rho)=0,$$
$$c_2f_0(\rho+2)+c_1f_1(\rho+1)+c_0f_2(\rho)=0,$$
$$.\quad .\quad .\quad .\quad .\quad .\quad .\quad .\quad .$$
$$c_\nu f_0(\rho+\nu)+c_{\nu-1}f_1(\rho+\nu-1)+\ .\ .\ .\ +c_0f_\nu(\rho)=0,$$
$$.\quad .\quad .\quad .\quad .\quad .\quad .\quad .\quad .$$

and these recurrence-relations determine $c_1, c_2, \ldots, c_\nu, \ldots$ when c_0 is given (cf. § 16·11).

* Floquet, *Ann. Éc. Norm.* (2), 8 (1879), *suppl.* p. 63.

Now the essential difference between the present case, and the case, treated in the preceding chapter, in which all solutions are regular at the origin, is that $f_0(\rho)$ is not of degree n but only of degree $n-r$ in ρ. On the other hand, among the functions $f_\nu(\rho)$ there are some whose degree is n; the first of these is $f_{g-n}(\rho)$.

If the process of evaluating the coefficients c_ν terminates, so that the expression for w contains only a finite number of terms, then the expression so found is a solution regular at the origin. In general, however, the series does not terminate; in this case it will be shown to diverge.

For certain values of k, for example $k=g-n$,

$$\lim_{\nu\to\infty}\left|\frac{f_k(\rho+\nu-k)}{f_0(\rho+\nu)}\right|=\infty,$$

for the numerator is of degree n, and the denominator of lower degree, in ν. Thus, in order that the recurrence-relation

$$\frac{c_{\nu-1}f_1(\rho+\nu-1)}{c_\nu f_0(\rho+\nu)}+\frac{c_{\nu-2}f_2(\rho+\nu-2)}{c_\nu f_0(\rho+\nu)}+\ldots+\frac{c_0 f_\nu(\rho)}{c_\nu f_0(\rho+\nu)}=-1$$

may be satisfied, it is necessary, in general, that

$$\lim_{\nu\to\infty}\left|\frac{c_{\nu-k}}{c_\nu}\right|=0;$$

the series therefore diverges.

17·4. The Adjoint Equation.—When the indicial equation relative to an irregular singularity is of degree $n-r$, there cannot be more than $n-r$ regular solutions. But since the number of regular solutions may fall short of the maximum, it is expedient to find a criterion for ascertaining whether or not the possible number of regular solutions is attained. This required criterion can be obtained by means of the adjoint equation.*

Let $\bar{L}_1$ be the differential operator adjoint to L_1. In the Lagrange identity (§ 5·3)

$$vL_1(u)-u\bar{L}_1(v)=\frac{d}{dz}\{P(u,v)\},$$

let $u=z^\rho$, $v=z^{-\rho-\nu-1}$, where ρ is arbitrary, but ν an integer, then

$$z^{-\rho-\nu-1}L_1(z^\rho)-z^\rho\bar{L}_1(z^{-\rho-\nu-1})=\frac{d}{dz}\{P(z^\rho,z^{-\rho-\nu-1})\}.$$

Now $P(z^\rho, z^{-\rho-\nu-1})$ is free from terms in z^ρ; from the assumption made concerning the coefficients of the operator L it follows that P has at the origin no singularity other than a pole. Consequently no term in z^{-1} can exist in

$$z^{-\rho-\nu-1}L_1(z^\rho)-z^\rho\bar{L}_1(z^{-\rho-\nu-1}).$$

As before, let

$$L_1(z^\rho)=\sum f_\lambda(\rho)z^{\lambda+\rho},$$

and now let

$$\bar{L}_1(z^\rho)=\sum\phi_\lambda(\rho)z^{\lambda+\rho}.$$

The coefficient of z^{-1} in $z^{-\rho-\nu-1}L_1(z^\rho)$ is $f_\nu(\rho)$ and that of z^{-1} in $z\ L_1(z^{-\rho-\nu-1})$ is $\phi_\nu(-\rho-\nu-1)$, hence

$$f_\nu(\rho)=\phi_\nu(-\rho-\nu-1)$$

and similarly,

$$\phi_\nu(\rho)=f_\nu(-\rho-\nu-1).$$

* Thomé, *J. für. Math.* 75 (1873), p. 276; Frobenius, *ibid.* 80 (1875), p. 320.

An immediate consequence is that the degrees of the two indicial equations $f_0(\rho)=0$, relating to $L_1(u)=0$, and $\phi_0(\rho)=0$, relating to $\overline{L}_1(v)=0$, are equal. Let this degree be $n-r$, then the class r is the same for both equations. In particular if one of $L_1(u)=0$ and $\overline{L}_1(v)=0$ has all or no solutions regular at a singular point, the same is true of the other.

It will now be supposed that $L_1(u)=0$ actually has $n-r$ solutions regular at the origin. Then

$$L_1=R_1M_1,$$

where $M_1(u)=0$ is the equation satisfied by the $n-r$ regular solutions, and R_1 is an operator of order r. But if $\overline{R}_1$ and $\overline{M}_1$ are the adjoint operators of R_1 and M_1 respectively,

$$\overline{L}_1=\overline{M}_1\overline{R}_1.$$

Now the indicial equations, relative to the origin, of both $\overline{L}_1$ and $\overline{M}_1$ are of degree $n-r$. Consequently the equation $\overline{R}_1(u)=0$ has no indicial equation. If, therefore, the equation $L(v)=0$ has $n-r$ regular solutions, it is necessary that the adjoint equation $\overline{L}(u)=0$ should be satisfied by all the solutions of an equation $\overline{R}(u)=0$, of order r, which has no indicial equation.

But this condition is also sufficient, for when it is satisfied the equation $R_1(u)=0$, adjoint to $\overline{R}_1(v)=0$, has also no indicial equation. Consequently the order of the equation $M_1(u)=0$ is equal to the degree of its indicial equation relative to the singularity considered, and all the solutions of $M_1(u)=0$ are regular at the origin. The equation $L_1(u)=0$ therefore has $n-r$ solutions regular at the origin.

Thus a necessary and sufficient condition that an equation of order n should have n−r solutions regular at a singular point at which the indicial equation is of degree n−r, is that the adjoint equation should be satisfied by all the solutions of an equation of order r, which has no indicial equation at the singular point in question.

When regular solutions exist, explicit expressions for them may be obtained by solving the set of recurrence-relations given in § 17·3. Any cases in which roots of the indicial equation are repeated, or differ from one another by integers, can be treated by applying the general method of § 16·3.

17·5. Normal Solutions.—The next problem which arises is that of obtaining, if possible, developments to represent those solutions which are not regular at a singular point. The case of an equation of the first order for which the origin is an irregular singular point will serve as an introduction to the more general case. Consider, then, the equation

$$z\frac{dw}{dz}+p(z)w=0,$$

in which

$$p(z)=\frac{ma_1}{z^m}+\frac{(m-1)a_2}{z^{m-1}}+\ldots+\frac{2a_{m-1}}{z^2}+\frac{a_m}{z}-\rho-\phi(z),$$

where $\phi(z)$ is analytic in the neighbourhood of the origin, and $\phi(0)=0$. The general solution is

$$w=Ae^{Q(z)}z^{\rho}\Phi(z),$$

where A is an arbitrary constant,

$$Q(z)=\frac{a_1}{z^m}+\frac{a_2}{z^{m-1}}+\ldots+\frac{a_{m-1}}{z^2}+\frac{a_m}{z},$$

and

$$\Phi(z)=\exp\int\frac{\phi(z)}{z}\,dz.$$

If the solution is written as

$$w=e^{Q(z)}v(z),$$

then $v(z)$ is the solution (regular at the origin) of the equation

$$z\frac{dv}{dz}=\{\rho+\phi(z)\}v.$$

The essential singularity of the solution is thus due to the presence of the factor $e^{Q(z)}$, which is known as the *determining factor* of the solution. When a solution of this form exists it is known as a *normal* solution* ; the number ρ is the *exponent.*

17·51. Equations in which the Point at Infinity is an Irregular Singularity.—In equations arising out of physical problems, when a point is an irregular singularity, that point is almost invariably the point at infinity. It is therefore expedient to suppose that any particular singular point, say z_0, has been transferred to infinity by the substitution

$$Z=\frac{1}{z-z_0}.$$

No loss in generality, and an appreciable gain in ease of manipulation results from this transformation.

Consider, then, the equation

$$\frac{d^nw}{dz^n}+p_1(z)\frac{d^{n-1}w}{dz^{n-1}}+\ldots+p_{n-1}(z)\frac{dw}{dz}+p_n(z)w=0,$$

in which the coefficients are developable in series of descending integral powers of z, thus

$$p_\nu(z)=z^{K_\nu}(a_{\nu_0}+a_{\nu_1}z^{-1}+a_{\nu_2}z^{-2}+\ldots).$$

If, as is supposed, the point at infinity is an irregular singular point, $K_\nu\geqslant 1-\nu$ for at least one value of ν. Suppose that

$$K_\nu+\nu<K_r+r \quad \text{when } \nu<r,$$
$$K_\nu+\nu\leqslant K_r+r \quad \text{when } \nu>r,$$

then the degree of the indicial equation relative to the point at infinity will be $n-r$, and r will be the class of the singularity.

It will now be shown that a necessary condition for the existence of a normal solution is that $K_\nu\geqslant 0$ for at least one value of ν. When a solution, normal at infinity, exists, it is of the form,

$$w=e^{Q(z)}u(z),$$

where $Q(z)$ is a determinate polynomial in z and $u(z)$ is of the form

$$z^\rho(c_0+c_1z^{-1}+c_2z^{-2}+\ldots).$$

Let

$$\frac{d^m}{dz^m}e^Q=t_me^Q,$$

so that

$$t_0=1, \quad t_1=Q'$$

and, in general,

$$t_{m+1}=t_m'+t_mQ'.$$

If $Q(z)$ is a polynomial of degree s, then at infinity

$$t_m=O(z^{ms-m}).$$

* Thomé, *J. für Math.* 95 (1883), p. 75.

Let the equation satisfied by $u(z)$ be

$$\frac{d^nu}{dz^n}+q_1(z)\frac{d^{n-1}u}{dz^{n-1}}+\ldots+q_{n-1}(z)\frac{du}{dz}+q_n(z)u=0,$$

then it may be verified that

$$q_\nu=p_\nu+(n-\nu+1)t_1p_{\nu-1}+{}_{n-\nu+2}C_2\,t_2p_{\nu-2}+\ldots+{}_nC_\nu t_\nu,$$

and in particular

$$q_n=p_n+t_1p_{n-1}+t_2p_{n-2}+\ldots+t_n.$$

Now if a normal solution exists, it will be possible to determine Q so that the equation in u has at least one solution regular at infinity. This condition limits the degree of the dominant term in q_n. The degrees of the dominant terms of the components of q_ν are, in order

$$K_\nu,\quad K_{\nu-1}+s-1,\quad K_{\nu-2}+2s-2,\quad \ldots,\quad \nu(s-1)$$

and therefore, when the polynomial Q is of degree s, but otherwise arbitrary, the degree of the leading term in q_ν exceeds that of the leading term in $q_{\nu-1}$ by at least $s-1$. In general, therefore, the dominant term in q_n will not be less than the dominant term of any other coefficient q_ν. The equation in u will therefore have no indicial equation, and consequently no regular solution, at infinity.

Thus when a normal solution exists, it must be possible, by a proper choice of the degree of $Q(z)$ and of its coefficients, to make the degree of the dominant term in q_n at least one unit lower than the degree of the dominant term in q_{n-1}, in which case only can the equation in u have a solution regular at infinity. In order that this may be possible, it is necessary that no one of the numbers

$$K_n,\quad K_{n-1}+s-1,\quad K_{n-2}+2(s-1),\quad \ldots,\quad n(s-1)$$

should exceed all the rest, that is, of the numbers

$$K_n-n(s-1),\quad K_{n-1}-(n-1)(s-1),\quad K_{n-2}-(n-2)(s-1),\quad \ldots,\quad 0$$

the two greatest should be equal. Let g be the greatest of the numbers

$$K_1,\quad \tfrac{1}{2}K_2,\quad \tfrac{1}{3}K_3,\quad \ldots,\quad \frac{1}{n}K_n,$$

then it is necessary that

$$K_\nu-\nu(s-1)\geqslant 0$$

for some value of ν, from which it follows that

$$g\geqslant s-1.$$

But since the solution is normal, and not regular,* $s\geqslant 1$, and therefore $g\geqslant 0$. It follows that $K_\nu\geqslant 0$ for at least one value of ν.

For instance, the equation

$$zw''+w'+w=0$$

has no solution, normal at infinity, because $K_1=K_2=-1$, and therefore $g=-\frac{1}{2}<0$.

17·52. Calculation of the Determining Factor.—The degree, s, of the polynomial $Q(z)$ is thus limited by the inequality

$$s\leqslant g+1.$$

When g is a positive integer or zero, it is clearly admissible to take $s=g+1$, because, in that case,

$$K_\nu=\nu(s-1)$$

* Note also that, when the point at infinity is an irregular singularity,

$$g\geqslant\frac{1}{\nu}K_\nu\geqslant\frac{1}{\nu}-1$$

for at least one value of ν, so that $g>-1$.

for at least one value of ν, and for all other values of ν

$$K_\nu < \nu(s-1),$$

and therefore, of the numbers

$$K_n, \quad K_{n-1}+s-1, \quad K_{n-2}+2(s-1), \ldots, \quad n(s-1),$$

the number $n(s-1)$ is equal to at least one other, and greater than the remaining, numbers of the set.

Now the *class* has been defined as the number r such that

$$K_\nu+\nu<K_r+r \qquad \text{when } \nu<r,$$
$$K_\nu+\nu\leqslant K_r+r \qquad \text{when } \nu>r,$$

and thus, when $\nu>r$,

$$K_r+(n-r)(s-1)\geqslant K_\nu+(n-\nu)(s-1)+s(\nu-r)$$
$$>K_\nu+(n-\nu)(s-1).$$

Consequently the equality

$$K_\nu+(n-\nu)(s-1)=n(s-1),$$

or

$$K_\nu=\nu(s-1)=\nu g,$$

which is certainly true for at least one value of ν, namely r, can only hold when $\nu\leqslant r$, and therefore

$$K_\nu\leqslant\nu g \quad \text{when } \nu<r,$$
$$K_\nu=\nu g \quad \text{when } \nu=r,$$
$$K_\nu<\nu g \quad \text{when } \nu>r.$$

Let m be the least value of ν for which $K_\nu=\nu g$, then

$$K_m+(n-m)(s-1)=K_r+(n-r)(s-1)=ng,$$

and

$$K_\nu+(n-\nu)(s-1)<ng \quad \text{when} \quad \nu<m \quad \text{or} \quad \nu>r.$$

The terms of highest order in $q_n(z)$ are therefore

$$t_n, \quad t_{n-m}p_m, \quad \ldots, \quad t_{n-r}p_r.$$

But

$$t_\nu=Q't_{\nu-1}+O\{z^{(\nu-1)(s-1)}\}$$
$$=Q'^\nu+O\{z^{(\nu-1)(s-1)}\},$$

whereas

$$t_\nu=O\{z^{\nu(s-1)}\}.$$

and therefore the dominant expression in $q_n(z)$ is

$$Q'^n+p_mQ'^{n-m}+\ldots+p_rQ'^{n-r}.$$

Let

$$Q(z)=A_1z+\tfrac{1}{2}A_2z^2+\ldots+\tfrac{1}{s}A_sz^s,$$

then since

$$p_\nu(z)=z^{K_\nu}(a_{\nu0}+a_{\nu1}z^{-1}+\ldots),$$

the condition for the vanishing of the term of highest order in $q_n(z)$ is

$$A_s^r+a_{m0}A_s^{r-m}+\ldots+a_{r0}=0.$$

There are therefore at most r distinct values of the constant A_s. When a value of A_s has been obtained, the remaining constants $A_{s-1}, \ldots, A_1$ can be calculated in succession. Thus, when $s=g+1$ the determining factor can be determined in one or more ways.

The assumption that $s=g+1$ is necessary when $g=0$, but when g is a positive fraction, and in general also when g is a positive integer, integral values of s less than $g+1$ will be admissible. To obtain the admissible values of s, use is made of the Puiseux diagram * which is constructed as follows. The points whose Cartesian coordinates (x, y) are

$$(0, r), \quad (K_1, r-1), \quad \ldots, \quad (K_r, 0),$$

are plotted, and a vector line is drawn through the point $(0, r)$ in the first quadrant and parallel to the x-axis. This vector is rotated about $(0, r)$ in the clockwise direction until it encounters one or more of the other points. It is then rotated in the same direction about that one of these points which is most remote from $(0, r)$ until it meets other points, and so on until it passes through the point $(K_r, 0)$. A polygonal line joining $(0, r)$ to $(K_r, 0)$ is thus formed such that none of the points lie, in the ordinary sense, above or to the right of that line.

Consider any one rectilinear segment of the line, and suppose, for instance, that it passes through the points

$$(K_\sigma, r-\sigma), \ \ldots, \ (K_\tau, r-\tau),$$

and let it make the angle θ with the negative direction of the y-axis. If $\mu=\tan\theta$, points on this segment satisfy the equation

$$x+\mu y=C,$$

where C is constant, and therefore

$$K_\sigma+\mu(r-\sigma)= \ \ldots \ =K_\tau+\mu(r-\tau),$$

and if $(K_\nu, r-\nu)$ is a point not on the segment

$$K_\sigma+\mu(r-\sigma)>K_\nu+\mu(r-\nu).$$

If, therefore, μ is a positive integer or zero, an admissible value of s will be

$$s=\mu+1,$$

and there will be as many admissible values of s as there are distinct rectilinear segments in the polygonal line, for which μ is a positive integer or zero.

When an admissible value of s has been obtained, the method of deriving the polynomial $Q(z)$ proceeds on the same lines as before. The next step is to obtain the differential equation in u, and to ascertain whether or not it has solutions regular at infinity, for it is only when $u(z)$ is regular at infinity that a normal solution $w(z)$ can be said to exist. The existence of the determining factor $e^{Q(z)}$ is not of itself sufficient for the existence of a normal solution; the convergence of the series in $u(z)$ is also necessary, and this, as has been seen, is exceptional. When, however, a normal solution exists, it is said to be of *grade* s, where s is the degree of the polynomial $Q(z)$. The *rank* of an equation, relative to the singular point considered, is the number h where

$$h=g+1.$$

When h is an integer, h may be equal to s, but in general

$$h>s.$$

When the polynomial $Q(z)$ has been determined, the next step is to obtain the indicial equation satisfied by ρ. When this equation has equal roots, or roots which differ from one another by integers, there may exist, in addition to a normal solution free from logarithmic terms, solutions of the form

$$e^{Q(z)}z^\rho\{\phi_0(z)+\phi_1(z)\log z+ \ \ldots \ +\phi_m(z)(\log z)^m\},$$

in which the functions $\phi(z)$ are analytic at infinity.

17·53. Subnormal Solutions.—For any rectilinear segment of the Puiseux diagram, the inclination μ is a rational fraction. In order to construct any

* Cf. § 12·61.

normal solutions which may exist, any zero or positive integral values of μ may be selected; non-integral values have to be discarded. These, however, are not altogether useless, for they may lead to solutions of a new type, known as *subnormal* solutions.*

Let the rational fraction μ, expressed in its lowest terms, be l/k, and transform the equation by writing

$$\zeta = z^k.$$

Then the Puiseux diagram of the transformed equation will possess a rectilinear segment inclined at an angle θ' to the negative direction of the y-axis, where

$$\tan \theta' = l.$$

If l is a *positive* integer, the transformed equation may possess a normal solution; if it does, the determining factor $Q(\zeta)$ will be a polynomial in ζ of degree s, where

$$s = l + 1.$$

Thus the original equation may possess a solution of normal type in the variable $z^{1/k}$; such a solution is said to be *subnormal*. Obviously, if one subnormal solution in $z^{1/k}$ exists, there will be $k-1$ other subnormal solutions of the same type. These solutions are said to form an aggregate of subnormal solutions.

For example, the equation

$$z\frac{d^2w}{dz^2} + 2\frac{dw}{dz} - \left\{\frac{1}{4} + \frac{5}{16z}\right\}w = 0$$

has two subnormal solutions. Its general solution is

$$w = Ae^{\sqrt{z}}\{z^{-\frac{3}{4}} - z^{-\frac{5}{4}}\} + Be^{-\sqrt{z}}\{z^{-\frac{3}{4}} + z^{-\frac{5}{4}}\},$$

where A and B are arbitrary constants.

When the determining factor $Q(z^{1/k})$ is of degree s in $z^{1/k}$, the subnormal solution is said to be of grade s/k; in this case

$$h = g + 1 \geqslant \frac{s}{k}.$$

Thus when a normal or subnormal solution exists, its grade does not exceed the rank of the equation.

17·54. Rank of the Equation satisfied by a given Fundamental Set of Normal and Subnormal Functions.—Let

$$w_1 = e^{Q_1}z^{\rho_1}u_1, \quad w_2 = e^{Q_2}z^{\rho_2}u_2, \quad \ldots, \quad w_n = e^{Q_n}z^{\rho_n}u_n$$

be n functions of normal or subnormal type arranged so that, if their grades are respectively $\gamma_1, \gamma_2, \ldots, \gamma_n$, then

$$\Gamma \geqslant \gamma_1 \geqslant \gamma_2 \geqslant \ldots \geqslant \gamma_n.$$

Then the differential equation of order n satisfied by these functions will be of rank h not exceeding Γ, with respect to the singular point at infinity.† Let

$$\Delta = \begin{vmatrix} w_1, & w_2, & \ldots, & w_n \\ w_1', & w_2', & \ldots, & w_n' \\ \cdot & \cdot & \cdot & \cdot \\ w_1^{(n-1)}, & w_2^{(n-1)}, & \ldots, & w_n^{(n-1)} \end{vmatrix}$$

be the Wronskian of the n given functions; it is assumed that Δ is not

* Fabry, *Thèse* (Faculté des Sciences, Paris, 1885).
† Poincaré, *Acta Math.* 8 (1886), p. 305.

identically zero. Let Δ_r be the determinant obtained from Δ by replacing $w_1^{(n-r)}$ by $w_1^{(n)}$, $w_2^{(n-r)}$ by $w_2^{(n)}$ and so on. Then, if

$$p_r = -\Delta_r/\Delta,$$

the differential equation satisfied by $w_1, w_2, \ldots, w_n$ will be

$$\frac{d^n w}{dz^n} + p_1 \frac{d^{n-1} w}{dz^{n-1}} + \ldots + p_{n-1}\frac{dw}{dz} + p_n w = 0.$$

The rank of this equation depends upon the order of the coefficients p_r at infinity. Now from

$$w_\mu = e^{Q_\mu} z^{\rho_\mu} u_\mu,$$

it follows that

$$w_\mu^{(\nu)} = z^{\nu(\gamma_\mu - 1)} e^{Q_\mu} z^{\rho_\mu} \phi_{\mu\nu}.$$

where $\phi_{\mu\nu}$ is analytic, and not zero, at infinity. Now

$$p_r = -\begin{vmatrix} \ldots, & u_\mu & , \ldots \\ \ldots, & z^{(\gamma_\mu-1)}\phi_{\mu 1} & , \ldots \\ \cdot & \cdot & \cdot \\ \ldots, & z^{n(\gamma_\mu-1)}\phi_{\mu n} & , \ldots \\ \cdot & \cdot & \cdot \\ \ldots, & z^{(n-1)(\gamma_\mu-1)}\phi_{\mu, n-1} & , \ldots \end{vmatrix} \div \begin{vmatrix} \ldots, & u_\mu & , \ldots \\ \ldots, & z^{(\gamma_\mu-1)}\phi_{\mu 1} & , \ldots \\ \cdot & \cdot & \cdot \\ \ldots, & z^{(n-r)(\gamma_\mu-1)}\phi_{\mu, n-r} & , \ldots \\ \cdot & \cdot & \cdot \\ \ldots, & z^{(n-1)(\gamma_\mu-1)}\phi_{\mu, n-1} & , \ldots \end{vmatrix}$$

When the determinants are expanded according to the elements of the $n-r+1^{\text{th}}$ row (which is the only row in which the determinants differ), p_r takes the form

$$p_r = \frac{\phi_{1n} U_1 z^{\alpha_1 + r(\gamma_1 - 1)} + \ldots + \phi_{nn} U_n z^{\alpha_n + r(\gamma_n - 1)}}{\phi_{1, n-r} U_1 z^{\alpha_1} + \ldots + \phi_{n, n-r} U_n z^{\alpha_n}}.$$

The functions $U_1, \ldots, U_n$ are analytic at infinity, and it will be supposed that the numbers $\alpha_1, \ldots, \alpha_n$ have been so chosen that $U_1, \ldots, U_n$ are not zero at infinity.

If, therefore,

$$p_r = O(z^{K_r}),$$

K_r will be the greatest of the numbers

$$r(\gamma_1 - 1) + \alpha_1 - \alpha_m, \ldots, r(\gamma_m - 1), \ldots, r(\gamma_n - 1) + \alpha_n - \alpha_m,$$

which are in turn not less than

$$r(\gamma_1 - 1), \ldots, r(\gamma_m - 1), \ldots r(\gamma_n - 1),$$

and of these the greatest is $r(\gamma_1 - 1)$.

Thus, for all values of r,

$$\frac{1}{r} K_r \leqslant \gamma_1 - 1 \leqslant \Gamma - 1.$$

and therefore

$$h = g + 1 \leqslant \Gamma.$$

When all of the given functions are normal functions, they are uniform in z, and consequently the coefficients p_r are also uniform in z. Consider the case in which, among the functions $w_1(z), \ldots, w_n(z)$, there occurs an aggregate of subnormal functions. Thus suppose, for definiteness, that $w_1(z), \ldots, w_k(z)$ form an aggregate of subnormal solutions. Then if $\zeta = z^k$ they may be written as

$$W_1(\zeta), \quad W_2(\zeta), \quad \ldots, \quad W_k(\zeta),$$

where $W_1, W_2, \ldots W_k$ are normal functions of ζ. But they may also be arranged in such an order that

$$W_2(\zeta) = W_1(\omega\zeta), \ldots, W_k(\zeta) = W_1(\omega^{k-1}\zeta),$$

where $\omega^k = 1$.

From this it follows that the effect of replacing $z^{1/k}$ by $\omega z^{1/k}$ is to leave p_r unaltered. That is to say, p_r is uniform in z. The same is clearly also true when two or more aggregates of subnormal solutions are present.

Consequently *a set of n functions which are normal or subnormal and of grade equal to or less than Γ satisfies a differential equation of order n and rank h not greater than Γ, with uniform coefficients, provided that when a subnormal function is present, the remaining members of the corresponding aggregate are likewise present.*

It follows from this theorem that when an equation $L(w)=0$, having uniform coefficients, possesses normal or subnormal solutions, it is reducible. For any number of the normal solutions, or of aggregates of the subnormal solutions will satisfy an equation

$$M(w)=0,$$

with uniform coefficients. If this equation has, as may be supposed, no solutions other than those which satisfy $L(w)=0$, the latter equation can be written in the form

$$RM(w)=0$$

and is therefore reducible.

17·6. Hamburger Equations.—No general set of explicit conditions is known which is sufficient to ensure that an equation of order n should admit of a normal solution. Only in one or two particular cases are explicit sets of conditions known; of these cases the most important is that of an equation of order n which is such that

(i) there are two and only two singular points, namely $x=0$ and $x=\infty$,
(ii) the origin is a regular singularity,
(iii) the point at infinity is an essential singularity for every solution.*

The equation may be written in the form

$$z^n\frac{d^nw}{dz^n}+z^{n-1}p_1\frac{d^{n-1}w}{dz^{n-1}}+\ .\ .\ .\ +zp_{n-1}\frac{dw}{dz}+p_nw=0,$$

where $p_1, p_2, \ldots, p_n$ are necessarily integral functions of z; for simplicity it will be assumed that they are polynomials in z.

Now since the origin is a regular singular point, there exists at least one solution of the form

$$w=z^\rho V(z),$$

where $V(z)$ is a power series convergent within any arbitrarily large circle $|z|=R$, and $V(0)\neq0$. This solution can be obtained by the methods of Chapter XVI.

A set of conditions will now be found sufficient to ensure that this solution is normal with respect to the singular point at infinity. Since any solution, normal at infinity is of the form

$$w=e^{Q(z)}z^\sigma U(z),$$

where $Q(z)$ is the polynomial

$$\frac{a_0z^s}{s}+\frac{a_1z^{s-1}}{s-1}+\ .\ .\ .\ +a_{s-1}z,$$

and $U(z)$ is analytic throughout any region which does not include the origin, and does not vanish at infinity, it follows that

$$\frac{w'}{w}=a_0z^{s-1}+a_1z^{s-2}+\ .\ .\ .\ +a_{s-1}+\sigma z^{-1}+U'/U,$$

where, for large values of $|z|$, $U'/U=O(z^{-2})$. But

$$\frac{w'}{w}=\rho z^{-1}+V'/V,$$

* Hamburger, *J. für Math.* 103 (1888), p. 238.

and therefore V'/V must be developable as a series of descending powers of z containing only a finite number of positive integral powers of z. In order that this may be possible it is necessary that V should have only a finite number of zeros within any circle $|z|=R$ however large; let V have k zeros apart from zeros at the essential singularity $z=\infty$. Then, by Weierstrass' theorem,

$$V(z)=P(z)e^{g(z)},$$

where $P(z)$ is a polynomial in z of degree k and $g(z)$ is an integral function of z. Hence

$$z^{\sigma}e^{Q(z)}U(z)=z^{\rho}e^{g(z)}P(z),$$

and consequently $g(z)$ on the one hand is a polynomial in z and $U(z)$ on the other is a polynomial in z^{-1}; also

$$\sigma=\rho+k.$$

Now let

$$\frac{w'}{w}=z^{s-1}P_1,$$

where

$$\begin{aligned}P_1&=a_0+a_1z^{-1}+\ldots+a_{s-1}z^{-s+1}+O(z^{-s})\\&=v+O(z^{-s}),\end{aligned}$$

then

$$\begin{aligned}\frac{w''}{w}&=\left(\frac{w'}{w}\right)^2+\frac{d}{dz}\left(\frac{w'}{w}\right)\\&=z^{2s-2}P_1^2+(s-1)z^{s-2}P_1+z^{s-1}P_1'\\&=z^{2s-2}P_2,\end{aligned}$$

where

$$P_2=v^2+O(z^{-s}).$$

In general

$$\frac{w^{(\kappa)}}{w}=z^{\kappa(s-1)}P_\kappa,$$

where

$$P_\kappa=v^\kappa+O(z^{-s}).$$

Substitute for $w'/w, w''/w, \ldots, w^{(n)}/w$ in the differential equation, then the resulting equation

$$z^{ns}P_n+z^{(n-1)s}p_1P_{n-1}+\ldots+z^sp_{n-1}P_1+p_n=0$$

is an identity. Now the positive integer s has not been restricted; let it be taken so large that each of the polynomials $z^{(n-\kappa)s}p_\kappa$ $(\kappa=1, 2, \ldots, n-1)$ is at most of degree ns, and let

$$p_\kappa=\sum_{\nu=0}^{\kappa s}a_{\kappa,\nu}\,z^\nu.$$

Now the determining factor is

$$Q(z)=\int_0^z z^{s-1}v(z)dz,$$

and since

$$P_\kappa=v^\kappa+O(z^{-s}),$$

$v(z)$ is obtained by taking the first s terms of a root of the equation

$$(z^sv)^n+p_1(z^sv)^{n-1}+\ldots+p_{n-1}z^sv+p_n=0.$$

In particular the equation which determines a_0, the leading term in v, is

$$A(a_0)\equiv a_0{}^n+a_{1,s}a_0{}^{n-1}+a_{2,2s}a_0{}^{n-2}+\ldots+a_{n,ns}=0;$$

it will be supposed that a_0 is a simple root of this equation.

17·61. Conditions for a Normal Solution.—Let

$$w=e^{Q(z)}u,$$

then if w is a normal solution, the equation satisfied by u will admit of at least one regular solution. Now

$$\frac{d^\kappa}{dz^\kappa}e^{Q(z)}=z^{\kappa(s-1)}v_\kappa e^{Q(z)},$$

where v_κ is identical with v^κ in the terms in $z^0, z^{-1}, \ldots, z^{-s+1}$. Consequently

$$\frac{d^\nu w}{dz^\nu}=e^{Q(z)}\sum_{\kappa=0}^{\nu}\left\{\frac{\nu\,!}{\kappa\,!\,(\nu-\kappa)\,!}z^{\kappa(s-1)}v_\kappa\frac{d^{\nu-\kappa}u}{dz^{\nu-\kappa}}\right\},$$

with $v_0=1$, $v_1=v$, and therefore the differential equation for u is

$$\sum_{r=0}^{n}z^{n-r}p_r\sum_{\kappa=0}^{n-r}\left\{\frac{(n-r)\,!}{\kappa\,!\,(n-r-\kappa)\,!}z^{\kappa(s-1)}v_\kappa\frac{d^{n-r-\kappa}u}{dz^{n-r-\kappa}}\right\}=0,$$

or, as it may be written,

$$\sum_{r=0}^{n}\sum_{\nu=0}^{n-r}\frac{(n-r)\,!}{\nu\,!\,(n-r-\nu)\,!}p_r z^{(n-r-\nu)s}v_{n-r-\nu}z^\nu\frac{d^\nu u}{dz^\nu}=0,$$

with $p_0=1$.

The coefficient of u in the differential equation is

$$\sum_{r=0}^{n}p_r z^{(n-r)s}v_{n-r}.$$

But since v is obtained by taking the s leading terms of a root of the equation

$$\sum_{r=0}^{n}p_r(z^s v)^{n-r},$$

and since the s leading terms of v_{n-r} and v^{n-r} agree, it follows that the s highest terms in the coefficient of u, namely the terms in $z^{ns}, \ldots, z^{(n-1)s+1}$ must vanish, and therefore

$$\sum_{r=0}^{n}p_r z^{(n-r)s}v_{n-r}=z^{(n-1)s}(\theta_0+\theta_1 z^{-1}+\ldots),$$

where, since v is known, θ_0 is a known constant.

Likewise the coefficient of $z\dfrac{du}{dz}$ in the differential equation is

$$\sum_{r=0}^{n-1}(n-r)p_r z^{(n-r-1)}v_{n-r-1},$$

but since a_0 is assumed to be a simple root of the equation

$$\sum_{r=0}^{n}a_{r,\,rs}\,a_0^{\,n-r}=0,$$

it follows that

$$\sum_{r=0}^{n-1}(n-r)a_{r,\,rs}\,a_0^{\,n-r-1}\neq 0.$$

Consequently the leading term in the coefficient of $z\dfrac{du}{dz}$, that is to say the term in $z^{(n-1)s}$ does not vanish identically, and therefore the coefficient of $z\dfrac{du}{dz}$ is of the form

$$z^{(n-1)s}(\eta_0+\eta_1 z^{-1}+\ldots),$$

where, since v is known, η_0 is a known constant which is not zero.

Now if $u=z^{\sigma}U(z)$, where $U(z)$ is a polynomial in z^{-1}, σ must satisfy the indicial equation

$$\eta_0\sigma+\theta_0=0.$$

But $\sigma=\rho+k$, where k is a positive integer or zero, and ρ satisfies the indicial equation

$$I(\rho)\equiv[\rho]_n+a_{10}[\rho]_{n-1}+\ldots+a_{n-1,0}\,\rho+a_{n0}=0.$$

It follows that *a necessary condition for the existence of a normal solution is that the equation*

$$I(-k-\theta_0/\eta_0)=0,$$

regarded as an equation in k, should have at least one root which is a positive integer or zero.

Let it be supposed that this condition is satisfied and that κ is the corresponding value of k. Then

$$\begin{aligned}u&=z^{\sigma}(c_0+c_1z^{-1}+\ldots+c_{\kappa}z^{-\kappa})\\&=z^{\rho}(c_{\kappa}+c_{\kappa-1}z+\ldots+c_0z^{\kappa}).\end{aligned}$$

When this expression is substituted in the differential equation for u, the set of recurrence-relations

$$\begin{aligned}&c_{\kappa}I(\rho)=0,\\&c_{\kappa-1}I(\rho+1)+c_{\kappa}G_1(\rho+1)=0,\\&c_{\kappa-2}I(\rho+2)+c_{\kappa-1}G_1(\rho+2)+c_{\kappa}G_2(\rho+2)=0,\\&\quad\cdot\quad\cdot\quad\cdot\quad\cdot\end{aligned}$$

where $G_1, G_2, \ldots$ are polynomials in their arguments, must be satisfied by the coefficients $c_{\kappa}, c_{\kappa-1}, \ldots$. The first equation is satisfied independently of c_{κ}; when the value of c_{κ} is assigned the succeeding κ equations determine $c_{\kappa-1}, \ldots, c_0$. In all there are $s(n-1)$ recurrence-relations of which $k+1$ have been used; the remaining equations must now be satisfied identically in virtue of the determined values of $c_{\kappa}, \ldots, c_0$. When the aggregate of these relations is satisfied a normal solution exists.

If the equation

$$A(\alpha_0)=0$$

has more than one simple root, in respect to which all the requisite conditions are satisfied, there will be a corresponding number of normal solutions of the differential equation. The possibility of the existence of n normal solutions will now be investigated. Let $\beta_1, \beta_2, \ldots, \beta_n$ be the n distinct roots of $A(\alpha_0)=0$, and let

$$Q_r(z)=\frac{\beta_r z^s}{s}+\ldots+\gamma_r z.$$

Then if normal solutions exist, they will be of the form

$$w_r=e^{Q_r(z)}z^{\sigma_r}U_r(z),$$

where $U_r(z)$ is polynomial in z^{-1}; if κ_r is its degree, then

$$\sigma_r=\rho_r+\kappa_r,$$

where $\rho_1, \rho_2, \ldots, \rho_n$ are the roots of $I(\rho)=0$. Now

$$\begin{aligned}\sum_{r=1}^{n}(\sigma_r-\kappa_r)&=\sum_{r=1}^{n}\rho_r\\&=\tfrac{1}{2}n(n-1)-a_{10},\end{aligned}$$

and $\sum\sigma_r$ can be evaluated as follows.

Let Δ be the Wronskian of the solutions $w_1, \ldots, w_n$, then since there is no loss of generality in writing

$$w_r = e^{Q_r} z^{\sigma_r}\{1+O(z^{-1})\},$$
$$w_r' = \beta_r e^{Q_r} z^{\sigma_r+(s-1)}\{1+O(z^{-1})\},$$
$$w_r'' = \beta_r^2 e^{Q_r} z^{\sigma_r+2(s-1)}\{1+O(z^{-1})\},$$
$$\cdots\cdots,$$

the first approximation to Δ is

$$\Delta = e^{Q_1+\cdots+Q_n}\begin{vmatrix} z^{\sigma_1}, & \ldots, & z^{\sigma_n} \\ \beta_1 z^{\sigma_1+(s-1)}, & \ldots, & \beta_n z^{\sigma_n+(s-1)} \\ \beta_1^2 z^{\sigma_1+2(s-1)}, & \ldots, & \beta_n^2 z^{\sigma_n+2(s-1)} \\ \cdot & \cdot & \cdot \end{vmatrix}$$

and more exactly

$$\Delta = e^{\Sigma Q_r} z^{\Sigma\sigma_r+\frac{1}{2}n(n-1)(s-1)}\{B+O(z^{-1})\},$$

where

$$B = \begin{vmatrix} 1, & \ldots, & 1 \\ \beta_1, & \ldots, & \beta_n \\ \cdot & \cdot & \cdot \\ \beta_1^{n-1}, & \ldots, & \beta_n^{n-1} \end{vmatrix} \neq 0,$$

On the other hand,

$$\Delta = Ae^{-\int^z p_1(z)dz}$$
$$= Az^{-a_{10}}e^{-a_{11}z-\frac{1}{2}a_{12}z^2-\frac{1}{3}a_{13}z^3}\cdots.$$

Thus it is found that

$$\sum Q_r(z) = -a_{11}z - \tfrac{1}{2}a_{12}z^2 - \cdots - \tfrac{1}{n}a_{1n}z^n,$$
$$\sum \sigma_r + \tfrac{1}{2}n(n-1)(s-1) = -a_{10},$$
$$B+O(z^{-1}) = A.$$

Also

$$\sum k_r = \sum \sigma_r + a_{10} - \tfrac{1}{2}n(n-1)$$
$$= -\tfrac{1}{2}sn(n-1).$$

But since $k_1, \ldots, k_n$ are positive integers, this equation is impossible. Thus *if the numbers $\beta_1, \ldots, \beta_n$ are unequal and the numbers $\sigma_1 \ldots, \sigma_n$ are unequal, and if the numbers σ are associated with n distinct roots of the equation $I(\rho)=0$, the differential equation cannot have n normal solutions.*

On the other hand, if the numbers σ are not unequal, or if each σ is not associated with a distinct ρ, the equation

$$\sum_{r=1}^{n}(\sigma_r - k_r) = \sum_{r=1}^{n}\rho_r$$

is no longer true, and the theorem is in default. It can in fact be shown by examples that in these cases n normal solutions may possibly exist.

Now consider the case in which α_0 is a multiple root of the equation

$$A(\alpha_0)=0,$$

then $\eta_0=0$; since

$$\sigma = -\theta_0/\eta_0$$

must be finite, it is necessary for the existence of a normal solution that $\theta_0=0$. When the factor $z^{(n-1)s-1}$ has been removed, the differential equation has the form

$$\theta_1\{1+O(z^{-1})\}u + \eta_1\{1+O(z^{-1})\}z\frac{du}{dz} + \zeta_0\{1+O(z^{-1})\}z^{-s+3}\frac{d^2u}{dz^2} + \cdots = 0$$

and in general the coefficient of $z^\nu \dfrac{d^\nu u}{dz^\nu}$ is $O(z^{(1-\nu)s+1})$. Thus the indicial equation is

$$\theta_1+\eta_1\sigma+\zeta_0\sigma(\sigma-1)=0 \qquad \text{when } s=1,$$

or

$$\theta_1+\eta_1\sigma=0 \qquad \text{when } s>1.$$

A set of conditions sufficient to ensure the existence of a normal solution is obtained by continuing the investigation on the same lines as before.*

It may happen that a zero value for v is obtained so that $Q(z)$ disappears. This would happen if the solution under consideration were regular; when this is the case the solution is developed by the methods of Chapter XVI. If, however, the solution is found not to be regular, the possibility that it is of subnormal type (§ 17·53) must then be considered.

17·62. The Hamburger Equation of the Second Order.—Consider the equation

$$\frac{d^2w}{dz^2}-(a+2bz^{-1}+cz^{-2})=0\,;$$

the origin is a regular singular point relative to which the indicial equation is

$$\rho(\rho-1)=c.$$

It will be assumed that the regular solution has only a finite number of zeros in the finite part of the plane, and that the normal solution

$$w=e^{Q(z)}u(z)$$

exists, where

$$Q(z)=\frac{\alpha_0 z^s}{s}+\ .\ .\ .\ +\alpha_{s-1}z.$$

Then the equation for u is

$$u''+2Q'u'+(Q''+Q'^2-a-2bz^{-1}-cz^{-2})u=0\,;$$

in order that this equation may admit of the solution

$$u=z^\sigma(1+c_1z^{-1}+c_2z^{-2}+\ .\ .\ .\ )$$

it is necessary that

$$s=1, \quad \alpha_0{}^2=a, \quad \alpha_0\sigma=b.$$

The coefficients c_r satisfy the recurrence-relations

$$2\alpha_0c_1=\sigma(\sigma-1)-c,.$$
$$2r\alpha_0c_r=\{(\sigma-r+1)(\sigma-r)-c\}c_{r-1} \qquad (r=2,\ 3,\ 4,\ .\ .\ .)\,;$$

if the series $\sum c_rz^{-r}$ did not terminate, it would diverge for all values of $|z|$ and the solution would be illusory. Let the series terminate with $c_\kappa z^{-\kappa}$; then

$$(\sigma-\kappa)(\sigma-\kappa-1)=c.$$

It is therefore necessary that the equation

$$\left(\kappa-\frac{b}{\alpha_0}\right)\left(\kappa+1-\frac{b}{\alpha_0}\right)=c$$

should, either for $\alpha_0=+\sqrt{a}$ or for $\alpha_0=-\sqrt{a}$, have a root κ which is a positive integer or zero, and this condition is manifestly sufficient for the existence of one normal solution.

Additional conditions are necessary to ensure the existence of two normal solutions. If the two values of σ, namely

$$\sigma_1=+b/\sqrt{a}, \quad \sigma_2=-b/\sqrt{a},$$

* Günther, *J. für Math.* 105 (1889), p. 1.

are not zero, and if they are associated with distinct values of ρ, thus

$$\sigma_1=\rho_1+\kappa_1, \quad \sigma_2=\rho_2+\kappa_2,$$

then

$$\begin{aligned} 0&=\sigma_1+\sigma_2 \\ &=\rho_1+\rho_2+\kappa_1+\kappa_2=1+\kappa_1+\kappa_2, \end{aligned}$$

which is impossible since κ_1 and κ_2 are positive integers or zero.

If, on the other hand,

$$\sigma_1=\sigma_2=0,$$

that is to say, if $b=0$, and if the equation

$$\kappa(\kappa+1)=c$$

has a positive integral root, there will exist two normal solutions

$$e^{z\sqrt{a}}(1+c_1z^{-1}+\ .\ .\ .\ +c_\kappa z^{-\kappa}),$$
$$e^{-z\sqrt{a}}(1-c_1z^{-1}+\ .\ .\ .\ \pm c_\kappa z^{-\kappa}).$$

Again, if σ_1 and σ_2 are unequal, but are associated with the same value ρ_1 of ρ so that

$$\sigma_1=\rho_1+\kappa_1, \quad \sigma_2=\rho_1+\kappa_2,$$

then since

$$\sigma_1-\sigma_2=\kappa_1-\kappa_2,$$
$$\sigma_1+\sigma_2=0,$$

$2\sigma_1$ and $2\sigma_2$ must be integers, that is $2b/\sqrt{a}$ must be an integer. Also

$$\kappa_1+\kappa_2+2\rho_1=0,$$

and therefore $2\rho_1$ is a negative integer, not zero. But

$$\begin{aligned} 4c+1&=4\rho_1(\rho_1-1)+1 \\ &=(2\rho_1-1)^2, \end{aligned}$$

that is $4c+1$ is the square of an integer, not zero. These conditions are necessary and sufficient for the existence of two normal solutions.

Miscellaneous Examples.

1. Prove that the equation

$$z\frac{d^2w}{dz^2}+2(1-z)\frac{dw}{dz}-w=0$$

has two solutions normal at infinity and obtain them.

2. Prove that the equations

(i) $z^6\dfrac{d^3w}{dz^3}+6z^5\dfrac{d^2w}{dz^2}-w=0,$

(ii) $z^2(z^3+6)\dfrac{d^3w}{dz^3}+(z^3+12)\left(3z\dfrac{d^2w}{dz^2}+3\dfrac{dw}{dz}-z^2w\right)=0,$

(iii) $z^2(2z+1)\dfrac{d^3w}{dz^3}+(2z^2+9z+5)z\dfrac{d^2w}{dz^2}+(-2z^3+3z^2+6z+4)\dfrac{dw}{dz}+(-2z^2-5z+3)zw=0$

have each three solutions normal at infinity and obtain them.

3. Prove that the equation

$$4z^2\frac{d^2w}{dz^2}+8z\frac{dw}{dz}-(4z^2+12z+3)w=0$$

has one solution normal at infinity.

4. Prove that the equation

$$\frac{d^2w}{dz^2}+\frac{a}{z}\cdot\frac{dw}{dz}+bw=0$$

has two solutions normal at infinity if a is an integer or zero.

5. Prove that the equation

$$\frac{d^2w}{dz^2}-(az^2+2b+cz^{-2})w=0$$

has a normal solution if the quadratic equation

$$\left(\kappa+\tfrac{1}{2}-\frac{b}{\sqrt{a}}\right)\left(\kappa+\tfrac{3}{2}-\frac{b}{\sqrt{a}}\right)=c$$

has a positive integral or zero root for either value of $\sqrt{a}$. Consider also the two cases: (i) both roots are integers for the same value of $\sqrt{a}$; (ii) the equation has a positive integral root for both values of $\sqrt{a}$.

6. Prove that the equation

$$z\frac{d^2w}{dz^2}+\mu\frac{dw}{dz}+\lambda w=0$$

possesses two solutions of subnormal type at infinity if 2μ is an odd integer.

7. Prove that the equation

$$\frac{d^2w}{dz^2}=(2bz^{-1}+cz^{-2})w$$

has two solutions of subnormal type at infinity. Express them in terms of the solutions regular at the origin.

8. Prove that the equation

$$z^p\frac{d^3w}{dz^3}=w$$

possesses three solutions of subnormal type at infinity when

$$p=3\left(1-\frac{1}{n}\right),$$

and n is an integer not divisible by 3. Obtain them. [Halphen.]

CHAPTER XVIII

THE SOLUTION OF LINEAR DIFFERENTIAL EQUATIONS BY METHODS OF CONTOUR INTEGRATION

18·1. Extension of the Scope of the Laplace Transformation.—The general principle of the Laplace transformation was explained in an earlier section (§ 8·1) of this treatise. Let

$$L_z \equiv \sum_{r=0}^{n} \sum_{s=0}^{m} a_{rs} z^s D_z^r$$

be a differential operator in z, whose coefficients are polynomials in z of degree m at most. Then the equation

$$L_z(w)=0$$

is satisfied by

$$w(z)=\int_C e^{z\zeta} v(\zeta) d\zeta,$$

where the function $v(\zeta)$ and the contour of integration C are defined as follows.

In the first place let M_ζ be the differential operator

$$\sum_{r=0}^{n} \sum_{s=0}^{m} a_{rs} \zeta^r D_\zeta^s,$$

and let $\overline{M}_\zeta$ be its adjoint. Then $v(\zeta)$ must satisfy the differential equation

$$\overline{M}_\zeta(v)=0,$$

whose order is equal to the degree of the polynomial coefficients in the operator L. Secondly, the contour C is to be so chosen that, if $P\{e^{z\zeta}, v\}$ is the bilinear concomitant of the transformation, then

$$\Big[P\{e^{z\zeta}, v\}\Big]_C=0$$

identically.

The advantage of replacing a definite integral by a contour integral lies partly in the increased liberty in the choice of a path of integration which is thereby gained. But this in itself would not justify a separate discussion of the expression of solutions of differential equations in terms of contour integrals. The real reason why this discussion is now taken up again is that the contour integral provides a powerful instrument for investigating those solutions which are irregular at infinity, and whose developments in series diverge, and are therefore illusory. The nature of the coefficients in the equation $L_z(w)=0$ shows that the point at infinity is an irregular singular point; by means of contour integral expressions for the solutions of the equation, the behaviour of the solutions in the neighbourhood of the singularity may be investigated.

18·11. Equations whose Coefficients are of the First Degree.—In the case in which the coefficients of the given equation are of the first degree, the equation satisfied by $v(\zeta)$ is of the first degree, and therefore completely soluble. Let the given equation be

$$(a_0z+b_0)\frac{d^nw}{dz^n}+(a_1z+b_1)\frac{d^{n-1}w}{dz^{n-1}}+\ .\ .\ .\ +(a_nz+b_n)w=0.$$

Then $v(\zeta)$ will satisfy an equation of the form

$$P(\zeta)\frac{dv}{d\zeta}-Q(\zeta)v=0,$$

where $P(\zeta)$ and $Q(\zeta)$ are polynomials of degree n; * this equation may be written as

$$v^{-1}\frac{dv}{d\zeta}=\mu+\frac{\lambda_1}{\zeta-\alpha_1}+\ .\ .\ .\ +\frac{\lambda_n}{\zeta-\alpha_n},$$

where $\alpha_1, \ldots, \alpha_n$ are the zeros of $P(\zeta)$, and are supposed, for the moment, to be distinct. Then

$$v(\zeta)=e^{\mu\zeta}(\zeta-\alpha_1)^{\lambda_1}\ .\ .\ .\ (\zeta-\alpha_n)^{\lambda_n}.$$

The bilinear concomitant is found to be

$$(a_0\zeta^n+a_1\zeta^{n-1}+\ .\ .\ .\ +a_n\zeta+a_n)v(\zeta)e^{z\zeta},$$

and therefore the contour integral

$$\int_C e^{z\zeta}v(\zeta)d\zeta$$

will satisfy the given differential equation provided that the contour C (which must be independent of z) is so chosen that

$$\left[(\zeta-\alpha_1)^{1+\lambda_1}\ .\ .\ .\ (\zeta-\alpha_n)^{1+\lambda_n}e^{(\mu+z)\zeta}\right]_C=0$$

identically with respect to z.

Let the real parts of λ_1 and λ_2 be greater than -1, then

$$w=\int_{\alpha_1}^{\alpha_2}e^{z\zeta}v(\zeta)d\zeta$$

will be a solution of the equation if the integration is taken over any simple curve of finite length joining α_1 to α_2, but remaining always at a finite distance from any point α_r for which the real part of the index $1+\lambda_r$ is negative or zero.

If the real parts of $\lambda_1, \ldots, \lambda_n$ are all greater than -1, there will be $n-1$, but no more, distinct integrals of the above type, each of which satisfies the given equation.

Now consider the case in which the numbers λ are unrestricted. For simplicity, each of the points $\alpha_1, \ldots, \alpha_n$ will be considered to be at a finite distance from the origin. Then the contour will be that formed by the aggregate of four loops described in succession such that each loop begins and ends at the origin and encloses one and only one singular point. For instance, let the first loop pass round α_1 in the positive direction, the second round α_2 in the positive direction, the third round α_1 in the negative direction, and the fourth round α_2 in the negative direction. The function

$$(\zeta-\alpha_1)^{1+\lambda_1}\ .\ .\ .\ (\zeta-\alpha_n)^{1+\lambda_n}e^{(\mu+z)\zeta}$$

returns to its initial value after this circuit has been described and therefore the contour is appropriate. In this way $n-1$ distinct integrals may be formed, which satisfy the given equation.

* Note that $P(\zeta)$ is a constant multiple of $a_0\zeta^n+a_1\zeta^{n-1}+\ldots+a_{n-1}\zeta+a_n$; in order that $P(\zeta)$ may not be of lower degree than n it will be supposed that $a_0\neq0$. The point at infinity is then an irregular singularity for w of rank unity.

A set of n distinct contour integrals which satisfy the equation cannot be obtained without some restriction on z. Suppose, for instance, that

$$\mathbf{R}(z+\mu)>0,$$

then a suitable contour is described when the point ζ moves from $-\infty$ along the line drawn through a_1 parallel to the real axis until it reaches a distance r from a_1, describes a circle about a_1 in the negative direction, and then retraces its rectilinear path. It is of course supposed that every point on the loop is at a finite distance from $a_2, \ldots, a_n$. In general, n integrals of this type will exist.

18·12. Discussion of the Integral when $\mathbf{R}(z)$ is large.—Consider the integral

$$\int_C (\zeta-a_1)^{\lambda_1} \ldots (\zeta-a_n)^{\lambda_n} e^{(z+\mu)\zeta} d\zeta,$$

for large values of $\mathbf{R}(z)$. There is no loss in generality in taking μ to be zero, which amounts to replacing $z+\mu$ by z, nor in taking a_1 to be zero which amounts to replacing $\zeta-a_1$ by ζ and putting aside the factor $e^{a_1 z}$ which will subsequently be restored. The contour is then composed of the three following parts:

(i) the real axis from $-\infty$ to $-r$,

(ii) the circle γ of radius r described in the negative direction about the origin, and

(iii) the real axis from $-r$ to $-\infty$.

Let I_1, I_2, I_3 be the respective contributions of these three paths to the integral. Now when ζ is real, and $\zeta<-r$ a positive number κ can be found such that

$$|\zeta^{\lambda_1}(\zeta-a_2)^{\lambda_2} \ldots (\zeta-a_n)^{\lambda_n}|<e^{-\kappa\zeta},$$

and therefore, when $\mathbf{R}(z)>\kappa$

$$|I_1|<\int_{-\infty}^{-r} e^{\zeta(z-\kappa)}d\zeta=\frac{e^{-r(z-\kappa)}}{z-\kappa}.$$

Consequently

$$I_1 \to 0 \quad \text{as} \quad \mathbf{R}(z)\to+\infty,$$

and similarly

$$I_3 \to 0 \quad \text{as} \quad \mathbf{R}(z)\to+\infty.$$

There remains the integral I_2, taken round the circle γ of arbitrarily small radius r. The first part of the integrand may be expanded thus:

$$\zeta^{\lambda_1}(\zeta-a_2)^{\lambda_2} \ldots (\zeta-a_n)^{\lambda_n}=\zeta^{\lambda_1}\{A_0+A_1\zeta+ \ldots +A_m\zeta^m+R\zeta^{m+1}\},$$

where $A_0, A_1, \ldots, A_m$ are constants. Let M be the upper bound of

$$(\zeta-a_2)^{\lambda_2} \ldots (\zeta-a_n)^{\lambda_n}$$

when $|\zeta|\leqslant r$. Then there will be a positive number ρ such that

$$|A_0|\leqslant M, \quad |A_1|\leqslant M\rho, \ldots, |A_m|\leqslant M\rho^m,$$

and consequently

$$|R|<M\frac{\rho^{m+1}}{1-r\rho}.$$

Now

$$I_2=\int_\gamma \{A_0+A_1\zeta+ \ldots +A_m\zeta^m\}\zeta^{\lambda_1}e^{z\zeta}d\zeta+\int_\gamma R\zeta^{\lambda_1+m+1}e^{z\zeta}d\zeta;$$

the problem is to determine the behaviour of I_2 as $\mathbf{R}(z)\to+\infty$. No essential point will, however, be lost by restricting z to be real.

Consider the integral

$$z^{p+1}\int_{\gamma} \zeta^p e^{z\zeta} d\zeta,$$

where z is large and real; let $z\zeta = -t$, then the integral becomes

$$\int_{\kappa} (-t)^p e^{-t} dt,$$

where κ is a circle of radius rz described in the positive direction about the origin in the t-plane, and the value of the integral therefore is *

$$2i \sin p\pi \Gamma(p+1).$$

This quantity is finite except when p is a positive integer.

Consequently, as $z \rightarrow +\infty$ along the real axis,

$$z^{\lambda_1+1}\int_{\gamma} \zeta^{\lambda_1} e^{z\zeta} d\zeta \rightarrow 2i \sin \lambda_1 \pi \Gamma(\lambda_1+1),$$

$$z^{\lambda_1+1}\int_{\gamma} \zeta^{\lambda_1+\nu} e^{z\zeta} d\zeta \rightarrow 0 \qquad (\nu=1, 2, \ldots, m),$$

and

$$\left| z^{\lambda_1+1}\int_{\gamma} R\zeta^{\lambda_1+m+1} e^{z\zeta} d\zeta \right| < 2\, |z^{-m-1}|\, M \frac{\rho^{m+1}}{1-r\rho} \Gamma(\lambda_1+m+2) \rightarrow 0,$$

since r can be so chosen that $r\rho<1$. Hence

$$I_2 = O(z^{-\lambda_1-1}),$$

except when λ_1 is a positive integer.

The factor $e^{a_1 z}$ was temporarily discarded, on restoring this factor it is seen that the integral considered approaches the limit

$$K_1 e^{a_1 z} z^{-\lambda_1-1},$$

where K_1 is finite and not zero, as z approaches $+\infty$ along the real axis.

18·13. Existence of n linearly Distinct Integrals.—Let it be supposed that the numbers

$$a_1, a_2, \ldots, a_n$$

are arranged so that their real parts form a decreasing sequence. In order that the loop corresponding to each point a may be drawn, it is necessary to suppose that the imaginary parts of these numbers are all unequal. When that is the case there will be an integral corresponding to each a; let these integrals be respectively

$$w_1, \quad w_2, \quad \ldots, \quad w_n.$$

These integrals are linearly distinct, for if this were not the case, there would be an identical relation of the form

$$C_1 w_1 + C_2 w_2 + \ldots + C_n w_n = 0.$$

But as $z \rightarrow +\infty$,

$$\lim w_1 e^{-a_1 z} z^{\lambda_1+1} = K_1,$$

$$\lim w_\nu e^{-a_1 z} z^{\lambda_1+1} = \lim K_\nu e^{-(a_1-a_\nu)z} z^{\lambda_1-\lambda_\nu}$$

$$= 0 \qquad (\nu=2, 3, \ldots, n),$$

since the real part of $a_1 - a_\nu$ is positive.† Consequently the relation cannot hold unless $C_1=0$. In the same way $C_2, \ldots, C_n$ are zero, and therefore no such linear relation exists.

* Whittaker and Watson, *Modern Analysis*, § 12·22.

† When the real parts of any two or more successive numbers a are equal the theorem is still true, but the proof of this fact is much more difficult.

18·14. The Case in which $P(\zeta)$ has Repeated Linear Factors.—This case will be illustrated by considering the effect of two of the numbers a, say a_1 and a_2, becoming equal. In this case

$$v^{-1}\frac{dv}{d\zeta}=\mu+\frac{\lambda_1}{\zeta-a_1}+\frac{\kappa_1}{(\zeta-a_1)^2}+\frac{\lambda_3}{\zeta-a_3}+\ldots+\frac{\lambda_n}{\zeta-a_n},$$

so that

$$v=e^{\mu\zeta-\frac{\kappa_1}{\zeta-a_1}}(\zeta-a_1)^{\lambda_1}\ldots(\zeta-a_n)^{\lambda_n}.$$

In order to obtain the full complement of integrals, two distinct contours relative to the point a_1 must be obtainable. One suitable contour is the loop which has been discussed; the real interest of this case lies in the second contour. It has been seen that when a_1 and a_2 are distinct there exists a suitable contour which enlaces these two points and does not proceed to infinity. The contour which now provides the second integral relative to the point a_1 is in reality a limiting case of the contour enlacing the two points a_1 and a_2 which now coincide.

In the present case the bilinear concomitant is

$$e^{-\frac{\kappa_1}{\zeta-a_1}}(\zeta-a_1)^{1+\lambda_1}\ldots(\zeta-a_n)^{1+\lambda_n}e^{(\mu+z)\zeta}.$$

An appropriate contour would be a closed curve starting from a_1 in a certain direction and returning to a_1 with a different direction. In other words it would be a contour whose gradient is discontinuous at a_1. Moreover, it must be such that as ζ approaches a_1, in either of these directions, the bilinear concomitant must tend to zero.

Let

$$\zeta-a_1=\rho e^{i\phi},\quad \kappa_1=ke^{i\beta},$$

so that

$$\frac{\zeta-a_1}{\kappa_1}=\frac{\rho}{k}e^{i(\phi-\beta)};$$

then

$$\left|e^{-\frac{\kappa_1}{\zeta-a_1}}\right|=e^{-\frac{k}{\rho}\cos(\phi-\beta)}.$$

Now $\rho\to 0$ as $\zeta\to a_1$. In order, therefore, that this exponential factor may tend to zero as ζ approaches a_1, it is necessary that $\cos(\phi-\beta)$ should be positive or that

$$\beta-\tfrac{1}{2}\pi<\phi<\beta+\tfrac{1}{2}\pi.$$

Thus all possible directions of approach to a_1 will lie on one side of the straight line drawn through a_1 in the direction β.

18·15. The Equation with Constant Coefficients.—When the method of Laplace is applied to the equation

$$a_0\frac{d^nw}{dz^n}+a_1\frac{d^{n-1}w}{dz^{n-1}}+\ldots+a_{n-1}\frac{dw}{dz}+a_nw=0,$$

in which the coefficients a are constant, it appears to break down. For if the equation is satisfied by an integral such as

$$w=\int_C e^{z\zeta}v(\zeta)d\zeta,$$

the condition to be satisfied is simply that

$$\int_C e^{z\zeta}v(\zeta)\phi(\zeta)d\zeta=0$$

identically, where

$$\phi(\zeta)=a_0\zeta^n+a_1\zeta^{n-1}+\ldots+a_{n-1}\zeta+a_n.$$

Thus there is no differential equation to be satisfied by $v(\zeta)$; the only condition to be fulfilled is that the function

$$f(\zeta)=v(\zeta)\phi(\zeta)$$

should be analytic in a region of the ζ-plane. The contour C may then be taken to be in that region.

Consider, for example, the case in which $\zeta=a$ is a root of multiplicity m of the characteristic equation

$$\phi(\zeta)=0,$$

and let the contour C enclose this root, but no other root of the equation. Choose $f(\zeta)$ so that

$$v(\zeta)=\frac{A_1}{(\zeta-a_1)^m}+\frac{A_2}{(\zeta-a_1)^{m-1}}+\ldots+\frac{A_m}{\zeta-a_1}+\psi(\zeta),$$

where $\psi(\zeta)$ is analytic within C; the constants A depend on the choice of $f(\zeta)$ and are therefore arbitrary.

Then

$$w=\int_C\left\{\frac{A_1}{(\zeta-a_1)^m}+\frac{A_2}{(\zeta-a_1)^{m-1}}+\ldots+\frac{A_m}{\zeta-a_1}+\psi(\zeta)\right\}e^{z\zeta}d\zeta.$$

When this integral is evaluated, w is found to be of the form (cf. § 6·12)

$$e^{a_1z}\{C_1+C_2z+\ldots+C_mz^{m-1}\}.$$

18·2. Discussion of the Laplace Transformation in the more general Case.—The restriction that the coefficients of the differential equation are of the first degree will now be abandoned. Let the equation be

$$L(w)\equiv P_0(z)\frac{d^nw}{dz^n}+P_1(z)\frac{d^{n-1}w}{dz^{n-1}}+\ldots+P_{n-1}(z)\frac{dw}{dz}+P_n(z)w=0,$$

in which $P_0(z)$ is a polynomial in z of degree p, and the remaining coefficients are polynomials of degrees not exceeding p.* Let

$$P_0(z)=a_0z^p+\ldots \qquad (a_0\neq 0),$$
$$P_r(z)=a_rz^p+\ldots \qquad (r=1, 2, \ldots, n).$$

Then if

$$W=\int_C e^{z\zeta}v(\zeta)d\zeta,$$

$$L(W)=\int_C\{a_0z^p\zeta^n+\ldots+a_{n-r}z^p\zeta^r+\ldots\}e^{z\zeta}v(\zeta)d\zeta.$$

By repeated integration by parts it may be verified that

$$\begin{aligned}\int_C z^p\zeta^rve^{z\zeta}d\zeta&=[z^{p-1}\zeta^rve^{z\zeta}]-\int_C z^{p-1}e^{z\zeta}\frac{d(\zeta^rv)}{d\zeta}d\zeta\\&=[z^{p-1}\zeta^rve^{z\zeta}]-\left[z^{p-2}\frac{d(\zeta^rve^{z\zeta})}{d\zeta}\right]+\int_C z^{p-2}e^{z\zeta}\frac{d^2(\zeta^rv)}{d\zeta^2}d\zeta,\\&=\ \ldots\\&=[R_p]+(-1)^p\int_C e^{z\zeta}\frac{d^p(\zeta^rv)}{d\zeta^p}d\zeta.\end{aligned}$$

Consequently

$$L(W)=[R]+(-1)^p\int_C\left\{a_0\frac{d^p(\zeta^nv)}{d\zeta^p}+a_1\frac{d^{p-1}(\zeta^{n-1}v)}{d\zeta^{p-1}}+\ldots\right\}e^{z\zeta}d\zeta,$$

* The rank of the singular point at infinity is therefore at most unity.

where R is a series of terms of the form

$$e^{z\zeta}\frac{d^s(\zeta^r v)}{d\zeta^s} \qquad \begin{pmatrix} s=0, 1, 2, \ldots, p-1 \\ r=0, 1, 2, \ldots, n \end{pmatrix},$$

whose coefficients are polynomials in z alone.

Thus if the integral W is a solution of the given differential equation it is necessary that $v(z)$ should satisfy the differential equation

$$(a_0\zeta^n+a_1\zeta^{n-1}+\ldots+a_n)\frac{d^pv}{d\zeta^p}+\ldots=0.$$

Consequently the determination of a contour integral which satisfies the given differential equation depends upon (i) the solution of the associated equation of order p, and (ii) the determination of the contour C so that $[R]$ is zero identically in z. It will now be proved that n distinct integrals of the type considered do in fact exist.*

Let $\alpha_1, \alpha_2, \ldots, \alpha_n$ be the roots of the equation

$$a_0\zeta^n+a_1\zeta^{n-1}+\ldots+a_n=0,$$

then $\zeta=\alpha_1, \alpha_2, \ldots, \alpha_n$ are the singular points of the differential equation in v. Now each of these singular points is regular and, moreover, relative to each singularity α_r there are $n-1$ solutions which are analytic in the neighbourhood of that singularity and one non-analytic solution of the form

$$v=(\zeta-\alpha_r)^{\lambda_r}\phi_r(\zeta),$$

where $\phi_r(\zeta)$ is analytic near α_r. Consider this non-analytic solution v. Let z tend to infinity along a straight line l drawn in the negative direction parallel to the axis of reals. Then, by an unimportant modification of the theorem of Liapounov (§ 6·6) it follows that a positive number μ exists such that

$$ve^{\mu\zeta}, \quad \frac{dv}{d\zeta}e^{\mu\zeta}, \quad \ldots, \quad \frac{d^nv}{d\zeta^n}e^{\mu\zeta}$$

tend to zero as $\mathbf{R}(\zeta)\to-\infty$. The same is evidently true. whatever ν may be, with regard to

$$v\zeta^\nu e^{\mu\zeta}, \quad \frac{dv}{d\zeta}\zeta^\nu e^{\mu\zeta}, \quad \ldots, \quad \frac{d^nv}{d\zeta^n}\zeta^\nu e^{\mu\zeta}.$$

If, therefore, $\mathbf{R}(z)$ is positive and sufficiently large and the contour C is a loop beginning and ending at the point at infinity on the line l and encircling the point α_r, $[R]$ will vanish independently of z.

Thus there will exist n integrals

$$W_1, \quad W_2, \quad \ldots, \quad W_n,$$

such that the integral W_r corresponds to the point α_r. Moreover, as in § 18·2, it follows that when $\lambda_1, \ldots, \lambda_n$ are not positive integers

$$W_1e^{-\alpha_1 z}z^{\lambda_1+1}, \quad \ldots, \quad W_ne^{-\alpha_n z}z^{\lambda_n+1}$$

tend to non-zero limits as z approaches $+\infty$ along the real axis. Thus the earlier discussion virtually also covers the more general case.

18·21. Asymptotic Representations.—The contour integrals obtained in the preceding section lead directly to asymptotic representations of the solutions which they represent.† It follows as in § 18·12 that if W represents the typical contour integral

$$\int_C e^{z\zeta}\zeta^\lambda\phi(\zeta)d\zeta,$$

* Poincaré, *Am. J. Math.* 7 (1885), p. 217.

† Poincaré, *Acta Math.* 8 (1886), p. 295.

then

$$We^{-az}z^{\lambda+1}=z^{\lambda+1}\int_C\{A_0+A_1\zeta+\ .\ .\ .\ +A_m\zeta^m\}\zeta^\lambda e^{z\zeta}d\zeta+z^{\lambda+1}\int_C R\zeta^{\lambda+m+1}e^{z\zeta}d\zeta.$$

Along the rectilinear parts of the contour the integral itself, and the product of the integral by any arbitrary power of z, tend to zero as $\mathbf{R}(z)\to+\infty$. The important part of the contour is the small circle γ encircling the origin in the negative direction. Now if r is a positive integer,

$$z^{\lambda+r+1}\int_\gamma \zeta^{\lambda+r}e^{z\zeta}d\zeta=(-1)^r 2i\sin\lambda\pi\,\Gamma(\lambda+r+1),$$

and

$$\left|z^{\lambda+1}\int_\gamma R\zeta^{\lambda+m+1}e^{z\zeta}d\zeta\right|<K|z^{-m-1}|,$$

where K is independent of z.

Let

$$S_m=2i\sin\lambda\pi\{A_0\Gamma(\lambda+1)-A_1\Gamma(\lambda+2)z^{-1}+\ .\ .\ .\ \pm A_m\Gamma(\lambda+m+1)z^{-m}\},$$

then

$$z^m(We^{-az}z^{\lambda+1}-S_m)\to 0$$

as $z\to\infty$ along the real axis. Consequently $e^{az}z^{-\lambda-1}S_m$ *is an asymptotic representation of the integral* W, that is

$$W\sim 2i\sin\lambda\pi\,e^{az}\left\{\frac{A_0\Gamma(\lambda+1)}{z^{\lambda+1}}-\frac{A_1\Gamma(\lambda+2)}{z^{\lambda+2}}+\ .\ .\ .\pm\frac{A_m\Gamma(\lambda+m+1)}{z^{\lambda+m+1}}\mp\ .\ .\ .\right\}.$$

The asymptotic series is formally identical with the series obtained in examining the equation for the presence of a normal solution. Thus when the normal series does not terminate and furnish a normal solution it furnishes an asymptotic representation of a solution.

In the preceding investigation it has been supposed that z tended to infinity along the real axis. This is a restriction adopted merely for the sake of simplicity; there is no essential difference in the case in which z tends to infinity along any ray of definite argument. The series S_m cannot be an asymptotic representation of the *same* function $We^{-az}z^{\lambda+1}$ for all values of the argument, for if

$$z^m\{We^{-az}z^{\lambda+1}-S_m\}$$

were to tend *uniformly* to zero for sufficiently large values of $|z|$, $We^{-az}z^{\lambda+1}$ would be analytic, and the series representation would converge, which, at least in the general case, is untrue. What actually happens is that as arg z increases, the solution which S_m asymptotically represents changes abruptly. Thus when a solution is developed asymptotically it is essential to specify the limits of arg z between which the representation is valid.*

18·3. Equations of Rank greater than Unity: Indirect Treatment.—In the preceding sections an explicit solution of equations of rank unity was obtained by means of the Laplace integral. The restriction that the rank should not exceed unity is essential; when the equation is of rank greater than unity the method breaks down entirely. It will now be shown that an equation of grade s greater than unity can be replaced by an equation of unit grade and rank which in turn lends itself to treatment by the Laplace integral.† A more direct method of procedure will be given in a later section.‡

* See the example of § 18·61 below and compare §§ 19·5, 19·6.

† Poincaré, *Acta Math.* 8 (1886), p. 328, originated the method and discussed in detail the case of an equation of grade 2. Horn, *Acta Math.* 23 (1900), p. 171, continued the discussion in the case of an equation of the second order and of rank p.

‡ § 18·31; see also §§ 19·41, 19·42.

Let the equation be

$$P_0 \frac{d^n w}{dz^n} + P_1 \frac{d^{n-1} w}{dz^{n-1}} + \dots + P_{n-1} \frac{dw}{dz} + P_n w = 0,$$

in which the coefficients are polynomials in z; let P_r be of degree K_r. Then if the equation possesses solutions which are normal and of grade s at infinity,

$$K_r \leqslant K_0 + r(s-1)$$

and the sign of equality holds at least once for $r \geqslant 1$.

Let

$$w_1(z), \quad w_2(z), \quad \dots, \quad w_n(z)$$

be n independent normal solutions and let

$$\omega = e^{\frac{2\pi i}{s}}.$$

Form all possible products, each of s factors, such as

$$v = w_\alpha(z) w_\beta(\omega z) \dots w_\mu(\omega^{s-1} z),$$

where the suffixes $\alpha, \beta, \dots, \mu$ may assume any of the values $1, 2, \dots, n$. The number N of distinct products is n^s, and the products satisfy a differential equation of the type

$$Q_0 \frac{d^N v}{dz^N} + Q_1 \frac{d^{N-1} v}{dz^{N-1}} + \dots + Q_{N-1} \frac{dv}{dz} + Q_N v = 0,$$

whose coefficients are polynomials in z. Now v is a normal solution of grade s, and therefore, if θ_r is the degree of Q_r,

$$\theta_r \leqslant \theta_0 + r(s-1).$$

If z is replaced by $\omega z, \omega^2 z, \dots$, or $\omega^{s-1} z$, the products v are permuted among themselves and therefore the equation remains unaltered.

Thus a number m can be found such that

$$Q_r(z) = z^{m-r} q_r(z^s) \qquad (r = 0, 1, \dots, N),$$

where $q_r(z^s)$ is a polynomial in z^s. The equation in v may therefore be written

$$q_0(z^s) z^N \frac{d^N v}{dz^N} + \dots + q_{N-1}(z^s) z \frac{dv}{dz} + q_N(z^s) v = 0.$$

Now let $z^s = \zeta$, then $z^r \frac{d^r v}{dz^r}$ is a linear expression in

$$\zeta^r \frac{d^r v}{dz^r}, \quad \zeta^{r-1} \frac{d^{r-1} v}{dz^{r-1}}, \quad \dots, \quad \zeta \frac{dv}{d\zeta}$$

with constant coefficients. The equation therefore becomes

$$R_0 \frac{d^N v}{d\zeta^N} + R_1 \frac{d^{N-1} v}{d\zeta^{N-1}} + \dots + R_{N-1} \frac{dv}{d\zeta} + R_N v = 0,$$

where the coefficients are polynomials in ζ.

If η_r is the degree of q_r in z^s,

$$\theta_r = m - r + s\eta_r,$$

and therefore

$$\eta_r \leqslant \eta_0 + r.$$

Now the degree of R_r is the degree of the highest term in

$$\zeta^N q_0(\zeta), \quad \zeta^{N-1} q_1(\zeta), \quad \dots, \quad \zeta^{N-r} q_r(\zeta),$$

and in general is the greatest of the numbers

$$N + \eta_0, \quad N - 1 + \eta_1, \quad \dots, \quad N - r + \eta_r,$$

that is $N + \eta_0$. Consequently the degree of each of the coefficients $R_1, \dots, R_N$

is at most equal to the degree of R_0, and therefore the equation is of unit rank, and v can be expressed in the form of a Laplace integral.

It remains to deduce w from v. Let

$$w=\phi_1(z)$$

be the solution aimed at and write

$$\phi_r(z)=\phi_1(z\omega^{r-1}) \qquad (r=1, 2, \ldots, s),$$

then the equation in v is satisfied by the product

$$v=\phi_1(z)\phi_2(z) \ldots \phi_s(z).$$

Form the first N derivatives of v; since $\phi_1^{(N)}(z)$ and higher derivatives are expressible in terms of the first $N-1$ derivatives, there will be in all $N+1$ equations of the form

$$\frac{d^r v}{dz^r}=\sum Z_{r, \alpha, \beta, \ldots, \mu} \frac{d^\alpha \phi_1}{dz^\alpha} \cdot \frac{d^\beta \phi_2}{dz^\beta} \ldots \frac{d^\mu \phi_s}{dz^\mu}$$

$(r=0, 1, \ldots, N)$, where the coefficients Z are rational functions of z. When the N products

$$\frac{d^\alpha \phi_1}{dz^\alpha} \cdot \frac{d^\beta \phi_2}{dz^\beta} \ldots \frac{d^\mu \phi_s}{dz^\mu}$$

are eliminated determinantally from these equations, the differential equation of order N in v is obtained.

Now consider only the first N equations, in which r has in succession the values $0, 1, \ldots, N-1$. From these equations any of the N products may, in general, be expressed in terms of $v, v', \ldots, v^{(N-1)}$. In particular

$$\phi_1(z)\phi_2(z) \ldots \phi_s(z)=v, \quad \phi_1'(z)\phi_2(z) \ldots \phi_s(z)=\Phi,$$

where Φ is a linear expression in $v, v', \ldots, v^{(N-1)}$ whose coefficients are rational functions of z.* Hence

$$\frac{\phi_1'(z)}{\phi_1(z)}=\frac{\Phi}{v},$$

and thus when v is known $w=\phi_1(z)$ is obtained by a quadrature.

18·301. An Example of the Reduction to Unit Rank.—Consider the equation

$$z\frac{d^2w}{dz^2}-(z^3+1)w=0,$$

which is of rank 2 with respect to the point at infinity. If $w=\phi(z)$ is a solution, $w_1=\phi(-z)$ satisfies the equation

$$z\frac{d^2w_1}{dz^2}-(z^3-1)w_1=0.$$

Let $v=ww_1$, then

$$v'=w'w_1+ww_1',$$

$$v''=w''w_1+2w'w_1'+ww_1''$$

$$=2z^3ww_1+2w'w_1',$$

$$v'''=4zww_1+\left\{4z^2+\frac{2}{z}\right\}ww_1'+\left\{4z^2-\frac{2}{z}\right\}w'w_1,$$

$$v^{\mathrm{IV}}=\left\{8z^4+4-\frac{4}{z^2}\right\}ww_1+\left\{12z-\frac{2}{z^2}\right\}ww_1'+\left\{12z+\frac{2}{z^2}\right\}w'w_1+8z^2w'w_1'.$$

* The case in which the determinant of the coefficients Z vanishes is dealt with by Poincaré in the memoir quoted. In this case Φ is not rational but algebraic in $z, v, v', \ldots, v^{(N-1)}$.

By eliminating ww_1, $w'w_1$, ww_1' and $w'w_1'$ it is found that the equation satisfied by v is

$$z^2\frac{d^4v}{dz^4}+z\frac{d^3v}{dz^3}-4z^4\frac{d^2v}{dz^2}-16z^3\frac{dv}{dz}-(8z^4-4)v=0,$$

and is of rank 2. But when it is transformed by the substitution $z^2=\zeta$ it becomes

$$4\zeta^3\frac{d^4v}{d\zeta^4}+14\zeta^2\frac{d^3v}{d\zeta^3}-(4\zeta^3+6\zeta)\frac{d^2v}{d\zeta^2}-10\zeta^2\frac{dv}{d\zeta}-(2\zeta-1)v=0,$$

and is of rank 1.

18·31. Equations of Rank greater than Unity : Direct Treatment.—When an equation is of rank p, greater than unity, the integral representation of solutions which replaces the Laplace integral is of the form *

$$w=\int\ldots\int e^{\zeta_1 z+\frac{1}{2}\zeta_2 z^2+\ldots+\zeta_p z^p/p}Zd\zeta_1\ldots d\zeta_p,$$

where Z is a function of $\zeta_1, \ldots, \zeta_p$. The problem of this representation will now be studied in the case of $p=2$; the more general case presents much complexity but no additional difficulty.

Let the equation be

$$L(w)\equiv p_0(z)\frac{d^nw}{dz^n}+p_1(z)\frac{d^{n-1}w}{dz^{n-1}}+\ldots+p_{n-1}(z)\frac{dw}{dz}+p_n(z)w=0,$$

where the coefficients are polynomials in z and the degree of $p_r(z)$ exceeds that of $p_0(z)$ by r. Let $p_0(z)$ be of even degree λ † and let $\lambda+2n=2m$.

Now consider the possibility of satisfying the equation by a double integral of the form

$$w=\iint e^{tz+\frac{1}{2}uz^2}Ududt,$$

in which U is a function, to be determined, of u and t, and the u- and t-contours are independent of z. Then

$$\frac{dw}{dz}=\iint e^{tz+\frac{1}{2}uz^2}(t+uz)Ududt,$$

$$\frac{d^2w}{dz^2}=\iint e^{tz+\frac{1}{2}uz^2}\{(t+uz)^2+u\}Ududt,$$

and, in general,

$$\frac{d^rw}{dz}=\iint e^{tz+\frac{1}{2}uz^2}\omega_rUdudt,$$

where

$$\omega_r=\frac{\partial\omega_{r-1}}{\partial z}+\omega_1\omega_{r-1}.$$

It will be observed that ω_r is a polynomial of the form

$$(t+uz)^r+\frac{r(r-1)}{2}u(t+uz)^{r-2}+\frac{r(r-1)(r-2)(r-3)}{8}u^2(t+uz)^{r-4}+\ldots;$$

in general the coefficient of $(t+uz)^{r-\nu}$ is 0 when ν is odd, and a constant multiple of $u^{\frac{1}{2}\nu}$ when ν is even.

Thus

$$z^nL(w)=\iint e^{tz+\frac{1}{2}uz^2}\Pi(t, u, z)Ududt,$$

where

$$\Pi(t, u, z)=z^n\{\omega_np_0+\omega_{n-1}p_1+\ldots+\omega_1p_{n-1}+p_n\}.$$

* Cunningham, *Proc. London Math. Soc.* (2), 4 (1906), p. 374. It should be noted that Cunningham's definition of *rank* differs from that now accepted.

† If $p_0(z)$ is of odd degree, multiply the left-hand member of the equation by z.

Now Π is a polynomial in z of degree $2m=\lambda+2n$. Let a_{rs} be the coefficient of $z^{\lambda+r-s}$ in p_r and let B_s be the coefficient of z^{2m-s} in Π. Then

$$B_0=a_{00}u^n+a_{10}u^{n-1}+\ .\ .\ .\ +a_{n0},$$
$$B_1=t\{na_{00}u^{n-1}+(n-1)a_{10}u^{n-2}+\ .\ .\ .\ +a_{n-1,\,0}\}$$
$$+\{a_{01}u^n+a_{11}u^{n-1}+\ .\ .\ .\ +a_{n1}\},$$

and in general

$$B_r=t^r\{{}_nC_ra_{00}u^{n-r}+{}_{n-1}C_ra_{10}u^{n-r-1}+\ .\ .\ .\}+\sum_{s=1}^{r}t^{r-s}B_{rs}(u),$$

where $B_{rs}(u)$ is a polynomial in u of degree $n-r+s$ at most.

Thus

$$B_r=\frac{t^r}{r\,!}\frac{\partial^r B_0}{\partial u^r}+\sum_{s=1}^{r}t^{r-s}B_{rs}(u).$$

Now single integrations with respect to u and t give

$$\int\!\!\int e^{tz+\frac{1}{2}uz^2}z^rU\,dudt=\int e^{tz}\Big[2e^{\frac{1}{2}uz^2}z^{r-2}U\Big]_u dt-2\int\!\!\int e^{tz+\frac{1}{2}uz^2}z^{r-2}\frac{\partial U}{\partial u}\,dudt$$
$$=\int e^{\frac{1}{2}uz^2}\Big[e^{tz}z^{r-1}U\Big]_t du-2\int\!\!\int e^{tz+\frac{1}{2}uz^2}z^{r-1}\frac{\partial U}{\partial t}\,dudt,$$

where the brackets denote the difference between the final and initial values after description of the u- or t- contour as the case may be. The single integrals containing these brackets will be referred to as the semi-integrated terms.

This reduction is repeatedly applied to $z^nL(w)$ so that the latter is reduced finally into the form

$$\int\!\!\int e^{tz+\frac{1}{2}uz^2}M(U,\,u,\,t)dudt+[R],$$

where $[R]$ denotes an aggregate of semi-integrated terms. Thus in order that the integral considered may satisfy the differential equation, it is necessary firstly, that $U(u,\,t)$ should satisfy the *partial differential equation*

$$M(U,\,u,\,t)=0,$$

and secondly, that the contours be so chosen that $[R]$ vanishes identically. When these conditions are satisfied and the integral exists, it furnishes a solution of the given equation.

The highest power of z in $\Pi(t,\,u,\,z)$ is z^{2m}, and this may be reduced by m successive integrations with respect to u, thus contributing to $M(U,\,u,\,t)$ the term

$$(-2)^m\frac{\partial^m(B_0U)}{\partial u^m}.$$

In the same way the term in z^{2m-1} is reduced by $m-1$ integrations with respect to u and one integration with respect to t and contributes the term

$$-(-2)^{m-1}\frac{\partial^m(B_1U)}{\partial u^{m-1}\partial t}.$$

The remaining terms may be reduced in the same manner; if a sufficient number of integrations with respect to u is made, no partial differential coefficient need be of order exceeding m.*

* It may be observed that the equation $M(U,\,u,\,t)=0$ is not uniquely determined, for in reducing the later terms there is a certain freedom of choice as to when integrations are made with respect to u and when with respect to t.

The partial differential equation satisfied by U is therefore of the form

$$B_0\frac{\partial^m U}{\partial u^m}=\sum A_{rs}\frac{\partial^{r+s}U}{\partial u^r dt^s}\quad (r=0,\ 1,\ .\ .\ .,\ m-1\ ;\ r+s\leqslant m),$$

where the coefficients A_{rs} are polynomials in u and t. Let $u=a$ be a non-repeated root * of the equation $B_0=0$. Then as in the case of an ordinary linear equation, the point $u=a$ is in general a singularity of the solution of the partial differential equation. It will now be shown that this equation admits of a solution expressible as a convergent double series.

18·32. Determination of the Function U.—Since $u=a$ is a simple root of

$$a_{00}u^n+a_{10}u^{n-1}+\ .\ .\ .\ a_{n-1,0}u+a_{n0}=0,$$

it follows that if

$$na_{00}a^n+(n-1)a_{10}a^{n-1}+\ .\ .\ .\ +a_{n-1,0}=\beta,$$

then $\beta\neq 0$. Let

$$a_{01}a^n+a_{11}a^{n-1}+\ .\ .\ .\ +a_{n1}=\gamma,$$

and write

$$v=u-a,\quad s=t+\gamma/\beta,$$

then the term in B_1 which does not involve v is βs.

Now

$$L(w)=e^{\frac{1}{2}az^2-\gamma z/\beta}\int\int e^{\frac{1}{2}vz^2+sz}U\Phi(s,\ v,\ z)dvds,$$

where

$$\Phi(s,\ v,\ z)\equiv\Pi(t,\ u,\ z).$$

A term in $z^\kappa v^\lambda s^\mu$ is reduced to a term independent of z by μ or $\mu+1$ integrations with respect to s together with $\frac{1}{2}(\kappa-\mu)$ or $\frac{1}{2}(\kappa-\mu-1)$ integrations with respect to v according as the integer $\kappa-\mu$ is even or odd. It will be observed that since Φ contains the factor z, κ is at least n ; μ is at most n and therefore $\kappa-\mu$ is a positive integer or zero.

Let $(s,\ v)_{r,n}$ denote a polynomial of degree r in s and n in v. Then the differential equation in U is of the form

$$(-2)^m\frac{\partial^m}{\partial v^m}\{B_0U\}-(-2)^{m-1}\frac{\partial^m}{\partial v^{m-1}\partial s}\{\beta sU+v(s,\ v)_{1,n-1}U\}$$
$$+(-2)^{m-2}\frac{\partial^m}{\partial v^{m-2}\partial s^2}\{(s,\ v)_{2,n}U\}-(-2)^{m-3}\frac{\partial^m}{\partial v^{m-3}\partial s^3}\{(s\ \ v)_{3,n}U\}$$
$$+\ .\ .\ .\ =0,$$

or, when expanded,

$$2B_0\frac{\partial^m U}{\partial v^m}+\{\beta s+v(s,\ v)_{1,n-1}\}\frac{\partial^m U}{\partial v^{m-1}\partial s}$$
$$+\sum_{\nu=0}^{n+1}\sum_{\mu=2}^{m}\{s^\nu(1,\ v)_n+s^{\nu-1}(1,\ v)_n\}\frac{\partial^{m-\mu+\nu}U}{\partial v^{m-\mu}\partial s^\nu}=0,$$

where the expressions $(1,\ v)_n$ denote polynomials in v of degree n. Assume a solution of the form

$$U=v\ \{f_0(s)+vf_1(s)+v^2f_2(s)+\ .\ .\ .\}\ ;$$

* The case of a repeated root involves a somewhat tedious analysis, and does not present any points of special interest.

then if $[\rho]_m=\rho(\rho-1)\ldots(\rho-m+1)$, the functions f satisfy the recurrence-relations

$$2[\rho]_m\beta f_0+[\rho]_{m-1}\beta s\frac{df_0}{ds}+[\rho]_{m-1}a_0f_0=0,$$

$$2[\rho+1]_m\beta f_1+[\rho+1]_{m-1}\beta s\frac{df_1}{ds}+[\rho+1]_{m-1}a_0f_1=(a_1s^2+b_1s)\frac{d^2f_0}{ds^2}$$
$$+(c_0s+d_0)\frac{df_0}{ds}+e_0f_0,$$

$$\cdot \quad \cdot \quad \cdot \quad \cdot$$

where a_0, a_1, b_1, c_0, ... are constants which occur in the differential equation.

The first recurrence-relation reduces to

$$s\frac{df_0}{ds}+\left\{2(\rho-m+1)+\frac{a_0}{\beta}\right\}f_0=0$$

and is satisfied by

$$f_0=s^{-\sigma},$$

where

$$\sigma=2(\rho-m+1)+\frac{a_0}{\beta}.$$

The second recurrence-relation then takes the form

$$s\frac{df_1}{ds}+(\sigma+2)f_1=A_1s^{-\sigma}+A_2s^{-\sigma-1},$$

where A_1 and A_2 are definite constants (dependent upon σ). Consequently,

$$f_1=s^{-\sigma}(\tfrac{1}{2}A_1+A_2s^{-1})$$

and, in general,

$$f_r=s^{-\sigma}g_r(s^{-1}),$$

where $g_r(s^{-1})$ is a polynomial in s^{-1} of degree r.

It will now be proved that the formal solution

$$U=v^\rho s^{-\sigma}\{1+vg_1(s^{-1})+\ldots+v^rg_r(s^{-1})+\ldots\}$$

is convergent within any finite circle $|v|=\gamma$ for all values of $|s|$ greater than a fixed positive number s_0. There will be no loss of generality in assuming that $p=0$, $a=0$, for the form of the series $\sum g_r(s)$ is the same in all cases. For simplicity let $s^{-1}=t$, then the partial differential equation for U becomes

$$2B_0\frac{\partial^mU}{\partial v^m}-\{\beta t+vt(t,v)_{1,n-1}\}\frac{\partial^mU}{\partial v^{m-1}\partial t}+\sum\sum\{t^\nu(1,v)_n+t^{\nu+1}(1,v)_n\}\frac{\partial^{m-\mu+\nu}U}{\partial v^{m-\mu}\partial t^\nu}=0.$$

Its solution may be developed as the series

$$U=1+vg_1(t)+\ldots+v^rg_r(t)+\ldots,$$

whose coefficients $g_r(t)$ are polynomials determined by relations of the form

$$-t\frac{dg_r}{dt}+2rg_r=\sum_{(h)}\sum_{k=0}^{r-1}(a_{hk}t^h+b_{hk}t^{h+1})\frac{d^hg_k}{dt^h},$$

where a_{hk} and b_{hk} are constants.

If polynomials $\phi_r(t)$ are defined by the relations

$$-t\frac{d\phi_r}{dt}+2r\phi_r=\sum_{(h)}\sum_{k=0}^{r-1}(|a_{hk}|t^h+|b_{hk}|t^{h+1})\frac{d^h\phi_k}{dt^h},$$

with $\phi_0=g_0=1$, the coefficients of $\phi_r(t)$ will be the moduli of the corresponding coefficients of $g_r(t)$ and therefore

$$|g_r^{(h)}(t)|\leqslant|\phi_r^{(h)}(t)|$$

for all values of t. Consider also the sequence of functions $\psi_r(t)=c_r t^r$, where

$$-t\frac{d\psi_r}{dt}+2r\psi_r=\sum_{(h)}\sum_{k=0}^{r-1}(|a_{hk}|+|b_{hk}|)t^{h+1}\frac{d^h\psi_k}{dt^h};$$

the coefficients c_r are positive if $c_0=1$. If $|t|>1$ and if

$$\left|\frac{d^h\psi_k}{dt^h}\right|\geqslant\left|\frac{d^h\phi_k}{dt^h}\right| \qquad (k=0, 1, \ldots, r-1),$$

then

$$|\psi_r|\geqslant|\phi_r|.$$

Therefore, by induction, for all values of r and for $|t|>1$,

$$|\psi_r|\geqslant|\phi_r|\geqslant|g_r|.$$

But $\psi_r=c_r t^r$ and ϕ_r is a polynomial of degree r with positive coefficients and therefore

$$\left|\frac{d^h\psi_r}{dt^h}\right|\geqslant\left|\frac{d^h\phi_r}{dt^h}\right|,$$

for $h=1, 2, \ldots, r$, and therefore

$$\left|\frac{d^h\psi_r}{dt^h}\right|\geqslant\left|\frac{d^h g_r}{dt^h}\right|.$$

Now consider the expression

$$\begin{aligned}V&=1+v\psi_1+v^2\psi_2+\ldots\\&=1+c_1vt+c_2v^2t^2+\ldots;\end{aligned}$$

it satisfies a partial differential equation of the form

$$\frac{\partial^m V}{\partial v^m}-2t\frac{\partial^m V}{\partial v^{m-1}\partial t}=\sum\sum t^{h+1}P_{hk}(v)\frac{\partial^{m-k+h}V}{\partial v^{m-k}\partial t^h},$$

in which $P_{hk}(v)$ is a power series in v which converges within the circle $|v|=\delta$, where δ is the modulus of the zero of $B_0(v)$ nearest the origin. Consequently, if $vt=\zeta$, $V(\zeta)$ satisfies an ordinary differential equation of the form

$$t^m\frac{d^mV}{d\zeta^m}=\sum_{r=1}^{m}Q_r(\zeta, t)\frac{d^{m-r}V}{d\zeta^{m-r}},$$

where Q_r is developable as a power series in ζ which converges for $|\zeta|<\delta t$ and therefore the series V converges for $|\zeta|<\delta t$, that is for $|v|<\delta$.

It follows that the series

$$1+vg_1(t)+v^2g_2(t)+\ldots$$

converges absolutely and uniformly if $|v|<\delta$ and if $|t|$ is finite and greater than unity. But since the coefficients $g_r(t)$ are polynomials in t, the series converges also when $|t|<1$.

The function $U(v, s)$ is thus represented by a double series which converges for all non-zero values of s, including $s=\infty$, and for $|v|<\delta$. It remains to prove that contours in the s- and v-planes can be assigned such that the double integral exists and the semi-integrated part $[R]$ vanishes.

18·33. Completion of the Proof.—The series for V satisfies a linear partial differential equation whose coefficients $P_{hk}(v)$ can be developed as power series in $(v-c)$, where c is not a zero of $B_0(v)$. Its solutions may similarly be developed and will converge within the circle $|v-c|=\eta$, where η is the distance from c of the nearest zero of $B_0(v)$. From this remark it follows that V admits of an analytic continuation throughout any closed region in the

v-plane which contains no zero of $B_0(v)$. The same is true of the differential coefficients of V with respect to v and t.*

But when $|t|>1$, the coefficients in the development of V are dominant functions for those in the development of U and therefore U and its derivatives admit in the same way of an analytic continuation.

Now by considering the source of the coefficients $P_{hk}(v)$ in the partial differential equation for V it will be seen that these coefficients, and therefore also the coefficients $Q_r(\zeta, t)$ in the ordinary equation for V are bounded for $v=\infty$. It follows that, if $|t|>1$, a number λ can be found such that as v tends to infinity in a definite direction,

$$e^{-\lambda vt}V \to 0$$

and therefore

$$e^{-\lambda vt}U \to 0.$$

Thus if $1<|t|<\tau$ and if v tends to infinity in such a manner that $\mathbf{R}(vz^2)$ is positive,

$$e^{-\frac{1}{2}vz^2}U \to 0.$$

But since U is an absolutely convergent series of positive powers of t, the restriction $1<t$ can be removed and the result is true for $0\leqslant t<\tau$. Under the same conditions

$$e^{-\frac{1}{2}vz^2}\frac{\partial^{h+k}U}{\partial v^h \partial t^k} \to 0.$$

Consequently if $|s|>s_0=\tau^{-1}$, as $|v|\to\infty$

$$e^{-\frac{1}{2}vz^2}\frac{\partial^{h+k}U}{\partial v^h \partial s^k} \to 0,$$

and similarly as $|s|\to\infty$

$$e^{-sz}\frac{\partial^{h+k}U}{\partial v^h \partial s^k} \to 0,$$

provided that ultimately,

$$\mathbf{R}(vz^2)>0, \quad \mathbf{R}(sz)\to 0.$$

Thus it is always possible to find contours in the v- and s-planes, encircling the points $v=0$ and $s=0$ and extending to infinity in appropriate directions such that the double integral

$$\int\int e^{sz+\frac{1}{2}vz^2}U\,dv\,ds$$

exists and such that the semi-integrated term $[R]$ vanishes at the infinite limits of integration.

The double integral

$$w=e^{\frac{1}{2}\alpha z^2-\beta z/\gamma}\int\int e^{sz+\frac{1}{2}vz^2}v^\rho s^{-\sigma}\{1+vg_1(s^{-1})+v^2g_2(s^{-1})+\ldots\}dv\,ds$$

is therefore a solution of the given differential equation of rank 2. Setting aside the exponential factor, the integral solution consists of terms such as

$$\int\int e^{sz+\frac{1}{2}vz^2}v^{\rho+h-1}s^{-\sigma-k+1}dv\,ds$$

$$=2^{\rho+k}z^{\sigma+k-2(\rho+h)}\int\int e^{\xi+\eta}\xi^{\rho+h-1}\eta^{-\sigma-k+1}d\xi\,d\eta \qquad (h=1, 2, \ldots; \; k\leqslant h).$$

Let the contours in the ξ- and η-planes be loops each encircling the origin and proceeding to infinity along the negative real axis. Then the term considered is seen to be a constant multiple of

$$z^{\sigma+k-2\rho-2h}\Gamma(\rho+h)\Gamma(-\sigma-k),$$

* The proof would be on the lines of § 12·3.

that is of

$$\frac{\Gamma(\rho+h)}{\Gamma(\sigma+k+1)} z^{-2m+a_0/\beta-2h+k+2} \qquad (k=1, \ldots, h).$$

Hence

$$w = e^{\frac{1}{2}az^2-\beta z/\gamma} z^{-2m+a_0/\beta+1} P(z^{-1}),$$

where $P(z^{-1})$ may formally be developed as an ascending series in z^{-1}. But since an infinite number of the coefficients

$$\frac{\Gamma(\rho+h)}{\Gamma(\sigma+k+1)}$$

increase without limit as $h \to 0$, the series will in general diverge. Thus unless $P(z^{-1})$ terminates, the development will not furnish a valid solution of the equation. It may, however, be proved that it does furnish an asymptotic representation of the solution.

18·4. Integrals of Jordan and Pochhammer.—The Euler transformation (§ 8·31) furnishes a powerful method of discussing equations of the type

$$L(w) \equiv Q(z)\frac{d^n w}{dz^n} - \mu Q'(z)\frac{d^{n-1}w}{dz^{n-1}} + \frac{\mu(\mu+1)}{1.2}Q''(z)\frac{d^{n-2}w}{dz^{n-2}} - \cdots$$
$$-R(z)\frac{d^{n-1}w}{dz^{n-1}} + (\mu+1)R'(z)\frac{d^{n-2}w}{dz^{n-2}} - \cdots = 0,$$

where $Q(z)$ and $R(z)$ are polynomials such that one of $Q(z)$ and $zR(z)$ is of degree n, whilst the degree of the other does not exceed n.

The complete discussion of the contour integrals which arise out of this transformation is due to Jordan and Pochhammer; * by considering the various possible contours of integration it is possible, in general, to obtain n distinct particular solutions which together compose the general solution.

The integral to be considered is

$$W(z) = \int_C (\xi - z)^{\mu+n-1} U d\zeta,$$

where U is a function of ζ alone, determined by the Euler-transform

$$\frac{d}{d\zeta}\{Q(\zeta)U\} = R(\zeta)U,$$

namely,

$$U = \frac{1}{Q(\zeta)} e^{\int \frac{R(\zeta)}{Q(\zeta)} d\zeta}.$$

Then

$$L\{W(z)\} = \int_C dV,$$

where

$$V = Z(\zeta)Q(\zeta)(\zeta - z)^{\mu} = (\zeta - z)^{\mu} e^{\int \frac{R(\zeta)}{Q(\zeta)} d\zeta},$$

and the contour C has to be so chosen that

$$\int_C dV = 0$$

independently of z. This condition will be satisfied if either

(i) C is a closed contour such that the initial and final values of V are the same,

or (ii) C is a curvilinear arc such that V vanishes at its end-points.

* Jordan, *Cours d'Analyse*, 3 (3rd ed. 1915), p. 251 ; Pochhammer, *Math. Ann.* 35 (1889), pp. 470, 495 ; 37 (1890), p. 500. Further applications of the method were made by Hobson, *Phil. Trans. Roy. Soc.* (A) 187 (1896), p. 493.

As a general principle it may be stated that when $Q(z)$ is a polynomial of degree n with unequal zeros, there are n contours of the first kind, one corresponding to each zero of $Q(z)$, which give rise to n distinct contour integral solutions. When, on the other hand, $Q(z)$ is of degree n but with repeated zeros, or of degree less than n the number of possible distinct contours of the first type falls short of n and the deficit is made up by contours of the second type.

18·41. Contours associated with zeros of $Q(z)$.—Let the zeros of $Q(z)$ be $a_1, \ldots, a_m$ $(m \leqslant n)$; then

$$\frac{R(\zeta)}{Q(\zeta)} = \sum_{r=1}^{m} \frac{\alpha_r}{\zeta - a_r} + S(\zeta),$$

where, in the most general case, $S(\zeta)$ consists of a polynomial in ζ with terms in $(\zeta - a_r)^{-2}$, $(\zeta - a_r)^{-3}$, etc. Consequently

$$V = Ke^{P(\zeta)}(\zeta - z)^{\mu} \prod_{r=1}^{m} (\zeta - a_r)^{\alpha_r},$$

where K is a constant and

$$P(\zeta) = \int S(\zeta) d\zeta$$

is meromorphic throughout the plane. Thus as ζ describes a simple closed contour in the positive direction around the point a_r, V returns to its initial value multiplied by $e^{2\pi i \alpha_r}$.

Let O be any point in the plane, and let A_r denote the loop beginning and ending at O and encircling the point a_r in the positive direction. A_r^{-1} will signify the same loop described in the reverse direction. Now consider the composite or *double-circuit* contour $A_r A_s A_r^{-1} A_s^{-1}$ consisting of the loop A_r followed in succession by the loop A_s, the loop A_r reversed and the loop A_s reversed. When ζ describes this contour, V evidently returns to O with its initial value. If O is taken on the line (a_r, a_s) the double-circuit contour is as shown diagrammatically (Fig. 13); the four parallel lines, drawn separately for clearness really coincide in the line (a_r, a_s).*

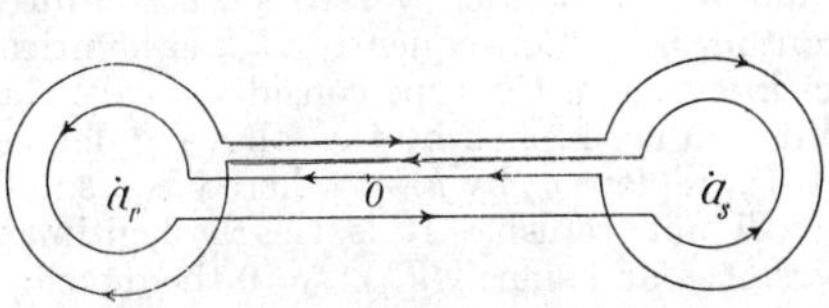

FIG. 13.

Let W_r denote the value of the integral

$$\int (\zeta - z)^{\mu + n - 1} U d\zeta,$$

where

$$U = Ke^{P(\zeta)} \frac{\prod_{r=1}^{m} (\zeta - a_r)^{\alpha_r}}{Q(\zeta)},$$

for the contour A_r and for a definite initial determination I_0 of the integrand. Also let W_{rs} be the value for the composite contour $A_r A_s A_r^{-1} A_s^{-1}$. Then W_{rs} is a solution of the differential equation.

Consider now the contribution of each of the four loops to the value of W_{rs}. The contribution of A_r is W_r, and after A_r has been described the

* It is assumed that no other singular point lies on this line.

final value of the integrand is $e^{2\pi i a_r} I_0$. This is the initial value for the loop A_s which therefore contributes the amount $e^{2\pi i a_r} W_s$ to the value of W_{rs} and leaves the integrand with the value $e^{2\pi i(a_r+a_s)} I_0$. Now if the loop A_r^{-1} were described with the initial value $e^{2\pi i a_r} I_0$ assigned to the integrand, the contribution would be $-W_r$ and the final value of the integrand would be I_0. But actually the integrand has, with regard to this loop, the initial value $e^{2\pi i(a_r+a_s)} I_0$; the contribution to the value of W_{rs} is therefore $-e^{2\pi i a_s} W_r$ and the final value of the integrand is $e^{2\pi i a_s} I_0$. Lastly, the loop A_s^{-1} contributes the amount $-W_s$ to the value of W_{rs} and the integrand returns to its initial value I_0.

The four loops together therefore give

$$W_{rs}=(1-e^{2\pi i a_s})W_r-(1-e^{2\pi i a_r})W_s.$$

Thus

$$W_{rs}=-W_{sr},$$

and it may readily be verified that

$$(1-e^{2\pi i a_t})W_{rs}=(1-e^{2\pi i a_r})W_{st}+(1-e^{2\pi i a_s})W_{rt}.$$

A similar contour with respect to the points a_r, z may be constructed; let W_{rz} be the value of the integral for this contour. Then

$$(1-e^{2\pi i\mu})W_{rs}=(1-e^{2\pi i a_r})W_{sz}+(1-e^{2\pi i a_s})W_{rz},$$

and therefore all integrals of the type W_{rs} may be expressed linearly in terms of the integrals W_{rz}. Consequently there are not more than m linearly distinct integrals of the type in question.

18·42. The Case of Integral Residues.—When any of the residues a_r in $R(\zeta)/Q(\zeta)$ is an integer, the method fails. Thus let

$$a_r=k,$$

where k is an integer. Then in the relation

$$W_{rz}=(1-e^{2\pi i\mu})W_r-(1-e^{2\pi i a_r})W_z,$$

$e^{2\pi i a_r}-1$ is zero and W_r is identically zero since the integrand is analytic throughout the contour A_r. Consequently W_{rz} is identically zero, and the number of distinct integrals of the type considered falls short of m. In this case the missing integral is supplied by the following device.

In the integral W_{rz} replace a_r by $k+\epsilon$, where ϵ is a small quantity; then the integral W_{rz} will not vanish. It is clearly legitimate to expand the integrand in powers of ϵ, and since $[W_{rz}]_{\epsilon=0}=0$ the development is

$$W_{rz}=\epsilon\left[\frac{dW_{rz}}{d\epsilon}\right]_{\epsilon=0}+O(\epsilon^2).$$

The differential equation is satisfied by

$$\lim \frac{W_{rz}}{\epsilon}=\left[\frac{dW_{rz}}{d\epsilon}\right]_{\epsilon=0}$$
$$=(1-e^{2\pi i\mu})\overline{W}_r+2\pi i e^{2\pi i k}W_z,$$

where $\overline{W}_r$ is the form which the integral W_r takes when the term $(\zeta-a_r)^{a_r}$ is replaced by

$$\left[\frac{d}{da_r}(\zeta-a_r)^{a_r}\right]_{a_r=k}=(\zeta-a_r)^k\log(\zeta-a_r).$$

When, for any reason the number of distinct integrals falls below m, this method may be employed to furnish integrals to bring the total number up to m.

18·43. Contours associated with Multiple Zeros of $Q(z)$.—Let a be a zero of $Q(z)$ of multiplicity k. Then the preceding methods furnish one and only one integral-solution relative to this point. By choosing a contour such that V vanishes at its end-points an additional set of $k-1$ distinct integrals may be obtained.

Let the principal part of $R(\zeta)/Q(\zeta)$ relative to $\zeta=a$ be

$$\frac{\beta_k}{(\zeta-a)^k}+\ .\ .\ .\ +\frac{\beta_1}{\zeta-a},$$

and write

$$\zeta-a=\rho\ (\cos\phi+i\sin\phi),$$
$$-\frac{\beta_k}{k-1}=r\ (\cos t+i\sin t).$$

Then

$$V=V_0(\zeta-a)^{\beta_1}e^{-\frac{\beta k}{(k-1)(\zeta-a)^{k-1}}-\cdots}$$
$$=V_0\rho^{\beta_1}(\cos\beta_1\phi+i\sin\beta_1\phi)e^{r\rho^{1-k}(\cos\omega+i\sin\omega)+\cdots},$$

where

$$\omega=t-(k-1)\phi \qquad (k>1),$$

and V_0 is finite (non-zero) in the neighbourhood of $\zeta=a$. The exponential term

$$e^{r\rho^{1-k}\cos\{t-(k-1)\phi\}}$$

dominates the function V, which tends to zero or infinity as ρ tends to zero according as $\cos\{t-(k-1)\phi\}$ is negative or positive.

The equation

$$\cos\{t-(k-1)\phi\}=0$$

gives rise to $2(k-1)$ equally-spaced values of ϕ in the interval $0\leqslant\phi<2\pi$. If through the point a rays are drawn in the corresponding directions, these rays divide the plane into $2(k-1)$ sectors of equal angle. As ζ tends to a in the various sectors V tends alternately to zero and to infinity. Let any sector in which V tends to zero be termed the first sector, and number the remaining sectors consecutively.

Consider a simple curve C issuing from a in the first sector, crossing the second sector at a finite distance from a and returning to a within the third sector (Fig. 14). Then, since V vanishes at the end-point of C, this curve

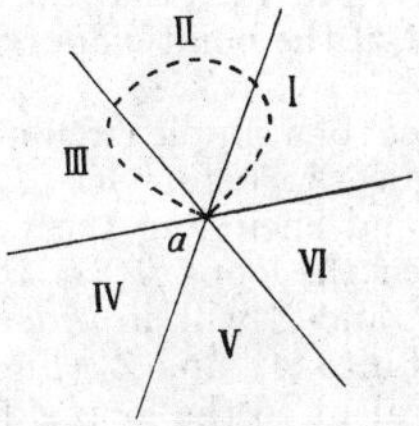

FIG. 14.

may be taken as the contour of integration. Without any loss in generality it may be assumed that the contour C is sufficiently small not to include any singular point of the equation. Another integral may be obtained by drawing another contour from the third sector to the fifth and so on and thus $k-1$ new integrals are finally obtained. Thus to a root of $Q(z)$ of multiplicity k there correspond k contour-integral solutions of the equation.*

* It is left to the reader to prove that the k integrals are linearly distinct.

18·44. $Q(z)$ of degree less than n.—The preceding discussion leads to distinct contour-integrals equal in number to the degree of $Q(z)$. When the degree is n the discussion is complete; when the degree is less than n further integrals must be sought to raise the total number to n. Let the degree of Q be $n-\lambda$, then since R is of degree $n-1$,

$$\frac{R(\zeta)}{Q(\zeta)}=\lambda\beta_{\lambda-1}\zeta^{\lambda-1}+\ .\ .\ .\ +2\beta_1\zeta+\beta_0+\sum_{r=1}^{m}\frac{a_r}{\zeta-a_r}+O(\zeta^{-2}) \qquad (\lambda\geqslant 1)$$

when ζ is large. Let

$$\zeta=\rho(\cos\phi+i\sin\phi), \qquad \beta^{\lambda-1}=r(\cos t+i\sin t),$$

then

$$V=V_0e^{\beta_{\lambda-1}\zeta^{\lambda}+\cdots}\prod_{r=1}^{m}(\zeta-a_r)^{\alpha_r}$$

$$=V_0\rho^{\alpha}(\cos\alpha\phi+i\sin\alpha\phi)e^{r\rho^{\lambda}(\cos\omega+i\sin\omega)+\ .\ .\ .},$$

where

$$\alpha=\alpha_1+\alpha_2+\ .\ .\ .\ +\alpha_m, \quad \omega=t+\lambda\phi,$$

and V_0 is finite at infinity.

V therefore tends to zero or to infinity as ρ tends to infinity according as $\cos(t+\lambda\phi)$ is negative or positive. If therefore the plane is divided into 2λ sectors by rays drawn from any convenient point in the directions

$$\cos(t+\lambda\phi)=0,$$

V will tend to zero and to infinity in alternate segments as ρ tends to infinity. A suitable contour of integration is therefore a curve starting from infinity in a segment in which the limiting value of V is zero, crossing a consecutive segment, and then returning to infinity in the next segment following. There are λ possible distinct curves of this character which do not enclose any singular points of the equation, and which give rise to the λ integrals necessary to make up the full complement of n contour-integral solutions.

18·45. The Group of the Equation.—For any fixed values of z the contours may be deformed in any continuous manner without altering the value of the integrals, provided that they do not encounter any of the points $a_1, \ldots, a_m, z$. In the same way, if z varies continuously the integrals will likewise vary continuously provided that the deformation of the contours consequent to the movement of the point z does not involve passage through any of the singular points.

Consider the resultant effect of a simple circulation in the positive direction around the singular point a_r. As before let A_r denote a loop proceeding from an arbitrary point O and encircling the point a_r; let Z be the loop encircling the point z. Then the loops A_s $(s\neq r)$ will be unaffected by the circulation, but the loops A_r and Z will, in order to avoid encountering the points z and a_r, be deformed into A_r' and Z' (Fig. 15).

The new loop Z' is equivalent to the loop Z followed by a double-circuit contour encircling a_r and z, that is, to $ZA_rZA_r^{-1}Z^{-1}$, and the new loop A_r' to a double-circuit encircling z and a_r followed by the loop A_r, that is to $ZA_rZ^{-1}A_r^{-1}A_r$, or ZA_rZ^{-1}.

Let W_z' and W_r' be the respective contributions of Z' and A_r' to the value of the integral taken round the corresponding double-circuit. Then

$$W_z'=W_z+e^{2\pi i\mu}W_{rz}, \quad W_r'=-W_{rz}+W_r,$$

and consequently the integral W_{rz} whose value for the undeformed contour is

$$(1-e^{2\pi i\mu})W_r-(1-e^{2\pi i\alpha_r})W_z$$

is transformed into

$$\begin{aligned}&(1-e^{2\pi i\mu})W_r'-(1-e^{2\pi i a_r})W_z'\\&=(1-e^{2\pi i\mu})W_r-(1-e^{2\pi i a_r})W_z-(1-e^{2\pi i(\mu+a_r)})W_{rz}\\&=e^{2\pi i(\mu+a_r)}W_{rz}.\end{aligned}$$

On the other hand, since the loop A_s is unaffected, the integral W_{sz} is transformed into

$$\begin{aligned}&(1-e^{2\pi i\mu})W_s-(1-e^{2\pi i a_r})W_z'\\&=W_{sz}-e^{2\pi i\mu}(1-e^{2\pi i a_r})W_{rz}.\end{aligned}$$

Now consider the effect on the integrals of § 18·43 of a circulation around the multiple zero a. The contours are simple closed curves beginning and ending at a and may be made arbitrarily small. Consequently a circulation of z around a has no effect upon this contour. The only effect is that which is due to the presence of the factor $(\zeta-z)^\mu$ in the integrand, for as z encircles the point a it also encircles the point ζ on the contour. The effect of the

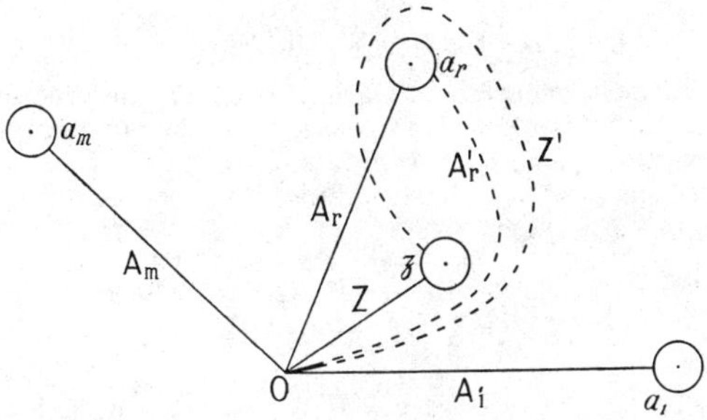

FIG. 15.

circulation therefore is to multiply the integral by the factor $e^{2\pi i\mu}$. The integrals of this type relative to multiple zeros other than a are unaffected by a circulation around a.

Finally, the effect of a circulation in the positive direction including all the singular points is to multiply the integrals of § 18·44 by the same factor $e^{2\pi i\mu}$.

Thus the fundamental substitutions of the group of the equation are known and therefore the group itself is known.

18·46. Recurrence-Relations and Contiguous Functions.—In order to emphasise the dependence of the integral-solution upon the parameters $\alpha_1, \ldots, \alpha_m, \mu$ it may be written in the form

$$W(\alpha_1, \ldots, \alpha_m, \mu;\ z).$$

In particular, let $Q(z)$ be of degree n and let the roots of $Q(z)=0$ be unequal, then

$$W(\alpha_1, \ldots, \alpha_n, \mu;\ z)=\int_C(\zeta-a_1)^{\alpha_1-1}\ldots(\zeta-a_n)^{\alpha_n-1}(\zeta-z)^{\mu+n-1}d\zeta,$$

where C is such that the initial and final values of the integrand are equal.

By differentiation under the integral sign it is found that

$$\frac{d^\kappa W(\alpha_1, \ldots, \alpha_n, \mu;\ z)}{dz^\kappa}=(-1)^\kappa(\mu+n-1)\ldots(\mu+n-\kappa)W(\alpha_1, \ldots, \alpha_n, \mu-\kappa;\ z).$$

By substituting this expression, with $\kappa=1, 2, \ldots, n$, in the differential equation a linear relation with polynomial coefficients between the $n+1$ functions

$$W(a_1, \ldots, a_n, \mu; z),\ W(a_1, \ldots, a_n, \mu-1; z), \ldots, W(a_1, \ldots, a_n, \mu-n; z)$$

is obtained.

Again, since

$$(\zeta-a)=(\zeta-z)+(z-a),$$

it follows that

$$W(a_1+1, a_2, \ldots, a_n, \mu; z)=W(a_1, a_2, \ldots, a_n, \mu+1; z) +(z-a)W(a_1, a_2, \ldots, a_n, \mu; z).$$

By considering all possible formulæ of these types it may be seen that all the functions

$$W(a_1+p_1, \ldots, a_n+p_n, \mu+q; z);$$

where $p_1, \ldots, p_n, q$ are integers or zero, may be expressed as linear combinations, with polynomial coefficients, in terms of any n of these functions, as for instance,

$$W(a_1, \ldots, a_n, \mu-1; z), \ldots, W(a_1, \ldots, a_n, \mu-n; z).$$

These relations are the recurrence-relations between the functions.

When one of the parameters is increased by unity and another diminished by unity a *contiguous function* * is produced. The relations which involve contiguous functions are particularly simple, thus by eliminating the function $W(a_1, \ldots, a_n, \mu-1; z)$ between

$$W(a_1+1, a_2, \ldots, a_n, \mu-1; z)=W(a_1, a_2, \ldots, a_n, \mu; z) +(z-a_1)W(a_1, \ldots, a_n, \mu-1; z)$$

and

$$W(a_1, a_2+1, \ldots, a_n, \mu-1; z)=W(a_1, a_2, \ldots, a_n, \mu; z) +(z-a_2)W(a_1, \ldots, a_n, \mu-1; z),$$

it is found that

$$(z-a_2)W(a_1+1, a_2, \ldots, a_n, \mu-1; z)-(z-a_1)W(a_1, a_2+1, \ldots, a_n, \mu-1; z) =(a_1-a_2)W(a_1, a_2, \ldots, a_n, \mu; z).$$

Other sets of recurrence-relations may be derived from formulæ similar to

$$(\mu+n-1)\frac{\partial W}{\partial a_1}-(a_1-1)\frac{\partial W}{\partial z}=(z-a_1)\frac{\partial^2 W}{\partial a_1 \partial z},$$

where $W=W(a_1, \ldots, a_n, \mu; z)$.

18·47. Contour-Integral Solutions of the Riemann *P*-Equation.—If, in the equation of the Riemann P-function (§ 15·93) the transformation

$$w=(z-a)^{\alpha}(z-b)^{\beta}(z-c)^{\gamma}u$$

is made, the resulting equation is

$$Q(z)\frac{d^2u}{dz^2}-\{\mu Q'(z)+R(z)\}\frac{du}{dz}+\{\tfrac{1}{2}\mu(\mu+1)Q''(z)+(\mu+1)R'(z)\}u=0,$$

where

$$\mu=-\alpha-\beta-\gamma-1=\alpha'+\beta'+\gamma'-2,$$
$$Q(z)=(z-a)(z-b)(z-c),$$
$$R(z)=\sum(\alpha'+\beta+\gamma)(z-b)(z-c).$$

* Riemann, *Gött. Abh.* 7 (1857); [*Math. Werke*, p. 67].

In this case

$$U(\zeta)=\frac{1}{Q(\zeta)}e^{\int\frac{R(\zeta)}{Q(\zeta)}d\zeta}$$
$$=(\zeta-a)^{\alpha'+\beta+\gamma-1}(\zeta-b)^{\alpha+\beta'+\gamma-1}(\zeta-c)^{\alpha+\beta+\gamma'-1}.$$

If, therefore, C is a double-loop contour encircling any two of the points a, b, c, the integral

$$\int_C U(\zeta)(\zeta-z)^{-\alpha-\beta-\gamma}d\zeta$$

multiplied by the factor $(z-a)^\alpha(z-b)^\beta(z-c)^\gamma$ represents a Riemann P-function.

In particular, let the double-loop contour encircle the points b and c. Let z lie in a circle Γ whose centre is a and which does not include either of the points b and c, then the contour C may be deformed, if necessary, so as to be wholly outside Γ. Then, for all points ζ on C

$$|z-a|<|\zeta-a|.$$

Let

$$|\arg(z-a)|<\pi,$$

also let $\arg(a-b)$ and $\arg(a-c)$ have their principal values, and let $\arg(\zeta-a)$, $\arg(\zeta-b)$ and $\arg(\zeta-c)$ be similarly made definite when ζ is at the initial point O. Then if $\arg(z-b)$, $\arg(z-c)$ and $\arg(\zeta-z)$ are so defined that they reduce respectively to $\arg(a-b)$, $\arg(a-c)$, $\arg(\zeta-a)$ when $z\to a$,

$$(z-b)^\beta=(a-b)^\beta\left\{1+\beta\frac{z-a}{a-b}+\ldots\right\},$$

$$(z-c)^\gamma=(a-c)^\gamma\left\{1+\gamma\frac{z-a}{a-c}+\ldots\right\},$$

$$(\zeta-z)^{-\alpha-\beta-\gamma}=(\zeta-a)^{-\alpha-\beta-\gamma}\left\{1-(\alpha+\beta+\gamma)\frac{a-z}{\zeta-a}+\ldots\right\},$$

and the series on the right converge absolutely and uniformly for all z in and on Γ and for all ζ on C.

If, therefore, $P^{(\alpha)}$ is that Riemann P-function which admits of the development

$$(z-a)^\alpha\{1+c_1(z-a)+c_2(z-a)^2+\ldots\},$$

then the integral solution *

$$(z-a)^\alpha(z-b)^\beta(z-c)^\gamma\int_0^{(b+,\,c+,\,b-,\,c-)}U(\zeta)(\zeta-z)^{-\alpha-\beta-\gamma}d\zeta$$

represents $P^{(\alpha)}$ multiplied by the factor

$$(a-b)^\beta(a-c)^\gamma\int_0^{(b+,\,c+,\,b-,\,c-)}(\zeta-a)^{\alpha'-\alpha-1}(\zeta-b)^{\gamma+\alpha+\beta'-1}(\zeta-c)^{\alpha+\beta+\gamma'-1}d\zeta.$$

In the same way the solutions $P^{(\alpha')}$, $P^{(\beta)}$, $P^{(\beta')}$, $P^{(\gamma)}$, $P^{(\gamma')}$ may be expressed as double-circuit integrals.†

18·471. The Periods of an Abelian Integral.—When the indices $\alpha_1, \ldots, \alpha_n, \nu$ are rational real numbers, the indefinite integral

$$\int(\zeta-a_1)^{\alpha_1-1}\ldots(\zeta-a_n)^{\alpha_n-1}(\zeta-z)^{\nu-1}d\zeta$$

is an Abelian integral. Its value for a closed contour such that the integrand returns

* The manner of writing this integral indicates the order and sense in which the loops composing the contour are described.

† The exceptional cases in which $\alpha-\alpha'$, $\beta-\beta'$ or $\gamma-\gamma'$ are integers or zero require the special treatment of § 18·42.

to its initial value is a *period* of the integral. From what has gone before it is not difficult to deduce the fact that the periods, which are functions of z, satisfy a linear differential equation with coefficients which are polynomial in z.

Consider in particular the elliptic integral

$$\int(1-t^2)(1-k^2t^2)dt$$

and let I be one of its periods. Then if

$$k^{-2}=z, \quad t^2=\zeta, \quad kI=w,$$

$$w=\tfrac{1}{2}\int\zeta^{-\frac{1}{2}}(\zeta-1)^{-\frac{1}{2}}(\zeta-z)^{-\frac{1}{2}}d\zeta,$$

and in the notation of the previous sections,

$$Q(\zeta)=\zeta(\zeta-1), \quad n=2, \quad \mu=-\tfrac{3}{2},$$

$$R(\zeta)=\tfrac{1}{2}\left\{\frac{1}{\zeta}+\frac{1}{\zeta-1}\right\}G(\zeta)=\zeta-\tfrac{1}{2},$$

and therefore w satisfies the hypergeometric equation

$$z(z-1)\frac{d^2w}{dz^2}+(2z-1)\frac{dw}{dz}+\tfrac{1}{4}w=0.$$

In fact, if K and K' are the quarter-periods of the Jacobian elliptic function then *

$$K=\tfrac{1}{2}\pi F(\tfrac{1}{2}, \tfrac{1}{2}\,;\, 1\,;\, k^2), \quad K'=\tfrac{1}{2}\pi F(\tfrac{1}{2}, \tfrac{1}{2}\,;\, 1\,;\, 1-k^2).$$

18·5. The Legendre Function $P_n(z)$.—A result obtained in an earlier section (§ 8·311) may now be restated in the following terms. The contour integral

$$\int_C(z-\zeta)^{-n-1}(1-\zeta^2)^n d\zeta$$

furnishes a solution of the Legendre equation

$$(1-z^2)\frac{d^2w}{dz^2}-2z\frac{dw}{dz}+n(n+1)w=0,$$

provided that the contour C is such that the expression

$$(\zeta-z)^{-n-2}(\zeta^2-1)^{n+1}$$

resumes its initial value after the contour has been described.

Let A be a point on the real axis to the right of $\zeta=1$ † and at A let

$$\arg(\zeta-1)=\arg(\zeta+1)=0\,; \quad |\arg(\zeta-z)|<\pi.$$

Now if ζ starts from A, describes a positive loop around the point $\zeta=1$ and returns to A, the expression $(\zeta-z)^{-n-2}(\zeta^2-1)^{n+1}$ assumes its initial value multiplied by $e^{2\pi i(n+1)}$; if a similar loop is made round $\zeta=z$, the expression returns to its initial value multiplied by $e^{2\pi i(-n-2)}$. If therefore the two loops are described, or what is the same thing if the contour of integration begins and ends at A and encircles $\zeta=1$ and $\zeta=z$ in the positive direction, but does not encircle $\zeta=-1$, the contour integral is a solution of the Legendre equation for all values of n.

Thus the contour integral ‡

$$\frac{1}{2\pi i}\int_A^{(1+,\, z+)}\frac{(\zeta^2-1)^n}{2^n(\zeta-z)^{n+1}}d\zeta$$

is a Legendre function, and since when n is a positive integer and $z=1$ it reduces to unity, it may consistently be represented by the symbol $P_n(z)$ which, when n is a positive integer, represents the Legendre polynomials.

* Whittaker and Watson, *Modern Analysis*, § 22·3, *et seq.*

† If z is real and greater than unity, A must be to the right of $\zeta=z$.

‡ Schläfli, *Über die zwei Heine'schen Kugelfunctionen*, Bern, 1881.

The contours C and C' (Fig. 16) both satisfy the requisite conditions, but the one cannot be transformed into the other without encountering the singular point $\zeta=-1$. Thus when n is not an integer, $P_n(z)$ will not be a

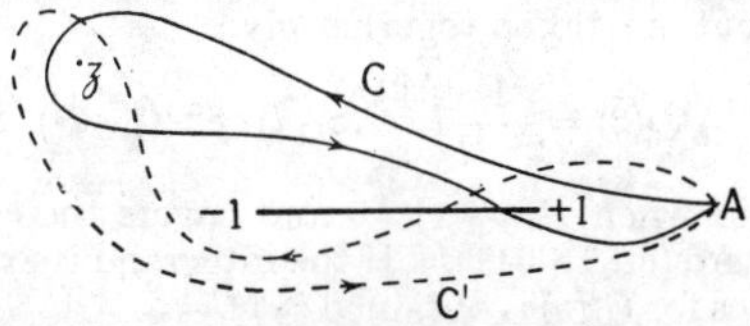

FIG. 16.

single-valued function of z. To render the function single-valued a cut along the real axis from -1 to $-\infty$ must be made in both the ζ- and the z-planes. Throughout the cut z-plane the function $P_n(z)$ is analytic.

18·51. The Legendre Function $Q_n(z)$.—The contour which leads to the Legendre function of the second kind $Q_n(z)$ is described as follows.* Let z be not a real number lying in the interval $(-1, +1)$ and describe an ellipse with the points ± 1 as foci such that z lies outside the ellipse. Then from A, the right-hand extremity of the major axis, describe a figure-of-eight contour C encircling the point $+1$ clockwise and the point -1 counter-clockwise, and lying within the ellipse (Fig. 17). Then the expression $(\zeta-z)^{-n-2}(\zeta^2-1)^{n+1}$

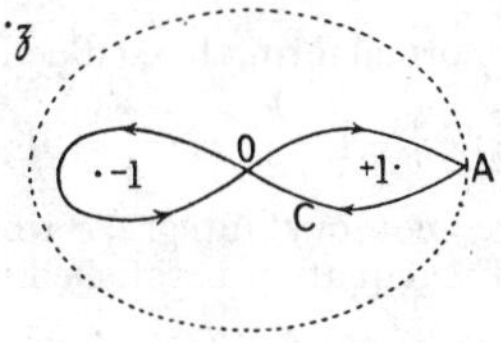

FIG. 17.

returns to its initial value as ζ returns to the starting point A after having described the contour.

Let $|\arg z| \leqslant \pi$, let $|\arg(z-\zeta)| \to \arg z$ as $\zeta \to 0$ on the contour, and at A let $\arg(\zeta-1)=\arg(\zeta+1)=0$. Then

$$Q_n(z)=\frac{1}{4i \sin n\pi}\int_C \frac{(\zeta^2-1)^n}{2^n(z-\zeta)^{n+1}}\,d\zeta$$

is a solution of the Legendre equation valid when n is not an integer, and is analytic throughout the z-plane cut along the real axis from 1 to $-\infty$.

Now let $\mathbf{R}\,(n+1)>0$ and consider the contour as composed of:

(i) a small circle described around $+1$ in the negative direction,

(ii) a small circle described around -1 in the positive direction,

(iii) the lines $(+1, -1)$ and $(-1, +1)$.

Since $\mathbf{R}\,(n+1)>0$ the contributions of (i) and (ii) tend to zero as the dimensions of the circles diminish.

The contribution of the line $(+1, -1)$ is

$$\frac{e^{-n\pi i}}{2^{n+2}\sin n}\int_{+1}^{-1}(z-t)^{-n-1}(1-t^2)^n dt,$$

* Whittaker and Watson, *Modern Analysis*, § 15·3.

and that of the line $(-1, +1)$ is

$$\frac{e^{n\pi i}}{2^{n+2}\sin n\pi}\int_{-1}^{+1}(z-t)^{-n-1}(1-t^2)^n dt,$$

and the two contributions taken together gives

$$Q_n(z)=\frac{1}{2^{n+1}}\int_{-1}^{+1}(z-t)^{-n-1}(1-t^2)^n dt.$$

This formula is valid when $\mathbf{R}(n+1)>0$ and covers the case in which n is a positive integer or zero (cf. § 8·311). If the integrand is expanded as a power-series in z^{-1} the series for $Q_n(z)$ is obtained (§ 7).

18·6. The Confluent Hypergeometric Functions.—The equation of the confluent hypergeometric functions of Whittaker * is derived from the Riemann P-equation, which is effectively the hypergeometric equation, by the following limiting process. In the equation of the P-function

$$w=P\left\{\begin{matrix} 0 & \infty & c & \\ \frac{1}{2}+m & -c & c-k & z \\ \frac{1}{2}-m & 0 & k & \end{matrix}\right\}$$

let $c\to\infty$, then the equation becomes

$$\frac{d^2w}{dz^2}+\frac{dw}{dz}+\left\{\frac{k}{z}+\frac{\frac{1}{4}-m^2}{z^2}\right\}w=0.$$

The substitution

$$w=e^{-\frac{1}{2}z}W$$

reduces this equation to its normal form, the confluent hypergeometric equation

$$\frac{d^2W}{dz^2}+\left\{-\tfrac{1}{4}+\frac{k}{z}+\frac{\frac{1}{4}-m^2}{z^2}\right\}W=0.$$

The limiting form of the contour integrals which represent the above P-function suggests that this equation is satisfied by an integral of the form

$$W=e^{-\frac{1}{2}z}z^k\int_C\left(1+\frac{\zeta}{z}\right)^{k+m-\frac{1}{2}}(-\zeta)^{-k+m-\frac{1}{2}}e^{-\zeta}d\zeta,$$

for a proper choice of the contour C.

It is readily found that this integral is a solution of the confluent hypergeometric equation if

$$\int_C\frac{d}{d\zeta}\left\{\left(1+\frac{\zeta}{z}\right)^{k+m-\frac{3}{2}}\zeta^{-k+m+\frac{1}{2}}e^{-\zeta}\right\}d\zeta=0,$$

and this condition is satisfied if the contour is a simple loop proceeding from infinity in a direction asymptotic to the positive real axis, encircling the origin in the positive direction but not encircling the point $\zeta=-z$, and returning to infinity on the positive real axis.

The standard solution of the confluent hypergeometric equation is defined as

$$W_{k,m}(z)=-\frac{1}{2\pi i}\Gamma(k-m+\tfrac{1}{2})e^{-\frac{1}{2}z}z^k\int_\infty^{(0+)}\left(1+\frac{\zeta}{z}\right)^{k+m-\frac{1}{2}}(-\zeta)^{-k+m-\frac{1}{2}}e^{-\zeta}d\zeta,$$

where, to make matters perfectly definite, it is supposed that $\arg z$ has its principal value, that $|\arg(-\zeta)|\leqslant\pi$ and that $\arg(1+\zeta/z)\to 0$ as $\zeta\to 0$ along a simple path inside the contour. The confluent hypergeometric function $W_{k,m}(z)$ is then analytic throughout the plane, cut along the negative real axis.

The above definition of $W_{k,m}(z)$ ceases to be valid when, and only when

* Whittaker and Watson, *Modern Analysis*, Chap. XVI.

$m-k+\frac{1}{2}$ is a positive integer. But when $\mathbf{R}(m-k+\frac{1}{2})\geqslant 0$ the contour integral may be transformed, as was done in the last paragraph, into the definite integral

$$W_{k,m}(z)=\frac{e^{-\frac{1}{2}z}z^k}{\Gamma(m-k+\frac{1}{2})}\int_0^\infty\left(1+\frac{t}{z}\right)^{k+m-\frac{1}{2}}t^{m-k-\frac{1}{2}}e^{-t}dt,$$

which is also valid when $m-k+\frac{1}{2}$ is a positive integer.

The function $W_{-k,m}(-z)$ is also a solution of the given equation valid when $|\arg(-z)|<\pi$. But since, in their respective regions of validity,

$$W_{k,m}(z)=e^{-\frac{1}{2}z}z^k\{1+O(z^{-1})\},$$
$$W_{-k,m}(-z)=e^{\frac{1}{2}z}(-z)^k\{1+O(z^{-1})\},$$

the ratio of these two solutions is not a constant and therefore, taken together they form a fundamental set.

18·61. The Asymptotic Expansion of $W_{k,m}(z)$.—In order to derive the asymptotic expansion from the contour integral for $W_{k,m}(z)$ use is made of the formula *

$$\left(1+\frac{\zeta}{z}\right)^\lambda=1+\frac{\lambda\zeta}{z}+\ldots+\frac{\lambda(\lambda-1)\ldots(\lambda-n+1)}{n!}\cdot\frac{\zeta^n}{z^n}+R_n(\zeta,z),$$

where

$$R_n(\zeta,z)=\frac{\lambda(\lambda-1)\ldots(\lambda-n)}{n!}\left(1+\frac{\zeta}{z}\right)^\lambda\int_0^{\zeta/z}u^n(1+n)^{-\lambda-1}du,$$

and $\lambda=k+m-\frac{1}{2}$.

Then by substituting this series in the contour integral for $W_{k,m}(z)$ and integrating term-by-term it is found that the $(r+1)^{\text{th}}$ term in the expansion is

$$(-1)^{r+1}\frac{\lambda(\lambda-1)\ldots(\lambda-r+1)}{r!}\cdot\frac{\Gamma(k-m+\frac{1}{2})}{2\pi i}e^{-\frac{1}{2}z}z^{k-r}\int_\infty^{(0+)}(-\zeta)^{r+m-k-\frac{1}{2}}e^{-\zeta}d\zeta,$$

and since

$$\frac{1}{2\pi i}\int_\infty^{(0+)}(-\zeta)^p e^{-\zeta}d\zeta=-\frac{1}{\Gamma(p)},$$

this reduces to

$$(-1)^r\frac{\Gamma(k+m+\frac{1}{2})\Gamma(k-m+\frac{1}{2})}{r!\,\Gamma(k+m-r+\frac{1}{2})\Gamma(k-m-r+\frac{1}{2})}e^{-\frac{1}{2}z}z^{k-r},$$

that is to

$$\frac{\{m^2-(k-\frac{1}{2})^2\}\{m^2-(k-\frac{3}{2})^2\}\ldots\{m^2-(k-r+\frac{1}{2})^2\}}{r!}e^{-\frac{1}{2}z}z^{k-r}.$$

When n is so large that $\mathbf{R}(n-k+m-\frac{1}{2})>0$, the remainder term may be expressed as the definite integral

$$\frac{1}{\Gamma(m-k+\frac{1}{2})}\int_0^\infty t^{m-k-\frac{1}{2}}R_n(t,z)e^{-t}dt.$$

Now suppose that $\lambda=k+m-\frac{1}{2}$ is real, that $|z|>1$ and that $|\arg z|\leqslant\pi-\alpha$, where $\alpha>0$. Then

$$1\leqslant|1+tz^{-1}|\leqslant 1+t \qquad \text{when } \mathbf{R}(z)\geqslant 0,$$
$$|1+tz^{-1}|\geqslant\sin\alpha \qquad \text{when } \mathbf{R}(z)\leqslant 0,$$

and consequently, in either case, if $\rho=|\lambda|$ and $r=|tz^{-1}|$,

$$|R_n(t,z)|\leqslant\left|\frac{\lambda(\lambda-1)\ldots(\lambda-n)}{n!}\right|\left(\frac{1+t}{\sin\alpha}\right)^\rho\int_0^r u^n(1+u)^\rho du$$
$$<\left|\frac{\lambda(\lambda-1)\ldots(\lambda-n)}{n!}\right|\left(\frac{1+t}{\sin\alpha}\right)^\rho\left|\frac{t}{z}\right|^{n+1}\frac{1+t^\rho}{n+1}.$$

* *See* Jacobi (Diss. Berlin, 1825), *Ges. Werke*, 3, pp. 1–44.

Therefore when $|z|>1$ the remainder term is in absolute magnitude less than

$$A \operatorname{cosec}^\rho \alpha\, |z|^{-n-1} \left| \int_0^\infty (1+t)^{2\rho} t^{m-k+n+\frac{1}{2}} e^{-t} dt \right|,$$

where A is independent of z, and since the integral converges, the remainder term is of the order of

$$\operatorname{cosec}^\rho \alpha\; z^{-n-1},$$

and in particular, when $\alpha \geqslant \alpha_0 > 0$, it is of the order of z^{-n-1}.

Therefore for $|z|>1$ and $|\arg z| \leqslant \pi - \alpha < \pi$,

$$W_{k,m}(z) \sim e^{-\frac{1}{2}z} z^k \left\{ 1 + \sum_{r=1}^{\infty} \frac{\{m^2-(k-\frac{1}{2})^2\}\{m^2-(k-\frac{3}{2})^2\} \ldots \{m^2-(k-r+\frac{1}{2})^2\}}{n!\, z^n} \right\}.$$

If $k-\frac{1}{2} \pm m$ is a positive integer, the series terminates and therefore furnishes an exact representation of the function.

18·7. The Bessel Functions.—The Bessel functions of integral order n may be defined * (cf. § 8·22) as the coefficient of ζ^n in the Laurent expansion of $e^{\frac{1}{2}z(\zeta-\zeta^{-1})}$. Consequently

$$J_n(z) = \frac{1}{2\pi i} \int^{(0+)} e^{\frac{1}{2}z(\zeta-\zeta^{-1})} \zeta^{-n-1} d\zeta,$$

where the contour is any simple closed curve encircling the origin in the positive direction.

The substitution $\zeta = 2t/z$ transforms the integral into

$$J_n(z) = \frac{1}{2\pi i} \left(\frac{z}{2}\right)^n \int^{(0+)} t^{-n-1} \exp\left\{t - \frac{z^2}{4t}\right\} dt;$$

the contour is again any closed curve encircling the origin in the positive direction, and may conveniently be taken to be the circle $|t|=1$ described counter-clockwise.

Now consider how the contour must be modified in order that the integral for $J_n(z)$ may, for any value of n, satisfy the Bessel equation

$$z^2 \frac{d^2 w}{dz^2} + z \frac{dw}{dz} + (z^2 - n^2) w = 0.$$

It is an easy matter to verify that the contour C must be such that

$$\int_C \frac{d}{dt} \left\{ t^{-n-1} \exp\left(t - \frac{z^2}{4t}\right) \right\} dt = 0$$

identically in z. When n is an integer, the function $t^{-n-1} \exp(t - z^2/4t)$ resumes its initial value after ζ has described the circle $|t|=1$, but when n is not an integer, this function is not one-valued on the circle. A suitable contour is one in which $t^{-n-1} \exp(t - z^2/4t)$ vanishes at the end-points and this is furnished by a loop beginning at a great distance along the negative real axis, encircling the origin positively, and returning to its starting point. Thus for all values of n, $J_n(z)$ is defined by the integral

$$\frac{1}{2\pi i} \left(\frac{z}{2}\right)^n \int_{-\infty}^{(0+)} t^{-n-1} \exp\left\{t - \frac{z^2}{4t}\right\} dt,$$

where $\arg z$ has its principal value and $|\arg t| \leqslant \pi$ on the contour.

The function thus defined is analytic for all values of z and admits of the series development

$$J_n(z) = \sum_{r=0}^{\infty} (-1)^r \frac{(\frac{1}{2}z)^{n+2r}}{r!\, \Gamma(n+r+1)}.$$

* Schlömilch, *Z. Math. Phys.* 2 (1857), p. 137.

The contour integral may, for all values of n, be transformed into a definite integral where $|\arg z|<\frac{1}{2}\pi$.* The formula

$$J_n(z)=\frac{1}{2\pi i}\int_{-\infty}^{(0+)} e^{\frac{1}{2}z(\zeta-\zeta^{-1})}\zeta^{-n-1}d\zeta$$

holds for all values of n when $|\arg z|<\frac{1}{2}\pi$. Let the contour be taken to be the circle $|\zeta|=1$ joined to the point at infinity by a double line lying along the negative real axis.

The contribution of the circle is (writing $\zeta=e^{i\theta}$)

$$\frac{1}{2\pi}\int_{-\pi}^{\pi} e^{-ni\theta+iz\sin\theta}d\theta,$$

and the contribution of the lines $(-\infty,\ -1)$ and $(-1,\ -\infty)$ together give, when ζ is replaced by $te^{-\pi i}$ in the first and by $te^{\pi i}$ in the second,

$$\left\{\frac{e^{(n+1)\pi i}-e^{-(n+1)\pi i}}{2\pi i}\right\}\int_1^{\infty} e^{\frac{1}{2}z(-t+t^{-1})}t^{-n-1}dt.$$

In the latter integral write $t=e^{i\theta}$, and then, taking the two integrals together,

$$J_n(z)=\frac{1}{\pi}\int_0^{\pi}\cos(n\theta-z\sin\theta)d\theta-\frac{\sin n\pi}{\pi}\int_0^{\infty} e^{-n\theta-z\sinh\theta}d\theta.$$

When n is a positive integer the second integral disappears and the result reduces to that of § 8·22.

Miscellaneous Examples.

1. Transform the Schlafli integral (§ 18·5) into the Laplace integral

$$P_n(z)=\frac{1}{\pi}\int_0^{\pi}\{z+(z^2-1)^{\frac{1}{2}}\cos\phi\}^n d\phi.$$

[Whittaker and Watson, *Modern Analysis*, § 15·23.]

Transform the corresponding integral for $Q_n(z)$ into

$$Q_n(z)=\int_0^{\infty}\{z+(z^2-1)^{\frac{1}{2}}\cosh\theta\}^{-n-1}d\theta.$$

[*Ibid.* § 15·33.]

2. Prove that the associated Legendre equation

$$(1-z^2)\frac{d^2w}{dz^2}-2z\frac{dw}{dz}+\left\{n(n+1)-\frac{m^2}{1-z^2}\right\}w=0$$

is satisfied by

$$P_n{}^m(z)=\frac{1}{\Gamma(1-m)}\left(\frac{z+1}{z-1}\right)^{\frac{1}{2}m}F(-n,\ n+1\ ;\ 1-m\ ;\ \tfrac{1}{2}-\tfrac{1}{2}z)$$

$$=\frac{(n+1)(n+2)\ .\ .\ .\ (n+m)}{2^n\pi i}(z^2-1)^{\frac{1}{2}m}\int_A^{(1+,\ z+)}(\zeta-z)^{-n-m-1}(\zeta^2-1)^n d\zeta,$$

and transform the last expression into

$$\frac{(n+1)(n+2)\ .\ .\ .\ (n+m)}{\pi}\int_0^{\pi}\{z+(z^2-1)^{\frac{1}{2}}\cos\phi\}^n\cos m\phi\, d\phi.$$

3. Show that the Weber-Hermite equation

$$\frac{d^2w}{dz^2}+\{n+\tfrac{1}{2}-\tfrac{1}{4}z^2\}w=0$$

is satisfied by the function

$$D_n(z)=2^{\frac{1}{2}n+\frac{1}{4}}z^{-\frac{1}{2}}W_{\frac{1}{2}n+\frac{1}{4},\ -\frac{1}{4}}(-\tfrac{1}{2}z^2),$$

that

$$D_n(z)=-\frac{\Gamma(n+1)}{2\pi i}e^{-\frac{1}{4}z^2}\int_{\infty}^{(0+)} e^{-z\zeta-\frac{1}{2}\zeta^2}(-\zeta)^{-n-1}d\zeta,$$

* Schläfli, *Math. Ann.* 3 (1871), p. 148. A similar result which holds when $\frac{1}{2}\pi<|\arg z|<\pi$ was given by Sonine, *ibid.* 16 (1880), p. 14.

and that when n is a positive integer

$$D_n(z)=(-1)^n e^{\frac{1}{4}z^2}\frac{d^n}{dz^n}\left(e^{-\frac{1}{4}z^2}\right).$$

[Whittaker and Watson, *Modern Analysis*, § 16·5.]

4. Prove that

$$\frac{1}{2\pi i}\int_{-\infty i}^{\infty i}\frac{\Gamma(\zeta+\alpha)\Gamma(\zeta+\beta)\Gamma(-\zeta)}{\Gamma(\zeta+\gamma)}(-z)^\zeta d\zeta=\frac{\Gamma(\alpha)\Gamma(\beta)}{\Gamma(\gamma)}F(\alpha,\beta\,;\ \gamma\,;\ z),$$

provided that $|\arg(-z)|<\pi$ and the contour is in general parallel to the imaginary axis but is curved where necessary to ensure that the poles of $\Gamma(\zeta+\alpha)\Gamma(\zeta+\beta)$ lie to the left and the poles of $\Gamma(-\zeta)$ lie to the right of the path.

[Barnes, *Proc. London Math. Soc.* (2), 6 (1908), p. 141.]

5. Prove that when $|\arg z|<\pi$

$$W_{k,m}(z)=\frac{e^{-\frac{1}{2}z}z^k}{2\pi i}\int_{-\infty i}^{\infty i}\frac{\Gamma(-\zeta-k-m+\frac{1}{2})\Gamma(-\zeta-k+m+\frac{1}{2})\Gamma(\zeta)}{\Gamma(-k-m+\frac{1}{2})\Gamma(-k+m+\frac{1}{2})}z^\zeta d\zeta,$$

and that this expression is a definition of $W_{k,m}(z)$ when $\pi\leqslant|\arg z|<\frac{3}{2}\pi$.

[Barnes.]

6. From the last result deduce that, when $|\arg z|<\frac{3}{2}\pi$,

$$W_{k,m}(z)=\frac{\Gamma(-2m)}{\Gamma(\frac{1}{2}-m-k)}M_{k,m}(z)+\frac{\Gamma(2m)}{\Gamma(\frac{1}{2}+m-k)}M_{k,m}(z),$$

where

$$M_{k,m}(z)=z^{\frac{1}{2}+m}e^{-\frac{1}{2}z}\left\{1+\frac{\frac{1}{2}+m-k}{1!\,(2m+1)}z+\frac{(\frac{1}{2}+m-k)(\frac{3}{2}+m-k)}{2!\,(2m+1)(2m+2)}z^2+\ldots\right\}.$$

[Whittaker and Watson, *Modern Analysis*, § 16·41.]

7. Prove that

$$J_n(z)=\frac{z^{-\frac{1}{2}}}{2^{2n+\frac{1}{2}}e^{\frac{1}{2}(n+\frac{1}{2})\pi i}\Gamma(n+1)}M_{0,n}(2iz)$$

and deduce the asymptotic expansion for $J_n(z)$.

8. Prove that

$$J_n(z)=\frac{\Gamma(\frac{1}{2}-n)}{2^{n+1}\pi i\,\Gamma(\frac{1}{2})}\int_C(\zeta^2-1)^{n-\frac{1}{2}}\cos(z\zeta)d\zeta,$$

where C is a figure-of-eight contour encircling $\zeta=1$ in the positive and $\zeta=-1$ in the negative directions. Deduce that when $\mathbf{R}(n+\frac{1}{2})>0$,

$$J_n(z)=\frac{z^n}{2^{n-1}\Gamma(n+\frac{1}{2})\Gamma(\frac{1}{2})}\int_0^{\frac{1}{2}\pi}\sin^{2n}\theta\cos(z\cos\theta)d\theta.$$

[Hankel, *Math. Ann.* 1 (1869), p. 467.]

9. Prove that when n is an integer,

$$\begin{aligned}Y_n(z)&=\lim_{\epsilon\to 0}\epsilon^{-1}\{J_{n+\epsilon}(z)-(-1)^nJ_{n-\epsilon}(z)\}\\&=\left(\frac{\pi}{2z}\right)^{\frac{1}{2}}\left\{e^{(\frac{1}{2}n+\frac{1}{4})\pi i}W_{0,n}(2iz)+e^{-(\frac{1}{2}n+\frac{1}{4})\pi i}W_{0,n}(-2iz)\right\}\end{aligned}$$

is a second solution of the Bessel equation, and deduce its asymptotic expansion.

[Hankel; Whittaker and Watson, *Modern Analysis*, § 17·6.]

CHAPTER XIX

SYSTEMS OF LINEAR EQUATIONS OF THE FIRST ORDER

19·1. Equivalent Singular Points.—In the system of n linear differential equations of the first order

$$\frac{dw_r}{dz} = \sum_{s=1}^{n} p_{rs}(z) w_s \qquad (r=1, 2, \ldots, n),$$

it will be supposed that the coefficients $p_{rs}(z)$ are analytic functions of the independent variable z, and have no singularities but poles even at infinity. Any finite point is an ordinary point of the system if the coefficients are analytic at that point; the point at infinity is an ordinary point if

$$p_{rs}(z) = O(z^{-2})$$

as $z \to \infty$. In studying the behaviour of the solutions at a singular point, it is a convenience, and no restriction, to transfer that point to infinity.

Outside a circle $|z|=R$, which includes all the finite singular points of the equation, the coefficients may be expanded in series of descending powers of z. If q is the greatest exponent of the leading term in any of these expansions, the number $q+1$ is, consistently with the previous definition, termed the *rank* of the singular point at infinity. Thus when $q \leqslant -2$, the point at infinity is an ordinary point; when $q=-1$ it is a regular singular point.

Let $q \geqslant 0$ and consider the possibility of satisfying the system of equations by a set of formal solutions of the normal type

$$w_r = e^{Q(z)} u_r(z),$$

where

$$Q(z) = \frac{\alpha z^{q+1}}{q+1} + \frac{\beta z^q}{q} + \ldots + \lambda z.$$

Then if

$$p_{rs}(z) = a_{rs} z^q + O(z^{q-1}),$$

α is determined by the *characteristic equation*

$$|a_{rs} - \delta_{rs}\alpha| = 0,$$

where

$$\delta_{rr}=1, \quad \delta_{rs}=0 \quad (r \neq s).$$

When $q=-1$, this same equation determines the exponent α in the regular solution

$$w_r = z^{\alpha}\{1+O(z^{-1})\}.$$

The nature of the formal solutions depends upon whether the roots

$$\alpha_1, \quad \alpha_2, \quad \ldots, \quad \alpha_n$$

of the characteristic equation are equal or unequal and, when $q=-1$, differ or do not differ by integers. But in any case, the fundamental existence theorem implies that there exists a set of n linearly independent solutions

$$w_1 = w_1^{(s)}, \quad w_2 = w_2^{(s)}, \quad \ldots, \quad w_n = w_n^{(s)} \quad (s=1, 2, \ldots, n),$$

such that each element $w_r^{(s)}$ is analytic for $|z| \geqslant R$, and the general solution may be expressed as a linear combination of these solutions, thus

$$w_1 = c_1 w_1^{(1)} + c_2 w_1^{(2)} + \ldots + c_n w_1^{(n)},$$
$$\cdot \quad \cdot \quad \cdot \quad \cdot \quad \cdot$$
$$w_n = c_1 w_n^{(1)} + c_2 w_n^{(2)} + \ldots + c_n w_n^{(n)}.$$

Now by any linear transformation of the form

$$w_r = \sum_{s=1}^{n} a_{rs}(z)\bar{w}_s \qquad (r=1, 2, \ldots, n),$$

where the coefficients $a_{rs}(z)$ are analytic at infinity and such that the determinant

$$\Delta = |a_{rs}(z)|$$

is not zero for $z=\infty$, the given linear differential system is transformed into a system of the same form, namely

$$\frac{d\bar{w}_r}{dz} = \sum_{s=1}^{n} \bar{p}_{rs}(z)\bar{w}_s \qquad (r=1, 2, \ldots, n).$$

The coefficients of this transformed equation are explicitly given by the formula

$$\bar{p}_{rs}(z) = \sum_{k, l=1}^{n} \bar{a}_{rk}(z) p_{kl}(z) a_{ls}(z) - \sum_{k=1}^{n} a_{rk}(z) \frac{d}{dz} a_{ks}(z) \qquad (r, s=1, 2, \ldots, n),$$

where $\{\bar{a}_{rs}(z)\}$ is the matrix of functions inverse to the matrix $\{a_{rs}(z)\}$, that is to say such that

$$\sum_{k=1}^{n} \bar{a}_{rk}(z) a_{ks}(z) = \delta_{rs}.$$

When the transformation is such that the coefficients are not only analytic at infinity but also satisfy the relations

$$a_{rs}(z) = \delta_{rs} \quad \text{for } z=\infty,$$

the original and the transformed systems are said to have an *equivalent singular point* at infinity. Since the inverse transformation has also this special property at infinity, the relation of equivalence is reciprocal. Moreover, since the product of two such transformations is also of this special form, the relation is transitive.

It is clear from the formulæ which express the coefficients $\bar{p}_{rs}(z)$ in terms of the coefficients $p_{rs}(z)$ that the rank of the transformed system cannot exceed that of the original system. But since the relation of equivalence is reciprocal, the converse is also true, and therefore the rank of all systems having an equivalent singular point is the same.

The conception of equivalent singular points suggests the problem of determining the simplest possible system which is equivalent, at infinity, to the given system. This problem is solved in the general case by a theorem which will be proved in the following section, namely that *every system of n linear differential equations, with a singular point of rank $q+1$ at infinity is equivalent at infinity to a canonical system of the form*

$$z\frac{dW_r}{dz} = \sum_{s=1}^{n} P_{rs}(z) W_s \qquad (r=1, 2, \ldots, n),$$

*in which the coefficients $P_{rs}(z)$ are polynomials of degree not exceeding $q+1$.**

* This theorem is due to Birkhoff, *Trans. Am. Math. Soc.* 10 (1909), p. 436. The simpler and more general proof here reproduced is also due to Birkhoff, *Math. Ann.* 74 (1913), p. 134.

Consider, for a moment, the implication of this theorem when the point at infinity is regular, and the roots $\alpha_1, \alpha_2, \ldots, \alpha_n$ of the characteristic equation are unequal and do not differ by integers. The canonical system is then, in its simplest form,

$$z\frac{dW_1}{dz}=\alpha_1 W_1, \ldots, z\frac{dW_n}{dz}=\alpha_n W.$$

It is soluble and has the fundamental set of n solutions

$$\begin{array}{llll} W_1^{(1)}=z^{\alpha_1}, & W_2^{(1)}=0, & \ldots, & W_n^{(1)}=0, \\ \cdot & \cdot & \cdot & \cdot \\ W_1^{(n)}=0, & W_2^{(n)}=0, & \ldots, & W_n^{(n)}=z^{\alpha_n}. \end{array}$$

Consequently the original equation has the fundamental set of solutions

$$w_1^{(s)}, \ldots, w_n^{(s)} \qquad (s=1, 2, \ldots, n),$$

where

$$w_r^{(s)}=\sum_{t=1}^{n} a_{rt}(z)W_t^{(s)}$$
$$=z^{\alpha_s}\, a_{rs}(z).$$

This is, in fact, the fundamental existence theorem for a regular singular point; in the same way the solutions of the canonical system lead to solutions of the original system when the point at infinity is an irregular singularity.

19·2. Reduction to a Canonical System.—The proof of the theorem enunciated in the preceding section depends upon a lemma in the theory of analytic functions which will be stated, without proof, in the following terms: *

Let $\{l_{rs}(z)\}$ be any matrix of functions, single-valued and analytic for $|z|\geqslant R$, but not necessarily analytic for $z=\infty$, and such that the determinant of this matrix does not vanish for $|z|\geqslant R$. Then there exists a matrix $\{a_{rs}(z)\}$ of functions analytic at infinity and reducing at infinity to the unit matrix (δ_{rs}), and also a matrix $\{e_{rs}(z)\}$ of integral functions, whose determinant is nowhere zero in the finite plane, such that

$$\{l_{rs}(z)\}=\{a_{rs}(z)\}\{e_{rs}(z)z^{k_s}\},$$

where $k_1, k_2, \ldots, k_n$ are integers.

The significance of the lemma may be illustrated by considering a single function $l(z)$ and taking R so large that $l(z)$ does not vanish for $|z|\geqslant R$. Then $\log l(z)$ is analytic for $|z|\geqslant R$, but not single-valued. But after a positive circuit around $z=\infty$, $\log l(z)$ becomes

$$\log l(z)-2\pi ki,$$

where k is an integer. Consequently

$$\log l(z)-k\log z$$

is both analytic and single-valued for $|z|\geqslant R$, and its expansion as a Laurent series shows it to be of the form

$$A(z)+E(z),$$

where $A(z)$ is analytic at infinity and $A(\infty)=0$, and $E(z)$ is an integral function. Let

$$a(z)=\exp A(z), \quad e(z)=\exp E(z),$$

then

$$l(z)=a(z)e(z)z^{-k},$$

* For a proof based upon the theory of linear integral equations see Birkhoff, *Bull. Am. Math. Soc.* 18 (1911), p. 64; *Math. Ann.* 74 (1913), p. 122. A proof in matrix notation of an equivalent theorem is given by Birkhoff in *Trans. Am. Math. Soc.* 10 (1909), p. 438, and generalised in *Proc. Am. Acad.* 49 (1913), p. 521. These theorems are included in more general theorems by Hilbert, *Gött. Nach.* 1905, p. 307, and Plemelj, *Monatsh. Math. Phys.* 19 (1908), p. 211.

where $a(z)$ is analytic at infinity and $a(\infty)=1$, $e(z)$ is an integral function and k is an integer.

Now let z describe in the negative sense a simple closed curve C, enclosing the circle $|z|=R$ within which lie all the finite singularities of the system. This curve is equivalent to a circuit described in the positive sense about the point at infinity. Since every finite point outside the circle $|z|=R$ is an ordinary point of the system, there exists at any point of the curve C, a fundamental set of n solutions

$$(w_1^{(1)}, \ldots, w_n^{(1)}), \ldots, (w_n^{(1)}, \ldots, w_n^{(n)}),$$

each element of which is analytic at all points of C. The elements of these solutions are not, however, single-valued, and thus when z has described a complete circuit along the curve C, the solutions are transformed into a new fundamental set

$$(\overline{w}_1^{(1)}, \ldots, \overline{w}_n^{(1)}), \ldots, (\overline{w}_1^{(n)}, \ldots, \overline{w}_n^{(n)}).$$

The two sets of solutions are connected by linear relations

$$\overline{w}_r^{(s)}=c_1^{(s)}w_r^{(1)}+ \ldots +c_n^{(s)}w_r^{(n)},$$

or in matrix notation

$$(\overline{w}_r^{(s)})=(w_r^{(s)})(c_r^{(s)}),$$

where $(c_r^{(s)})$ is a matrix of constants of non-zero determinant.

In the general case, that is to say, when the roots $\rho_1, \rho_2, \ldots, \rho_n$ of the equation

$$|c_r^{(s)}-\delta_{rs}\rho|=0$$

are unequal,* the initial fundamental set of solutions may be so chosen that the matrix $(c_r^{(s)})$ has the simple form $(\delta_{rs}\rho_s)$. Thus the substitution relative to a circuit in the positive direction around $z=\infty$ is

$$\overline{w}_1^{(1)}=\rho_1 w_1^{(1)}, \ldots, \overline{w}_n^{(1)}=\rho_1 w_n^{(1)},$$

$$\cdot \quad \cdot \quad \cdot \quad \cdot$$

$$\overline{w}_1^{(n)}=\rho_n w_1^{(n)}, \ldots, \overline{w}_n^{(n)}=\rho_n w_n^{(n)}.$$

Now let $\lambda_1, \lambda_2, \ldots, \lambda_n$ be numbers which satisfy the equations

$$\lambda_s=\frac{-1}{2\pi i}\log \rho_s \qquad (s=1, 2, \ldots, n).$$

These equations leave $\lambda_1, \lambda_2, \ldots, \lambda_n$ undetermined to the extent of additive integers. For any chosen determination of λ_s let

$$w_r^{(s)}(z)=z^{\lambda_s}l_{rs}(z),$$

then each function $l_{rs}(z)$ is single-valued and analytic for $|z|\geqslant R$ and the determinant of these functions has the value

$$\begin{aligned}|l_{rs}(z)|&=z^{-(\lambda_1+\ldots+\lambda_n)}|w_r^{(s)}(z)|\\ &=cz^{-(\lambda_1+\ldots+\lambda_n)}\exp\left\{\int[p_{11}(z)+p_{22}(z)+\ldots+p_{nn}(z)]dz\right\},\end{aligned}$$

where c is a constant, and is not zero for $|z|\geqslant R$.

The matrix of functions $\{l_{rs}(z)\}$ thus satisfies the conditions of the lemma and can therefore be decomposed into the product of matrices

$$\{l_{rs}(z)\}=\{a_{rs}(z)\}\{e_{rs}(z)z^{k_s}\}.$$

Let

$$W_{rs}^{(s)}=e_{rs}(z)z^{k_s+\lambda_s}=e_{rs}(z)z^{\mu_s},$$

* Strictly speaking, it is not necessary to assume the inequality of $\rho_1, \rho_2, \ldots, \rho_n$; the correct assumption to make is that the elementary divisors of the matrix $(c_r^{(s)}-\delta_{rs}\rho)$ are distinct. *Vide* Kowalewski, *Determinantentheorie*, Chap. XIII.

then with the functions $a_{rs}(z)$ so defined, the transformation

$$\{w_r^{(s)}\}=\{a_{rs}(z)\}\{W_r^{(s)}\}$$

connects each particular set

$$w_1^{(s)}, \ldots, w_n^{(s)}$$

of solutions of the original equation

$$\frac{dw_r}{dz}=\sum_{s=1}^{n} p_{rs}(z)w_s \qquad (r=1, 2, \ldots, n)$$

with the corresponding set

$$W_1^{(s)}, \ldots, W_n^{(s)}$$

of solutions of the transformed equation

$$\frac{dW_r}{dz}=\sum_{s=1}^{n} \bar{p}_{rs}(z)W_s \qquad (r=1, 2, \ldots, n).$$

The n^2 equations satisfied by the elements $W_r^{(s)}$ may be combined into the matrix equation

$$\left\{\frac{dW_r^{(s)}}{dz}\right\}=\{\bar{p}_{rs}(z)\}\{W_r^{(s)}\},$$

whence, if $\{W_r^{(s)}\}^{-1}$ is the matrix inverse to $\{W_r^{(s)}\}$,

$$\{\bar{p}_{rs}(z)\}=\left\{\frac{dW_r^{(s)}}{dz}\right\}\{W_r^{(s)}\}^{-1}.$$

Now

$$\{W_r^{(s)}\}=\{e_{rs}(z)z^{-\mu_s}\}=\{e_{rs}(z)\}\{\delta_{rs}z^{\mu_s}\},$$

and therefore

$$\{W_r^{(s)}\}^{-1}=\{\delta_{rs}z^{-\mu_s}\}\{e_{rs}(z)\}^{-1}.$$

Also

$$\left\{\frac{dW_r^{(s)}}{dz}\right\}=\left\{z^{\mu_s}\frac{d}{dz}e_{rs}(z)+\mu_s z^{\mu_s-1}e_{rs}(z)\right\}$$
$$=\{f_{rs}(z)z^{\mu_s-1}\}=\{f_{rs}(z)\}\{\delta_{rs}z^{\mu_s-1}\},$$

where the functions $f_{rs}(z)$ are integral functions. Consequently

$$\{\bar{p}_{rs}(z)\}=\{f_{rs}(z)\}\{\delta_{rs}z^{\mu_s-1}\}\{\delta_{rs}z^{-\mu_s}\}\{e_{rs}(z)\}^{-1}$$
$$=z^{-1}\{f_{rs}(z)\}\{e_{rs}(z)\}^{-1}.$$

Since the determinant $|l_{rs}(z)|$ is nowhere zero in the finite plane, the matrix $\{e_{rs}(z)\}^{-1}$ is a matrix of integral functions. Consequently each function

$$z\,\bar{p}_{rs}(z)$$

is an integral function.

But since the rank of the singular point at infinity is $q+1$,

$$\bar{p}_{rs}(z)=O(z^q)$$

as $z\to\infty$. Thus $z\,\bar{p}_{rs}(z)$ is an integral function which has a pole of order $q+1$ at most at infinity and is therefore a polynomial of degree not greater than $q+1$.

The given system is therefore equivalent at infinity to the canonical system

$$z\frac{dW_r}{dz}=\sum_{s=1}^{n} P_{rs}(z)W_s \qquad (r=1, 2, \ldots, n),$$

where the coefficients $P_{rs}(z)$ are polynomials of degree $q+1$ at most.

The canonical system may be still further simplified by a substitution of the form

$$\overline{W}_r=\sum_{s=1}^{n} c_{rs}W_s \qquad (r=1, 2, \ldots, n).$$

In particular, when the roots $a_1, a_2, \ldots, a_n$ of the characteristic equation

$$|a_{rs}-\delta_{rs}a|=0$$

are unequal, the constants c_{rs} may be so chosen that the polynomials $P_{rs}(z)$ are of the form *

$$P_{rs}=p_{rs}^{(0)}+p_{rs}^{(1)}z+\ \ldots\ +p_{rs}^{(q)}z^q \qquad (r\neq s),$$
$$P_{rr}=p_{rr}^{(0)}+p_{rr}^{(1)}z+\ \ldots\ +p_{rr}^{(q)}q^q+a_r z^{q+1}.$$

When the polynomials $P_{rs}(z)$ are thus simplified the system is said to be in the *standard canonical form.*

19·21. Modification of the Proof in the Degenerate Case.—To illustrate how the argument is modified in the degenerate case in which two or more of the multipliers ρ, corresponding to a positive circuit around the point at infinity, are equal, consider the particular case $\rho_1=\rho_2$. If, as in the general case, there is a fundamental set of solutions such that

$$\overline{w}_r^{(s)}=\rho_s w_r^{(s)},$$

for $s=1, 2, \ldots, n$, no modification is necessary. When this is not the case,† a fundamental set of solutions exists such that (cf. § 15·22) for $r=1, 2, \ldots, n$,

$$\overline{w}_r^{(1)}=\rho_1 w_r^{(1)},$$
$$\overline{w}_r^{(2)}=\rho_1 w_r^{(2)}+w_r^{(1)},$$
$$\overline{w}_r^{(s)}=\rho_s w_r^{(s)} \qquad (s=3, 4, \ldots, n).$$

As before, let

$$\lambda_s=\frac{-1}{2\pi i}\log\rho_s \qquad (s=1, 3, \ldots, n),$$

and write

$$w_r^{(1)}=z^{\lambda_1}l_{r1}(z),$$
$$w_r^{(2)}=z^{\lambda_1}\left\{l_{r2}(z)+\frac{1}{2\pi\rho_1 i}l_{r1}(z)\log z\right\},$$
$$w_r^{(s)}=z^{\lambda_s}l_{rs}(z) \qquad (s=3, 4, \ldots, n).$$

In this way there is defined a matrix $\{l_{rs}(z)\}$ of functions which are single-valued and analytic for $|z|\geqslant R$ and whose determinant

$$z^{-2\lambda_1-\lambda_3-\ \ldots\ -\lambda_n}|w_r^{(s)}|$$

is not zero for $|z|\geqslant R$.

Then, as before, the transformation

$$\{w_r^{(s)}\}=\{a_{rs}(z)\}\{W_r^{(s)}\}$$

changes the given system into an equivalent canonical system

$$z\frac{dW_r}{dz}=\sum_{s=1}^{n}P_{rs}(z)W_s \qquad (r=1, 2, \ldots, n),$$

in which the coefficients $P_{rs}(z)$ are polynomials of maximum degree $q+1$, and which has the fundamental set of solutions

$$(W_r^{(1)}, \ldots, W_r^{(n)}) \qquad (r=1, 2, \ldots, n),$$

* The coefficients c_{rs} are such that the operations

$$\text{new col. } r=c_{r1}\text{ (col. 1)}+\ \ldots\ +c_{rn}\text{ (col. } n) \qquad (r=1, 2, \ldots, n)$$

transforms the determinant $|a_{rs}-\delta_{rs}a|$ into $|\delta_{rs}(a_r-a)|$. The corresponding theorem when $a_1, a_2, \ldots, a_n$ are not all distinct may be supplied by the reader.

† That is to say, when the elementary divisors of the matrix $(c_r^{(s)}-\delta_{rs}\rho)$ corresponding to $\rho_1=\rho_2$ are equal.

where

$$W_r^{(1)}=z^{k_1+\lambda_1}e_{r1}(z)=z^{\mu_1}e_{r1}(z),$$

$$W_r^{(2)}=z^{\lambda_1}\left\{z^{k_2}e_{r2}(z)+\frac{1}{2\pi\rho_1 i}z^{k_1}e_{r1}(z)\log z\right\},$$

$$W_r^{(s)}=z^{k_s+\lambda_s}e_{rs}(z)=z^{\mu_s}e_{rs}(z) \qquad (s=3, 4, \ldots, n).$$

As before $\{l_{rs}(z)\}$ is a matrix of integral functions whose determinant does not vanish anywhere in the finite plane and $k_1, \ldots, k_n$ are integers. The standard canonical form is reached as before.

Cases of further degeneracy may be disposed of in the same manner; and thus the possibility of reduction to the canonical form is established in all cases.

19·22. A simple Example of the Reduction to Standard Canonical Form.—Consider the linear differential equation of the second order *

$$\frac{d^2w}{dz^2}+p(z)\frac{dw}{dz}+q(z)w=0,$$

in which $p(z)$ and $q(z)$ are analytic for $|z|\geqslant R$ and, at infinity,

$$p(z)=p_0+O(z^{-1}), \quad q(z)=q_0+O(z^{-1}).$$

In the most general case the point at infinity is an irregular singularity of rank unity. If b_1 and b_2 are the roots of the quadratic equation

$$b^2+p_0b+q_0=0,$$

and are distinct, and if the constant c is properly chosen, the change of variables

$$z=(b_2-b_1)\bar{z}, \quad w=e^{b_1z}z^c\bar{w}$$

will transform the given equation into an equation of the same form but with

$$p(z)=-1+p_1z^{-1}+O(z^{-2}), \quad q(z)=O(z^{-2}).$$

It will therefore be supposed that $p(z)$ and $q(z)$ are of these forms.

Now if $v=z\dfrac{dw}{dz}$, the single equation of the second order may be replaced by the pair of equations of the first order

$$\frac{dw}{dz}=\frac{v}{z}, \quad \frac{dv}{dz}=-zq(z)w+\left\{-p(z)+\frac{1}{z}\right\}v.$$

A pair of solutions w_1, w_2 of the original equation can always be found such that if the point z describes a positive circuit about the point at infinity, then either

$$\overline{w}_1=\rho_1w_1, \quad \overline{w}_2=\rho_2w_2$$

or

$$\overline{w}_1=\rho_1w_1, \quad \overline{w}_2=\rho_1w_2+w_1.$$

The first case will be dealt with in detail; the modifications which the second case involves will be indicated subsequently.

Thus the linear system admits of the solutions

$$w_1=z^{\lambda_1}l_{11}(z), \quad w_2=z^{\lambda_2}l_{12}(z),$$

$$v_1=zw_1'=z^{\lambda_1}l_{21}(z), \quad v_2=zw_2'=z^{\lambda_2}l_{22}(z),$$

where the exponents λ_1, λ_2 satisfy the equations

$$\lambda_1=\frac{-1}{2\pi i}\log\rho_1, \quad \lambda_2=\frac{-1}{2\pi i}\log\rho_2,$$

and are thus arbitrary as to additive integers, and the functions $l_{11}(z)$, $l_{12}(z)$,

* Birkhoff, *Trans. Am. Math. Soc.* 14 (1913), p. 462.

$l_{21}(z)$ and $l_{22}(z)$ are single-valued and analytic for $|z|\geqslant R$. Moreover the determinant has the value

$$l_{11}(z)l_{22}(z)-l_{12}(z)l_{21}(z)=z^{1-\lambda_1-\lambda_2}(w_1w_2'-w_2w_1')$$
$$=z^{1-\lambda_1-\lambda_2}e^{-\int p(z)dz},$$

and is not zero for $|z|\geqslant R$.

In order to carry out explicitly the reduction to canonical form, it is convenient to restate the lemma of § 19·2 for the particular case $n=2$, as follows : *Let* $l_{11}(z)$, $l_{12}(z)$, $l_{21}(z)$, $l_{22}(z)$ *be functions single-valued and analytic for* $|z|\geqslant R$ *(but not necessarily analytic at infinity), and such that their determinant* $l_{11}(z)l_{22}(z)-l_{12}(z)l_{21}(z)$ *does not vanish for* $|z|\geqslant R$. *Then there exist a set of functions* $a_{11}(z)$, $a_{12}(z)$, $a_{21}(z)$, $a_{22}(z)$ *analytic at infinity and reducing respectively to* 1, 0, 0, 1 *at infinity, and a set of integral functions* $e_{11}(z)$, $e_{12}(z)$, $e_{21}(z)$, $e_{22}(z)$ *whose determinant does not vanish at any point in the finite plane, such that*

$$l_{11}(z)=\{a_{11}(z)e_{11}(z)+a_{12}(z)e_{21}(z)\}z^{k_1},$$
$$l_{12}(z)=\{a_{11}(z)e_{12}(z)+a_{12}(z)e_{22}(z)\}z^{k_2},$$
$$l_{21}(z)=\{a_{21}(z)e_{11}(z)+a_{22}(z)e_{21}(z)\}z^{k_1},$$
$$l_{22}(z)=\{a_{21}(z)e_{12}(z)+a_{22}(z)e_{22}(z)\}z^{k_2},$$

where k_1 *and* k_2 *are integers.*

Now four functions $l_{11}(z)$, $l_{12}(z)$, $l_{21}(z)$, $l_{22}(z)$, satisfying these conditions, have been defined by means of the relations

$$w_1=z^{\lambda_1}l_{11}(z), \quad w_2=z^{\lambda_2}l_{12}(z),$$
$$v_1=z^{\lambda_1}l_{21}(z), \quad v_2=z^{\lambda_2}l_{22}(z),$$

and their definition depends upon the actual choice of λ_1 and λ_2. By properly choosing these exponents, the integers k_1 and k_2 can be made zero, and it will be supposed that this definite choice of λ_1 and λ_2 has been made.

Now make the transformation

$$w=a_{11}(z)W+a_{12}(z)V, \quad v=a_{21}(z)W+a_{22}(z)V\,;$$

then the transformed system is

$$\frac{dW}{dz}=P_{11}(z)W+P_{12}(z)V, \quad \frac{dV}{dz}=P_{21}(z)W+P_{22}(z)V,$$

where

$$P_{11}=\frac{1}{\Delta}\left[a_{22}\left\{\frac{a_{21}}{z}-a'_{11}\right\}-a_{12}\left\{-zqa_{11}+\left(-p+\frac{1}{z}\right)a_{21}-a'_{21}\right\}\right],$$
$$P_{12}=\frac{1}{\Delta}\left[a_{22}\left\{\frac{a_{22}}{z}-a'_{12}\right\}-a_{12}\left\{-zqa_{12}+\left(-p+\frac{1}{z}\right)a_{22}-a'_{22}\right\}\right],$$
$$P_{21}=\frac{-1}{\Delta}\left[a_{21}\left\{\frac{a_{21}}{z}-a'_{11}\right\}-a_{11}\left\{-zqa_{11}+\left(-p+\frac{1}{z}\right)a_{21}-a'_{21}\right\}\right],$$
$$P_{22}=\frac{-1}{\Delta}\left[a_{21}\left\{\frac{a_{22}}{z}-a'_{12}\right\}-a_{11}\left\{-zqa_{12}+\left(-p+\frac{1}{z}\right)a_{22}-a'_{22}\right\}\right],$$

and the determinant

$$\Delta=a_{11}a_{22}-a_{12}a_{21}$$

is not identically zero.

Since, at infinity,

$$a_{11}=a_{22}=1, \quad a_{12}=a_{21}=0,$$

these expressions admit of developments of the form

$$P_{11}=O(z^{-2}), \qquad P_{12}=rz^{-1}+O(z^{-2}),$$
$$P_{21}=sz^{-1}+O(z^{-2}), \quad P_{22}=1+(1+p_1)z^{-1}+O(z^{-2}),$$

where r and s are constants whose values will be determined later.

The solutions of the transformed system are

$$W_1=z^{\lambda_1}e_{11}(z), \quad W_2=z^{\lambda_2}e_{12}(z),$$
$$V_1=z^{\lambda_1}e_{21}(z), \quad V_2=z^{\lambda_2}e_{22}(z)\,;$$

on substituting these expressions in the first equation of the transformed system it is found that

$$\frac{d}{dz}e_{11}(z)+\frac{\lambda_1}{z}e_{11}(z)=P_{11}(z)e_{11}(z)+P_{12}(z)e_{21}(z),$$
$$\frac{d}{dz}e_{12}(z)+\frac{\lambda_2}{z}e_{12}(z)=P_{11}(z)e_{12}(z)+P_{12}(z)e_{22}(z).$$

Since $e_{11}(z)$, $e_{12}(z)$, $e_{21}(z)$, $e_{22}(z)$ are integral functions and their determinant

$$e_{11}(z)e_{22}(z)-e_{12}(z)e_{21}(z)$$

is not zero for any finite value of z, the functions $P_{11}(z)$ and $P_{12}(z)$ are analytic throughout the finite plane except for a possible simple pole at the origin. By considering the second equation of the transformed system it may be proved that the same is true with regard to the functions $P_{21}(z)$ and $P_{22}(z)$. But the four functions $P_{rs}(z)$ are analytic at infinity; they are therefore linear in z^{-1}. Thus the terms $O(z^{-2})$ in the developments of these functions disappear and the transformed system has the simple canonical form *

$$z\frac{dW}{dz}=rV, \quad z\frac{dV}{dz}=sW+(z+1-p_1)V.$$

This leads to the theorem: *If $w(z)$ is a solution of the equation*

$$\frac{d^2w}{dz^2}+p(z)\frac{dw}{dz}+q(z)w=0,$$

where

$$p(z)=-1+p_1z^{-1}+O(z^{-2}), \quad q(z)=O(z^{-2}),$$

then $w(z)$ and $zw'(z)$ may be represented in the forms

$$w(z)=a_{11}(z)W+a_{12}(z)\frac{z}{r}\cdot\frac{dW}{dz}, \quad z\frac{dw(z)}{dz}=a_{21}(z)W+a_{22}(z)\frac{z}{r}\cdot\frac{dW}{dz},$$

where W is a particular solution of

$$\frac{d^2W}{dz^2}+\left\{-1+\frac{p_1}{z}\right\}\frac{dW}{dz}-\frac{rs}{z^2}W=0,$$

and $a_{11}(z)$, $a_{12}(z)$, $a_{21}(z)$, $a_{22}(z)$ are analytic at infinity and reduce when $z=\infty$ to 1, 0, 0, 1 respectively.†

The constants r and s will now be identified. The origin is a regular singular point of the transformed equation with exponents λ_1 and λ_2. But the indicial equation relative to this singularity is

$$\lambda^2+(p_1-1)\lambda-rs=0,$$

and therefore

$$\lambda_1+\lambda_2=1-p_1, \quad \lambda_1\lambda_2=-rs.$$

In the exceptional case when

$$\overline{w}_1=\rho_1w_1, \quad \overline{w}_2=\rho_1w_2+w_1,$$

* The system is integrable by quadratures when either r or s is zero.

† When $r=0$ it is necessary to replace

$$\frac{z}{r}\cdot\frac{dW}{dz} \text{ by } \lim_{r\to 0}\frac{z}{r}\cdot\frac{dW}{dz}.$$

the functions $l_{11}(z)$, $l_{12}(z)$, $l_{21}(z)$, $l_{22}(z)$ are defined, so as to satisfy the conditions of the lemma, by means of the relations

$$w_1 = z^{\lambda_1} l_{11}(z), \quad w_2 = z^{\lambda_1}\left\{l_{12}(z) + \frac{1}{2\pi\rho_1 i} l_{11}(z) \log z\right\},$$
$$v_1 = zw_1' = z^{\lambda_1} l_{21}(z), \quad v_2 = zw_2' = z^{\lambda_1}\left\{l_{22}(z) + \frac{1}{2\pi\rho_1 i} l_{21}(z) \log z\right\},$$

where

$$\lambda_1 = \frac{-1}{2\pi i} \log \rho_1,$$

and λ_1 is so determined that, in the lemma, $k_1 = 0$. The argument then proceeds on the main lines as before, and ends with precisely the same theorem.

19·3. Formal Solutions.—It will now be supposed that all the roots, $\alpha_1, \ldots, \alpha_n$ of the characteristic equation of the given system are unequal and that $q \geqslant 0$. The equivalent standard canonical system is therefore

$$z\frac{dW_r}{dz} = \sum_{s=1}^{n} P_{rs}(z) W_s \qquad (r=1, 2, \ldots, n),$$

where

$$P_{rs}(z) = p_{rs}^{(0)} + p_{rs}^{(1)} z + \ldots + p_{rs}^{(q)} z^q \qquad (r \neq s),$$
$$P_{ss}(z) = p_{ss}^{(0)} + p_{ss}^{(1)} z + \ldots + p_{ss}^{(q)} z^q + \alpha_s z^{q+1}.$$

Then for each value of s there will arise a formal solution

$$W_1 = T_1^{(s)}, \quad \ldots, \quad W_n = T_n^{(s)}$$

of the normal type in which

$$T_r^{(s)} = e^{Q_s(z)} z^{\mu_s} B_{rs}(z),$$

where

$$Q_s(z) = \frac{\alpha_s z^{q+1}}{q+1} + \frac{\beta_s z^q}{q} + \ldots + \lambda_s z,$$
$$B_{rs}(z) = B_{rs} + B_{rs}^{(1)} z^{-1} + \ldots,$$

and μ_s is so chosen that the constant B_{ss} is not zero. To make the formal solutions definite, B_{ss} will be given the value unity.

By direct substitution it may be verified that

$$B_{rs} = 0 \qquad (r \neq s),$$

and that

$$\beta_s = p_{rr}^{(q)} \qquad (q > 0),$$
$$\mu_s = p_{ss}^{(0)} \qquad (q = 0).$$

The remaining coefficients are then determined in the order

$$B_{rs}^{(1)}, \gamma_s, B_{rs}^{(2)}, \ldots, \lambda_s, B_{rs}^{(q)}, \mu_s.$$

The determinant of the formal solutions is

$$|T_r^{(s)}| = e^{Q_1(z) + \ldots + Q_n(z)} z^{\mu_1 + \ldots + \mu_n}\{D + D^{(1)} z^{-1} + \ldots\},$$

where

$$D = |B_{rs}| = 1.$$

The formal determinant therefore does not vanish identically.

Since solutions of the original system

$$\frac{dw_r}{dz} = \sum_{s=1}^{n} p_{rs}(z) w_s \qquad (s=1, 2, \ldots, n)$$

are connected with the solutions of the canonical system by the relations

$$w_r^{(s)} = \sum_{t=1}^{n} a_{rt}(z) W_t^{(s)},$$

where each function $a_{rt}(z)$ is analytic at infinity and reduces to δ_{rt} for $z=\infty$, it follows that the system admits of precisely n formal solutions

$$w_1 = S_1^{(s)}, \quad . \; . \; ., \quad w_n = S_n^{(s)} \qquad (s=1, 2, \; . \; . \; ., n),$$

in which

$$S_r^{(s)} = e^{Q_s(z)} z^{\mu_s} A_{rs}(z),$$

where $Q_s(z)$ is the same polynomial as for the canonical system, and $A_{rs}(z)$ is a series of descending powers of z and has the value δ_{rs} for $z=\infty$.

19·4. Solution of the Standard Canonical System of Rank Unity by Laplace Integrals.—When $q=0$ the standard canonical system is of the form

$$z\frac{dW_1}{dz} = \{p_{11}^{(0)} + a_1 z\} W_1 + p_{12}^{(0)} W_2 + \; . \; . \; . \; + p_{1n}^{(0)} W_n,$$

$$z\frac{dW_2}{dz} = p_{21}^{(0)} W_1 + \{p_{22}^{(0)} + a_2 z\} W_2 + \; . \; . \; . \; + p_{2n}^{(0)} W_n,$$

$$. \quad . \quad . \quad . \quad . \quad . \quad . \quad . \quad . \quad . \quad . \quad .$$

$$z\frac{dW_n}{dz} = p_{n1}^{(0)} W_1 + p_{n2}^{(0)} W_2 + \; . \; . \; . \; + \{p_{nn}^{(0)} + a_n z\} W_n.$$

The formal solutions

$$W_1 = T_1^{(s)}, \quad . \; . \; ., \quad W_n = T_n^{(s)} \qquad (s=1, 2, \; . \; . \; ., n)$$

are given by expressions such as

$$T_r^{(s)} = e^{a_s z} z^{\mu_s} B_{rs}(z),$$

where

$$\mu_s = p_{ss}^{(0)}$$

and

$$B_{rs}(z) = B_{rs} + B_{rs}^{(1)} z^{-1} + \; . \; . \; . \; .$$

Now consider the possibility of satisfying the system by the set of Laplace integrals

$$W_r = \int e^{z\zeta} v_r(\zeta) d\zeta \qquad (r=1, 2, \; . \; . \; ., n).$$

By direct substitution in the differential system it is found that the condition to be satisfied is

$$\sum_{s=1}^{n} p_{rs}^{(0)} \int e^{z\zeta} v_s(\zeta) d\zeta = \int z e^{z\zeta} (\zeta - a_r) v_r(\zeta) d\zeta$$

$$= [e^{z\zeta}(\zeta - a_r) v_r(\zeta)] - \int e^{z\zeta} \left\{ v_r(\zeta) + (\zeta - a_r) \frac{dv_r(\zeta)}{d\zeta} \right\} d\zeta \quad (r=1, 2, \; . \; . \; ., n).$$

Consequently the functions $v_1(\zeta)$, $v_2(\zeta)$, . . ., $v_n(\zeta)$ must satisfy the Laplace transformed system

$$-(\zeta - a_1) \frac{dv_1}{d\zeta} = \{p_{11}^{(0)} + 1\} v_1 + p_{12}^{(0)} v_2 + \; . \; . \; . \; + p_{1n}^{(0)} v_n,$$

$$-(\zeta - a_2) \frac{dv_2}{d\zeta} = p_{21}^{(0)} v_1 + \{p_{22}^{(0)} + 1\} v_2 + \; . \; . \; . \; + p_{2n}^{(0)} v_n,$$

$$. \quad . \quad . \quad . \quad . \quad . \quad . \quad . \quad . \quad . \quad . \quad .$$

$$-(\zeta - a_n) \frac{dv_n}{d\zeta} = p_{n1}^{(0)} v_1 + p_{n2}^{(0)} v_2 + \; . \; . \; . \; + \{p_{nn}^{(0)} + 1\} v_n,$$

and the corresponding contour of integration must be such that every one of the terms

$$[e^{z\zeta}(\zeta-a_r)v_r(\zeta)] \qquad (r=1, 2, \ldots, n)$$

vanishes identically in z.

Now the Laplace transformed system has regular singular points at $\zeta=a_1, a_2, \ldots, a_n$ and at infinity. The exponents relative to $\zeta=a_s$ are all zero except one which has the value

$$-p_{ss}^{(0)}-1=-\mu_s-1.$$

It will, for the moment, be supposed that this exponent is not a negative integer. Then the corresponding solutions of the transformed system, namely

$$v_1^{(s)}=(\zeta-a_s)^{-\mu_s-1}\phi_1^{(s)}(\zeta-a_s), \ldots, v_n^{(s)}=(\zeta-a_s)^{-\mu_s-1}\phi_n^{(s)}(\zeta-a_s),$$

where the functions $\phi_r^{(s)}(\zeta-a_s)$ are analytic in the neighbourhood of $\zeta=a_s$, lead to a set of integral solutions if the corresponding contour is a loop C_s from infinity in the ζ-plane along a suitable ray, encircling the point a_r in the negative direction, and returning to infinity along the ray. The conditions which must be imposed upon the ray are that it does not meet any singular point other than a_s, and that $\mathbf{R}\{z(\zeta-a_s)\}$ is negative along the ray.

Then a set of solutions is represented by the formulæ

$$W_r^{(s)}=\int_{C_s} e^{z\zeta}(\zeta-a_s)^{-\mu_s-1}\phi_r^{(s)}(\zeta-a_s)d\zeta \quad (r=1, 2, \ldots, n).$$

To each finite singular point a_s corresponds a set of solutions, that is n sets in all.

When $-\mu_s-1$ is a positive integer, the contour degenerates into a rectilinear path extending in an appropriate direction from a_s to infinity. When $-\mu_s-1$ is a negative integer or zero, the logarithmic case arises but does not present any special difficulty. Thus *each set of integrals*

$$W_1^{(s)}, \quad \ldots, \quad W_n^{(s)}$$

represents a solution of the standard canonical system of rank unity, which is valid in certain sectors of the z-plane.

19·41. Solution of the System of Rank Two.—It will now be shown that the foregoing process may be modified and extended so as to cover systems of rank greater than unity. Consider first of all the system of rank two ($q=1$):

$$z\frac{dW_1}{dz}=\{p_{11}^{(0)}+p_{11}^{(1)}z+a_1z^2\}W_1+\{p_{12}^{(0)}+p_{12}^{(1)}z\}W_2+\ldots+\{p_{1n}^{(0)}+p_{1n}^{(1)}z\}W_n,$$

$$z\frac{dW_2}{dz}=\{p_{21}^{(0)}+p_{21}^{(1)}z\}W_1+\{p_{22}^{(0)}+p_{22}^{(1)}z+a_2z^2\}W_2+\ldots+\{p_{2n}^{(0)}+p_{2n}^{(1)}z\}W_n,$$

$$\cdot \quad \cdot \quad \cdot \quad \cdot \quad \cdot \quad \cdot \quad \cdot \quad \cdot \quad \cdot \quad \cdot \quad \cdot \quad \cdot$$

$$z\frac{dW_n}{dz}=\{p_{n1}^{(0)}+p_{n1}^{(1)}z\}W_1+\{p_{n2}^{(0)}+p_{n2}^{(1)}z\}W_2+\ldots+\{p_{nn}^{(0)}+p_{nn}^{(1)}z+a_nz^2\}W_n.$$

The formal solutions are

$$W_1=T_1^{(s)}, \quad \ldots, \quad W_n=T_n^{(s)} \quad (s=1, 2, \ldots, n),$$

where

$$T_r^{(s)}=e^{\frac{1}{2}a_sz^2+\beta_sz}z^{\mu_s}\{B_{rs}+B_{rs}^{(1)}z^{-1}+\ldots\}$$

and

$$\beta_s=p_{ss}^{(1)}.$$

Now in this case the Laplace integral is replaced by an integral ot more general form, namely

$$W_r=\int e^{z^2\zeta}\{v_{r0}(\zeta)+zv_{r1}(\zeta)\}d\zeta \qquad (r=1, 2, \ldots, n).$$

When this expression for W_r is substituted in the system of equations it is found that

$$\sum_{s=1}^{n}\{p_{rs}^{(0)}+p_{rs}^{(1)}z\}\int e^{z^2\zeta}\{v_{s0}(\zeta)+zv_{s1}(\zeta)\}d\zeta$$

$$=\int z^2e^{z^2\zeta}(2\zeta-a_r)\{v_{r0}(\zeta)+zv_{r1}(\zeta)\}d\zeta+\int ze^{z^2\zeta}v_{r1}(\zeta)d\zeta,$$

or, transferring the terms which involve z^2 from the left-hand to the right-hand member,

$$\sum_{s=1}^{n}\{p_{rs}^{(0)}+p_{rs}^{(1)}z\}\int e^{z^2\zeta}v_{s0}(\zeta)d\zeta+\sum_{s=1}^{n}p_{rs}^{(0)}z\int e^{z^2\zeta}v_{s1}(\zeta)d\zeta$$

$$=\int z^2e^{z^2\zeta}(2\zeta-a_r)\{v_{r0}(\zeta)+zv_{r1}(\zeta)\}d\zeta+\int ze^{z^2\zeta}v_{r1}(\zeta)d\zeta-\sum_{s=1}^{n}p_{rs}^{(1)}\int z^2e^{z^2\zeta}v_{s1}(\zeta)d\zeta$$

$$=[e^{z^2\zeta}(2\zeta-a_r)\{v_{r0}(\zeta)+zv_{r1}(\zeta)\}-e^{z^2\zeta}\sum p_{rs}^{(1)}v_{s1}(\zeta)]$$

$$-\int e^{z^2\zeta}\left\{2v_{r0}(\zeta)+(2\zeta-a_r)\frac{dv_{r0}(\zeta)}{d\zeta}\right\}d\zeta-z\int e^{z^2\zeta}\left\{v_{r1}(\zeta)+(2\zeta-a_r)\frac{dv_{r1}(\zeta)}{d\zeta}\right\}d\zeta$$

$$+\sum_{s=1}^{n}p_{rs}^{(1)}\int e^{z^2\zeta}\frac{dv_{s1}(\zeta)}{d\zeta}d\zeta \qquad (r=1, 2, \ldots, n).$$

The integrals on the two sides of this equation cancel one another if the $2n$ functions $v_{r0}(\zeta)$ and $v_{r1}(\zeta)$ satisfy the $2n$ simultaneous equations

$$-(2\zeta-a_r)\frac{dv_{r1}}{d\zeta}=v_{r1}+\sum_{s=1}^{n}\{p_{rs}^{(0)}v_{s1}+p_{rs}^{(1)}v_{s0}\},$$

$$-(2\zeta-a_r)\frac{dv_{r0}}{d\zeta}+\sum_{s=1}^{n}p_{rs}^{(1)}\frac{dv_{s1}}{d\zeta}=2v_{r0}+\sum_{s=1}^{n}p_{rs}^{(0)}v_{s0}$$

$$(r=1, 2, \ldots, n).$$

The finite singular points of this system are $\zeta=\frac{1}{2}a_1, \frac{1}{2}a_2, \ldots, \frac{1}{2}a_n$; they are not regular but irregular singularities of rank unity. The point at infinity is a regular singularity. If, in the original system, the transformation

$$\overline{W}_r=e^{-\beta_m z}W_r \qquad (r=1, 2, \ldots, n)$$

were made, the effect would be to replace $p_{rr}^{(1)}$ by $p_{rr}^{(1)}-\beta_m$ throughout; in particular $p_{mm}^{(1)}$ would be reduced to zero.

Now the system of equations by which $v_{r0}(\zeta)$ and $v_{r1}(\zeta)$ are defined may be written

$$\frac{dv_{r1}}{d\zeta}=\frac{-1}{2\zeta-a_r}\left[v_{r1}+\sum_{s=1}^{n}\left\{p_{rs}^{(0)}v_{s1}+p_{rs}^{(1)}v_{s0}\right\}\right],$$

$$\frac{dv_{r0}}{d\zeta}=\frac{-1}{2\zeta-a_r}\left[2v_{r0}+\sum_{s=1}^{n}p_{rs}^{(0)}v_{s0}+\sum_{s=1}^{n}\frac{p_{rs}^{(1)}}{2\zeta-a_s}\left\{2v_{s1}+\sum_{t=1}^{n}(p_{st}^{(0)}v_{t1}+p_{st}^{(1)}v_{t0})\right\}\right]$$

$$(r=1, 2, \ldots, n).$$

The singularity $\zeta=\frac{1}{2}a_m$ is irregular when poles of the second order at $\zeta=\frac{1}{2}a_m$ occur in the coefficients of the system, and this can only happen if $p_{mm}^{(1)}\neq 0$. But since by the transformation just mentioned $p_{mm}^{(1)}$ may be reduced to zero, that transformation renders the singularity at $\zeta=\frac{1}{2}a_m$ regular. It will be supposed that this transformation has been effected.

The exponents relative to the regular singularity $\zeta=\frac{1}{2}a_m$ are all zero except two, which are

$$-\tfrac{1}{2}(\mu_m+1), \quad -\tfrac{1}{2}(\mu_m+2).$$

Hence there exists a solution of the system in $v_{r0}(\zeta)$ and $v_{r1}(\zeta)$ which is of the form

$$\left.\begin{aligned} v_{r0}^{(m)}(\zeta)&=(\zeta-\tfrac{1}{2}a_m)^{-\frac{1}{2}(\mu_m+1)}\phi_{r0}^{(m)}(\zeta-\tfrac{1}{2}a_m) \\ v_{r1}^{(m)}(\zeta)&=(\zeta-\tfrac{1}{2}a_m)^{-\frac{1}{2}(\mu_m+1)}\phi_{r1}^{(m)}(\zeta-\tfrac{1}{2}a_m) \end{aligned}\right\} \quad (r=1, 2, \ldots, n),$$

where $\phi_{r0}^{(m)}(\zeta-\frac{1}{2}a_m)$ and $\phi_{r1}^{(m)}(\zeta-\frac{1}{2}a_m)$ are analytic in the neighbourhood of $\zeta=\frac{1}{2}a_m$.

Each individual singularity $\zeta=\frac{1}{2}a_s$ is dealt with separately, and is made regular by the appropriate transformation. To each singularity $\zeta=\frac{1}{2}a_s$ corresponds the set of $2n$ functions

$$v_{10}^{(s)}(\zeta), \quad \ldots, \quad v_{n0}^{(s)}(\zeta), \quad v_{11}^{(s)}(\zeta), \quad \ldots, \quad v_{n1}^{(s)}(\zeta).$$

The corresponding contour C_s is a loop-circuit encircling the point $\zeta=\frac{1}{2}a_s$ in the negative sense and proceeding to infinity along a ray such that $\mathbf{R}\{z^2(\zeta-\frac{1}{2}a_s)\}$ is negative. Then *each set of integrals*

$$W_r^{(s)}=e^{-\beta_s z}\int_{C_s} e^{z^2\zeta}(\zeta-\tfrac{1}{2}a_s)^{-\frac{1}{2}(\mu_s+1)}\{\phi_{r0}^{(s)}(\zeta-\tfrac{1}{2}a_s)+z\phi_{r1}^{(s)}(\zeta-\tfrac{1}{2}a_s)\}d\zeta$$

($r=1, 2, \ldots, n$) *represents a solution of the standard canonical system of rank two.*

The case in which the exponent $-\frac{1}{2}(\mu_s+1)$ is an integer is easily disposed of; the other exponent $-\frac{1}{2}(\mu_s+2)$, which is then not an integer, simply takes its place.

19·42. Solution of the System of general Rank $q+1$.—In the general case the formal solutions are given by

$$T_r^{(s)}=e^{Q_s(z)}z^{\mu_s}(B_{rs}+B_{rs}^{(1)}z^{-1}+\ldots),$$

where

$$Q_s(z)=a_s\frac{z^{q+1}}{q+1}+\beta_s\frac{z^q}{q}+\ldots+\lambda_s z.$$

The generalised Laplace integrals

$$W_r=\int \exp(\zeta z^{q+1})\{v_{r0}(\zeta)+zv_{r1}(\zeta)+\ldots+z^q v_{rq}(\zeta)\}d\zeta \quad (r=1, 2, \ldots, n)$$

satisfy the system of rank $q+1$ if

$$\int \exp(\zeta z^{q+i})\sum_{j=1}^{q} z^j\{(q+1)z^{q+1}\zeta+j\}v_{rj}(\zeta)d\zeta$$

$$=\sum_{s=1}^{n}\int \exp(\zeta z^{q+1})\left\{\sum_{k=1}^{q} p_{rs}^{(k)}z^k\right\}\left\{\sum_{l=1}^{q} z^l v_{sl}(\zeta)\right\}d\zeta$$

$$+a_r\int \exp(\zeta z^{q+1})z^{q+1}\sum_{l=1}^{q} z^l v_{rl}(\zeta)d\zeta \qquad (r=1, 2, \ldots, n).$$

All integral powers of z up to z^{q+1} are involved. By equating to zero the aggregate of terms in z and $z^{\nu+q+1}$ for $\nu=0, 1, \ldots, q$, a set of $q+1$

equations in the $q+1$ unknown functions $v_{r0}(\zeta)$, $v_{r1}(\zeta)$, . . ., $v_{rq}(\zeta)$ is obtained. The typical equation is (after the factor z^ν has been suppressed):

$$\int \exp(\zeta z^{q+1})\{(q+1)z^{q+1}\zeta+\nu\}v_{r\nu}(\zeta)d\zeta$$
$$=\sum_{s=1}^{n}\int \exp(\zeta z^{q+1})\Big\{\sum_{k+l=\nu} p_{rs}{}^{(k)}v_{sl}(\zeta)+z^{q+1}\sum_{k+l=q+\nu+1} p_{rs}{}^{(k)}v_{sl}(\zeta)\Big\}d\zeta$$
$$+a_r\int \exp(\zeta z^{q+1})z^{q+1}v_{r\nu}(\zeta)d\zeta.$$

But since, by integration by parts,

$$\int \exp(\zeta z^{q+1})z^{q+1}u(\zeta)d\zeta=[\exp(\zeta z^{q+1})u(\zeta)]-\int \exp(\zeta z^{q+1})du(\zeta),$$

each of the $q+1$ equations is found to be satisfied, for an appropriate choice of the contour, if the functions

$$v_{r0}(\zeta),\quad v_{r1}(\zeta),\quad . . .,\quad v_{rq}(\zeta) \qquad (r=1, 2, . . ., n)$$

satisfy the set of $n(q+1)$ transformed equations

$$-\{(q+1)\zeta-a_r\}\frac{dv_{r\nu}}{d\zeta}+\sum_{s=1}^{n}\sum_{k+l=q+\nu+1} p_{rs}{}^{(k)}\frac{dv_{sl}}{d\zeta}=(q+1-\nu)v_{r\nu}+\sum_{s=1}^{n}\sum_{k+l=\nu} p_{rs}{}^{(k)}v_{sl},$$

where $r=1, 2, . . ., n$; $\nu=0, 1, . . ., q$.

Thus, taking in succession $\nu=q, q-1, . . ., 0$, the complete set of transformed equations may be written:

$$-\{(q+1)\zeta-a_r\}\frac{dv_{rq}}{d\zeta}=v_{rq}+\sum_{s=1}^{n}\sum_{k+l=q} p_{rs}{}^{(k)}v_{sl},$$
$$-\{(q+1)\zeta-a_r\}\frac{dv_{r,\,q-1}}{d\zeta}+\sum_{s=1}^{n} p_{rs}{}^{(q)}\frac{dv_{sq}}{d\zeta}=2v_{r,\,q-1}+\sum_{s=1}^{n}\sum_{k+l=q-1} p_{rs}{}^{(k)}v_{sl},$$
$$. \quad . \quad . \quad . \quad . \quad . \quad . \quad . \quad . \quad . \quad . \quad .$$
$$-\{(q+1)\zeta-a_r\}\frac{dv_{r0}}{d\zeta}+\sum_{s=1}^{n} p_{rs}{}^{(q)}\frac{dv_{s1}}{d\zeta}+ . . . +\sum_{s=1}^{n} p_{rs}{}^{(0)}\frac{dv_{sq}}{d\zeta}$$
$$=(q+1)v_{r0}+\sum_{s=1}^{n} p_{rs}{}^{(0)}v_{s0}.$$

In each equation $r=1, 2, . . ., n$.

The finite singularities of this system are

$$\zeta=\frac{a_1}{q+1},\quad \frac{a_2}{q+1},\quad . . .,\quad \frac{a_n}{q+1},$$

and are irregular singularities of rank q at most. The point at infinity is regular. The transformation

$$\overline{W}=W_r\exp\Big\{-\beta_m\frac{z^q}{q}-\gamma_m\frac{z^{q-1}}{q-1}- . . . -\lambda_m z\Big\}$$

has, for fixed m, the effect of changing

$$p_{rr}{}^{(q)},\quad p_{rr}{}^{(q-1)},\quad . . .,\quad p_{rr}{}^{(0)}$$

respectively into

$$p_{rr}{}^{(q)}-\beta_m,\quad p_{rr}{}^{(q-1)}-\gamma_m,\quad . . .,\quad p_{rr}{}^{(0)}-\lambda_m$$

for $r=1, 2, . . ., n$. The equations for $v_{r0}, v_{r1}, . . ., v_{rq}$ then have a regular singular point * at $\zeta=a_m/(q+1)$, relative to which all $n(q+1)$ exponents are

* For a proof of this fact see Birkhoff, *Trans. Am. Math. Soc.* 10, p. 460. The statement concerning the exponents admits of an indirect proof by the principle of continuity; no direct proof appears to be known.

zero except $q+1$, namely,

$$-(\mu_m+1)/(q+1), \quad -(\mu_m+2)/(q+1), \quad \ldots, \quad -(\mu_m+q+1)/(q+1),$$

and a solution of the form

$$v_{rk}^{(m)}=\left(\zeta-\frac{a_m}{q+1}\right)^{-\frac{\mu_m+1}{q+1}}\phi_{rk}^{(m)}\left(\zeta-\frac{a_m}{q+1}\right) \quad (r=1, 2, \ldots, n;\ k=0, 1, \ldots, q)$$

exists, where the functions $\phi_{rk}^{(m)}$ are analytic near $\zeta=a_m/(q+1)$.

Thus if C_s is a loop-circuit about $\zeta=a_s/(q+1)$ such that $\mathbf{R}[z^{q+1}\{\zeta-a_s/(q+1)\}]$ is negative along its ray, *each set of integrals*

$$W_r^{(s)}=e^{\beta_s\frac{z^q}{q}+\ldots+\lambda_s z}\int_{C_s} e^{z^{q+1}}\left(\zeta-\frac{a_s}{q+1}\right)^{-\frac{\mu_s+1}{q+1}}\sum_{k=0}^{q} z^k\phi_{rk}^{(s)}\left(\zeta-\frac{a_s}{q+1}\right)d\zeta$$

represents a solution of the standard canonical system of rank q.

If $(\mu_s+1)/(q+1)$ is an integer, it may be replaced by any one of the other (non-integer) exponents. The sectors in which this integral representation of the solution is valid will be specified more particularly in the following section.

19·5. Asymptotic Representations.—In the integral representation of $W_r^{(s)}$ make, for each s, the substitution

$$t=z^{q+1}\left\{\zeta-\frac{a_s}{q+1}\right\},$$

then

$$W_r^{(s)}=e^{Q_s(z)}z^{\mu_s}\int_{\Gamma_s} e^t t^{-\frac{\mu_s+1}{q+1}}\sum_{k=0}^{q} z^k\phi_{rk}^{(s)}(z^{-q-1}t)dt,$$

where Γ_s is a loop-circuit enlacing the point $t=0$. Then by a suitable modification of the reasoning of § 18·21 it may be proved, on expanding the integrand, that if $\arg z=\phi$ is a ray for which

$$\mathbf{R}[z^{q+1}\{\zeta-a_s/(q+1)\}]<0,$$

there will be a sector for which $\arg z=\phi$ is an interior ray, and for which

$$W_r^{(s)}=e^{Q_s(z)}z^{\mu_s}\left\{B_{rs}+\frac{B_{rs}^{(1)}}{z}+\ldots+\frac{B_{rs}^{(m)}}{z^m}+\frac{\epsilon_m}{z^{m+1}}\right\},$$

where $\epsilon_m\to 0$ for all m as $z\to\infty$. So far as the first $m+1$ terms are concerned, this development coincides with the formal solution $T_r^{(s)}$. Thus $W_r^{(s)}$ *is asymptotically represented by* $T_r^{(s)}$ *along the ray* $\arg z=\phi$ or symbolically

$$W'^{(s)}_r \sim T_r^{(s)} \qquad (\arg z=\phi).$$

The ray in the ζ-plane along which the loop-circuit C_s proceeds to infinity is such that $\mathbf{R}z^{q+1}\{\zeta-a_s/(q+1)\}$ is negative; subject to this condition it may vary so long as it does not pass through any finite singular point other than $\zeta=a_s/(q+1)$. It is not difficult to determine the exact sectors in the z-plane for which the corresponding formulæ are valid, and this is the question which will now be considered.

In all there are $N=n(n-1)(q+1)$ rays for which

$$\mathbf{R}\{(a_s-a_t)z^{q+1}\}=0 \qquad (t\neq s),$$

and these rays are given by the formula

$$\tan(q+1)\phi=\cot\arg(a_s-a_r).$$

Assuming that these rays are distinct, let them be denoted, in increasing angular order, by

$$\arg z=\tau_1, \quad \tau_2, \quad \ldots, \quad \tau_N;$$

let $\tau_{N+1}=\tau_1+2\pi$.

As the point z passes from any sector (τ_{m-1}, τ_m) into the consecutive sector (τ_m, τ_{m+1}), the real part of a particular one of the differences $(\alpha_s-\alpha_t)z^{q+1}$ changes from positive to negative. Let this particular difference be denoted by

$$(\alpha_{s_m}-\alpha_{t_m})z^{q+1}.$$

Consider any one of the $q+1$ values of m for which $\tau_m=s$, and let $\tau_{m'}$ be the ray next in increasing angular order to τ_m on which the real part of another difference, say

$$(\alpha_{s_{m'}}-\alpha_{t_{m'}})z^{q+1},$$

for which $t_{m'}=s$, changes from positive to negative. Then the argument of the loop circuit C_s is intermediate between the consecutive pair of arguments

$$\arg(\alpha_{s_m}-\alpha_s), \quad \arg(\alpha_{s_{m'}}-\alpha_s)$$

and $\mathbf{R}[z^{q+1}\{\zeta-\alpha_s/(q+1)\}]$ remains negative for

$$\tau_m < \arg z < \tau_{m'}+\frac{\pi}{q+1}.$$

The integrals

$$W_1^{(s)}, \quad W_2^{(s)}, \quad \ldots, \quad W_n^{(s)}$$

furnish a set of $q+1$ *solutions of the canonical system, fixed by assigning the sector in which the ray of the loop circuit* C_s *is to lie. Each set is valid for any one of the* $q+1$ *corresponding sections*

$$\tau_m<\arg z<\tau_{m'}+\frac{\pi}{q+1}.$$

For every ray $\arg z=\phi$ *which lies within any one of these sectors, there exists a fundamental set of solutions*

$$W_1=W_1^{(s)}, \quad W_2=W_2^{(s)}, \quad \ldots, \quad W_n=W_n^{(s)} \qquad (s=1, 2, \ldots, n)$$

such that

$$W_1^{(s)} \sim T_1^{(s)}, \quad W_2^{(s)} \sim T_2^{(s)}, \quad \ldots, \quad W_n^{(s)} \sim T_n^{(s)}.$$

The corresponding theorem for the original system * is that *there exist solutions* $w_1, w_2, \ldots, w_n$ *such that*

$$w_r \sim S_r^{(s)} \qquad (r=1, 2, \ldots, n)$$

within any given sector

$$\tau_m<\arg z<\tau_{m'}+\pi/(q+1).$$

19·6. Characterisation of the Solutions in the Neighbourhood of Infinity.—The solutions of the canonical system are characterised by the following theorem: *There exist* $N=n(n-1)(q+1)$ *fundamental sets of solutions of the standard canonical system, namely*

$$\begin{matrix} W_{1m}^{(1)}, & W_{2m}^{(1)}, & \ldots, & W_{nm}^{(1)}, \\ \cdot & \cdot & \cdot & \cdot \\ W_{1m}^{(n)}, & W_{2m}^{(n)}, & \ldots, & W_{nm}^{(n)}, \end{matrix} \qquad (m=1, 2, \ldots, N)$$

such that

$$W_{rm}^{(s)} \sim T_r^{(s)}, \qquad \tau_m \leqslant \arg z<\tau_{m+1},$$

and such that

$$W_{r,m+1}^{(s)} = W_{rm}^{(s)} \qquad (s\neq s_m),$$

$$W_{r,m+1}^{(s_m)}=W_{rm}^{(s_m)}+A_m W_{rm}^{(t_m)}$$

and finally

$$W_{r,N+1}^{(s)}=e^{2\pi i\mu_s}W_{r1}^{(s)}.$$

* This theorem generalises a result given by Horn, *J. für Math.* 133 (1907), p. 19.

The present section will be devoted to a proof of this theorem.*

By virtue of the theorem in the preceding section it is possible to divide the z-plane into a finite number of closed abutting sectors σ in each of which there exists a fundamental set of solutions

$$W_1 = W_1^{(s)}, \quad W_2 = W_2^{(s)}, \quad \ldots, \quad W_n = W_n^{(s)} \qquad (s = 1, 2, \ldots, n)$$

such that, in the sector considered,

$$W_r^{(s)} \sim T_r^{(s)}.$$

The sectors may be chosen so that the rays $\arg z = \tau_m$ are internal rays and so that at most one ray lies within each sector. Now if σ is a sector not containing any ray τ, every solution of the canonical system of equations will have a definite asymptotic representation throughout σ. For the general solution is

$$W_r = c_1 W_r^{(1)} + c_2 W_r^{(2)} + \ldots + c_n W_r^{(n)} \qquad (r = 1, 2, \ldots, w),$$

and this leads to the asymptotic relationship

$$W_r \sim c_1 T_r^{(1)} + c_2 T_r^{(2)} + \ldots + c_n T_r^{(n)}.$$

For large values of $|z|$ the relative magnitudes of the terms of this expression are respectively the relative magnitudes of

$$\mathbf{R}(a_1 z^{q+1}), \quad \mathbf{R}(a_2 z^{q+1}), \quad \ldots, \quad \mathbf{R}(a_n z^{q+1}),$$

and the relative order of magnitude does not change except at the rays

$$\arg z = \tau_1, \quad \tau_2, \quad \ldots, \quad \tau_N,$$

and therefore does not alter in any sector σ not containing a ray τ. Let it be supposed that, for the sector under consideration, the suffixes $1, 2, \ldots, n$ are so chosen that

$$\mathbf{R}(a_1 z^{q+1}) > \mathbf{R}(a_2 z^{q+1}) > \ldots > \mathbf{R}(a_n z^{q+1})$$

and let

$$c_1 = c_2 = \ldots = c_{k-1} = 0, \quad c_k \neq 0.$$

Then for the sector σ

$$W_r \sim c_k T_r^{(k)}.$$

But since consecutive sectors abut on one another, every solution W_1, $W_2, \ldots, W_n$ has the same asymptotic representation in successive sectors until a sector which contains a ray τ is reached. Thus if the sector σ and consecutive sectors up to and including that which contains the ray τ_{m+1} are amalgamated into a single sector σ_m it follows that *there exists a fundamental set of solutions*

$$W_1^{(s)}, \quad W_2^{(s)}, \quad \ldots, \quad W_n^{(s)} \qquad (s = 1, 2, \ldots, n)$$

such that throughout the sector σ_m

$$W_r \sim T_r^{(s)}.$$

Now consider the character of the general solution

$$W_1, \quad W_2, \ldots, \quad W_m$$

in the sector σ_m. As the point z crosses the ray $\arg z = \tau_m$, the order of magnitude of $\mathbf{R}\{Q_{s_m}(z)\}$ and $\mathbf{R}\{Q_{t_m}(z)\}$ is inverted, and moreover

$$\mathbf{R}\{(a'_{s_m} - a_{t_m}) z^{q+1}\} > 0 \qquad (\arg z < \tau_m),$$
$$< 0 \qquad (\arg z > \tau_m).$$

Suppose that on the initial bounding ray of the sector σ_m

$$\mathbf{R}(a_1 z^{q+1}), \quad \mathbf{R}(a_2 z^{q+1}), \quad \ldots, \mathbf{R}(a_n z^{q+1})$$

* The solutions referred to in this theorem are not, in general, the integral solutions of §§ 19·4–19·42; the Laplace integral solutions retain their asymptotic form throughout maximum sectors; the sectors of the present theorem are minimum sectors.

are in descending order of magnitude. Then as z crosses the ray $\arg z=\tau_m$, two particular consecutive terms say

$$\mathbf{R}\{a_k z^{q+1}\}, \quad \mathbf{R}\{a_{k+1} z^{q+1}\}$$

will change their relative order. But the general solution

$$W_r=c_1W_r^{(1)}+c_2W_r^{(2)}+\ .\ .\ .\ +c_nW_r^{(n)} \qquad (r=1, 2, \ldots, n)$$

will nevertheless preserve its asymptotic form unless

$$c_1=c_2=\ .\ .\ .\ =c_{k-1}=0, \quad c_k\neq 0, \quad c_{k+1}\neq 0.$$

When this is the case the solution is of the asymptotic form

$$W_r \sim c_k T_r^{(k)}$$

on the initial bounding ray, and of the asymptotic form

$$W_r \sim c_{k+1} T_r^{k+1}$$

on the terminal boundary ray of the sector σ_m.

If, however, two particular solutions W_r and W_r' can be found such that, on the initial bounding ray of σ_m

$$W_r \sim c_k T_r^{(k)}, \quad W_r' \sim c'_{k+1} T_r^{(k+1)} \qquad (r=1, 2, \ldots, n),$$

then a linear combination W_r+AW_r' of these solutions can be chosen which will preserve its asymptotic form throughout the sector σ_m. For since

$$W_r=c_rW_r^{(k)}+c_{k+1}W_r^{(k+1)}+\ .\ .\ .\ +c_nW_r^{(n)},$$
$$W_r'=c'_{k+1}W_r^{(k+1)}+\ .\ .\ .\ +c_n'W_r^{(n)},$$

it is only necessary to assign to A the value $-c_{k+1}/c'_{k+1}$.

Now let

$$W_{11}^{(s)}, \quad W_{21}^{(s)}, \ldots, \quad W_{n1}^{(s)} \qquad (s=1, 2, \ldots, n)$$

be any fundamental set of solutions such that, in the sector σ_1,

$$W_{r1}^{(s)} \sim T_r^{(s)}.$$

Each of the n distinct solutions of the set will preserve its asymptotic form throughout the consecutive sector σ_2 except possibly the solution

$$W_{11}^{(s_1)}, \quad W_{21}^{(s_1)}, \quad \ldots, \quad W_{n1}^{(s_1)},$$

but when this exceptional case does arise,* a constant A_1 can be so chosen that the solution

$$W_{11}^{(s_1)}+A_1W_{11}^{(t_1)}, \quad W_{21}^{(s_1)}+A_1W_{21}^{(t_1)}, \quad \ldots, \quad W_{n1}^{(s_1)}+A_1W_{n1}^{(t_1)}$$

preserves its asymptotic form

$$T_1^{(s_1)}, \quad T_2^{(s_1)}, \quad \ldots, \quad T_n^{(s_1)}$$

throughout the sector σ_2. Therefore the new fundamental set of solutions

$$W_{12}^{(s)}, \quad W_{22}^{(s)}, \quad \ldots, \quad W_{n2}^{(s)},$$

where

$$W_{r2}^{(s)}=W_{r1}^{(s)} \qquad (s\neq s_1),$$
$$W_{r2}^{(s_1)}=W_{r1}^{(s_1)}+A_1W_{r1}^{(t_1)},$$

preserves its asymptotic form throughout the sector σ_2.

In the same way fundamental sets of solutions

$$\begin{matrix} W_{13}^{(s)}, & W_{23}^{(s)}, & \ldots, & W_{n3}^{(s)} \\ \cdot & \cdot & \cdot & \cdot \\ W_{1N}^{(s)}, & W_{2N}^{(s)}, & \ldots, & W_{nN}^{(s)} \end{matrix} \qquad (s=1, 2, \ldots, n)$$

are determined in succession, which respectively preserve their asymptotic

* It is important to note that this exceptional case arises only when $\mathbf{R}(a_{s_1}z^{q+1})$ changes its order relative to the other expressions $\mathbf{R}(a_s z^{q+1})$ and goes into a lower rank.

forms throughout the sectors $\sigma_3, \ldots, \sigma_N$. From the last set the same process leads to a new set

$$W_{1,N+1}^{(s)}, \quad W_{2,N+1}^{(s)}, \quad \ldots, \quad W_{n,N+1}^{(s)} \qquad (s=1, 2, \ldots, n)$$

which preserves its asymptotic character throughout the sector σ_1. It now remains to prove that a choice of the initial fundamental set of solutions made be made so that

$$W_{r,N+1}^{(s)}=e^{2\pi i\mu_s}W_{r1}^{(s)} \qquad (r, s=1, 2, \ldots, n).$$

The first step is to show that the final fundamental set of solutions is entirely independent of the choice of the initial set

$$W_{11}^{(s)}, \quad W_{21}^{(s)}, \quad \ldots, \quad W_{n1}^{(s)} \qquad (s=1, 2, \ldots, n).$$

Let

$$U_{11}^{(s)}, \quad U_{21}^{(s)}, \quad \ldots, \quad U_{n1}^{(s)} \qquad (s=1, 2, \ldots, n)$$

be a new initial set of fundamental solutions; let

$$U_{1m}^{(s)}, U_{2m}^{(s)}, \ldots, U_{nm}^{(s)} \qquad (s=1, 2, \ldots, n; \ m=2, 3, \ldots, N)$$

be the successive fundamental sets derived therefrom, and let the constant which corresponds to A_m be denoted by B_m.

In the sector (τ_1, τ_2) let

$$\mathbf{R}(a_iz^{q+1})<\mathbf{R}(a_jz^{q+1})<\mathbf{R}(a_kz^{q+1})< \ldots,$$

then since $\mathbf{R}(a_iz^{q+1})$ is the expression of lowest order there can be but one solution asymptotic in (τ_1, τ_2) to

$$T_1^{(i)}, \quad T_2^{(i)}, \quad \ldots, \quad T_n^{(i)},$$

and therefore

$$U_{r1}^{(i)}=\mathrm{W}_{r1}^{(i)} \qquad (r=1, 2, \ldots, n).$$

Now every two of the expressions $\mathbf{R}(a_sz^{q+1})$ become equal $2(q+1)$ times as z describes a complete circuit about the origin; if they become equal on the ray $\arg z=\tau'$, they also become equal on the rays

$$\arg z=\tau'+\frac{\kappa\pi}{q+1} \qquad (\kappa=1, 2, \ldots, 2q+1),$$

and nowhere else. Consequently in the sector

$$\tau_1 \leqslant \arg z \leqslant \tau_1+\frac{\pi}{q+1}=\tau_\nu,$$

where $\nu=\frac{1}{2}n(n-1)$, every two of the expressions $\mathbf{R}(a_sz^{q+1})$ become equal on one and only one ray. In particular, as $\arg z$ increases from τ_1 to τ_ν, $\mathbf{R}(a_iz^{q+1})$ steadily increases and finally surpasses all the remaining expressions $\mathbf{R}(a_sz^{q+1})$, and therefore

$$W_{r1}^{(i)}=W_{r2}^{(i)}= \ldots =W_{r\nu}^{(i)},$$
$$U_{r1}^{(i)}=U_{r2}^{(i)}= \ldots =U_{r\nu}^{(i)}.$$

Thus since

$$U_{r1}^{(i)}=W_{r1}^{(i)},$$

it follows that

$$U_{rm}^{(i)}=W_{rm}^{(i)} \qquad (r=1, 2, \ldots, n)$$

for $m\leqslant\nu$.

Now since, in (τ_1, τ_2), $\mathbf{R}(a_jz^{q+1})$ is second in increasing order of magnitude, there will be a relation of the form

$$U_{r1}^{(j)}=W_{r1}^{(j)}+cW_1^{(i)} \qquad (r=1, 2, \ldots, n),$$

and from this there follows the relation *

$$U_{rm}^{(j)}=W_{rm}^{(j)}+cW_{rm}^{(i)},$$

* Note that $\mathbf{R}(a_jz^{q+1})$ cannot fall below $\mathbf{R}(a_sz^{q+1})$ except for $s=i$.

for $m=2, 3, \ldots, \theta$, where θ is the value of m for which the magnitude of $\mathbf{R}(a_j z^{q+1})$ falls below that of $\mathbf{R}(a_i z^{q+1})$. Now since

$$W_{r,\,\theta+1}^{(j)}=W_{r\theta}^{(j)}+A_\theta W_{r\theta}^{(i)}, \quad W_{r,\,\theta+1}^{(i)}=W_{r\theta}^{(i)},$$
$$U_{r,\,\theta+1}^{(j)}=U_{r\theta}^{(j)}+B_\theta U_{r\theta}^{(i)}, \quad U_{r,\,\theta+1}^{(i)}=U_{r\theta}^{(i)},$$

the relation

$$U_{r\theta}^{(j)}=W_{r\theta}^{(j)}+cW_{r\theta}^{(i)}$$

may be written

$$U_{r,\,\theta+1}^{(j)}-B_\theta U_{r,\,\theta+1}^{(i)}=W_{r,\,\theta+1}^{(j)}-(A_\theta-c)W_{r,\,\theta+1}^{(i)}.$$

In the sector $(\tau_\theta, \tau_{\theta+1})$, $\mathbf{R}(a_i z^{q+1})>\mathbf{R}(a_j z^{q+1})$, and since it has been proved that

$$U_{r,\,\theta+1}^{(i)}=W_{r,\,\theta+1}^{(i)},$$

it follows that

$$B_\theta=A_\theta-c,$$

and consequently that

$$U_{r,\,\theta+1}^{(j)}=W_{r,\,\theta+1}^{(j)}.$$

But for $m=\theta+1, \theta+2, \ldots, \nu$, the order of $\mathbf{R}(a_j z^{q+1})$ does not fall below that of any other expression $\mathbf{R}(a_s z^{q+1})$ and therefore

$$U_{rm}^{(j)}=W_{rm}^{(j)} \qquad (r=1, 2, \ldots, n)$$

for $m=\theta+1, \theta+2, \ldots, \nu$.

In the same way a relation of the form

$$U_{rm}^{(k)}=W_{rm}^{(k)}+cW_{rm}^{(j)}+dW_{rm}^{(i)}$$

holds for successive values of m. The constants c and d in this relation alter their values only for values of m such that the relative order of the three expressions

$$\mathbf{R}(a_i z^{q+1}), \quad \mathbf{R}(a_j z^{q+1}), \quad \mathbf{R}(a_k z^{q+1})$$

is changed at the ray $\arg z=\tau_m$. If the first expression, which is initially lowest in order, increases over the second, the value of d may change; when the second increases over the third, c becomes zero; when the first increases over the third, d becomes zero. Thus if θ' is the value of m for which the first expression increases over the third,

$$U_{rm}^{(k)}=W_{rm}^{(k)} \qquad (r=1, 2, \ldots, n)$$

for $m=\theta'+1, \theta'+2, \ldots, \nu$.

By continuing the argument on these lines it may be proved that on and after a fixed value of $m\leqslant\nu$, the relation

$$U_{rm}^{(s)}=W_{rm}^{(s)} \qquad (r=1, 2, \ldots, n)$$

holds for every value of s. In particular the final fundamental system

$$U_{1,\,N+1}^{(s)}, \quad U_{2,\,N+1}^{(s)}, \ldots, \quad U_{n,\,N+1}^{(s)} \quad (s=1, 2, \ldots, n)$$

is identical with the system

$$W_{1,\,N+1}^{(s)}, \quad W_{2,\,N+1}^{(s)}, \ldots, \quad W_{n,\,N+1}^{(s)} \quad (s=1, 2, \ldots, n).$$

The final fundamental system is therefore independent of the choice of the initial fundamental system, provided, of course, that the initial choice is consistent with the conditions of the theorem. Let the initial system be defined in terms of the invariant final system by the relations

$$W_{r1}^{(s)}=e^{-2\pi i\mu_s}W_{r,\,N+1}^{(s)} \qquad (r, s=1, 2, \ldots, n).$$

This definition is self-consistent for, since $T_r^{(s)}$ is multiplied by the factor $e^{2\pi i\mu_s}$ when the point z has described a complete positive circuit about the point at infinity, the asymptotic relationship

$$W_{r,\,N+1}^{(s)}\sim e^{2\pi i\mu_s}T_r^{(s)}$$

holds for the sector (τ_1, τ_2).

Thus the theorem proposed has been completely proved. Its extension to the original system is immediate and may be formulated as follows: *There exist $N=n(n-1)(q+1)$ fundamental solutions of the system*

$$\frac{dw_r}{dz}=\sum_{s=1}^{n} p_{rs}(z)w_s, \qquad (r=1, 2, \ldots, n)$$

whose rank at infinity is q, namely

$$\left.\begin{matrix} w_{1m}^{(1)}, & w_{2m}^{(1)}, & \ldots, & w_{nm}^{(1)}, \\ \cdot & \cdot & \cdot & \cdot \\ w_{1m}^{(n)}, & w_{1m}^{(n)}, & \ldots, & w_{nm}^{(n)}, \end{matrix}\right\} \quad (m=1, 2, \ldots, N)$$

such that if the formal fundamental solution is

$$\left.\begin{matrix} S_1^{(1)}, & S_2^{(1)}, & \ldots, & S_n^{(1)}, \\ \cdot & \cdot & \cdot & \cdot \\ S_1^{(n)}, & S_2^{(n)}, & \ldots, & S_n^{(n)}, \end{matrix}\right\}$$

then

$$w_{rm}^{(s)} \sim S_r^{(s)}, \qquad \tau_m \leqslant \arg z < \tau_{m+1}.$$

The N fundamental solutions are linked up by the relations

$$w_{r,\, m+1}^{(s)}=w_{rm}^{(s)} \qquad (s \neq s_m),$$

$$w_{r,\, m+1}^{(s_m)}=w_{rm}^{(s_m)}+A_m w_{rm}^{(t_m)},$$

where, for $m=N$,

$$w_{r,\, N+1}^{(s)}=e^{2\pi i\mu_s} w_{r1}^{(s)}.$$

Any set of functions $w_{rm}^{(s)}$ which satisfies all these conditions furnishes a solution of the differential system. The theorem is therefore said to give a *complete characterisation* of the solutions of the system with reference to the point at infinity.

The constants which determine the nature of the standard canonical system are known as the *characteristic constants* and fall into two classes. The exponential constants are the $q+1$ constants $\alpha_s, \beta_s, \ldots, \lambda_s$ of each polynomial $Q_s(z)$ and the exponents μ_s; altogether they are $n(q+2)$ in number and are independent of one another. The transformation constants $A_1, A_2, \ldots, A_N$ are not all independent, for $n-1$ of them may be disposed of by the transformation

$$\overline{W}_r=c_r W_r \qquad (r=1, 2, \ldots, n),$$

where the constants c_r are properly chosen. The number of essential characteristic constants is therefore

$$n(q+2)+n(n-1)(q+1)-(n-1)=n^2(q+1)+1.$$

The coefficients in the standard canonical system involve, in all, $n^2(q+1)+n$ constants which may be reduced to $n^2(q+1)+1$ by multiplying $W_1, W_2, \ldots, W_n$ by suitable constants. In the general case the number of constants in the equation cannot further be reduced; these constants are therefore said to be the *irreducible constants* of the system. Since the number of characteristic constants and the number of irreducible constants is the same, it follows that the *characteristic constants are not connected by any necessary relation.*

19·7. The Generalised Riemann Problem.—The Riemann problem which, in its original form (§ 15·92), referred to three singular points, all of which were regular, has been generalised by Birkhoff in the following terms: *To* *construct a system of n linear differential equations with prescribed singular points*

$$z_1, \quad z_2, \quad . \; . \; ., \quad z_m, \quad z_{m+1}=\infty$$

of respective rank

$$q_1, \quad q_2, \quad . \; . \; ., \quad q_m, \quad q_{m+1},$$

and with a given monodromic group, the characteristic constants being assigned for each singular point.

To show that the problem thus postulated is self-consistent consider the simultaneous system of equations

$$\frac{dw_r}{dz}=\sum_{s=1}^{n}\left\{\sum_{k=1}^{m}\sum_{l=1}^{2+q_k}\frac{A_{rskl}}{(z-z_k)^l}+\sum_{k=0}^{q_{m+1}}B_{rsk}z^k\right\}w_s \qquad (r=1, 2, \ldots, n),$$

which is the most general equation whose singular points $z_1, z_2, \ldots, z_m, \infty$ are of the prescribed ranks $q_1, q_2, \ldots, q_m, q_{m+1}$. The number of arbitrary constants A_{rskl} and B_{rsk} to be disposed of is

$$n^2\left\{\sum_{l=1}^{m+1} q_l+2m+1\right\}.$$

Now let

$$w_1^{(s)}, \quad w_2^{(s)}, \quad \ldots, \quad w_n^{(s)} \qquad (s=1, 2, \ldots, n)$$

be a fundamental set of solutions fixed by assigning the condition that at some particular ordinary point a,

$$w_r^{(s)}=\delta_{rs};$$

the group of this particular fundamental set will be regarded as assigned.

Now the monodromic group possesses m fundamental substitutions, one corresponding to each finite singular point.* Each substitution is defined by a matrix of n^2 constants, and therefore the group involves, altogether, mn^2 arbitrary constants.

The characteristic constants relative to the singularity z_k are $n^2(q_k+1)+1$ in number, in all there are

$$\sum_{k=1}^{m}\{n^2(q_k+1)+1\}$$

characteristic constants. But the exponents μ are determined both by the group and by the characteristic constants, and are $n(m+1)$ in number. Thus between the constants of the group and the characteristic constants there are $n(m+1)$ relations.

Finally a correspondence must be set up, at each singular point, between the chosen fundamental set of solutions and the canonical fundamental sets defined by the theorem of § 19·6. This correspondence is determined by the group (which fixes the exponents at each singularity) except for n multiplicative constants. Thus $n-1$ additional conditions are imposed at each singularity; in all $(n-1)(m+1)$ further conditions.

The total number of conditions to be satisfied is therefore

$$nm^2+\sum_{k=1}^{m+1}\{n^2(q_k+1)+1\}-n(m+1)+(n-1)(m+1)$$
$$=n^2\left\{\sum_{k=1}^{m+1} q_k+2m+1\right\},$$

* If $S_1, S_2, \ldots, S_m$ are these substitutions, and S_{m+1} is the substitution corresponding to $z=\infty$, then

$$S_{m+1}S_m \ldots S_2 S_1=I,$$

where I is the identical substitution.

and is equal to the number of constants to be disposed of. *The problem is therefore self-consistent.*

The problem thus formulated was virtually solved by Birkhoff (*Proc. Am. Acad.* 49 (1913), p. 536). When obvious conditions of consistency are satisfied, either a solution of the problem as stated, or a solution of the problem modified by replacing the exponents $\mu_1, \ldots, \mu_n$ relative to any one of the singular points by $\mu_1+k_1, \ldots, \mu_n+k_n$, where $k_1, \ldots, k_n$ are integers, will exist.

Miscellaneous Examples.

[These examples are all taken from Birkhoff, *Trans. Am. Math. Soc.* 14 (1913), pp. 462–476.]

1. The system

$$z\frac{dW}{dz}=rV, \quad z\frac{dV}{dz}=sW+(z+1-p_1)V$$

has the formal solutions

$$W=S_1(z), \quad V=\frac{z}{r}\frac{dS_1(z)}{dz},$$

$$W=S_2(z), \quad V=\frac{z}{r}\frac{dS_2(z)}{dz},$$

where, if $\lambda_1+\lambda_2=1-p_1$, $\lambda_1\lambda_2=-rs$,

$$S_1(z)=\left\{1-\frac{\lambda_1\lambda_2}{1}z^{-1}+\frac{\lambda_1(\lambda_1+1)\lambda_2(\lambda_2+1)}{1.2}z^{-2}-\ldots\right\},$$

$$S_2(z)=e^z z^{-p_1}\left\{1+\frac{(\lambda_1-1)(\lambda_2-1)}{1}z^{-1}+\frac{(\lambda_1-1)(\lambda_1-2)(\lambda_2-1)(\lambda_2-2)}{1.2}z^{-2}+\ldots\right\}.$$

2. Let

$$\rho_1=e^{-2\pi i\lambda_1}, \quad \rho_2=e^{-2\pi i\lambda_2},$$

then if $\rho_1\neq 1$, $\rho_2\neq 1$ neither r nor s is zero and the formal solutions diverge. If $\rho_1=1$ either r or s may be taken to be zero and at least one of the formal solutions terminates.

The two formal solutions are, when $r=0$

$$W=1, \quad V=-\frac{s}{z}\left\{1+\frac{p_1-2}{z}+\frac{(p_1-2)(p_1-3)}{z^2}+\ldots\right\},$$

$$W=0, \quad V=e^z z^{1-p_1},$$

and when $s=0$

$$W=1, \quad V=0,$$

$$W=re^z z^{-p_1}\left\{1+\frac{p_1}{z}+\frac{p_1(p_1+1)}{z^2}+\ldots\right\}, \quad V=e^z z^{1-p_1}.$$

When both the formal series terminate both r and s may be taken to be zero.

3. By determining the formal solutions $s_1(z)$ and $s_2(z)$ of the equation

$$\frac{d^2w}{dz^2}+p(z)\frac{dw}{dz}+q(z)w=0,$$

where

$$p(z)=-1+\frac{p_1}{z}+\frac{p_2}{z^2}+\ldots, \qquad q(z)=\frac{q_2}{z^2}+\frac{q_3}{z^3}+\ldots,$$

and using the formal solutions $S_1(z)$, $S_2(z)$, show that the coefficients in the transformation

$$w=a_{11}(z)W+a_{12}(z)V,$$
$$v=a_{21}(z)W+a_{22}(z)V,$$

can be developed in power series in z when $r\neq 0$.

4. Two linearly independent solutions of the equation of Ex. 3 may be represented in the form

$$w_i = z^{\lambda_1}\left\{A_i(z)\int_{C_i} e^{z\zeta}\zeta^{\lambda_1-1}(1-\zeta)^{-\lambda_2}d\zeta + B_i(z)\int_{C_i} e^{z\zeta}\zeta^{\lambda_1}(1-\zeta)^{-\lambda_2}d\zeta\right\} \qquad (i=1, 2),$$

where $A_i(z)$ and $B_i(z)$ are analytic at infinity and reduce to 1 and 0 respectively for $z=\infty$. This representation breaks down when one or other of ρ_1 and ρ_2 reduces to zero, when it may be replaced by one of the following:

$$w_1 = A(z) + B(z)e^z z^{1-p_1}\int e^{-z}z^{p_1-2}d\zeta, \quad w_2 = B(z)e^z z^{1-p_1},$$

$$w_1 = A(z)\int e^z z^{-p_1}dz + B(z)e^z z^{1-p_1}, \quad w_2 = A(z).$$

5. If the multipliers ρ_1 and ρ_2 are distinct from one another and from unity the coefficients in the Laurent series

$$w_1 = z^{\lambda_1}\sum_{\nu=-\infty}^{\infty} l_1^{(\nu)}z^\nu, \quad w_2 = z^{\lambda_2}\sum_{\nu=-\infty}^{\infty} l_2^{(\nu)}z^\nu,$$

in which two particular linearly independent solutions of the equation of Ex. 3 may be expanded have the form

$$l_i^{(\nu)} = L_i^{(\nu)} + \{a_1 + b_1(\lambda_i+\nu+1)\}L_i^{(\nu+1)} + \{a_2 + b_2(\lambda_i+\nu+2)\}L_i^{(\nu+2)} + \ldots \qquad (i=1, 2),$$

where

$$L_1^{(\nu)} = \frac{\lambda_1(\lambda_1+1)\ldots(\lambda_1+\nu-1)}{\nu!(\lambda_1-\lambda_2+1)\ldots(\lambda_1-\lambda_2+\nu)},$$

$$L_2^{(\nu)} = \frac{\lambda_2(\lambda_2+1)\ldots(\lambda_2+\nu-1)}{\nu!(\lambda_2-\lambda_1+1)\ldots(\lambda_2-\lambda_1+\nu)},$$

and a_ν and b_ν are numbers such that $|a_\nu|^{1/\nu}$, $|b_\nu|^{1/\nu}$ are finite for all values of ν.

6. If ρ_1 and ρ_2 are distinct from one another and from unity and if $\psi(z)$ is analytic at infinity, then for every solution $W(z)$ of the equation

$$\frac{d^2W}{dz^2} + \left\{-1 + \frac{p_1}{z}\right\}\frac{dW}{dz} - \frac{rs}{z^2}W = 0,$$

there is a relation of the form

$$W\{z+\psi(z)\} = a(z)W(z) + b(z)\frac{dW(z)}{dz},$$

where $a(z)$ and $b(z)$ are analytic at infinity.

CHAPTER XX

CLASSIFICATION OF LINEAR DIFFERENTIAL EQUATIONS OF THE SECOND ORDER WITH RATIONAL COEFFICIENTS

20·1. The Necessity for a Systematic Classification.—The foundations of the abstract theory of ordinary linear differential equations are firmly placed upon the classical theorems which assert the existence and specify the nature of solutions in the neighbourhood of an ordinary point. The nature of the solutions in the neighbourhood of a regular singularity is known with equal exactitude, and the behaviour of solutions with regard to irregular singular points has been revealed. On the other hand, the information available regarding the functions defined by particular equations or classes of equations is very scanty. Apart from simple equations, whose solutions are elementary functions, the only equation which has been exhaustively studied is the hypergeometric equation in its general form or under a particular guise such as the Legendre equation, that of Bessel or that of Weber or the equation of the confluent hypergeometric functions. The equations of Mathieu and of Lamé have been studied to some extent, but the knowledge of the functions defined by these equations is, even now, far from complete.

It would thus seem desirable that the study of linear differential equations should be resumed from a point of view intermediate between the most general on the one hand and the highly particularised on the other. In this intermediate aspect, any given equation appears as the common member of a number of specific classes whose salient properties it possesses. Thus what is inherent and essential in any given equation is readily discriminated from what is purely accidental.

In the present chapter a systematic classification of linear differential equations with rational coefficients is carried out by grouping the equations into types according to the number and the nature of their singular points. This classification is of value in that it not only indicates those properties which are common to the members of a particular class, but also suggests the existence of relationships between the individual members of one class and the corresponding members of another.

This systematisation was suggested by the discovery of Klein and Bôcher * that the chief linear differential equations which arise out of problems of mathematical physics can be derived from a single equation with five distinct regular singular points in which the difference between the two exponents relative to each singular point is $\frac{1}{2}$. The coalescence of two such singular points produces a regular singularity whose exponent-difference is arbitrary; the coalescence of three or more in one point generates an irregular singularity.

Every linear differential equation of the second order with rational coefficients has associated with it a definite number of regular and irregular

* Klein, *Vorlesungen über lineare Differentialgleichungen der zweiten Ordnung* (1894), p. 40; Bôcher, *Über die Reihenentwickelungen der Potentialtheorie* (1894), p. 193.

singular points. By regarding each of these singularities as generated by the confluence of the appropriate number of regular singularities with exponent-difference $\frac{1}{2}$, it is possible to consider the equation as derived, by definite processes, from one of a standard set of equations. The whole ground can be covered in this way, and those characteristic features which may be attributed to the presence of certain singularities of a certain definite type may be brought to light.

20·2. The Confluence of Singular Points.—It is convenient to introduce a term signifying a regular singular point with exponent-difference $\frac{1}{2}$; such a singular point will be called *elementary*. When a regular singular point is not so qualified, it is to be assumed that the exponent-difference is arbitrary.

The most general equation which has p elementary singularities, situated at the points

$$a_1, \quad a_2, \quad . \; . \; ., \quad a_{p-1}, \infty$$

is (§ 15·4)

$$\frac{d^2w}{dz^2}+\left\{\sum_{r=1}^{p-1}\frac{\frac{1}{2}-2\alpha_r}{z-a_r}\right\}\frac{dw}{dz}+\left\{\sum_{r=1}^{p-1}\frac{\alpha_r(\alpha_r+\frac{1}{2})}{(z-a_r)^2}+\frac{A_0+A_1z+\;.\;.\;.\;+A_{p-3}z^{p-3}}{\prod_{r=1}^{p-1}(z-a_r)}\right\}w=0,$$

where the exponents relative to a_r are α_r and $\alpha_r+\frac{1}{2}$. Since the exponents relative to the singular point at infinity also differ by $\frac{1}{2}$,

$$A_{p-3}=\left\{\sum_{r=1}^{p-1}\alpha_r\right\}^2-\sum_{r=1}^{p-1}\alpha_r^{\,2}-\frac{p-2}{2}\sum_{r=1}^{p-1}\alpha_r+\frac{(p-2)(p-4)}{16}.$$

The constant A_{p-3} is therefore definite; the remaining $p-3$ constants $A_0, A_1, \ldots, A_{p-4}$ are, on the other hand, entirely arbitrary.

Now suppose that two of the elementary singularities are caused to coalesce; thus let $a_2=a_1$. Then the indicial equation relative to the singular point $z=a_1$ becomes

$$\rho^2-2(\alpha_1+\alpha_2)\rho+\alpha_1(\alpha_1+\tfrac{1}{2})+\alpha_2(\alpha_2+\tfrac{1}{2})+A=0,$$

where

$$A=\frac{A_0+A_1a_1+\;.\;.\;.\;+A_{p-3}a_1^{\,p-3}}{(a_1-a_3)(a_1-a_4)\;.\;.\;.\;(a_1-a_{p-3})}.$$

The exponent-difference relative to the singularity $z=a_1$ is now dependent upon A, that is upon the arbitrary constants $A_0, \ldots, A_{p-4}$, and is therefore arbitrary if $p\geqslant 4$. The singularity, however, remains regular.

If the coalescence is not between a_1 and a_2 but between say a_{p-1} and ∞, let the arbitrary constants $A_0, \ldots, A_{p-4}$ be such that

$$\lim\frac{A_0}{a_{p-1}}=-A'_0, \quad . \; . \; ., \quad \lim\frac{A_{p-4}}{a_{p-1}}=-A'_{p-4};$$

where $A'_0, \ldots, A'_{p-4}$ are finite but otherwise arbitrary; since A_{p-3} is necessarily finite,

$$\lim\frac{A_{p-3}}{a_{p-1}}=0.$$

Then the equation takes the form

$$\frac{d^2w}{d^2z}+\left\{\sum_{r=1}^{p-2}\frac{\frac{1}{2}-2\alpha_r}{z-a_r}\right\}\frac{dw}{dz}+\left\{\sum_{r=1}^{p-2}\frac{\alpha_r(\alpha_r+\frac{1}{2})}{(z-a_r)^2}+\frac{A'_0+A'_1z+\;.\;.\;.\;+A'_{p-4}z^{p-4}}{\prod_{r=1}^{p-2}(z-a_r)}\right\}w=0,$$

and the singular point at infinity is regular but, since A'_{p-4} is arbitrary, with arbitrary exponent-difference.

Again, suppose that any q elementary singular points coalesce, then if

$q > 2$, the resulting singularity does not admit of an indicial equation and is consequently an irregular singularity.

The nature of an irregular singular point depends entirely upon the number of elementary singularities by whose coalescence it was generated. An irregular singularity generated by the coalescence of three elementary singularities will be said to be of *the first species*, and in general an irregular singularity of *the r-th species* will be defined as one which arises out of the coalescence of $r+2$ elementary singularities. It is evident that the order in which the singularities coalesce has no influence upon the nature of the resulting singularity.

20·21. Standard Forms: Transformations.—By multiplying the dependent variable by an appropriate factor it is possible, without altering the exponent-difference, to give to one exponent at any regular singular point any chosen value. Thus if the equation with dependent variable u has an elementary singularity a_r with exponents α_r and $\alpha_r+\frac{1}{2}$ the transformation

$$u=(z-a_r)^{\alpha_r}v$$

gives rise to an equation in v with a singularity at a_r with exponents 0 and $\frac{1}{2}$.

More generally if the equation in u is defined by the scheme

$$u=P\left\{\begin{matrix} a_1 & a_2 & \cdots & a_m & \infty & \\ \alpha_1 & \alpha_2 & \cdots & \alpha_m & * & z \\ \alpha_1+\frac{1}{2} & \alpha_2+\frac{1}{2} & \cdots & \alpha_m+\frac{1}{2} & * & \end{matrix}\right\}$$

where the asterisks denote that the point at infinity is any singularity, regular or irregular, the transformation

$$u=v\prod_{r=1}^{m}(z-a_r)^{\alpha_r}$$

leads to the equation

$$v=P\left\{\begin{matrix} a_1 & a_2 & \cdots & a_m & \infty & \\ 0 & 0 & \cdots & 0 & * & z \\ \frac{1}{2} & \frac{1}{2} & \cdots & \frac{1}{2} & * & \end{matrix}\right\}$$

in which the *nature* of the singularity at infinity has not been altered.

Thus there is no loss in generality in taking as the standard equation with p elementary singularities $a_1, a_2, \ldots, a_{p-1}, \infty$ the following

$$\frac{d^2w}{dz^2}+\left\{\sum_{r=1}^{p-1}\frac{\frac{1}{2}}{z-a_r}\right\}\frac{dw}{dz}+\frac{A_0+A_1z+\ldots+A_{p-3}z^{p-3}}{\prod_{r=1}^{p-1}(z-a_r)}w=0,$$

where, since the point at infinity is also elementary,

$$A_{p-3}=\frac{(p-2)(p-4)}{16}.$$

This equation is known as *the generalised Lamé equation.* There is occasionally an advantage in taking the exponents at the finite singularities to be $\frac{1}{4}$ and $\frac{3}{4}$, for then the equation assumes its normal form

$$\frac{d^2w}{dz^2}+\left\{\frac{3}{16}\sum_{r=1}^{p-1}\frac{1}{(z-a_r)^2}+\frac{A_0+A_1z+\ldots+A_{p-3}z^{p-3}}{\prod_{r=1}^{p-1}(z-a_r)}\right\}w=0,$$

where

$$A_{p-3}=-\tfrac{3}{16}(p-2).$$

There are two algebraic transformations in the independent variable which will occasionally be made. The *projective transformation*

$$z' = \frac{a_j - a_k}{a_j - a_i} \cdot \frac{z - a_i}{z - a_k}$$

transforms the singular points a_i, a_j, and a_k into 0, 1 and ∞ respectively, without altering the exponents relative to these points.* Thus there is no loss in generality in fixing three singularities at the three points 0, 1, ∞ ; if there are more than three singularities the distribution of the remainder is arbitrary.

Next in importance is the *quadratic transformation*

$$z'^2 = z$$

with two fixed points 0 and ∞. An elementary singularity at either of these two points becomes an ordinary point, a regular singularity remains regular, and an irregular singularity has its species doubled. A singularity at any other point $z=a$ is replaced by two precisely similar singularities at $z' = \pm\sqrt{a}$ and thus in general complicates the equation.

Finally, transcendental transformations are used to reduce the equation to a known form, *e.g.* to the Mathieu equation. Their general effect is to replace a number of elementary singularities by an irregular singularity of transfinite species.

20·22. The Formula of an Equation : the Irreducible Constants.—Any given equation is, in the first place, characterised by

(α) the number a of its elementary singularities,
(β) the number b of its non-elementary regular singularities,
(γ) the number c of its essential singularities of all species.

In the second place the c irregular singularities may be subdivided into

(i) c_1 singularities of the first species,
(ii) c_2 ,, ,, second ,, ,
(iii) c_3 ,, ,, third ,, ,

and so on. The equation will then be said to have the *formula* †

$$[a, \quad b, \quad c_1, \quad c_2, \quad c_3, \quad \ldots].$$

Equations which have the same formula may differ from one another firstly as to the actual location of the singular points, secondly as to the actual exponents relative to the regular singularities, and thirdly as to certain arbitrary constants. An equation, whose formula is given, is determinate except as to these three variants, the arbitrary nature of which introduces three categories of constants into the equation. Of the constants in the first category, which determine the position of the singularities, all but three must be regarded as arbitrary. Secondly, to each non-elementary regular singularity corresponds an arbitrary constant which represents the exponent-difference. These arbitrary constants, together with the constants of the third category, are the *irreducible constants* of the general equation with the given formula. Thus the first equation of § 20·2 whose formula is $[p, 0, 0]$ has $p-1$ constants of the first category $(a_1, a_2, \ldots, a_{p-1})$, which are reducible to $p-3$; it has $p-1$ constants of the second category $(\alpha_1, \alpha_2, \ldots, \alpha_{p-1})$ all of which are removable, and $p-3$ arbitrary constants of the third category $(A_0, A_1, \ldots, A_{p-4})$. It has thus in all $2p-6$ irreducible constants.

The coalescence of singularities is a process affecting the constants of the

* Alternatively a transformation into $+1$, -1, ∞ is occasionally used.

† When $c=0$ the formula $[a, b, 0]$ is used. When there is only one irregular singularity the formula is shortened to $[a, b, 1_s]$, where s is the species.

first category alone; each individual coalescence of two singular points diminishes the number of irreducible constants by one and only one provided that at least three singularities remain. When, however, the number of distinct singularities is reduced to two (0 and ∞), a transformation $z'=Cz$, where C is a constant properly chosen, can be applied which reduces one of the constants in the third category to a predetermined numerical value. If now further coalescence takes place, and but one singularity (at ∞) remains, a linear transformation can be applied which again diminishes the number of constants in the third category by unity. Hence the equation $[p, 0, 0]$ and all others derived from it by coalescence have at most $2p-6$ and at least $p-5$ irreducible constants.

20·221. The Number of Distinct Types of Equation which can be derived from the Equation $[p, 0, 0]$.—It may easily be verified that the number of distinct types of equation, having only regular singularities, which can be derived from the equation $[p, 0, 0]$ is $\frac{1}{2}p$ or $\frac{1}{2}(p-1)$ according as p is even or odd. Any such equation is in fact of type $[p-2r, r, 0]$.

Similarly the number of types of equation possessing one irregular singularity of the first species is $\frac{1}{2}p-1$ or $\frac{1}{2}(p-1)$ according as p is even or odd. More generally the total number of types of equation having one irregular singular point of any possible species is

$$(\tfrac{1}{2}p-1)+(\tfrac{1}{2}p-1)+(\tfrac{1}{2}p-2)+(\tfrac{1}{2}p-2)+\ \ldots\ +2+2+1+1=\tfrac{1}{4}p(p-2)$$

when p is even, or

$$\tfrac{1}{2}(p-1)+\tfrac{1}{2}(p-3)+\tfrac{1}{2}(p-3)\ +\ \ldots\ +2+2+1+1=\tfrac{1}{4}(p-1)^2$$

when p is odd. The typical equations having two or more irregular singularities may be enumerated in the same way.

If each regular singularity is counted once or twice according as the exponent-difference is $\frac{1}{2}$ or arbitrary, and each irregular singularity of the rth species is counted $r+2$ times, the sum of the numbers thus obtained will be p. Conversely the number N of distinct types of equations which may be derived from the equation $[p, 0, 0]$ is the same as the number of partitions of the integer p into any number of integral parts each less than p. The results are summarised, for particular values of p, in the following table in which N_r denotes the number of distinct types of equation with r irregular singularities and N the total number theoretically possible. The equation $[p, 0, 0]$ itself is not included.

$p=$	4	5	6	7	8	9	10	11	12
N_0	2	2	3	3	4	4	5	5	6
N_1	2	4	6	9	12	16	20	25	30
N_2	—	—	1	2	5	8	14	20	30
N_3	—	—	—	—	—	1	2	5	9
N_4	—	—	—	—	—	—	—	—	1
N	4	6	10	14	21	29	41	55	76

20·3. Equations derived from the Equation with four Elementary Singularities.—The equations which have two or three elementary, and no other singularities are trivial; the present section deals with the equation having four elementary singularities and its coalescent cases.

Let the four elementary singularities be $z=a_1, a_2, a_3, \infty$; since the sum of the eight exponents is 2, the exponents relative to each singularity can be chosen to be 0 and $\frac{1}{2}$. A_1 is then zero and the standard form of $[4, 0, 0]$ can therefore be taken as

$$\frac{d^2w}{dz^2}+\left\{\frac{\frac{1}{2}}{z-a_1}+\frac{\frac{1}{2}}{z-a_2}+\frac{\frac{1}{2}}{z-a_3}\right\}\frac{dw}{dz}+\frac{A_0}{(z-a_1)(z-a_2)(z-a_3)}w=0.$$

There are two irreducible constants, namely $\dfrac{a_3-a_2}{a_3-a_1}$ and A_0. The equation is a particular case of the Lamé equation and is of no importance in itself.

Now let the singular point $z=a_3$ coalesce with the singular point at infinity, and let

$$\lim A_0/a_3=n^2.$$

If the points a_1 and a_2 are transferred to -1 and $+1$ respectively, the equation becomes [2, 1, 0]:

$$\text{I.}\qquad \frac{d^2w}{dz^2}+\left\{\frac{\frac{1}{2}}{z+1}+\frac{\frac{1}{2}}{z-1}\right\}\frac{dw}{dz}-\frac{n^2}{z^2-1}w=0,$$

and contains one irreducible constant, n. It is the equation of the Gegenbauer function * $C_n{}^0(z)$.

If $a_3\to\infty$ and also $a_2\to a_1\to 0$, and n^2 is as before, the equation becomes [0, 2, 0]:

$$\frac{d^2w}{dz^2}+\frac{1}{z}\frac{dw}{dz}-\frac{n^2}{z}=0$$

with one irreducible constant, n. Multiplication of the dependent variable by z^n reduces the equation to its standard form:

$$\text{II.}\qquad \frac{d^2w}{dz^2}+\frac{1-2n}{z}\frac{dw}{dz}=0.$$

The equation [1, 0, 1] is obtained by the coalescence of a_2 and a_3 with ∞, producing at infinity an irregular singularity of the first species. Let $a_1\to 0$. Since A_0 is arbitrary, it may be so chosen that

$$\lim A_0/a_2a_3=-m^2,$$

where m is finite. This gives rise to the equation:

$$\frac{d^2w}{dz^2}+\frac{\frac{1}{2}}{z}\frac{dw}{dz}-\frac{m^2}{z}w=0.$$

The constant m^2 is not irreducible; if the independent variable is multiplied by m^{-2} the equation reduces to its standard form:

$$\text{III.}\qquad \frac{d^2w}{dz^2}+\frac{\frac{1}{2}}{z}\frac{dw}{dz}-\frac{w}{z}=0.$$

Finally let

$$a_3\to a_2\to a_1\to\infty$$

and let

$$\lim A_0/a_1a_2a_3=m^2,$$

then the equation has an irregular singularity of the second species at infinity and is

$$\frac{d^2w}{dz^2}-m^2w=0.$$

The constant m^2 is removable, and therefore the standard form of [0, 0, 0, 1] or [0, 0, 1_2] is

$$\text{IV.}\qquad \frac{d^2w}{dz^2}-w=0.$$

Thus there can be derived from [4, 0, 0] the four types:

I. [2, 1, 0] with one irreducible constant,
II. [0, 2, 0] ,, one ,, ,, ,
III. [1, 0, 1] ,, no ,, ,, ,
IV. [0, 0, 0, 1] ,, no ,, ,, .

It may be noted that the quadratic transformation changes III. into IV. in accordance with § 20·21.

* Whittaker and Watson, *Modern Analysis*, § 15·8.

20·31. Equations derived from the Equation with Five Elementary Singularities.—The standard form of [5, 0, 0] is

$$\frac{d^2w}{dz^2}+\left\{\sum_{r=1}^{4}\frac{\frac{1}{2}}{z-a_r}\right\}\frac{dw}{dz}+\left\{\frac{A_0+A_1z+\frac{3}{16}z^2}{\prod_{r=1}^{4}(z-a_r)}\right\}w=0,$$

and contains four irreducible constants.

Let $a_4 \to 0$ and let

$$\lim A_0/a_4=\tfrac{1}{4}h, \qquad \lim A_1/a_4=\tfrac{1}{4}n(n+1).$$

Then the equation which arises is [3, 1, 0] :

$$\text{I.}\quad \frac{d^2w}{d^2z}+\left\{\frac{\frac{1}{2}}{z-a_1}+\frac{\frac{1}{2}}{z-a_2}+\frac{\frac{1}{2}}{z-a_3}\right\}\frac{dw}{dz}\frac{h+n(n+1)z}{4(z-a_1)(z-a_2)(z-a_3)}w=0,$$

and has three irreducible constants, $\dfrac{a_3+a_2}{a_3-a_1}$, h and n. It is the Lamé equation in its algebraic form.*

Now in I. let $a_2 \to a_3 \to 1$, $a_1 \to 0$, then equation [1, 2, 0] arises in the form :

$$\text{II.}\qquad \frac{d^2w}{dz^2}+\left\{\frac{\frac{1}{2}}{z}+\frac{1}{z-1}\right\}\frac{dw}{dz}-\frac{h+n(n+1)z}{4z(z-1)^2}w=0,$$

and contains two irreducible constants.

It is transformed by the quadratic substitution $z=x^2$ into the Associated Legendre equation :

$$\text{IIa.}\qquad (1-x^2)\frac{d^2w}{dx^2}-2x\frac{dw}{dx}+\left\{n(n+1)-\frac{m^2}{1-x^2}\right\}w=0.$$

Equation IIa. has the formula [0, 3, 0] but is particularised in that the exponents at $z=-1$ are the same as those at $z=+1$. It has only two irreducible constants whereas the general equation of type [0, 3, 0] has three (Equation III. of the following section).

The first of the two possible equations having the point at infinity as an irregular singularity of the first species is obtained by the process :

$$a_1 \to 0, \quad a_2 \to 1, \quad a_3 \to a_4 \to \infty,$$
$$\lim A_0/a_3a_4=\tfrac{1}{4}a, \quad \lim A_1/a_3a_4 \to \tfrac{1}{4}k^2.$$

Thus the typical equation [2, 0, 1] is :

$$\text{III.}\qquad \frac{d^2w}{dz^2}+\left\{\frac{\frac{1}{2}}{z}+\frac{\frac{1}{2}}{z-1}\right\}\frac{dw}{dz}-\frac{a+k^2z}{4z(z-1)}w=0,$$

and contains two irreducible constants. By means of the transcendental substitution $z=\cos^2x$ it is transformed into the Mathieu equation

$$\text{IIIa.}\qquad \frac{d^2w}{dx^2}+(a+k^2\cos^2x)w=0.$$

Now let

$$a_1 \to a_2 \to 0, \qquad a_3 \to a_4 \to \infty,$$
$$\lim A_0/a_3a_4=\tfrac{1}{4}n^2, \quad \lim A_1/a_3a_4=\tfrac{1}{4}k^2,$$

then there arises the equation

$$\frac{d^2w}{dz^2}+\frac{1}{z}\frac{dw}{dz}-\frac{n^2+k^2z}{4z^2}w=0.$$

* Whittaker and Watson, *Modern Analysis*, § 23·4.

The constant k is removable by multiplying the independent variable by $-k^{-2}$; thus the typical equation [0, 1, 1] is

$$\text{IV.}\qquad \frac{d^2w}{dz^2}+\frac{1}{z}\frac{dw}{dz}+\frac{z-n^2}{4z}w=0$$

and involves one irreducible constant. The quadratic transformation $z=x^2$ reduces it to the Bessel equation:

$$\text{IVa.}\qquad x^2\frac{d^2w}{dx^2}+x\frac{dw}{dx}+(x^2-n^2)w=0.$$

The Bessel equation is a particular case of [0, 1, 0, 1], the general case of which involves two arbitrary constants (Equation VIII. of the following section).

An irregular singularity at infinity of the second species is obtained by the operations

$$a_2\to a_3\to a_4\to\infty,\quad a_1\to 0.$$

The equation [1, 0, 0, 1] or $[1, 0, 1_2]$ thus generated reduces to

$$\text{V.}\qquad z\frac{d^2w}{dz^2}+\tfrac{1}{2}\frac{dw}{dz}+\tfrac{1}{4}(n+\tfrac{1}{2}-\tfrac{1}{4}z)w=0,$$

and contains one irreducible constant. Under the transformation $z=x^2$ this equation becomes the Weber equation:

$$\text{Va.}\qquad \frac{d^2w}{dx^2}+(n+\tfrac{1}{2}-\tfrac{1}{4}x^2)w=0,$$

which has the formula $[0, 0, 1_4]$ (Equation X. of the following section).

Lastly, if $a_1\to a_2\to a_3\to a_4\to\infty$, the equation $[0, 0, 1_3]$ arises, which may be reduced to the standard form:

$$\text{VI.}\qquad \frac{d^2w}{dz^2}+zw=0,$$

and contains no irreducible constant.

Equation VI. is transformed by the substitutions

$$w=z^{\frac{1}{2}}y,\quad z=(\tfrac{3}{2}x)^{\frac{2}{3}}$$

into a particular case of the Bessel equation, namely

$$x^2\frac{d^2y}{dx^2}+x\frac{dy}{dx}(x^2-\tfrac{1}{9})y=0.$$

Thus the six types of equation which can be derived, by coalescence of singularities, from the equation [5, 0, 0] are as follows:

I. [3, 1, 0] with three irreducible constants,
II. [1, 2, 0] ,, two ,, ,, ,
III. [2, 0, 1] ,, two ,, ,, ,
IV. [0, 1, 1] ,, one ,, ,, ,
V. $[1, 0, 1_2]$,, one ,, ,, ,
VI. $[0, 0, 1_3]$,, no ,, ,, .

20·32. Equations derived from the Equation with six Elementary Singularities.—It is convenient to take [6, 0, 0] in its most general form:

$$\frac{d^2w}{dz^2}+\left\{\sum_{r=1}^{5}\frac{\frac{1}{2}-2a_r}{z-a_r}\right\}\frac{dw}{dz}+\left\{\sum_{r=1}^{5}\frac{a_r(a_r+\frac{1}{2})}{(z-a_r)^2}+\frac{A_0+A_1z+A_2z^2+A_3z^3}{\prod_{r=1}^{5}(z-a_r)}\right\}w=0,$$

where

$$A_3=\left(\sum_{r=1}^{5}a_r\right)^2-\sum_{r=1}^{5}a_r{}^2-2\sum_{r=1}^{5}a_r+\tfrac{1}{2}.$$

There are six irreducible constants, namely A_0, A_1, A_2 and the anharmonic ratios of three tetrads of the numbers $a_1, a_2, \ldots, a_5$.

Let $a_5 \to \infty$ and let

$$\lim A_0/a_5 = -\tfrac{1}{4}C_0, \quad \lim A_1/a_5 = -\tfrac{1}{2}C_1, \quad \lim A_2/a_5 = \tfrac{1}{4}n(n+1);$$

also let $a_1=a_2=a_3=a_4=0$. Then the equation which arises is [4, 1, 0]:

$$\text{I.} \qquad \frac{d^2w}{dz^2} + \left\{ \sum_{r=1}^{4} \frac{\frac{1}{2}}{z-a_r} \right\} \frac{dw}{dz} + \left\{ \frac{C_0+2C_1z-n(n+1)z^2}{4\prod_{r=1}^{4}(z-a_r)} \right\} w = 0.$$

This equation is a generalised form of the Lamé equation; it has four elementary singularities and a regular singularity at infinity with exponent-difference $n+\frac{1}{2}$, and involves five irreducible constants.

The next equation [2, 2, 0] is obtained from I. by the operations

$$a_1 \to 0, \quad a_2 \to a_3 \to a, \quad a_4 \to 1$$

and is of the form:

$$\text{II.} \qquad \frac{d^2w}{dz^2} + \left\{ \frac{\frac{1}{2}}{z} + \frac{1}{z-a} + \frac{\frac{1}{2}}{z-1} \right\} \frac{dw}{dz} + \frac{C_0+2C_1z-n(n+1)z^2}{4z(z-a)^2(z-1)} w = 0,$$

with four irreducible constants a, C_0, C_1, n.

Let $a=k^{-2}$ and make the transformation $z=\mathrm{sn}^2(x, k)$, then with a little manipulation the equation may be brought into the form: *

$$\text{IIa.} \qquad \frac{d^2y}{dx^2} - \left\{ h + n(n+1)k^2 \,\mathrm{sn}^2 x + \frac{m(m+1)k^2\,\mathrm{cn}^2 x}{\mathrm{dn}^2 x} \right\} y = 0.$$

Equation [0, 3, 0] is most conveniently obtained directly from [6, 0, 0]. Let $a_1 \to a_2 \to 0$, $a_3 \to a_4 \to 1$, $a_5 \to \infty$, and let C_0, C_1, and n be as above. The four exponents $\alpha_1, \alpha_2, \alpha_3, \alpha_4$ may be assigned in any arbitrary manner; there remain three irreducible constants, let them be α, β, γ defined as follows:

$$\begin{aligned}
1-\gamma &= 2(\alpha_1+\alpha_2), \\
0 &= \alpha_1(\alpha_1+\tfrac{1}{2}) + \alpha_2(\alpha_2+\tfrac{1}{2}) + \tfrac{1}{4}C_0, \\
\gamma-\alpha-\beta &= 2(\alpha_3+\alpha_4), \\
0 &= \alpha_3(\alpha_3+\tfrac{1}{2}) + \alpha_4(\alpha_4+\tfrac{1}{2}) + \tfrac{1}{4}\{C_0+2C_1-n(n+1)\}, \\
\alpha\beta &= -\tfrac{1}{2}(C_0+C_1).
\end{aligned}$$

The equation then reduces to the ordinary hypergeometric equation:

$$\text{III.} \qquad z(1-z)\frac{d^2w}{dz^2} + \{\gamma-(\alpha+\beta+1)z\}\frac{dw}{dz} - \alpha\beta w = 0.$$

The equation [3, 0, 1] is obtained by the operations

$$a_1 \to 0, \quad a_2 \to a, \quad a_3 \to 1, \quad a_4 \to a_5 \to \infty,$$

$$\lim A_0/a_4a_5 = \tfrac{1}{4}C_0, \quad \lim A_1/a_4a_5 = \tfrac{1}{2}C_1, \quad \lim A_2/a_4a_5 = \tfrac{1}{2}C_2.$$

Let

$$\alpha_1 = \alpha_2 = \alpha_3 = 0,$$

and the equation becomes:

$$\text{IV.} \qquad \frac{d^2w}{dz^2} + \left\{ \frac{\frac{1}{2}}{z} + \frac{\frac{1}{2}}{z-a} + \frac{\frac{1}{2}}{z-1} \right\} \frac{dw}{dz} + \frac{C_0+2C_1z+C_2z^2}{4z(z-a)(z-1)} w = 0,$$

with four irreducible constants. If $a=k^{-2}$ and $z=\mathrm{sn}^2(x, k)$ the equation becomes

$$\frac{d^2w}{dx^2} - k^2(C_0 + 2C_1\,\mathrm{sn}^2 x + 2C_2\,\mathrm{sn}^4 x)w = 0.$$

This equation is thus an extension of the Lamé equation.

* Hermite, *J. für Math.* 89 (1880), p. 9 [*Œuvres*, 4, p. 8]; Darboux, *C. R. Acad. Sc. Paris*, 94 (1882), p. 1645.

The equation [1, 1, 1] is a confluent case of IV. It is, however, more convenient to derive it from [6, 0, 0] as follows. Let $a_4 \to a_5 \to \infty$ and let C_0, C_1 and C_2 be as above. Let $a_1 \to 0$ with $\alpha_1 = 0$ and let $a_2 \to a_3 \to 1$, forming a regular singularity with exponents 0 and r. The conditions for this are

$$r = 2(\alpha_2 + \alpha_3),$$
$$0 = \alpha_2(\alpha_2 + \tfrac{1}{2}) + \alpha_3(\alpha_3 + \tfrac{1}{2}) + \tfrac{1}{4}(C_0 + 2C_1 + 2C_2).$$

The equation then assumes the form :

$$\text{V.} \qquad \frac{d^2w}{dz^2} + \left\{\frac{\frac{1}{2}}{z} + \frac{1-r}{z-1}\right\}\frac{dw}{dz} - \frac{a+k^2z}{4z(z-1)}w = 0,$$

where $a = C_0$, $k^2 = -2C_2$; it has three irreducible constants. The substitution $z = \cos^2 x$ transforms it into the Associated Mathieu equation : *

$$\text{Va.} \qquad \frac{d^2w}{dx^2} + (1-2r)\cot x\,\frac{dw}{dx} + (a + k^2\cos^2 x)w = 0.$$

The equation [0, 0, 2] having two irregular singularities of the first species, the one at the origin and the other at infinity, arises as follows. Let $a_4 \to a_5 \to \infty$ and let C_0, C_1 and C_2 be as before. Let $a_1 \to a_2 \to a_3 \to 0$ with $\alpha_1 = \frac{1}{4}$, $\alpha_2 = 0$, $\alpha_3 = 0$. Then the equation becomes :

$$z^3\frac{d^2w}{dz^2} + z^2\frac{dw}{dz} + \{C_0 + 2C_1z + 2C_2z^2\}w = 0.$$

There are only two irreducible constants ; if the independent variable is multiplied by an appropriate constant, the equation can be reduced to its standard form :

$$\text{VI.} \qquad z^2\frac{d^2w}{dz^2} + z\frac{dw}{dz} - \tfrac{1}{4}\{a + \tfrac{1}{2}k^2 + \tfrac{1}{4}k^2(z + z^{-1})\}w = 0.$$

The transcendental substitution $z = e^{2ix}$ now transforms it into the Mathieu equation :

$$\frac{d^2w}{dx^2} + (a + k^2\cos^2 x) = 0.$$

Two equations for which the point at infinity is an irregular singularity of the second species are obtainable. In the first place, let $a_3 \to a_4 \to a_5 \to \infty$ and let

$$\lim A_0/a_3a_4a_5 = -\tfrac{1}{4}C_0, \quad \lim A_1/a_3a_4a_5 = -\tfrac{1}{2}C_1, \quad \lim A_2/a_3a_4a_5 = -\tfrac{1}{2}C_2.$$

Let $a_1 \to 0$, $a_2 \to 1$ with $\alpha_1 = \alpha_2 = 0$. There arises the equation $[2, 0, 1_2]$:

$$\text{VII.} \qquad \frac{d^2w}{dz^2} + \left\{\frac{\frac{1}{2}}{z} + \frac{\frac{1}{2}}{z-1}\right\}\frac{dw}{dz} + \frac{C_0 + 2C_1z + 2C_2z^2}{4z(z-1)}w = 0,$$

which contains three irreducible constants. The transformation $z = \cos^2 x$ followed by a modification of the constants brings the equation into the form : †

$$\text{VIIa.} \qquad \frac{d^2w}{dx^2} + \{a - (n+1)l\cos 2x + \tfrac{1}{8}l^2\cos 4x\}w = 0.$$

Secondly, let $a_3 \to a_4 \to a_5 \to \infty$ as in the previous case, and let $a_1 \to a_2 \to 0$ with $\alpha_1 = \alpha_2 = \frac{1}{4}$. If $C_0 = -\frac{1}{2} - 4m^2$ the exponents relative to $z = 0$ are $\frac{1}{2} - m$

* Ince, *Proc. Edin. Math. Soc.* 41 (1923), p. 94.

† This equation was first obtained by Whittaker, *Proc. Edin. Math. Soc.* 33 (1914), p. 22, by confluence from [2, 2, 0]. It was investigated in detail by Ince, *Proc. London Math. Soc.* (2), 23 (1924), p. 56 ; *ibid.* (2), 25 (1926), p. 53 ; for its physical significance see Ince, *Proc. Roy. Soc. Edin.* 45 (1925), p. 106.

and $\frac{1}{2}+m$. No loss of generality is involved in taking $C_1=2k$, $C_2=-\frac{1}{2}$. The equation is now reduced to its standard form :

VIII. $$\frac{d^2w}{dz^2}+\left\{-\frac{1}{4}+\frac{k}{z}+\frac{\frac{1}{4}-m^2}{z^2}\right\}w=0,$$

and involves two irreducible constants. It is the equation of the confluent hypergeometric functions * $W_{k,m}(z)$.

Let $a_2\to a_3\to a_4\to a_5\to\infty$ and let $a_1\to 0$ with $\alpha_1=0$. If

$$\lim A_0/a_2a_3a_4a_5=\tfrac{1}{4}C_0,\quad \lim A_1/a_2a_3a_4a_5=\tfrac{1}{2}C_1,\quad \lim A_2/a_2a_3a_4a_5=\tfrac{1}{2}C_2,$$

the equation becomes $[1, 0, 1_3]$:

IX. $$\frac{d^2w}{dz^2}+\frac{1}{2z}\cdot\frac{dw}{dz}+\frac{C_0+2C_1z+2C_2z^2}{4z}w=0.$$

The equation has only two irreducible constants, $C_0{}^2/C_1$ and $C_0{}^3/C_2$. The quadratic transformation $z=x^2$ brings it into the form :

IXa. $$\frac{d^2w}{dx^2}+\{C_0+2C_1x^2+2C_2x^4\}w=0,$$

which is a particular case of $[0, 0, 1_6]$.

Finally, let all the elementary singularities coalesce in the point at infinity, then the equation $[0, 0, 1_4]$ which arises can easily be reduced to the Weber equation :

X. $$\frac{d^2w}{dz^2}+(n+\tfrac{1}{2}-\tfrac{1}{4}z^2)w=0.$$

It involves one irreducible constant.

Thus the ten distinct equations which arise out of the equation [6, 0, 0] by coalescence of its singularities are : †

I.	[4, 1, 0]	with	five	irreducible	constants,
II.	[2, 2, 0]	,,	four	,,	,, ,
III.	[0, 3, 0]	,,	three	,,	,, ,
IV.	[3, 0, 1]	,,	four	,,	,, ,
V.	[1, 1, 1]	,,	three	,,	,, ,
VI.	[0, 0, 2]	,,	two	,,	,, ,
VII.	$[2, 0, 1_2]$	,,	three	,,	,, ,
VIII.	$[0, 1, 1_2]$	,,	two	,,	,, ,
IX.	$[1, 0, 1_3]$	,,	two	,,	,, ,
X.	$[0, 0, 1_4]$	,,	one	,,	,, .

20·4. Constants-in-Excess.—It may be noted that in the set of equations derived from [6, 0, 0] the number of irreducible constants is equal to the number of singularities ; in the set derived from [5, 0, 0] the number of singularities exceeds the number of irreducible constants by unity. In general the number of irreducible constants in an equation derived from $[p, 0, 0]$ exceeds the number of singularities by $p-6$. It is interesting to inquire how these constants are to be accounted for.

The typical equation $[p, 0, 0]$ involves altogether $2p-6$ irreducible constants, of which $p-3$ are accounted for by the arbitrary position of $p-3$ singularities, and $p-3$ remain unspecified. Similarly in the equation $[p, q, 0]$ there are $p+q-3$ arbitrary constants which are not accounted for by the positions of the singular points or by the arbitrary exponent-differences relative to the q regular singularities. These constants are termed the *constants-in-excess*.

* Whittaker, *Bull. Am. Math. Soc.* 10 (1903), p. 125 ; Whittaker and Watson, *Modern Analysis*, Chap. XVI. It is essentially equivalent to the Hamburger equation, § 17·62.

† The types I., IV., V. and IX. have not yet been investigated in detail.

Now consider the class of equations which have one irregular singularity of the first species. Any such equation $[p, q, 1]$ may be regarded as generated from $[p+1, q+1, 0]$ by coalescence of an elementary with a regular singularity. In this process one constant is lost, but it is not a constant-in-excess. Therefore $[p, q, 1]$ has the same number of constants-in-excess as $[p+1, q+1, 0]$, namely $p+q-1$. Similarly by considering $[p, q, 1_2]$ to be derived from $[p, q+2, 0]$ by the coalescence of two regular singularities it may be proved that the number of constants-in-excess in $[p, q, 1_2]$ is $p+q-1$. In general the equation $[p, q, 1_s]$ has $2p+3q+s-3$ irreducible constants, of which $p+q-2$ are accounted for by the arbitrary positions of that number of singularities, q are accounted for by the exponent differences relative to the q regular singularities, and s by the constants in the determining factor relative to the irregular singularity. There remain $p+q-1$ constants-in-excess.

The constants-in-excess are involved in the group of the equation; by a proper choice of these constants the group may be simplified. An example is furnished by the Mathieu equation (§ 20·31, IIIa.), in which the constant k occurs in the determining factor relative to the irregular singularity at infinity,* and the constant a is the constant-in-excess.

20·5. Sequences of Equations with Regular Singularities.—The equations of formulæ

$$[3, 1, 0], \quad [4, 1, 0], \quad \ldots, \quad [p, 1, 0], \quad \ldots$$

form an important sequence. The first is the Lamé equation

$$\frac{d^2w}{dz^2}+\left\{\sum_{r=1}^{3}\frac{\frac{1}{2}}{z-a_r}\right\}\frac{dw}{dz}-\frac{n(n+1)z+h}{4\prod\limits_{r=1}^{3}(z-a_r)}w=0.$$

Only one of the constants a_r is reducible; there is therefore no loss in generality in supposing that the singularities are so disposed that

$$a_1+a_2+a_3=0.$$

Let z be transformed by the substitution

$$x=\tfrac{1}{2}\int_z^\infty \{(t-a_1)(t-a_2)(t-a_3)\}^{-\frac{1}{2}}dt,$$

so that $z=\wp(x)$, and the equation becomes

$$\frac{d^2w}{dx^2}-\{h+n(n+1)\wp(x)\}w=0.$$

The generalised Lamé equation is

$$\frac{d^2w}{dz^2}+\left\{\sum_{r=1}^{p}\frac{\frac{1}{2}}{z-a_r}\right\}\frac{dw}{dz}+\left\{\frac{A_0+A_1z+\ \ldots\ +A_{p-2}z^{p-2}}{4\prod\limits_{r=1}^{p}(z-a_r)}\right\}w=0$$

with $p-2$ constants-in-excess, namely $A_0, A_1, \ldots, A_{p-3}$. Under the transformation

$$x=\tfrac{1}{2}\int_z^\infty\left\{\prod_{r=1}^{p}(t-a_r)\right\}^{-\frac{1}{2}}dt$$

the equation becomes

$$\frac{d^2w}{dx^2}+\{A_0+A_1z+\ \ldots\ +A_{p-2}z^{p-2}\}w=0.$$

* Ince, *Proc. Roy. Soc. Edin.* 46 (1926), p. 386.

Another important set of Fuchsian equations is the set having p singularities. The distinct types are

$$[p, 0, 0], \quad [p-1, 1, 0], \quad \ldots, \quad [0, p, 0],$$

and each equation has $p-3$ constants in excess. The equations $p=3$ are equations of the Riemann P-function; the equations $p=4$ are the Lamé equation, and the associated equations derived from the Lamé equation by generalising its elementary singularities. The equations for $p=5, 6, 7, \ldots$ have not yet been studied.

20·51. Sequences of Equations with one Irregular Singularity.—The Weber equation $[0, 0, 1_4]$

$$\frac{d^2w}{dz^2}+(n+\tfrac{1}{2}-\tfrac{1}{4}z^2)w=0,$$

may be regarded as a particular case of $[0, 0, 1_p]$:

$$\frac{d^2w}{dz^2}+\{A_0+A_2z^2+\ \ldots\ +A_{p-2}z^{p-2}\}w=0,$$

which has $p-3$ irreducible constants.

It was seen that the equation $[1, 0, 1_2]$ (§ 20·31, V.) is transformed by the quadratic substitution into $[0, 0, 1_4]$. Now the more general equation $[1, 0, 1_r]$ is

$$z\frac{d^2w}{dz^2}+\frac{1}{2}\frac{dw}{dz}+\tfrac{1}{4}(B_0+B_1z+B_2z^2+\ \ldots\ +B_{r-1}z^{r-1})w=0$$

with $r-1$ irreducible constants. It is transformed by the substitution $z=x^2$ into

$$\frac{d^2w}{dx^2}+(B_0+B_1x^2+B_2x^4+\ \ldots\ +B_{r-1}x^{2r-2})w=0$$

and now has the formula $[0, 0, 1_{2r}]$. But it is not the typical equation of that formula since it contains only $r-1$ instead of the full number $2r-3$ of irreducible constants. The sequence of equations $[1, 0, 1_r]$ can therefore be ignored; they are effectively included in the sequence $[0, 0, 1_p]$.

The equation $[0, 1, 1_1]$ is transformed by the quadratic substitution into Bessel's equation which is a particular case of $[0, 1, 1_2]$, the confluent hypergeometric equation. Similarly, the more general equation $[0, 1, 1_p]$:

$$z^2\frac{d^2w}{dz^2}+z\frac{dw}{dz}+\tfrac{1}{4}(A_pz^p+A_{p-1}z^{p-1}+\ \ldots\ +A_1z-n^2)w=0$$

with p irreducible constants, is transformed by the substitution $z=x^2$ into

$$x^2\frac{d^2w}{dx^2}+x\frac{dw}{dx}+(A_px^{2p}+A_{p-1}x^{2p-2}+\ \ldots\ +A_1x^2-n^2)w=0,$$

which is a particular case of $[0, 1, 1_{2p}]$.

20·52. Equations with Periodic Coefficients.—Just as the equation $[2, 0, 1_1]$ is transformed by the substitution $z=\cos^2 x$ into the Mathieu equation, so also is the more general equation $[2, 0, 1_p]$:

$$\frac{d^2w}{dz^2}+\left\{\frac{\frac{1}{2}}{z}+\frac{\frac{1}{2}}{z-1}\right\}\frac{dw}{dz}-\frac{A_0+A_1z+\ \ldots\ +A_pz^p}{4z(z-1)}w=0,$$

with $p+1$ irreducible constants, transformed into

$$\frac{d^2w}{dx^2}+\{A_0+A_1z\cos^2 x+\ \ldots\ +A_p\cos^{2p} x\}w=0,$$

which may be written in the form

$$\frac{d^2w}{dx^2}+\{\Theta_0+2\Theta_1\cos 2x+\ .\ .\ .\ +2\Theta_p\cos 2px\}w=0,$$

and is virtually the equation of G. W. Hill in the Lunar Theory. When $p=1$ it reduces to the Mathieu equation, when $p=2$ to Equation VIIa. of § 20·32: no particular properties of equations for which $p>2$ are known.

If the two elementary singularities $z=0$ and $z=1$ of $[2, 0, 1_p]$ are caused to coalesce in the origin, the equation becomes $[0, 1, 1_p]$.

The Lamé equation may be generalised in a somewhat similar manner by replacing the regular singular point at infinity by an irregular singularity of species $p-1$. The equation $[3, 0, 1_{p-1}]$ is

$$\frac{d^2w}{dz^2}+\left\{\frac{\frac{1}{2}}{z-a_1}+\frac{\frac{1}{2}}{z-a_2}+\frac{\frac{1}{2}}{z-a_3}\right\}\frac{dw}{dz}+\left\{\frac{A_0+A_1z+\ .\ .\ .\ +A_pz^p}{4(z-a_1)(z-a_2)(z-a_3)}\right\}w=0,$$

with $p+2$ irreducible constants. If $a_1=0$, $a_2=k^{-2}$, $a_3=1$, the substitution $z=sn^2(x, k)$ brings the equation into the form

$$\frac{d^2w}{dx^2}-k^2(A_0+A_1\text{sn}^2 x+\ .\ .\ .\ +A_p\text{sn}^{2p} x)w=0.$$

By means of the operations

$$a_1\rightarrow 0,\quad a_2\rightarrow\infty,\quad a_3\rightarrow 1$$

$[3, 0, 1_{p-1}]$ becomes $[2, 0, 1_p]$ and the generalised Lamé equation degenerates into the Hill equation.

20·6. Asymptotic Behaviour of Solutions at an Irregular Singularity.—Since any equation which has an irregular singular point at infinity of odd species can be converted by the quadratic transformation into an equation with a singularity of even species, it will be sufficient to consider the latter type. The equation $[0, 0, 1_{2p}]$ may be written

$$\frac{d^2w}{dz^2}+\{A_0+A_2z^2+A_3z^3+\ .\ .\ .\ +A_{2p-3}z^{2p-3}-m^2z^{2p-2}\}w=0,$$

where $m\neq 0$. If a normal solution exists, the determining factor is of the form

$$e^{\pm\frac{m}{p}z^p+\frac{a_1}{p-1}z^{p-1}+\ .\ .\ .\ +a_{p-1}z},$$

and therefore the equation is of rank p. The same is true even when other singularities are present.

Miscellaneous Examples.

1. Find conditions sufficient to ensure that $[2, 0, 1_2]$ should possess a normal solution. Examine the possibility of two normal solutions. Express the results in terms of Equation VIIa. (§ 20·32).

2. Illustrate in tabular form the statement that equation $[2, 0, 1_2]$ bear the same relation to $[2, 0, 1_1]$ as $[0, 1, 1_2]$ bears to $[0, 1, 1_1]$.

3. Write down the formulæ of the 14 typical equations which can be derived from $[7, 0, 0]$.

CHAPTER XXI

OSCILLATION THEOREMS IN THE COMPLEX DOMAIN

21·1. Statement of the Problem.—In Chapters X. and XI. a series of theorems was developed, whose aim was to specify the number and the distribution of the real zeros of functions of the Sturm-Liouville type. The complex zeros of such particular functions as the hypergeometric function, Bessel functions * and Legendre functions † have been investigated by modern writers, but until quite recently no general theorems covering the whole field of Sturm-Lionville functions were known. This gap was filled up by Hille,‡ who, in turn, applied his results to such well-known functions as those of Legendre § and Mathieu. || Hille's methods will be expounded in the present chapter, and illustrated by the special example of the equation

$$\frac{d^2w}{dz^2}-\frac{w}{z}=0,$$

whose solutions may be expressed in terms of Bessel functions of the first order.

The method depends upon the study of certain integral equalities, known as the *Green's transforms*, which are derived from the differential equation of the problem. The behaviour of the zeros of a particular solution of the equation is reflected in the behaviour of the corresponding Green's transform. It will be found that there exist certain regions of the plane of the complex independent variable, known as *zero-free* regions, throughout which the particular solution does not vanish. In the more important cases, the zero-free regions will be found to extend over the greater part of the plane, thus confining the zeros of the solution to a comparatively small region.

21·2. The Green's Transform.—In the self-adjoint linear differential equation of the second order,

$$\text{(A)} \qquad \frac{d}{dz}\left\{K(z)\frac{dw}{dz}\right\}+G(z)w=0,$$

it will be supposed that $K(z)$ and $G(z)$ are analytic in a domain D throughout which $K(z)$ does not vanish. If

$$w_1=w, \quad w_2=K(z)\frac{dw}{dz},$$

* See especially Hurwitz, *Math. Ann.* 33 (1889), p. 246.

† Hille, *Arkiv för Mat.* 13 (1918), No. 17.

‡ *Arkiv för Mat.* 16 (1921), No. 17; *Bull. Am. Math. Soc.* 28 (1922), pp. 261, 462; *Trans. Am. Math. Soc.* 23 (1922), p. 350.

§ *Arkiv för Mat.* 17 (1922), No. 22.

|| *Proc. London Math. Soc.* (2), 23 (1924), p. 185.

the single equation (A) is replaced by the pair of simultaneous equations of the first order

$$\text{(B)} \qquad \begin{cases} \dfrac{dw_1}{dz} = w_2/K(z), \\ \dfrac{dw_2}{dz} = -G(z)w_1. \end{cases}$$

The first equation of this system is also true if each term is replaced by its conjugate, thus

$$d\bar{w}_1 = \bar{w}_2 d\bar{z}/\overline{K(z)}.$$

It follows that

$$w_2 d\bar{w}_1 + \bar{w}_1 dw_2 = w_2\bar{w}_2 d\bar{z}/\overline{K(z)} - w_1\bar{w}_1 G(z)dz,$$

and, on integrating between limits z_1 and z_2, and assuming that every point on the path of integration lies in D,

$$\Big[\bar{w}_1 w_2\Big]_{z_1}^{z_2} - \int_{z_1}^{z_2} \frac{|w_2|^2}{\overline{K(z)}} d\bar{z} + \int_{z_1}^{z_2} |w_1|^2 G(z)dz = 0.$$

This equation is known as the *Green's transform* of the given equation; it plays a part in the investigation of the complex zeros analogous to that played by the original Green's formula in the case of the real variable.

Let

$$dz/K(z) = d\mathbf{K} = d\mathbf{K}_1 + id\mathbf{K}_2,$$
$$G(z)dz = d\mathbf{G} = d\mathbf{G}_1 + id\mathbf{G}_2,$$

$\mathbf{K}_1$, $\mathbf{K}_2$, $\mathbf{G}_1$, $\mathbf{G}_2$ being real, then the Green's transform becomes

$$\text{(C)} \qquad \Big[\bar{w}_1 w_2\Big]_{z_1}^{z_2} - \int_{z_1}^{z_2} |w_2|^2 d\overline{\mathbf{K}} + \int_{z_1}^{z_2} |w_1|^2 d\mathbf{G} = 0,$$

and when the real and imaginary parts are separated,

$$\text{(E)} \qquad \begin{aligned} &\mathbf{R}\Big[\bar{w}_1 w_2\Big]_{z_1}^{z_2} - \int_{z_1}^{z_2} |w_2|^2 d\mathbf{K}_1 + \int_{z_1}^{z_2} |w_1|^2 d\mathbf{G}_1 = 0, \\ &\mathbf{I}\Big[\bar{w}_1 w_2\Big]_{z_1}^{z_2} + \int_{z_1}^{z_2} |w_2|^2 d\mathbf{K}_2 + \int_{z_1}^{z_2} |w_1|^2 d\mathbf{G}_2 = 0. \end{aligned}$$

21·21. Invariance of the Green's Transform.—Let Z be a new independent variable, defined by the relation

$$dz = f(Z)dZ,$$

where $f(Z)$ is any analytic function. In the new variable, the system (21·2, B) becomes

$$dw_1 = w_2 dZ/k(Z), \quad dw_2 = -g(Z)w_1 dZ,$$

where

$$k(Z) = K(z)/f(Z), \quad g(Z) = G(z)f(Z).$$

If

$$dZ/k(Z) = d\mathfrak{K}, \quad g(Z)dZ = d\mathfrak{G}$$

then the Green's transform becomes

$$\Big[\bar{w}_1 w_2\Big]_{Z_1}^{Z_2} - \int_{Z_1}^{Z_2} |w_2|^2 d\overline{\mathfrak{K}} + \int_{Z_1}^{Z_2} |w_1|^2 d\mathfrak{G} = 0,$$

as (C) above, and therefore *the Green's transform is invariant under a transformation of the independent variable.*

Three special cases of the transformation deserve special mention.

(i) Let $Z=\mathbf{K}(z)$ and write $J(Z)=G(z)K(z)$,

then the differential equation and the Green's transform become respectively

$$\frac{d^2w}{dZ^2}+J(Z)w=0,$$

$$\left[\bar{w}_1w_2\right]_{Z_1}^{Z_2}-\int_{Z_1}^{Z_2}|w_2|^2d\bar{Z}+\int_{Z_1}^{Z_2}|w_1|^2J(Z)dZ=0.$$

(ii) Let $Z=\mathbf{G}(z)$ and write $H(Z)=G(z)K(z)$, then

$$\frac{d}{dZ}\left\{H(Z)\frac{dw}{dZ}\right\}+w=0,$$

$$\left[\bar{w}_1w_2\right]_{Z_1}^{Z_2}-\int_{Z_1}^{Z_2}\frac{|w_2|^2}{H(Z)}d\bar{Z}+\int_{Z_1}^{Z_2}|w_1|^2dZ=0.$$

(iii) To obtain a symmetrical form, let

$$dZ=\sqrt{\left\{\frac{G(z)}{K(z)}\right\}}dz,\quad S(Z)=\sqrt{\{G(z)K(z)\}},$$

then

$$\frac{d}{dZ}\left\{S(Z)\frac{dw}{dZ}\right\}+S(Z)w=0,$$

$$\left[\bar{w}_1w_2\right]_{Z_1}^{Z_2}-\int_{Z_1}^{Z_2}\frac{|w_2|^2}{S(Z)}d\bar{Z}+\int_{Z_1}^{Z_2}|w_1|^2S(Z)dZ=0.$$

21·3. Selection of an appropriate Path of Integration.—The path of integration (z_1, z_2) has not yet been specified; by choosing the path to be such that one or other of the conditions

$$d\mathbf{K}_1=0,\quad d\mathbf{K}_2=0,$$
$$d\mathbf{G}_1=0,\quad d\mathbf{G}_2=0$$

is satisfied, the formulæ (21·2, E), derived from the Green's transform, may be simplified.

The curves $\mathbf{K}_1=$const., $\mathbf{K}_2=$const. are mutually-orthogonal families of curves in the z-plane, and will be known as the **K**-net. In the particular case $K=1$, the **K**-net consists of the network of straight lines parallel to the x- and y-axes. Similarly the curves $\mathbf{G}_1=$const., $\mathbf{G}_2=$const. constitute a pair of mutually-orthogonal families, known as the **G**-net.

Now consider the **G**-net,* and write

$$G(z)=g_1(z)+ig_2(z),$$

where g_1 and g_2 are real. Let Δ be a region of the z-plane for which $G(z)$ is meromorphic, and let a be an interior point of Δ for which $G(a)\neq 0$. Through a there passes one and only one curve of each family $\mathbf{G}_1=$const., $\mathbf{G}_2=$const. The slopes of these two curves at a are respectively

$$g_1(a)/g_2(a),\quad -g_2(a)/g_1(a).$$

Thus the curves of the family $\mathbf{G}_1=$const. have tangents parallel to the x-axis at points where they meet the curve $g_1(z)=0$, and tangents parallel to the y-axis at points where they meet the curve $g_2(z)=0$. The reverse is true in the case of the family $\mathbf{G}_2=$const.

* Since the K-net becomes trivial in the most important case, namely $K=1$, it is advantageous to concentrate on the G-net. The corresponding results for the K-net will be stated at the end of the section.

The only exceptional points of Δ are zeros and poles of $G(z)$. In the first place let $z=a$ be a zero of multiplicity k. Then, if

$$G(z)=a_k(z-a)^k+O\{(z-a)^{k+1}\},$$

$$\int_a^z G(z)dz=\frac{a_k}{k+1}(z-a)^{k+1}+O\{(z-a)^{k+2}\}.$$

Write

$$z-a=re^{i\theta}, \quad a_k=\rho e^{i\phi}.$$

then, separating real and imaginary parts,

$$\mathbf{G}_1(z)-\mathbf{G}_1(a)\equiv\mathbf{R}\int_a^z G(z)dz=\frac{\rho}{k+1}r^{k+1}\cos\{(k+1)\theta+\phi\}+O(r^{k+2}),$$

$$\mathbf{G}_2(z)-\mathbf{G}_2(a)\equiv\mathbf{I}\int_a^z G(z)dz=\frac{\rho}{k+1}r^{k+1}\sin\{(k+1)\theta+\phi\}+O(r^{k+2}).$$

Thus through the point $z=a$ there pass $k+1$ curves of each of the families $\mathbf{G}_1=\text{const.}$, $\mathbf{G}_2=\text{const.}$ The curves of the two families alternate with one another and consecutive tangents intersect at the constant angle $\pi/(k+1)$.

In the second place, let $z=a$ be a pole of order k, where $k>1$. If

$$G(z)=a_k(z-a)^{-k}+O\{(z-a)^{1-k}\},$$

and if it is assumed that the term in $(z-a)^{-1}$ is absent from the expansion of $G(z)$, then

$$\int^z G(z)dz=\gamma+i\delta-\frac{a_k}{k-1}(z-a)^{1-k}+O\{(z-a)^{2-k}\},$$

where $\gamma+i\delta$ is the complex constant of integration. It follows that

$$\mathbf{G}_1(z)\equiv\mathbf{R}\int^z G(z)dz=\gamma-\frac{\rho}{k-1}r^{1-k}\cos\{(k-1)\theta-\phi\}+O(r^{2-k}),$$

$$\mathbf{G}_2(z)\equiv\mathbf{I}\int^z G(z)dz=\delta+\frac{\rho}{k-1}r^{1-k}\sin\{(k-1)\theta-\phi\}+O(r^{2-k}).$$

Since, under the assumption made, $\mathbf{G}_1$ and $\mathbf{G}_2$ involve no logarithmic terms, every curve of either family which has points in the neighbourhood of $z=a$ actually passes through $z=a$. The curves in question which belong to the $\mathbf{G}_1$-family are tangent to the lines

$$\arg(z-a)=\{\phi+(\nu+\tfrac{1}{2})\pi\}/(k-1)$$

and those of the $\mathbf{G}_2$-family are tangent to the lines

$$\arg(z-a)=\{\phi+\nu\pi\}/(k-1),$$

where, in each case, $\nu=0, 1, 2, \ldots, k-2$.

Lastly, consider the case in which $z=a$ is a simple pole of $G(z)$, and let

$$G(z)=a_1(z-a)^{-1}+O(1).$$

Then

$$\int^z G(z)dz=\gamma+i\delta+a_1\log(z-a)+O(z-a),$$

and, if $a_1=\alpha+i\beta$,

$$\mathbf{G}_1(z)\equiv\mathbf{R}\int^z G(z)dz=\gamma+\alpha\log r-\beta\theta+O(r),$$

$$\mathbf{G}_2(z)\equiv\mathbf{I}\int^z G(z)dz=\delta+\beta\log r+\alpha\theta+O(r).$$

When $\alpha\neq0$, $\beta\neq0$, the point $z=a$ is a spiral point for the curves of both families; when $\alpha\neq0$, $\beta=0$, the curves of the $\mathbf{G}_1$-family near $z=a$ are approximately circular ovals enclosing this point, and those of the $\mathbf{G}_2$-family have the point $z=a$ for a point of ramification of infinite order. When $\alpha=0, \beta\neq 0$, the reverse is true.

The corresponding results in the case of the **K**-net are as follows. When $z=a$ is a pole of $K(z)$, the **K**-net behaves at the point a as the **G**-net behaved when $z=a$ was a zero of $G(z)$. Similarly the behaviour of the **K**-net when $z=a$ is a zero of $K(z)$ of order greater than unity, or of order unity, is similar to that of the **G**-net when $z=a$ is a pole, of the same order, of $G(z)$.

21·31. Special Case : $G(z)$ a Polynomial.—The case in which $G(z)$ is a polynomial of degree n is of prime importance ; let

$$G(z)=A_0(z-a_1)^{\nu_1}(z-a_2)^{\nu_2} \dots (z-a_m)^{\nu_m} \qquad (\nu_1+\nu_2+ \dots +\nu_m=n),$$

then the following deductions may be made from the theory of the preceding section. Any general curve of either of the **G**-families which does not pass through any of the points $a_1, a_2, \dots, a_m$ has no multiple points in the z-plane. On the other hand one curve of each family has a multiple point of order ν_k+1 at a_k, and therefore there are at most m singular curves of each family.

Every curve intersects the line at infinity in $n+1$ distinct points, but these intersections are the same for all curves of the same family. The asymptotes of all the curves are real and distinct and intersect in one point, namely the point

$$z=\frac{\nu_1 a_1+\nu_2 a_2+ \dots +\nu_m a_m}{n}.$$

If arg $A_0=\phi_0$, the asymptotic directions of the $\mathbf{G}_1$-curves are

$$\frac{(k+\frac{1}{2})\pi-\phi_0}{n+1}$$

and those of the $\mathbf{G}_2$-curves are

$$\frac{k\pi-\phi_0}{n+1},$$

where $k=0, 1, \dots, n$. The asymptotes of each family therefore make equal angles with one another, and bisect the angles between the asymptotes of the other family.

$\mathbf{G}_1(z)$ and $\mathbf{G}_2(z)$ are functions harmonic throughout the finite part of the z-plane. Therefore they can have neither maxima nor minima for any finite value of z.* It follows that a **G**-curve cannot begin or end at a finite point, nor can a **G**-curve be closed. Thus a path can be drawn from infinity to any chosen point in the z-plane without crossing the curve in question.

21·4 Zero-Free Intervals on the Real Axis.—Let x_1 and x_2 $(x_1<x_2)$ be two arbitrary points of any interval (a, b) of the real axis throughout which $J(z)$ is analytic, then if w is any solution of the equation

$$\text{(A)} \qquad \frac{d^2w}{dz^2}+J(z)w=0$$

and

$$w=w_1, \quad w'=w_2, \quad J(z)=g_1(z)+ig_2(z),$$

the formulæ deduced from the Green's transform become

$$\text{(B)} \qquad \mathbf{R}\Big[\bar{w}_1 w_2\Big]_{x_1}^{x_2}-\int_{x_1}^{x_2} |\, w_2\,|^2 dx+\int_{x_1}^{x_2} g_1(x)|\, w_1\,|^2 dx=0,$$

$$\text{(C)} \qquad \mathbf{I}\Big[\bar{w}_1 w_2\Big]_{x_1}^{x_2}+\int_{x_1}^{x_2} g_2(x)|\, w_1\,|^2 dx=0.$$

* Forsyth, *Theory of Functions*, p. 475, IV.

It will now be proved that *if throughout the interval* (a, b) *either* $\mathbf{R}J(z)\leqslant 0$ *or* $\mathbf{I}J(z)$ *does not change sign, then there can be at most one zero of* $w\,dw/dz$ *in that interval.*

For let there be more zeros than one in the interval, and let x_1 and x_2 be consecutive zeros. Then, in equation (B), $\left[\overline{w}_1 w_2\right]_{x_1}^{x_2}$ is zero, whereas if $g_1(x)\leqslant 0$ the sum of the remaining two terms is definitely negative. Again in the second equation (C), $\left[\overline{w}_1 w_2\right]_{x_1}^{x_2}$ is zero, whereas if $g_2(x)$ is of one sign and vanishes only at discrete points, the second term is not zero. Thus under either hypothesis the supposition that there is more than one zero leads to a contradiction, which proves the theorem. As a corollary it follows that two necessary conditions for oscillation in an interval of the real axis free from singular points are that

(*a*) $\mathbf{R}J(z)\geqslant 0$,

(*b*) $\mathbf{I}J(z)$ changes sign or vanishes identically.

The above theorem will require modification if any singular point occurs within or at an end-point of the interval (a, b). To take a particular instance, let $z=a$ be a regular singular point of (A) with exponents λ_1 and λ_2 $(\lambda_1+\lambda_2=1)$. Assume also that $\mathbf{R}(\lambda_1)>\frac{1}{2}$ and let $w=w_1$ be the solution corresponding to the exponent λ_1. Then since $z=a$ is at most a pole of order 2 for $J(z)$, the integrals in (B) and (C) are finite provided that no singular point other than $z=a$ occurs in (a, b). If, then, the conditions previously imposed upon $J(z)$ hold, and if also $w_1w_2\to 0$ as $z\to a$, then w_1w_2 will not vanish in the interval $a<x<b$.

21·41. Zero-Free Regions.—Let z_1z_2 be any rectilinear segment in the z-plane along which $J(z)$ is analytic. Write

$$z=z_1+re^{i\theta},$$

then θ is constant along the segment chosen. If $w(z)$ is any solution of equation (21·4, A), the Green's transform becomes

$$\left[\overline{w}\frac{dw}{dr}\right]_0^r-\int_0^r\left|\frac{dw}{dr}\right|^2dr+e^{2i\theta}\int_0^r|w|^2J(z)dr=0.$$

Let

$$\overline{w}\frac{dw}{dr}=f_1(r)+if_2(r),$$

$$g_1(z)\cos 2\theta-g_2(z)\sin 2\theta=P(z, \theta),$$
$$g_2(z)\cos 2\theta+g_1(z)\sin 2\theta=Q(z, \theta),$$

where f_1, f_2, P and Q are real, and separate the real and imaginary parts of the Green's transform. Then

$$f_1(r)-f_1(0)-\int_0^r\left|\frac{dw}{dr}\right|^2dr+\int_0^r P(z, \theta)|w|^2dr=0,$$

$$f_2(r)-f_2(0)+\int_0^r Q(z, \theta)|w|^2dr=0.$$

A line of reasoning similar to that followed in the previous section now leads to the following theorem.

There is at most one zero of $w\,dw/dz$ *on the segment* z_1z_2 *provided that along that segment either*

(i) $P(z, \theta)\leqslant 0$, *or* (ii) $Q(z, \theta)$ *does not change sign.*

If, in addition to (i), $f_1(0)\geqslant 0$, *or if, in addition to* (ii), $f_2(0)$ *has the opposite* *sign to that which $Q(z, \theta)$ has on the segment, then the product wdw/dz has no zero at all on the segment.*

This theorem will now be modified in such a way as to lead to a lemma which, in its turn, provides an important theorem on the distribution of the complex zeros. Consider the pencil of parallel lines (l)

$$z=z_0+re^{i\theta},$$

in which z_0 is regarded as a variable parameter. Let T be a simply-connected region in the z-plane throughout which $J(z)$ is analytic, and which is such that every line of the pencil which cuts the boundary cuts it in two points. Two lines of the pencil each meet the boundary in coincident points; let these points be α and β. The boundary is thus divided into two distinct arcs, one of which will be regarded as the locus of z_0 and termed *the arc C*. Then there follows the lemma.

There is at most one zero of wdw/dz on that part of each line l which lies within T, provided that throughout T either

$$\text{(i) } P(z, \theta)\leqslant 0, \quad \textit{or} \quad \text{(ii) } Q(z, \theta)\neq 0.$$

If, in addition to (i), $\mathbf{R}\{\bar{w}dw/dr\}\geqslant 0$ *along C, or if, in addition to* (ii) $\mathbf{I}\{\bar{w}dw/dr\}$ *has along C the opposite sign to that which $Q(z, \theta)$ has throughout T, then wdw/dz has no zero in T.*

Now let C be a segment of the real axis, let $w(z)$ be real for all points of C, and let $\theta=\frac{1}{2}\pi$. Then

$$\mathbf{R}\left\{\bar{w}\frac{dw}{dr}\right\}=\mathbf{R}\left\{e^{i\theta}\bar{w}\frac{dw}{dz}\right\}=0, \qquad P(z, \theta)=-g_1(z)=-\mathbf{R}\{J(z)\}.$$

This leads to the important theorem which follows.

If $w(z)$ is a solution which is real on a segment (a, b) of the real axis; if, further, T is a region symmetrically situated with respect to the real axis, and such that every line perpendicular to the real axis which cuts the region cuts its boundary in two points and meets (a, b) in an interior point; and if finally $\mathbf{R}\{J(z)\}\geqslant 0$ *throughout T, then $w(z)$ can have no complex zero or extremum* * *in T.*

In the statement of this theorem the words *real axis* may be replaced by *imaginary axis* and the condition $\mathbf{R}\{J(z)\}\geqslant 0$ by $\mathbf{R}\{J(z)\}\leqslant 0$.

If the equation considered is $w''+w=0$ and $w(z)$ is taken to be $\sin z$, the above theorem shows that $\sin z$ and $\cos z$ have no complex zeros.

The following theorem and a similar theorem for the imaginary axis may be deduced in a similar manner.

Let the region T be as before, and let $w(z)$ be a solution, real on the segment (a, b) and such that in (a, b) wdw/dz has a fixed sign; let $\mathbf{I}\{J(z)\}$ *have this sign throughout that part of the region T which lies above the real axis, then $w(z)$ can have no complex zero or extremum in T.*

21·411. Application.—Consider the differential equation

$$\frac{d^2w}{dz^2}-\frac{w}{z}=0;$$

it has a regular singular point at the origin with exponents 0 and 1, and an irregular singularity at infinity. One solution is finite at the origin, and this solution may be written as $E(z)=iz^{\frac{1}{2}}J_1(2iz^{\frac{1}{2}})$, where J_1 is the Bessel function of order 1. This solution is real for all real values of z, and has an infinite number of real negative

* An *extremum* (point for which $w'(z)=0$) corresponds in the theory of the complex variable to a stationary point in the theory of the real variable.

zeros.* Any other solution, which is not a mere multiple of $E(z)$, necessarily involves $\log z$. Such a solution can be real on a half axis at most; if it is real on the negative half of the real axis, it must oscillate there. A solution which is real for positive real values of x can have at most one positive zero or extremum. In general, when w is real,

$$\lim_{x \to +\infty} \frac{x^{\frac{1}{2}} w'(x)}{w(x)} \to +1,$$

and therefore w increases in absolute magnitude without limit, but there is one exceptional solution † for which the limiting ratio is -1, and for this solution $w \to 0$ as $x \to +\infty$.

Now consider the distribution of the complex zeros. When $\mathbf{R}(z)<0$, $\mathbf{R}(-1/z)>0$ and therefore by the main theorem of the preceding section, no solution which is real for negative values of the real variable x can have any complex zeros in the half-plane $\mathbf{R}(z)<0$. Moreover, $\mathbf{I}(-1/z)>0$ when $\mathbf{I}(z)>0$. Let $w(z)$ be any solution; if it has a positive real zero or extremum let this be x_0, otherwise let x_0 be any positive number. Then, by the last theorem of § 21·41, $w(z)$ will have no zero in the half-plane $\mathbf{R}(z)>x_0$. In the case of the exceptional solution, there is no zero in the half-plane $\mathbf{R}(z)\geqslant 0$.

21·42. The Zero-Free Star.—Consider now a pencil of lines radiating out from a point $z=a$ at which $J(z)$ is regular but not zero. Write

$$z=a+re^{i\theta},$$

$$(z-a)^2 J(z)=P(z)+iQ(z).$$

The curves

$$P(z)=0, \quad Q(z)=0$$

intersect at the point a, where each curve has a double point. The directions of the tangents to these curves at the point a are given by

$$g_1(a) \cos 2\theta - g_2(a) \sin 2\theta=0, \quad g_2(a) \cos 2\theta + g_1(a) \sin 2\theta=0$$

respectively.

On the ray through a, of vectorial angle θ, mark the point p_θ, which is arrived at as follows. A moving point starts from a and traverses the ray until Q changes sign. If P has been positive or changed sign, then the point at which Q changes sign is p_θ. If, on the other hand, P has been constantly negative, then the moving point continues still further until P changes sign, and then that point is p_θ.

The process is repeated for all the rays of the pencil, and the aggregate of segments ap_θ is termed the *star* belonging to a. If a singular point of $J(z)$ falls within the star, it is excluded by a rectilinear cut drawn in the direction away from a.

In the neighbourhood of the point a the boundary of the cut consists of that branch of the curve $Q=0$ which lies in the region $P>0$, together with the tangent to that branch at $z=a$. (Fig. 18.)

Now it follows from the first theorem of § 21·41 that *if $z=a$ is a zero of $w dw/dz$, then this product does not vanish at any point of the star belonging to a, including the non-singular points of its boundary.*

This theorem can be applied to the solution

$$w=z^{\frac{1}{2}}J(2iz^{\frac{1}{2}})$$

of the equation

$$\frac{d^2w}{dz^2} - \frac{w}{z}=0.$$

* Concerning the zeros of Bessel and allied functions, see Watson, *Bessel Functions* Chap. XV.

† *Cf.* Wiman, *Arkiv för Mat.* 12 (1917), No. 14.

This solution has a simple zero at $z=0$. The star corresponding to this point covers the whole plane except for the negative half of the real axis. It follows that the

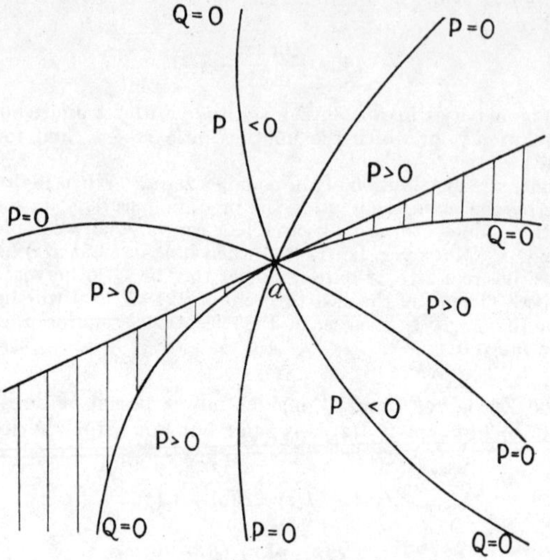

FIG. 18.

[The region near $z=a$ which does not belong to the star is shaded.]

solution in question has no zeros except those which lie upon the negative half of the real axis.

21·43. The Standard Domain.—A zero-free region which in general is more extensive than the star will now be obtained. Consider the differential system in its more general form

$$\frac{dw_1}{dz}=\frac{w_2}{K(z)}, \quad \frac{dw_2}{dz}=-G(z)w_1,$$

and let $K(z)$ and $G(z)$ be analytic throughout the whole plane except at a number of isolated points. These singular points, together with the zeros of $K(z)$ are the singularities of the differential system; let them be excluded from the plane by a number of appropriately-drawn cuts. In the cut-plane the functions $\mathbf{K}$ and $\mathbf{G}$, which define the two networks of curves of § 21·3, are one-valued.

A *standard path* will now be defined as a curve issuing from any ordinary point of the plane and satisfying the following conditions.

(i) It does not encounter a cut, except possibly at its end point.

(ii) It is composed of a finite number of arcs belonging to the two networks.

(iii) Throughout the path a particular one of the four following pairs of inequalities is satisfied, namely

$$(\alpha)\ \begin{matrix} d\mathbf{G}_1\leqslant 0, \\ d\mathbf{K}_1\geqslant 0; \end{matrix} \quad (\beta)\ \begin{matrix} d\mathbf{G}_1\geqslant 0, \\ d\mathbf{K}_1\leqslant 0; \end{matrix} \quad (\gamma)\ \begin{matrix} d\mathbf{G}_2\geqslant 0, \\ d\mathbf{K}_2\geqslant 0; \end{matrix} \quad (\delta)\ \begin{matrix} d\mathbf{G}_2\leqslant 0, \\ d\mathbf{K}_2\leqslant 0. \end{matrix}$$

In order to avoid the possibility of discontinuous tangents it will also be supposed that at a point where two different arcs meet, the angular point is replaced by a small arc having a continuous tangent. This can always be carried out in such a way that the characteristic pair of inequalities of the curve is not violated.

Now let the point a be such that, if $W(z)=w_1(z)w_2(z)$, then $W(a)=0$. If b is any other point on a standard path issuing from a, then it follows immediately from the equalities

$$\mathbf{R}\Big[\overline{w}_1 w_2\Big]_a^b - \int_a^b |w_2|^2 d\mathbf{K}_1 + \int_a^b |w_1|^2 d\mathbf{G}_1 = 0,$$

$$\mathbf{I}\Big[\overline{w}_1 w_2\Big]_a^b + \int_a^b |w_2|^2 d\mathbf{K}_2 + \int_a^b |w_1|^2 d\mathbf{G}_2 = 0,$$

that $W(z)$ will have no zero except a on the standard path.

If the aggregate of standard paths issuing from the point a is called the *standard domain* of a, then follows the theorem. *If $W(a)=0$, $W(z)$ has no zero, other than $z=a$, in the standard domain of a.*

Similarly there may be constructed the standard paths of all the points of a continuous curve C upon which the variation of the sign of $K(z)w_1(z)w_2(z)$ is known. The aggregate of these standard paths is known as the standard domain of C with respect to the solution considered.

21·431. Example of a Standard Domain.—In the case of the equation

$$\frac{d^2w}{dz^2} - \frac{w}{z} = 0$$

it is found that

$$\mathbf{K}(z)=z, \qquad \mathbf{G}(z)=-\log z.$$

To make $\mathbf{G}(z)$ single-valued, the plane is cut along the negative half of the real axis.

Now, if $z=re^{i\theta}$, the standard curves are made up of arcs of the networks of curves

$$x=\text{const.},\quad y=\text{const.}\,;\quad r=\text{const.},\quad \theta=\text{const.},$$

and the four characteristic pairs of inequalities are effectively

$$(\alpha)\ \begin{matrix} dr\geqslant 0, \\ dx\geqslant 0\,; \end{matrix} \qquad (\beta)\ \begin{matrix} dr\leqslant 0, \\ dx\leqslant 0\,; \end{matrix} \qquad (\gamma)\ \begin{matrix} d\theta\leqslant 0, \\ dy\geqslant 0\,; \end{matrix} \qquad (\delta)\ \begin{matrix} d\theta\geqslant 0, \\ dy\leqslant 0. \end{matrix}$$

Let there be a solution such that $W(x_0)=0$, where x_0 is a point on the negative half of the real axis. Then standard curves issuing from x_0 can be made to cover the following regions.

When (α) is satisfied the region is $\mathbf{R}(z)\geqslant x_0$, $|z|\geqslant|x_0|$.

There is no region in which (β) is satisfied.

When (γ) is satisfied, the region is $\mathbf{I}(z)>0$.

When (δ) is satisfied, the region is $\mathbf{I}(z)<0$.

Thus the standard domain covers the whole of the plane with the exception of that part of the real axis for which $\mathbf{R}(z)<|x_0|$, and it follows that no solution of the equation considered which has a negative real zero $z=x_0$ has a complex zero or a real zero $z>|x_0|$.

21·5. Asymptotic Distribution of the Zeros.—The theorems which have been developed in the preceding sections have as their aim a more or less complete solution of the problem of determining extensive regions of the z-plane which are free from zeros of a particular solution of the differential equation in question. A complementary problem will now be taken up, namely to investigate the distribution of the zeros in the neighbourhood of an irregular singular point.*

* Similar problems have been studied in connection with the Painlevé transcendents by Boutroux, *Ann. Éc. Norm.* (3), 30 (1913), p. 255 ; (3), 31 (1914), p. 99, and in connection with the solutions of linear differential equations by Garnier, *J. de Math.* (8), 2 (1919), p. 99.

The differential equation

(A) $$\frac{d}{dz}\left\{K(z)\frac{dw}{dz}\right\}+G(z)w=0$$

is transformed, by the change of independent variable

$$Z=\int_{z_0}^{z}\sqrt{\left\{\frac{G(z)}{K(z)}\right\}}dz,$$

into

(B) $$\frac{d}{dZ}\left\{S(Z)\frac{dw}{dZ}\right\}+S(Z)w=0,$$

where

$$S(Z)=\sqrt{\{G(z)K(z)\}}.$$

It may also be written in the form

(C) $$\frac{d^2w}{dZ^2}+F(Z)\frac{dw}{dZ}+w=0,$$

where

$$F(Z)=\frac{d}{dZ}\{\log S(Z)\}.$$

The change of dependent variable

$$W=\sqrt{\{S(Z)\}}w$$

now transforms the equation into

(D) $$\frac{d^2W}{dZ^2}+\{1-\Phi(Z)\}W=0,$$

where

$$\Phi(Z)=\tfrac{1}{2}\frac{dF}{dZ}+\tfrac{1}{4}F^2.$$

The new variable Z, regarded as a function of z, is infinitely many-valued. It will be assumed that Z may be so determined that $\Phi(Z)$ is analytic throughout an infinite region Δ of the Z-plane having the following properties :

(**A**1) Δ is simply connected and smooth.

(**A**2) Every line parallel to the real axis cuts the boundary Γ of the region (i) in a line segment, or (ii) in a point, or (iii) not at all.

(**A**3) Δ lies wholly within a sector

$$-\pi+\delta<\arg Z<\pi-\delta, \qquad |Z|\geqslant R_0>0.$$

A region which satisfies these conditions is said to be of type **A**. When Γ is cut by every parallel to the real axis the region is said to be of type **Aa**, otherwise it is of type **Ab**. The conditions **A** ensure that Δ contains a strip Δ_0 of finite width defined by inequalities such as

$$\mathbf{R}(Z)\geqslant A>R_0, \quad B_1\geqslant\mathbf{I}(Z)\geqslant B_2.$$

It is also assumed that at every point of Δ, $\Phi(Z)$ satisfies the condition

(**B**) $$\Phi(Z)<\frac{M}{|Z|^{1+\nu}},$$

where M and ν are positive numbers.

It follows from the existence-theorems that any solution $W(Z)$ of (D) is bounded in the strip Δ_0. Consider the expression

$$f(Z)=W_0(Z)+\int^{\infty}\sin(T-Z)\,\Phi(T)\,W(T)\,dT,$$

where $W_0(Z)$ is a solution of

$$W_0''+W_0=0,$$

and the path of integration is parallel to the real axis. It is found that

$$f''(Z)+f(Z)-\Phi(Z)W(Z)=0,$$

and therefore, if $f(Z)$ is a solution of the integral equation

$$\text{(E)} \qquad f(Z)=W_0(Z)+\int_Z^\infty \sin(T-Z)\,\Phi(T)f(T)\,dT,$$

then $f(Z)$ is also a solution of the differential equation (D). In this sense (E) may be spoken of as the *equivalent integral equation.**

21·51. Discussion of the Integral Equation.—Before considering the integral equation itself, it is necessary to obtain an expression for the upper bound of the integral

$$\text{(F)} \qquad I(z\,;\,\mu)=\int_z^\infty \frac{dt}{|t|^\mu},$$

where μ is real and

$$\mu \geqslant \mu_0 > 1\,;$$

where z is not a negative real number, and the path of integration is parallel to the real axis. Let

$$z=re^{i\theta}, \quad t=r(v+e^{i\theta}),$$

then

$$I(re^{i\theta}\,;\,\mu)=r^{1-\mu}\int_0^\infty \frac{dv}{|v+e^{i\theta}|^\mu}$$
$$=r^{1-\mu}J(\theta\,;\,\mu).$$

Now

$$|v+e^{i\theta}|^{-\mu}=(1+2v\cos\theta+v^2)^{-\mu/2}$$
$$=(1+v)^{-\mu}\left\{1-\frac{4v\sin^2\frac{1}{2}\theta}{(1+v)^2}\right\}^{-\mu/2}$$

When $|\theta|<\pi$, the second factor may be expanded as a series in $v/(1+v)^2$; the resulting series for $|v+e^{i\theta}|^{-\mu}$ is uniformly convergent for $0\leqslant v\leqslant\infty$. By integrating this series term-by-term, employing the formula

$$\int_0^\infty \frac{v^k dv}{(1+v)^{\mu+2k}}=B(k+1,\,\mu+k-1)$$
$$=\frac{\Gamma(k+1)\Gamma(\mu+k-1)}{\Gamma(\mu+2k)},$$

it is found that

$$J(\theta\,;\,\mu)=\frac{1}{\Gamma(\frac{1}{2}\mu)}\sum_{k=0}^\infty \frac{\Gamma(\frac{1}{2}\mu+k)\Gamma(\mu+k-1)}{\Gamma(\mu+2k)}(4\sin^2\tfrac{1}{2}\theta)^k$$
$$=\frac{\Gamma(\frac{1}{2}\mu+\frac{1}{2})}{\Gamma(\mu)}\sum_{k=0}^\infty \frac{\Gamma(\mu+k-1)}{\Gamma(\frac{1}{2}\mu+\frac{1}{2}+k)}\sin^{2k}\tfrac{1}{2}\theta$$
$$=\frac{1}{\mu-1}F(\mu-1,\,1\,;\,\tfrac{1}{2}\mu+\tfrac{1}{2}\,;\,\sin^2\tfrac{1}{2}\theta),$$

and consequently †

$$I(re^{i\theta}\,;\,\mu)=\frac{F(\mu-1,\,1\,;\,\frac{1}{2}\mu+\frac{1}{2}\,;\,\sin^2\frac{1}{2}\theta)}{(\mu-1)r^{\mu-1}}.$$

* It is a singular integral equation of the Volterra type. The following discussion of the integral equation is due to Hille, *Trans. Am. Math. Soc.* 26 (1924), p. 241.

† In particular, $I(re^{i\theta}\,;\,2)=\dfrac{\theta}{r\sin\theta}$.

Now it may be proved that,* when $0<|\theta|<\pi$,

$$F(\mu-1, 1;\ \tfrac{1}{2}\mu+\tfrac{1}{2};\ \sin^2 \tfrac{1}{2}\theta)=-F(\mu-1, 1;\ \tfrac{1}{2}\mu+\tfrac{1}{2};\ \cos^2 \tfrac{1}{2}\theta) +2\sqrt{\pi}\,\frac{\Gamma(\frac{1}{2}\mu+\frac{1}{2})}{\Gamma(\frac{1}{2}\mu)}|\sin\theta|^{1-\mu}.$$

Since each of the two hypergeometric functions in this equation has a positive sum when $\mu>1$, $0\leqslant|\theta|<\pi$, it follows that

$$I(re^{i\theta};\ \mu)\leqslant\frac{2\sqrt{\pi}\,\Gamma(\frac{1}{2}\mu+\frac{1}{2})}{(\mu-1)\Gamma(\frac{1}{2}\mu)}|r\sin\theta|^{1-\mu}.$$

Now the hypergeometric function in the expression for $I(re^{i\theta};\ \mu)$ increases with $|\theta|$ when $0\leqslant|\theta|<\pi$, and therefore if $|\theta|\leqslant\frac{1}{2}\pi$,

$$I(re^{i\theta};\ \mu)\leqslant I(re^{\frac{1}{2}i\pi};\ \mu) =\frac{2\sqrt{\pi}\,\Gamma(\frac{1}{2}\mu+\frac{1}{2})}{(\mu-1)\Gamma(\frac{1}{2}\mu)}r^{1-\mu}.$$

On the other hand, if $\frac{1}{2}\pi<|\theta|<\pi$,

$$I(re^{i\theta};\ \mu)<\frac{2\sqrt{\pi}\,\Gamma(\frac{1}{2}\mu+\frac{1}{2})}{(\mu-1)\Gamma(\frac{1}{2}\mu)}|y|^{1-\mu}.$$

Now when $\mu\geqslant\mu_0>1$, the expression

$$\frac{2\sqrt{\pi}\,\Gamma(\frac{1}{2}\mu+\frac{1}{2})}{\sqrt{(\mu-1)}\Gamma(\frac{1}{2}\mu)}$$

is bounded, let its upper bound be C. Then finally,

$$\text{(G)}\qquad I(z;\ \mu)<C\frac{R^{1-\mu}}{\sqrt{(\mu-1)}},$$

where, if $z=x+iy$,

$$R=|z|\quad\text{when}\quad 0\leqslant|\arg z|\leqslant\tfrac{1}{2}\pi,$$
$$R=|y|\quad\text{when}\quad \tfrac{1}{2}\pi\leqslant|\arg z|<\pi.$$

Now consider the integral equation

$$W(Z)=W_0(Z)+\int_Z^\infty \sin(T-Z)\,\Phi(T)\,W(T)\,dT,$$

and write

$$K(Z, T)=\sin(T-Z)\,\Phi(T).$$

It will be shown, by a method of successive approximation, that a solution of the integral equation exists. Define the sequence of functions $W_1(Z), \ldots, W_n(Z), \ldots$ where

$$W_1(Z)=W_0(Z)+\int_Z^\infty K(Z, T)W_0(T)dT,$$

and, in general,

$$W_n(Z)=W_0(Z)+\int_Z^\infty K(Z, T)W_{n-1}(T)dT \qquad (n=1, 2, 3, \ldots).$$

Then

$$W_{n+1}(Z)-W_n(Z)=\int_Z^\infty K(Z, T)\{W_n(T)-W_{n-1}(T)\}dT.$$

* The proof follows from the formula (§ 7·231) expressing $F(a, \beta;\ \gamma;\ x)$ in terms of $F(a, \beta;\ a+\beta-\gamma+1;\ 1-x)$ and $(1-x)^{\gamma-a-\beta}F(\gamma-a, \gamma-\beta;\ \gamma-a-\beta+1;\ 1-x)$ and from the fact that $F(a, \beta;\ a;\ x)=(1-x)^{-\beta}$.

Let L be the upper bound of $|W_0(Z)|$ in Δ_0; since $T-Z$ is real on the path of integration,

$$|K(T, Z)| \leqslant |\phi(T)| < \frac{M}{|Z|^{1+\nu}}.$$

Now let it be supposed that for some value of n

$$|W_n(Z)-W_{n-1}(Z)| < \left(\frac{CM}{\nu^{\frac{1}{2}}|Z|^{\nu}}\right)^n \cdot \frac{L}{\{n!\}^{\frac{1}{2}}},$$

then

$$|W_{n+1}(Z)-W_n(Z)| < \left(\frac{CM}{\nu^{\frac{1}{2}}}\right)^n \cdot \frac{L}{\{n!\}^{\frac{1}{2}}} \int_Z^{\infty} \frac{MdT}{|T|^{n\nu+\nu+1}}.$$

$$< \left(\frac{CM}{\nu^{\frac{1}{2}}|Z|^{\nu}}\right)^{n+1} \frac{L}{\{(n+1)!\}^{\frac{1}{2}}},$$

that is, the inequality holds for the next value of n. But since

$$|W_1(Z)-W_0(Z)| \leqslant \left| \int_Z^{\infty} K(Z, T) W_0(T) dT \right|$$

$$< \int_Z^{\infty} \frac{MLdT}{|T|^{1+\nu}}$$

$$< \frac{CML}{\nu^{\frac{1}{2}}|Z|^{\nu}},$$

the inequality holds for $n=1$, and the proof by induction follows.

Consequently $W_n(Z)$ converges uniformly in Δ_0 to a limit-function $W(Z)$ which is analytic throughout Δ_0 and satisfies the integral equation. Moreover $W(Z)$ is the only bounded solution of the integral equation, for if a second bounded solution existed, the difference $D(Z)$ would satisfy the homogeneous integral equation

$$D(Z) = \int_Z^{\infty} K(Z, T) D(T) dT.$$

Let Δ_a be that part of Δ_0 for which $\mathbf{R}(Z) \geqslant a$, where a is to be determined, then it may be verified that, if μ_a is the upper bound of $|D(Z)|$ in Δ_a, then

$$\mu_a \leqslant \frac{CM}{\nu^{\frac{1}{2}} a^{\nu}} \mu_a.$$

Thus if a is so chosen that $a^{\nu} > CM/\nu^{\frac{1}{2}}$, this inequality will lead to a contradiction unless $\mu_a = 0$, which proves that $D(Z)$ must be identically zero.

The proof holds for any strip of type Δ_0 which Δ may contain. It follows that the integral equation possesses, in that part of Δ which lies in the half-plane $\mathbf{R}(Z) \geqslant 0$, a unique analytic solution. Now consider the half-plane $\mathbf{R}(Z) < 0$, and let b be an arbitrarily large positive number. Then a positive number M_b exists such that, in Δ,

$$|\Phi(Z)| < \frac{M_b}{|Z+b|^{1+\nu}}.$$

The only modification necessary to complete the proof in this case is that due to the altered form of the inequality (G). It follows that a unique solution also exists in that part of Δ which lies on the negative side of the imaginary axis provided that

$$\mathbf{R}(Z) > -b, \quad |\mathbf{I}(Z)| \geqslant \rho.$$

Let D be a part of Δ in which $\mathbf{I}(Z)$ is bounded, $\mathbf{R}(Z)$ is bounded below, and $|\mathbf{I}(Z)| \geqslant \rho$ when $\mathbf{R}(Z) < 0$. Let Λ be the upper bound of $|W(Z)|$ in D

and let $z_1=x_1+iy_1$ be a point at which this upper bound is attained, then if L is the upper bound of $|W_0(Z)|$ in D,

$$\Lambda<L+\Lambda M\int_0^\infty \frac{dv}{|z_1+v|^{1+\nu}}<L+\Lambda\frac{CM}{\nu^{\frac{1}{2}}R_1^{\nu}},$$

where

$$R_1=|z_1| \quad \text{when} \quad x_1>0,$$
$$R_1=|y_1| \quad \text{when} \quad x_1<0.$$

Now if D is so chosen that $R_1^{\nu}>2CM/\nu^{\frac{1}{2}}$, then $\Lambda<2L$ and

$$|W(Z)-W_0(Z)|<\frac{2CLM}{\nu^{\frac{1}{2}}R^{\nu}},$$

where $R=|Z|$ or $|Y|$ according as $X>0$ or $X<0$.

It is not difficult to obtain similar equalities which are valid in that part of Δ in which $\mathbf{I}(Z)>B_2$. For the integral equation

$$W^+(Z)=W_0^+(Z)+\int_Z^\infty K^+(Z, T)W^+(T)dT,$$

in which

$$W_0^+(Z)=e^{iZ}W_0(Z), \quad K^+(Z, T)=e^{i(Z-T)}K(Z, T),$$

is satisfied by

$$W^+(Z)=e^{iZ}W(Z),$$

and it may be proved that $|W^+(Z)|$ is bounded for $\mathbf{I}(Z)>B_2$. By an appropriate choice of B_2, the upper bound of $|W^+(Z)|$ in the region considered may be made less than twice the upper bound L^+ of $|W_0^+(Z)|$ in that region. It follows that

$$|e^{iZ}\{W(Z)-W_0(Z)\}|<\frac{2CL^+M}{\nu^{\frac{1}{2}}R^{\nu}},$$

where R is as before. An analogous formula may be obtained for that part of Δ for which $\mathbf{I}(Z)<B_1$.

On account of these inequalities $W(Z)$ is said to be *asymptotic to* $W_0(Z)$; in the same way it may be proved that $W'(Z)$ is asymptotic to $W'_0(Z)$.

21·52. Truncated Solutions.—Now let $W_1(Z)$ be the solution asymptotic to e^{iZ}. Then the integral equation

$$U(Z)=1+\frac{1}{2i}\int_Z^\infty \{e^{2i(T-Z)}-1\}\Phi(T)U(T)dT$$

is satisfied by $U(Z)=e^{-iZ}W_1(Z)$. From this integral equation it may be shown * that $W_1(Z)$ is analytic in the sector

$$-\pi+\epsilon\leqslant \arg Z\leqslant 2\pi+\epsilon, \quad |z|\geqslant\rho,$$

and that

$$e^{-iZ}W_1(Z)=1+\frac{\Theta_1(Z)}{Z^{\nu}},$$

where $|\Theta_1(Z)|$ is bounded in the sector. Similarly, if $W_2(Z)$ is the solution asymptotic to e^{-iZ},

$$e^{iZ}W_2(Z)=1+\frac{\Theta_2(Z)}{Z^{\nu}}.$$

It follows from these formulæ that $W_1(Z)$ and $W_2(Z)$ have no zeros outside a sufficiently large circle; they are said to be *truncated in* Δ. The same is true of the derivatives $W_1'(Z)$ and $W_2'(Z)$.

Now when the region Δ is of type **Aa**, in which case every line parallel

* For a proof valid when $\nu=1$, see Hille, *Proc. London Math. Soc.* 23 (1924), § 2·24.

to the real axis intersects the boundary, $W_1(Z)$ and $W_2(Z)$ are the only solutions truncated in Δ. For any other solution may be written in the form

$$W(Z)=C_1W_1(Z)+C_2W_2(Z) \qquad (C_1C_2\neq 0),$$

and is asymptotic to

$$W_0(Z)=C_1e^{iZ}+C_2e^{-iZ}.$$

Without any loss of generality $W_0(Z)$ may be assumed to be $\sin(Z-a)$; its zeros are then $a_n=a+n\pi$. Now since the region is of type **Aa**, the strip Δ_0 may be so chosen as to contain all the zeros of $W_0(Z)$ on and after a certain value of n, say N_0. Let the parts of Δ which lie above and below Δ_0 be denoted by Δ_1 and Δ_{-1} respectively. Then, in Δ_0,

$$W(Z) = \sin(Z-a)+\frac{\theta_0(Z)}{Z^\nu};$$

in Δ_1,

$$e^{iZ}W(Z) = e^{iZ}\sin(Z-a)+\frac{\theta_1(Z)}{Z^\nu};$$

and in Δ_2,

$$e^{-iZ}W(Z) = e^{-iZ}\sin(Z-a)+\frac{\theta_{-1}(Z)}{Z^\nu}.$$

In each case, in Δ_λ,

$$|\theta_\lambda(Z)|<\frac{2CL_\lambda M}{\nu^{\frac{1}{2}}} \qquad (\lambda=-1,0,+1)$$

when

$$|Z|^\nu\geqslant 2CM/\nu^{\frac{1}{2}},$$

where L_λ denotes the upper bound of $|e^{\lambda iZ}\sin(Z-a)|$ in Δ_λ.

Now let Γ_n be the circle of small radius ϵ surrounding the point a_n, then on Γ_n,

$$|\sin(Z-a)|>\frac{2}{\pi}\epsilon,$$

and if

$$|Z|\geqslant r=\left\{\frac{CL_0M\pi}{\nu^{\frac{1}{2}}\epsilon}\right\}^{1/\nu}$$

then

$$\left|\frac{\theta_0(Z)}{Z^\nu}\right|<\frac{2}{\pi}\epsilon.$$

This proves that $\sin(Z-a)$ is the dominant term for $W(Z)$ on any circle Γ_n which lies in Δ and without the circle $|Z|=\gamma$.

Let Δ^+ be that part of Δ which lies outside the circle $|Z|=\gamma$. Then within each circle Γ_n in Δ^+ lies one and only one zero of $W(Z)$* Let Δ^* be what remains of Δ^+ when the interior of each circle Γ_n in Δ^+ is removed. Then $W(Z)$ has no zero in Δ^*. In the same manner it may be proved that the zeros of $W'(Z)$ in Δ^+ lie one within each of the circles

$$|Z-a'_n|=\epsilon,$$

where $a_n'=a+\frac{1}{2}\pi$.

Thus the zeros of $W(Z)$ and $W'(Z)$ may be denoted respectively by A_n and A'_n where

$$\lim(A_n-a_n)=0, \quad \lim(A'_n-a'_n)=0,$$

and the set of points A_n is said to form a *string of zeros* of the oscillatory solution $W_n(Z)$. The two truncated solutions, and these only, have no string of zeros.

* Cf. Rouché, *J. Éc. Polyt.* cah. 39 (1862), p. 217.

Where the region Δ is of type **Ab**, an infinite number of solutions exist which are truncated in Δ, namely those which are asymptotic to a function $W_0(Z)$ whose zeros lie outside of Δ. On the other hand if, on and after a certain value of n, the set of points $a_n=a+n\pi$ lies in Δ, a solution can be constructed whose string of zeros is approximated by (a_n), and this solution is asymptotic to $W_0(Z)=C\sin(Z-a)$.

Thus whatever be the type of the region, a solution can always be found whose zeros approximate to the set $(a+n\pi)$ if this set ultimately lies in Δ. To indicate the dependence of the zeros upon a, write $A_n(a)$ instead of A_n and $W(Z, a)$ for $W(Z)$. The question now arises as to how $A_n(a)$ varies with a. Let $a=\sigma+i\tau$ and assigning to τ the constant value τ_0, let σ vary from σ_0 to $\sigma_0+\pi$. Then $A_n(a)$ describes a continuous curve between the points $A_n(a_0)$, where $a_0=\sigma_0+i\tau_0$, and $A_{n+1}(a_0)$. As σ continues to increase, $A_n(a)$ describes a curve joining the zeros of the string and approaching its asymptote $\mathbf{I}(Z)=\tau_0$. This curve is called the *zero-curve* of the differential equation. It is evident that through every point in Δ^+ there passes one and only one zero-curve.

21·53. Distribution of Zeros in the z-plane.—The foregoing results are referred back to the z-plane by means of the substitution

$$Z=\int_{z_0}^{z}\sqrt{\left\{\frac{G(z)}{K(z)}\right\}}dz.$$

This substitution sets up a conformal transformation between the Z- and the z-plane. The simply connected domain Δ on the Z-plane will transform into a simply-connected, but in general overlapping domain D in the z-plane; the transformed domain lies in the most general case upon an infinitely-many leaved Riemann surface. Any solution $w(z)$ is analytic throughout the domain D, but on the boundary of this domain there may be one or more singular points corresponding to $z=\infty$.

The results which have been obtained concerning the distribution of zeros upon the Z-plane may now be re-stated in regard to the z-plane. The circle $|Z|=\gamma$ corresponds to a curve dividing the region D into two parts; let D^+ be the part corresponding to Δ^+. The points a_n become points a_n and the circles Γ_n become closed contours C_n enclosing the points a_n. D^* is defined to be that part of D^+ which is left when the interior of the contours C_n are removed. If, on and after a certain value of n the points a_n all lie in Δ, the corresponding points a_n will lie in D. To the solution $W(Z, a)$ corresponds the solution $w(z, a)$, where

$$w(z, a)=\{S(Z)\}^{-\frac{1}{2}}W(Z, a),$$

and one and only one zero of $w(z, a)$ lies within each of the contours C_n in D^+ whereas no zero at all lies in D^*.

The zero-curves in the Z-plane are represented by zero-curves $\mathfrak{C}$ on the Riemann surface which are asymptotic to the curves

$$\mathbf{I}\left\{\sqrt{\frac{G(z)}{K(z)}}\right\}=0.$$

Through every point a in D passes one and only one zero-curve $\mathfrak{C}(a)$. Let the points a_n be marked upon $\mathfrak{C}(a)$ in the direction of increasing values of $\mathbf{R}(Z)$, where

$$n\pi=\mathbf{R}\left\{\int_{a}^{a_n}\sqrt{\left(\frac{G(z)}{K(z)}\right)}dz\right\} \qquad (n=1, 2, 3, \ldots)$$

and the path of integration is the curve $\mathfrak{C}(a)$. Then there exists a solution $w(z, a)$ such that its zeros A_n can be so ordered that

$$\lim (A_n-a_n)=0.$$

Consider two circles Σ_1 and Σ_2 drawn in the Z-plane with radii R_1 and R_2 respectively where $R_2>R_1>R$, and suppose for simplicity that each of these circles cuts Γ, the boundary of Δ, in two and only two points. In the z-plane the circular arcs Σ_1 and Σ_2 transform into curves S_1 and S_2 which, together with the transformed portions of Γ enclosed between Σ_1 and Σ_2 form a curvilinear quadrilateral $[D]$. This quadrilateral is cut by $\mathfrak{C}(a)$ in two points, say z_1 on S_1 and z_2 on S_2. Then the number of zeros of $w(z, a)$ in $[D]$ is given by the formula

$$N[D]=\frac{1}{\pi}R\left\{\int_{z_1}^{z_2}\sqrt{\left\{\frac{G(z)}{K(z)}\right\}}dz\right\}+\theta$$

where the path of integration lies along $\mathfrak{C}(a)$ and $-1\leqslant\theta\leqslant+1$.

Similar results may be obtained with regard to $w'(z, a)$ by considering the formula

$$\frac{dw}{dz}=\left\{\frac{dW}{dZ}-\tfrac{1}{2}F(Z)W(Z)\right\}\{S(Z)\}^{-\frac{1}{4}}\frac{dZ}{dz},$$

in which the first factor alone is relevant.

21·54. Equations with Polynomial Coefficients.—A definite and important example is provided by the case in which $K(z)$ and $G(z)$ are polynomials in z; let

$$K(z)=z^k+\ldots,\quad G(z)=g_0z^g+\ldots\quad(g_0\neq 0).$$

In order that the point at infinity may be an irregular singular point it will be supposed that $g\geqslant k-1$; let $m=g-k+2$ so that $m\geqslant 1$.

Since

$$Z=\int_{z_0}^{z}\sqrt{\left\{\frac{g_0z^g+\ldots}{z^k+\ldots}\right\}}dz,$$

Z is in general an Abelian integral of the third kind. For large values of z, Z has the form

$$Z=\frac{2}{m}g_0^{\frac{1}{2}}z^{\frac{1}{2}m}\{1+C_{\frac{1}{2}m}z^{-\frac{1}{2}m}\log z+O(z^{-1})\},$$

in which the logarithmic term occurs only if m is an even number. Conversely

$$z=\left\{\frac{m^2Z^2}{4g_0}\right\}^{\frac{1}{m}}\left\{1+\sum\right\},$$

where $\sum$ is a double series of ascending powers of $Z^{-\frac{2}{m}}$ and $\log Z$, which is convergent for sufficiently large values of $|Z|$.

Again, since

$$F(Z)=\frac{G(z)K'(z)+G'(z)K(z)}{2\{G(z)\}^{\frac{3}{2}}\{K(z)\}^{\frac{1}{2}}},$$

it follows that

$$F(Z)=\frac{g+k}{2g_0^{\frac{1}{2}}}z^{-\frac{m}{2}}\{1+O(z^{-1})\},$$

and therefore

$$F(Z)=\frac{g+k}{m}Z^{-1}\left\{1+\sum\right\},$$

where $\sum$ is a double series of the same type as above.

Similarly

$$\Phi(Z)=\frac{3k-g+4}{4m^2}Z^{-2}\left\{1+\sum\right\},$$

and therefore $\Phi(Z)$ satisfies a condition **B** with $\nu=1$ in any region outside a sufficiently large circle $|Z|=R$ in which arg Z is bounded. Let Δ be the region

$$|Z|>R, \quad \mathbf{R}(Z)\geqslant 0.$$

For a sufficiently large value of R this region is conformally represented on the z-plane by a sectorial region D_μ in which

$$\frac{1}{m}\{(2\mu-1)\pi-\theta_0\}-\delta<\arg z<\frac{1}{m}\{(2\mu+1)\pi-\theta_0\}+\delta,$$

where δ is a small positive number, θ_0 is arg g_0, and μ has the value 0, 1, . . ., or $m-1$ corresponding to the chosen determination of $Z^{1/m}$.

If $z=re^{i\theta}$, the asymptotic zero curves in the z-plane are of the form

$$r^{\frac{1}{2}m}\sin\tfrac{1}{2}(m\theta+\theta_0)+\text{ lower terms}=\text{const.},$$

and their asymptotic directions are

$$\theta_\mu=\frac{1}{m}(2\mu\pi-\theta_0).$$

The solution $w(z, a)$ is not, in general, one-valued in the neighbourhood of infinity, but if $\mathfrak{D}$ represents that part of the Riemann surface of log z which lies outside a sufficiently large circle, $w(z, a)$ is one-valued on $\mathfrak{D}$. The zeros of $w(z, a)$ thus form m strings which are asymptotic to the directions θ_μ in each leaf of $\mathfrak{D}$. If $N(r)$ denotes the number of zeros in a string within the circle $|z|=r$ then as $r\to\infty$,

$$N(r)\to\frac{2}{m\pi}|g_0|^{\frac{1}{2}}r^{\frac{1}{2}m}.$$

The results of § 21·52 show that there are two solutions which are truncated in the direction θ_μ, from which it follows that the total number of truncated solutions does not exceed $2m$. It will now be shown that the actual number of truncated solutions is m.

Consider the region Δ' whose boundary is the large circular arc

$$|Z|=R, \quad -\tfrac{1}{2}\pi+\delta\leqslant\arg Z\leqslant\tfrac{1}{2}\pi-\delta,$$

and the tangents drawn to the extremities of this arc and extending to infinity in the half plane $\mathbf{I}(Z)<0$. The region thus defined is of type **A**, and in it $\Phi(Z)$ satisfies a condition **B**. Let $W_1(Z)$ and $W_2(Z)$ be the truncated solutions asymptotic to e^{iZ} and e^{-iZ} respectively. Now $W_1(Z)$ is asymptotic to e^{iZ} in the more extended region

$$-\pi<\arg Z<2\pi,$$

and, as may be seen by considering a region symmetrical to Δ' with respect to the imaginary axis, $W_2(Z)$ is asymptotic to e^{-iZ} in the more extended region

$$-2\pi<\arg Z<\pi.$$

If therefore $|\mathbf{I}(Z)|\to\infty$,

$$W_1(Z)\to 0 \text{ in the upper half of } \Delta',$$
$$W_2(Z)\to 0 \text{ in the lower half of } \Delta',$$

and in view of the properties of the integral equation satisfied by $W(Z)$, these conditions suffice uniquely to determine $W_1(Z)$ and $W_2(Z)$ respectively.

In the z-plane there are m distinct regions D'_μ which correspond to Δ', and D'_μ is such that

$$\theta_{\mu-1}+\epsilon\leqslant\arg z\leqslant\theta_{\mu+1}-\epsilon \qquad (\mu=0, 1, \ldots, m-1).$$

Consecutive regions D'_μ and $D'_{\mu+1}$ have a common part namely a region U_μ where

$$\theta_\mu+\epsilon\leqslant\arg z\leqslant\theta_{\mu+1}-\epsilon.$$

Now there is one solution truncated in D'_μ which tends to zero in U_μ and a solution truncated in $D'_{\mu+1}$ which also tends to zero in U_μ. But as only one such solution can tend to zero in U_μ, the two solutions in question must be identical. The number of truncated solutions thus reduces to m; that this number is actually attained may be seen by considering the equation

$$\frac{d^2w}{dz^2}+z^g w=0.$$

If $w_\mu(z)$ is the solution which tends to zero in U_μ, this solution is truncated in the adjacent directions θ_μ and $\theta_{\mu+1}$, and moreover, it preserves the same asymptotic representation in the three adjacent regions $U_{\mu-1}$, U_μ and $U_{\mu+1}$.

Miscellaneous Examples.

1. Prove the formula

$$\Big[w_1\bar{w}_2\Big]_{z_1}^{z_2}-\int_{z_1}^{z_2}\frac{|w_2|^2}{K(z)}dz+\int_{z_1}^{z_2}|w_1|^2\bar{G}(z)d\bar{z}=0.$$

2. Considering the dynamical system

$$\frac{d^2x}{dt^2}=-g_1x+g_2y,$$
$$\frac{d^2y}{dt^2}=-g_1y-g_2x,$$

where g_1 and g_2 are functions of t, employ the results of § 21·4 to prove that a particle starting from the origin at time t_1 with a given velocity will continue to move away from the origin so long as $g_1(t)\leqslant 0$ and the sign of $g_2(t)$ remains unchanged.

3. Extend the results of §§ 21·4, 21·41 to the general self-adjoint equation of the second order.

4. Let $F(z)$ be real and positive when z is real and greater than x_1, analytic throughout a region D including the real axis for $\mathbf{R}(z)>x_1$, and such that either

$$\mathbf{R}\{F(z)\}>0 \quad \text{or} \quad \mathbf{I}\{F(z)\}\neq 0$$

in D; let $W(z)$ be a solution of

$$\frac{d^2w}{dz^2}-F(z)w=0$$

such that $W(z)\to 0$ as $z\to\infty$ in D along a parallel to the real axis, then under very general assumptions, $W(z)$ has no zero nor extremum in D.

5. Construct the standard domain for a solution of

$$\frac{d^2w}{dz^2}-\frac{w}{z}=0$$

which has a complex zero $z=a+ib$.

6. Prove that when $\Phi(Z)$ is analytic and satisfies a condition B in the half-plane Υ, $\mathbf{I}(Z)>B_2$, every solution is asymptotic to one sine-function in $\Upsilon+$, the extreme right-hand part of the region, and asymptotic to another sine-function in $\Upsilon-$, the extreme left-hand part of the region. Discuss the zeros of this solution.

7. Given the function $\sin(Z-a)$, there exsts one solution $W+(Z)$ asymptotic to it in $\Upsilon+$, and another solution $W-(Z)$ asymptotic to it in $\Upsilon-$. But if $\tau=\mathbf{I}(a)$ is large, there exists a solution $W(Z)$ asymptotic to $\sin(Z-a)$ throughout Υ, and the strings in $\Upsilon+$ and $\Upsilon-$ join into a single string. There are no zeros above this string and only a finite number of zeros in Υ below it.

8. The asymptotic zero-curves of the equation

$$\frac{d^2w}{dz^2}-\frac{w}{z}=0$$

are parabolas with focus at the origin and the negative real axis as axis. Work out the distribution of the zeros in the neighbourhood of the asymptotic parabola.

9. The general solution of the equation

$$\frac{d^2w}{dz^2}+G(z)w=0,$$

in which $G(z)$ is a polynomial in z, can be represented in the form

$$w=w_1(z)-\lambda w_2(z),$$

where $w_1(z)$ and $w_2(z)$ are linearly independent solutions and λ is a complex parameter. Show that a necessary and sufficient condition that the solution be truncated is that λ is one of the asymptotic values of the meromorphic function $\lambda(z)=w_1(z)/w_2(z)$.

[NOTE.—A number a is said to be an asymptotic value of an integral or meromorphic function $f(z)$ if there is a simple curve tending to infinity along which $f(z)\rightarrow a$.]

APPENDIX A

HISTORICAL NOTE ON FORMAL METHODS OF INTEGRATION

A·1. Differential Equations to the End of the Seventeenth Century.—The early history of a branch of mathematics which has enjoyed two and a half centuries of vigorous life naturally tends more and more to be masked by the density of its later growth. Yet our hazy knowledge of the birth and infancy of the science of differential equations condenses upon a remarkable date, the eleventh day of November, 1675, when Leibniz first set down on paper the equation

$$\int y\,dy = \tfrac{1}{2}y^2,$$

thereby not merely solving a simple differential equation, which was in itself a trivial matter, but what was an act of great moment, forging a powerful tool, the integral sign.

The early history of the infinitesimal calculus abounds in instances of problems solved through the agency of what were virtually differential equations ; it is even true to say that the problem of integration, which may be regarded as the solution of the simplest of all types of differential equations, was a practical problem even in the middle of the sixteenth century. Particular cases of the inverse problem of tangents, that is the problem of determining a curve whose tangents are subjected to a particular law, were successfully dealt with before the invention of the calculus.*

But the historical value of a science depends not upon the number of particular phenomena it can present but rather upon the power it has of coordinating diverse facts and subjecting them to one simple code.

A·11. Newton and Leibniz.—Thus it was that the first step of moment was that which Newton took when he classified differential equations of the first order, then known as fluxional equations, into three classes.†

The first class was composed of those equations in which two fluxions $\dot{x}$ and $\dot{y}$ and one fluent x or y, are related, as for example,

$$\text{(i)}\ \frac{\dot{y}}{\dot{x}} = f(x)\ ;\quad \text{(ii)}\ \frac{\dot{y}}{\dot{x}} = f(y),$$

or as they would to-day be written

$$\text{(i)}\ \frac{dy}{dx} = f(x),\quad \text{(ii)}\ \frac{dy}{dx} = f(y).$$

The second class embraced those equations which involve two fluxions and two fluents thus

$$\frac{\dot{y}}{\dot{x}} = f(x, y).$$

The third class was made up of equations which involve more than two fluxions ; these are now known as partial differential equations.

Newton's general method was to develop the right-hand member of the equation

* For example, by Isaac Barrow (1630–1677).

† *Methodus Fluxionum et serierum infinitarum*, written about the year 1671, published in 1736 [*Opuscula*, 1744, Vol. I. p. 66].

in powers of the fluents and to assume as a solution an infinite series whose coefficients were to be determined in succession. For example, if the equation to be solved was

$$\frac{\dot{y}}{\dot{x}}=2+3x-2y+x^2+x^2y,$$

a solution of the form

$$y=A_0+A_1x+A_2x^2+A_3x^3+\ldots$$

was assumed. Then

$$\frac{\dot{y}}{\dot{x}}=A_1+2A_2x+3A_3x^2+\ldots,$$

and by substituting in the equation it was found that

$$A_1=2-2A_0,\quad 2A_2=3-2A_1,\quad 3A_3=1+A_0-2A_2,\ldots.$$

It was noted that A_0 could be chosen in an arbitrary manner, and it was concluded that the equation possessed an infinite number of particular solutions. Yet the real significance of this fact that the general solution of an equation of the first order depends upon an arbitrary constant remained hidden until the middle of the eighteenth century. Newton did, however, observe that any solution of the equation

$$y^{(n)}=f(x)$$

remains a solution after the addition thereto of an arbitrary polynomial of degree $n-1$.*

One of the earliest discoveries in the integral calculus was that the integral of a given function could only in very special cases be finitely expressed in terms of known functions. So is it also in the theory of differential equations. That any particular equation should be integrable in a finite form is to be regarded as a happy accident; in the general case the investigator has to fall back, as in the example just quoted, upon solutions expressed in infinite series whose coefficients are determined by recurrence-formulæ.

The general statement of the problem of integrating a given differential equation was first formulated by Newton in the following anagram: †

6a, 2c, d, ae, 13e, 2f, 7i, 3l, 9n, 4o, 4q, 2r, 4s, 8t, 12v, x,

which was subsequently deciphered thus: *Data aequatione quotcumque fluentes quantitates involvente, fluxiones invenire et vice versa.* Two methods of solution are stated in a second anagram which when unravelled runs as follows: *Una methodus consistit in extractione fluentis quantitatis ex aequatione simul involvente fluxionem ejus; altera tantum in assumptione seriei pro quantitate qualibet incognita, ex qua cetera commode derivari possint, et in collatione terminorum homologorum aequationis resultantis ad eruendos terminos assumptae seriei.*

The inverse problem of tangents led Leibniz on to many important developments. Thus, in 1691, he implicitly discovered the method of separation of variables by proving that a differential equation of the form

$$y\frac{dx}{dy}=X(x)Y(y)$$

is integrable by quadratures.‡ A year later he made known the method of integrating the homogeneous differential equation of the first order, and not long afterwards reduced to quadratures the problem of integrating a linear equation of the first order.

To Leibniz is due the modern differential notation and the use of the sign of integration. The notorious controversy § which centres round Newton and Leibniz had the effect of depriving English mathematicians of the use of this powerful

* *Tractatus de quadratura curvarum*, written about 1676, published for the first time as an appendix to the *Opticks* (1704) [*Opuscula*, 1744, Vol. I. p. 244].

† It occurs in a letter to Leibniz (through the intermediary of Oldenburg) dated the 26th October, 1676.

‡ This theorem was communicated to Huygens towards the end of the year 1691, *Briefwechsel von Leibniz*, 1, p. 680.

§ See Gibson, *Proc. Edin. Math. Soc.* 14 (1896), p. 148.

system of notation, and for more than a century England was barren, whereas the Continent flourished in the field of analysis.

A·12. The Elder Bernoullis.—In May, 1690, James Bernoulli published his solution of the problem of the isochrone,* of which a solution had already been given by Leibniz. This problem leads to the differential equation

$$dy\sqrt{(b^2y-a^3)}=dx\sqrt{(a^3)}.$$

In this form the equation expresses the equality of two differentials from which, in the words *ergo et horum Integralia aequuntur*, Bernoulli concludes the equality of the integrals of the two members of the equation and uses the word *integral* for the first time on record. From this beginning also sprang the idea of obtaining the equation of a curve which has a kinematical or a dynamical definition by expressing the mode of its description in the guise of a differential equation and integrating this equation under certain initial conditions. Instances of such curves are the *spira mirabilis* or logarithmic spiral, the elastica and the lemniscate.

To John Bernoulli (a younger brother of James) is due the term and the explicit process of *seperatio indeterminatarum* or separation of variables.† But it was noticed that in one particular yet important case this process broke down; for although the variables in the equation

$$axdy-ydx=0$$

are separable, yet the equation could not be integrated by this particular method. The reason was that the differential dx/x had not at that time been integrated; in fact Bernoulli, assuming that the formula

$$ax^p dx=d\left(\frac{a}{p+1}x^{p+1}\right)$$

holds when $p=-1$, comes to the conclusion *neutrius habetur integrale*. In this particular instance the difficulty was overcome by the introduction of the *integrating factor* ‡ y^{a-1}/x^2 which brings the equation into the form

$$\frac{ay^{a-1}}{x}dy-\frac{y^a}{x^2}dx=0,$$

when it is immediately integrable and has the solution

$$\frac{y^a}{x}=b,$$

where b is any constant.

In the same year, however, the true interpretation of $\int dx/x$ as $\log x$ became known,§ and the scope of the method of separation of variables was vastly extended.

The equation known as the Bernoulli equation,

$$ady=ypdx+by^nqdx,$$

in which a and b are constants, and p and q are functions of x alone, was proposed for solution by James Bernoulli in December, 1695.‖ As was pointed out by Leibniz ¶ it may be reduced to a linear equation by taking y^{1-n} as the dependent variable. John Bernoulli chose a different line of attack, making use of the process

* *Acta Erud.* May, 1690 [*Opera*, 1, p. 421].

† *Acta Erud.*, November, 1694; given in a letter to Leibniz, May 9, 1694.

‡ From a letter to Huygens dated 14/24 June, 1687, it appears that Fatio de Duillier applied this process to the equation $3xdy-2ydx=0$. No earlier instance of an integrating factor seems to be known.

§ It may have been known to Nicolaus Mercator (N. Kaufmann) in 1668. It was certainly known to Leibniz, through the problem of the quadrature of the hyperbola, in 1694. Napier's *Mirifici Logarithmorum Canonis Descriptio* was published in 1614, some fifty years before the invention of the infinitesimal calculus.

‖ *Acta Erud.* (1695), p. 553 [*Opera* I. p. 663].

¶ *Acta Erud.* (1696), p. 145.

by which the homogeneous equation was reduced to an integrable form; he made the substitution

$$y=mz, \quad dy=mdz+zdm,$$

where m and z are new variables, and thus obtained the relation

$$amdz+azdm=mzpdx+bm^nz^nqdx.$$

The fact that one unknown y has been replaced by two unknown m and z introduces an element of choice which is exercised in writing

$$amdz=mzpdx,$$

whence

$$\frac{adz}{z}=pdx.$$

This auxiliary equation can be integrated, giving z as a function of x. Then in the remaining equation

$$azdm=bm^nz^nqdx$$

the variables are separable; the equation can be integrated and thus m and therefore y are explicitly found in terms of x.

A·2. The Early Years of the Eighteenth Century.—By the end of the seventeenth century practically all the known elementary methods of solving equations of the first order had been brought to light. The problem of determining the orthogonal trajectories of a one-parameter family of curves was solved by John Bernoulli in 1698; the problem of oblique trajectories presented no further difficulties.

The early years of the eighteenth century are remarkable for a number of problems which led to differential equations of the second or third orders. In 1696 James Bernoulli formulated the isoperimetric problem, or the problem of determining curves of a given perimeter which shall under given conditions, enclose a maximum area. Five years later he published his solution,* which depends upon a differential equation of the third order.

Attention was now turned to trajectories in a general sense and in particular to trajectories defined by the knowledge of how the curvature varies from point to point; these gave rise to differential equations of the second order. Thus, for example, John Bernoulli, in a letter to Leibniz dated May 20, 1716, discussed an equation which would now be written

$$\frac{d^2y}{dx^2}=\frac{2y}{x^2}$$

and stated that it gave rise to three types of curves, parabolæ, hyperbolæ and a class of curves of the third order.†

A·21. Riccati and the Younger Bernoullis.—An Italian mathematician, Count Jacopo Riccati, was destined to play an important part in furthering the theory of differential equations. In investigating those curves whose radii of curvature were dependent solely upon the corresponding ordinates, he was led to a differential equation of the general form

$$f(y, y', y'')=0,$$

that is to say to an equation explicitly involving y, y' and y'' but not x. By regarding y as an independent variable and p or y' as the dependent variable, and making use of the relationship

$$y''=p\frac{dp}{dy},$$

* *Acta Erud.*, May, 1701 [*Opera* 2, p. 895].

† The general solution may be written

$$y=\frac{x^2}{a}+\frac{b^2}{3x},$$

where a and b are constants of integration. When $b=0$ the curves are parabolæ, when $a=\infty$ they are rectangular hyperbolæ, in other cases they are of the third order.

Riccati brought the equation into the form *

$$f\left(y, p, p\frac{dp}{dy}\right)=0,$$

and thus reduced it from the second order in y to the first order in p.

The particular equation to which the name of Riccati is attached was first exhibited in the form †

$$x^m\frac{dq}{dx}=\frac{du}{dx}+\frac{u^2}{q}.$$

Before the equation can be dealt with some restrictive hypothesis as to u or q must be made. Riccati chose to suppose that q was a power of x, say x^n, and thus reduced the equation to the form

$$nx^{m+n-1}=\frac{du}{dx}+u^2x^{-n}.$$

The problem now became one of choosing values of n such that the equation could be integrated, if possible, in a finite form.

This problem attracted the attention of the Bernoulli family. Following immediately upon Riccati's paper is a note by Daniel Bernoulli, who claimed that he and three of his kinsmen had independently discovered the value of n by means of which the variables became separable.‡ What these solutions may have been is not known; Daniel Bernoulli concealed his own solution under the form of an anagram which has not yet been deciphered. §

Daniel Bernoulli published the conditions under which the equation, written in a form equivalent to

$$\frac{dy}{dx}+ay^2=bx^m,$$

is integrable in a finite form, namely that m must be of the form $-4k/(2k\pm1)$ where k is a positive integer. ||

A·3. Euler.—The next important advance was made by Euler, who proposed and solved the problem of reducing a particular class of equations of the second order to equations of the first order.¶ The germ of Euler's method lies in replacing x and y by new variables v and t by the substitution

$$x=e^{\alpha v}, \quad y=e^{v}t,$$

where α is a constant subsequently to be determined. In modern symbolism the formulæ of transformation are

$$\frac{dy}{dx}=\frac{1}{\alpha}e^{v(1-\alpha)}\left\{\frac{dt}{dv}+t\right\},$$

$$\frac{d^2y}{dx^2}=\frac{1}{\alpha^2}e^{v(1-2\alpha)}\left\{\frac{d^2t}{dv^2}+2\frac{dt}{dv}+(1-\alpha)t\right\}.$$

The idea of the method is to choose α, if possible, in such a way that no exponential terms shall appear in the transformed equation, which implies a certain degree of

* *Giornale de' Letterati d'Italia*, 11 (1712). The device by which the lowering of the order is effected had already been used by James Bernoulli.

† *Acta Erud. Supp.* 8 (1723), pp. 66–73. The equation arose as the result of reducing the equation $x^md^2x=d^2y+dy^2$ to the first order through the substitution $dx/dy=q/u$, where u and q are, in the first place, supposed to depend upon x and y.

‡ *Ibid.* p. 74: *Praescribit frater meus, se illud solvisse; sed praeter illum alii quoque existunt solutores, solutionem enim eruerunt Pater et Patruelis Nicolaus Bernoulli pariter ac egomet.* Daniel was the second son of John Bernoulli; Nicholas the younger was his elder brother, and Nicholas the elder his cousin.

§ The anagram is reproduced in Watson, *Bessel Functions*, p. 2.

|| *Exercitationes quaedam mathematicae* (Venice, 1724), pp. 77–80; *Acta Erud.*, November, 1725, pp. 473–475.

¶ *Comm. Acad. Petrop.* 3 (1728), pp. 124–137.

homogeneity in the original equation. Thus consider, as a particular instance, the equation

$$y^n\left(\frac{dy}{dx}\right)^{p-2}\cdot\frac{d^2y}{dx^2}=ax^m,$$

which is transformed into

$$\alpha^{-p}e^{(n+p-p\alpha-1)v}t^n\left\{\frac{dt}{dv}+t\right\}^{p-2}\left\{\frac{d^2t}{dv^2}+2\frac{dt}{dv}+(1-\alpha)t\right\}=ae^{m\alpha v}.$$

The exponential term cancels out if

$$\alpha=\frac{n+p-1}{m+p},$$

and with this choice of α the equation takes the form

$$t^n\left\{\frac{dt}{dv}+t\right\}^{p-2}\left\{\frac{d^2t}{dv^2}+2\frac{dt}{dv}+\frac{m-n+1}{m+p}t\right\}=a\left\{\frac{n+p-1}{m+p}\right\}^p.$$

It is now simpler than the original equation in the sense that the independent variable v is not explicitly involved; let v be replaced by a new variable z defined by the relation

$$\frac{dv}{dt}=z,$$

then the equation is reduced to the first order in z and t. Several types of equations of order higher than the second may be reduced to a lower order by similar methods.

The fundamental conception of an integrating factor is also due to Euler,* for although instances of its use in the integration of a differential equation of the first order had already been given, Euler went further and set up classes of equations which admit of integrating factors of given types. He also proved that if two distinct integrating factors of any equation of the first order can be found. then their ratio is a solution of the equation. In the development of the theory of the integrating factor an important part was played by Clairaut.†

A·4. Linear Equations.—With a letter from Euler to John Bernoulli, dated September 15, 1739, begins the general treatment of the homogeneous linear differential equation with constant coefficients.‡ It appears from Bernoulli's replies that before the year 1700 he had studied the differential equation

$$0=y+ax\frac{dy}{dx}+bx^2\frac{d^2y}{dx^2}+cx^3\frac{d^3y}{dx^3}+\ .\ .\ .\ +kx^n\frac{d^ny}{dx^n}.$$

He first multiplied it throughout by the factor x^p, then defining z by the relation

$$z=\frac{1}{p+1}\cdot\frac{d(x^{p+1}y)}{dx}=x^py+\frac{x^{p+1}}{p+1}\cdot\frac{dy}{dx},$$

and making use of the formulæ

$$ax^{p+1}\frac{dy}{dx}=a(p+1)z-a(p+1)x^py,$$

$$bx^{p+2}\frac{d^2y}{dx^2}=-2b(p+1)^2z+b(p+1)x\frac{dz}{dx}+b(p+1)(p+2)x^py,$$

etc., he transformed the equation into one of the form

$$0=\alpha x^py+a'z+b'x\frac{dz}{dx}+c'x^2\frac{d^2z}{dx^2}+\ .\ .\ .\ +k'x^{n-1}\frac{d^{n-1}z}{dx^{n-1}}.$$

Now α depends upon p, and p may be so chosen as to reduce α to zero, by which means the order of the equation is reduced by unity. This process of reduction can be repeated as often as is necessary.

* *Comm. Acad. Petrop.* 7 (1734), p. 168; *Novi Comm. Acad. Petrop.* 8 (1760), p. 3.

† *Hist. Acad. Paris*, 1739, p. 425; 1740, p. 293.

‡ *Bibl. Math.* (3), 6 (1905), p. 37. On the discovery of the general solution of this equation, see Eneström, *Bibl. Math.* (2), 11 (1897), p. 43.

Euler's method of dealing with the linear equation with constant coefficients was as follows.* If $y=u$ is any solution of the differential equation

$$0=Ay+\frac{Bdy}{dx}+\frac{Cd^2y}{dx^2}+\frac{Dd^3y}{dx^3}+\ldots+\frac{Nd^ny}{dx^n},$$

then $y=\alpha u$ is a solution, where α is any constant. Moreover, if n particular solutions (*valores particuliares*) $y=u$, $y=v$, . . . are obtainable, then the complete or general solution (*aequatio integralis completa*) of the differential equation will be

$$y=\alpha u+\beta v+\ldots,$$

where α, β, . . . are constants.

Now if the root $z=\frac{q}{p}$ of the algebraic equation of the first degree $q-pz=0$ satisfies the algebraic equation of the nth degree,

$$A+Bz+Cz^2+\ldots+Nz^n=0,$$

then the solution $y=\alpha e^{\frac{qx}{p}}$ of the differential equation

$$qy-p\frac{dy}{dx}=0$$

will satisfy the differential equation of order n. Thus there are as many particular solutions of this form as there are distinct real linear factors in

$$A+Bz+Cz^2+\ldots+Nz^n.$$

The complication introduced by a multiple factor $(q-pz)^k$ is met by the substitution

$$y=e^{\frac{qx}{p}}u,$$

whereby a particular solution involving k constants is found:

$$y=e^{\frac{qx}{p}}(\alpha+\beta x+\gamma x^2+\ldots+\kappa x^{k-1}).$$

When a pair of complex linear factors arise they are united in a real quadratic factor $p-qz+rz^2$ or

$$p-2z\sqrt{pr}\cos\phi+rz^2,\quad \text{where } \cos\phi=\frac{q}{2\sqrt{pr}}.$$

To this factor corresponds the differential equation

$$0=py-2z\sqrt{pr}\cos\phi\frac{dy}{dx}+r\frac{d^2y}{dz^2}.$$

The transformation

$$y=e^{fx\cos\phi}u,\quad f=\sqrt{pr}$$

reduces the equation to

$$r\frac{d^2u}{dx^2}+(rf^2\cos^2\phi-2f^2\cos^2\phi+p)u=0,$$

which is of the form

$$\frac{d^2y}{dx^2}+ky=0,$$

an equation which had already been solved.† Repeated quadratic factors were next dealt with and the discussion of the homogeneous linear equation with constant coefficients was complete.

Euler next turned his attention to the non-homogeneous linear equation

$$X=Ay+\frac{Bdy}{dx}+\frac{Cd^2y}{dx^2}+\ldots+\frac{Nd^ny}{dx^n},$$

* Published in *Misc. Berol.* 7 (1743), pp. 193–242.

† Euler, *Inquisitio physica in causam fluxus ac refluxus maris*, 1740. Daniel Bernoulli had solved it independently, *Comm. Acad. Petrop.* 13 (1741), p. 5.

a particular case of which, namely

$$\frac{d^2y}{dx^2}+ky=X,$$

he had also discussed in 1740. The method now adopted was that of a successive reduction of the order of the equation by the aid of integrating factors of the form $e^{\alpha x}dx$. Thus, in the case of the equation of the second order,

$$\int e^{\alpha x}Xdx=\int\left[e^{\alpha x}Aydx+e^{\alpha x}Bdy+e^{\alpha x}\frac{Cd^2y}{dx}\right]$$
$$=e^{\alpha x}\left(A'y+\frac{B'dy}{dx}\right).$$

By differentiating and comparing like terms it is found that

$$B'=C,\quad A'=B-\alpha C=\frac{A}{\alpha},\quad \text{whence}\quad A-B\alpha+C\alpha^2=0.$$

Thus α, A' and B' are found, and the equation is reduced to an equation of the same form as before, but of lower order, namely

$$e^{-\alpha x}\int e^{\alpha x}Xdx=A'y+\frac{B'dy}{dx}.$$

An integrating factor for this equation is $e^{\beta x}dx$ where $\alpha+\beta=\frac{B}{C}$, and therefore α and β are the two roots of $A-B\alpha+C\alpha^2=0$.

In the case of the equation of order n, there are n integrating factors of the type $e^{\alpha x}dx$, by means of which the equation is reduced in order step by step and finally integrated. The complications due to equal or complex roots of the equation in α were also disposed of by Euler.

To Euler is also due the process of integrating by series equations which were not integrable in a finite form. Thus, for example, he integrated the equation *

$$\frac{d^2y}{dx^2}+\frac{c^2y}{x^{2m+2}}=0$$

in the form

$$y=kx^{\frac{m+1}{2}}(1-Bx^{2m}+Dx^{4m}-\ .\ .\ .\ )\sin\left(\frac{c}{mx^m}+\vartheta\right)$$
$$-kx^{\frac{m+1}{2}}(Ax^m-Cx^{3m}+\ .\ .\ .\ )\cos\left(\frac{c}{mx^m}+\vartheta\right).$$

A·41. Lagrange and Laplace.—The problem of determining an integrating factor for the general linear equation

$$Ly+M\frac{dy}{dt}+N\frac{d^2y}{dt^2}+\ .\ .\ .\ =T,$$

where $L, M, N, \ldots, T$ are functions of t, led Lagrange to the conception of the adjoint equation.† If the equation is multiplied throughout by zdt, where z is a function of t, then the equation can be integrated once if z is a solution of the adjoint equation

$$Lz-\frac{d\,.\,Mz}{dt}+\frac{d^2\,.\,Nz}{dt^2}+\ .\ .\ .\ =0.$$

In this way Lagrange solved the equation ‡

$$Ay+B(h+kt)\frac{dy}{dt}+C(h+kt)^2\frac{d^2y}{dt^2}+\ .\ .\ .\ =T,$$

where $A, B, C, \ldots, h$ and k are constants and T is a function of t. He formed

* *Novi Comm. Acad. Petrop.* 9 (1762/63), p. 298. It is virtually the Bessel equation.

† *Misc. Taur.* 3 (1762/5), pp. 179–186 [*Œuvres*, 1, pp. 471–478]. See also Euler, *Novi Comm. Acad. Petrop.* 10 (1764), p. 134.

‡ *Ibid.* pp. 190, 199 [*Œuvres*, 1, pp. 481–490].

the adjoint equation and assumed that it was satisfied by $z=(h+kt)^r$. The index r was then found to satisfy the equation

$$A-Bk(r+1)+Ck^2(r+1)(r+2)+\ldots=0.$$

Lagrange also proved * that the general solution of a homogeneous linear equation of order n is of the form

$$y=c_1y_1+c_2y_2+\ldots+c_ny_n,$$

where $y_1, y_2, \ldots, y_n$ are a set of linearly independent solutions and $c_1, c_2, \ldots, c_n$ are arbitrary constants.

Laplace generalised Lagrange's methods † by considering not a single integrating factor but a system of multipliers. In the equation

$$X=y+H\frac{dy}{dx}+H'\frac{d^2y}{dx^2}+\ldots+H^{(n-1)}\frac{d^ny}{dx^n},$$

where $X, H, H', \ldots$ are functions of x, Laplace made the substitution

$$\omega\frac{dy}{dx}+y=T,$$

where ω and T were functions of x to be determined. The equation then became

$$X=T+\omega'\frac{dT}{dx}+\omega''\frac{d^2T}{dx^2}+\ldots+\omega^{(n-1)}\frac{d^{n-1}T}{dx^{n-1}},$$

where

$$\omega\omega^{(n-1)}=H^{(n-1)},$$

$$\omega\omega^{(n-2)}+\omega^{(n-1)}+(n-1)\omega^{(n-1)}\frac{d\omega}{dx}=H^{(n-2)},$$

$$\cdot\quad\cdot\quad\cdot\quad\cdot\quad\cdot\quad\cdot$$

$$\omega+\omega'+\omega'\frac{d\omega}{dx}+\ldots=H.$$

The first $n-1$ equations determine $\omega', \omega'' \ldots$ in terms of $\omega, H', H'', \ldots$. The last equation then becomes an equation of order $n-1$ for ω; the equation for T is also of order $n-1$. Thus the given equation of order n has been replaced by a pair of equations of order $n-1$, which are not, in general, linear. If, however, $n-1$ particular solutions of the equations in ω and in T are known, the general solution of the linear equation in y can be obtained by quadratures.

In particular, if the given equation is of the second order:

$$X=y+H\frac{dy}{dx}+H'\frac{d^2y}{dx^2},$$

then ω is determined by the Riccati equation

$$\frac{d\omega}{dx}=-1+\frac{H\omega}{H'}-\frac{\omega^2}{H'};$$

let β and β' be two independent solutions. T is determined by a similar equation, let two solutions be T and T'. Then the given equation has the general solution

$$y=e^{-\int\frac{dx}{\beta}}\left(C+\int\frac{T}{\beta}e^{\int\frac{dx}{\beta}}dx\right)+e^{-\int\frac{dx}{\beta'}}\left(C'+\int\frac{T'}{\beta'}e^{\int\frac{dx}{\beta'}}dx\right).$$

Lagrange also discovered in its general form the method of variation of parameters ‡ by means of which. if a linear equation can be solved when the term

* *Ibid.* p. 181 [*Œuvres*, 1, p. 473].

† *Misc. Taur.* 4 (1766/9), p. 173.

‡ *Nouv. Mém. Acad. Berlin*, 5 (1774), p. 201 ; 6 (1775), p. 190 [*Œuvres*, 4, pp. 9, 159]. The method had been used by Euler in 1739 in his investigations on the equation

$$\frac{d^2y}{dx^2}+ky=X.$$

It was also known to Daniel Bernoulli (*Comm. Acad. Petrop.* 13 (1741), p. 5).

independent of y and its derivatives is made zero, its solution when that term is restored can be obtained by quadratures.

On the basis of Lagrange's work d'Alembert considered the conditions under which the order of a linear differential equation could be lowered.* D'Alembert also derived a special method of dealing with the exceptional cases of the solutions of linear equations with constant coefficients, and initiated the study of linear differential systems.† His main work lies, however, in the field of partial differential equations.

A·5. Singular Solutions.—Singular solutions were discovered in a rather surprising manner. Brook Taylor ‡ set out to discover the solution of a certain differential equation which, in modern symbolism, would be written

$$(1+x^2)^2\left(\frac{dy}{dx}\right)^2=4y^3-4y^2.$$

He made the substitution

$$y=u^\lambda v^\vartheta,$$

where u and v were new variables, and λ and ϑ constants to be determined, and so transformed the equation into

$$(1+x^2)^2\left(\vartheta u\frac{dv}{dx}+\lambda v\frac{du}{dx}\right)^2=4u^{\lambda+2}v^{\vartheta+2}-4u^2v^2.$$

In this equation there are three elements whose choice is unrestricted, namely λ, ϑ and v; u is then the new dependent variable.

Firstly let

$$v=1+x^2,$$

then, after division by $(1+x^2)^2$, the equation becomes

$$\left(2\vartheta xu+\lambda v\frac{du}{dx}\right)^2=4u^{\lambda+2}v^\vartheta-4u^2.$$

Now let $\lambda=-2$, $\vartheta=1$ and the equation reduces to

$$\left(2xu-2v\frac{du}{dx}\right)^2=4v-4u^2,$$

that is

$$(1+x^2)u^2-2xuv\frac{du}{dx}+v^2\left(\frac{du}{dx}\right)^2=v,$$

or, since $v=1+x^2$,

$$u^2-2xu\frac{du}{dx}+v\left(\frac{du}{dx}\right)^2=1.$$

Now, if this equation is differentiated with respect to x, the derived equation is

$$2\frac{d^2u}{dx^2}\left(v\frac{du}{dx}-xu\right)=0$$

and breaks up into two equations namely

$$\frac{d^2u}{dx^2}=0, \qquad v\frac{du}{dx}-xu=0.$$

The first gives $\frac{du}{dx}=a$, a constant; when this value is substituted in the differential equation for u, the latter degenerates into the algebraic equation

$$(u-ax)^2=1-a^2.$$

The general solution of the original equation is therefore

$$y=\frac{v}{u^2}=\frac{1+x^2}{\{ax+\sqrt{(1-a^2)}\}^2}.$$

* *Misc. Taur.* 3 (1762/5), p. 381.
† *Hist. Acad. Berlin*, 4 (1748), p. 283.
‡ *Methodus Incrementorum* (1715), p. 26.

The second equation,

$$v\frac{du}{dx}-xu=0,$$

taken in conjunction with

$$u^2-2xu\frac{du}{dx}+v\left(\frac{du}{dx}\right)^2=1$$

gives

$$1=u^2-\frac{2x^2u^2}{v}+\frac{x^2u^2}{v}$$

or

$$v=u^2(v-x^2)=u^2,$$

and therefore

$$y=\frac{v}{u^2}=1.$$

This is truly a solution of the original equation, but it cannot be derived from the general solution by attributing a particular value to a. It is therefore a singular solution.

Nearly twenty years later Clairaut published his researches * on the class of equations with which his name is now associated. Here, also, the general and the singular solutions were arrived at by differentiation and elimination, and the fact that the singular solution was not included in the general solution was made clear. Geometrically the general solution represents a one-parameter family of straight-lines ; the singular solution represents their envelope. Closely allied to the work of Clairaut are the researches of d'Alembert † on the more general class of equations of the form

$$y=x\phi\left(\frac{dy}{dx}\right)+\psi\left(\frac{dy}{dx}\right).$$

A·6. The Equations of Mathematical Physics.—The history of formal methods of integration practically ends at the middle of the eighteenth century. In conclusion it remains but to mention the Laplace partial differential equation ‡

$$\frac{\partial^2 V}{\partial x^2}+\frac{\partial^2 V}{\partial y^2}+\frac{\partial^2 V}{\partial z^2}=0.$$

This and allied equations associated with various types of boundary conditions led to the ordinary differential equations, such as those of Legendre and Bessel which together with the hypergeometric equation suggested much of the modern analytical theory. As the power of analytical methods grew, the problem of formal integration dropped into comparative insignificance in comparison with the wider problems of the existence and validity of solutions.

* *Hist. Acad. Paris*, 1734, pp. 196–215.
† *Hist. Acad. Berlin*, 4 (1748), pp. 275–291.
‡ *Hist. Acad. Paris*, 1787, p. 252.

APPENDIX B

NUMERICAL INTEGRATION OF ORDINARY DIFFERENTIAL EQUATIONS

B·1. The Principle of the Method.—Of all ordinary differential equations of the first order only certain very special types admit of explicit integration, and when an equation which is not of one or other of these types arises in a practical problem the investigator has to fall back upon purely numerical methods of approximating to the required solution.

It will be supposed that the equation to be considered has been reduced to the first degree, and can therefore be expressed in the form

$$\frac{dy}{dx}=f(x, y).$$

It will also be supposed that the initial pair of values (x_0, y_0) is not singular with respect to the equation, and that, therefore, a solution exists which can be developed in the Taylor series,

$$k=h\left(\frac{dy}{dx}\right)_0+\frac{h^2}{2!}\left(\frac{d^2y}{dx^2}\right)_0+\frac{h^3}{3!}\left(\frac{d^3y}{dx^3}\right)_0+\frac{h^4}{4!}\left(\frac{d^4y}{dx^4}\right)_0+\ldots,$$

where

$$h=x-x_0, \quad k=y-y_0$$

and h is sufficiently small.

Now the coefficients in the Taylor series may be calculated as follows:

$$\frac{dy}{dx}=f(x, y),$$

$$\frac{d^2y}{dx^2}=\frac{\partial f}{\partial x}+f\frac{\partial f}{\partial y},$$

$$\frac{d^3y}{dx^3}=\frac{\partial^2 f}{dx^2}+2f\frac{\partial^2 f}{\partial x\partial y}+f^2\frac{\partial^2 f}{\partial y^2}+\left(\frac{\partial f}{\partial x}+f\frac{\partial f}{\partial y}\right)\frac{\partial f}{\partial y},$$

$$\cdot \quad \cdot \quad \cdot \quad \cdot \quad \cdot \quad \cdot \quad \cdot \quad \cdot \quad \cdot$$

but the increasing complexity of these expressions renders the process impracticable.

The actual method adopted in practice * is an adaptation of Gauss' method of numerical integration.† Four numbers k_1, k_2, k_3, k_4, are defined as follows:

$$k_1=hf(x_0, y_0),$$

$$k_2=hf(x_0+\alpha h, y_0+\beta k_1),$$

$$k_3=hf(x_0+\alpha_1 h, y_0+\beta_1 k_1+\gamma_1 k_2),$$

$$k_4=hf(x_0+\alpha_2 h, y_0+\beta_2 k_1+\gamma_2 k_2+\delta_2 k_3),$$

* In its original form the method is due to Runge, *Math. Ann.* 46 (1895), p. 167; later modifications are due, among others to Kutta, *Z. Math. Phys.* 46 (1901), p. 435. A detailed exposition of this and other methods is given by Runge and König, *Numeriches Rechnung* (1924), Chap. X.

† Whittaker and Robinson, *Calculus of Observations* (1924), p. 159.

where the nine constants $\alpha, \beta, \ldots, \delta_2$, and four weights R_1, R_2, R_3, R_4 are to be determined so that the expression

$$R_1k_1+R_2k_2+R_3k_3+R_4k_4$$

agrees with the Taylor series up to and including the term in h^4.

B·2. Equations connecting the Constants.—The expressions k_2, k_3, k_4 are developed in powers of h by making use of the Taylor expansion in two variables

$$f(x+p, y+q)=f(x, y)+Df(x, y)+\frac{1}{2!}D^2f(x, y)+\frac{1}{3!}D^3f(x, y)+\ldots,$$

where

$$D^rf(x, y)=\left(p\frac{\partial}{\partial x}+q\frac{\partial}{\partial y}\right)^r f(x, y).$$

Thus, to evaluate k_2, let

$$D_1=\alpha\frac{\partial}{\partial x}+\beta f_0\frac{\partial}{\partial y},$$

where $f_0=f(x_0, y_0)$, then

$$\alpha h\frac{\partial}{\partial x}+\beta k_1\frac{\partial}{\partial y}=h\left(\alpha\frac{\partial}{\partial x}+\beta f_0\frac{\partial}{\partial y}\right)=hD_1,$$

and therefore

$$k_2=h\left[f+hD_1f+\frac{h^2}{2!}D_1^2f+\frac{h^3}{3!}D_1^3f+\ldots\right]_0.$$

To evaluate k_3, let

$$D_2=\alpha_1\frac{\partial}{\partial x}+(\beta_1+\gamma_1)f_0\frac{\partial}{\partial y},$$

then

$$\begin{aligned}\alpha_1h\frac{\partial}{\partial x}+(\beta_1k_1+\gamma_1k_2)\frac{\partial}{\partial y}&=h\left(\alpha_1\frac{\partial}{\partial x}+\beta_1f_0\frac{\partial}{\partial y}\right)+\gamma_1k_2\frac{\partial}{\partial y}\\&=hD_2+\gamma_1(k_2-hf_0)\frac{\partial}{\partial y}\\&=hD_2+h^2\gamma_1\left[D_1f+\frac{h}{2!}D_1^2f+\ldots\right]_0\frac{\partial}{\partial y},\end{aligned}$$

and therefore

$$\begin{aligned}k_3=h\Big[f+hD_2f+\frac{h^2}{2!}D_2^2f+\frac{h^3}{3!}D_2^3f+\ldots\\+h^2\gamma_1\Big\{f_yD_1f+\frac{h}{2!}f_yD_1^2f+hD_1f.D_2f_y+\ldots\Big\}\Big]_0,\end{aligned}$$

where

$$f_y=\frac{\partial f}{\partial y}.$$

Lastly, to evaluate k_4, let

$$D_3=\alpha_2\frac{\partial}{\partial x}+(\beta_2+\gamma_2+\delta_2)f_0\frac{\partial}{\partial y},$$

then

$$\alpha_2h\frac{\partial}{\partial x}+(\beta_2k_1+\gamma_2k_2+\delta_2k_3)\frac{\partial}{\partial y}=hD_3+\Big\{\gamma_2(k_2-hf_0)+\delta_2(k_3-hf_0)\Big\}\frac{\partial}{\partial y}$$

$$=hD_3+h^2\left[\gamma_2\Big\{D_1f+\frac{h}{2!}D_1^2f+\ldots\Big\}+\delta_2\Big\{D_2f+\frac{h}{2!}(D_2^2f+2\gamma_1f_yD_1f)+\ldots\Big\}\right]_0\frac{\partial}{\partial y},$$

and therefore

$$\begin{aligned}k_4=h\Big[f+hD_3f+\frac{h^2}{2!}D_3^2f+\frac{h^3}{3!}D_3^3f+\ldots\\+h^2(\gamma_2D_1f+\delta_2D_2f)f_y+\frac{h^3}{2!}(\gamma_2D_1^2f+\delta_2D_2^2f+2\gamma_1\delta_2f_yD_1f)f_y\\+h^3(\gamma_2D_1f+\delta_2D_2f)D_3f_y+\ldots\Big]_0.\end{aligned}$$

Now k itself has the development

$$k=\left[hf+\frac{h^2}{2!}Df+\frac{h^3}{3!}(D^2f+f_yDf)+\frac{h^4}{4!}(D^3f+f_yD^2f+f_y{}^2Df+3DfDf_y)+\ .\ .\ .\right]_0,$$

where

$$D=\frac{\partial}{\partial x}+f_0\frac{\partial}{\partial y}$$

and this development is to agree, as far as the terms in k^4 inclusive, with that of

$$R_1k_1+R_2k_2+R_3k_3+R_4k_4,$$

whatever may be the function $f(x, y)$.

Now, to the order in question there are eight terms in the development of k; if each of these terms is equated to the corresponding term in the development of $R_1k_1+\ .\ .\ .\ +R_4k_4$ the following eight relations must hold:

$$\begin{aligned}
R_1+R_2\quad +R_3\qquad\qquad +R_4\qquad\qquad &= 1,\\
R_2D_1f+R_3D_2f\qquad +R_4D_3f\qquad &= \frac{1}{2!}Df,\\
R_2D_1{}^2f+R_3D_2{}^2f\qquad +R_4D_3{}^2f\qquad &= \frac{2!}{3!}D^2f,\\
R_2D_1{}^3f+R_3D_2{}^3f\qquad +R_4D_3{}^3f\qquad &= \frac{3!}{4!}D^3f,\\
R_3\gamma_1D_1f\qquad +R_4(\gamma_2D_1f+\delta_2D_2f) &= \frac{1}{3!}Df,\\
R_3\gamma_1D_1{}^2f\qquad +R_4(\gamma_2D_1{}^2f+\delta_2D_2{}^2f) &= \frac{2!}{4!}D^2f,\\
R_3\gamma_1D_1fD_2f_y+R_4(\gamma_2D_1f+\delta_2D_2f)D_3f_y &= \frac{3}{4!}DfDf_y,\\
R_4\gamma_1\delta_2D_1f\qquad &= \frac{1}{4!}Df.
\end{aligned}$$

These equations are homogeneous in the operators D_1, D_2, D_3, D with constant coefficients. These operators must therefore bear a constant ratio to one another which can only be the case if

$$\begin{aligned}
a&=\beta,\\
a_1&=\beta_1+\gamma_1,\\
a_2&=\beta_2+\gamma_2+\delta_2,
\end{aligned}$$

and consequently

$$D_1=aD,\quad D_2=a_1D,\quad D_3=a_2D.$$

In view of these relations the eight equations assume a form independent of the function $f(x, y)$, namely

$$\text{(A)}\quad\left\{\begin{aligned}
R_1+R_2\quad +R_3\qquad +R_4\qquad &= 1,\\
R_2a\ +R_3a_1\qquad +R_4a_2\qquad &= \tfrac{1}{2},\\
R_2a^2+R_3a_1{}^2\qquad +R_4a_2{}^2\qquad &= \tfrac{1}{3},\\
R_2a^3+R_3a_1{}^3\qquad +R_4a_2{}^3\qquad &= \tfrac{1}{4},\\
R_3a\gamma_1\qquad +R_4(a\gamma_2+a_1\delta_2) &= \tfrac{1}{6},\\
R_3a^2\gamma_1\qquad +R_4(a^2\gamma_2+a_1{}^2\delta_2) &= \tfrac{1}{12},\\
R_3aa_1\gamma_1+R_4(a\gamma_2+a_1\delta_2)a_2 &= \tfrac{1}{8},\\
R_4a\gamma_1\delta_2\qquad &= \tfrac{1}{24}.
\end{aligned}\right.$$

Thus between the thirteen unknowns R_1, . . ., R_4, a, . . ., δ_2 there are eleven equations, so that two further consistent relations may be set up between the unknowns.

B·3. Determination of the Constants.—To the fourth of the equations (A), add the second multiplied by aa_2 and the third multiplied by $-(a+a_2)$, then

(B) $$R_3 a_1(a-a_1)(a_2-a_1) = \frac{aa_2}{2} - \frac{a+a_2}{3} + \tfrac{1}{4}.$$

From the fifth and seventh equations it follows that

(C) $$R_3 a \gamma_1 (a_2-a_1) = \frac{a_2}{6} - \tfrac{1}{8},$$

and from the fifth and sixth :

$$R_4 a_1 \delta_2 (a_1-a) = \tfrac{1}{12} - \frac{a}{6}.$$

When R_4 is eliminated between this equation and the eighth of the set (A), it is found that

$$a\gamma_1 = \frac{a_1(a-a_1)}{2(2a-1)},$$

and, substituting this expression in (C),

$$R_3 a_1(a-a_1)(a_2-a_1) = (2a-1)\left(\frac{a_2}{3} - \tfrac{1}{4}\right).$$

Finally, comparing this equation with (B),

$$\frac{aa_2}{2} - \frac{a+a_2}{3} + \tfrac{1}{4} = (2a-1)\left(\frac{a_2}{3} - \tfrac{1}{4}\right) = \frac{2aa_2}{3} - \frac{a}{2} - \frac{a_2}{3} + \tfrac{1}{4},$$

whence

$$aa_2 = a.$$

Now it is clear from the eighth of the equations (A) that a cannot be zero, it therefore follows that

$$a_2 = 1.$$

The same equation shows that R_4 cannot be zero, and it is now evident from equation (C) that R_3 cannot be zero.

The first four equations of the set (A) determine R_1, R_2, R_3, R_4 uniquely in terms of a and a_1 provided that their determinant, which, since $a_2=1$, has the value

(D) $$aa_1(a-a_1)(a_1-1)(1-a),$$

does not vanish. The values found are

(E) $$\begin{cases} R_1 = \tfrac{1}{2} + \dfrac{1-2(a+a_1)}{12aa_1}, & R_2 = \dfrac{2a_1-1}{12a(a_1-a)(1-a)}, \\ R_3 = \dfrac{1-2a}{12a_1(a_1-a)(1-a_1)}, & R_4 = \tfrac{1}{2} + \dfrac{2(a+a_1)-3}{12(1-a)(1-a_1)}. \end{cases}$$

The fifth, sixth and seventh of equations (A) now determine γ_1, γ_2 and δ_2 in terms of a and a_1 provided that their determinant

(F) $$\begin{vmatrix} R_3 a & , & R_4 a & , & R_4 a_1 \\ R_3 a^2 & , & R_4 a^2 & , & R_4 a_1{}^2 \\ R_3 a a_1 & , & R_4 a & , & R_4 a_1 \end{vmatrix} = R_3 R_4{}^2 a^2 a_1(a_1-a)(a_1-1)$$

does not vanish. The values obtained are

(G) $$\begin{cases} \gamma_1 = \dfrac{a_1(a_1-a)}{2a(1-2a)}, \\ \gamma_2 = \dfrac{(1-a)\{a+a_1-1-(2a_1-1)^2\}}{2a(a_1-a)\{6aa_1-4(a+a_1)+3\}}, \\ \delta_2 = \dfrac{(1-2a)(1-a)(1-a_1)}{a_1(a_1-a)\{6aa_1-4(a+a_1)+3\}}. \end{cases}$$

Finally β, β_1 and β_2 are obtained from the equations

$$\text{(H)} \qquad \begin{cases} \beta = a, \\ \beta_1 = a_1 - \gamma_1, \\ \beta_2 = 1 - \gamma_2 - \delta_2. \end{cases}$$

Thus the six coefficients β, β_1, β_2, γ_1, γ_2, δ, and the four weights R_1, R_2, R_3, R_4 are expressed in terms of a and a_1 which may be regarded as arbitrary.

B·4. Particular Values of the Coefficients and Weights.—Any two conditions, consistent with the previous equations, may be imposed. For example, a symmetrical expression for k is obtained if

$$R_1 = R_4, \quad R_2 = R_3.$$

This is, however, equivalent to the single condition

$$a + a_1 = 1,$$

under which the weights and coefficients take the simple form

$$\text{(K)} \qquad \begin{cases} 12R_1 = 12R_4 = 6 - \dfrac{1}{aa_1}, \quad 12R_2 = 12R_3 = \dfrac{1}{aa_1}, \\ a + a_1 = 1, \quad a_2 = 1, \\ \gamma_1 = \dfrac{a_1}{2a}, \quad \gamma_2 = \dfrac{a_1(a - a_1)}{2a(6aa_1 - 1)}, \quad \delta_2 = \dfrac{a}{6aa_1 - 1}. \end{cases}$$

The second condition may be imposed by supposing the range $(x_0, x_0 + h)$ to be divided into three equal parts so that $a = \frac{1}{3}$, $a_1 = \frac{2}{3}$. Then

$$\begin{aligned} &R_1 = \tfrac{1}{8}, \\ &R_2 = \tfrac{3}{8}, \quad a = \tfrac{1}{3}, \quad \beta = \tfrac{1}{3}, \\ &R_3 = \tfrac{3}{8}, \quad a_1 = \tfrac{2}{3}, \quad \beta_1 = -\tfrac{1}{3}, \quad \gamma_1 = 1, \\ &R_4 = \tfrac{1}{8}, \quad a_2 = 1, \quad \beta_2 = 1, \quad \gamma_2 = -1, \quad \delta_2 = 1. \end{aligned}$$

This gives the formulæ due to Kutta :

$$\begin{aligned} k_1 &= hf(x_0, y_0), \\ k_2 &= hf(x_0 + \tfrac{1}{3}h, y_0 + \tfrac{1}{3}k_1), \\ k_3 &= hf(x_0 + \tfrac{2}{3}h, y_0 - \tfrac{1}{3}k_1 + k_2), \\ k_4 &= hf(x_0 + h, y_0 + k_1 - k_2 + k_3), \\ k &= \tfrac{1}{8}(k_1 + 3k_2 + 3k_3 + k_4). \end{aligned}$$

It is interesting and important to examine the cases which arise when the determinants (D) and (F) vanish. There are three, and only three, possible cases in which the solutions are finite, namely

(i) $a = a_1$, (ii) $a = 1$, (iii) $a_1 = 0$.

The first case, for the finiteness of R_2 and R_3, implies the further condition

$$a = a_1 = \tfrac{1}{2};$$

either R_2 or R_3 may now be regarded as arbitrary, but

$$R_2 + R_2 = \tfrac{2}{3}.$$

Let $R_2 = R_3 = \frac{1}{3}$, then

$$\begin{aligned} &R_1 = \tfrac{1}{6}, \\ &R_2 = \tfrac{1}{3}, \quad a = \tfrac{1}{2}, \quad \beta = \tfrac{1}{2}, \\ &R_3 = \tfrac{1}{3}, \quad a_1 = \tfrac{1}{2}, \quad \beta_1 = 0 \quad \gamma_1 = \tfrac{1}{2}, \\ &R_4 = \tfrac{1}{6}, \quad a_2 = 1, \quad \beta_2 = 0, \quad \gamma_2 = 0, \quad \delta_2 = 1. \end{aligned}$$

This gives rise to a very convenient set of formulæ, due to Runge :

$$\begin{aligned} k_1 &= hf(x_0, y_0), \\ k_2 &= hf(x_0 + \tfrac{1}{2}h, y_0 + \tfrac{1}{2}k_1), \\ k_2 &= hf(x_0 + \tfrac{1}{2}h, y_0 + \tfrac{1}{2}k_2), \\ k_3 &= hf(x_0 + h, y_0 + k_3), \\ k &= \tfrac{1}{6}(k_1 + 2k_2 + 2k_3 + k_4). \end{aligned}$$

When the equation to be integrated takes the special form

$$\frac{dy}{dx}=f(x),$$

Runge's formula reduces to Simpson's rule :

$$k=\int_{x_0}^{x_0+h} f(x)dx=\tfrac{1}{6}h\{f(x_0)+4f(x_0+\tfrac{1}{2}h)+f(x_0+h)\}.$$

The second and third cases do not lead to formulæ of any particular importance.

B·5. Arrangement of the Work.—The practical problem may be stated as follows : To tabulate the solution of the differential equation

$$\frac{dy}{dx}=f(x, y)$$

which reduces to $y=y_0$ when $x=x_0$, the tabular interval being h. Let $x_r=x_0+rh$, and let y_r be the corresponding value of y

Runge's formula is, on account of its particular simplicity, adopted as the standard, and the work of evaluating y_1 is carried out in the following self-explanatory scheme :—

x	y	f	$f\times h$	
x_0	y_0	$f(x_0, y_0)$	k_1	$\tfrac{1}{2}(k_1+k_4)$
$x_0+\tfrac{1}{2}h$	$y_0+\tfrac{1}{2}k_1$	$f(x_0+\tfrac{1}{2}h, y_0+\tfrac{1}{2}k_1)$	k_2	k_2+k_3
$x_0+\tfrac{1}{2}h$	$y_0+\tfrac{1}{2}k_2$	$f(x_0+\tfrac{1}{2}h, y_0+\tfrac{1}{2}k_2)$	k_3	sum.
x_0+h	y_0+k_3	$f(x_0+h, y_0+k_3)$	k_4	$k=\tfrac{1}{3}$ sum.
x_1	y_1			

The work is repeated with (x_1, y_1) as the pair of initial values, giving y_2, and so on.

So far no estimate of the error due to the neglecting of terms in h^5 and higher terms has been made. An estimate of the error, when h is reasonably small, may be made by repeating the working with the double interval $2h$. Let ε be the error in y_1, so that approximately

$$\varepsilon=ch^5,$$

where c is a constant. Then the error in y_2, calculated in two stages, is $2\varepsilon=2ch^5$. On the other hand the error in y_2, calculated in one stage, is

$$\varepsilon'=32ch^5,$$

and therefore

$$\begin{aligned}2\varepsilon&=\tfrac{1}{15}(\varepsilon'-2\varepsilon)\\&=\tfrac{1}{15}(y_2'-y_2),\end{aligned}$$

where y_2 is the value determined by two stages, and y_2' the value determined in a single stage.

The process will be illustrated by calculating the value for $x=0{\cdot}4$ of the solution of

$$\frac{dy}{dx}=y^2+x$$

which reduces to zero for $x=0$. When the calculation is performed first by two steps and then in a single step, the working is as follows :

Tabular Interval : $h=0{\cdot}2$.

x	y	f	hf	
0	0	0	0	·020040
0·1	0	·100000	·020000	·040020
0·1	·010000	·100100	·020020	·060060
0·2	·020020	·200401	·040080	·020020
0·2	·020020	·200401	·040080	·060688
0·3	·040060	·301605	·060321	·120825
0·3	·050180	·302518	·060504	·181513
0·4	·080524	·406484	·081297	·060504
0·4	·080524			

Tabular Interval : $h=0{\cdot}4$.

x	y	f	hf	
0	0	0	0	·081301
0·2	0	·200000	·080000	·160640
0·2	·040000	·201600	·080640	·241941
0·4	·080640	·406503	·162601	·080647
0·4	·080647			

The difference between the two determinations is ·000125, which points to an error of roughly ·000008 in the first determination. The errors are both in excess, and therefore the corrected value is

$$0{\cdot}080516.$$

It may easily be verified that the solution in question may be developed as follows :

$$y=\tfrac{1}{2}x^2+\tfrac{1}{20}x^5+\tfrac{1}{160}x^8+\tfrac{1}{4400}x^{11}+\ .\ .\ .$$

and that the true value of y, for $x=0{\cdot}4$, is

$$0{\cdot}0805161\ .\ .\ ..$$

B·6. Extension to Systems of Equations.—The foregoing processes of numerical integration may be extended to systems of any number of equations of the first order, and therefore to equations of order higher than the first. For a system of two equations

$$\frac{dy}{dx}=f(x, y, z), \quad \frac{dz}{dx}=g(x, y, z),$$

if the initial conditions are that

$$y=y_0, \quad z=z_0, \quad \text{when} \quad x=x_0$$

then Runge's formulæ for the increments k and l which y_0 and z_0 receive when x_0 is increased by h are

$$k_1=hf(x_0, y_0, z_0), \qquad l_1=hg(x_0, y_0, z_0),$$
$$k_2=hf(x_0+\tfrac{1}{2}h, y_0+\tfrac{1}{2}k_1, z_0+\tfrac{1}{2}l_1), \qquad l_2=hg(x_0+\tfrac{1}{2}h, y_0+\tfrac{1}{2}k_1, z_0+\tfrac{1}{2}l_1),$$
$$k_3=hf(x_0+\tfrac{1}{2}h, y_0+\tfrac{1}{2}k_2, z_0+\tfrac{1}{2}l_2), \qquad l_3=hg(x_0+\tfrac{1}{2}h, y_0+\tfrac{1}{2}k_2, z_0+\tfrac{1}{2}k_2),$$
$$k_4=hf(x_0+h, y_0+k_3, y_0+l_3), \qquad l_4=hg(x_0+h, y_0+k_3, z_0+l_3),$$
$$k=\tfrac{1}{6}(k_1+2k_2+2k_3+k_4),$$
$$l=\tfrac{1}{6}(l_1+2l_2+2l_3+l_4).$$

Examples for Solution.

1. Given the differential equation

$$\frac{dy}{dx}=\frac{y-x}{y+x},$$

with the initial conditions $x_0=0$, $y_0=1$, tabulate y for $x=0{\cdot}2, 0{\cdot}4, \ldots, 1{\cdot}2$ to six places of decimals. [The accurate solution

$$\tfrac{1}{2}\log(x^2+y^2)+\text{arc}\tan\frac{y}{x}=\frac{\pi}{2}$$

gives $y=1{\cdot}1678417$ when $x=1{\cdot}2$.]

2. For the above range of values, and initial conditions tabulate the solution of

$$\frac{dy}{dx}=\frac{1}{10}y^2-xy.$$

[When $x=1{\cdot}2$ the value of y correct to seven places of decimals is $0{\cdot}5387334$. This example is treated in detail in Runge-König, *Numerisches Rechnung*.]

3. The equation of the Bessel functions of order zero

$$x\frac{d^2y}{dx^2}+\frac{dy}{dx}+xy=0$$

is equivalent to the system

$$\frac{dy}{dx}=\frac{z}{x}, \quad \frac{dz}{dx}=-xy,$$

and the solution

$$y=J_0(x)$$

corresponds to the initial conditions

$$x_0=0, \quad y_0=1, \quad z_0=0.$$

Tabulate $J_0(x)$, to five places of decimals, at intervals of $0{\cdot}2$ from $x=0$ to $x=1{\cdot}2$.

APPENDIX C

LIST OF JOURNALS QUOTED IN FOOTNOTES TO THE TEXT

[For fuller information concerning these Journals, and for a list of the libraries in which they may be found, the *Catalogue of Current Mathematical Journals, etc.*, published by the Mathematical Association, may be consulted.]

Abh. Akad. Wiss. Berlin . . . Abhandlungen der königlichen Akademie der Wissenschaften in Berlin.

Abh. Ges. Wiss. Gött. . . . . Abhandlungen der königlichen Gesellschaft der Wissenschaften zu Göttingen [continuation of *Comm. Gott.*].

Acta Erud. . . . . . . Acta Eruditorum publicata Lipsiæ.

Acta Erud. Suppl. . . . . . Acta Eruditorum quæ Lipsiæ publicantur Supplementa.

Acta Math. . . . . . . Acta Mathematica, Stockholm.

Acta Soc. Sc. Fenn. . . . . . Acta Societatis Scientiarum Fennicæ, Helsingfors.

Am. J. Math. . . . . The American Journal of Mathematics, Baltimore, Md.

Ann. di Mat. . . . . . . Annali di Matematica pura ed applicata, Rome and Milan.

Ann. Éc. Norm. . . . . . Annales scientifiques de l'École Normale supérieure, Paris.

Ann. Fac. Sc. Toulouse . . . Annales de la Faculté des Sciences de Toulouse.

Ann. of Math. . . . . . . Annals of Mathematics, Princeton, N.J.

Ann. Scuola Norm. Pisa . . . Annali della R. Scuola Normale superiore di Pisa.

Archiv d. Math. u. Phys. . . . Archiv der Mathematik und Physik (Grunert's Archiv), Greifswald and Leipzig.

Archiv for Math. . . . . . Archiv for Mathematik og Naturvidenskab, Christiania (Oslo).

Arkiv för Mat. . . . . . . Arkiv för Matematik, Astronomi och Fysik, Stockholm.

Bibl. Math. . . . . . . Bibliotheca Mathematica, Stockholm and Leipzig.

Bull. Am. Math. Soc. . . . . Bulletin of the American Mathematical Society, Lancaster, Pa., and New York.

Bull. Acad. Sc. Belg. . . . . Bulletins de l'Académie royale des Sciences de Belgique, Brussels.

Bull. Sc. Math. . . . . . . Bulletin des Sciences mathématiques, Paris.

Bull. Soc. Math. France . . . Bulletin de la Société mathématique de France, Paris.

Bull. Soc. Philomath. Paris . . Bulletin de la Société philomathique de Paris.

Camb. Math. J. . . . . . The Cambridge Mathematical Journal.

Comm. Acad. Petrop. . . . . Commentarii Academiæ Scientiarum Imperialis Petropolitanæ [Continued as *Novi Comm.*].

Comm. Gott.	Commentarii Societatis Regiæ Scientiarum Gottingensis. [Continued successively as Novi Commentarii, Commentationes and Commentationes recentiores.]
Comm. Math. Soc. Kharkov . .	Communications and Proceedings of the Mathematical Society of the Imperial University of Kharkov.
C.R. Acad. Sc. Paris. . . .	Comptes Rendus hebdomadaires des Séances de l'Académie des Sciences, Paris.
Forhand. Vid.-Selsk. Christiania .	Forhandlinger i Videnskabs-Selskabet i Christiania (Oslo).
Gött. Nach.	Nachrichten von der königlichen Gesellschaft der Wissenschaften zu Göttingen.
Hist. Acad. Berlin	Histoire de l'Académie royale des Sciences et des Belles-Lettres de Berlin.
Hist. Acad. Paris	Histoire de l'Académie royale des Sciences, Paris.
J. de Math.	Journal de Mathématiques pures et appliquées (Liouville), Paris.
J. Éc. Polyt.	Journal de l'École Polytechnique, Paris. [Reference is made to the *Cahier*, each of which is separately paged. The number of *cahiers* to the volume is irregular.]
J. für Math.	Journal für die reine und angewandte Mathematik (Crelle's Journal), Berlin.
Math. Ann.	Mathematische Annalen, Leipzig.
Mathésis	Mathésis, Recueil Mathématique, Gand and Paris.
Mém. Acad. Sc. Paris . . .	Mémoires de l'Académie des Sciences de l'Institut de France ; since 1805, Mémoires présentés par divers savants . . ..
Mess. Math.	The Messenger of Mathematics, London and Cambridge.
Misc. Berol.	Miscellanea Berolinensia, Berlin.
Misc. Taur.	Miscellanea Taurinensia, Turin.
Monatsh. Math. Phys. . . .	Monatshefte für Mathematik und Physik, Vienna.
Nouv. Mém. Acad. Berlin . .	Nouveaux Mémoires de l'Académie royale des Sciences et Belles-Lettres, Berlin. [Continuation of *Hist. Acad. Berlin.*]
Öfv. Vet.-Akad. Stockholm . .	Öfversigt af Kongliga Vetenskaps-Akademiens Förhandlingar, Stockholm.
Phil. Trans. R.S.	Philosophical Transactions of the Royal Society of London.
Proc. Am. Acad.	Proceedings of the American Academy of Arts and Sciences, Boston, Mass.
Proc. Camb. Phil. Soc. . . .	Proceedings of the Cambridge Philosophical Society.
Proc. Edin. Math. Soc. . . .	Proceedings of the Edinburgh Mathematical Society.
Proc. London Math. Soc. . .	Proceedings of the London Mathematical Society.
Proc. Roy. Soc. Edin. . . .	Proceedings of the Royal Society of Edinburgh.

Quart. J. Math.	The Quarterly Journal of Pure and Applied Mathematics, London.
Rend. Accad. Lincei	Atti della R. Accademia dei Lincei, Rendiconti, Rome.
Rend. Circ. Mat. Palermo . .	Rendiconti del Circolo Matematico di Palermo.
Rend. Ist. Lombard.	Reale Istituto Lombardo di Scienze e Lettere, Rendiconti, Milan.
Sitz. Akad. Wiss. Berlin . . .	Sitzungsberichte der königlichen preussischen Akademie der Wissenschaften, Berlin.
Trans. Am. Math. Soc. . . .	Transactions of the American Mathematical Society, Lancaster, Pa. and New York.
Trans. Camb. Phil. Soc. . .	Transactions of the Cambridge Philosophical Society.
Trans. Roy. Soc. Edin. . . .	Transactions of the Royal Society of Edinburgh.
Z. Math. Phys.	Zeitschrift für Mathematik und Physik, Leipzig.

APPENDIX D

BIBLIOGRAPHY

I. Treatises.

(1) Forsyth, A. R., *Theory of Differential Equations*, Cambridge, 1900–1902, six volumes, of which the first four deal with ordinary differential equations, namely :

Vol. I. *Exact Equations and Pfaff's Problem.*
Vols. II., III. *Ordinary Equations, not Linear.*
Vol. IV. *Ordinary Linear Equations.*

(2) Craig, T., *Treatise on Linear Differential Equations*, New York, 1889.

(3) Page, J. M., *Ordinary Differential Equations, with an Introduction to Lie's Theory of Groups of One Parameter*, London, 1897.

(4) Bateman, H., *Differential Equations*, London, 1918.

(5) Goursat, E. *Cours d'Analyse mathématique*, Paris, Tome II. (4th ed. 1924) and Tome III. (3rd ed. 1922).

(5a) *A Course in Mathematical Analysis*, translated by E. R. Hendrick and O. Dunkel, Vol. II., part 2, Boston, 1917.

(6) Jordan, C., *Cours d'Analyse de l'École Polytechnique*, Paris, Tome III. (3rd ed. 1915).

(7) Picard, E., *Traité d'Analyse*, Paris, Tome II. (3rd ed. 1926), Tome III. (2nd ed. 1908).

(8) Schlesinger, L., *Einfühung in die Theorie der Differentialgleichungen auf funktiontheoretischer Grundlage*, Berlin and Leipzig (3rd ed. 1922 ; a revised version of Sammlung Schubert XIII.).

(9) Schlesinger, L., *Handbuch der Theorie der linearen Differentialgleichungen*, Leipzig, Band I. 1895, Band II_1. 1897, Band II_2. 1898.

(10) Schlesinger, L., *Vorlesungen über lineare Differentialgleichungen*, Leipzig and Berlin, 1908.

(11) Kœnigsberger, L., *Lehrbuch der Theorie der Differentialgleichungen*, Leipzig, 1889.

(12) Heffter, L., *Einleitung in die Theorie der linearen Differentialgleichungen mit einer unabhängigen Variabeln*, Leipzig, 1894.

(13) Horn, J., *Gewöhnliche Differentialgleichungen beliebiger Ordnung*, Leipzig, 1905 (Sammlung Schubert L.).

(14) Bieberbach, L., *Theorie der Differentialgleichungen*, Berlin, 1923.

II. Monographs.

(1) *Enzyklopädie der Mathematischen Wissenschaften*, Leipzig:

II. A 4a. Painlevé, P., *Gewöhnliche Differentialgleichungen; Existenz der Lösungen*, 1900.

II. A 4b. Vessiot, E., *Gewöhnliche Differentialgleichungen; Elementare Integrationsmethoden*, 1900.

[These are reproduced, in an improved form, in the *Encyclopédie des Sciences mathématiques*, Paris and Leipzig, Tome II. vol. 3, fasc. 1, 1910.]

II. A 7a. Bôcher, M., *Randwertaufgaben bei gewöhnlichen Differentialgleichungen*, 1900.

II. B 5. Hilb, E., *Lineare Differentialgleichungen im komplexen Gebiet*, 1915.

II. B 6. Hilb, E., *Nichtlineare Differentialgleichungen*, 1921.

III. D 8. Liebmann, H., *Geometrische Theorie der Differentialgleichungen*, 1916.

(2) Klein, F., *Über lineare Differentialgleichungen der zweiten Ordnung*. Göttingen, 1914 (autographed; a printed edition is said to be in preparation).

(3) Bôcher, M., *Über die Reihenentwickelungen der Potentialtheorie*, Leipzig, 1894.

(4) Painlevé, P., *Leçons sur la théorie analytique des équations différentielles, professées à Stockholm*, Paris, 1897 (lithographed).

(5) Boutroux, P., *Leçons sur les fonctions définies par les équations differentielles de premier ordre*, Paris, 1908.

(6) Bôcher, M. *Leçons sur les méthodes de Sturm dans la théorie des équations différentielle linéaires et leurs développements modernes*, Paris, 1917.

[Several monographs in preparation for the Series: *Mémorial des Sciences Mathématiques* deal with various aspects of the theory of ordinary differential equations.]

INDEX OF AUTHORS

[*As a rule no reference is made to authors of current text-books quoted in the text.*]

GENERAL INDEX

[*Numbers refer to pages; those in italics relate to special examples.*]

THE END

A CATALOGUE OF SELECTED DOVER SCIENCE BOOKS

A CATALOGUE OF SELECTED DOVER BOOKS IN ALL FIELDS OF INTEREST

CELESTIAL OBJECTS FOR COMMON TELESCOPES, T. W. Webb. The most used book in amateur astronomy: inestimable aid for locating and identifying nearly 4,000 celestial objects. Edited, updated by Margaret W. Mayall. 77 illustrations. Total of 645pp. 5⅜ x 8½.
20917-2, 20918-0 Pa., Two-vol. set $8.00

HISTORICAL STUDIES IN THE LANGUAGE OF CHEMISTRY, M. P. Crosland. The important part language has played in the development of chemistry from the symbolism of alchemy to the adoption of systematic nomenclature in 1892. ". . . wholeheartedly recommended,"—Science. 15 illustrations. 416pp. of text. 5⅝ x 8¼. 63702-6 Pa. $6.00

BURNHAM'S CELESTIAL HANDBOOK, Robert Burnham, Jr. Thorough, readable guide to the stars beyond our solar system. Exhaustive treatment, fully illustrated. Breakdown is alphabetical by constellation: Andromeda to Cetus in Vol. 1; Chamaeleon to Orion in Vol. 2; and Pavo to Vulpecula in Vol. 3. Hundreds of illustrations. Total of about 2000pp. 6⅛ x 9¼.
23567-X, 23568-8, 23673-0 Pa., Three-vol. set $26.85

THEORY OF WING SECTIONS: INCLUDING A SUMMARY OF AIRFOIL DATA, Ira H. Abbott and A. E. von Doenhoff. Concise compilation of subatomic aerodynamic characteristics of modern NASA wing sections, plus description of theory. 350pp. of tables. 693pp. 5⅜ x 8½.
60586-8 Pa. $6.50

DE RE METALLICA, Georgius Agricola. Translated by Herbert C. Hoover and Lou H. Hoover. The famous Hoover translation of greatest treatise on technological chemistry, engineering, geology, mining of early modern times (1556). All 289 original woodcuts. 638pp. 6¾ x 11.
60006-8 Clothbd. $17.50

THE ORIGIN OF CONTINENTS AND OCEANS, Alfred Wegener. One of the most influential, most controversial books in science, the classic statement for continental drift. Full 1966 translation of Wegener's final (1929) version. 64 illustrations. 246pp. 5⅜ x 8½. 61708-4 Pa. $3.00

THE PRINCIPLES OF PSYCHOLOGY, William James. Famous long course complete, unabridged. Stream of thought, time perception, memory, experimental methods; great work decades ahead of its time. Still valid, useful; read in many classes. 94 figures. Total of 1391pp. 5⅜ x 8½.
20381-6, 20382-4 Pa., Two-vol. set $13.00

TONE POEMS, SERIES II: TILL EULENSPIEGELS LUSTIGE STREICHE, ALSO SPRACH ZARATHUSTRA, AND EIN HELDENLEBEN, Richard Strauss. Three important orchestral works, including very popular *Till Eulenspiegel's Marry Pranks*, reproduced in full score from original editions. Study score. 315pp. 9⅜ x 12¼. (Available in U.S. only)
23755-9 Pa. $7.50

TONE POEMS, SERIES I: DON JUAN, TOD UND VERKLARUNG AND DON QUIXOTE, Richard Strauss. Three of the most often performed and recorded works in entire orchestral repertoire, reproduced in full score from original editions. Study score. 286pp. 9⅜ x 12¼. (Available in U.S. only) 23754-0 Pa. $7.50

11 LATE STRING QUARTETS, Franz Joseph Haydn. The form which Haydn defined and "brought to perfection." *(Grove's)*. 11 string quartets in complete score, his last and his best. The first in a projected series of the complete Haydn string quartets. Reliable modern Eulenberg edition, otherwise difficult to obtain. 320pp. 8⅜ x 11¼. (Available in U.S. only)
23753-2 Pa. $6.95

FOURTH, FIFTH AND SIXTH SYMPHONIES IN FULL SCORE, Peter Ilyitch Tchaikovsky. Complete orchestral scores of Symphony No. 4 in F Minor, Op. 36; Symphony No. 5 in E Minor, Op. 64; Symphony No. 6 in B Minor, "Pathetique," Op. 74. Bretikopf & Hartel eds. Study score. 480pp. 9⅜ x 12¼. 23861-X Pa. $10.95

THE MARRIAGE OF FIGARO: COMPLETE SCORE, Wolfgang A. Mozart. Finest comic opera ever written. Full score, not to be confused with piano renderings. Peters edition. Study score. 448pp. 9⅜ x 12¼. (Available in U.S. only) 23751-6 Pa. $11.95

"IMAGE" ON THE ART AND EVOLUTION OF THE FILM, edited by Marshall Deutelbaum. Pioneering book brings together for first time 38 groundbreaking articles on early silent films from *Image* and 263 illustrations newly shot from rare prints in the collection of the International Museum of Photography. A landmark work. Index. 256pp. 8¼ x 11.
23777-X Pa. $8.95

AROUND-THE-WORLD COOKY BOOK, Lois Lintner Sumption and Marguerite Lintner Ashbrook. 373 cooky and frosting recipes from 28 countries (America, Austria, China, Russia, Italy, etc.) include Viennese kisses, rice wafers, London strips, lady fingers, hony, sugar spice, maple cookies, etc. Clear instructions. All tested. 38 drawings. 182pp. 5⅜ x 8.
23802-4 Pa. $2.50

THE ART NOUVEAU STYLE, edited by Roberta Waddell. 579 rare photographs, not available elsewhere, of works in jewelry, metalwork, glass, ceramics, textiles, architecture and furniture by 175 artists—Mucha, Seguy, Lalique, Tiffany, Gaudin, Hohlwein, Saarinen, and many others. 288pp. 8⅜ x 11¼. 23515-7 Pa. $6.95

THE DEPRESSION YEARS AS PHOTOGRAPHED BY ARTHUR ROTHSTEIN, Arthur Rothstein. First collection devoted entirely to the work of outstanding 1930s photographer: famous dust storm photo, ragged children, unemployed, etc. 120 photographs. Captions. 119pp. 9¼ x 10¾.
23590-4 Pa. $5.00

CAMERA WORK: A PICTORIAL GUIDE, Alfred Stieglitz. All 559 illustrations and plates from the most important periodical in the history of art photography, *Camera Work* (1903-17). Presented four to a page, reduced in size but still clear, in strict chronological order, with complete captions. Three indexes. Glossary. Bibliography. 176pp. 8⅜ x 11¼.
23591-2 Pa. $6.95

ALVIN LANGDON COBURN, PHOTOGRAPHER, Alvin L. Coburn. Revealing autobiography by one of greatest photographers of 20th century gives insider's version of Photo-Secession, plus comments on his own work. 77 photographs by Coburn. Edited by Helmut and Alison Gernsheim. 160pp. 8⅛ x 11. 23685-4 Pa. $6.00

NEW YORK IN THE FORTIES, Andreas Feininger. 162 brilliant photographs by the well-known photographer, formerly with *Life* magazine, show commuters, shoppers, Times Square at night, Harlem nightclub, Lower East Side, etc. Introduction and full captions by John von Hartz. 181pp. 9¼ x 10¾. 23585-8 Pa. $6.00

GREAT NEWS PHOTOS AND THE STORIES BEHIND THEM, John Faber. Dramatic volume of 140 great news photos, 1855 through 1976, and revealing stories behind them, with both historical and technical information. Hindenburg disaster, shooting of Oswald, nomination of Jimmy Carter, etc. 160pp. 8¼ x 11. 23667-6 Pa. $5.00

THE ART OF THE CINEMATOGRAPHER, Leonard Maltin. Survey of American cinematography history and anecdotal interviews with 5 masters—Arthur Miller, Hal Mohr, Hal Rosson, Lucien Ballard, and Conrad Hall. Very large selection of behind-the-scenes production photos. 105 photographs. Filmographies. Index. Originally *Behind the Camera.* 144pp. 8¼ x 11. 23686-2 Pa. $5.00

DESIGNS FOR THE THREE-CORNERED HAT (LE TRICORNE), Pablo Picasso. 32 fabulously rare drawings—including 31 color illustrations of costumes and accessories—for 1919 production of famous ballet. Edited by Parmenia Migel, who has written new introduction. 48pp. 9⅜ x 12¼. (Available in U.S. only) 23709-5 Pa. $5.00

NOTES OF A FILM DIRECTOR, Sergei Eisenstein. Greatest Russian filmmaker explains montage, making of *Alexander Nevsky,* aesthetics; comments on self, associates, great rivals (Chaplin), similar material. 78 illustrations. 240pp. 5⅜ x 8½. 22392-2 Pa. $4.50